Organic Reactions

Organic Reactions

VOLUME 72

A JOHN WILEY & SONS, INC., PUBLICATION

Published by John Wiley & Sons, Inc., Hoboken, New Jersey

Published simultaneously in Canada.

Library of Congress Catalog Card Number: 42-20265
ISBN 978-0-470-42374-5

Printed in the United States of America

10 9 8 7 6 5 4 3 2 1

INTRODUCTION TO THE SERIES
ROGER ADAMS, 1942

In the course of nearly every program of research in organic chemistry, the investigator finds it necessary to use several of the better-known synthetic reactions. To discover the optimum conditions for the application of even the most familiar one to a compound not previously subjected to the reaction often requires an extensive search of the literature; even then a series of experiments may be necessary. When the results of the investigation are published, the synthesis, which may have required months of work, is usually described without comment. The background of knowledge and experience gained in the literature search and experimentation is thus lost to those who subsequently have occasion to apply the general method. The student of preparative organic chemistry faces similar difficulties. The textbooks and laboratory manuals furnish numerous examples of the application of various syntheses, but only rarely do they convey an accurate conception of the scope and usefulness of the processes.

For many years American organic chemists have discussed these problems. The plan of compiling critical discussions of the more important reactions thus was evolved. The volumes of *Organic Reactions* are collections of chapters each devoted to a single reaction, or a definite phase of a reaction, of wide applicability. The authors have had experience with the processes surveyed. The subjects are presented from the preparative viewpoint, and particular attention is given to limitations, interfering influences, effects of structure, and the selection of experimental techniques. Each chapter includes several detailed procedures illustrating the significant modifications of the method. Most of these procedures have been found satisfactory by the author or one of the editors, but unlike those in *Organic Syntheses*, they have not been subjected to careful testing in two or more laboratories. Each chapter contains tables that include all the examples of the reaction under consideration that the author has been able to find. It is inevitable, however, that in the search of the literature some examples will be missed, especially when the reaction is used as one step in an extended synthesis. Nevertheless, the investigator will be able to use the tables and their accompanying bibliographies in place of most or all of the literature search so often required. Because of the systematic arrangement of the material in the chapters and the entries in the tables, users of the books will be able to find information desired by reference to the table of contents of the appropriate chapter. In the interest of economy, the entries in the indices have been kept to a minimum, and, in particular, the compounds listed in the tables are not repeated in the indices.

The success of this publication, which will appear periodically, depends upon the cooperation of organic chemists and their willingness to devote time and effort to the preparation of the chapters. They have manifested their interest already by the almost unanimous acceptance of invitations to contribute to the work. The editors will welcome their continued interest and their suggestions for improvements in *Organic Reactions*.

INTRODUCTION TO THE SERIES
SCOTT E. DENMARK, 2008

In the intervening years since "The Chief" wrote this introduction to the second of his publishing creations, much in the world of chemistry has changed. In particular, the last decade has witnessed a revolution in the generation, dissemination, and availability of the chemical literature with the advent of electronic publication and database services. Although the exponential growth in the chemical literature was one of the motivations for the creation of *Organic Reactions*, Adams could never have anticipated the impact of electronic access to the literature. Yet, as often happens with visionary advances, the value of this critical resource is now even greater than at its inception.

From 1942 to the 1980's the challenge that *Organic Reactions* successfully addressed was the difficulty in compiling an authoritative summary of a preparatively useful organic reaction from the primary literature. Practitioners interested in executing such a reaction (or simply learning about the features, advantages, and limitations of this process) would have a valuable resource to guide their experimentation. As abstracting services, in particular *Chemical Abstracts* and later *Beilstein*, entered the electronic age, the challenge for the practitioner was no longer to locate all of the literature on the subject, but rather how to critically and efficiently digest it. *Organic Reactions* chapters are much more than a surfeit of primary references; they constitute a distillation of an avalanche of information into the knowledge needed to correctly implement a reaction. It is in this capacity, namely to provide focused, scholarly, and comprehensive overviews of a given transformation, that *Organic Reactions* takes on even greater significance for the practice of chemical experimentation in the 21st century.

Adams' description of the content of the intended chapters is still remarkably relevant today. The development of new chemical reactions over the past decades has greatly accelerated and has embraced more sophisticated reagents derived from elements representing all reaches of the Periodic Table. Accordingly, the successful implementation of these transformations requires more stringent adherence to experimental details and conditions. The suitability of a given reaction for an unknown application is best judged from the informed vantage point provided by precedent and guidelines offered by a knowledgeable author.

As Adams clearly understood, the ultimate success of the enterprise depends on the willingness of organic chemists to devote their time and efforts to the preparation of chapters. The fact that, well into the 21st century, the series continues to thrive is fitting testimony to those chemists whose contributions serve as the foundation of this edifice. Chemists who are considering the preparation of a manuscript for submission to *Organic Reactions* are urged to contact the Editor-in-Chief.

PREFACE

Beginning with Volume 72, a new feature for the *Organic Reactions* series will be introduced by providing a preface that summarizes the chapters included in each volume.

Volume 72 comprises two chapters that can be broadly aligned under the general heading of functionalization and de-functionalization reactions. The first chapter by Engelbert Ciganek (a long-time member of the *Organic Reactions* family and author or coauthor of three previous chapters in the series) covers the introduction of nitrogen-containing functional groups by amination of carbon-based nucleophiles. The range of reagents capable of transferring a nitrogen-containing moiety to a nucleophile is enormous and includes inter alia, haloamines, hydroxylamines, diazonium salts, azo compounds and azides. The scope of carbon nucleophiles is also very broad and ranges from highly reactive organometallic species ("carbanions") to nucleophiles stabilized by heteroatomic groups (sulfur and phosphorus) and more common substrates such as enolates, metalloenamines, enol ethers, and dicarbonyl compounds. In view of the very large substrate scope, this chapter provides invaluable guidance for the selection of an appropriate aminating agent for each type of nucleophile.

The second chapter by Diego A. Alonso and Carmen Nájera covers the removal of a specific functional group, the sulfone. Because of their ability to stabilize directly bound carbanions, sulfones have evolved as useful workhorses in organic synthesis. These anions are powerful nucleophiles and can form carbon-carbon bonds through alkylation, carbonyl and conjugate addition reactions. However, few desirable end products contain the sulfonyl group such that efficient desulfonylation is central to the overall utility of sulfonyl-mediated reactions. Alonso and Nájera present two different kinds of desulfonylation processes, reductive desulfonylation (replacement with hydrogen) and reductive eliminations (formation of multiple bonds). Reductive desulfonylation can be effected by a wide range of reducing agents that operate primarily by single-electron transfer mechanisms. Reductive elimination is the second step of the well-known Julia olefination process and its descendants. Both of these transformations have significant applications in the total synthesis of complex natural products and are thus key components in retrosynthetic strategy. Phantom sulfonyl groups serve in much the same way that phantom carboxyl groups serve in classic carbonyl condensation reactions.

It is appropriate here to acknowledge the expert assistance of the entire editorial board, in particular Stuart McCombie and T. V. RajanBabu who shepherded the first and second chapters in this volume, respectively, and Jeffery B. Press, the responsible secretary. In addition, the *Organic Reactions* enterprise could not maintain the quality of production without the dedicated efforts of its editorial staff, Dr. Linda S. Press and Dr. Danielle Soenen. Insofar as the essence of *Organic Reactions* chapters resides in the massive tables of examples, the author's and editorial coordinator's painstaking efforts are highly prized.

Scott E. Denmark
Urbana, Illinois

CONTENTS

CHAPTER 1

ELECTROPHILIC AMINATION OF CARBANIONS, ENOLATES, AND THEIR SURROGATES

Engelbert Ciganek

121 Spring House Way, Kennett Square, PA, 19348, USA

CONTENTS

eciganek@verizon.net
Organic Reactions, Vol. 72, Edited by Scott E. Denmark et al.

ACKNOWLEDGEMENTS

I am indebted to E. I. du Pont de Nemours & Co., Inc. and Dr. Pat Confalone for permission to use the company libraries and especially to Ms. Susan Titter of the Agricultural Products Department for valuable assistance. Professor Scott Denmark and Ms. Donna Whitehill of the University of Illinois and Professor

Peter Wipf and Ms. Michelle Woodring of the University of Pittsburgh graciously provided copies of less common journals. I also thank the many colleagues who answered questions or provided copies of their papers. My editor, Dr. Stuart McCombie, is thanked for his guidance and advice and for painstakingly proof-reading the manuscript. Last, but not least, I owe a large debt of gratitude to Dr. Linda Press for valuable help during the preparation of this chapter and for patiently answering my many questions regarding the mysteries of computer software.

INTRODUCTION

Nitrogen-containing organic compounds are ubiquitous in nature and essential to life. They are also important intermediates and products of the chemical and pharmaceutical industries. As a consequence, chemists have developed a plethora of methods for their generation, starting with the first organic synthesis, Wöhler's preparation of urea from ammonium cyanate in 1828.[1] There are many reports of the formation of carbon-nitrogen bonds by electrophilic amination of carbanions and enolates in the early literature, but development of this method as a useful synthetic tool, especially for asymmetric synthesis, is of more recent date.

Most electrophilic aminations can be divided into two types: substitutions (e.g. Eq. 1) and additions (e.g. Eq. 2) to give products that in many cases are not amines. A detailed discussion of the conversion of these intermediates into amines is beyond the scope of this chapter, but references to relevant methods are given in the section on Experimental Conditions.

$$R^1M + (R^2)_2NX \longrightarrow R^1N(R^2)_2 + MX \qquad \text{(Eq. 1)}$$

R^1M = Grignard or organolithium reagent, etc.

$$(MO)(R^1)C{=}C(R^2)(R^3) \xrightarrow[\text{2. } H_2O]{\text{1. } R^4N{=}NR^5} R^1C(O)C(R^2)(R^3)N(R^4)NHR^5 \qquad \text{(Eq. 2)}$$

M = metal

The initial intent to cover the subject exhaustively had to be abandoned because of the overwhelming amount of relevant literature. The following reactions are not covered but are briefly discussed, with references to reviews and seminal papers, in the section on Comparison with Other Methods: reactions of carbanions and enolates and their surrogates with nitrogen oxides, nitrite and nitrate esters, and nitroso and nitro compounds; reactions of enolates with diazonium salts, including the Japp-Klingemann reaction; the diazo transfer reaction except as it interferes with the synthesis of azides; the amination of boranes; and the Neber rearrangement.

The large number of reagents that are available for amination necessitated a deviation from the standard *Organic Reactions* format. The section on Reagents and Mechanisms includes discussion and exemplification of each reagent or reagent class as well as comments on mechanism, particularly in context of reagent-substrate combinations that can lead to more than one product. Stereochemistry is discussed in the relevant sections of Scope and Limitations.

There is only one previous comprehensive review of the electrophilic amination of carbanions;[2] shorter reviews[3–9] and reviews limited to particular reagents, substrates, or products have appeared: amination with haloamines,[10] sulfonylhydroxylamines,[11] oxaziridines,[12] oximes,[13] diazonium salts,[14,15] diazo compounds,[16] activated azo compounds,[17] azides,[18–23] and nitridomanganese(V) reagents;[24,25] amination of enolates;[26–30] and the preparation of α-amino acids by electrophilic amination.[31–34]

REAGENTS AND MECHANISMS

Preparation of Carbanions, Enolates, and Their Surrogates

The preparation of carbanions,[35] organolithium reagents,[36,37] Grignard reagents,[38,39] and organozinc reagents[40,41] has been reviewed. For reviews on the generation of enolates see refs. 42–45. The synthesis of silyl enol ethers is reviewed in refs. 46–49, that of silyl ketene acetals in ref. 50. The term "carbanion" is used loosely without regard to aggregation or solvation.

Aminating Reagents

All aminating reagents dealt with in this chapter are listed here; references to their preparation are found in the section on Experimental Conditions. Stereochemistry is discussed in the relevant sections of Scope and Limitations. The term amination refers to the formation of a carbon-nitrogen bond, not just to the introduction of an amine group. For a quantum Monte Carlo study of electrophilic amination reagents see ref. 51.

Metal Amides. Amidocuprates, when treated with molecular oxygen at low temperatures, give secondary or tertiary amines (Eq. 3). The substrates may be generated from disubstituted lithium cuprates and a primary or secondary amine (method A);[52] one equivalent of the cuprate may be used but yields are higher with three to five equivalents. Only one of the two R^1 groups enters into the product; it may be, among others, an aryl or *tert*-butyl group. Acyl and hydroxy groups in the amine are tolerated. Method B involves the reaction of an organolithium reagent with an excess of a copper amide, which in turn is generated from a lithium amide with copper(II) iodide.[52] The copper amide may be replaced by an anilido cuprate $ArN(R^3)Cu(X)Li$ where X is Cl or CN.[53] The third method (C) employs a lower-order cuprate and a lithium amide. R^1 may be alkyl, aryl, heteroaryl, or styryl. Yields in the three methods are moderate to good. Substituted hydrazines are obtained by replacing the lithium amides in method C with a lithium hydrazide, but yields are only in the 20–40% range.[54] THF is the preferred solvent in these reactions, which fail with Grignard or organolithium reagents. An eight-membered planar complex has been suggested[54] as the intermediate, which reacts with oxygen to give the product via an aminyl radical.

Yields are improved in method C when zinc cyanocuprates and co-oxidants (*o*-dinitrobenzene or copper(II) nitrate) are employed.[55]

$$\left.\begin{array}{l} (R^1)_2CuLi + R^2R^3NH \xrightarrow{A} \\ R^1Li + R^2R^3NCu \xrightarrow{B} \\ R^1Cu(CN)Li + R^2R^3NLi \xrightarrow{C} \end{array}\right\} \longrightarrow \text{amidocuprate} \xrightarrow{O_2} R^1NR^2R^3 \qquad \text{(Eq. 3)}$$

Haloamines. Chloramine was one of the earliest reagents investigated for the amination of Grignard reagents and organolithium compounds.[56–59] An excess of the latter is usually required because of the acidic nature of the haloamine hydrogens. Replacement of one of these by lithium to give a nitrenoid has been suggested as the first step (Eq. 4).[60] Bromamine offers no advantage over chloramine.[61] In the reactions of haloamines with Grignard reagents, yields decrease in the order of RMgCl > RMgBr > RMgI.[61] Chloramine aminates sodio malonates.[62–64] With sodium phenolates, ring-expanded products are obtained.[65] The mechanism of these reactions is unknown[62] but a nitrenoid intermediate is unlikely because of the lower basicity of the substrates. No reaction occurs between 2-lithio *N*-methylimidazole and chloramine.[66]

$$RLi + ClNH_2 \longrightarrow ClNHLi \xrightarrow[-LiCl]{RLi} RNHLi \xrightarrow{H_2O} RNH_2 \qquad \text{(Eq. 4)}$$

Monosubstituted chloramines have not received much attention. The reaction of *N*-chloro-*tert*-butylamine with di(*tert*-butyl)magnesium gives di(*tert*-butyl) amine in 10% yield.[67] Butylmagnesium chloride and *N*-chloromethylamine produce mostly methylamine by reduction and only 14% of *N*-methylbutylamine.[68]

Disubstituted chloramines are claimed to not react with phenylmagnesium bromide[69] and with only very poor yields with *n*-butyl- or benzylmagnesium chloride.[68] *N*-Chlorodiisopropylamine reacts with isopropylpotassium to give triisopropylamine in 3% yield.[70] Similar low yields are obtained in the reactions of phenylethynyllithium,[71] phenylethynylmagnesium bromide,[71] or diethylzinc[72] with *N*-chlorodiethylamine. Chloramines of type ClNRCHRAr, prepared from the secondary amines with *N*-chlorosuccinimide, react with arylmagnesium chlorides to give the corresponding tertiary amines (see Eq. 62).[73] N,N-Disubstituted *N*-chloroamines react with enamines to give mixtures of α-amino aldehydes in moderate to excellent yields where the α-amino group is derived from the chloro amine in one product and from the enamine in the other (see Eq. 86). A mechanism involving aziridinium intermediates has been suggested.[74]

N,*N*-Dibromoamine,[75] *N*,*N*-dichloroalkylamines,[68,72,76] and even trichloroamine[58,77] react with Grignard or dialkylzinc reagents to give amines by reduction of the excess halogen. Yields are low and these reagents are currently of no value in synthesis.

Chloramine-T, the sodium salt of *N*-chloro-*p*-toluenesulfonamide, tosylaminates a number of in situ generated enamines of α-substituted propionaldehydes (see Eq. 78), α-substituted arylacetaldehydes, and methyl arylmethyl ketones.[78]

Hydroxylamines. A number of O-substituted hydroxylamines are electrophilic aminating reagents for introduction of unsubstituted as well as mono- and disubstituted amino groups.

N-Unsubstituted O-*Alkylhydroxylamines.* The most widely used in this category are *O*-methylhydroxylamine, and, to a lesser extent, *O*-benzylhydroxylamine. In the amination of the dianion of 3-methylbutanoic acid with $RONH_2$,[79] yields decrease in the order R = Me > Et = *i*-Pr > *t*-Bu > Bn and range from 34% for $MeONH_2$ to a trace for $BnONH_2$. However, the latter aminates organolithium and Grignard reagents (two equivalents) in fair to good yields.[80] The mechanism of the amination of organolithium reagents with *O*-alkylhydroxylamines involves the nitrenoid intermediate **1** (Eq. 5) and eventual displacement of the methoxy group by R in a counterintuitive reaction between two negatively charged species that is sterically akin to an S_N2 reaction. The mechanism is based on extensive experimental[81–85] and computational work[60,86–90] and also applies to Grignard, organozinc, and organocopper reagents.[91] However, it should be kept in mind that other mechanisms are, at least in principle, available, in view of the fact that N,N-disusbstituted *O*-alkylhydroxylamines are also aminating reagents even though a process involving a nitrenoid is impossible with these reagents. By generating the nitrenoid **1** with methyllithium only one equivalent of RLi is required. Application of this method to aminations with *O*-alkylhydroxylamines reported in the earlier literature should increase the efficiency of these reactions. An excess of the nitrenoid MeONHLi is recommended; in the reaction with *n*-butyllithium the yields of *n*-butylamine are 51% with one equivalent, 71% with two (see also Eq. 63), and 85% with four.[92]

$$MeLi + MeONH_2 \longrightarrow \underset{\mathbf{1}}{MeONHLi} \xrightarrow{RLi} R(\mu\text{-}Li)_2N(H)OMe \longrightarrow$$

$$\left[R\text{----}\underset{H}{N}\text{----}OMe\right]^{2-} 2\,Li^+ \xrightarrow{-\,LiOMe} RNHLi \xrightarrow{H_2O} RNH_2 \qquad \text{(Eq. 5)}$$

N-Unsubstituted O-*Arylhydroxylamines.* Amination of malonic and cyanoacetic ester enolates[93] and of methyl 9-fluorenecarboxylate[94] may be carried out in fair to good yields with *O*-(2,4-dinitrophenyl)hydroxylamine. Yields are low with the more basic phenylacetic ester enolates and the anion of phenylacetonitrile, both of which partially decompose the reagent with formation of diimide.[93] This reagent provides much poorer yields than $Ph_2P(O)ONH_2$ in the amination of the anion of tetraethyl methylenebis(phosphonate).[95] The corresponding *N*-methyl derivative is unreactive in an N-amination.[94] Various analogs of the highly explosive *O*-(2,4-dinitrophenyl)hydroxylamine have been tested in N-aminations only [94,96] and *O*-(4-nitrophenyl)hydroxylamine was found to provide the highest yields and to have the highest onset temperature of explosive decomposition.[96]

N-Monosubstituted O-*Alkylhydroxylamines.* Various *O*-methylhydroxylamine derivatives (MeONHR) aminate aliphatic and aromatic organolithium compounds: R = Me,[82,83,97] *n*-Pr and *i*-Pr,[83] benzyl,[83,85] α-methylbenzyl,[82,83,85,97] and 2-phenylethyl.[83] The order of reactivity of BnNLiOMe toward butyllithium reagents is n-Bu < s-Bu < t-Bu.[85] BnNLiOMe reacts much more rapidly with these three alkyllithium reagents than its α-methyl derivative PhCHMeNLiOMe;[85] the latter is about equal in reactivity to MeNLiOMe.[97] Reagents of type $RCH_2NLiOBn$ may be prepared by addition of an organolithium reagent RLi to formaldehyde *O*-benzyl oxime (Eq. 6).[98] A nitrenoid of this class is also formed in the reaction of phenyllithium with nitrosobenzene (Eq. 7),[99] but it reacts so rapidly with unreacted phenyllithium that the possibility of trapping it with another organolithium reagent seems remote.

$CH_2{=}N{-}OBn$ —(n-BuLi, THF, –40°)→ n-BuCH_2N(Li)OBn —(1. PhLi, 0-40°; 2. 4-PhC_6H_4COCl)→ n-BuCH_2N(Ph)C(O)C_6H_4-4-Ph (47%) (Eq. 6)

PhNO —(PhLi (1.1 eq), THF, –100°, 70 min)→ [PhN(Li)OPh —(PhLi)→ Ph_2NLi + PhOLi] —(H_2O)→ Ph_2NH (41%) + PhOH (41%) (Eq. 7)

O-Trimethylsilylhydroxylamine reagents (RNHOTMS where R is TMS or alkyl), aminate organocuprates of type $R^1_2Cu(CN)Li_2$ (see Eqs. 64 and 73), but not organolithium reagents.[100–102] Small amounts of alcohols R^1OH are formed in some reactions as a consequence of the nitrenoid **2**/oxenoid **3** equilibrium (Eq. 8), with the latter acting as a hydroxylating agent.[60,103]

$$\mathrm{TMS\bar{N}(M^+)OTMS}\ (\mathbf{2}) \rightleftharpoons \mathrm{(TMS)_2NO^-M^+}\ (\mathbf{3}) \quad \text{(Eq. 8)}$$

Amination with an N-monosubstituted cyclic hydroxylamine is shown in Eq. 9.[104]

Cyclic N-H hydroxylamine (acetonide) —(PhMgBr (3 eq), –78° to 0°, 1 h)→ OH / NHPh product (~100%) (Eq. 9)

N,N-Disubstituted O-*Alkylhydroxylamines.* In the amination with a series of N,N-disubstituted *O*-methylhydroxylamines, more bulky alkyllithium compounds react more readily (product **4**, Eq. 10).[85] The small amounts of products **5** are the result of elimination of methanol from the substrate to give the imine followed by addition of R^1Li to the latter. Reagents where R^2, R^2 is H, Me or Me, Me do not react. A single-electron-transfer process involving a nitrogen radical has been proposed,[85] but no cyclized product is formed when R^3 is a dimethylvinyl group.

Bn–CH₂–CH(R^2)–N(OMe)–CH(R^2)–CH₂CH₂–R^3 →[R^1Li, –78°, 3 h; rt, 1-2 d] Bn–CH₂–CH(R^2)–N(R^1)–CH(R^2)–CH₂CH₂–R^3 (**4**) + Bn–CH₂–CH(R^2)–NH–C(R^2)(R^1)–CH₂CH₂–R^3 (**5**)

R^1	R^2	R^3	**4**	**5**
n-Bu	H	Ph	(5%)	(5%)
s-Bu	H	Ph	(47%)	(5%)
t-Bu	H	Ph	(72%)	(5%)
t-Bu	H	CH=CMe$_2$	(67%)	(—)

(Eq. 10)

Silyl ketene acetals are aminated by the ethoxycarbonylnitrene precursor $EtO_2CN(TMS)OTMS$ to give α-ethoxycarbonylamino esters via aziridines in fair to good yields (see Eq. 124).[105]

O-Acyl Hydroxylamines. *O*-Acyl N-unsubstituted hydroxylamines have been used occasionally in the amination of enolates.[79,106] In the amination of the sodium salt of diethyl phenylmalonate, *O*-(4-nitrobenzoyl)hydroxylamine is somewhat more efficient than $(4\text{-}MeOC_6H_4)_2P(O)ONH_2$ (99% vs 92% yields).[106] This reagent also gives the highest yield in the N-amination of oxazolidinone anions.[107] A series of N,N-disubstituted *O*-benzoylhydroxylamines is used in the amination of alkyl- and arylzinc chlorides in the presence of a catalytic amount of $(Ph_3P)_2NiCl_2$[108] and of dialkyl-, diaryl-, and di(heteroaryl)zinc reagents in the presence of a catalytic amount of a copper(II) salt (see Eq. 36).[109–112] The disubstituted zinc reagents may be prepared in situ by reaction of Grignard reagents with a catalytic amount of zinc chloride because transmetalation is faster than the reaction of the Grignard reagent with *O*-benzoylhydroxylamine. Functional groups on the aryl ring, such as NO_2, CO_2R, and CN are tolerated and 0.6 equivalent of the disubstituted zinc reagent may be employed with a slight reduction of the yield. Arylmagnesium reagents may be aminated in this way without the intervention of the corresponding zinc reagents.[113] An S_N2 mechanism has been advanced.[113]

N-Unsubstituted O-Sulfonylhydroxylamines. The acidic nature of hydroxylamine *O*-sulfonic acid makes it essentially useless in electrophilic aminations of carbanions. One of the few exceptions is shown in Eq. 161. The explosive[114,115] *O*-(mesitylenesulfonyl)hydroxylamine aminates alkylzirconium complexes (see Eqs. 41 and 51),[116] acid dianions,[115] and ester enolates.[117] *O*-Arenesulfonylhydroxylamines with no ortho substituents are thermally unstable at room temperature.[11]

N-Monosubstituted O-Sulfonylhydroxylamines. *N*-Ethoxycarbonyl-*O*-(*p*-toluenesulfonyl)hydroxylamine (**6**) is used in the amination of enamines.[118,119] The more reactive *N*-ethoxycarbonyl-*O*-(4-nitrobenzenesulfonyl)hydroxylamine

(**7**) aminates enamines[120,121] and enol ethers[122] derived from ketones (see Eq. 96), as well as metalloimines,[123] enolates of β-dicarbonyl compounds,[124] and enamines derived from β-dicarbonyl compounds.[125] The lithium salt of *N*-(*tert*-butoxycarbonyl)-*O*-(*p*-toluenesulfonyl)hydroxylamine (**8**) aminates alkyl- and aryllithium and -copper reagents (see Eq. 69),[126–128] esters and *N*-acyloxazolidinone enolates,[126] and α-alkylphosphonamides.[129] The allyloxycarbonyl analogs **10** and **11** are similarly used.[130] The structure of the mesityl analog **9** (dimer, crystallizing with three molecules of THF) has been determined by single crystal X-ray crystallography.[131] Because this class of reagents offers a much better leaving group, the possibility exists that the nitrenoids lose the elements of $ArSO_3M$ to give nitrenes NCO_2R.[60] The involvement of these reactive intermediates has been proposed in a number of examples.

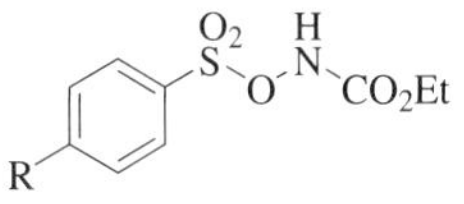

O_2 H, S, O, N, CO_2Et, R
6 R = Me
7 R = O_2N

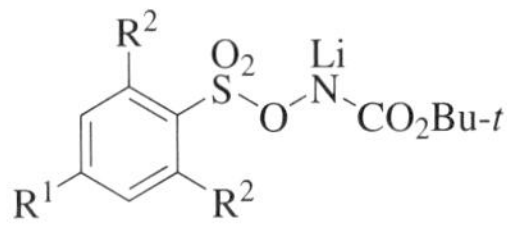

R^2, O_2 Li, S, O, N, CO_2Bu-*t*, R^1, R^2
8 R^1 = Me R^2 = H
9 R^1, R^2 = Me

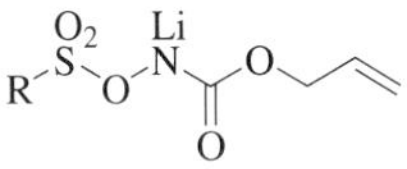

O_2 Li, R, S, O, N, O, O
10 R = Me
11 R = 4-MeC_6H_4

N,N-Disubstituted O-*Sulfonylhydroxylamines.* Compounds of type $R^1SO_2ON(R^2)_2$ (R^1 = Me, Ph, *p*-tolyl, mesityl; R^2 = Me, Et) are versatile aminating reagents for a wide variety of substrates: aliphatic (see Eq. 35),[132,133] allylic,[133,134] olefinic (see Eq. 56),[133] acetylenic (see Eq. 60),[135] benzylic (see Eq. 53),[133,136] and aromatic[132,133] metal derivatives and enolates (see Eq. 89).[133,134] Reactions of $MeSO_2ONMe_2$ (and probably other similar reagents) with RMgI should be avoided because iodide reduces the reagent.[137] Both an electron-transfer and an S_N2-type substitution mechanism have been considered for these transformations.[136]

O-*Phosphinoylhydroxylamines.* The non-explosive[138] *O*-diphenylphosphinoylhydroxylamine, $Ph_2P(O)ONH_2$, aminates alkyl,[139,140] aryl,[139] ethynyl (see Eq. 60),[135] cyanomethyl, and phosphinoylmethyl (see Eq. 152)[95,141] metal derivatives and enolates of esters,[139,142] lactams (see Eq. 137),[143] α,β-unsaturated carbonyl compounds (see Eq. 153),[144] and β-dicarbonyl compounds.[139] The equally stable methoxy analog $(4\text{-}MeOC_6H_4)_2P(O)ONH_2$ has been recommended[106] as a better reagent because of its increased solubility in organic solvents at low temperatures but there is a report of a low yield and formation of a hydroxylation product in the amination of a malonic ester enolate.[145] Amination with the disubstituted analog $Ph_2P(O)ONMe_2$[146] and the chiral, non-racemic cyclic derivative **12** (see Eqs. 109 and 143)[147] has also been reported. There appear to be no mechanistic studies of these reagents but it is relevant that equimolar amounts of the substrate and the reagent or a slight excess of the latter are usually employed.

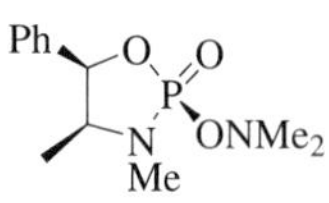

12

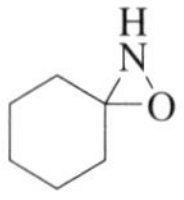

13a

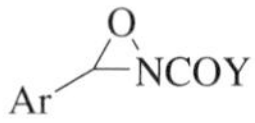

Y = *t*-Bu, NEt_2, (–)-menthoxide

14

Oxaziridines. The readily synthesized 1-oxa-2-azaspiro[2,5]octane (**13a**)[148] aminates[12] enolates of β-dicarbonyl compounds,[149,150] α-cyano carbonyl compounds,[149,150] and anions derived from cyanomethyl derivatives further activated by aryl or heteroaryl groups.[150] The products are either amines, *N*-cyclohexylidene derivatives, or more complex structures (see Eq. 162). The camphor-derived oxaziridine **13b** aminates enolates of esters, β-dicarbonyl and α-cyano carbonyl compounds,[151] and anions derived from various cyanomethyl compounds.[151] Esters are aminated only if they carry an additional aryl group.[151] The products resulting from β-dicarbonyl and α-cyano carbonyl compounds are camphorimines that have lost the ester group by hydrolysis and decarboxylation. Camphorimines derived from aminations of esters retain the ester group. The cyano group in all substrates is converted into an amide group and the mechanism shown in Eq. 11 has been proposed. The first step is analogous to that of the mechanistically fairly well-established hydroxylation of enolates with *N*-sulfonyl oxaziridines[152] except that attack by the anion is on nitrogen rather than oxygen. When R is methyl or ethyl, only rearrangement products of the aminating reagent are isolated.[151]

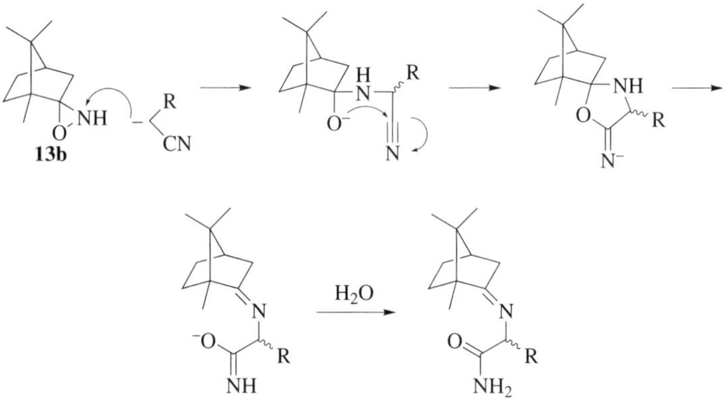

(Eq. 11)

R = CH=CH_2, Ar (45-80%)

R = Me, Et (0%)

Oxaziridines **14** transfer the NCOY group to enolates of ketones (see Eq. 90),[153–156] esters (see Eq. 110),[153,155,157,158] amides,[158] *N*-acyloxazolidinones,[153,157] and β-dicarbonyl compounds,[155] anions stabilized by cyano (see Eq. 141),[155] sulfonyl (see Eq. 145),[158] and phosphinoyl[154] groups, and ketone enol ethers.[155] Yields are in the 20–60% range. The first step in these reactions is presumably attack of the enolate on nitrogen as in Eq. 11, followed by elimination of an aldehyde ArCHO and formation of the amination product. With esters, the aldehyde may undergo an aldol reaction with the substrate enolate when LiHMDS, KHMDS, LDA, or *t*-BuLi are used as the bases to generate the enolates. This undesired side reaction is not observed with NaHMDS provided that two equivalents of the reagent are used, but yields are low.[155]

Imines. Organometallic compounds normally attack imines at the carbon atom. Predominant or exclusive attack on nitrogen may be forced by attaching one or two electron-withdrawing groups to the imine carbon atom.[159–167] In the examples of Eq. 12[161] involving a substrate with a fairly bulky group on nitrogen, the ratios of product **15** to **16** demonstrate that only the *tert*-butyl and allyl Grignard reagents attack on carbon, the former presumably for steric reasons. All cadmium reagents RCdX tested (R = Me, *n*-Pr, *i*-Pr, Bn) add normally on carbon.

R^2MX (X not specified), Et_2O → **15** + **16** (Eq. 12)

R^1 =

R^2	M	**15 + 16**	**15:16**
Et	Mg	(45-55%)	95:5
n-Pr	Mg	(44-55%)	96:4
i-Pr	Mg	(44-55%)	60:40
$CH_2CH=CH_2$	Mg	(45-55%)	0:100
i-Bu	Mg	(45-55%)	96:4
t-Bu	Mg	(45-55%)	0:100
Bn	Mg	(45-55%)	100:0
Bn	Cd	(55-70%)	0:100

A second method of favoring attack on nitrogen involves systems where the imine carbon is surrounded by fairly bulky substituents and where placing a negative charge on this carbon is favored by formation of a cyclopentadienyl anion (Eq. 13).[168] A phenyl group on nitrogen reverses this trend, with product **18** now predominating over **17**.

NR^1 → R^2Li, THF, hexane, –78°, 2 h; to rt → R^1R^2N H (**17**) + R^1NH R^2 (**18**) (Eq. 13)

R^1	R^2	**17**	**18**
Me	*n*-Bu	(71%)	(0%)
n-Bu	Et	(65%)	(5%)
Ph	*n*-Bu	(15%)	(50%)

Attack of isopropylmagnesium bromide on the hindered imine in Eq. 14 surprisingly occurs on nitrogen whereas the less bulky ethylmagnesium bromide adds to the carbonyl group.[169] Organozinc reagents react with anthranil under $Ni(acac)_2$ catalysis to give α-aminobenzaldehyde derivatives by a proposed single-electron transfer mechanism (Eq. 15).[170] Diethyl zinc adds to 1,4-diaza-1,3-butadienes in a net 1,4-fashion (Eq. 16).[171]

t-Bu, HN=(cyclohexadienone)=O, t-Bu → (i-PrMgBr, Et_2O) → i-PrNH=(cyclohexadienone)=O (35%) (Eq. 14)

RZnCl + (2,1-benzisoxazole) → ($Ni(acac)_2$, THF, 0° to rt, 2 h) → 2-(NHR)benzaldehyde (4-86%) (Eq. 15)

R = Me, 2-thienyl, Ph, 2-, 3-, and 4-$MeOC_6H_4$

t-BuN=CH–CH=NBu-t → (Et_2Zn, toluene, −70°) → $ZnEt_2$ complex → (−50°) → N-Et ZnEt complex → (t-BuOH, pentane, rt) → Et–N(t-Bu)–CH=CH–NHBu-t (76%) + Et–N(t-Bu)–CH₂–CH=NBu-t (12%) (Eq. 16)

(*N*-Arenesulfonylimino)phenyliodinanes. [*N*-(*p*-tolylsulfonyl)imino]phenyliodinane (TsN=IPh) and its pentafluoro analog $C_6F_5SO_2N{=}IPh$ react readily on warming in acetonitrile with silyl enol ethers derived from acetophenones to give the α-tosylamino derivatives in high yields. The reaction is less efficient in methylene chloride, gives low yields with the trimethylsilyl ether of 3-pentanone and with 1-trimethylsilyloxybutadiene, and fails completely with 1-trimethylsilyloxycyclohexene and a ketene acetal, 1-phenoxy-1-(trimethylsilyloxy)ethylene.[172] The latter two types of substrates do react when a copper catalyst is employed, but yields do not exceed 50% (see also Eq. 92).[173] With chiral (ligand **19** or **20**) copper catalysts, modest to fair enantiomeric excesses are achieved (Eq. 17).[174] The proposed mechanism involves a slightly favored front-side attack of the enol derivative on the initially formed ligand–copper nitrene complex with formation of an aziridine, which is converted directly into the α-tosylamino product during isolation when methyl or trimethylsilyl enol ethers are used.

Ph(AcO)C=CHCH₃ + TsN=IPh → ($[Cu(MeCN)_4]PF_6$, **19** or **20**, CH_2Cl_2, −40°) → aziridine (AcO, Ph, NTs) → (HCl, MeOH) → PhCOCH(CH₃)NHTs

19 Ar = $C_6H_3Cl_2$-2,6 **20**

Ligand	Conversion [a]	ee
19	(>95%)	28% R
20	(61%)	52% R

[a] based on TsN=IPh reacted

(Eq. 17)

Oximes. Reaction of alkyl- or arylmagnesium reagents with two equivalents of acetone oxime in toluene gives alkyl or arylamines, respectively, in low yields. The yields are improved by converting the oxime into the salt with ethylmagnesium halide followed by addition of the desired Grignard reagent. A mechanism involving a four-membered cyclic transition state is postulated (Eq. 18).[174a] Similar reactions with the lithium salt or methyl ether of benzaldoxime have also been reported.[175] Among the *O*-sulfonyloxime derivatives **21**[176–178] (see Eq. 61), **22**[178,179] (see Eq. 40), **23**,[180] **24**,[181] and **25**,[181,182] the dioxolane **25b** combines the advantages of high product yields in reactions with alkyl-, vinyl-, aryl-, and heteroarylmagnesium reagents with ease of hydrolysis of the initially formed imine to the amine (see Eq. 37).[182] Reactions with other types of anions do not seem to have been investigated except that phenolates (Eq. 176) and enolates of β-dicarbonyl (Eq. 175) and α-sulfonyl carbonyl compounds undergo an intramolecular version of this amination reaction. The mechanism is believed to involve direct S_N2 substitution on the sp^2 nitrogen of the oxime[13,183] rather than addition/elimination or electron transfer.

$$\text{Me}_2\text{C{=}NOH} \xrightarrow{\text{EtMgX}} \text{Me}_2\text{C{=}NO}^-\ \text{MgX}^+ \xrightarrow{\text{Ph(CH}_2\text{)}_2\text{MgX (2 eq)}}$$

$$\left[\ \text{Me}_2\text{C{=}N(O{-}MgX)(MgX)(CH}_2\text{Bn)} \longrightarrow \text{Me}_2\text{C{=}N(CH}_2\text{)}_2\text{Ph}\ \right] \longrightarrow \text{H}_2\text{N(CH}_2\text{)}_2\text{Ph}\ \ (48\%) \qquad \text{(Eq. 18)}$$

21 R = Me, Ph ($R(Me)C{=}N{-}OSO_2C_6H_2Me_3$-2,4,6)

22 Ar = Ph, 4-$CF_3C_6H_4$, 3,5-$(CF_3)_2C_6H_3$; R = Me, 4-MeC_6H_4 ($Ar_2C{=}N{-}OSO_2R$)

23 (tetraphenylcyclopentadienone O-tosyloxime, C=N–OTs)

24 ($(EtO)_2C{=}N{-}OSO_2Ph$)

25 (cyclic R^1_4-substituted Y,Z-heterocycle C=N–OSO_2Ph)

	R^1	Y	Z
25a	H	O	O
25b	Me	O	O
25c	H	NMe	O
25d	H	O	NMe
25e	H	NMe	NMe

Diazonium Salts. *Diazonium salts are potentially explosive. See the cautionary note in Experimental Conditions.* Aryldiazonium salts **26** react with alkyl- and arylmagnesium reagents,[184–191] arylzinc,[190,192,193] and aryltin reagents[194] to give azo compounds. Yields vary considerably; the best are achieved with the diazonium salt **26e**[191] (see Eq. 48). Aryldiazonium salts also react with enolates, enol derivatives, or enamines of aldehydes (see Eq. 85),[195] ketones (see Eq. 95),[185] and with silyl ketene acetals (see Eq. 121).[196,197]

$ArN_2^+ X^-$

	X
26a	Cl
26b	$ZnCl_3$
26c	BF_4
26d	$Zn(BF_4)Cl_2$

26e X = (benzo-fused cyclic disulfonylimide: O_2S–N–SO_2)

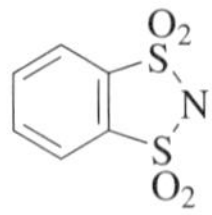

Diazo Compounds.[198] Alkyl- and arylmagnesium[199–204] and alkyllithium reagents[205] add to diazo compounds in a little-used reaction to give hydrazones. Diazo compounds add to enolates to give azines.[206] With enamines, diazo compounds give hydrazones of α-diketones.[207]

Azo Compounds. *Alkyl Azo Compounds.* The only aminations with alkyl azo compounds found in the literature involve the cyclic derivatives **27**,[208] **28**,[209] and **29**.[210] Reaction of **29** with phenyllithium followed by in situ arylation of the anion (Eq. 19)[210] is one of the few examples of tandem reactions in aminations reported thus far. Azo compounds **27** add to cyclohexyl- and phenylmagnesium reagents at −78° with fair to excellent yields,[208] and the bicyclic azo compound **28** gives an adduct with *t*-BuLi at −78° in almost quantitative yield.[209] Relief of strain no doubt is one of the driving forces for these reactions but the low temperatures involved may indicate that they could be extended to acyclic alkyl azo compounds.

27 R, R
Me, Me
n-Pr, *n*-Pr
—$(CH_2)_5$—

28

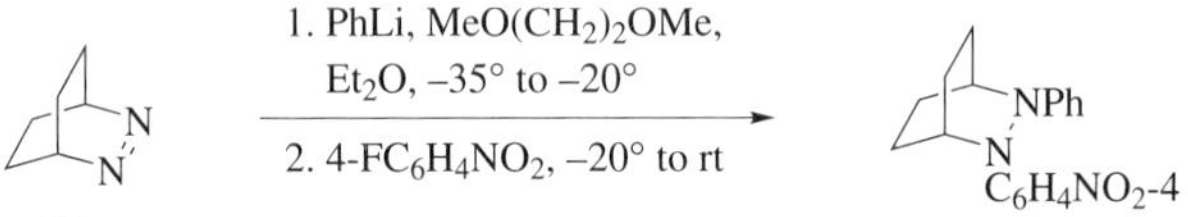

29 —[1. PhLi, $MeO(CH_2)_2OMe$, Et_2O, −35° to −20°; 2. 4-$FC_6H_4NO_2$, −20° to rt]→ NPh / N–$C_6H_4NO_2$-4 (34%) (Eq. 19)

Aryl Azo Compounds. Alkyl- (including *tert*-butyl) and aryllithium reagents add to azo benzene to give trisubstituted hydrazines in fair to excellent yields (see Eqs. 44 and 45); alkylation of the intermediate anion in situ leads to tetrasubstituted hydrazines.[211] Benzyl and heteroarylmethyl (see Eq. 54) anions and the enolate of phenylacetamide add to azo benzene in fair to excellent yields.[212] Aromatic Grignard reagents are reported to reduce azo benzene and its derivatives to the hydrazo compounds (cf. also Eq. 20).[213] The only other aryl azo compound investigated in aminations appears to be benzo[*c*]cinnoline.[214]

Esters of Azodicarboxylic Acid. These compounds are versatile aminating reagents for alkyl- (see Eq. 46), allenyl- (see Eq. 59), aryl- and heteroarylmetal (see Eq. 75) derivatives, and especially enolates (see Eqs. 87, 88, 115–117, and

119) and metalloimines (see Eqs. 104–106). An important new reaction involves addition of azo esters to alkenes,[215] dienes,[216] and enynes[216] in the presence of silanes catalyzed by cobalt and manganese complexes to give the more highly substituted hydrazino esters (see Eqs. 49, 52, and 55). Based on preliminary mechanistic studies of this hydrohydrazination reaction, rate-limiting addition of a metal hydride species to the double bond is followed by a fast amination step.[215]

Benzyl and *tert*-butyl esters are widely used because of their ready conversion into the hydrazines after the amination step and the presence of an aromatic chromophore in the former. Addition of the organometallic species to the ester carbonyl group does not appear to be a problem, although *tert*-butyl esters often provide higher yields. Formation of substantial amounts of an α,β-unsaturated carbonyl compound by elimination of the hydrazino ester from the desired product has been reported in the reaction of dibenzyl azodicarboxylate with the enolate of a sugar ketone.[217] Esters derived from azodicarboxylic acid and chiral alcohols have been prepared[218,219] and a chiral amide has been used in the amination of an achiral enolate (see Eq. 134).[219] The failure of a secondary Grignard reagent to add to diisopropyl azodicarboxylate is shown in Eq. 20.[220] The asymmetric amination of aldehydes (see Eqs. 76 and 77)[221–227] and ketones (see Eq. 91)[228,229] by azo esters is catalyzed by proline and its derivatives. The proposed mechanism involving a hydrogen bond from the catalyst to the N=N double bond in the transition structure is shown in Eq. 21[221] (see also ref. 224). The amination of β-keto esters by azo esters proceeds at room temperature neat or in polar solvents such as alcohols[230,231] or, as with β-aminocrotonic ester, even in petroleum ether.[230] The former reaction may be carried out enantioselectively with catalysts such as cinchona alkaloids (see Eq. 163),[231,232] chiral urea and thiourea derivatives,[233] chiral copper(bis)oxazoline complexes[234] (see Eqs. 103, 151, and 164),[235–237] and chiral palladium BINAP complexes (see Eqs. 150 and 165).[238,239]

MgCl, Ph + $i\text{-PrO}_2\text{C-N=N-CO}_2\text{Pr-}i$ → Ph (82%) + $i\text{-PrO}_2\text{C-NH-NH-CO}_2\text{Pr-}i$ (82%)

(Eq. 20)

R^1CHO → (proline, CO_2H, N H) → N+ CO_2H, OH^-, R^1 → N CO_2H, R^1 → ($R^2O_2CN{=}NCO_2R^2$)

R^2O_2C, N, N, H, O, O, N, CO_2R^2, R^1 → H, N+, CO_2^-, R^1, N, $NHCO_2R^2$, CO_2R^2 → H, CHO, R^1, N, $NHCO_2R^2$, CO_2R^2 + N H CO_2H

(Eq. 21)

Azo esters also aminate enol ethers (see Eq. 82),[240–245] enamines (see Eq. 147),[118,246–250] ketene acetals (see Eqs. 112 and 113),[251] ketene aminals (see Eqs. 125 and 126),[251,252] and ketene thioacetals.[253]

Other Acyl Azo Compounds. Various azo derivatives [$R^1N{=}NCOR^2$: R^1 = aryl, $R^2 = CO_2R$, $CONR_2$, or COAryl; and $R^1CON{=}NCOR^2$: $R^1 = R^3O$, $(R^3)_2N$, Ar, $R^2 = (R^3)_2N$, Ar] have been used as aminating agents. The site selectivity is governed by the degree to which a substituent stabilizes the negative charge on nitrogen, which increases in the order Aryl < $CONR_2$ < CO_2R < COAr. *N*-Phenyltriazolinedione has been used to aminate acetone[254] and a silyl enol ether.[245]

Sulfonyl Azo Compounds. Aryl and cyclopropyl Grignard reagents add to ArN=NTs to give diaryl or cyclopropylarylamines after allylation and reduction (Eq. 22).[255] For a similar reaction involving organozinc reagents see Eq. 38.

$$Ar^1I \xrightarrow[-20^\circ]{i\text{-PrMgCl, THF}} Ar^1MgI \xrightarrow[\text{2. ICH}_2\text{CH=CH}_2\text{, }N\text{-methylpyrrolidinone, rt, 2 h}]{\text{1. Ar}^2\text{N=NTs, THF, }-20^\circ}$$

$$Ar^1N(Ar^2)N(Ts)CH_2CH{=}CH_2 \xrightarrow[\text{2. Zn, HOAc, CF}_3\text{CO}_2\text{H, }75^\circ]{\text{1. Remove solvents}} \underset{(63\text{-}86\%)}{Ar^1NHAr^2} \quad \text{(Eq. 22)}$$

Azides. *Alkyl Azides.* A variety of alkyl azides react with alkyl- and aryl-metal species to give triazenes (Eq. 23) (*see cautionary note with regard to both azides and triazenes in Experimental Conditions*): methyl azide,[256–258] ethyl azide,[258] isopropyl azide,[259] *n*-butyl azide,[260–262] cyclopropylmethyl azide,[262] allyl azide,[263] trimethylsilylmethyl azide,[264–267] a protected 2-hydroxyethyl azide,[268] *n*-hexyl and cyclohexyl azide,[269] benzyl azide,[261,269,270] and polymethylene diazides $N_3(CH_2)_nN_3$ (n = 2,3).[271,272] Protolysis of the intermediate metal salts of the triazenes may give rise to two different triazenes (Eq. 23) and their structures have not always been determined with certainty. The product of the reaction of benzyl azide with phenylmagnesium bromide is identical to that obtained from phenyl azide and benzylmagnesium chloride and was assigned structure **30** with the extended conjugation (Eq. 24)[270] on the basis of the product obtained with phenyl cyanate. Protolysis of triazene **30** with 1 N HCl gives aniline hydrochloride and benzyl chloride (Eq. 24);[270] similarly, *N*-methyl- and *N*-ethyl-*N′*-phenyltriazenes, on treatment with HCl, give aniline hydrochloride and methyl or ethyl chloride, respectively.[270] The intermediate triazenes obtained from trimethylsilylmethyl azide and aryllithium or arylmagnesium reagents decompose to arylamines on aqueous workup.[264] Triazenes are also not isolated from the reaction of allylindium species, generated in situ from the bromides and indium metal, with alkyl and aryl azides in DMF; however, *N*-alkyl and *N*-aryl allylamines, respectively, are obtained (Eq. 25).[269] This example appears to be one of only two instances where, in a reaction of an organometallic species with an azide, both substituents on the intermediate triazene appear in the product. The other is the addition of alkylmagnesium species to aryl azides mentioned below.

By contrast, allyl azide, and aryllithium or arylmagnesium species react to give arylamines after acidic workup (Eq. 26).[263] The triazene intermediate should be the same, except for the counter ion and the solvent, as the one in Eq. 25. No explanation for these differing results has been advanced.

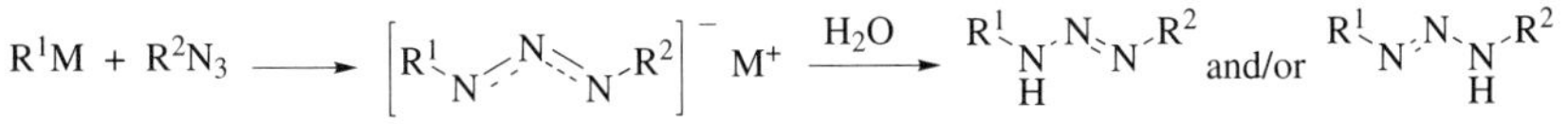

(Eq. 23)

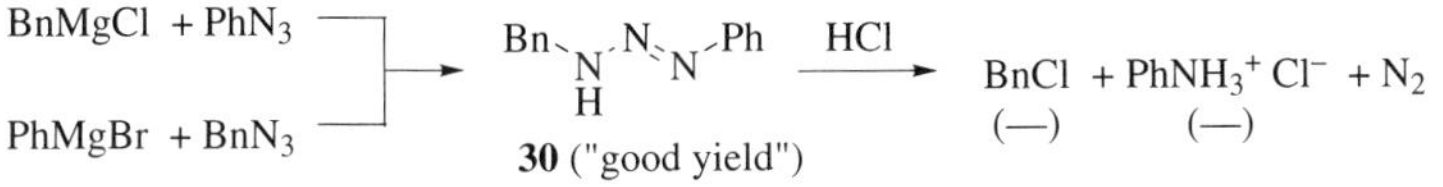

(Eq. 24)

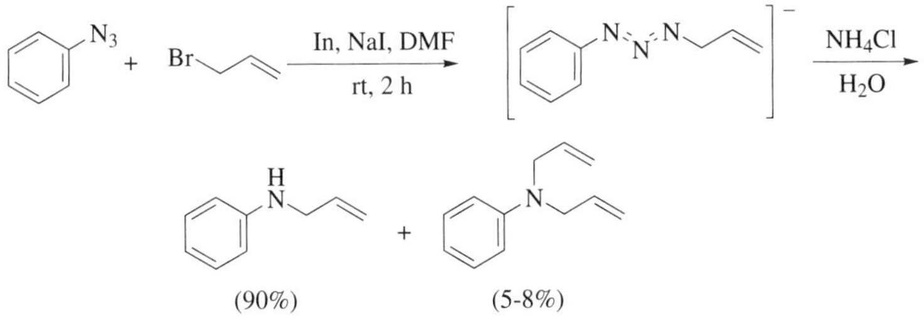

(Eq. 25)

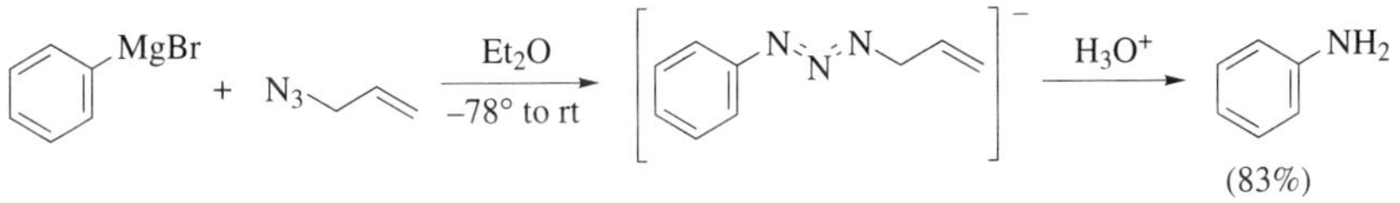

(Eq. 26)

Both N,N'-di(n-butyl) and N,N'-di(cyclopropylmethyl)triazenes react differently with dilute HCl (0.1% in acetone) to give nitrogen gas and nitrogen-free products (n-BuOH, s-BuOH, 1-butene, and 2-butene with the former triazene) via alkyldiazonium species.[262]

Reaction of the α-heteroatom-substituted azides **31** and **32** with 2-phenethylmagnesium bromide proceeds with equal rates at −78°; analog **33** only reacts at 0°, whereas both azides **34** and **35** are essentially unreactive at this temperature.[273] Both aliphatic (see Eq. 40) and aromatic Grignard reagents, but not aromatic lithium reagents, may be used with azide **32**, which has a low steric requirement as evidenced by its reaction with the exo and endo isomers of 2-norbornylmagnesium bromide at about equal rates[274] (see also Eq. 39).

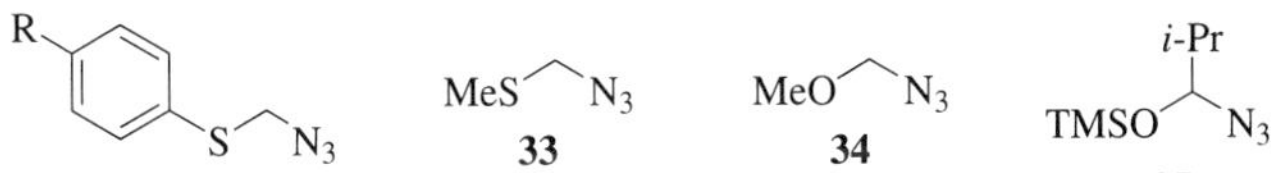

Hydrolysis of the triazenes so obtained from aromatic Grignard reagents to give aromatic amines may be carried out with either aqueous formic acid or aqueous potassium hydroxide.[275] Triazene anions derived from aliphatic Grignard reagents are quenched with acetic anhydride (or benzoyl chloride) and the acetates **36** are then converted into the aliphatic amines using the conditions shown in Eq. 27.[273] The scope of this method is somewhat limited, however: the unstable triazenes, obtained in almost quantitative yields from *tert*-butylmagnesium chloride and *n*-octylmagnesium bromide, could not be converted into the amines and quenching the triazene anion obtained from azide **32** and 1-octenylmagnesium bromide with acetic anhydride gives the regioisomer of acetate **36**, which is unsuitable for further manipulation.[274] The 2-anions of furan, thiophene, *N*-methylpyrrole, and *N*-methylindole do not react with azide **32**.[274]

$$[\text{R-N=N-N-CH}_2\text{SPh}]^- \text{MgBr}^+ \xrightarrow[-60^\circ \text{ to } -30^\circ,\ 1.5\text{ h}]{\text{Ac}_2\text{O}} \text{R-N(Ac)-N=N-CH}_2\text{SPh}\ (\mathbf{36}) \xrightarrow[\text{or: KOH, Me}_2\text{SO, }0^\circ]{n\text{-Bu}_4\text{NH}^+\ \text{HCO}_2^-,\ \text{DMF, }45^\circ} \text{RNH}_2 \quad \text{(Eq. 27)}$$

Azide **32** aminates ester enolates (see Eq. 114)[275] and a sugar-derived azide aminates the anion derived from cyanoacetamide[276] (see Eq. 167).

Vinyl Azides. Vinyl azides such as **37** or **38** react with alkyl-, aryl-, and heteroaryllithium reagents like other azides to give the corresponding triazenes. Hydrolysis of the latter leads to nitrogen-free carbonyl compounds when aliphatic lithium reagents are used (path A, Eq. 28),[277] but when benzyl, aromatic, and heteroaromatic lithium reagents are used, amines are formed in fair to good yields (path B).[278]

$$\mathbf{37}\ \text{or}\ \mathbf{38} \xrightarrow[2.\ \text{H}_2\text{O}]{1.\ \text{RLi, THF }-78^\circ} \text{R-N=N-NH-CH=CH-Bu-}t \xrightarrow{\text{HCl}} \text{A: } t\text{-BuCH(R)CHO};\ \text{B: RNH}_2\ (45\text{-}70\%) \quad \text{(Eq. 28)}$$

Aryl Azides. The triazenes formed by addition of alkylmagnesium halides to aryl azides lose nitrogen and give *N*-alkylaniline derivatives on workup with aqueous ammonium chloride (Eq. 29).[279] This is unusual in two respects: earlier reports[270,280–283] state that triazenes are isolated under these conditions (see also Eq. 58) and that anilines, rather than *N*-alkylanilines, are formed on treatment with acid at room temperature (see discussion under Alkyl Azides, above).

$$4\text{-FC}_6\text{H}_4\text{N}_3 \xrightarrow[\text{Et}_2\text{O, rt, 1 h}]{c\text{-C}_6\text{H}_{11}\text{MgBr}} [4\text{-FC}_6\text{H}_4\text{N=N-N}^-\text{C}_6\text{H}_{11}\text{-}c\ \text{MgBr}^+ \xrightarrow{-\text{N}_2} 4\text{-FC}_6\text{H}_4\text{N}^-\text{C}_6\text{H}_{11}\text{-}c\ \text{MgBr}^+] \xrightarrow{\text{NH}_4\text{Cl, H}_2\text{O}} 4\text{-FC}_6\text{H}_4\text{NHC}_6\text{H}_{11}\text{-}c\ (85\%) \quad \text{(Eq. 29)}$$

Aromatic Grignard reagents react normally with aryl azides to give triazenes[280,281,284,285] as do vinylmagnesium halides.[286–288] Grignard reagents also add to a variety of aromatic diazides to give the corresponding bis(triazenes).[272,289,290] Phenylmagnesium bromide adds preferentially to an azide group in the presence of a diaryl azo group.[290] Addition of N-protected imidazole anions to phenyl azide gives the corresponding 2-amino derivatives after acid hydrolysis.[66] Addition of phenyl azide to ketene dimethyl acetals and decomposition of the intermediate triazolines gives α-anilino esters in low yields.[291] The formation of diazomalonamide in addition to aniline from the enolate of malonamide and phenyl azide is the earliest example of a diazo transfer reaction.[292] Aryl azides undergo net reduction to arylamines and *N*-formyl arylamines on reaction with the enolate of acetaldehyde.[293]

Acyl Azides. The only additions of a Grignard reagent to acyl azides appear to be those of phenylmagnesium bromide to carbonyl azide (N_3CON_3) and methyl and ethyl azidoformates (N_3CO_2R) to give triazenes in low or unstated yields with retention of the carbonyl group.[284] However, the same Grignard reagent reacts faster with the carbonyl than the azide group in azido acetone.[294] Ethyl and *tert*-butyl azidoformates aminate tetrahydropyrans,[295] ketone silyl enol ethers (see Eq. 98),[296,297] ketene acetals,[298–301] and enamines.[302,303] A camphorsulfone-derived acyl azide has also been used.[304] Either irradiation or thermolysis or a combination of the two is used and the reactions proceed either via the triazoline and aziridine or directly via the latter. Yields vary widely from poor to good.

Sulfonyl Azides. Alkyl- and arylmagnesium halides,[305,306] as well as alkyl-[307], aryl- (see Eq. 70),[308–312] and heteroaryllithium[313] reagents add to sulfonyl azides to give triazene salts which may be reduced to amines [305,310–312] or converted into azides. The latter reaction has been accomplished by an aqueous workup with the highly hindered 2,6-dimesitylphenyl azide,[314] whereas quenching with aqueous potassium hydroxide (see Eq. 72)[305,315] sodium bicarbonate,[313] or sodium pyrophosphate[305,316] (see Eqs. 67 and 74) is necessary with other arenesulfonyl azide adducts. Thermolysis of the dry triazene salts also leads to azides,[307,308] but because of the hazards involved, this procedure is not recommended.

Azidations of certain phosphorus-stabilized anions with 2,4,6-triisopropylbenzenesulfonyl azide ("trisyl azide," **41a**) may be reversible.[317]

The most widely used application of sulfonyl azides is in the azidation of enolates and other stabilized carbanions. The main challenge here is the avoidance of the diazo transfer reaction, which leads to diazo compounds and thus makes a diastereoselective amination impossible. Addition of the enolates to the sulfonyl azide proceeds rapidly at low temperatures ($-78°$ or lower) to give the mesomeric ion **42** (Eq. 30).[318] Reagents **41**, the counter ion M^+, the solvent, and the quenching reagent all influence the subsequent partition between azide and diazo compound. For enolates of esters (**39**) and *N*-acyloxazolidinones (**40**) the preferred reagent is trisyl azide (**41a**); 4-nitrobenzenesulfonyl azide (**41c**) promotes diazo transfer, and tosyl azide (**41b**) usually leads to mixtures of the two types of products. For ester enolates **39**, either lithium or potassium as the counter ion in combination with trisyl azide favors azidation (see Eqs. 118, 120, 122, and 123), whereas for *N*-acyloxazolidinone enolates **40** the potassium enolates are usually employed. Diazidation may occur with ester enolates (but not with *N*-acyloxazolidinone enolates) as a consequence of proton transfer from the initial adduct **42** to the enolate **39** (see Eq. 122); it can be avoided or minimized by use of the lithium enolate or by inverse addition of the enolate to the sulfonyl azide. Quenching agents are added after short reaction times (about one minute). Acetic acid is the reagent of choice for azidation, whereas trifluoroacetic acid promotes diazo transfer.[318] In the triethylamine-promoted reaction of a β-keto ester with trisyl azide, use of THF or acetonitrile as the solvent leads to the azide exclusively, whereas in methylene chloride only diazo transfer and other products are formed.[319] The use of TMSCl as the quenching agent gives considerably higher yields than acetic acid in the azidation of a lactone enolate.[320] The reasons for the above experimental observations do not appear to be clear. In the azidation of cyclic β-keto esters, where trisyl azide also promotes azidation,[319,321] the bulky and less electrophilic trisyl azide may inhibit formation of the triazoline precursor to the diazo compound. However, trisyl azide is the only reagent that promotes diazo transfer to a number of simple ketone enolates, which do not normally react with sulfonyl azides.[322–324] One of the few exceptions is the azidation of a taxane-derived ketone enolate where reaction with tosyl azide followed by quenching with acetic acid gives the diazo compound, whereas quenching with aqueous ammonium chloride leads to the azide.[325] In another example, a lactone lithium enolate reacts with 4-nitrobenzenesulfonyl azide (**41c**) to give exclusively the azide.[326] These examples underscore the fact that exceptions exist to the above-mentioned rules. Other factors that affect yields and azide/diazo compound partitioning in specific reactions are discussed in the relevant sections of Scope and Limitations. A reaction in which the *N*-arenesulfonylamide rather than the azide is obtained on quenching with aqueous ammonium chloride is shown in Eq. 106.[327]

39 $R^2 = OR^3$

40 $R^2 =$ (oxazolidinone, R^4)

$ArSO_2N_3$

41a Ar = 2,4,6-(i-Pr)$_3C_6H_2$
41b Ar = 4-MeC_6H_4
41c Ar = 4-$O_2NC_6H_4$

42

quench

A → azide (N_3) + $ArSO_2H$

B → diazo (N_2) + $ArSO_2NH_2$

(Eq. 30)

Trifluoromethanesulfonyl azide, prepared in situ from trifluoromethanesulfonyl chloride and sodium azide in dimethylformamide, is reported to azidate phosphonoacetic esters and β-dicarbonyl compounds in the presence of triethylamine,[309] whereas the same, but preformed, reagent gives the diazo compounds with α-nitro[328] and α-cyano[329] carbonyl compounds in the presence of pyridine. The reason for this dichotomy is not clear but because the former reaction was carried out under typical diazo transfer conditions the products may have been misidentified.[330]

Sodium Azide/Ammonium Cerium(IV) Nitrate. Silyl enol ethers give α-azido ketones on treament with sodium azide and anhydrous ammonium cerium(IV) nitrate in anhydrous acetonitrile (see Eq. 97).[297,325,331] With a glycal, the 2-azido-1-hydroxy nitrate derivative is formed.[332] Low yields due to hydrolysis of the silyl enol ether may be improved by use of the triisopropylsilyl (TIPS) derivatives,[331] although with a sterically encumbered taxane-derived enol ether the TMS derivative gives higher yields than the TIPS derivative.[325] The mechanism is believed to involve addition of an azide radical to the double bond.

Diphenyl Phosphorazidate. $(PhO)_2P(O)N_3$, reacts with aromatic Grignard and lithium reagents to give aromatic amines after in situ reduction of the initially formed triazene salt.[333,334] Reaction of a lithiated poly(phenylsulfone) with this reagent is not as clean as the corresponding reaction with tosyl azide.[335] Addition of lithium amide enolates to $(PhO)_2P(O)N_3$ at low temperature and trapping the triazene salt with di-*tert*-butyl dicarbonate gives protected α-amino amides in high yields (Eq. 31).[336] When the initial addition is carried out at 0°, the α-diazo amides are formed exclusively.[337] Similarly, reaction of $(PhO)_2P(O)N_3$ with an ester enolate gives exclusively the diazo ester whereas azidation only occurs with trisyl azide.[338]

(PhO)$_2$P(O⁻ Li⁺)–N=N–N... R–CH–C(O)N(Me)Ph ⟷ N⁻ Li⁺ ... $\xrightarrow[-78°]{(t\text{-}BuO_2C)_2O}$ NCO$_2$Bu-t ... $\xrightarrow{H_2O}$ ... ⟶ R–CH(NHCO$_2$Bu-t)–C(O)N(Me)Ph

(Eq. 31)

Miscellaneous Azides. Ethyl (*N*-methanesulfonyl)azidoformimidate [$N_3C(OEt)=NSO_2Me$] has been used to aminate chiral cyclopentanone enamines but the yields are low and the reaction could not be extended to the corresponding cyclohexanone enamines.[303] Trimethylsilyl azide ($TMSN_3$) transfers the TMS rather than the azide group to a lactam enolate.[339]

Miscellaneous Reagents. *Chloramine-T/Osmium Tetroxide.* The Sharpless asymmetric aminohydroxylation system for olefins ($4\text{-MeC}_6\text{H}_4\text{SO}_2\text{N(Na)Cl}/\text{OsO}_4$/cinchona alkaloid derived catalysts)[340,341] converts silyl enol ethers into α-(*p*-tosylamino) ketones in 34–40% yield and 76–92% ee (see Eq. 99).[342]

N-*Chlorocarbamate/Chromium(II) Chloride.* Enol ethers (see Eq. 80) and glycals (see Eq. 84) react with N-chlorocarbamates in the presence of chromous chloride to produce α-amino carbonyl derivatives.[343] Trimethylsilyl enol ethers give low yields because of their ease of hydrolysis. A radical chain mechanism has been proposed with the N-haloamide acting as the transfer agent (Eq. 32).[344]

$$\text{ClNHCOR}^5 + \text{CrCl}_2 \longrightarrow \cdot\text{NHCOR}^5 + \text{CrCl}_3$$

$$\cdot\text{NHCOR}^5 + \text{R}^3\text{R}^4\text{C}{=}\text{C(OR}^1)\text{R}^2 \longrightarrow \text{R}^5\text{CONH}{-}\text{CR}^3\text{R}^4{-}\dot{\text{C}}(\text{OR}^1)\text{R}^2$$

$$\text{R}^5\text{CONH}{-}\text{CR}^3\text{R}^4{-}\dot{\text{C}}(\text{OR}^1)\text{R}^2 + \text{ClNHCOR}^5 \longrightarrow \text{R}^5\text{CONH}{-}\text{CR}^3\text{R}^4{-}\text{C(OR}^1)(\text{Cl})\text{R}^2 + \cdot\text{NHCOR}^5 \qquad \text{(Eq. 32)}$$

$$\text{R}^5\text{CONH}{-}\text{CR}^3\text{R}^4{-}\text{C(OR}^1)(\text{Cl})\text{R}^2 \xrightarrow{\text{H}_3\text{O}^+} \text{R}^5\text{CONH}{-}\text{CR}^3\text{R}^4{-}\text{C(O)R}^2$$

*Bis[*N*-(p-Toluenesulfonyl)]selenodiimide.* The reagent obtained from the reaction of chloramine-T with selenium metal, proposed to have structure TsN=Se=NTs, reacts with TIPS enol ethers in an ene-like reaction to give the corresponding α-tosylamino enol ethers (see Eq. 100).[345–349]

Nitridomanganese Complexes. Stoichiometric amounts of chiral complexes of type **43** react with silyl enol ethers in the presence of trifluoroacetic or *p*-toluenesulfonic anhydride to give α-(*N*-trifluroacetyl)amino- and α-(*N*-*p*-tosylamino) ketones, respectively (see Eq. 160).[350–353] With glycals, the 1-hydroxy-2-(*N*-trifluoroacetyl)amino derivatives are formed (see Eq. 83).[354] A mechanism involving approach of the enol ether from the least hindered side of the **43**•TFA complex has been proposed.[353]

43

SCOPE AND LIMITATIONS

Amination of Aliphatic Carbanions

Preparation of Alkyl Amines. The main application of the electrophilic amination of aliphatic carbanions is in the preparation of hindered amines. These are not usually accessible by nucleophilic displacement involving an alkyl halide and ammonia or an amine and have been prepared by alternate methods such as the Curtius rearrangement or the Ritter reaction. Examples are shown in Eqs. 10, 12, 33,[52] 34,[355] 35,[133] 36,[112] 37,[182] and 38.[356]

t-BuCuMeLi + NH_2 (1-naphthylamine) → 1. Et_2O, –20°, 2 h; 2. O_2, –20° → NHBu-t (35%) (Eq. 33)

Ph_3C–Li → $ClNH_2$, Et_2O, sonication → Ph_3C–NH_2 (67%) (Eq. 34)

Li → 2,4,6-$Me_3C_6H_2SO_2ONMe_2$, Et_2O; –10° to –15°; to rt, 15 h → NMe_2 (54%) (Eq. 35)

$(t\text{-Bu})_2Zn$ + Bn_2NOBz → THF, $(CuOTf)_2$•C_6H_6 (1 mol%); 15-60 min → t-$BuNBn_2$ (98%) (Eq. 36)

MgBr + N-OSO_2Ph (O, O) → 1. Et_2O, CH_2Cl_2, 0°, 30 min; 2. HCl, EtOH, H_2O, reflux, 10 h → NH_2 (89%) (Eq. 37)

$(\ )_2Zn$ + EtO_2C–C_6H_4–N=NTs → 1. THF, –20°, 30 min; 2. RaNi, EtOH, reflux, 90 min → N–H–C_6H_4–CO_2Et (50%) (Eq. 38)

Preparation of *N*-alkylanilines from aliphatic Grignard reagents and aryl azides was discussed previously (Eq. 29). The net insertion of a methylene group between the alkyl or aryl group of an organolithium reagent and the nitrogen as part of an amination was also mentioned earlier (Eq. 6).

Both lithium and Grignard reagents are aminated with retention of configuration (Eqs. 39[274] and 40[220]). On the other hand, preparation of an organozinc reagent from a chiral, non-racemic bromide with highly reactive zinc, subsequent amination with an azo ester, and reduction of the adduct gives the racemic amine; racemization is believed to have occurred during preparation of the zinc reagent.[357]

H, Br, H, H → 1. t-BuLi (2 eq), pentane, –78°, 30 min; to rt; 2. $PhSCH_2N_3$, THF, pentane, –78°; to rt, 1.5 h; 3. NH_4Cl, H_2O; 4. KOH, DMSO, rt, 1 h → H, NH_2, H, H (45%) (Eq. 39)

Ph–CH₂–CH(Cl)–S(O)–C₆H₄-4-Cl $\xrightarrow[\text{THF, –78° to –30°}]{\text{EtMgCl (5 eq)}}$ Ph–CH₂–CH(MgCl)–CH₂CH₃ $\xrightarrow{\text{A or B}}$ Ph–CH₂–CH(NHAc)–CH₂CH₃ (Eq. 40)

A: 1. $[3,5\text{-}(CF_3)_2C_6H_3]_2C{=}NOTs$, toluene, Et_2O, –70°, 10 d (25%) 90% ee
2. Ac_2O, Et_3N

B: 1. $PhSCH_2N_3$, THF, –78°, 1 h (82%) 92-95% ee
2. Ac_2O, –60° to –30°
3. KOH, DMSO, 0° to rt, 3 h

Zirconium complexes, generated in situ by addition of $HZrCp_2Cl$ to alkenes, can be aminated with *O*-(mesitylenesulfonyl)hydroxylamine; an example is shown in Eq. 41.[116] When the initial hydrozirconation is not regioselective, as with styrene, mixtures of amines are formed. A reaction that permits amination at the tertiary carbon in a similar substrate is discussed below (Eq. 49).

$\xrightarrow[\text{2. 2,4,6-}Me_3C_6H_2SO_2ONH_2\text{, 0°, 10 min}]{\text{1. }HZrCp_2Cl\text{ (inverse addition), THF, rt}}$ NH₂ (88%) (Eq. 41)

Chiral ligands of type **44** may be prepared from chiral amines via amidocuprates (Eq. 42).[54]

n-BuCu(CN)Li + 1-(1-naphthyl)ethyl-NHLi $\xrightarrow[\text{2. }O_2\text{, –78°, 20 min; to rt}]{\text{1. THF, –40°, 15 min}}$ 1-(1-naphthyl)ethyl-NHBu-*n* **44** (60%) (Eq. 42)

Preparation of Alkyl Hydrazines. As mentioned previously (Eq. 19), additions of aliphatic carbanions to unactivated azo compounds are rare. Another example is shown in Eq. 43.[208] On the other hand, additions to diaryl azo compounds (Eq. 44)[211] and esters of azodicarboxylic acids (Eq. 46)[358] are well documented. The intermediate anion in Eq. 44 can be trapped with alkyl halides to give tetrasubstituted hydrazines. An extension of the reaction of Eq. 44 exploits the ready displacement of the benzotriazole functionality by Grignard reagents (Eq. 45).[359] Because of the instability of the intermediate **45**, the Grignard reagent is added before the azobenzene in the actual experiments.

c-C₆H₁₁MgBr + pentamethylenediazirine $\xrightarrow{Et_2O\text{, 0° to rt, 1 h}}$ HN–N(c-C₆H₁₁) diaziridine (86%) (Eq. 43)

t-BuLi + Ph–N=N–Ph $\xrightarrow[\text{rt, 10 h}]{\text{hexane, THF, –78°, 2 h}}$ Ph–N(*t*-Bu)–NH–Ph (47%) (Eq. 44)

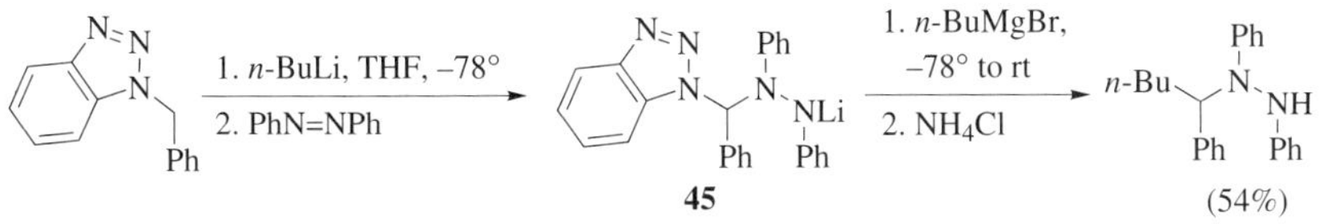

(Eq. 45)

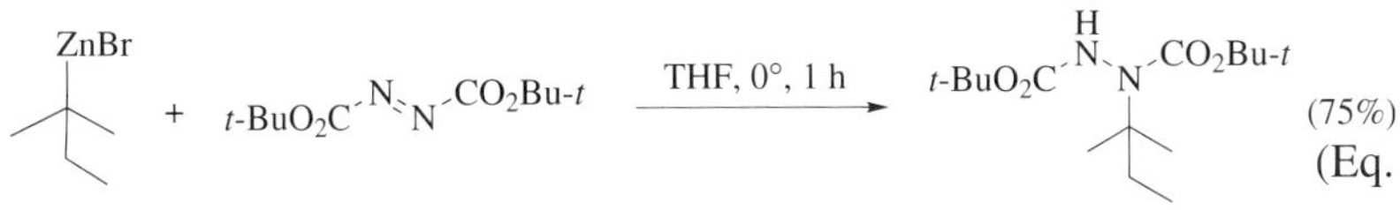

(Eq. 46)

Hydrazines may also be obtained via amidocuprates (Eq. 47)[54] but the yields are low. Addition of Grignard reagents to diazonium salts provides azo compounds, which may be reduced to hydrazines. Yields in the former reaction are often low and the requirement to use dry diazonium salts adds a potential hazard. The best yields are obtained with *o*-benzenedisulfonimide salts (Eq. 48).[191]

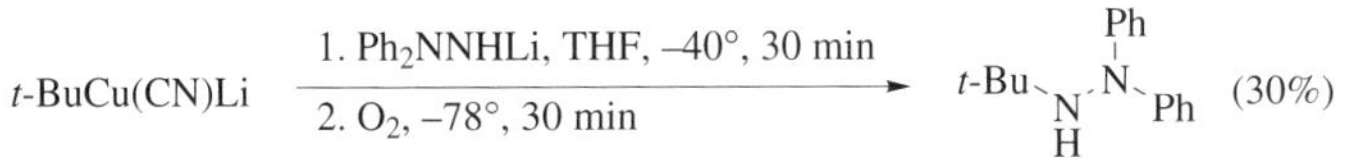

(Eq. 47)

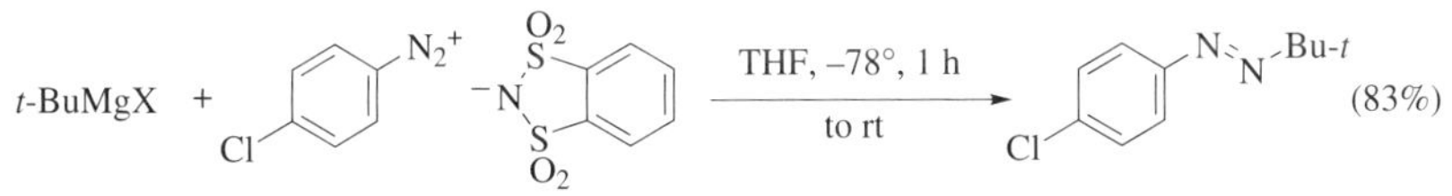

(Eq. 48)

A wide variety of *N*-alkyl hydrazinedicarboxylic esters may be obtained in excellent yields by the hydrohydrazination reaction depicted in Eq. 49.[215] Use of cobalt complexes results in more highly regioselective reactions at the cost of lower reaction rates as compared to additions where manganese complexes are employed. Di(*tert*-butyl) azodicarboxylate is the preferred azo ester; reduction of the N=N double bond becomes more prominent when less hindered azo esters are used. Alcoholic solvents are essential; the reaction fails when methylene chloride or THF is used.

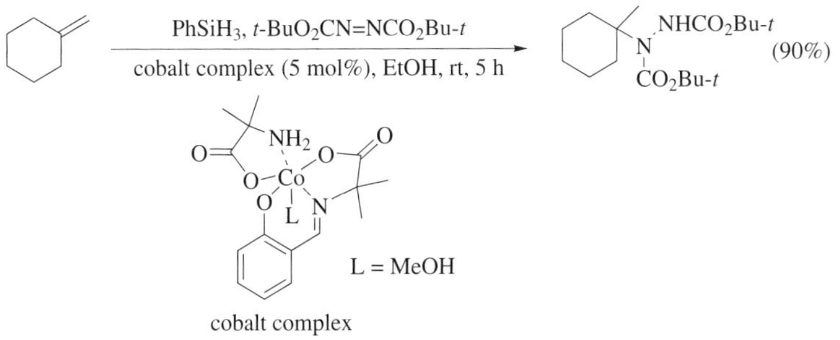

(Eq. 49)

Preparation of Alkyl Azides. A hydroazidation reaction similar to the reaction of Eq. 49 permits preparation of alkyl azides (Eq. 50).[215]

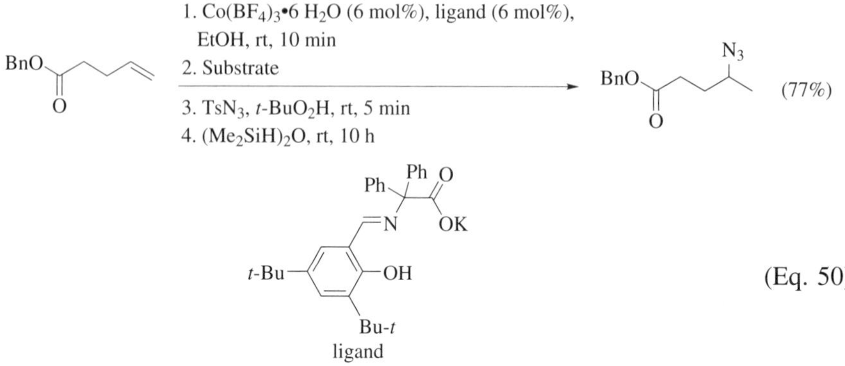

1. $Co(BF_4)_3$•6 H_2O (6 mol%), ligand (6 mol%), EtOH, rt, 10 min
2. Substrate
3. TsN_3, t-BuO_2H, rt, 5 min
4. $(Me_2SiH)_2O$, rt, 10 h

(77%)

ligand

(Eq. 50)

Amination of Allylic and Propargylic Carbanions

The literature in this area is fairly sparse, presumably because of the ease of preparation of allyl- and propargylamines by nucleophilic amination. The reaction of allylindium species with aryl azides to give *N*-allylarylamines was mentioned earlier (Eq. 25). It has also been applied to the indium species derived from methyl 2-(bromomethyl)acrylate.[269] The amination of alkylzirconium species mentioned above (Eq. 41) can also be applied to allenes (Eq. 51).[116]

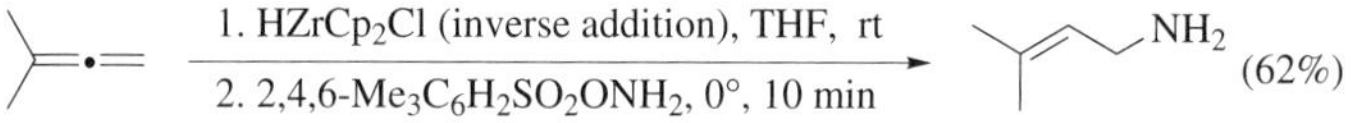

1. $HZrCp_2Cl$ (inverse addition), THF, rt
2. 2,4,6-$Me_3C_6H_2SO_2ONH_2$, 0°, 10 min

(62%) (Eq. 51)

Application of the hydrohydrazination mentioned above (Eq. 49) to dienes and enynes gives *N*-allyl- and *N*-propargyl- (Eq. 52)[216] hydrazinedicarboxylic esters in generally good yields. Serious competition from the Diels-Alder reaction is a problem only with very reactive dienes such as cyclopentadiene.

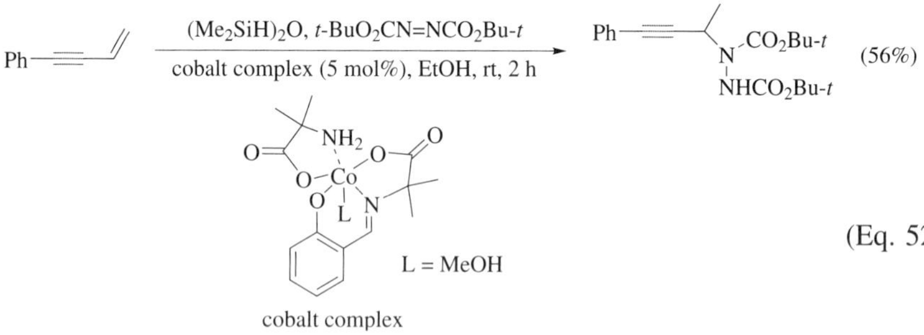

$(Me_2SiH)_2O$, t-$BuO_2CN{=}NCO_2Bu$-t

cobalt complex (5 mol%), EtOH, rt, 2 h

(56%)

L = MeOH

cobalt complex

(Eq. 52)

Amination of Arylmethyl and Heteroarylmethyl Carbanions

Arylmethyl carbanions such as benzyl carbanions in general undergo most of the amination reactions discussed for aliphatic carbanions. The difference is that they may often be generated directly by metalation of the arylmethyl compounds as shown in Eq. 53.[136] Heteroarylmethyl carbanions frequently are also accessible by direct metalation but they have been used in electrophilic aminations much less frequently, although the method shown in Eq. 54[212] should be applicable to other aminating reagents.

1. n-BuLi (2.1 eq), THF, hexane, 0°
2. 2,4,6-$Me_3C_6H_2SO_2ONEt_2$, –78°, to rt; rt, overnight

NEt_2 (43%) (Eq. 53)

1. LDA (2 eq), THF, hexane, 1 h
2. PhN=NPh, –78°, 10 min

(77%) (Eq. 54)

Catalytic hydrohydrazination of vinylarenes and vinylheteroarenes proceeds regioselectively and with often excellent yields (Eq. 55).[215]

1. manganese complex (1 mol%), i-PrOH, rt to 0°
2. $PhSiH_3$, t-$BuO_2CN{=}NCO_2Bu$-t, 0°
3. substrate, 0°, 5 h

$NHCO_2Bu$-t, CO_2Bu-t (88%) (Eq. 55)

manganese complex

Amination of Vinyl and Allenyl Carbanions

Amination of vinyl carbanions gives enamines (Eqs. 56[133] and 57[55]) or their derivatives (Eq. 58).[286] Only arylamines are isolated when products of type **46** are hydrolyzed with acid.

2,4,6-$Me_3C_6H_2SO_2ONEt_2$, Et_2O or Et_2O/THF; –10° to –20°; to rt, 14 h

NEt_2 (28%) (Eq. 56)

Ph–CH=CH–Cu(CN)ZnCl

1. $(i\text{-}Pr)_2NLi$, THF, –78° to –40°, 40 min
2. 1,2-$(O_2N)_2C_6H_4$, THF, –78°
3. O_2, –78°, 30 min

Ph–CH=CH–$N(Pr\text{-}i)_2$ (60%) (Eq. 57)

$Ph_2C{=}CHMgBr$ + $N_3C_6H_4OMe$ —THF, rt, 2 h→ (Eq. 58)

46 (55%)

In situ generated allenyltitanium complexes of type **47** are aminated by azodicarboxylic esters and the products may be degraded to α-hydrazino acids (Eq. 59).[360] High α-symmetric induction is achieved only when R is a methyl group; when it is *n*-butyl or isobutyl, the enantiomeric excess in the product decreases to 55% and 27%, respectively.

$$\text{TMS-C}\equiv\text{C-CH(R)OP(O)(OEt)}_2 + \text{Ti(OPr-}i)_4 \xrightarrow[-50^\circ,\ 2\ h]{i\text{-PrMgBr (3 eq), Et}_2\text{O}} \mathbf{47}$$

R = Me (94% ee)

$$\xrightarrow[-78^\circ;\ \text{to } 0^\circ,\ 1\ h]{t\text{-BuO}_2\text{CN=NCO}_2\text{Bu-}t} \text{TMS-C}\equiv\text{C-CH(R)N(CO}_2\text{Bu-}t\text{)NHCO}_2\text{Bu-}t\ (77\%)\ 81\%\ \text{ee} \xrightarrow[\text{NaIO}_4]{\text{RuCl}_3} \text{HO}_2\text{C-CH(R)N(CO}_2\text{Bu-}t\text{)NHCO}_2\text{Bu-}t\ (80\text{-}83\%)$$

(Eq. 59)

Amination of Ethynyl Carbanions

Amination of alkynylcuprates gives ynamines (Eq. 60);[135] the yields are based on two of the three ethynyl groups reacting. Yields are very low with organolithium and Grignard reagents.[135] Amination of lithium bis(phenylethynyl)cuprate with $Ph_2P(O)ONH_2$ gives phenylacetonitrile by rearrangement of the initially formed primary ynamine.[139] Imines of primary ynamines, however, can be isolated (Eq. 61).[178] Phenylethynylsodium and tosyl azide react to give the triazoline by cycloaddition rather than the ethynyl azide.[361]

$$(\text{RC}\equiv\text{C})_3\text{CuLi}_2 \xrightarrow{\text{Me}_2\text{NX, Et}_2\text{O}} \text{RC}\equiv\text{CNMe}_2$$

R	X	
t-Bu	$Ph_2P(O)O$	(71%)
Ph	MsO	(52%)

(Eq. 60)

$$(\text{PhC}\equiv\text{C})_3\text{CuLi}_2 + \text{Ph(Me)C=NOSO}_2\text{Ph} \xrightarrow{\text{Et}_2\text{O, rt, 20 h}} \text{Ph(Me)C=N-C}\equiv\text{C-Ph}\quad (39\%)$$

(Eq. 61)

Amination of Aryl Carbanions

Preparation of Arylamines. Many methods to prepare arylamines by electrophilic amination are available. Some have been mentioned previously (Eqs. 13, 15, 22, 24, 25, 26, 28, and 29) and some of the methods described for the preparation of alkylamines (Eqs. 33, 35–37) can also be used to synthesize arylamines. Additional methods are shown in Eqs. 62,[73] 63,[82] 64,[101] 65[264] 66,[333,334] and 67.[305] The recently developed direct catalytic amination of aryl halides and aryl sulfonates,[362–373] and arylboronic acids,[374] however, has the advantage over these methods of requiring one or more fewer steps. The approach that merits consideration will need to be decided based on each individual objective.

(Eq. 62): 3-NC-C6H4-MgCl•LiCl + Cl-N(Me)-CH2-C6H4-Br-p, THF, –45°, 15 min → 3-NC-C6H4-N(Me)-CH2-C6H4-Br-p (70%)

(Eq. 63): 2-MeC6H4Li + MeONHLi (2 eq); 1. Et_2O, hexane, –78° to –15°, 2 h; 2. BzCl → 2-MeC6H4NHBz (96%)

(Eq. 64): (4-F-C6H4)2Cu(CN)Li_2 + t-BuNHOTMS; THF, –50° to rt, 2 h → 4-F-C6H4-NHBu-t (45%)

(Eq. 65): 2,6-Me2C6H3MgBr + $TMSCH_2N_3$; 1. Et_2O, rt, 3 h; 2. H_2O → 2,6-Me2C6H3NH2 (79%)

(Eq. 66): 2,4,6-Me3C6H2MgBr + $(PhO)_2P(O)N_3$; 1. Et_2O, –73° to –69°, 2 h; 2. $NaAlH_2(OCH_2CH_2OMe)_2$, toluene, –70° to 0°, 1 h → 2,4,6-Me3C6H2NH2 (67%)

(Eq. 67): 2-BnC6H4MgBr; 1. TsN_3, THF, 0°; 2. $Na_4P_2O_7$, H_2O → 2-BnC6H4N3 (49%); or 2. RaNi, NaOH; 3. HCl → 2-BnC6H4NH_3^+ Cl^- (71%)

The situation is more favorable when the aryl carbanion can be prepared directly from the arene by ortho lithiation. Examples are shown in Eqs. 68,[53] 69 (the copper reagent gives higher yields than the lithium reagent),[128] and 70.[311] Phenylthiomethyl azide (**32**) does not react with aryllithium reagents but this failure can be remedied by converting them into magnesium reagents (Eq. 71).[274,275,375] Trimethylsilylmethyl azide (Eq. 65) aminates aryllithium reagents but the yields are lower than for Grignard reagents. On the other hand, the reactions of diphenyl phosphorazidate, illustrated in Eq. 66, work equally well with organolithium reagents.

(Eq. 68): 2-MeO-C6H4-CONEt2; 1. s-BuLi, TMEDA, THF, –78°, 50 min; 2. PhNHCu(CN)Li, –78°, 2 h; 3. O_2, –78°, 30 min → 2-OMe-6-NHPh-C6H3-CONEt2 (63%)

(Eq. 69): spiro[1,3-benzodioxole-2,1'-cyclohexane]; 1. n-BuLi, THF, Et_2O, 0°, 2 h; rt, 22 h; 2. CuI, 0°, 15 min; 3. TsON(Li)CO_2Bu-t, –78°, 30 min; 0°, 2 h → 4-($NHCO_2$Bu-t) derivative (45%)

1. s-BuLi, TMEDA, THF, –78°, 1 h
2. TsN_3
3. $NaBH_4$, n-$Bu_4N^+ HSO_4^-$

(50%)

(Eq. 70)

1. s-BuLi, THF, hexane
2. $MgBr_2$
3. $PhSCH_2N_3$, –78° to 0°; 0°, 1 h
4. NH_4Cl, H_2O
5. 50% KOH in H_2O, MeOH, THF, rt, 16 h

(71%)

(Eq. 71)

The direct amination of arenes with chloramines in the presence of redox catalysts is another alternative that usually proceeds with good yields.[376]

Preparation of Aryl Hydrazines. All methods mentioned above for the hydrazination of alkyl carbanions may also be applied to aryl carbanions. Addition of phenyllithium to a cyclic azo compound followed by in situ arylation to give a tetrasubstituted hydrazine was mentioned earlier (Eq. 19). An alternate hydrazination method, not involving aryl anions, is the reaction of electron-rich arenes with azodicarboxylic esters and aroylazocarboxylic esters under the influence of various catalysts.[230,377–384]

Preparation of Aryl Azides. Aryl azides may be prepared by reaction of aryl carbanions with tosyl azide followed by treatment of the triazene salt with sodium pyrophosphate (Eq. 67)[305] or aqueous base (Eq. 72).[315]

1. n-BuLi (5.4 eq), Et_2O, rt, 5 h
2. TsN_3, rt, overnight
3. 10% KOH in H_2O

(28%) + (6%)

(Eq. 72)

Amination of Heterocyclic Carbanions

Aminations in this area involve anions of both π-excessive and π-deficient heterocycles, which are generated from the halo compounds or by direct metalation. Most of the aminating reagents seem to be applicable except that phenylthiomethyl azide (**32**) fails with the 2-lithium or 2-copper derivatives of furan, thiophene, N-methylpyrrole, and N-methylindole.[274] Similarly, chloramine and O-methylhydroxylamine, but not phenyl azide, fail to aminate 2-lithio-1-methylimidazole[66] and the MeN(Li)OMe nitrenoid does not react with 2-lithiothiophene.[97] The reactions that appear to be most widely applicable to heterocyclic carbanions are shown in Eqs. 73,[100,101] 74,[316] and 75.[358]

$$R^1Li \xrightarrow[\text{2. } R^2NHOTMS,\ -50° \text{ to rt, 2 h}]{\text{1. CuCN, THF, } -40°, \text{ 20 min}} R^1NHR^2$$

(Eq. 73)

R^1	R^2	
2-thienyl	*i*-Pr	(65%)
3-pyridyl	TMS	(58%)
2-benzo[*b*]thienyl	TMS	(58%)

R^2 = H after hydrolytic workup (3-pyridyl, 2-benzo[*b*]thienyl)

Li; O; S; O

1. TsN_3, Et_2O, –78°
2. $Na_4P_2O_7$, H_2O, rt, overnight

N_3; O; S; O (65%) (Eq. 74)

S; ZnBr

t-$BuO_2CN{=}NCO_2Bu$-*t*
THF, 0°, 1 h

S; N; $NHCO_2Bu$-*t*; CO_2Bu-*t* (80%) (Eq. 75)

Amination of Aldehyde Enolates, Enol Ethers, and Enamines

There appear to be no reports of aminations of aldehyde enolates in the literature, presumably because of their instability and their tendency to undergo aldol self-condensations. Since electrophilic *hydroxylations* of sterically hindered aldehyde enolates have been reported,[152] these should also be amenable to electrophilic amination. α-Amino aldehydes or their derivatives, however, can be generated by the use of aldehyde enol ethers or enamines, either as substrates, or as in situ generated intermediates. An example of the latter is shown in Eq. 76[223] where the aldehyde product is isolated.[222,226,229,385] The mechanism of this reaction was discussed earlier (Eq. 21). D-Proline gives the enantiomeric product.[224,227] Derivatives of proline[385,386] and L-azetidinecarboxylic acid[222,223,229] are also used as catalysts. In other applications of this method the products are reduced in situ to the α-amino alcohols[221,223,227] or their cyclization products.[222–225,386] An example of the latter reaction sequence involves a diastereoselctive Michael addition to an α,β-unsaturated aldehyde to generate the precursor aldehyde enolate (Eq. 77).[225] L-Proline in this reaction gives lower diastereo- and enantioselectivities. Reaction of α-branched aldehydes with chloramine-T in the presence of L-proline gives the racemic α-tosylamino aldehydes in high yield (Eq. 78).[78] A similar reaction with sulfonyl azides also produces α-tosylamino aldehydes, but with modest yields and enantioselectivities (Eq. 79).[386a]

CHO

1. L-proline (0.5 eq), CH_2Cl_2, 0°; rt, 1 h
2. $EtO_2CN{=}NCO_2Et$, rt, 2.5 d
3. H_2O

CHO; N–$NHCO_2Et$; CO_2Et

(54%) 86% ee

(Eq. 76)

(1.5 eq) + BnSH → [Ar = 3,5-$(CF_3)_2C_6H_3$ catalyst (0.1 eq), BzOH, toluene, –15°, 16 h] → $EtO_2CN{=}NCO_2Et$, –15°, 16 h → 1. $NaBH_4$, MeOH, 0°, 10 min; 2. NaOH

(63%) 90% de, >99% ee

(Eq. 77)

$TsN(Cl)Na \cdot x\ H_2O$, L-proline (2 mol%), MeCN, rt, 2 d → NHTs (86%) (Eq. 78)

4-$O_2NC_6H_4SO_2N_3$, L-proline (1 eq), EtOH, rt, 1 d → $NHSO_2C_6H_4NO_2$-4 (49%) 69% ee (Eq. 79)

Examples where enol ethers of aldehydes are used as starting materials are shown in Eqs. 80,[343] 81,[387] and 82.[241] Glycals may also serve as substrates (Eqs. 83[354] and 84[343]).

OEt + $ClNHCO_2Bn$ → 1. $CHCl_3$, MeOH, –78°; 2. $CrCl_2$, –78°; 3. NaOMe, –78° to rt → BnO_2CHN (OMe, OEt) (81%) (Eq. 80)

OBu-*n* + O_2N–C_6H_4–N_3 → $CHCl_3$, 40°, 70 h → (96%) → AcOH, PhH, 50°, 10 min → *n*-BuO (OAc) (88%) (Eq. 81)

1. $MeO_2CN{=}NCO_2Me$; 2. HCl, MeOH → MeO_2CNH (MeO_2C, OMe) (85%) (Eq. 82)

BnO, BnO + (salen)Mn≡N (2 eq) — 1. 2,6-(t-Bu)$_2$-4-Me-pyridine, CH_2Cl_2; 2. $(CF_3CO)_2O$, –78° to rt, 5-6 h → BnO, BnO, $NHCOCF_3$, OH, 2 (80%) C2 de 86%

(Eq. 83)

AcO, OAc, AcO — 1. $ClNHCO_2CH_2CH_2Cl$, $CHCl_3$, MeOH, –78°; 2. $CrCl_2$; 3. MeOH, $AgNO_3$ → AcO, OMe, OAc, AcO, $NHCO_2CH_2CH_2Cl$ (55%) (Eq. 84)

Enamines may serve as precursors as well (Eqs. 85[195] and 86[74]). The latter reaction is of interest for the formation of rearrangement product **48**, which apparently has not been followed up as a means of preparing α-amino aldehydes. A mechanism involving an aziridinium intermediate has been proposed.[74]

Ph, NEt_2 + HO_2C, N_2^+ Cl^- — NaOAc, H_2O; pH 5-6 → Ph, CHO, N, N, H, CO_2H (Eq. 85)

(89%)

+ Me_2NCl — dioxane, Et_2O, 0°, 2 h; rt, overnight; reflux, 5 h → CHO, NMe_2 (53%) + CHO, N **48** (24%) (Eq. 86)

Amination of Ketone Enolates, Enol Ethers, and Enamines

With ketone enolates, issues of site selectivity arise. Generation of enolates under conditions of kinetic control results in preferential amination at the less substituted α-carbon (product **49**, Eq. 87;[388] Eq. 88[217]) unless one of the α-positions is benzylic (Eq. 89).[134] Trialkylsilyl groups may also be used to direct aminations (Eq. 90).[156] On the other hand, in reactions involving ketone enamine intermediates under thermodynamic control, amination at the more highly substituted α-carbon predominates, but as the bulk at that position increases, reaction times increase and selectivity decreases (products **51** and **52**, Eq. 91).[228] A potential solution to this problem that apparently has not been explored extensively is to selectively generate silyl enol ethers and treat them with one of the reagents that are known to aminate these derivatives. The lone example of this approach is shown in Eq. 92.[173]

1. $Ph(n\text{-}Bu)NMnMe \cdot 4\ LiBr$, THF, rt, 1 h
2. $R^4O_2CN{=}NCO_2R^4$, –30°; rt, 2.5 h

49 **50**

$E = N(CO_2R^4)NHCO_2R^4$

R^1	R^2	R^3	R^4	**49 + 50**	**49**:**50**	**49** dr
H	H	Me	Et	(50%)	1:1	—
$n\text{-}C_5H_{11}$	Me	Me	*t*-Bu	(60%)	98:2	—
Et	Et	Bn	Et	(93%)	98:2	3:1

(Eq. 87)

1. LDA, THF, –78°
2. $BnO_2CN{=}NCO_2Bn$, –78°

(74%)

(Eq. 88)

1. unspecified Li base, Et_2O or THF
2. Me_2NOMs, –30° to 0°

(52%) (Eq. 89)

1. LDA, THF, 0°
2. $4\text{-}O_2NC_6H_4$-oxaziridine-$NCO_2Bu\text{-}t$, –100° to rt

(29%) 88% de

(Eq. 90)

$EtO_2CN{=}NCO_2Et$, L-proline (0.1 eq)
MeCN, rt, time

51 **52**

$E = N(CO_2Et)NHCO_2Et$

R	Time	**51 + 52**	**51**:**52**	**51** ee
Me	10 h	(80%)	10:1	95%
Et	20 h	(77%)	4:1	98%
i-Pr	95 h	(69%)	3:1	99%

(Eq. 91)

TsN=IPh (0.67 eq),
$CuClO_4$ (8 mol%)
MeCN, 0°, 1.5 h

(53%)

(Eq. 92)

Ketone silyl enol ethers react with derivatives of diacyl azo compounds at room temperature[245] or on heating[242,243] (see also Eq. 82) as well as enantioselectively under the influence of silver trifluoromethanesulfonate and BINAP (Eq. 93)[244] or copper bis(oxazoline) complexes (Eq. 94). The latter is proposed to proceed via a formal hetero Diels-Alder adduct.[252] Ketones themselves react with azodicarboxylic esters either thermally[246,389,390] or in the presence of potassium carbonate[390] but yields are low. Higher yields can be achieved with LDA,[391–394] (see also Eq. 88), LiHMDS,[395,396] or KOBu-*t*[325] as the bases. Aryl diazonium salts also aminate lithium enolates (Eq. 95) but yields can be low.[185] Better yields could potentially be achieved with arenediazonium *o*-benzenedisulfonimides (**26d**), which are very efficient in the amination of Grignard reagents (see Eq. 48).[191]

OTMS

1. $BnO_2CN{=}NCO_2Bn$, AgOTf, (*R*)-BINAP (12 mol%), THF, 2,4,6-$Me_3C_6H_3$, –45°, 1 h
2. HF, THF

O; BnO_2C, N, H, N, CO_2Bn

(95%) 86% ee (Eq. 93)

TMSO

$Cl_3CCH_2O_2C$–N=N–C(O)–N (oxazolidinone)

copper complex (10 mol%), CF_3CH_2OH (1 eq), THF, –78°, overnight

O, N, Cu, $(OTf)_2$, *t*-Bu, *t*-Bu

copper complex

O, N, O, CuL_2, N, O, TMSO, $N\text{–}CO_2CH_2CCl_3$ $]^{2+}$

(Eq. 94)

CF_3CH_2OH

$CO_2CH_2CCl_3$, O, N–NH, O, N, O, O

(94%) 99% ee

OLi + $PhN_2^+\ BF_4^-$ —THF, –78°→ O, N=N, Ph (72%) (Eq. 95)

Hypervalent iodine reagents aminate ketone enol ethers.[172–174] Yields are often high but enantioselectivities in catalyzed reactions are generally considerably lower than the 52% achieved in Eq. 17.[174] Other reagents that aminate ketone enol ethers include *N*-arenesulfonyloxy carbamates[119,122,397] (Eq. 96),[122] the sodium azide/ammonium cerium(IV) nitrate reagent[297,331] (Eq. 97),[297] ethyl azidoformate,[296,397] (Eq. 98),[296] the *N*-chlorocarbamate/chromium(II) chloride reagent (Eq. 32),[343] the chloramine-T/osmium tetroxide system (Eq. 99),[342] and bis[*N*-(*p*-toluenesulfonyl)]selenodiimide (Eq. 100).[345] Nitridomanganese complexes (cf. Eq. 83) can also be applied to the amination of silyl enol ethers.[352,353,398]

OTMS + 4-$O_2NC_6H_4SO_2ONHCO_2Et$ —CH_2Cl_2, CaO, rt, 3.5 h→ O, $NHCO_2Et$ (Eq. 96)

(67%)

NaN_3, $Ce(NH_4)_2(NO_3)_6$ / MeCN, –15°, 2 h; to rt — OTBS → N_3, H, O (70%) (Eq. 97)

OTMS + EtO_2CN_3 — 1. 100°, 15 h; 2. SiO_2 → $NHCO_2Et$ (40%) (Eq. 98)

OTMS, t-Bu + TsN(Cl)Na — $(DHQD)_2CLB^a$ (0.008 eq), OsO_4 (0.004 eq), t-BuOH/H_2O (1:1), rt, 15 min → NHTs, t-Bu (34%) 76% ee (Eq. 99)

[a] see List of Abbreviations

OSi(Pr-i)$_3$ + "TsN=Se=NTs" — CH_2Cl_2, 0° → TsNH, OSi(Pr-i)$_3$ (62%) (Eq. 100)

Enamines derived from ketones undergo some of the same reactions described for enol ethers, for example with arenesulfonyloxy carbamates as in Eq. 96[120,121,399] and with ethyl azidoformate as in Eq. 98.[302,303] The reaction with activated azo compounds occurs readily at room temperature or below and diamination often cannot be avoided with the more electrophilic reagents (Eq. 101).[400,401] The proline-catalyzed reaction of ketones with azodicarboxylic esters, which proceeds by way of the enamines, has been mentioned above (Eq. 91).

1. $PhCON{=}NCO_2Me$ (1 eq), Et_2O, –30°, 3 h; 2. HCl, Me_2CO, 5°, 48 h → E (20%) + E, E (26%)

E = $C(CO_2Me)NHCOPh$

1. PhN=NCOPh, Et_2O, 0°, 15 min; 2. HCl, Me_2CO, 0°, few min → O, Ph, N, NHCOPh (55%)

(Eq. 101)

In the Morita-Baylis-Hillman reaction, enolate intermediates are formed by addition of a nucleophilic catalyst to an α,β-unsaturated carbonyl compound. These intermediates can be trapped with a variety of electrophiles,[402] including azodicarboxylic esters (Eq. 102).[403] The reaction fails with ethyl acrylate.

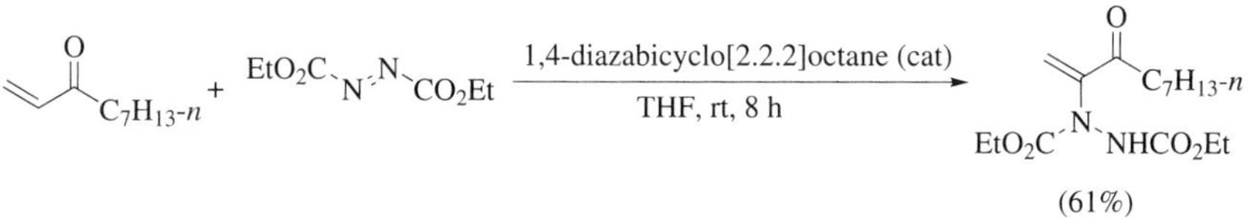

(Eq. 102)

α-Keto esters can be aminated enantioselectively with azodicarboxylic esters under the influence of copper bis(oxazoline) catalysts (Eq. 103);[404] the initial products were not isolated but were reduced and cyclized to give derivatives of syn-β-amino-α-hydroxy esters.

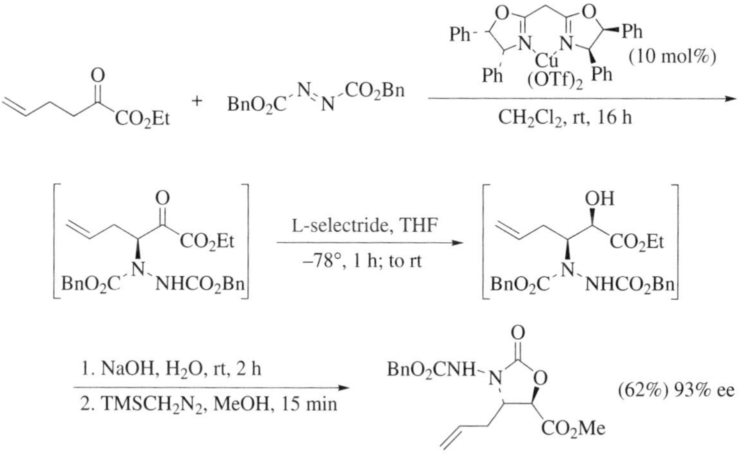

(Eq. 103)

Enamines derived from cyclohexane-1,2-dione react readily with azodicarboxylic esters but the enamine products are very resistant to hydrolysis.[249]

Amination of Imine and Hydrazone Anions

Imines have the advantage over ketones of permitting the introduction of a chiral auxiliary on the imine nitrogen, which is then removed when the imine is hydrolyzed to the ketone. An example involving a manganese enamine is shown in Eq. 104.[388] Amination occurs selectively at the less substituted α-carbon, as shown by the distribution of products **53** and **54**; the configuration at the newly created stereogenic center was not reported. Reaction of imines with azodicarboxylic esters proceeds slowly at room temperature (Eq. 105a), and yields and diastereoselectivities are comparable to those achieved via the aza enolate (Eq. 105b).[405]

t-BuS(O)(*)N=C(CH$_2$R^1)CH(R^2)R^3 → Ph(Me)NMnMe•4 LiBr, THF, rt, 1 h → [t-BuS(O)(*)N(MnN(Me)Ph)C(=CHR^1)CH(R^2)R^3] → 1. R^{4}O$_2$CN=NCO$_2$R^4, –30°; rt, 2.5 h; 2. HCl → **53** + **54** (Eq. 104)

E = N(CO$_2$R^4)NHCO$_2$R^4

R^1	R^2	R^3	*	R^4	**53 + 54**	**53** ee	**53**:**54**
Me	Me	Me	R	Et	(50%)	40%	90:10
n-C$_5$H$_{11}$	Me	Me	R	t-Bu	(65%)	68%	98:2
Et	Et	Bn	R,S	t-Bu	(50%)	—	99:1

t-BuO$_2$CN=NCO$_2$Bu-t, rt, 24 h; (85%) 64% ee (Eq. 105a)

1. LDA, hexane, THF, –45°, 75 min; 2. t-BuO$_2$CN=NCO$_2$Bu-t, –78°, 5 min; (82%) 76% ee (Eq. 105b)

E = N(CO$_2$Bu-t)NHCO$_2$Bu-t

Hydrazone anions have also been subjected to electrophilic amination. They react very rapidly at −78° but the overall yields of the α-aminated ketones are only fair (Eq. 106).[327] Interestingly, the N-arylsulfonamides rather than the azides are obtained in the attempted azidations.

Me$_2$NN=C(Ph)CH$_2$CH$_3$ → 1. LDA, THF, 0°, 4-6 h

→ 2. t-BuO$_2$CN=NCO$_2$Bu-t, –78°, 2-5 min → Me$_2$NN=C(Ph)CH(E)CH$_3$ (66%) → O$_3$, CH$_2$Cl$_2$, –30° → PhCOCH(E)CH$_3$ (65%)

E = N(CO$_2$Bu-t)NHCO$_2$Bu-t

→ 2. 2,4,6-(i-Pr)$_3$C$_6$H$_2$SO$_2$N$_3$, –78°, 2-5 min; 3. NH$_4$Cl, H$_2$O → Me$_2$NN=C(Ph)CH(NHR)CH$_3$ (69%) → O$_3$, CH$_2$Cl$_2$, –30° → PhCOCH(NHR)CH$_3$ (48%)

R = SO$_2$C$_6$H$_2$(Pr-i)$_3$-2,4,6

(Eq. 106)

Amination of Carboxylic Acid Dianions[406]

Although electrophilic amination of carboxylic acid dianions is potentially a very short route to α-amino acids and their derivatives, little work has been published and yields achieved so far, with few exceptions (Eq. 107),[407] are low. Aminations of the dianions of α,β-unsaturated acids are discussed in the section on α,β-unsaturated carbonyl compounds (see below).

Ph⌒CO_2H —— 1. LDA (2.2 eq), THF, HMPA; 2. $MeONH_2$, –15° to –10°, 2 h, rt, overnight ——> Ph–CH(NH_2)–CO_2H (55%) (Eq. 107)

Amination of Ester Enolates and Ketene Acetals

Efforts to introduce the amino or substituted amino group directly into ester enolates by electrophilic amination have met with limited success. *O*-[Di(*p*-methoxyphenyl)]phosphinoylhydroxylamine aminates the enolate of ethyl phenylacetate (Eq.108),[106] but the reaction has not been applied to enolates that do not contain a second activating group such as phenyl or carbonyl. The chiral phosphinoyl reagent **12** also has been applied only to phenylacetates and the products are obtained with low diastereoselectivities (Eq. 109).[147] *O*-Mesitylenesulfonylhydroxylamine aminates a simple ester in low yield[117] and *N*,*N*-dimethyl-*O*-methanesulfonylhydroxylamine converts the lithium enolate of ethyl phenylacetate into ethyl (α-dimethylamino)phenylacetate in 48% yield.[134] The amination with oxaziridines,[151,154,155,157,158] including chiral, non-racemic ones such as **55** (Eq. 110),[154] is often plagued by low yields and generally poor diastereoselectivities and sometimes[154] side reactions involving the aldehyde that is a product of the reaction.

Ph⌒CO_2Et —— 1. KOBu-*t*, –78°, 15 min; 2. $(4\text{-}MeOC_6H_4)_2P(O)ONH_2$, –78°, 6 h; to rt; 3. Ac_2O, Et_3N ——> Ph–CH(NHAc)–CO_2Et (76%) (Eq. 108)

Ph–CH(Li)–CO_2Et —— **12** (Ph, P=O, $ONMe_2$); THF, –15° ——> Ph–CH(NMe_2)–CO_2Et (50%) 23% ee (Eq. 109)

i-Pr branched CO_2Et —— 1. LDA, THF, –78°, 1 h; 2. **55**, –78°; to rt, 2-3 h ——> carbamate (HN, O, O, *i*-Pr, CO_2Et) (49%) 17% de + NC–C_6H_4–CHO (Eq. 110)

55 (NC, N, O, O, O, *i*-Pr)

Silyl ketene acetals are aminated by the hypervalent iodine reagent TsN=IPh (Eq. 111),[173] and by $EtO_2CN(TMS)(OTMS)$ (see Eq. 124 in the section on amination of lactones).[105]

OPh/OTMS ketene acetal + TsN=IPh (0.67 eq) —— MeCN, –20° ——> α-NHTs phenyl ester (O, OPh, NHTs) (50%) (Eq. 111)

Aminations of ester enolates with azodicarboxylic esters and arenesulfonyl azides are more successful but the most widely used method for the preparation of chiral, non-racemic α-amino acids involves *N*-acyloxazolidinones which are discussed in a separate section (see below). Ester enolates in general react rapidly with azodicarboxylic esters at low temperature as illustrated below in connection with β-substituted ester enolates (Eqs. 115–117 and 119). Esters of azodicarboxylic acid derived from borneol, menthol, and isoborneol aminate ester enolates with no or low diastereoselectivity.[408] Similarly, an ester enolate where the alcohol portion is derived from a camphorsulfonamide reacts with di(*tert*-butyl) azodicarboxylate with only moderate diastereoselectivity.[409] Ketene acetals react with azodicarboxylic esters either slowly at room temperature (Eq. 112),[251] or at low temperatures catalyzed by $TiCl_4$,[410,411] Ti(OPr-*i*)$_4$,[409,412] AgOTf,[244] or $AgClO_4$. The latter catalyst together with (*R*)-BINAP furnishes the amination product with moderate enantioselectivity.[244] Much higher diastereoselectivities are achieved with enol ethers derived from chiral alcohols[409,411,412] (Eq. 113).[409,412]

Ph, OMe, OMe + $EtO_2CN{=}NCO_2Et$ —(PhH, rt, 6 d)→ Ph, OMe, $EtO_2CNH{-}N$, OMe, CO_2Et (86%) (Eq. 112)

OTMs, $SO_2N(C_6H_{11}\text{-}c)_2$ —(1. Ti(OPr-*i*)$_4$, CH_2Cl_2, –78°; 2. *t*-$BuO_2CN{=}NCO_2Bu$-*t*, –78°, 5 min; 3. $TiCl_4$; 4. Add substrate, –78°, 1 h)→ H, O, N-$NHCO_2Bu$-*t*, CO_2Bu-*t*, $SO_2N(C_6H_{11}\text{-}c)_2$ (65%) >99.5% de (Eq. 113)

Reaction of ester enolates with trisyl azide and short reaction times at −78° gives the α-azido esters in 50–70% yields;[318,413,414] with 4-nitrobenzenesulfonyl azide, the diazo esters are formed almost exclusively.[318] Azidomethyl phenyl sulfide and ester enolates give α-amino amides[274,275] (Eq. 114),[274] but the scope of this reaction has not been determined.

CO_2Me —(1. LDA, THF, –78°; 2. $PhSCH_2N_3$, –78° to –20°, 45 min)→ O, N, N, N, SPh —(1. NH_4Cl, H_2O; 2. NH_4OH, THF, rt, 12-24 h)→ O, NH, SPh, NH_2 (79%) (Eq. 114)

Ketene dimethyl acetals react with phenyl azide to give α-anilido esters after acid hydrolysis of the intermediate triazolines, but yields are low.[291] The reaction of ketene acetals with arenesulfonyl azides does not appear to have been investigated.

A considerable amount of work has been carried out on the amination with azodicarboxylic esters of β-hydroxy esters, a class of compounds where both enantiomers are readily available by asymmetric reduction of β-keto esters. Yields are in the range of 50-70% for lithium[115,415–417] (Eq. 115),[417] magnesium,[416] zinc,[416,418–424] and titanium enolates,[416] but diastereoselectivities are highest with zinc enolates (Eq. 116).[416] Attack from the less hindered side of zinc enolate **57** accounts for the observed anti selectivity. Similar results are obtained with the other enantiomer.[416] The lithium enolate of the rigidized derivative of ester **56** gives higher yields with a somewhat reduced anti selectivity (Eq. 117).[416]

MeO OH, MeO, CO_2Et — 1. LDA (4 eq), THF, −78°; 2. t-$BuO_2CN{=}NCO_2Bu$-t → MeO OH, MeO, CO_2Et, N, t-BuO_2C, $NHCO_2Bu$-t (55%) 89% de (Eq. 115)

OH O, OEt, **56** — 1. MeZnBr (1.1 eq), THF, Et_2O, 0°; 2. LDA (2.2 eq), −78°, 1 h → Zn, O, O, OEt, **57** — 1. t-$BuO_2CN{=}NCO_2Bu$-t −78°, 10 min; 2. NH_4Cl, H_2O → OH O, OEt, N, t-BuO_2C, $NHCO_2Bu$-t (63%) >90% de

(Eq. 116)

O O, O — 1. LDA; 2. t-$BuO_2CN{=}NCO_2Bu$-t, −78° → O O, O, N, t-BuO_2C, $NHCO_2Bu$-t (90%), 90% de (Eq. 117)

Reaction of the lithium enolate of ethyl β-hydroxybutyrate with trisyl azide furnishes the azide in 77% yield but with only 64% anti diastereomeric excess; the diazo ester (10%) and the diazide (1%) are also formed.[318]

Other β-substituents also promote anti selectivity with both azo esters and trisyl azide. Examples are given in Eqs. 118,[425] and 119.[426] Use of trisyl azide in the latter reaction gives the two diastereomeric azides as a 1 : 1 mixture in 90% yield.[426] More remote substituents, however, may reverse the trend (Eq. 120).[427]

CO_2Me, $SiMe_2Ph$ — 1. LDA, THF, −78°; 2. 2,4,6-(i-$Pr)_3C_6H_2SO_2N_3$, −78° to rt, 10 h → N_3, CO_2Me, $SiMe_2Ph$ (73%) 95% de (Eq. 118)

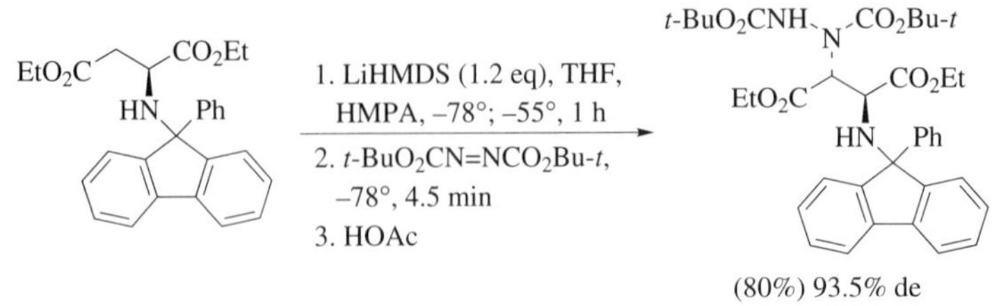

1. LiHMDS (1.2 eq), THF, HMPA, –78°; –55°, 1 h
2. t-$BuO_2CN{=}NCO_2Bu$-t, –78°, 4.5 min
3. HOAc

(80%) 93.5% de

(Eq. 119)

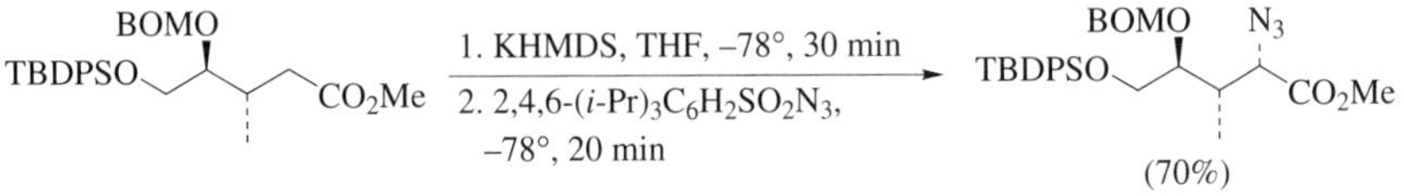

1. KHMDS, THF, –78°, 30 min
2. 2,4,6-(i-$Pr)_3C_6H_2SO_2N_3$, –78°, 20 min

(70%)

(Eq. 120)

Reaction of silyl ketene acetals with aryldiazonium salts produces α-keto ester hydrazones[196,197] by rearrangement of the initially formed azo compounds (Eq. 121). The latter are obtained with disubstituted ketene acetals.[197]

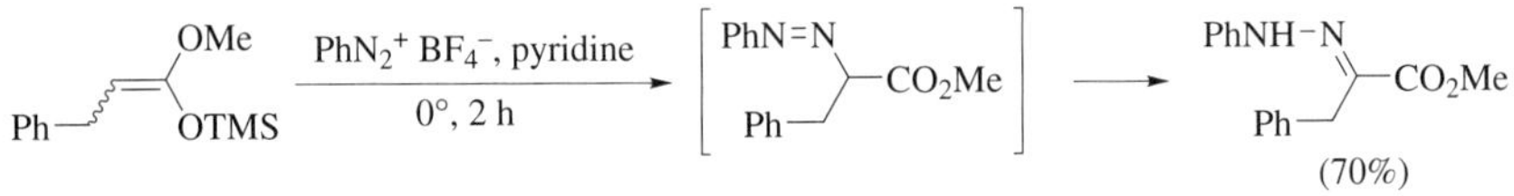

$PhN_2^+ BF_4^-$, pyridine, 0°, 2 h

(70%)

(Eq. 121)

Amination of Thioester Enolates and Ketene Thioacetals

Only a few examples in this category were found in the literature and azodicarboxylic esters are the only aminating reagents that have been used. The reactivities appear to be similar to those described above for ester enolates and ketene acetals. The catalyzed enantioselective amination of ketone silyl enol ethers described in Eq. 94 has also been applied to ketene thioacetals.[252]

Amination of Lactone Enolates

Lactone enolates behave similarly to ester enolates in electrophilic aminations. Examples are shown in Eqs. 122[428] and 123;[429] attack on the less-hindered side to give the equatorial azide is illustrated by the distribution of products **58** and **59** in Eq. 123.

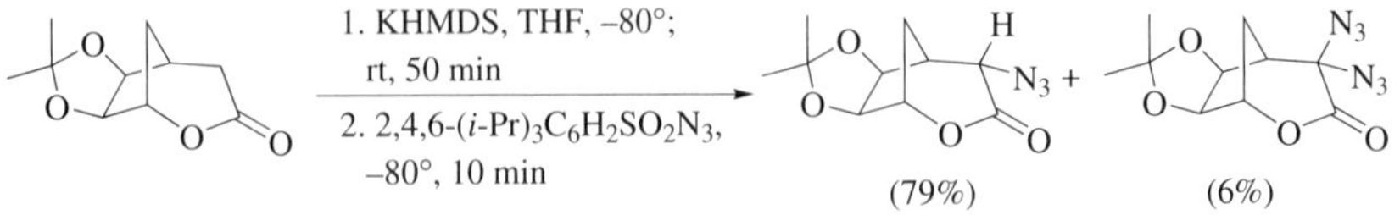

1. KHMDS, THF, –80°; rt, 50 min
2. 2,4,6-(i-$Pr)_3C_6H_2SO_2N_3$, –80°, 10 min

(79%) (6%)

(Eq. 122)

1. KHMDS, toluene, THF, –90°, 15 min
2. 2,4,6-(i-Pr)$_3C_6H_3SO_2N_3$, –90°, 2 min

R^1	R^2	**58**	**59**
BnO	H	(0%)	(70%)
H	BnO	(50%)	(0%)
H	(BnO, BnO, OBn, OBn sugar substituent)	(0%)	(60%)

58 **59**

(Eq. 123)

Ester- and lactone-derived silyl enol ethers are aminated by the $Et_2OCN(TMS)$ OTMS reagent (Eq. 124).[105]

OTMS + $EtO_2C-N(TMS)OTMS$ —(90°, 5 d)→ NHCO$_2$Et (44%) (Eq. 124)

Amination of Amide Enolates and Ketene Aminals

Amide enolates mirror ester enolates in their amination reactions. Secondary amides can be used by employing two equivalents of the base, but yields in the only example found in the literature are low to fair.[212] Ketene aminals react with azodicarboxylic esters at room temperature, but yields are low (Eq. 125).[251] Eq. 126 shows the application of the copper-catalyzed enantioselective addition of mixed ketene acetal/aminals to azodicarboxylic esters previously described for silyl enol ethers in Eq. 94.[252] Increasing bulk of the R substituent in the substrate causes partial or complete amination on the pyrrole, as evidenced by the yields of products **60** and **61** as R is varied.

+ $RO_2C-N=N-CO_2R$ —(PhH, rt, 3 d)→ (Eq. 125)

Y	R	
O	Et	(18%)
CH_2	Me	(41%)

R	Temp	Time	60	61
Me	–78°	30 min	(96%) 99% ee	(0%)
t-Bu	–20°	5 min	(0%)	(80%)

(Eq. 126)

Amination of *N*-Acyloxazolidinone Enolates

This reaction is arguably the most useful and certainly the most widely used application of the electrophilic C-amination of enolates in organic synthesis. A number of 4- and 4,5-substituted 2-oxazolidinones are commercially available in both enantiomeric forms and the chiral auxiliary is easily recovered.[430] Reactions of *N*-acyloxazolidinone enolates with azo esters[431,432] and arenesulfonyl azides[433] are rapid even at very low temperatures (-100°) and the diastereochemical outcome is reliably predictable. The facile removal of the chiral auxiliary and ready conversion of the azide or hydrazino ester functionalities into amines makes these reactions a standard method for the preparation of D- and L-α-amino acids.

The optimum conditions have been thoroughly worked out,[318,431] although a direct comparison of the diastereodirecting efficiency of various oxazolidinones does not appear to have been made for aminations. However, they all direct the incoming electrophile to the less hindered side of the Z-enolate as illustrated in Eqs. 127[431] and 128.[434] The diastereomer with the opposite configuration at the amination site can be obtained using the enantiomeric chiral auxiliary or from the same *N*-acyloxazolidinone by a bromination/S_N2 displacement sequence (Eq. 129)[431] or a hydroxylation/Mitsunobu reaction protocol.[427]

(Eq. 127)

1. KHMDS, THF, –78°, 30 min
2. 2,4,6-(i-Pr)$_3$C$_6$H$_2$SO$_2$N$_3$, –78°, 15 min
3. HOAc

(85%) >90% de

(Eq. 128)

1. i-Pr$_2$NEt, CH$_2$Cl$_2$, rt
2. (n-Bu)$_2$BOTf, –78°; to rt, 1 h
3. Add to NBS, CH$_2$Cl$_2$, –78°, 3 h
4. NaHSO$_3$, H$_2$O

(98%) 80% de

1. (Me$_2$N)$_2$C=NH$_2^+$ N$_3^-$, CH$_2$Cl$_2$, 0°, 3 h
2. NaHCO$_3$, H$_2$O

(85%) >98% de

(Eq. 129)

Lithium diisopropylamide (LDA) or KHMDS is used as the base although the former seems to be preferred for reactions with azodicarboxylic esters and the latter with trisyl azide. In one report[435] a mixture of KHMDS and sodium hydride (one equivalent of each) gave much-improved yields in an azidation. As little as 5 mol% of sodium *tert*-butoxide, lanthanum tri(*tert*-butoxide), or the conjugate base **62** (Eq. 127) effect the amination, indicating that the external base serves as initiator whereas anion **62** is the base in the catalytic cycle.[436] No yields were reported in this investigation. Most procedures call for slightly more than one equivalent of the base except when other acidic protons are present in the molecule (see below). In one azidation, 1.2 equivalents of KHMDS gave a mixture of the diazo compound and the azide in low yields, whereas the latter was formed exclusively in 78% yield with 1.5 equivalents of the base.[437] Trisyl azide is the electrophile of choice for the azidation; 4-nitrobenzenesulfonyl azide and tosyl azide lead to the diazo compounds either exclusively or in admixture with the azides. The benzyl and *tert*-butyl esters of azodicarboxylic acid are the most widely used members of that class of electrophiles because the products are easily cleaved to the hydrazines and the former has an aromatic chromophore for UV detection in chromatography. Azo esters and trisyl azide usually work equally well although there is one report where the former gives a cleaner product,[438] and one instance involving an N-acyloxazolidinone with a sugar attached to the γ-position where di(*tert*-butyl) azodicarboxylate reacted (Eq. 135), but trisyl azide did not.[439] Addition of the pre-cooled electrophile solution to the enolate (or vice versa) is often carried out by means of a cooled or insulated cannula although one report finds that addition of the solid trisyl azide to the cold enolate solution gives the highest yield.[440] The reaction is usually quenched with acetic acid after a short period. The effect of other quenching reagents was discussed in the section on Reagents and Mechanisms.

The following functional groups are tolerated in electrophilic aminations of N-acyloxazolidinones: Br [441–443] (but see below), CH$_2$CO$_2$Bu-t (with one equivalent of base),[444] NH (with two equivalents of base),[445,446] $NRCO_2Bu$-*t* (with two equivalents of base when R = H),[440,447–452] NHAc (with two equivalents of base),[453] $RNCO_2Bn$,[454] aliphatic alcohols protected by trialkylsilyl or tosyl,[455] protected phenols, phenylselenyl,[456] $(t\text{-BuO})_2P(O)CH_2$ (with one equivalent of base),[457] and $Ph_2P(S)CH_2$ (the amount of base was not reported).[458]

A few problems have been reported. Cleavage of the *N*-acyloxazolidinone occurs to a considerable extent in the reaction of Eq. 130.[445,446] A bromine atom at a distance of five carbons from the carbonyl group causes the enolate to cyclize under normal procedures (product **64**, Eq. 131); azide **63** (n = 3) is obtained in 40% yield only by adding an excess of trisyl azide early in the enolization step.[443] The α,β-unsaturated *N*-acyloxazolidinone **65** does not undergo amination under conditions where its isomer **66** does (Eq. 132).[453] However, product **67** epimerized on attempted removal of the auxiliary.

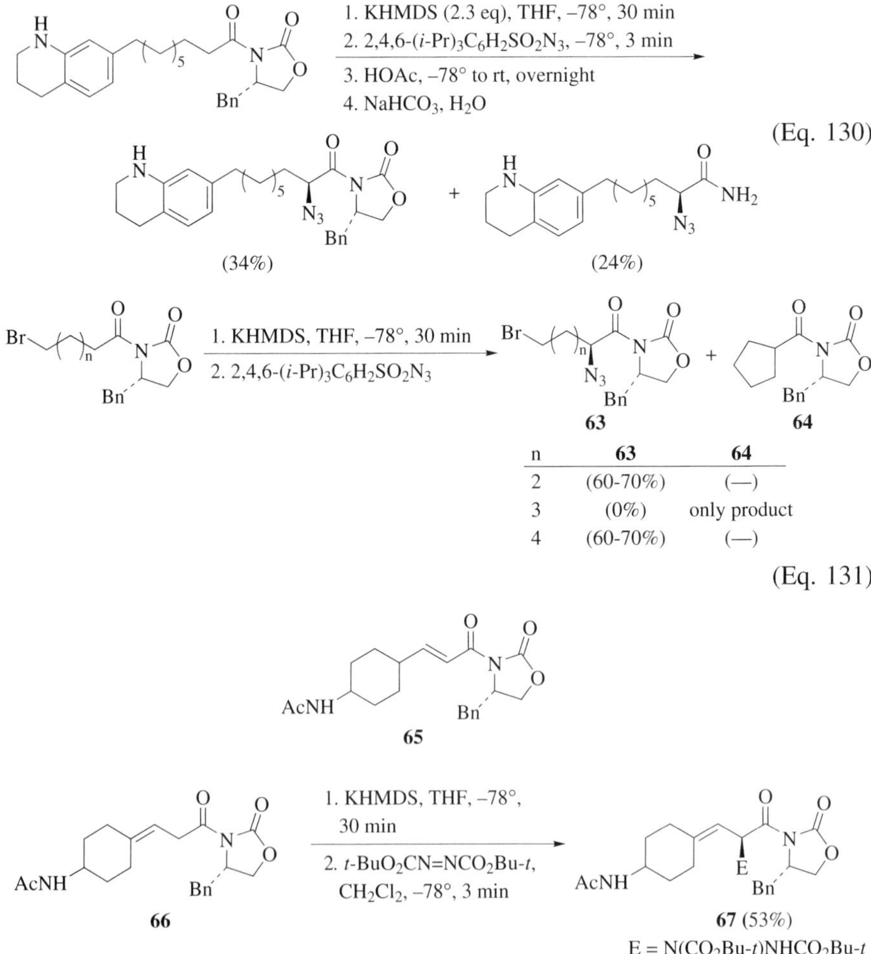

(Eq. 130)

n	63	64
2	(60-70%)	(—)
3	(0%)	only product
4	(60-70%)	(—)

(Eq. 131)

(Eq. 132)

In the attempted double diastereoselection shown in Eq. 133, amination of a pair of enantiomeric *N*-acyloxazolidinones with (−)-diisobornyl azodicarboxylate furnishes a single product for each. The same reactions with dibenzyl azodicarboxylate as the electrophile proceed with only 9:1 diasteromeric ratio. These experiments indicate that the only effect of the bulky isobornyl group is to enhance the diastereoselectivity, which is controlled by the enolate geometry.[408]

1. LDA, THF, –78°, 40 min
2. $RO_2CN{=}NCO_2R$, –78°, 4 min

E = $N(CO_2R)NHCO_2R$

C4 Config.	R	Yield	% de
S	Bn	(—)	80 (S)
S	(–)-isobornyl	(56%)	100 (S)
R	Bn	(—)	80 (R)
R	(–)-isobornyl	(88%)	100 (R)

(Eq. 133)

Alternate routes to chiral α-amino acids and α-amino alcohols that apparently proceed with somewhat higher diasteroselectivity involve the reactions of achiral α-chloronitroso compounds with chiral enolates or of chiral α-chloronitroso compounds with achiral enolates (see section on the amination with nitroso compounds in Comparison with Other Methods), but they have not been applied nearly as frequently as the aminations described above.

Chiral azo amide **68** reacts with an achiral oxazolidinone enolate to give a single product with the configuration indicated in Eq. 134, but the hydrazino amide could not be hydrolyzed.[219] A remote chiral group attached to an achiral *N*-acyloxazolidinone directs a diastereoselctive amination as shown in Eq. 135.[459]

1. LDA, THF, –78°, 20 min
2. **68**, CH_2Cl_2, –78°, 7 min

68

(85%)

(Eq. 134)

R = OBn

1. LDA, THF, –78°, 30 min
2. t-$BuO_2CN{=}NCO_2Bu$-t, –78°, 5 min

(70%) 73% de

(Eq. 135)

The amination of an achiral *N*-acyloxazolidinone with azo esters may also be carried out catalytically with magnesium complex **69** as the base (Eq. 136).[436] The role of *N*-methyl-*p*-toluenesulfonamide, which accelerates the reaction, is not clear. Enantiomeric excesses are in the range of 82–90% but the catalytic amination has only been carried out with *N*-arylmethylcarbonyloxazolidinones.

N-acyloxazolidinones are cleaved to the acid salts by lithium hydroxide/hydrogen peroxide.[318] The chiral auxiliary is recovered by extraction into an organic solvent; the acid is obtained by acidification of the aqueous phase.

t-$BuO_2CN{=}NCO_2Bu$-*t*, **69** (10 mol%), TsNHMe (20 mol%)

Et_2O, CH_2Cl_2, –65°, 2 d

(97%) 90% ee

69

(Eq. 136)

Amination of Lactam Enolates

O-(Diphenylphosphinoyl)hydroxylamine (Eq. 137),[143] azo esters (Eq. 138),[460] and arenesulfonyl azides (Eq. 139)[339] have been used to aminate lactam enolates. In the azidation of the lactam **70**,[461] the diazo compound **73** predominates over azide **72** even though trisyl azide is used as the aminating agent; amination with di(*tert*-butyl) azodicarboxylate was unsuccessful. The closely related lactam **71**[462] reacts normally with trisyl azide (Eq. 140).

1. LiHMDS, THF, –78°
2. $(Ph_2P(O)ONH_2$

(47%) >82% de (Eq. 137)

1. LDA, THF, hexane, –78°, 2 h
2. *t*-$BuO_2CN{=}NCO_2Bu$-*t*, –78°, 6 h; to rt, 3-6 h

(67%) >98% de (Eq. 138)

1. LiHMDS, THF, –78°, 4 h
2. TsN_3, –70°
3. TMSCl

(68%) (Eq. 139)

R = TBDPS
70 or **71**

1. LDA, THF, –78°, 20 min
2. 2,4,6-$(i\text{-Pr})_3C_6H_2SO_2N_3$, –78°, time

72 + **73**

	m	n	Time	**72**	**73**
70	2	1	10 min	(20%)	(64%)
71	1	2	3 h	(43%)	(0%)

(Eq. 140)

Amination of Nitrile-Stabilized Carbanions

Little work could be found on the electrophilic amination of simple nitrile-stabilized carbanions. The lithium anion of propionitrile reacts normally with an N-substituted oxaziridine (Eq. 141).[158] The amination of nitriles with a camphor-derived N-unsubstituted oxaziridine was discussed earlier (Eq. 11).[151] Aminomalononitrile is formed from malononitrile anion and O-(mesitylenesulfonyl)hydroxylamine (Eq. 142).[463]

1. n-BuLi, hexane, THF, 0°, 30 min
2. 2-NCC_6H_4-oxaziridine-$NCONEt_2$, –78°, 3 h; to rt, 1.5 h

Product: $NHCONEt_2$ (56%) (Eq. 141)

1. unspecified Na base
2. 2,4,6-$Me_3C_6H_2SO_2ONH_2$, THF, 0°, 2.5 h
3. TsOH

Product: NC, CN, NH_3^+ TsO^- (55%) (Eq. 142)

The anion of phenylacetonitrile has been aminated with a variety of reagents; examples are shown in Eq. 143.[106,147] In the reaction involving the chiral O-phosphinoylhydroxylamine, epimerization is believed to have occurred during isolation of the product.

Ph CN

1. unspecified Li base, THF
2. Ph, O, O, P, N, $ONMe_2$, –15°, Me
3. H_3O^+ (pH 4.5)

Product: NMe_2, Ph, CN (62%) 8% ee

1. KOBu-t, THF, –78°, 15 min
2. $(4\text{-MeOC}_6H_4)_2P(O)NH_2$, –78° to rt; rt, overnight
3. Ac_2O, Et_3N

Product: NHAc, Ph, CN (67%)

(Eq. 143)

Amination of Nitronates

Only one example involving a number of substituted nitromethane anions was found in the literature and the reaction with *p*-toluenesulfonyl azide proceeds with loss of the nitro group (Eq. 144). Nitromethane itself failed to react under these conditions.[464]

1. KH, THF, rt; 40°, 15 min
2. TsN_3, −10° to 0°; 0°, 1 h
(56%)

(Eq. 144)

Amination of Sulfone-Stabilized Carbanions

The few examples indicate that sulfone-stabilized carbanions should react normally with electrophilic aminating reagents (Eqs. 145[158] and 146[465]) with the caveat that free α-amino sulfones are unstable.[158,465] The β,γ-unsaturated sulfone **74** is aminated at the γ-position (Eq. 147),[250] presumably by an ene reaction. The preparation of α-tosyl azides from nitronates was shown above in Eq. 144. The scope of this reaction does not seem to have been determined. Reaction of the anions of nitrobenzyl aryl sulfones with 1-oxa-2-azaspiro[2.5]octane (**13a**) gives nitrobenzaldehydes by cleavage of the initially formed amination products.[466] Similarly, reaction of the lithium salt of benzyl phenyl sulfone with phenyl azide gives benzilydeneaniline and phenyl sulfinate.[467] No reports on aminations of sulfoxide-stabilized carbanions were found.

1. *n*-BuLi, THF, hexane, 0°, 30 min
2. 2-NCC_6H_4-oxaziridine-$NCONEt_2$, −78°, 3 h; to rt, 1.5 h
$NHCONEt_2$ (43%)

(Eq. 145)

R = TBDPS
1. *n*-BuLi, THF, pentane, −78°, 55 min
2. 2,4,6-$Me_3C_6H_2SO_2N_3$, −78°, 6 h
(40%) + (24%)

(Eq. 146)

74
$EtO_2CN{=}NCO_2Et$, MeCN
reflux, 3 h
(65%)

(Eq. 147)

Amination of Phosphorus-Stabilized Carbanions

Only one report on the amination of a phosphine oxide anion (Eq. 148) is known;[467] the product is claimed to have the structure shown but no spectral data excluding the isomer where the N=N double bond is conjugated with the phenyl group were provided.

1. $LiNEt_2$, PhH, rt, 1 h
2. PhN_3, rt, 18 h
3. H_2O

(26%) (Eq. 148)

All other reactions involve derivatives of methanephosphonic acid and a range of aminating reagents has been applied, including hydroxylamine derivatives, oxaziridines, azo esters, and sulfonyl azides. The products are α-amino phosphonic acids or derivatives that can be converted into these biologically interesting analogs of α-amino acids. The best results with methanephosphonic acid derivatives not containing an additional activating group have been obtained so far with phosphorinanes of type **75** (Eq. 149).[317] The diastereoselectivity using the standard acetic acid quench to generate the azide is disappointing, and yields from analogous compounds are low, possibly because here the addition of trisyl azide is reversible. Trapping the triazene salt with acetic anhydride resolves the problem. Cleavage of the product and removal of the chiral auxiliary gives the phosphono analog of (*S*)-phenylglycine.

75 (2*S*,6*S*)

1. LDA, Et_2O, –78°, 30 min
2. 2,4,6-$Me_3C_6H_2SO_2N_3$, –78°, 5 h
3. Ac_2O

(75%) 86% de

(Eq. 149)

Two catalytic enantioselective methods have been developed for β-keto phosphonic acid derivatives (Eqs. 150[238] and 151[468]).

$EtO_2CN{=}NCO_2Et$

$[(P)_2Pd(\mu\text{-}O)_2Pd(P)_2]^{+2}$ 2 BF_4^- (2.5 mol%),

Me_2CO, rt, 20 h

(91%) 99% ee

P–P = 2,2'-bis[$P(C_6H_3Me_2\text{-}3,5)_2$]-1,1'-binaphthyl

(Eq. 150)

1. (Ph-box)Zn(OTf)$_2$ (10 mol%), CH_2Cl_2
2. add substrate, then $BnO_2CN{=}NCO_2Bn$, rt, 140 h

(85%) 98% ee

(Eq. 151)

Standard amination methods may be used for the synthesis of racemic α-phosphono α-amino carboxylic esters (Eq. 152).[141] No diastereo- or enantioselective syntheses appear to have been reported.

$(EtO)_2P(O)CH_2CO_2Bn$ → 1. NaH, THF, rt, 1 h; 2. $Ph_2P(O)ONH_2$, THF, –78°, 2 h; 3. HO_2CCO_2H → $(EtO)_2P(O)CH(NH_3^+ \; {}^-O_2CCO_2H)CO_2Bn$ (60%) (Eq. 152)

Amination of Enolates of α,β-Unsaturated Carbonyl Compounds

Enolates of α,β-unsaturated carbonyl compounds can react at either the α- or γ-position and α,β-unsaturated ketones can react at the α′-position as well. On the basis of limited evidence, NH_2^+ synthons react at the α-position,[64,144] whereas azo esters aminate preferentially at the γ-position[144,469] (Eq. 153),[144] both by kinetic control, although there are exceptions (product **77** vs. **76**, Eq. 154).[469] With an α,β-unsaturated *N*-acyloxazolidinone, the two constitutiona isomers are formed in equal amounts (Eq. 155).[431] The catalytic method shown in Eq. 156[470] is believed to involve a hetero Diels-Alder reaction of the intermediate dienamine. Allyltin and allylgermanium reagents give mostly or exclusively the products of an S_E2' reaction (Eqs. 157 and 158).[469] The substrates for these reactions are prepared by addition of tin tetrachloride and trimethylgermanium chloride, respectively, to the lithium enolates of the corresponding α,β-unsaturated esters. The generation of the tin substrate can be carried out in situ. Silyl ketene acetal **78**, the only example of this type of derivative whose amination was found in the literature, reacts predominantly at the γ-position (Eq. 159).[469]

1. $LiNEt_2$ (2.0 eq), THF, –70°, 15 min; 2. $Ph_2P(O)ONH_2$, –70°, 25 min; rt, 2 h → α-amino product (CO_2H, NH_2) (64%)

1. $LiNEt_2$ (2.2 eq), THF, –70°, 30 min; 2. $EtO_2CN{=}NCO_2Et$, –70°, 55 min → γ-product (EtO_2CNH–N(CO_2Et)–, CO_2H) (50%)

(Eq. 153)

1. LDA, HMPA, THF, –78°, 70 min
2. $EtO_2CN{=}NCO_2Et$, –78°, 3 min
3. MeOH

76 **77**

E = $N(CO_2Et)NHCO_2Et$

n	**76**	**77**	**77** E:Z
1	(22%)	(65%)	1:2
2	(55%)	(14%)	1:1.5

(Eq. 154)

1. LDA, THF, –78°, 30 min
2. $t\text{-}BuO_2CN{=}NCO_2Bu\text{-}t$, CH_2Cl_2, –78°, 0.5 to 3 min

(42%) E:Z = 3:2 (51%) 96% de

E = $N(CO_2Bu\text{-}t)NHCO_2Bu\text{-}t$

(Eq. 155)

1. OTMS, Ar, Ar (10 mol%), toluene, rt, 15 min
2. $EtO_2CN{=}NCO_2Et$, 1.5 h

(43%), 88% ee (Eq. 156)

1. $EtO_2CN{=}NCO_2Et$, THF, –10°, 30 min
2. to –78°; MeOH, –78° to rt

(53%) (5%)

E = $N(CO_2Et)NHCO_2Et$

(Eq. 157)

1. $EtO_2CN{=}NCO_2Et$, $ZnCl_2$, CH_2Cl_2, –78°
2. add substrate, –78° to 0°, 30 min
3. MeOH

(55%) E:Z = 6:1 (Eq. 158)

78

1. $EtO_2CN{=}NCO_2Et$, $TiCl_4$, CH_2Cl_2, –78°
2. add substrate, –78°, 30 min
3. MeOH

(68%) (17%)

E = $N(CO_2Et)NHCO_2Et$

(Eq. 159)

The dianion of *trans,trans*-hepta-2,4-dienoic acid (sorbic acid) is aminated in the α-position by $Ph_2P(O)ONH_2$ and in the γ-position by diethyl azodicarboxylate.[144]

The amination of the only derivative of an α,β-unsaturated ketone is shown in Eq. 160.[471]

OTMS

CH_2Cl_2, TFAA, pyridine, –78° to rt, 3-4 h

$NHCOCF_3$ (50%)

(Eq. 160)

Amination of Enolates of α-Cyanocarbonyl and β-Dicarbonyl Compounds

The electrophilic amination of the sodium salts of α-unsubstituted β-dicarbonyl compounds is one of the few examples of an amination where hydroxylamine *O*-sulfonic acid gives useful yields (Eq. 161);[472] with two equivalents of the substrate, pyrroles are formed.[472,473] Chloramine,[62,64] *O*-arylhydroxylamines,[93,124,474] *O*-sulfonylhydroxylamines,[134] and *O*-(diarylphosphinoyl)hydroxylamines[106,139,475] have also been employed, although a low yield and formation of the hydroxylation product as a side product have been reported in one instance with $(4\text{-MeOC}_6H_4)_2P(O)ONH_2$.[145] Some of these aminations use chiral auxiliaries in the substrates with modest diastereoselectivities,[124,475,476] but these have been superseded by the catalytic methods discussed below.

NH_2OSO_3H, NaOH, H_2O, rt, few minutes → NH_2 (100%)

(Eq. 161)

The oxaziridine **13a** reacts with a variety of β-dicarbonyl and α-cyanocarbonyl compounds under base catalysis (Eq. 162).[149]

13a

R = H; toluene, NaOH, H_2O, 0°, 10 min → (56%)

R = C_6H_4OMe-4; toluene, NaOH, H_2O, rt, 12 h → (48%) + (39%)

EtOH, reflux, 5 min (95%)

(Eq. 162)

Lithium[477] and potassium[478] enolates of β-dicarbonyl compounds are aminated by azodicarboxylic esters in good to excellent yields. Diethyl malonate, ethyl acetoacetate, *N*,*N*-diethyl acetoacetamide, and acetylacetone have also been aminated with diethyl azodicarboxylate under nickel acetylacetonate catalysis,[479] and nickel salicylideneimine complexes catalyze the analogous amination of acetylacetone and its 2-methyl derivative.[480]

A number of catalytic, enantioselective reactions of azodicarboxylic esters with β-dicarbonyl and α-cyanocarbonyl compounds have been recently developed, using cinchona alkaloids or their derivatives,[231,232,481] BINAP-derived palladium complexes,[239] chiral copper bis(oxazoline) complexes,[235,237] and chiral amidines and amines[233] as catalysts. With cinchona alkaloid-derived catalysts, cyanoacetic esters carrying aryl substituents in the α-position give higher selectivities than those with alkyl groups in that position[232,481] (Eq. 163).[232] With the quinidine-derived catalyst **79**, the newly created stereogenic center has the S-configuration; the quinine-derived enantiomer furnishes the R-isomer. Cinchonine and cinchonidine catalyze the reaction of dibenzyl azodicarboxylate with ethyl α-ethylacetoacetate but the enantioselectivity is low (47 and 27% ee, respectively).[231] However, α-fluoro-[237] and α-alkyl [235] acyl and aroylacetates respond well to catalysis by chiral copper bis(oxazoline) complexes (Eq. 164).[235] The reaction of ethyl α-methylacetoacetate with dibenzyl azodicarboxylate is also catalyzed by a BINAP-derived palladium complex (95% ee).[239] This catalyst also induces good enantioselectivity in the amination of two cyclic β-dicarbonyl compounds (Eq. 165).[239]

79 (10 mol%), toluene, –78°

R	Time		ee
Me	30 min	(75%)	35%
4-BrC_6H_4	1 min	(86%)	91%

79

(Eq. 163)

(0.5 mol %), CH_2Cl_2, rt, 16 h

(89%) 98% ee

(Eq. 164)

i-$PrO_2CN{=}NCO_2Pr$-i

$2\ PF_6^-$ (5 mol%), MeOH, rt

(Eq. 165)

R	Y	Time		ee
Me	O	9 h	(93%)	93%
OEt	CH_2	31 h	(89%)	97%

Enolates of β-dicarbonyl and α-cyanocarbonyl compounds have a strong tendency to form diazo compounds with arenesulfonyl azides. α-Substituted substrates react normally to give azides[482,483] but even then a diazo transfer (Eq. 166)[484] or other transformations[319,321,484] may occur as side reactions.

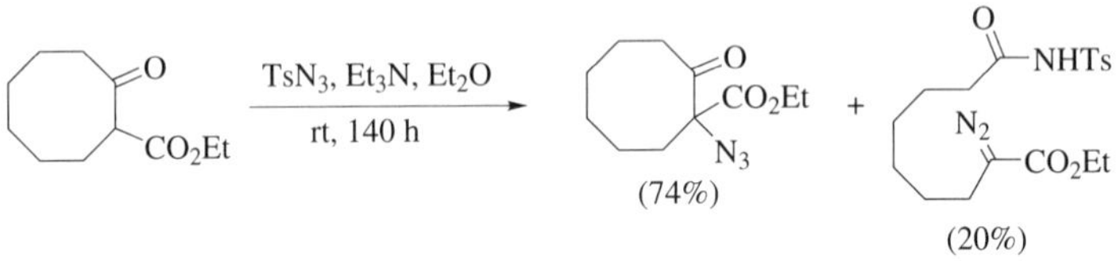

(Eq. 166)

An interesting addition of a sugar azide to the enolate of cyanoacetamide is shown in Eq. 167.[276]

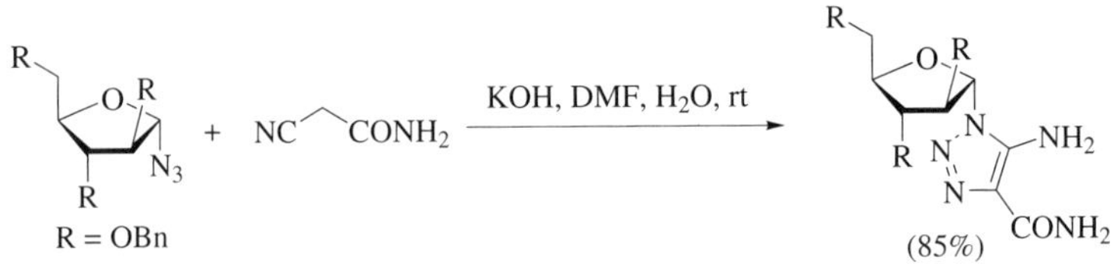

(Eq. 167)

Intramolecular Aminations

Formation of Aziridines. The addition of *O*-methylhydroxylamine to α,β-unsaturated carbonyl compounds gives β-methoxyamino derivatives which on treatment with sodium methoxide at elevated temperatures give aziridines (Eq. 168).[485,486] The products were initially considered to be the isomeric primary enamines.[485] The reaction has been carried out with other leaving groups: benzyloxy,[487–489] OCOBu-*t*,[490] TMSO,[491] arenesulfonyloxy[492,493] (Eq. 169),[492] and trimethylammonium (with formation of an azirine; Eq. 170).[494] An example involving a chiral auxiliary is shown in Eq. 171.[487,488]

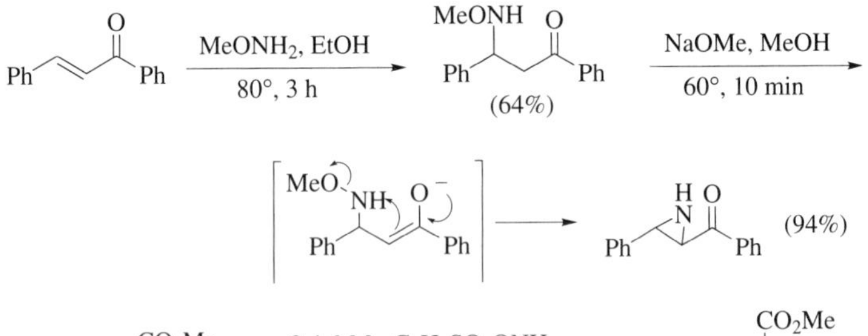

(Eq. 168)

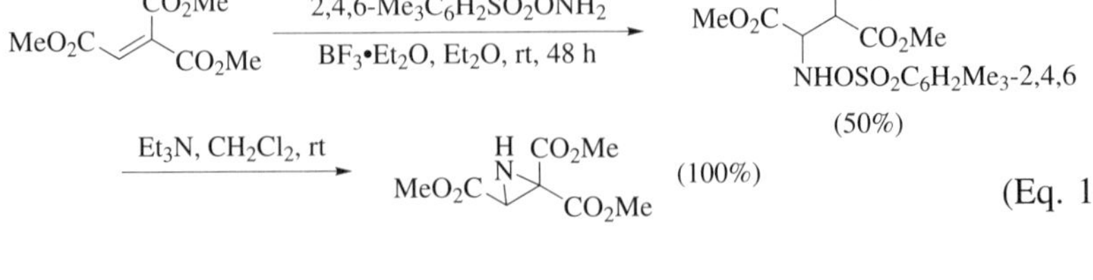

(Eq. 169)

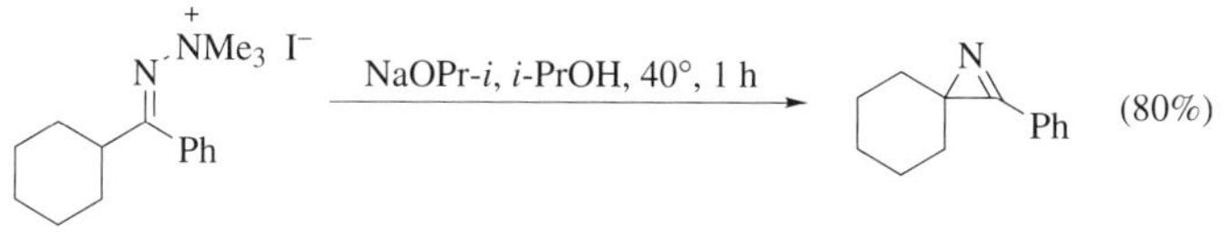

(80%) (Eq. 170)

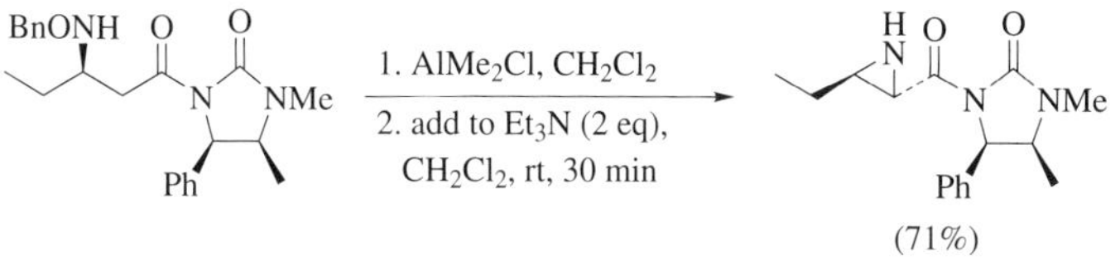

(71%) (Eq. 171)

Hydroxylamine derivatives add to activated double bonds in the presence of a base to give aziridines where intermediates of the type illustrated above have not been isolated or observed. These reactions may proceed via stereospecific addition of nitrenoid intermediates to the double bonds. However, in some instances, both isomeric aziridines are produced and these are included in Table 21 of the Tabular Survey since the possibility exists that they are formed by a Michael addition/cyclization process. An example is shown in Eq. 172.[495]

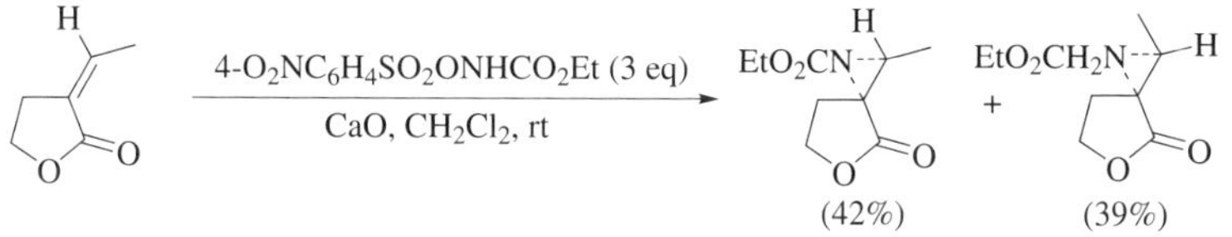

(42%) (39%) (Eq. 172)

Formation of Higher-Membered Rings. Intramolecular displacement of a methoxy group by an aryl carbanion by way of a nitrenoid intermediate (Eq. 5) produces 4- (Eq. 173),[83] 5-,[82] 6-,[83] and 7-membered[83] benzannulated ring systems. The 8-membered benzazocine cannot be prepared by this method.[83] The diphenylphosphinoyloxy functionality has also been employed as the leaving group (Eq. 174).[496] Five-, 6-, and 7-membered unsaturated nitrogen-containing rings are obtained from substituted oximes (Eqs. 175[497] and 176[498,499]). The former reaction is postulated to proceed by an S_N2 displacement on sp^2 nitrogen rather than an addition/elimination process. The intermediate **80** in the latter reaction is air sensitive and is either reduced to the tetrahydroquinoline or oxidized to the quinoline. The cyclization in this case is believed to involve a single-electron transfer.

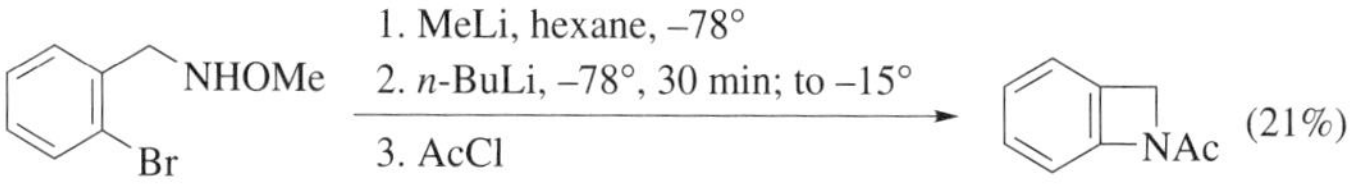

(21%) (Eq. 173)

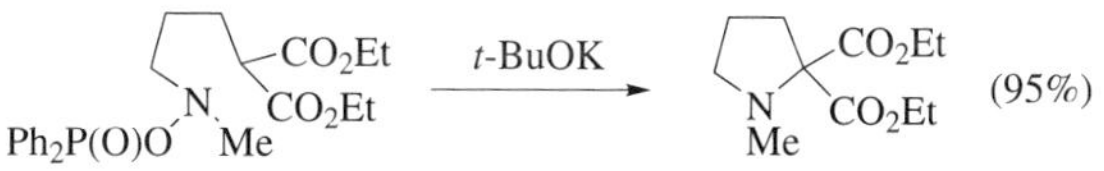

(95%) (Eq. 174)

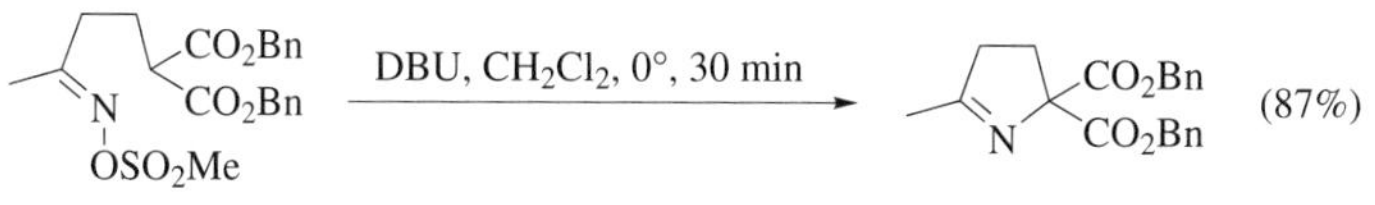

(87%) (Eq. 175)

NaH, dioxane; N–$OC_6H_3(NO_2)_2$-2,4; OH; **80**; $Na(CN)BH_3$, 50°, 10 h (78%); DDQ, HOAc, reflux, 2 h (70%)

(Eq. 176)

COMPARISON WITH OTHER METHODS

The number of methods for the formation of carbon-nitrogen bonds[500–505] is too large to permit a meaningful comparison. A few other methods were mentioned where appropriate in Scope and Limitations. The following is a brief discussion of the reagents for electrophilic amination that were excluded from the scope of this chapter.

Amination with Nitrogen Oxides

Nitrous oxide (N_2O) reacts with phenyllithium to give complex mixtures containing azobenzene, hydrazobenzene, and biphenyl, among others.[506] With 9-fluorenyllithium, fluorenone azine is formed in 60% yield; the analogous product is obtained with the sodium salt of phenylacetonitrile.[506] With *n*-butyllithium, the *N*-butylhydrazone of butyraldehyde is formed in low yield.[507] Nitric oxide (NO) reacts with alkyl- and arylmagnesium reagents to give *N*-nitrosohydroxylamines in low to fair yields.[508,509] α,β-Unsaturated amides react with nitric oxide and triethylsilane in the presence of cobalt complexes to give α-nitroso amides.[510] Dinitrogen tetroxide (N_2O_4) and ethylmagnesium bromide[511] or triethylaluminum etherate[512] give *N*,*N*-diethylhydroxylamine.

Amination with Nitrosyl Chloride, Nitryl Chloride, and Nitronium Tetrafluoroborate

Combination of arylmagnesium halides and nitrosyl chloride (NOCl) gives mixtures of arylnitroso compounds and diarylamines.[513–515] With alkylmagnesium halides, *N*-nitrosohydroxylamines[509] or dialkylhydroxylamines[516,517] are formed. Aldehyde and ketone trimethylsilyl enol ethers and ketene acetals react with nitrosyl chloride to give the α-oximino aldehydes, ketones, or esters, respectively, by rearrangement of the intermediate nitroso compounds (Eq. 177).[518] The latter are isolated from enol ethers of α, α-disubstituted aldehydes. α-Nitro aldehydes and α-nitro ketones are formed in low yields by reaction of enol acetates with nitryl chloride (NO_2Cl).[519] Alkyl- and allylsilanes react with nitronium tetrafluoroborate ($NO_2^+\ BF_4^-$) to give the corresponding nitro compounds.[520] The reaction of ketone enol ethers with nitronium tetrafluoroborate gives α-nitro ketones.[521,522]

OTMS; 1. NOCl (excess), CH_2Cl_2, −10° to −15°, 30 s; 2. 0°, vacuum; O; NO; O; NOH; (82%) (Eq. 177)

Amination with Alkyl Nitrites

Alkyl nitrites, of which the most commonly used representative is isopentyl nitrite,[523] react with a wide variety of compounds containing active methylene groups to give oximes.[524–526] Activating functionalities include carbonyl, cyano, nitro, and aryl. For the latter, the presence of two aryl groups is usually required but by using chromium-complexed arenes, one aryl group suffices (Eq. 178).[527] The oximes can then be reduced to hydroxylamines or amines.[528,529] Alkyl nitrites react with dialkylzinc[530] and alkylmagnesium[516] reagents to give dialkylhydroxylamines, whereas with arylmagnesium reagents, diarylnitroxyls are formed.[531] Activated olefins react with triphenylsilane and *n*-butyl nitrite in the presence of cobalt complexes to give the corresponding α-hydroxyimino derivatives.[532,533] A similar reaction of unactivated olefins in the presence of iron complexes gives nitrosoalkane dimers.[534]

$(CO)_3Cr$; MeO; 1. *t*-BuONO, DMSO, rt; 2. KOBu-*t*, 30 min; $(CO)_3Cr$; NOH; MeO; (60%) (Eq. 178)

Amination with Alkyl Nitrates

Alkyl nitrates[535] give *N*,*N*-dialkylhydroxylamines with alkylmagnesium reagents.[536,537] The reaction of 9-fluorenylpotassium with isopentyl nitrate forms the 9-nitro derivative in unspecified yield.[538] The main application of alkyl nitrates, however, has been in the nitration of ketone enolates[539–544] to give mono- or dinitro ketones. Many steroid nitro ketones have been prepared in this way but yields are variable.[542,545–548] α-Nitro amides,[549] α-nitro lactams,[540] and α-nitro nitriles[550] may also be prepared in this manner. Aza enolates give nitro enamines[551,552] (Eq. 179).[552] Acetyl nitrate, prepared in situ from acetic anhydride and nitric acid, nitrates enol acetates.[522,553–555] Similarly, α-nitro ketones are formed from the reaction of enol ethers and esters with trifluoroacetyl nitrate, prepared in situ from ammonium nitrate and trifluoroacetic anhydride.[522,548,556,557] Cyclohexanone triisopropylsilyl enol ethers and a mixture of tetra-*n*-butylammonium nitrate and trifluoroacetic anhydride give α-nitro enol ethers.[558]

N; 1. KNH_2, NH_3 (liq), −33°, 1 h; 2. *n*-$PrONO_2$, <−40°; −33°, 25 min; H N; NO_2; (50%) (Eq. 179)

Amination with Nitroso Compounds

Nitroso compounds[559–561] are versatile electrophiles that undergo a number of different amination reactions. Arylnitroso compounds and aryl Grignard reagents are variously reported to give diarylhydroxylamines,[99,562–568] diarylamines,[99,514,569] or diaryl azo compounds.[567] The reaction has been developed into a general diarylamine synthesis (Eq. 180).[570] Nitrosotrifluoromethane undergoes a nitroso aldol reaction with the anions of pentane-2,4-dione (Eq. 181)[571] and bis(trifluoromethyl)acetonitrile[572] as does nitrosobenzene with ketone lithium and tin enolates[573] and with aldehydes in the presence of a proline-derived catalyst.[574] The reaction of tin enolates with nitrosobenzene catalyzed by Lewis acids gives mostly the hydroxylation products.[575] Ketone trimethylsilyl enol ethers react with nitrosobenzene to give adducts of type **81**,[242,243,576,577] which on reaction with triethylamine give imines of α-keto aldehydes.[576] Oxidation of intermediates **81** leads to nitrones,[577] and reduction to amino alcohol derivatives (Eq. 182).[578] Similarly, ketene bis(trimethylsilyl)acetals give *N*-phenyl α-amino acids on reduction of the intermediate adducts.[578] Enamines react with nitroso arenes to give α-(*N*-arylhydroxylamino) ketones.[579,580]

Me_2N–C_6H_4–NO + ClMg–C_6H_4–CO_2Et → (1. THF, –20°, 1 h; 2. $FeCl_2$, $NaBH_4$, rt, 2 h) → Me_2N–C_6H_4–NH–C_6H_4–CO_2Et (72%)

(Eq. 180)

pentane-2,4-dione + CF_3NO → (Et_3N, Et_2O, –10°) → 3-[N(CF_3)OH]-pentane-2,4-dione (65%) (Eq. 181)

Ph(TMSO)C=CH_2 → (PhNO, $CHCl_3$, rt, 4-6 h) → PhC(O)CH_2N(OTMS)Ph **81** (80%)

81 → (Et_3N, $CHCl_3$, rt, 1 d) → PhC(O)CH=NPh (100%)

81 → (1. HCl, THF, rt; 2. Ag_2O, C_6H_6, rt) → PhC(O)CH=N^+(O^-)Ph (80%)

81 → ($LiAlH_4$, Et_2O, rt or Pd/C, H_2, THF, rt) → PhCH(OH)CH_2NHPh (75-78%)

(Eq. 182)

α-Chloronitroso compounds[581] react with alkyl- and arylmagnesium reagents[582,583] and with trialkylaluminum reagents[584] to give nitrones. In contrast, allylzinc reagents and α-chloronitroso compounds furnish mostly *O*-allyloximes.[585] An important application of these reagents is in the amination of enolates[586–594] (Eqs. 183[587] and 184[595]). Using these methods, the reactions apparently proceed with somewhat higher diastereoselectivity than aminations of *N*-acyloxazolidinones. However, amination of a β-lactam enolate with chloronitroso reagent **82**, while completely trans selective, occurs with poor discrimination between the two enantiomers of the enolate (products **83** and **84**, Eq. 185).[592] A mannose-derived α-chloronitroso compound has been prepared[596] but apparently not yet applied in amination reactions.

1. KHMDS, THF, –78°, 30 min
2. Cl–C(NO)(cyclohexyl), –78°, 30 min

HCl, H_2O, rt → NHOH product (78%) >99% de

1. Zn, HCl, HOAc, rt
2. LiOH

→ NH_2 amino acid (87%) >99% ee

(Eq. 183)

1. LiHMDS
2. $ZnCl_2$
3. **82**

1. HCl
2. $NaBH_4$, MeOH
3. Zn, HCl, HOAc

(68%) 90% de, 96% ee

82 R = $SO_2N(C_6H_{11}\text{-}c)_2$

(Eq. 184)

Ar = C_6H_4OMe-4

1. LDA, –78°, 10 min
2. **82**, –78° to –30°, 2 h
3. HCl

83 + **84**

(58%) **83**:**84** = 3:2

(Eq. 185)

Amination With Nitro Compounds

The reaction of Grignard reagents with nitro compounds is complex and the products depend on the nature of both reactants, but a number of useful synthetic schemes have been developed in recent years. Alkylmagnesium reagents undergo 1,2- or 1,4-addition to aromatic nitro compounds to give ring-alkylated intermediates that may be converted into ring-alkylated arylnitro compounds or anilines.[597] The less basic organocerium reagents react with nitroalkanes to give N,N-disusbstituted hydroxylamines.[598] *N*-Allyl-[599] and *N*-allenylmagnesium[600] halides react with nitroalkanes and nitroarenes to give *N*-allyl- and *N*-propargyl-, *N*-alkyl- and *N*-arylhydroxylamines after reduction of the intermediate hydroxylamine *N*-oxides. Nitrones can be isolated from the reaction of allyl- and benzylmagnesium reagents with nitroalkanes[601–603] (Eq. 186).[603] Arylmagnesium reagents react with nitroarenes to give nitroso arenes which rapidly react with another molecule of the arylmagnesium halide to give diarylhydroxylamines in low to good yields;[604] the formation of diarylamines has also been reported.[605] By reducing the unstable diarylhydroxylamines in situ, diarylamines are accessible in good yields (compare to Eq. 180).[606]

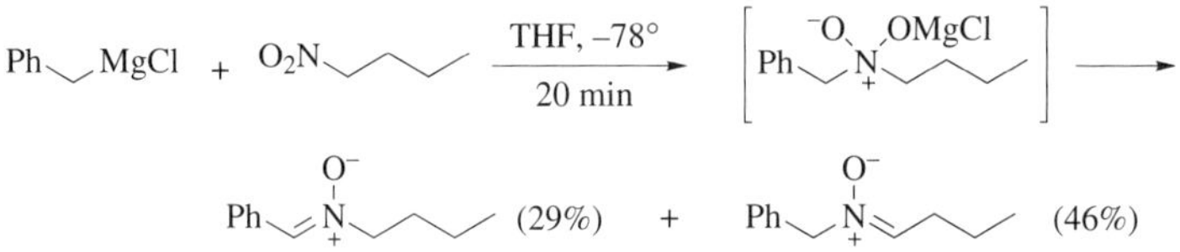

(Eq. 186)

Reactions of nitroarenes with vinylmagnesium halides give indoles (the Bartoli reaction).[607] Site selectivity problems may be avoided by temporarily installing a bromine ortho to the nitro group (Eq. 187).[608]

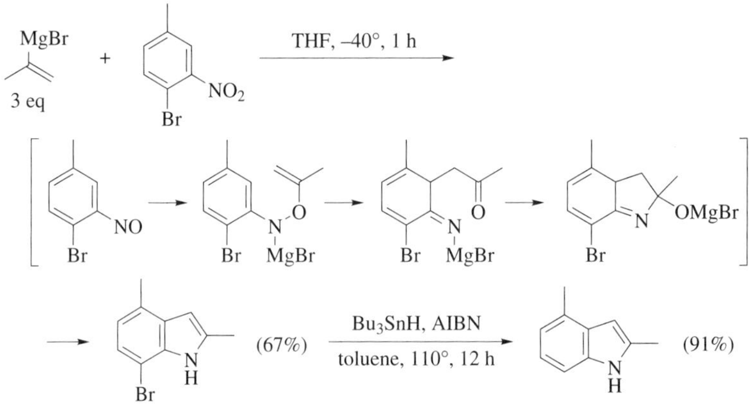

(Eq. 187)

Reaction of the highly explosive fluorotrinitromethane with the anion of 2,4,6-trinitrotoluene, prepared with potassium hydroxide, gives the highly explosive α,2,4,6-tetranitrotoluene in 89% yield.[609] Ketone enol silyl ethers and the equally highly explosive tetranitromethane react to give α-nitro ketones in low to very high yields (Eq. 188).[610]

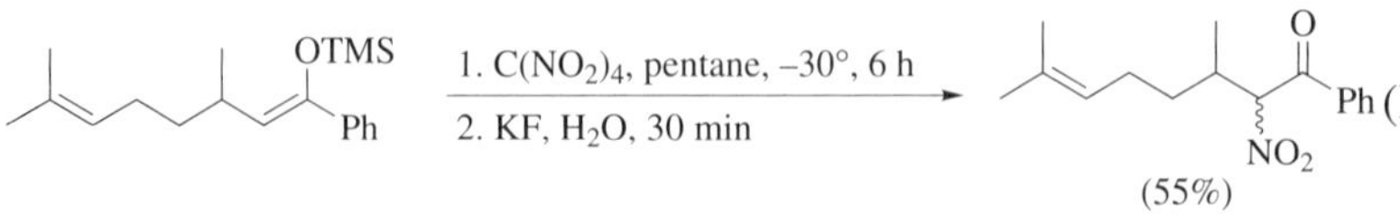

(Eq. 188)

N,N-Disubstituted nitroxides are formed in the reaction of *tert*-butylmagnesium chloride with 1,1-dimethylnitroethane[611] and nitroarenes,[612,613] and by reaction of 1,1-dimethylnitroethane with arylsodium or aryllithium reagents.[614]

Amination of Enolates with Diazonium Salts

Enolates of β-dicarbonyl and similar doubly activated compounds are aminated by aryldiazonium salts to give hydrazones by rearrangement of the intermediate azo compounds.[14,15,615,616] The Japp-Klingemann reaction[617,618] is a variation in which either acyl cleavage or decarboxylation occurs in situ after the amination. The hydrazones may be reduced to amines.[619]

The Diazo Transfer Reaction

Stabilized carbanions react with certain azides to give diazo compounds (Eq. 30, path A)[620–624] Substrates include enolates with one additional activating group and cyclopentadienide anions.[625] Simple ketones only rarely[322,324] undergo the diazo transfer reaction unless a formyl group is installed temporarily in the α-position. Only one example of an alkylcarbanion leading to a diazo compound was found in the literature.[626] The most widely used azide is tosyl azide but less dangerous sulfonyl azides have been proposed as alternatives.[627–629] The vast majority of diazo compounds preparared in this manner is used as precursors to carbenes or carbenoids although methods exist for their reduction to hydrazones, hydrazines, or amines.[198,205,630] Diastereo- or enantioselective reductions of this kind do not seem to have been reported although the carbenoid NH insertion[631] reaction shown in Eq. 189[632] indicates that they may be feasible.

Ph, O, O, P, Ph, N, Me, N_2 → ($BnOCONH_2$, $Rh_2(OAc)_4$ (cat), toluene, 80°, 5 h) → Ph, O, O, P, Ph, N, Me, $NHCO_2Bn$ (56%) 15% de (Eq. 189)

Amination of Boranes

Organoboranes, which are readily accessible by hydroboration of olefins,[633] undergo many of the amination reactions also observed with alkyl carbanions but often afford higher yields with fewer complications.[634] Thus organoboranes give amines by reaction with chloramine[635] and its dialkyl derivatives, *N*-chloro *O*-(2,4-dinitrophenyl)hydroxylamine,[636] hydroxylamine *O*-sulfonic acid,[635,637,638] *O*-(2,4-dinitrophenyl)hydroxylamine,[639] the lithium or potassium salts of *tert*-butyl *N*-tosyloxycarbamate,[640] chloramine-T,[641] and azides.[642–644] Enantiomerically enriched amines are formed using chiral, non-racemic borane[645,646] or boronic esters.[220,647,648] Reaction of triphenylborane with hydroxylamine *O*-sulfonic acid gives aniline.[649]

The Neber Rearrangement

The Neber rearrangement[650–653] is a method for preparing α-amino ketones by base-catalyzed intramolecular rearrangement of ketoxime *O*-sulfonates. The intermediate azirine,[654–656] which can be isolated, can also lead to aziridine derivatives when the base is lithium aluminum hydride[657] or a Grignard reagent (the Hoch-Campbell reaction)[658,659] (Eq. 190).

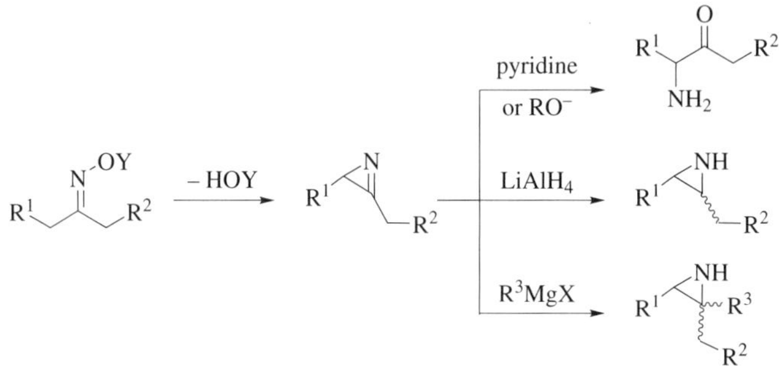

(Eq. 190)

EXPERIMENTAL CONDITIONS

A number of reagent and product classes discussed in this chapter require special handling. *Haloamines are toxic and explosive; the experimental hazards are eliminated or greatly reduced by using solutions in inert solvents.*[10] *Some* O-*sulfonylhydroxylamines are explosive:* O-*(2,4-dinitrobenzenesulfonyl)hydroxylamine is flammable, highly toxic, and highly explosive; an explosion occurred when brought in contact with potassium hydride.*[93] *An explosion of* O-*mesitylenesulfonylhydroxylamine occurred on storage below 0°.*[114] *Dry aryldiazonium salts are explosive. Hydrazoic acid and its salts are toxic. Organic azides are explosive. Distillation should be avoided or carried out at low temperatures behind a shield. Tosyl azide has the exposive power of TNT.*[627,660,661] *Triazenes, the products of the reaction of azides with carbanions, are potent chemical carcinogens*[258] *and vesicants.*[259] *Low-molecular-weight triazenes have high vapor pressures. Some are explosive and cause headaches.*[662] *Chromium and cadmium salts are toxic.*

The great variey of reagents and substrates dealt with in this chapter does not permit a detailed discussion of conditions for each experiment. Most of the reactions require flame-dried glassware, anhydrous solvents, and an inert atmosphere of nitrogen or argon.

Preparation of Electrophilic Aminating Reagents

References to the preparation of electrophilic aminating reagents are given in Table A.

Table A

References to the Preparation of Amination Reagents

1. Haloamines. Reviews: refs.10, 663.
   $ClNH_2$: refs. 664-668; method of analysis: ref. 64.
   $BrNH_2$: ref. 669.
   Cl_2NH: refs. 663, 667.
   Cl_3N: ref. 670.
   RNHCl, R_2NCl, $RNCl_2$; R = alkyl: ref. 68.
   R_2NCl from R_2NH and *N*-chlorosuccinimide: refs. 671-673.
   $ClNHCO_2R$: ref. 674.
2. O-Substituted Hydroxylamines: review ref. 675.
   a. *O*-Alkyl-Substituted Hydroxylamines: refs. 676-680.
      $MeONH_2$: refs. 681-683.
      $EtONH_2$: ref. 682.
      $BnONH_2$: refs. 85, 680, 684, 685.
      MeONHR [R = Bn, 2-MeC_6H_4, $Ph(CH_2)_3$]: ref. 85.
      $MeONR_2$ (R = alkyl): ref. 85.
      $RONMe_2$ (R = alkyl): ref. 677.
   b. *O*-Arylhydroxylamines
      $PhONH_2$: ref. 686.
      2-$O_2NC_6H_4ONH_2$: ref. 94.
      4-$O_2NC_6H_4ONH_2$: refs. 94, 96, 107.
      2,4-$(O_2N)_2C_6H_3ONH_2$: refs. 93, 94, 96, 687-689.
      various substituted 2-$O_2NC_6H_3ONH_2$ and 5-$O_2NC_6H_3ONH_2$: ref. 96.
      2,4,6-$(O_2N)_3C_6H_4ONH_2$: ref. 94.
   c. *O*-Acylhydroxylamines: refs. 679, 699.
      $Me_3CCO_2NH_2$: ref. 690.
      $BzONH_2$: refs. 690, 691.
      3-$ClC_6H_4CO_2NH_2$: ref. 690.
      4-$O_2NC_6H_4CO_2NH_2$: ref. 690.
      2,4,6-$Me_3C_6H_2CO_2NH_2$: refs. 690, 692, 693.
      BzONHR (R = alkyl): ref. 694.
      RCO_2NHCO_2Bu-*t* (R = *t*-Bu, aryl: ref. 690.
      $BzONR_2$: refs. 109, 695.
   d. *O*-Sulfonylhydroxylamines
      HSO_3ONH_2: refs. 696-698.
      $MeSO_2ONH_2$: ref. 137.
      $PhSO_2ONH_2$: ref. 133.
      4-$MeC_6H_4SO_2ONH_2$: refs. 132, 699.
      2,4,6-$Me_2C_6H_2SO_2ONH_2$: refs. 700 (review), 116, 133, 693, 699-703.
         Hazards: refs. 114, 116, 395, 703, 704.
      2-$O_2NC_6H_4SO_2ONH_2$: ref. 705.
      2,4-$(O_2N)_2C_6H_3SO_2ONH_2$: ref. 705.
      2,4,6-$(O_2N)_3C_6H_2SO_2ONH_2$: ref. 705.
      $ArSO_2ONEt_2$ (Ar = Ph, 2,4,6-$Me_3C_6H_2$): ref. 133.
      $ArSO_2ONR_2$ (Ar = Ph, 2,4,6-$Me_3C_6H_2$; NR_2 = 1-piperidinyl): ref. 134.
      $TsONHCO_2Et$: ref. 119.
      $TsON(M)CO_2Bu$-*t* (M = Li, MgCl): refs. 127.
      $ArSO_2ON(Li)CO_2CH_2CH{=}CH_2$ (Ar = 4-MeC_6H_4; 2,4,6-$Me_3C_6H_2$): ref. 130.
      4-$O_2NC_6H_4SO_2ONHP(O)NHSO_2C_6H_4NO_2$-4: ref. 706.

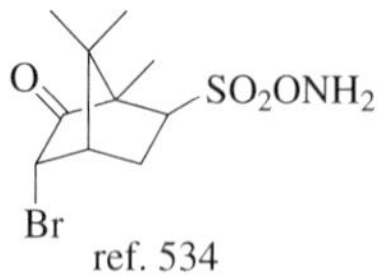

ref. 534 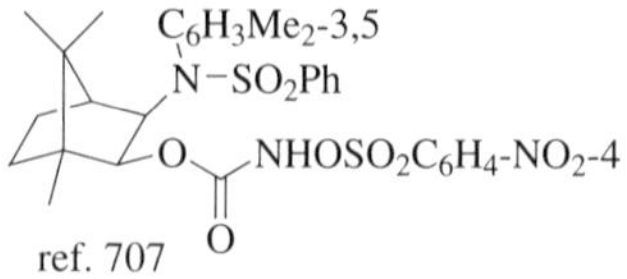 ref. 707 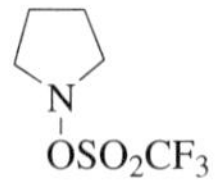 ref. 708

e. OSi-Substituted Hydroxylamines

TMSONHOTMS: refs. 699, 709.

TMSONHR (R = alkyl): refs. 101, 709.

TMSONHBn: ref. 709.

f. *O*-Phosphinylhydroxylamines

$Ph_2P(O)ONH_2$: refs 138, 141, 710-713.

$(4\text{-}MeOC_6H_4)_2P(O)ONH_2$: refs. 106, 141, 712, 713.

$(4\text{-}MeC_6H_4)_2P(O)ONH_2$: refs. 141, 712, 713.

$Ph_2P(O)ONMe_2$: refs. 85, 714.

$Ph_2P(O)ONR^1R^2$ (R^1 = alkyl; R^2 = alkyl, allyl, Bn): ref. 715.

$Ph_2P(O)ONR^1R^2$ ($R^1R^2 = (CH_2)_4$, CH=CH-CH=CH): ref. 715.

$Ph_2P(O)ONR_2$ (R = alkyl, $c\text{-}C_6H_{11}$, Bn): ref. 716.

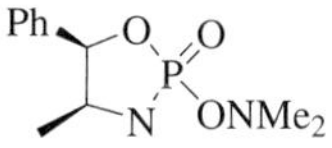

ref. 147

3. Oxaziridines

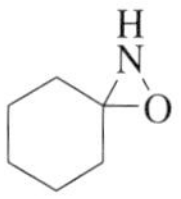

ref. 148 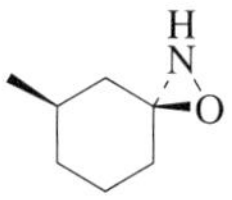 ref. 717 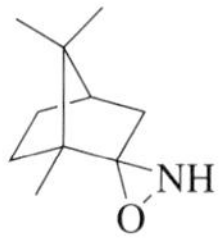 ref. 718  ref. 718

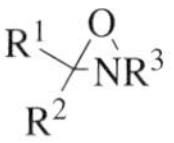

R^1	R^2	R^3	Refs.
CCl_3	H	$CO_2Bu\text{-}t$	719
Me	Me	COMe (chiral)	717
CO_2Et	CO_2Et	$CO_2Bu\text{-}t$	155
CO_2Me	Ph	$CO_2Bu\text{-}t$	155
Ph	CF_3	$CO_2Bu\text{-}t$	155
Ar	H	CO_2Me	153
Ar	H	$CO_2Bu\text{-}t$	157, 720
Ar	H	$CONEt_2$	158
Ar	H	CONHCH(Me)CH(ODBDPS)Ph (derived from pseudoephedrine)	155
Ar	H	9-fluorenylmethoxycarbonyl	720
Ph	Ph	CO_2Me	153

4. Imines

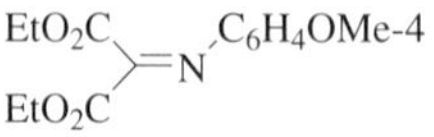

ref. 167 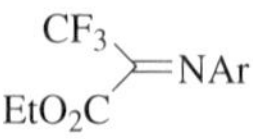 ref. 721

5. (*N*-Arylsulfonylimino)phenyliodinanes

TsN=IPh: ref. 722.

6. Oximes

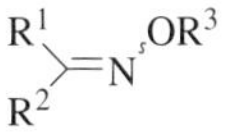

$R^1R^2C{=}N{-}OR^3$

R^1	R^2	R^3	Refs.
H	H	Bn (see Eq. 6)	723
Me	Me	$SO_2C_6H_2Me_3$-2,4,6	724
Ph	Me, Ph	SO_2Ph, 4-$BrC_6H_4SO_2$, 4-$MeC_6H_4SO_2$	725
4-$CF_3C_6H_4$	4-$CF_3C_6H_4$	SO_2Me	179, 726
4-$CF_3C_6H_4$	4-$CF_3C_6H_4$	$SO_2C_6H_4Me$-4	179
3,5-$(CF_3)_2C_6H_3$	3,5-$(CF_3)_2C_6H_3$	$SO_2C_6H_4Me$-4	179

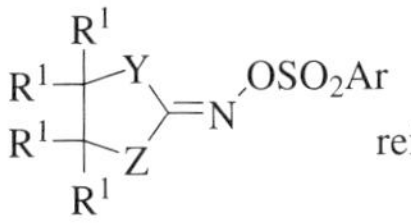

refs. 181, 182

R^1	Y	Z
H	O	O
Me	O	O
H	NMe	O
H	O	NMe
H	NMe	NMe

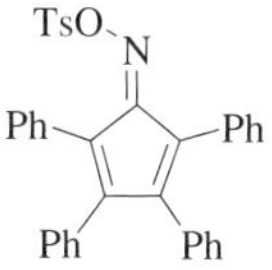

ref. 727

7. Diazonium Salts

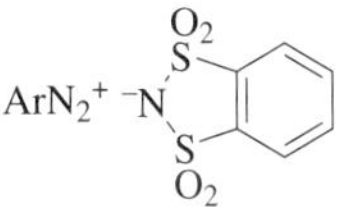

ArN_2^+ ref. 728

8. Azo Compounds

$R^1N{=}NR^2$

R^1	R^2	Refs.
Ph	CO_2R	401, 729, 730
Ph	COPh	729
Ar	$SO_2C_6H_4Me$-4	255
Ar	COAr	389, 729
ArNHCO	CO_2R	383, 731

$R^1O_2CN{=}NCO_2R^2$

R^1	R^2	Refs.
Et	Et	732
Cl_3CCH_2	Cl_3CCH_2	733, 734
Cl_3CCH_2	$(CH_2)_2TMS$	380
allyl	allyl	735
t-Bu	*t*-Bu	736
Ph	*t*-Bu	410
Bn	Bn	737
(+)-menthyl	(+)-menthyl	408
(−)-menthyl	(−)-menthyl	218
(−)-bornyl	(−)-bornyl	406
(−)-isobornyl	(−)-isobornyl	408

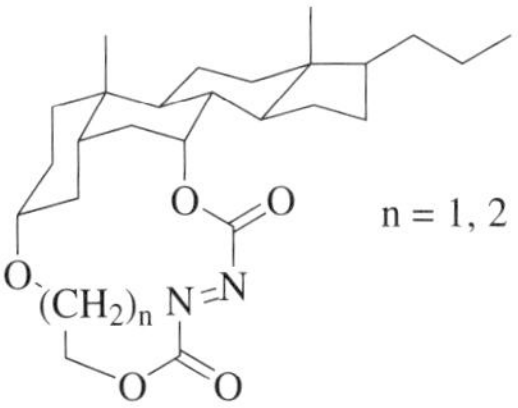

n = 1, 2 ref. 738

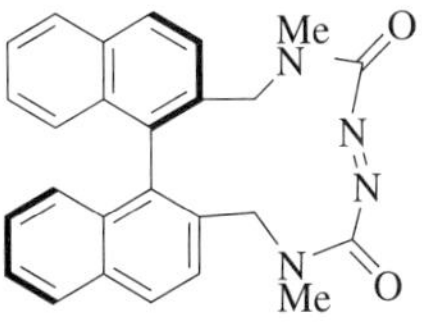

ref. 219

9. Azides

RN_3

R	Refs.
TMS	739
$TMSCH_2$	see Experimental Procedures
EtO_2C	302, 740
t-BuCH=CH	741
$PhSCH_2$	see Experimental Procedures
$(PhO)_2P(O)$	742

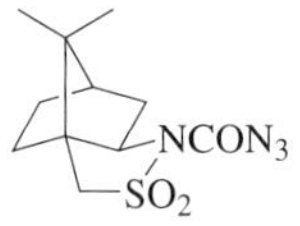

ref. 304

RSO_2N_3

R	Refs.
Me	743
CF_3	309, 744, 745
$4\text{-}MeC_6H_4$	335, 746, 747
	hazards: 627, 660, 661
	safer analogs: 627-629,
RPh R = H, 2-I, $2\text{-}NO_2$	748
2,4,6-(*i*-Pr)$_3C_6H_2$	448, 748-750
$4\text{-}AcNHCOC_6H_4$	751
polymer-bound	752, 753

10. "TSN=Se=NTs": refs. 348, 349, 754.

11. Nitridomanganese complexes reviews: 24, 25

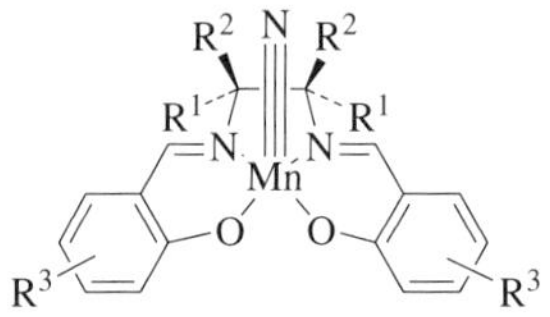

refs. 471, 755, 756

Conversions of Amination Products

The following is a selection of procedures for the conversion of non-amine amination products into amines and other nitrogen-containing compounds. Relevant information may also be found in reviews of protecting groups.[757,758]

N-Tosylamines into Amines: Bu_3SnH,[349] Na/liquid ammonia.[348]

N-Tosylamines into *N*-Tosylimines: SeO_2.[345]

Azo Compounds into Hydrazines: Al/Hg.[205]

Azo Compounds into Amines:[759] H_2/Pd.[196,197,415]

Hydrazides into Amides or Amines:[759] TFA-SmI_2;[233,481,760] peracids;[761] Raney nickel;[405,432,460,762–765] sodium in liquid ammonia;[762,766] N_2O_3 or $NaNO_2$/HOAc;[767,768] H_2/Pt.[411,769,770]

Triazene Salts into Amines: $NaBH_4$;[311] Ac_2O-Al/Hg;[317] sodium bis(2-methoxyethoxy)aluminum hydride.[333]

Azides into Amines:[20,23,771,772] H_2/Pd or H_2/Pt;[317,318,339,450,773] H_2/Pd-$(Boc)_2O$ in *N*-acyloxazolidinones to prevent reaction of the amine with the chiral auxiliary;[774] Raney nickel;[444] $SnCl_2$;[444,450,458,775,776] Zn;[777] Al/Hg;[777] sodium borohydride under phase-transfer conditions;[778] lithium aluminum hydride;[779] H_2S;[780] triphenylphosphine.[325,781,782]

Azides into Imines: base.[783–785]

Azides into Enamines: $NaReO_4$.[331]

EXPERIMENTAL PROCEDURES

Procedures are listed by type of reagent in the same order as in the section on Reagents and Mechanisms.

PhLi + CuCN —(THF, –40°)→ PhCu(CN)Li —(1. i-Pr_2NLi, –40°; 2. O_2, –78°, 20 min)→ $PhN(Pr$-$i)_2$ (60%)

***N,N*-Diisopropylaniline (Amination of an Arylcopper Reagent with a Lithium Dialkylamide).**[54] Copper(I) cyanide (2 mmol) was added at –40° to a solution of phenyllithium (2 mmol) in THF (10 mL) and the mixture was stirred for 20 minutes. A THF solution of LDA was added and after 15 minutes at –40° the mixture was cooled to –78° and a vigorous stream of oxygen was introduced for 20 minutes. The mixture was allowed to warm to room temperature and passed through a pad of celite. Concentration and kugelrohr distillation of the residue (100° bath temperature, 20 mmHg) gave 0.21 g (60%) of the title product as an oil: ^1H NMR ($CDCl_3$) δ 7.69–7.33 (m, 5H), 3.80 (m. 2H) and 1.24 (d, $J = 6.8$ Hz, 12H). Anal. Calcd for $C_{12}H_{19}N$: C, 81.29; H, 10.81; N, 7.90. Found: C, 82.00; H, 10.81; N, 8.92.

EtO_2C–CH_2–CO_2Et —(1. NaH, benzene; 2. $ClNH_2$, morpholine, Et_2O, rt, overnight; reflux, 5 h)→ $(EtO_2C)_2CHNH_2$ (89%)

Diethyl Aminomalonate (Amination of a β-Dicarbonyl Compound with Chloramine).[62] Diethyl malonate was converted into the sodium salt with sodium hydride in benzene and the solvent was removed. To a suspension of the salt (11.3 g, 0.06 mol) in Et_2O (100 mL) was added with cooling a solution of chloramine in Et_2O (0.12 mol) followed by morpholine (5.22 g, 0.06 mol). The mixture was stirred with cooling for 2 hours and at room temperature over night and then refluxed for 5 hours. The filtered mixture was concentrated and the residue was distilled to give 10.9 g (89%) of the title compound, bp 116–117° (18 mmHg). The product gave the correct elemental analysis and the physical properties were those reported in the literature.

MeLi + MeONHCH$_2$Ph —(ether, hexanes, –78°)→ MeON(Li)CH$_2$Ph —(t-BuLi, pentane, –78° to –10°)→ t-BuNHCH$_2$Ph (99%)

***N-tert*-Butylbenzylamine (Amination of an Alkyllithium Compound with a Lithium Nitrenoid).**[85] To a solution of MeLi in Et_2O (1.40 mL, 1.54 mmol) was added at –78° a solution of *N*-benzyl-*O*-methylhydroxylamine (0.21 g, 1.53 mmol) in hexanes (5 mL). After stirring for 5 minutes, a solution of *t*-BuLi in pentane (1.2 mL, 1.28 M, 1.53 mmol) was added, the mixture was allowed to warm to –10° and kept at that temperature for 2 hours. Water and Et_2O were added and the dried (Na_2SO_4) Et_2O solution was concentrated. The residue was distilled (kugelrohr) to give 0.28 g (99%) of the title product, bp 70° (0.5 mmHg): ^1H NMR ($CDCl_3$) δ 7.33 (s, 5H), 3.73 (s, 2H) and 1.18 (s, 9H); ^{13}C NMR δ 141.4, 128.4, 128.2, 126.7, 50.6, 47.2, 29.1. Anal. Calcd for $C_{11}H_{17}N$: C, 80.93; H, 10.50; N, 8.58. Found: C, 80.59; H, 10.91; N. 8.67.

Br, F (4-fluorobromobenzene) → 1. *n*-BuLi, hexanes, –78° 2. $CuBr{\bullet}Me_2S$ 3. $TsON(Li)CO_2Bu$-*t*, THF, –78° → $NHCO_2Bu$-*t*, F (50%)

***tert*-Butyl 4-Fluorophenylcarbamate (Amination of an Arylcopper Reagent with Lithium *tert*-Butyl *N*-Tosyloxycarbamate).**[127] A solution of *n*-BuLi in hexane (0.4 mL, 2.5 M, 1 mmol) was added dropwise at −78° to a solution of *tert*-butyl *N*-tosyloxycarbamate (0.287 g, 1 mmol) in THF. The mixture was stirred at −78° for one hour. In a separate vessel, a solution of 4-fluoro-1-bromobenzene (1 mmol) was treated with one equivalent of *n*-BuLi in hexane at −78° for 30 minutes and then cannulated into a suspension of $CuBr{\bullet}Me_2S$ (1 mmol) in THF (2 mL). The mixture was stirred at −60° to −78° for one hour, cooled to −78°, treated dropwise with the solution of lithium *tert*-butyl *N*-tosyloxycarbamate, and stirred at −78° for 30 minutes. A saturated aqueous solution of NH_4Cl and ammonia (5 mL) was added and the aqueous phase was extracted with Et_2O. The combined organic phases were washed with brine, dried ($MgSO_4$), and concentrated. Flash chromatography of the residue (1:5 EtOAc/cyclohexane) gave 0.105 g (50%) of the title product, mp. 111°: IR (KBr) 2255, 1690 cm^{-1}; ^1H NMR ($CDCl_3$) δ 7.31 (m, 2H), 6.97 (m, 2H), 6.6 (s, 1H); 1.51 (s, 9H); ^{13}C NMR δ 158.6, 156.2, 134.2, 120.2, 115.4, 80.5, 28.2. Anal. Calcd for $C_{11}H_{14}FO_2N$: C, 62.55; H, 6.68; N, 6.62. Found: C, 62.45; H, 6.69; N, 6.47.

Ph_2Zn + (O-Bz, N, O morpholine) → $CuCl_2$ (2.5 mol%) THF, 0-5°, 90 min → (Ph, N, O morpholine) (67%)

***N*-Phenylmorpholine (Amination of an Arylzinc Derivative with an *O*-Acylhydroxylamine).** This procedure is found in *Organic Syntheses*.[111]

(N, H) → 1. *n*-BuLi (2.1 eq), THF, hexane, TMEDA, 0° 2. 2,4,6-$Me_3C_6H_2SO_2ONEt_2$, –78°; to rt, overnight → Et_2N, (N, H) (43%)

N,N-Diethyl-5,10-dihydroindeno[1,2-*b*]indol-10-amine (Amination of a Benzylic Anion with an *N,N*-Disubstituted *O*-Arenesulfonylhydroxylamine).[136] A solution of *n*-BuLi (30 mL, 2.25 M in hexane, 67.5 mmol) was added with ice cooling to a solution of 5,10-dihydroindeno[1,2-*b*]indol (6.6 g, 32.1 mmol) and TMEDA (20 mL) in THF (200 mL), the mixture was stirred at room temperature for 45 minutes, and cooled to −78°. Solid *N,N*-diethyl *O*-mesitylenesulfonylhydroxylamine (8.7 g, 39.7 mmol) (*caution, the N,N-unsubstituted analog is explosive*) was added in one portion and the mixture was left to warm to room temperature and stirred overnight. Et_2O (150 mL) was added and the organic phase was washed with water (2 × 100 mL) and then extracted with 2 N HCl (2 × 60 mL). The precipitate that formed in the acid extracts was collected by filtration and suspended in 2 N NaOH solution (100 mL). The mixture was extracted with Et_2O (150 mL), which was then washed with water (3 × 100 mL). Concentration of the dried ($MgSO_4$) Et_2O solution gave 3.8 g (43%) of the title product as a brownish-pink solid, mp 126.0–126.5°, unchanged on crystallization from petrol ether: ^{1}H NMR ($CDCl_3$) δ 8.2 (br s, 1H), 7.91–7.03 (m, 8H), 4.87 (s, 1H), 2.58 (q, $J = 7$ Hz, 4H), 1.08 (t, $J = 7$ Hz, 6H). Anal. Calcd for $C_{19}H_{20}N_2$: C, 82.56; H, 7.28; N, 10.13. Found: C, 82.81; H, 7.29; N, 9.93.

CO_2Et → 1. KOBu-*t*, THF, −78°; 2. $(4\text{-}MeOC_6H_4)_2P(O)ONH_2$, −78° to rt; 3. Ac_2O, Et_3N → NHAc, CO_2Et (67%)

Ethyl (*N*-Acetylamino)phenylacetate (Amination of an Ester Enolate with an *O*-Phosphinoylhydroxylamine).[106] A freshly prepared solution of KOBu-*t* (31 mg, 0.28 mmol) in THF (2 mL) was added slowly to a solution of ethyl phenylacetate (41 mg; 0.25 mmol) in THF (3 mL) cooled to −78° and the mixture was stirred at −78° for 15 minutes. *O*-[Di(*p*-methoxyphenyl)]phosphinoylhydroxylamine (*caution, related hydroxylamine derivatives are explosive*) (81 mg, 0.28 mmol) was added as a solid in one portion, and the mixture was left to warm to room temperature and stirred overnight. Acetic anhydride (71 μL, 0.75 mmol) and triethylamine (210 μL, 1.5 mmol) were added and the mixture was stirred at room temperature for one hour. Et_2O (20 mL) and saturated aqueous NH_4Cl solution (30 mL) were added, and the aqueous layer was extracted with Et_2O (2 x 30 mL). The dried ($MgSO_4$) extracts were concentrated and the residue was purified by flash chromatography (1:1 EtOAc:cyclohexane) to give 37 mg (67%) of the title product as a colorless oil, R_f 0.20 (1:1 EtOAc:cyclohexane). No other data were reported.

CONHPh / CONHPh → (oxaziridine) NH (2.5 eq), DABCO, toluene, rt, 12 h → H_2N, H_2N CONHPh, CONHPh (91%) → EtOH, reflux → HN= CONHPh, CONHPh (96%)

Diamino-*N*,*N*′-diphenylmalonamide and Imino-*N*,*N*′-diphenylmalonamide (Diamination of a Malonamide with 1-Oxa-2-azaspiro[2.5]octane and Conversion of the Product into an Imine).[149] A suspension of *N*,*N*′-diphenylmalonamide in a mixture of toluene and 2.2–2.5 equivalents of 1-oxa-2-azaspiro[2.5]octane was treated with a solution of 1,4-diazabicyclo[2.2.2]octane (5–10 mol%) in toluene (1 mL). The solid was collected by filtration after 12 hours at room temperature, washed with a small amount of EtOH, and air dried to give diamino-*N*,*N*′-diphenylmalonamide in 91% yield, mp 130–132°: ^{1}H NMR (DMSO-d$_6$) δ 6.9-7.9 (m, 10H), 3.2–3.6 (br, 6H); ^{13}C NMR (DMSO-d$_6$) δ 170.6, 138.3, 128.6, 123.6, 119.4, 73.9. Anal. Calcd for $C_{15}H_{16}N_4O_2$: C, 63.37; H, 5.67; N, 19.71. Found: C, 62.60; H, 5.85; N, 19.74.

A 10% solution of diamino-*N*,*N*′-diphenylmalonamide in EtOH was heated under reflux for 15 minutes. The solid was collected after 12 hours at room temperature and air-dried to give imino-*N*,*N*′-diphenylmalonamide in 96% yield, mp 158–162°. ^{1}H NMR (DMSO-d$_6$) δ 12.33 (br, 1H), 10.5 (br, 2H), 6.9–7.9 (m, 10H); ^{13}C NMR (DMSO-d$_6$) δ 159.8, 162.3, 164.1, 137.3, 137.3, 119.7, 120.4, 128.6, 128.9, 124.3, 124.4. Anal. Calcd. for $C_{15}H_{13}N_3O_2$: C, 67.40; H, 4.90; N, 15.72. Found: C, 66.70; H, 5.05; N, 16.34.

CN, CO_2Et — 1. LiHMDS, THF, –78°; 2. EtO_2C, EtO_2C, O, NCO_2Bu-*t*, –78°, 20 h → CN, $NHCO_2Bu$-*t*, CO_2Et (70%)

Ethyl *tert*-Butoxycarbonylamino(cyano)phenylacetate (Amination of a Cyanoacetic Ester Enolate with an *N*-Acyloxaziridine).[155] Ethyl phenylcyanoacetate (0.22 mmol) was added to a solution of LiHMDS (0.22 mL, 1.0 M in hexane, 0.22 mmol) at −78°. After 30 minutes, a solution of *N*-*tert*-butoxycarbonyl-3,3-bis(ethoxycarbonyl)oxaziridine (95 mg, 0.33 mmol) in THF (1 mL) was added, the mixture was stirred at −78° for 12 hours and then left to warm to room temperature. CH_2Cl_2 and saturated aqueous NH_4Cl were added, and the organic layer was washed twice with saturated NH_4Cl. Removal of the solvent from the dried (Na_2SO_4) solution and flash chromatography of the residue (10:1 petrol ether/EtOAc) gave 47 mg (70%) of the title product as an oil: IR (film) 2253, 1754, 1721 cm^{-1}; ^{1}H NMR ($CDCl_3$) δ 7.67 (br 2H), 7.46-7.45 (m, 3H), 5.75 (br, 1H), 4.25 (q, $J = 7.2$ Hz, 2H), 1.46 (s, 9H), 1.25 (d, $J = 7.2$ Hz, 3H). MS-CI (m/z): $[M + H]^+$ calcd for $C_{16}H_{20}N_2O_4$: 305.1501; found: 305.1511.

i-PrMgBr + MeO, N, CO_2Me, CO_2Me — THF, –95° → Pr-*i*, N, CO_2Me, CO_2Me, MeO — KOH, EtOH / air, rt, 48 h → NHPr-*i*, MeO (57%)

***N*-Isopropyl-*p*-anisidine (Amination of a Grignard Reagent with an Imine).**[167] Isopropylmagnesium bromide (0.83 M in THF, 0.54 mL, 0.45 mmol) was added slowly to a solution of diethyl 2-[*N*-(*p*-methoxyphenyl)imino]malonate (84 mg, 0.30 mmol) in THF (5 mL) at −95°. After 30 minutes saturated aqueous $NaHCO_3$ was added and the mixture was extracted with EtOAc (3 × 10 mL). The combined extracts were washed with brine, dried (Na_2SO_4), and the solvent was removed. The residue was stirred vigorously with 1 M aqueous KOH (0.11 mL) and EtOH (3.3 mL) at room temperature for 48 hours. The EtOH was removed after addition of aqueous Na_2SO_3 and the residue was extracted with EtOAc (3 × 10 mL). The extracts were washed with brine, dried (Na_2SO_4), and the solvent was removed. Preparative TLC of the residue (silica gel, 1:15 EtOAc:hexane) gave 28 mg (57%) of the title product: ^{1}H NMR ($CDCl_3$) δ 6.57 (d, J = 8.9 Hz, 2H), 6.77 (d, J = 8.9 Hz, 2H), 1.19 (d, J = 6.3 Hz, 6H), 3.74 (s, 3H), 3.61–3.48 (m, 1H); ^{13}C NMR ($CDCl_3$) δ 23.07, 45.24, 55.79, 114.93, 141.73, 151.95.

OTMS + TsN=IPh —(MeCN, warm)→ O, NHTs (95%)

2-[*N*-(*p*-Toluenesulfonyl)amino]acetophenone (Amination of a Ketone Silyl Enol Ether with [*N*-(*p*-tolylsulfonyl)imino] phenyliodinane).[172] A solution of 1-(trimethylsilyloxy)styrene (0.5 mmol) in dry MeCN (7 mL) was treated with TsN=IPh (0.6 mmol). The mixture was warmed and the solvent was removed after the reagent had dissolved. The residue was purified by chromatography on silica gel followed by crystallization from Et_2O to give the title product in 95% yield. No analytical or spectroscopic data were reported.

MgBr + N, O, SO_2Ph, O, O —(PhCl, 0°)→ N, O, O —(HCl, EtOH, reflux, 6 h)→ NH_3^+ Cl^- (89%)

1-Aminoadamantane Hydrochloride (Amination of a Grignard Reagent with an *O*-Arenesulfonyloxime).[182] To a solution of 4,4,5,5-tetramethyl-1,3-dioxolane-2-one *O*-benzenesulfonyloxime (602 mg, 2.01 mmol) in chlorobenzene (14 mL) was added dropwise at 0° 1-adamantylmagnesium bromide (0.63 M in Et_2O, 3.5 mL, 2.2 mmol) and the mixture was stirred at 0° for 30 minutes. The reaction was quenched with pH 9 buffer at 0° and the mixture was extracted three times with EtOAc. The combined extracts were washed with brine, dried (Na_2SO_4), and concentrated. The crude imine was refluxed with 10 mL of EtOH and 1.3 mL of 6 M HCl for 10 hours. The ethanol was removed, the residue was made basic with 5 mL of 5 M NaOH, and the mixture was extracted three times with CH_2Cl_2. The combined extracts were washed with brine, dried (Na_2SO_4), and concentrated. The residue was dissolved in MeOH, HCl in Et_2O was added, and all volatiles were removed under vacuum. The residue was stirred with Et_2O and the solids were collected by filtration and dried to give 334 mg (89%) of the title product: ^{1}H NMR (DMSO-d_6) δ 8.18 (br, 3H), 2.05 (s, 3H), 1.79 (s, 6H), 1.62 (d, $J = 12.2$ Hz, 3H), 1.54 (d, $J = 12.2$ Hz, 3H); ^{13}C NMR (DMSO-d_6) δ 51.1, 40.1, 35.4, 28.5.

t-BuMgCl + [4-ClC$_6$H$_4$N$_2^+$ o-benzenedisulfonimide] —(THF, −78°)→ 4-ClC$_6$H$_4$N=NBu-t (86%)

***E*-(*tert*-Butyl)(4-chlorophenyl)diazene (Reaction of a Grignard Reagent with an Aryldiazonium Salt).**[191] A suspension of 4-chlorobenzenediazonium *o*-benzenedisulfonimide (1.77 g, 5 mmol) in anhydrous THF (15 mL) was stirred vigorously at −78°, a solution of *t*-BuMgCl (5 mmol) was added over a period of 10 minutes, and stirring at −78° was continued for one hour. The mixture was poured into 30 mL of water and extracted with Et_2O (2 × 30 mL). The washed (H_2O, 30 mL) and dried (Na_2SO_4) extracts were heated in a 70° water bath to remove the Et_2O and heating was continued for 1 hour to ensure conversion of any Z into the E isomer. Purification by column chromatography gave the title product in 83% yield, bp 57–58°/0.25 mm: ^{1}H NMR ($CDCl_3$) δ 7.60 (d, $J = 8.9$ Hz, 2H), 7.38 (d, $J = 8.9$ Hz, 2H), 1.32 (s, 9H); ^{13}C NMR ($CDCl_3$) δ 150.5, 135.5, 128.8, 122.9, 67.5, 26.4.

1-(4-methylbenzyl)benzotriazole —(1. *n*-BuLi, THF, −78°; 2. PhN=NPh)→ Bt–CH(p-tolyl)–N(Ph)–N(Li)Ph —(1. *n*-BuMgBr, −78° to rt; 2. NH_4Cl)→ *n*-Bu–CH(p-tolyl)–N(Ph)–NHPh (57%)

1,2-Diphenyl-1-(1-*p*-tolylpentyl)hydrazine (Amination of a Benzotriazolylmethyl Anion with an Azo Compound Followed by Displacement of the Benzotriazole Functionality by a Grignard Reagent).[359] To a solution of 1-(4-methylbenzyl)benzotriazole (2 mmol) in THF (7 mL) was added *n*-BuLi (2 mmol) at −78° and the mixture was stirred at −78° for 10 minutes. *n*-BuMgBr (4 mmol) in Et_2O was added followed by the azobenzene (2 mmol), and the mixture was left to warm to room temperature overnight. It was washed with 30 mL of 10% NH_4Cl solution and the washing was extracted with EtOAc (2 × 10 mL). Removal of the solvents from the dried ($MgSO_4$) organic phase and column chromatography of the residue (SiO_2, 1:1 toluene/hexane) gave the title product in 57% yield, mp 97–99°: ^{1}H NMR δ 7.20–7.04 (m, 8H), 6.93 (d, $J = 8$ Hz, 2H), 6.81–6.68 (m, 4H), 5.07 (br, s, 1H), 4.94 (t, $J = 7$ Hz, 1H), 2.28 (s, 3H), 2.18–2.06 (m, 1H), 2.01–1.86 (m, 1H), 1.57–1.26 (m, 4H), 0.88 (t, $J = 7$ Hz,

3H); ^{13}C NMR δ 150.3, 148.5, 137.1, 129.1, 128.9, 128.1, 119.5, 115.2, 112.2, 66.0, 31.2, 29.4, 22.7, 21.1, 14.1. Anal. Calcd for $C_{24}H_{28}N_2$: C, 83.68; H, 7.93; N, 8.48. Found: C, 83.68; H, 8.06; N, 8.48.

Br⌒⌒= —[$PhSiH_3$, t-$BuO_2CN{=}NCO_2Bu$-t, cobalt complex, EtOH, rt, 5 h]→ Br⌒⌒CH(CH_3)N(CO_2Bu-t)$NHCO_2Bu$-t (90%)

cobalt complex (L = MeOH)

***tert*-Butyl *N*-(3-Bromo-1-methylpropyl)-*N'*-(*tert*-butoxycarbonyl)hydrazinecarboxylic Acid (Catalyzed Hydrohydrazination of an Olefin with an Azo Ester).[215]** The Co catalyst (10 mg, 0.025 mmol) was dissolved in EtOH (2.5 mL) at room temperature under argon. To the brown-red solution were added 4-bromo-1-butene (68 mg, 0.50 mmol) and phenylsilane (65 μL, 0.52 mmol), followed by di(*tert*-butyl) azodicarboxylate (0.17 g, 0.75 mmol) in one portion. The resulting solution was stirred at room temperature for 5 hours. Water (1 mL) and brine (5 mL) were added and the reaction mixture was extracted with EtOAc (3 × 10 mL). The combined organic layers were dried over Na_2SO_4, filtered, and the solvents were removed under reduced pressure. The residue was purified by column chromatography (1:15 EtOAc:hexane) to give 166 mg (90%) of the title product, mp 88–90°: ^1H NMR ($CDCl_3$, 300 MHz, 52°) δ 6.06 (br s, 1H), 4.38 (m, 1H), 3.45 (m, 2H), 2.13 (m, 1H), 1.82 (m, 1H), 1.46 (s, 18H), 1.12 (d, $J = 6.5$ Hz, 3H); ^{13}C NMR ($CDCl_3$, 75 MHz, 52°) δ 156.0, 154.7, 81.3, 52.2, 37.5, 30.5, 28.3, 28.2, 18.0. Anal. Calcd for $C_{14}H_{27}N_2O_4Br$: C, 45.78; H, 7.41; N, 7.63. Found: C, 45.98; H, 7.48; N, 7.63.

2-Bromothiophene —[1. Zn*, THF, rt; 2. t-$BuO_2CN{=}NCO_2Bu$-t, 0° to rt]→ 2-thienyl-N(CO_2Bu-t)$NHCO_2Bu$-t (80%)

2-[*N*,*N'*-bis(*tert*-Butoxycarbonyl)hydrazino]thiophene (Amination of a Heterocyclic Zinc Reagent with an Azo Ester).[358] To 1.5 equivalents of active zinc in THF, contained in a 50-mL centrifuge tube, was added 2-bromothiophene (0.163 g, 1 mmol) with stirring at room temperature. The mixture was stirred for 30 minutes, then centrifuged. The supernatant was cannulated into another flask and di(*tert*-butyl) azodicarboxylate (1 mmol in THF) was added over 5 minutes at 0°. After stirring for one hour the reaction was quenched with saturated aqueous $NaHCO_3$, the mixture was extracted with Et_2O, the solvent was removed, and the residue was purified by flash chromatography (silica, hexanes/EtOAc) to give

1.2 g (80%) of the title product, mp. 82–84° (Et_2O): ^{1}H NMR (DMSO-d_6, 100°) δ 9.54 (br s, 1H), 7.03 (dd, $J = 5.5$, 1.6 Hz, 1H), 6.82 (dd, $J = 5.5$, 3.8 Hz, 1H), 6.70 (dd, $J = 3.8$, 1.6 Hz, 1H), 1.47 (s, 9H), 1.44 (s, 9H). Anal. Calcd for $C_{14}H_{22}N_2O_4S$: C, 53.43; H, 7.05; N, 8.91; S, 10.20. Found: C, 53.8; H, 7.0; N, 8.7; S, 10.2.

CHO → 1. $BnO_2CN{=}NCO_2Bn$, L-proline, MeCN, 0° to rt, 3 h; 2. $NaBH_4$, EtOH; 3. NaOH → BnO_2C–NH–N(CO_2Bn)–CH(Bu)–CH_2OH (94%) 97% ee

(*R*)-Dibenzyl 1-(1-Hydroxyhexan-2-yl)hydrazine-1,2-dicarboxylate (Catalytic Asymmetric Amination of an Aldehyde with an Azo Ester).[221] Hexanal (1.5 mmol) was added to a solution of dibenzyl azodicarboxylate (330 mg, 1 mmol) and L-proline (12 mg, 0.1 mmol) in MeCN (10 mL) at 0°. The mixture was stirred at 0° for 2 hours, warmed to room temperature during one hour, and cooled back to 0°. EtOH (10 mL) and $NaBH_4$ (40 mg) were added and the mixture was stirred at 0° for 5 minutes. Addition of aqueous NH_4Cl and EtOAc and removal of the solvent from the dried ($MgSO_4$) organic phase gave the crude title product, which was purified by column chromatography (EtOAc/hexanes) to give 376 mg (94%) of the title compound as a colorless solid: ^{1}H NMR ($CDCl_3$) δ 7.35 (m, 10H), 6.45 (s, 1H), 5.10 (m, 4H), 4.60–3.90 (m, 2H), 3.34 (m, 2H), 1.25 (m, 6H), 0.83 (m, 3H); ^{13}C NMR ($CDCl_3$) δ 136.2, 135.8, 129.5, 129.1, 128.7, 128.5, 69.1, 68.9, 62.9, 61.2, 28.8, 28.2, 22.3, 14.3; HRMS-MALDI (m/z): $[M + Na]^+$ calcd for $C_{22}H_{28}N_2O_5$, 423.1890; found 423.1889. The enantiomeric excess (97%) was determined by conversion into the oxazolidinone (K_2CO_3, toluene, reflux, 1 hour) and HPLC on a Chiralpak AD-RH column.

$OSiMe_3$ (dihydronaphthalene silyl enol ether) → 1. $AgClO_4$, (*R*)-BINAP, $BnO_2CN{=}NCO_2Bn$, THF, –45°; 2. add substrate, –45°, 5 h → tetralone with N(CO_2Bn)$NHCO_2Bn$ (82%) 65% ee

(*S*)-Dibenzyl 1-(1-Oxo-1,2,3,4-tetrahydronaphthalen-2-yl)hydrazine-1,2-dicarboxylate (Catalyzed Asymmetric Amination of a Ketone Silyl Enol Ether with an Azo Ester).[244] A solution of silver perchlorate (0.040 mmol) and (*R*)-BINAP (0.048 mmol, 12 mol%) in THF (1 mL) was stirred at room temperature for 30 minutes, cooled to −45°, and treated with dibenzyl azodicarboxylate (0.44 mmol). After stirring for 10 minutes, (3,4-dihydronaphthalen-1-yloxy)trimethylsilane (0.4 mmol) in THF (0.5 mL) was added and the mixture was stirred at −45° for 5 hours. Aqueous HF (20%) and THF (1:1) were added and the mixture was stirred at room temperature for one hour after which time it was made basic with aqueous $NaHCO_3$ solution and extracted with CH_2Cl_2.

Removal of the solvent from the dried ($MgSO_4$) extracts and preparative thin-layer chromatography of the residue gave the title product in 82% yield, mp 141°. ^{1}H NMR (DMSO-d_6, 70°) δ 2.2–2.4 (m, 2 H), 2.9–3.5 (m, 2 H), 4.92 (br s, 1 H), 5.09 (s, 2 H), 5.13 (s, 2 H), 7.2–7.5 (m, 12 H), 7.56 (t, $J = 7.5$ Hz, 1 H), 7.89 (d, $J = 7.9$ Hz, 1 H), 9.35 (br s, 1 H). Anal. Calcd for $C_{26}H_{24}N_2O_5$: C, 70.26; H, 5.44; N, 6.30. Found: C, 70.52; H, 5.57; N, 6.13. The enantiomeric excess (65%) was determined by HPLC analysis (DAICEL, CHIRALCEL OD or AS).

N3 + Br CO2Me —In, NaI, DMF / rt, 3.5 h→ H N CO2Me (75%)

Methyl 2-(Naphthalen-2-ylamino)methylacrylate (Amination of an Allylindium Species with an Azide).[269] A mixture of 2-azidonaphthalene (5 mmol), methyl 2-(bromomethyl)acrylate (7.5 mmol), indium powder (7.5 mmol), sodium iodide, (7.5 mmol), and DMF (15 mL) was stirred at room temperature for 3.5 hours. Saturated aqueous NH_4Cl (15 mL) was added and the mixture was extracted with Et_2O (2 × 15 mL). The solvent was removed from the extracts and the residue was purified by silica gel chromatography (0.5:9.5 EtOAc/hexane) to give the title product in 75% yield: IR (KBr) 1605 cm^{-1}; NMR ($CDCl_3$) δ 7.87–7.80 (m, 2H), 7.50–7.43 (m, 2H), 7.30 (d, $J = 8$ Hz, 1H), 7.25 (d, $J = 8.0$ Hz, 1H), 6.58 (d, $J = 7.8$ Hz, 1H), 6.35 (s, 1H), 5.85 (s, 1H), 4.25 (s, 2H), 3.83 (s, 3H); EIMS (m/z): 241 (M^+), 209, 180.

N3 + EtMgBr —Et2O, rt, 30 min→ NHEt (90%)

***N*-Ethylaniline (Preparation of an *N*-Substituted Aniline by Reaction of a Grignard Reagent with an Aromatic Azide).**[279] A solution of ethylmagnesium bromide (15 mmol) in Et_2O (20 mL) was added to a solution of phenyl azide (1.19 g, 10 mmol) in Et_2O (5 mL) at room temperature and the mixture was stirred another 30 minutes. Saturated aqueous NH_4Cl (15 mL) was added and the mixture was extracted with ethyl acetate (2 × 25 mL). The extracts were washed with water and brine, dried (Na_2SO_4), and concentrated. The residue was purified by column chromatography (silica, 1:9 EtOAc:hexane) to give 1.09 g (90%) of the title product as a pale yellow liquid. ^{1}H NMR δ 6.8-6.5 (m, 5H), 3.25 (br s, 1H), 3.15 (q, $J = 8.0$ Hz, 2H), 1.25 (t, $J = 8.0$ Hz, 3H); ^{13}C NMR δ 148.2, 128.9, 116.8, 112.5, 38.1, 14.5; MS (m/z): 121 (M^+), 106, 77, 51.

TMS Cl —NaN3, DMF / 80°, 44 h→ TMS N3 (98%) —MgBr / Et2O, rt, 3 h→ NH2 (79%)

2,4-Dimethylaniline (Preparation of Trimethylsilylmethyl Azide and Its Reaction with an Arylmagnesium Reagent to Give an Aniline).[264] A mixture of trimethylsilylmethyl chloride (0.2 mol), sodium azide (0.24 mol), and dry DMF was heated at 80° for 44 hours. Distillation gave trimethylsilylmethyl azide, bp 43° (43 mmHg) in 97% yield. ^{1}H NMR δ 2.75 (s, 2H), 0.12 (s, 9H). The product is stable and can be stored in a refrigerator for at least 6 months *but like all azides it is potentially explosive and should be handled with care.*

Trimethylsilylmethyl azide (1.2 eq) was added dropwise at room temperature to a solution of 2,4-dimethylphenylmagnesium bromide (1 eq) in ether and the mixture was stirred at room temperature for 3 hours. After conventional workup the low-boiling substances were removed under reduced pressure, leaving the title product in 79% yield. It was identified by comparison of its properties with those of an authentic sample.

NaN_3, NaI (cat)
MeCN, reflux, 4.4 h
(99%)

1. *n*-BuLi, –75°
2. MgBr, CuI, 0°
3. $PhSCH_2N_3$, 0° to rt
4. KOH, MeOH, rt
NH_2 (59%)

2-Aminobenzothiazole (Preparation of Azidomethyl Phenyl Sulfide and Its Reaction with a Heterocyclic Grignard Reagent to Give a Heterocyclic Amine).[274] A mixture of chloromethyl phenyl sulfide (40.0 g, 0.25 mol), sodium azide (32.5 g, 0.50 mol), dry MeCN (167 mL), and sodium iodide (100 mg) was stirred and heated under reflux for 4.4 hours, cooled, diluted with Et_2O, and filtered through celite. Removal of the solvents and distillation of the residue gave 40.8 g (99%) of azidomethyl phenyl sulfide as a colorless oil, bp 55–58° (0.23 mmHg): ^{1}H NMR ($CDCl_3$) δ 7.64–7.34 (m, 5H), 4.58 (s, 2H); ^{13}C NMR (acetone-d_6) δ 134.5, 131.2, 129.8, 128.0, 55.9.

A solution of benzothiazole (75 mg, 0.55 mmol) in Et_2O (0.75 mL) was added to a solution of *n*-BuLi in hexane (0.32 mL, 1.75 M, 0.55 mmol) and Et_2O (0.75 mL) at −75°. After 10 minutes, a solution of $MgBr_2$ (0.26 mL, 2.24 M in benzene/Et_2O, 0.58 mmol) was added, followed by THF (0.75 mL). Azidomethyl phenyl sulfide (96 mg, 0.58 mmol) was added and the solution was warmed to 0°. Cuprous iodide (5.0 mg, 0.026 mmol) was added and after 1 hour the mixture was warmed to room temperature, stirred for another 2 hours, and poured into saturated aqueous NH_4Cl. The mixture was extracted twice with Et_2O, the extracts were washed with brine, dried (Na_2SO_4), and concentrated. The residue was stirred with THF (1 mL), methanol (1 mL), and 50% KOH in H_2O (0.25 mL) at room temperature for 3 hours and the mixture was poured into water and extracted three times with Et_2O. Acid-base purification and crystallization of the crude product from water gave 49 mg (59%) of 2-aminobenzothiazole, mp 129–131° (lit. mp 129°): ^{1}H NMR ($CDCl_3$) δ 7.55 (t, $J = 9.0$ Hz, 2H), 7.30 (dt, $J = 7.6$, 1.2 Hz, 1H), 7.11 (dt, $J = 7.6$, 1.2 Hz), 5.35 (br s, 2H).

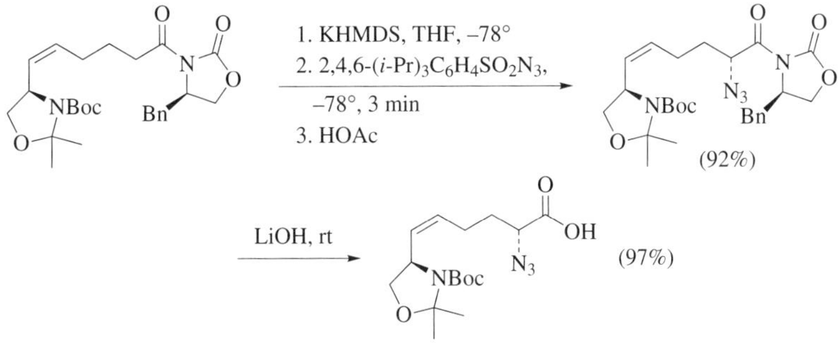

(4*R*)-3{(*Z*,2*R*)-2-Azido-6-[(4*R*)-3-*tert*-butoxycarbonyl-2,2-dimethyl-1, 3-oxazolidin-4-yl]-1-oxohex-5-enyl}-4-phenylmethyl-1,3-oxazolidinone and (4*R*)-4[(1*Z*,5*R*)-5-Azido-5-carboxypent-1-enyl]-3-*tert*-butoxycarbonyl-2,2-dimethyl-1,3-oxazolidine (Diastereoselective Azidation of an *N*-Acyloxazolidinone with Trisyl Azide and Removal of the Chiral Auxiliary).[440] KHMDS in toluene (2.85 mL, 0.5 M, 1.43 mmol) was added at −78° to THF (7.5 mL) followed by a pre-cooled (−78°) solution of the substrate (601 mg, 1.27 mmol; E/Z = 1:13) in THF (9.5 mL) by insulated steel cannula. The mixture was stirred at −78° for 80 minutes. Solid 2,4,6-triisopropylbenzenesulfonyl azide (591 mg, 1.91 mmol) was added in one portion with vigorous stirring and the reaction was quenched with AcOH/THF (1:1, 0.7 mL) after 3 minutes. The flask was immediately placed in a 28° water bath, the mixture was stirred for 30 minutes, and then partitioned between 50 mL of half-saturated aqueous NH_4Cl and 50 mL of EtOAc, and the aqueous phase was extracted with 2 × 50 mL of EtOAc. The combined extracts were dried ($MgSO_4$) and concentrated, and the residue was purified by flash chromatography (EtOAc/hexane) to give 602 mg (92%) of the title product as an oil, E/Z = 1:13. Crystallization (Et_2O/hexane) gave the pure Z-isomer, mp 87–88°: $[\alpha]_D^{22}$ −6.3° (*c* 1.15, $CHCl_3$); IR ($CHCl_3$)2108, 1783, 1690 cm^{-1}; ^1H NMR ($CDCl_3$) δ 7.12-6.82 (m, 5H), 5.51 (ddt, J = 10.7, 9.2, 1.4 Hz, 1H), 5.38 (br dt, J = 10.7, 7.5 Hz, 1H), 5.18 (br q, 1H), 4.62 (m, 1H), 4.15 (ddt, J = 9.2, 8.2, 3.2 Hz, 1H), 3.85 (dd, J = 8.6, 6.3 Hz, 1H), 3.54 (dd, J = 8.6, 3.1 Hz, 1H), 3.35 (t, J = 9.1 Hz, 1H), 2.93 (dd, J = 13.6, 3.2 Hz, 1H), 2.44–2.28 (m, 2H), 2.33 (dd, J = 13.6, 9.2 Hz, 1H), 2.08–1.94 (m, 1H), 1.92–1.78 (m, 1H), 1.67 (s, 3H), 1.56 (s, 3H), 1.43 (s, 9H). Anal. Calcd for $C_{26}H_{35}N_5O_6$: C, 60.80; H, 6.87; N, 13.64. Found: C, 60.8; H, 6.9; N. 13.0.

A solution of the above product (150 mg, 0.29 mmol) in 3:1 THF/water was treated at 0° with lithium hydroxide hydrate (25 mg, 0.59 mmol) and the mixture was stirred at 0–2° for 45 minutes. Aqueous $NaHCO_3$ (2 mL, 0.5 M) was added at 0–2° and the THF was removed under reduced pressure. The aqueous phase was extracted with CH_2Cl_2 (4 × 30 mL) to recover the chiral auxiliary (51 mg, 98%). The aqueous phase and aqueous back-extracts were acidified (2 mL of 2 N HCl) and the product was extracted into EtOAc (4 × 40 mL). The dried ($MgSO_4$) extracts were concentrated to give 101 mg (97%) of the title acid, mp 95.5–96.5°: $[\alpha]_D^{22}$ + 54.5° (*c* 0.53, $CHCl_3$); IR ($CHCl_3$) 2109, 1719, 1698 cm^{-1};

^1H NMR (C_6D_6) δ 1.3–1.9 (m, 2 H), 1.41 (s, 9 H), 1.53 (s, 3 H), 1.63 (s, 3 H), 2.07–2.33 (m, 2 H), 3.48 (dd, $J = 8.7$, 3.3 Hz, 1 H), 3.80 (br m, 1 H), 5.22 (dt, $J = 10.2$, 7.5 Hz, 1 H), 5.45 (dd, $J = 10.7$, 9.2 Hz, 1 H), 8.49 (br s 1 H). Anal. Calcd for $C_{16}H_{26}N_4O_5$: C, 54.22; H, 7.39; N, 15.81. Found, C, 54.0; H, 7.3; N, 15.8.

MgBr
4-MeC$_6$H$_4$SO$_2$N$_3$
Et$_2$O, 0°
N=N SO$_2$C$_6$H$_4$Me-4 ⁻ MgBr⁺
Na$_4$P$_2$O$_7$, H$_2$O
Et$_2$O, rt, overnight
N$_3$
(63%)

2-Azido-1,3,5-trimethylbenzene (Preparation of an Azide from a Grignard Reagent and Tosyl Azide).[305] A solution of 2,4,6-trimethylphenylmagnesium bromide in Et_2O, prepared from 39.8 g (0.2 mol) of 2-bromo-1,3,5-trimethylbenzene, was added with ice cooling to a solution of 19.7 g (0.1 mol) of tosyl azide (*caution; tosyl azide has the explosive power of TNT*) in Et_2O (500 mL). The mixture was stirred for 30 minutes and the tan precipitate was collected by filtration, washed with Et_2O and petrol ether, and dried to give 50.8 g of the triazene salt (*caution: triazenes are potential carcinogens*). It was suspended in 250 mL of Et_2O and a solution of tetrasodium pyrophosphate decahydrate (44.6 g) in H_2O (500 mL) was added dropwise with ice cooling. The mixture was stirred overnight, the layers were separated, and the aqueous layer was extracted with petrol ether (2 × 100 mL). The solvents were removed from the dried ($CaCl_2$) organic phases to leave 16.7 g of a red oil, which was passed through a column of 300 g of alumina and eluted with petrol ether to give 10.16 g (63%) of the title product as a colorless oil. An analytical sample was distilled at 65° (0.2 mm): IR (neat) 2130 cm^{-1}; ^1H NMR (CCl_4) δ 6.60 (s, 2H), 2.21 (s, 6H), 2.17 (s, 3H). Anal. Calcd for $C_9H_{11}N_3$: C, 67.05; H, 6.88; N, 26.07. Found: C, 66.98; H, 6.82; N, 26.03.

O
NMePh
S
1. LDA, THF, –78°, 1 h
2. (PhO)$_2$P(O)N$_3$, –78°, 5 min
3. (t-BuO$_2$C)$_2$O, –78° to rt, 6 h
O
NMePh (70%)
S
NHCO$_2$Bu-t

α-[(*tert*-Butoxycarbonyl)amino]-*N*-methyl-*N*-phenyl-2-thiopheneacetamide (Amination of an Amide Enolate with Diphenyl Phosphorazidate).[336] To a solution of *N*-methyl-*N*-phenyl-2-thiopheneacetamide (3 mmol) in THF (6 mL) was added LDA (1.5 M in cyclohexane, 3.3 mmol) at −78° and the mixture was stirred at −78° for one hour. Diphenyl phosphorazidate (3.3 mmol) was added, the mixture was stirred for 5 minutes, (t-BuO$_2$C)$_2$O (6 mmol) in THF

(3 mL) was added, and the mixture left to warm to room temperature during 6 hours. The solvents were removed and the residue was purified by chromatography (SiO_2, hexane/EtOAc) to give 725 mg (70%) of the title product as yellow crystals, mp 104–106°: IR 1705, 1655 cm^{-1}; ^{1}H NMR δ 7.45–7.40 (m, 3H), 7.35–7.25 (d-like, 1H), 7.20–7.00 (m, 2H), 6.85–6.80 (t-like, 1H), 6.70-6.65 (d-like, 1H), 5.74 (d, $J = 8$ Hz, 1H), 3.30 (s, 3H), 1.40 (s, 3H); MS–CI (m/z): $[M + 1]^+$ 293.

OSi(Pr-*i*)$_3$ — NaN_3, $(NH_4)_2Ce(NO_2)_6$ / MeCN, –20° → O, N_3 (49%)

2-Azido-2-methylcyclohexanone (Preparation of an α-Azido Ketone by Reaction of a Ketone Triisopropylsilyl Enol Ether with Sodium Azide and Ammonium Cerium(IV) Nitrate).[331] To a solution of 1-methyl-2-(triisopropylsilyloxy)cyclohexene in MeCN (0.4M, 1.99 mmol) was added at −20° sodium azide (8.86 mmol, 4.5 eq) followed dropwise by a solution of ammonium cerium-(IV) nitrate in MeCN (0.4M, 5.90 mmol, 3 eq). When the reaction was complete (TLC), ice-cold water was added and the mixture was extracted with ice-cold Et_2O. The combined extracts were washed with ice-cold water, dried (Na_2SO_4), and concentrated. The residue was purified by silica gel chromatography (1:3 ether/pentanes) to give the title product in 49% yield as a pale yellow oil: IR ($CHCl_3$) 2102, 1722 cm^{-1}; ^{1}H NMR ($CDCl_3$) δ 2.61–2.51 (m, 1H); 2.37–2.28 (m, 1H), 1.91-1.56 (m, 6H), 1.35 (s, 3H); ^{13}C NMR δ 207.7, 67.9, 39.1, 36.2, 26.9, 21.1, 20.2; HRMS (m/z): calcd for $C_7H_{11}NO$, 153.090; found, 153.090.

OMe — 1. $ClNHCO_2CH_2CCl_3$ (inverse addition), $CHCl_3$, MeOH, –78° / 2. CrO_2, MeOH, –78° to rt → O, $NHCO_2CH_2CCl_3$ (86%)

2,2,2-Trichloroethyl 2-Oxocyclohexylcarbamate (Amination of a Ketone Enol Ether with the Chromium(II) Chloride/Chlorocarbamate Reagent).[343] A solution of *N*-chloro 2,2,2-trichloroethyl carbamate (1.33 g, 5.74 mmol) in $CHCl_3$ (4 mL) and MeOH (1 mL) was cooled to −78° and treated with a precooled solution of 1-methoxycyclohexene (1.5 mL, 12 mmol) in $CHCl_3$ (2 mL). During 1 hour, a 1 M solution of $CrCl_2$ (about 5 mL, 5 mmol) in MeOH was added dropwise until a starch-iodide paper test was negative. The cooling bath was removed and air was admitted. Sulfuric acid (1 mL of a 1 N solution) was added and the mixture was stirred at room temperature for 4 hours, poured into 50 mL of water, and extracted with CH_2Cl_2 (3 × 100 mL). The combined extracts were washed twice with water, dried, and concentrated. The residue was separated by chromatography (1:4 ether/hexane) into 2,2,2-trichloroethyl carbamate

(0.185 g) and the less polar title product (1.302 g, 86%), mp 75–78°. Crystallization from hexane gave an analytical sample, mp 80–80.5°: IR (CCl_4)1745, 1720 cm^{-1}; ^{1}H NMR (CCl_4) δ 5.92 (m, 1H), 4.63 (s, 2H), 4.22 (dt, $J = 6, 12$ Hz; dd, $J = 6, 11.5$ Hz after D_2O exchange, 1H). Anal. Calcd for $C_9H_{12}Cl_3NO_3$: C, 37.44; H, 4.19; Cl, 36.89; N, 4.85. Found: C, 37.43, H, 4.17; Cl, 37.10; N, 5.02.

TABULAR SURVEY

An effort was made to include all relevant reactions that appeared in the literature up to the middle of 2007. However, in view of the difficulties in searching the subject, omissions are inevitable. The tables are arranged according to substrates and follow the organization of the section on Scope and Limitations. The titles of the individual tables are listed in the Table of Contents and are not repeated here.

Substrates are listed in the order of increasing carbon count. To group similar substrates together, protecting groups and chiral auxiliaries are not counted nor are groups on heteroatoms such as N, O, S, and P. This includes alcohol portions of esters and groups such as methyl or ethyl in ethers, amides, and amines. Ligands in metal complexes are excluded from the count but ferrocene is listed in Table 4 (Aromatic Carbanions) under C_{10}. However, all ring carbons in heterocycles are included in the carbon count. Within each carbon count or range of carbon counts entries are listed in the order in which reagents are discussed in the section on Reagents and Mechanisms: amines, haloamines, hydroxylamines, oxaziridines, imines, oximes, diazonium salts, diazo compounds, azo compounds, azides, and miscellaneous other reagents. This order is not followed in Table 5 (Heterocyclic Anions) where like heterocycles are grouped together. Only substrates where the carbanionic center is in the heterocyclic ring are listed here. Heterocyclic substrates where the carbanionic center is on a side chain are listed in Table 1A (Arylmethyl and Heteroarylmethyl Carbanions). Substrates where the carbanionic center is on an aromatic ring fused to, or attached to, a heterocycle are listed Table 4 (Aryl Carbanions). Table 10A (Esters) does not include lactones and Table 12 (amides) does not include lactams which are in separate tables (11 and 14, respectively) and which are not listed in Table 5 (Heterocyclic Carbanions). Surrogates of carbonyl compounds, such as enol ethers or enamines, are listed together with their parent carbonyl compounds.

A dash enclosed in parentheses [(–)] next to a product signifies that the product was isolated but no yield was reported. When a reaction involving the same aminating reagent has been reported in more than one publication, the conditions producing the highest yield are shown and the reference to that paper is given first. Extensive variations of catalysts, solvents, and conditions are not included in the tables; instead, one or two sets of conditions that produce the highest yield and best selectivity are given.

The following abbreviations are used in the tables:

Ac	acetyl
acac	2,4-pentadionato (acetylacetonato)
BINAP	2,2-bis(diphenylphosphino)-1-binaphthyl
[bmim][BF_4]	*N*-butyl-*N′*-methylimidazolium tetrafluoroborate
Bn	benzyl
Boc	*tert*-butoxycarbonyl
BOM	benzyloxymethyl
Bu	butyl
Bz	benzoyl
[capemim][BF_4]	*N*-5-carboxypentyl-*N′*-methylimidazolium tetrafluoroborate
Cbz	benzyloxycarbonyl
Cp	η^5-cyclopentadienyl
DABCO	1,4-diazabicyclo[2.2.2]octane
DBU	1,8-diazabicyclo[5.4.0]undec-7-ene
DDQ	2,3-dichloro-5,6-dicyanobenzoquinone
$(DHQD)_2$CLB	dihydroquinidinyl *p*-chlorobenzoate (see Chart 1)
$(DHQD)_2$PYR	dihydroquinidinyl pyrimidine (see Chart 1)
DMF	dimethylformamide
DME	dimethoxyethane
DMPU	1,3-dimethyl-3,4,5,6-tetrahydro-2(1*H*)-pyrimidinone
DMSO	dimethylsulfoxide
Et	ethyl
Fmoc	9-fluorenylmethoxycarbonyl
HMPA	hexamethylphosphoric triamide
ia	inverse addition
KHMDS	potassium hexamethyldisilazide
LDA	lithium diisopropylamide
LiHMDS	lithium hexamethyldisilazide
Me	methyl
MEM	(2-methoxyethoxy)methyl
Ms	methanesulfonyl
NaHMDS	sodium hexamethyldisilazide
Ph	phenyl
Piv	pivaloyl
PMB	*p*-methoxybenzyl
Pr	propyl
Py	pyridine
(saltmen)Mn(N)	nitrido[*N*,*N′*-(1,1,2,2-tetramethyl) bis(salicylideneaminato)]manganese (see Chart 1)
TMS	trimethylsilyl
TBS	*tert*-butyldimethylsilyl
TBDPS	*tert*-butyldiphenylsilyl

TEMPO	2,2,6,6-tetramethylpiperidinyl-1-oxyl
Tf	trifluoromethanesulfonyl
TFA	trifluoroacetic acid
TFAA	trifluoroacetic anhydride
THF	tetrahydrofuran
TMEDA	tetramethylethylenediamine
Tr	triphenylmethyl
Ts	tosyl; 4-methylbenzenesulfonyl

CHART 1. STRUCTURES OF REAGENTS AND CATALYSTS

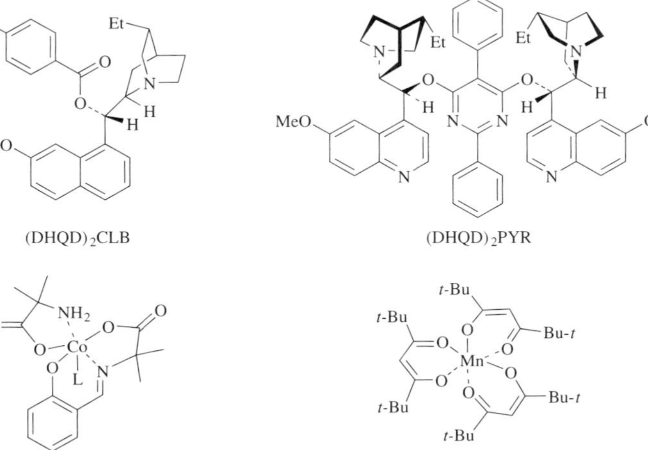

(saltmen)Mn(N)

Nitrido[*N,N'*-(1,1,2,2-tetramethyl)bis(salicylideneaminato)]manganese

$(DHQD)_2CLB$

$(DHQD)_2PYR$

catalyst A

L = MeOH, 2:1 mixture of isomers

catalyst B

TABLE 1A. ACYCLIC ALIPHATIC CARBANIONS

Substrate	Conditions	Product(s) and Yield(s) (%)	Refs.
C_{1-4}			
R^1M (1-5 eq)	1. R^2R^3NH, solvent, temp 1, 2 h 2. O_2, temp 2, time	$R^1NR^2R^3$	52

R^1	M	R^2	R^3	Solvent	Temp 1	Temp 2	Time	
Me	CuMeLi	Ph	Me	toluene	0°	–78°	—	(33)
Me	Cu	Ph	Me	THF, HMPA	rt	rt	—	(46)
Me	CuMeLi	n-C_7H_{15}	n-Bu	Et_2O	rt	rt	—	(39)
Me	Cu	n-C_7H_{15}	n-Bu	THF	rt	rt	—	(52)
Me	CuMeLi	$(CH_2)_2Ph$	$(CH_2)_2Ph$	Et_2O	rt	rt	—	(76)
n-Bu	CuMeLi	Ph	Me	Et_2O	–20°	–20°	—	(57)
n-Bu	CuMeLi	Ph	$(CH_2)_2OH$	Et_2O	–20°	–20°	—	(37)
n-Bu	CuMeLi	n-C_7H_{15}	n-Bu	Et_2O	–20°	–20°	3 min	(73)
n-Bu	CuMeLi	c-C_6H_{11}	c-C_6H_{11}	Et_2O	–20°	–20°	—	(38)
n-Bu	CuMeLi	$(CH_2)_2Ph$	$(CH_2)_2Ph$	Et_2O	–20°	–20°	—	(62)
t-Bu	CuMeLi	Ph	H	THF	–20°	–20°	—	(46)
t-Bu	CuMeLi	3-AcC_6H_4	H	Et_2O	–20°	–20°	—	(32)
t-Bu	CuMeLi	3-(MeCHOH)C_6H_4	H	Et_2O	–20°	–20°	—	(39)
t-Bu	CuMeLi	1-naphthyl	H	THF	–20°	–20°	—	(35)
t-Bu	CuMeLi	n-$C_{10}H_{21}$	H	THF	–20°	–20°	—	(23)
t-Bu	CuMeLi	$(CH_2)_2Ph$	$(CH_2)_2Ph$	THF	–20°	–20°	—	(26)

Substrate	Conditions	Product(s) and Yield(s) (%)	Refs.
$R^1Cu(CN)Li$	1. R^2R^3NLi, THF, –40°, 15 min 2. O_2, –78°. 20 min; to rt	$R^1NR^2R^3$	54

R^1	R^2	R^3	
Me	Ph	Bn	(50)
n-Bu	Me	Bn	(45)
n-Bu	i-Pr	i-Pr	(50)
n-Bu	Ph	H	(62)
n-Bu	(R)-1-(1-naphthyl)ethyl	H	(60)

$C_{1\text{-}4}$ $R^1Cu(CN)X$

1. R^2R^3NLi, THF, temp, time
2. Addend, THF, –78°
3. O_2, –78°, 30 min

$R^1NR^2R^3$

55

R^1	X	R^2	R^2	Addend	Temp	Time	
Me	ZnCl	Ph	Bn	1,2-$(O_2N)_2C_6H_4$	–78° to –40°	40 min	(60)
n-Bu	ZnCl	H	Ph	none	–78° to –40°	40 min	(70)
n-Bu	ZnCl	Me	Bn	1,2-$(O_2N)_2C_6H_4$	–78° to –40°	40 min	(57)
n-Bu	ZnCl	*i*-Pr	*i*-Pr	none	–78° to –40°	40 min	(85)
n-Bu	Li	Ph	Ph	$Cu(NO_3)_2$	–40°	20 min	(60)

$C_{1\text{-}8}$

RM (x eq)

See table.

RNH_2 + R_2NH

R	M	x	Reagent	Conditions			
Me	Li	1	NH_2Cl	Et_2O, 0°	(4)	(—)	58
Me	MgBr	excess	NH_2Cl	Et_2O, 0°	(26)	(—)	56
Et	MgCl	excess	NH_2Cl	Et_2O, 0°	(57)	(—)	56, 58
Et	MgCl	4	NCl_3	Et_2O	(29)	(6)	77, 58
Et	ZnEt	excess	NH_2Cl	petrol ether, –30°; rt, overnight	(46)	(—)	58
Et	ZnEt	excess	NCl_3	petrol ether, –30°; rt, overnight	(17)	(—)	58
n-Pr	MgCl	excess	NH_2Cl	Et_2O, 0°	(58)	(—)	58,56
n-Pr	ZnPr-*n*	excess	NH_2Cl	petrol ether, –30°; rt, overnight	(57)	(—)	58
n-Pr	ZnPr-*n*	excess	NCl_3	petrol ether, –30°; rt, overnight	(8)	(—)	58
i-Pr	MgCl	excess	NH_2Cl	Et_2O, 0°	(66)	(—)	57
i-Pr	MgCl	4	NCl_3	Et_2O	(23)	(2)	77
n-Bu	Li	3	NH_2Cl	Et_2O, –50°	(39)	(—)	58
n-Bu	MgCl	1	NH_2Cl	petrol ether, 0°; rt, overnight	(57)	(—)	59, 56, 58
n-Bu	MgBu-*n*	3 or more	NH_2Cl	Et_2O, dioxane, –60°	(97)	(—)	59
n-Bu	MgCl	4	NCl_3	Et_2O	(37)	(5)	77, 58
n-Bu	MgCl	excess	NH_2Br	E_2O, 2-3°	(29)	(—)	61
n-Bu	MgCl	excess	$NHBr_2$	Et_2O, 0°	(15)	(5)	75
s-Bu	MgCl	excess	NH_2Cl	Et_2O, 0°	(70)	(—)	57
s-Bu	MgCl	4	NCl_3	Et_2O	(23)	(3)	77

TABLE 1A. ACYCLIC ALIPHATIC CARBANIONS (*Continued*)

Substrate	Conditions	Product(s) and Yield(s) (%)	Refs.
$C_{1\text{-}8}$ RM, x eq	See table.	RNH_2 + R_2NH	

R	M	x	Reagent	Conditions	RNH_2	R_2NH	Refs.
s-Bu	MgCl	excess	NH_2Br	Et_2O, 2-3°	(46)	(—)	61
s-Bu	MgCl	excess	$NHBr_2$	Et_2O, 2-3°	(21)	(5)	75
t-Bu	MgCl	excess	NH_2Cl	Et_2O, 0°	(60)	(—)	57
t-Bu	MgCl	4	NCl_3	Et_2O	(30)	(2)	77
t-Bu	MgCl	excess	NH_2Br	Et_2O, 2-3°	(45)	(—)	61
n-C_5H_{11}	MgCl	4	NCl_3	Et_2O	(21)	(5)	77
i-C_5H_{11}	MgCl	excess	NH_2Cl	Et_2O, 0°	(55)	(—)	56
s-C_5H_{11}	MgCl	excess	NH_2Cl	Et_2O, 0°	(72)	(—)	57
t-C_5H_{11}	MgCl	excess	NH_2Cl	Et_2O, 0°	(66)	(—)	57
$Ph(CH_2)_2$	MgCl	excess	NH_2Cl	Et_2O, 0°	(74)	(—)	56
$Ph(CH_2)_2$	MgCl	4	NCl_3	Et_2O	(20)	(2)	77
$Ph(CH_2)_2$	MgCl	excess	$NHBr_2$	Et_2O, 2-3°	(18)	(3)	75

Substrate	Conditions	Product(s) and Yield(s) (%)	Refs.
$C_{1\text{-}4}$ R^1M	1. R^2ONHLi (2 eq, ia), Et_2O, –78° to –15°, 2 h 2. BzCl	R^1NHBz	

R^1	M	R^2	R^1NHBz	Refs.
Me	Li	Me	(80)	82, 786
Et	Li	Me	(78)	82, 786
n-Bu	Li	Me	(77)[a]	82, 786
n-Bu	Li	$CH_2CH{=}CH_2$	(95)	82
n-Bu	MgBr	Me	(16)	83, 786
i-Bu	Li	Me	(67)	82, 786
s-Bu	Li	Me	(71)	82, 83
s-Bu	MgBr	Me	(19)	83
s-Bu	CuLi	Me	(58)	83
s-Bu	Me_2ZnLi	Me	(18)	83
t-Bu	Li	Me	(80)	82, 786

C_{1-4} RM^1

$TsO\bar{N}CO_2Bu$-*t* $(M^2)^+$ → $RNHCO_2Bu$-*t*

R	M^1	M^2	Temp	Time	Yield (%)	Refs.
Me	Li	Li	–78° to –30°	2 h	(60)	126
n-Bu	Li	Li	–78° to –40°	3 h	(71)	127, 126
n-Bu	$(CuLi)_{0.5}$	Li	–78°	1.5 h	(62)	127
n-Bu	$(CuLi)_{0.5}$	MgCl	–78°	1.5 h	(70)	127
s-Bu	Li	Li	0°	3 h	(42)	127, 126
s-Bu	$(CuLi)_{0.5}$	Li	–78°	1.5 h	(32)	127
s-Bu	$(CuLi)_{0.5}$	MgCl	–78°	1.5 h	(57)	127

C_{1-8} R^1M

$R^2SO_2ON(R^3)_2$ (ia), Et_2O or Et_2O/THF → $R^1N(R^3)_2$

R^1	M	R^2	R^3	Temp, Time	Yield (%)	Refs.
Me	Li	2,4,6-$Me_3C_6H_2$	Me	–10° to –20°; to rt, 15 h	(45)	133
Me	Li	Ph	Et	–10° to –20°; to rt, 15 h	(39)	133
n-Bu	Li	Ph	Me	–10° to –20°; to rt, 15 h	(47)	133
$Ph(CH_2)_2$	MgBr	Me	Me	–30° to 0°	(70)	134

C_{1-4} RMX

X not specified

[structure: chiral (1-phenylethyl)imino acetate ester of menthyl (H, Ph, N=CH–CO_2–cyclohexyl, Pr-*i*)], (ia), Et_2O

→ [structure I: H, Ph, N(R)–CH_2–CO_2–] + [structure II: H, Ph, NH–CH(R)–CO_2–] 161

I II

R	M	I + II	I:II
Me	Cd	(55-70)	0:100[b]
Et	Mg	(45-55)	95:5
n-Pr	Mg	(45-55)	96:4
n-Pr	Cd	(55-70)	0:100[b]
i-Pr	Mg	(45-55)	60:40
i-Pr	Cd	(55-70)	0:100[b]
i-Bu	Mg	(45-55)	96:4
t-Bu	Mg	(45-55)	0:100

TABLE 1A. ACYCLIC ALIPHATIC CARBANIONS (*Continued*)

Substrate	Conditions	Product(s) and Yield(s) (%)	Refs.
C_{1-14}			
RMgBr	1. $4\text{-MeOC}_6H_4N{=}C(CO_2Et)_2$, THF, temp, 30 min (forms **I**) 2. Air, KOH, H_2O, EtOH (forms **II**)	$4\text{-MeOC}_6H_4N(R)CH(CO_2Et)_2$ **I** 4-MeOC_6H_4NHR **II**	167, 166

R	Temp	I	II[d]
Me	–78°	(98)[c]	(63)
Et	–78°	(91)[c]	(93)
n-Pr	–78°	(81)[c]	(79)
i-Pr	–95°	(86)[c]	(68)
n-Bu	–78°	(98)[c]	(98)
t-Bu	–95°	(56)	(0)
c-$C_6H_{11}CH_2$	–78°	(93)[c]	(91)
$Ph(CH_2)_2$	–78°	(86)[c]	(89)
n-$C_{10}H_{21}$	–78°	(78)[c]	(79)
n-$C_{12}H_{25}$	–78°	(94)[c]	(84)
n-$C_{14}H_{29}$	–78°	(79)[c]	(71)

Substrate	Conditions	Product(s) and Yield(s) (%)	Refs.
C_1			
MeLi	1. $ZnCl_2$, THF, 0° to rt 2. 2,1-benzisoxazole (ia), $Ni(acac)_2$ (cat), 2 h	2-(methylamino)benzaldehyde (CHO, NHMe) (70)	170
C_{1-2}			
R^1MgI	$R^2R^3C{=}N_2$, Et_2O	$R^2R^3C{=}NNHR^1$	

R^1	R^2	R^3		Refs.
Me	EtO_2C	H	(30)	199
Me	CN	CN	(60)	203
Me	Ph	Ph	(84)	202
Me	Ph	Bz	(—)	201
Et	EtO_2C	H	(—)	199

C_1

MeLi	[diazo camphor], Et_2O, cooling	[camphor NNHMe hydrazone] (—)	201
MeLi	1. [diazocyclopentadiene] (ia), Et_2O 2. $FeCl_3$	[1,1'-bis(N=NMe)ferrocene] (—)	787

C_{1-6}

RMX^1 + $ArN_2^+(X^2)^-$ → $PhN{=}NR$

R	M	X^1	Ar	X^2	Solvent	Temp	Time		Refs.
Me	Mg	I	Ph	Cl	Et_2O	reflux	15 min	(15)[e]	184
Me	Mg	I	1-naphthyl	Cl	Et_2O	reflux	15 min	(7)[e]	184
Me	Mg	I	2-naphthyl	Cl	Et_2O	reflux	15 min	(5)[e]	184
Et	Mg	Br	Ph	Cl	Et_2O	reflux	15 min	(15)[e]	184
Et	Mg	Br	1-naphthyl	Cl	Et_2O	reflux	15 min	(5)[e]	184
Et	Mg	Br	2-naphthyl	Cl	Et_2O	reflux	15 min	(10)[e]	184
t-Bu	Zn	Cl	Ph	BF_4	E_2O	−10°	22 h	(40)	192
t-Bu	Mg	—	Ph	[f]	THF	−78°	1 h[g]	(71)	191
t-Bu	Mg	—	$4\text{-}ClC_6H_4$	[f]	THF	−78°	1 h[g]	(83)	191
t-Bu	Mg	—	$4\text{-}MeOC_6H_4$	[f]	THF	−78°	1 h[g]	(78)	191
t-Bu	Mg	Cl	Ph	BF_4	THF	−78°	—	(40)	185
$n\text{-}C_6H_{11}$	$Zn(C_6H_{11}\text{-}n)_2$	Br	Ph	BF_4	THF	0°	1 h	(0)	190

TABLE 1A. ACYCLIC ALIPHATIC CARBANIONS (*Continued*)

Substrate	Conditions					Product(s) and Yield(s) (%)	Refs.
C_{1-10}							
R^1M	R^2N_3					$R^1NHN{=}NR^2$ and/or[h] $R^1N{=}N\text{-}NHR^2$	
	R^1	M	R^2	Solvent	Temp, Time		
	Me	Li	Me	pentane	0°, 1 h	(60)	258, 256
	Me	MgCl	Me	Et_2O	–40°; –10°, 30 min; to rt	(—)	257
	Me	MgBr	Me	Et_2O	–40°; 0°, 30 min; to rt	(—)	257
	Me	MgI	Me	Et_2O	–20°; to rt; reflux, 20 min	(—)	257
	Me	MgBr	*n*-Bu	Et_2O	0° to rt, 2 h	(85)	261
	Me	Li	$(CH_2)_2OTBDMS$ (ia)	Et_2O	–20°; to rt, 2 h	(70)	268
	Me	MgI	$4\text{-}BrC_6H_4$	Et_2O	reflux, 0.5 h	(—)	280
	Me	MgI	Ph	Et_2O	reflux, 0.5 h	(75)	270, 285
	Me	MgI	$4\text{-}MeC_6H_4$	Et_2O	reflux, 30 min	(—)	280
	Me	MgI	$4\text{-}EtC_6H_4$	Et_2O	reflux, 30 min	(—)	280

R^1	M	R^2	Solvent	Temp, Time		
Me	MgBr	Bn	Et_2O	reflux, 25 min	(95)	261, 270
Me	MgBr	1-naphthyl	Et_2O	0° to rt, 2 h	(80)	281
Me	MgI	$PhSO_2$ (ia)	Et_2O	—	(—)	306
Et	MgBr	Et	Et_2O	0°, 1 h	(45)	258
Et	$Al_{1/3}$	*n*-Bu	petrol ether	40°, 10 h	(60)	260
Et	$Al_{1/3}$	Ph	petrol ether	rt, 1 d	(50)	260
Et	MgI	Ph	Et_2O	reflux, 30 min	(55)	270
Et	MgBr	1-naphthyl	Et_2O	reflux, 25 min	(53)	281
n-Pr	MgBr	1-naphthyl	Et_2O	reflux, 25 min	(54)	281
i-Pr	MgBr	*i*-Pr	pentane	0° to rt, 2 h	(12)	259
n-Bu	Li	*n*-Bu	THF	0°, 1 h	(>88)	262
n-Bu	Li	Bn	pentane	0° to rt, 2 h	(91)	261
n-Bu	MgBr	1-naphthyl	Et_2O	reflux, 25 min	(65)	281
t-Bu	MgCl	$PhSO_2$ (ia)	Et_2O	—	(—)	306
t-Bu	MgCl	$PhSCH_2$ (ia)	THF	–78°; to –20°, 2.5 h	(96)	274
c-$C_3H_5CH_2$	Li	*c*-$C_3H_5CH_2$ (ia)	pentane	–70°, 30 min	(>55)	262
i-C_5H_{11}	MgBr	Ph	Et_2O	reflux, 25 min	(—)	281
i-C_5H_{11}	MgBr	1-naphthyl	Et_2O	reflux, 25 min	(50-55)	281
n-C_6H_{13}	MgBr	$PhSO_2$ (ia)	Et_2O	—	(—)	306
4-cyclohexenylmethyl	MgBr	$PhSCH_2$ (ia)	THF	–78°	(86)	273
n-C_8H_{17}	MgBr	$PhSCH_2$ (ia)	THF	–78°, 1 h; to rt	(100)	274
$Ph(CH_2)_2$	MgBr	$MeSCH_2$ (ia)	THF	0°, 2 h	(>86)	274
$Ph(CH_2)_2$	MgBr	$PhSCH_2$ (ia)	THF	–78°, 1.75 h; to 0°, 30 min	(>90)	274, 273
$Ph(CH_2)_2$	MgBr	4-$MeOC_6H_4SCH_2$ (ia)	THF	–78°, 1.75 h; to 0°, 30 min	(90)	274
$Ph(CH_2)_2$	MgBr	4-$MeC_6H_4SO_2$	THF	0°	(25)[i]	305
4-$MeOC_6H_4(CH_2)_2$	MgBr	$PhSCH_2$	Et_2O	–78°, 2 h	(>73)[j]	274, 273
$PhC(Me)_2CH_2$	MgBr	4-$MeC_6H_4SO_2$	THF	0°	(25)[i]	305

TABLE 1A. ACYCLIC ALIPHATIC CARBANIONS (*Continued*)

Substrate	Conditions	Product(s) and Yield(s) (%)	Refs.
C_1			
1,3-dithian-2-yl–Li	1. N_3(Ph)C=CH_2, THF, –78°; rt, 2 h 2. HCl	2-amino-1,3-dithiane (NH_2) (64)	278
C_{1-4}			
RLi	N_3CH=CHBu-*t*, Et_2O, –78°; to rt, 2-12 h	RHN–N=N–CH=CH–Bu-*t* R: Me (91); Et (—); *n*-Bu (—); *t*-Bu (—)	277
RMgX X = Cl or Br	triazido-triazine/pyrimidine (N_3 ×3; ring N, Y), Et_2O, 1 h	tris(N=NNHR) ring (RNHN=N, N=NNHR, N=NNHR; Y) R, Y: Me, CH (60); Et, CH (70); *n*-Bu, CH (52); *n*-Bu, N (—)	788
RMgBr	$PhCOCH_2N_3$, Et_2O	HO(Ph)(R)C–CH_2–N=N–NHR R: Me (93); Et (95); *n*-Bu (64)	789
C_1			
MeLi	N_3–C_6H_4–Y–C_6H_4–N_3, Et_2O	R–C_6H_4–Y–C_6H_4–R (—) Y = O, CH_2 ; R = N=NNHMe	290

$C_{1\text{-}5}$

RMgX	$N_3\text{–}Y\text{–}N_3$, Et_2O	$RHN\text{–}N{=}N\text{–}Y\text{–}N{=}N\text{–}NHR$[h]		272

R	X	Y	
Me	I	1,4-phenylene	(92)
Me	I	4,4'-biphenylene	(80)
Et	Br	$(CH_2)_2$	(41)
Et	Br	4,4'-biphenylene	(80)
n-Pr	Br	$(CH_2)_2$	(—)
n-Pr	Br	4,4'-biphenylene	(76)
i-Pr	Br	$(CH_2)_2$	(72)
i-Pr	Br	4,4'-biphenylene	(70)
n-Bu	Br	$(CH_2)_2$	(71)
n-Bu	Br	CH_2CHMe	(87)
n-Bu	Br	$(CH_2)_3$	(65)
n-Bu	Br	4,4'-biphenylene	(72)
i-C_5H_{11}	Br	$(CH_2)_2$	(61)
i-C_5H_{11}	Br	4,4'-biphenylene	(78)

TABLE 1A. ACYCLIC ALIPHATIC CARBANIONS (*Continued*)

Substrate	Conditions	Product(s) and Yield(s) (%)	Refs.
C_{1-5}			
RMgX X not specified	6-azido-2-methylbenzothiazole, Et_2O, 10-12 h	RHNN=N-(2-methylbenzothiazol-6-yl) R Me (78) Et (68) *n*-Pr (23) *i*-Pr (57) *n*-Bu (28) *i*-Bu (40) *n*-C_5H_{11} (38) *i*-C_5H_{11} (62)	790
C_{2-3}			
$(R^1)_2Zn$	R^2NCl_2, petrol ether, cold	R^2NH_2 + R^2NHR^1 R^1 R^2 Et Me (44) (46) Et Et (49) (42) Et *n*-Bu (57) (43) Et *i*-C_5H_{11} (52) (42) *n*-Pr *n*-Bu (61) (24)	72
C_2			
Et_2Zn	$EtNCl_2$, Et_2O, 0°	Et_3N (35) + $EtNH_2$ (—)	76
	Et_2NCl, petrol ether, cold	Et_2NH (70) + Et_3N (2)	72

C_{2-8}

R^1MgX^1

1. R^2ONH_2, solvent, temp (forms **I**)
2. HX^2 (forms **II**)

R^1NH_2 **I**
$R^1NH_3X^2$ **II**

R^1	X^1	X^2	R^2	Solvent	Temp	**I**	**II**	Refs.
Et	Br	Cl	Me	Et_2O	–10° to –15°	(—)	(81)	791, 792
Et	Br	Cl	Bn	Et_2O	–10° to –15°	(—)	(46)	80
n-Pr	Br	Cl	Me	Et_2O	–10° to –15°	(—)	(85)	791
n-Bu	Cl	—	Me	Et_2O	–10° to –15°	(58)	(—)	791
n-Bu	Br	—	Me	Et_2O	–10° to –15°	(63)	(—)	791
i-Bu	Br	—	Me	Et_2O	–10° to –15°	(90)	(—)	791
s-Bu	Cl	Cl	Me	Et_2O	–10° to –15°	(—)	(73)	792
t-Bu	Cl	Cl	Me	Et_2O	–10° to –15°	(—)	(74)	791, 792
n-C_5H_{11}	Br	—	Me	Et_2O	–10° to –15°	(65)	(—)	791
i-C_5H_{11}	Cl	Cl	Me	Et_2O	–10° to –15°	(—)	(80)	791
i-C_5H_{11}	Br	Cl	Me	Et_2O	–10° to –15°	(—)	(71)	791
i-C_5H_{11}	I	Cl	Me	Et_2O	–10° to –15°	(—)	(5)	792
i-C_5H_{11}	Br	Cl	Bn	Et_2O	–10° to –15°	(—)	(61)	80
t-C_5H_{11}	Br	—	Me	Et_2O	–10° to –15°	(48)	(—)	791
4-(-pentenyl)	Cl	—	Me	—	—	(—)	(—)	793
Ph$(CH_2)_2$	Cl	—	Me	Et_2O	–10° to –15°	(68)	(—)	791

C_{2-10}

$R^1CH(R^2)CH_2ZrCpCl$ [k]

2,4,6-$Me_3C_6H_2SO_2ONH_2$, Et_2O, 0°, 10 min

$R^1CH(R^2)CH_2NH_2$

116

R^1	R^2	
TMS	H	(81)
$TMSCH_2$	H	(76)
$Cl(CH_2)_3$	H	(78)
Et	*i*-Pr	(85)
n-C_6H_{13}	H	(77)
Ph	H	(60)[l]
Ph$(CH_2)_2$	H	(80)

TABLE 1A. ACYCLIC ALIPHATIC CARBANIONS (*Continued*)

Substrate	Conditions	Product(s) and Yield(s) (%)	Refs.
C_{2-4}			
RMgX	$Bn_2NOBz + CuCl_2$ (2.5 mol%, ia, slow addition), THF, rt, 15 min	$RNBn_2$ R / X: Et / Br (95); t-Bu / Cl (88)	113
$(R^1)_2Zn$ 1.1 eq	R^2R^3NOBz, catalyst, THF, rt	$R^1R^2R^3N$	
	R¹ Et, R² Bn, R³ Bn, Catalyst $(CuOTf)_2$•PhH, Time 1 h	(91)	109, 112
	R¹ *i*-Pr, R² Bn, R³ Bn, Catalyst $(CuOTf)_2$•PhH, Time 1 h	(77)	109, 112
	R¹ *t*-Bu, R² *t*-$BuCH_2CMe_2$, R³ H, Catalyst $CuCl_2$, Time 15-60 min	(43)	112
	R¹ *t*-Bu, R² Bn, R³ Bn, Catalyst $CuCl_2$, Time 15 min	(99)	109
	R¹ *t*-Bu, R² Bn, R³ Bn, Catalyst $(CuOTf)_2$•PhH, Time 1 h	(98)	109, 112
C_2			
Et_2Zn	$R^1N{=}CHCO_2R^2$, pentane, –80° to rt R^1 = *i*-Pr, *t*-Bu; R^2 = Me, Et	β-lactam: Et(R^1)N–, CO_2R^2, NR^1, O (80-90)	163
	1. *t*-BuN=CHCOMe, hexane, –100° 2. CH_2Cl_2, H_2O	*t*-Bu(Et)N–CH_2COMe (80)	162
	1. $R^1N{=}C(R^2)C(R^3){=}NR^1$, hexane, rt 2. *t*-BuOH	$R^1EtN(R^2)C{=}C(R^3)NEtR^1$	171

R¹	R²	R³	
i-Pr	H	H	(—)
i-Pr	H	Me	(—)
t-Bu	H	H	(88)
t-$BuCH_2$	H	H	(—)
c-C_6H_{11}	H	H	(—)
c-C_6H_{11}	Me	Me	(—)
(*i*-Pr)$_2$CH	H	H	(—)

C_{2-4}

R^1M | $R^2CF_2C(=NR^3)CO_2Et$ → I + II + III (165)

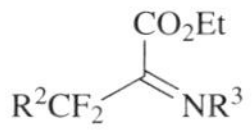

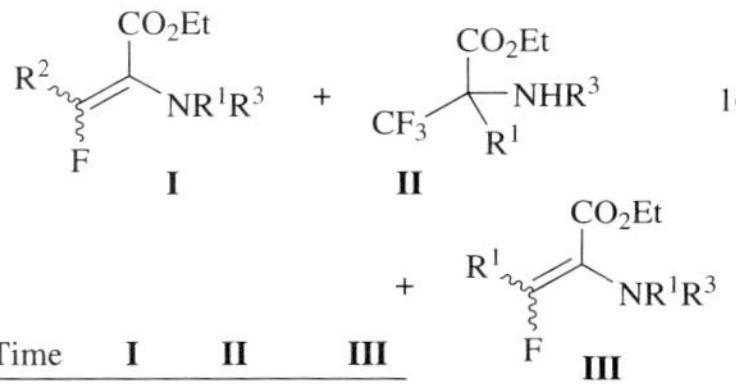

R^1	M	R^2	R^3	Solvent	Temp	Time	**I**	**II**	**III**
Et	MgBr	F	4-$MeOC_6H_4$	Et_2O	rt	30 s	(50)	(0)	(15)
Et	ZnEt	F	4-$MeOC_6H_4$	toluene	rt	30 s	(88)	(1)	(0)
Et	ZnEt	F	Ph	toluene	rt	30 s	(80)	(—)	(—)
Et	ZnEt	F	4-ClC_6H_4	toluene	rt	30 s	(84)	(—)	(—)
Et	ZnEt	F	2-EtC_6H_4	toluene	rt	30 s	(65)	(—)	(—)
Et	ZnEt	F	2,6-$Me_2C_6H_3$	toluene	rt	30 s	(1)	(—)	(—)
Et	ZnEt	F	PhCHMe	toluene	rt	30 s	(85)	(—)	(—)
Et	ZnEt	C_2F_5	4-$MeOC_6H_4$	toluene	100°	2 min	(77)	(—)	(—)
n-Bu	Li	F	4-$MeOC_6H_4$	toluene	80°	30 s	(0)	(80)	(0)

R^1Li | 9-(NR^2)fluorene (1-R^3), THF, hexane, –78°, 2 h; to rt → 9-(R^1R^2N)-9-H-fluorene + 9-(R^2HN)-9-R^1-fluorene (168)

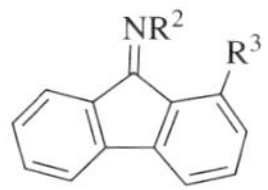

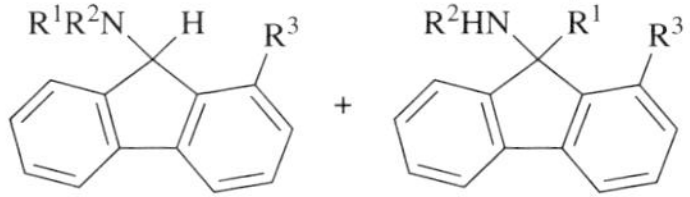

R^1	R^2	R^3		
Et	*n*-Bu	H	(65)	(5)
n-Bu	Me	H	(71)	(0)
n-Bu	*n*-Bu	H	(70)	(0)
n-Bu	*n*-Bu	Me	(0)	(—)
n-Bu	*i*-Bu	H	(41)	(16)
n-Bu	Ph	H	(15)	(50)
n-Bu	4-MeC_6H_4	H	(17)	(52)
n-Bu	4-$MeOC_6H_4$	H	(19)	(47)
n-Bu	$Ph(CH_2)_2$	H	(80)	(5)
n-Bu	$Ph(CH_2)_4$	H	(62)	(15)

TABLE 1A. ACYCLIC ALIPHATIC CARBANIONS (*Continued*)

Substrate	Conditions	Product(s) and Yield(s) (%)	Refs.
C_{2-9} RMgX	1. $[3,5\text{-}(CF_3)_2C_6H_3]_2C{=}NOTs$ (ia), Et_2O, toluene, rt, 0.5 h 2. HCl, Me_2CO 3. BzCl, Et_3N	RNHBz R = Et, X = Br (87) R = *t*-Bu, X = Cl (35) R = $Ph(CH_2)_3$, X = Br (96)[m]	179
C_{2-8} R^1MBr	$R^2CON{=}NR^3$	$R^2CO(H)N{-}N(R^1)R^3$	

R^1	M	R^2	R^3	Solvent	Temp, Time	Yield (%)	Refs.
Et	Mg	Ph	Ph	Et_2O	rt	(25)	794
Et	Mg	Ph	Bz	Et_2O	rt	(—)	794
n-Pr	Mg	Ph	Bz	Et_2O	rt	(—)	794
i-Pr	Mg	Ph	Bz	Et_2O	rt	(poor)	794
i-Bu	Mg	Ph	Bz	Et_2O	rt	(—)	794
t-Bu	Mg[n]	*t*-BuO	CO_2Bu-*t* (ia)	Et_2O	0°, 15 min; rt, 10 h	(26)	795
i-C_5H_{11}	Mg	Ph	Bz	Et_2O	rt	(40)	794
$Me_2(Et)C$	Zn	*t*-BuO	CO_2Bu-*t*	THF	rt, 30 min	(75)	358
$TMSC{\equiv}C(CH_2)_2$	Mg	$CH_2{=}CHCH_2O$	$CO_2CH_2CH{=}CH_2$ (ia)	THF	–78° to rt	(55)	735
$MeCO_2(CH_2)_4$	Zn	*t*-BuO	CO_2Bu-*t*	THF	rt, 3 h	(90)	358
5-$Cl(CH_2)_2$	Zn	*t*-BuO	CO_2Bu-*t*	THF	rt, 3 h	(81)	358
4-$NC(CH_2)_4$	Zn	*t*-BuO	CO_2Bu-*t*	THF	rt, 1 h	(90)	358
$EtO_2C(CH_2)_5$	Zn	*t*-BuO	CO_2Bu-*t*	THF	rt, 3 h	(90)	358
2-octyl	Zn	*t*-BuO	CO_2Bu-*t*	THF	rt, 3 h	(94)	358

Substrate	Conditions	Product(s) and Yield(s) (%)	Refs.
C_{2-4} RM	*t*-$BuO_2CN{=}NCON$(piperidine), –78°, solvent, 1 h	*t*-$BuO_2C(R)N{-}NHCON$(piperidine) **I** + *t*-$BuO_2CNH{-}N(R)CON$(piperidine) **II**	410

RM	Solvent	**I**	**II**
Et_2AlCl	CH_2Cl_2	(100)	(0)
$Et_2Zn{\bullet}TiCl_4$	CH_2Cl_2	(—)	(64)
n-BuLi	Et_2O	(65)	(23)

$C_{2\text{-}12}$

RMgX

X not specified

ArN_3 (ia), Et_2O, rt

ArNHR

279

R	Ar	Time	
Et	Ph	30 min	(90)
Et	MeC_6H_4	30 min	(89)
n-Pr	4-FC_6H_4	1 h	(82)
n-Pr	(3-OCH_2-4)C_6H_3	30 min	(88)
i-Pr	3,4-$Cl_2C_6H_3$	1 h	(85)
i-Pr	4-$MeOC_6H_4$	1 h	(88)
i-Pr	3,4-$(MeO)_2C_6H_3$	30 min	(87)
i-Pr	2-naphthyl	1 h	(88)
n-Bu	MeC_6H_4	30 min	(87)
n-C_5H_{11}	2-naphthyl	1 h	(87)
n-C_7H_{15}	4-ClC_6H_4	1 h	(84)
n-C_8H_{17}	4-$MeOC_6H_4$	30 min	(90)
n-$C_{10}H_{21}$	2-naphthyl	1 h	(88)
n-$C_{12}H_{25}$	4-ClC_6H_4	1.5 h	(85)

C_2

EtMgBr

4-$N_3C_6H_4COMe$, Et_2O

EtNHN=N–(4-C₆H₄)–C(Me)(Et)OH (—) 283

1,3-$(N_3)_2C_6H_4$, Et_2O, –20°

EtNHN=N–(1,3-C₆H₄)–N=NNHEt[h] (58) 289

TABLE 1A. ACYCLIC ALIPHATIC CARBANIONS (*Continued*)

Substrate	Conditions	Product(s) and Yield(s) (%)	Refs.
C_{3-5}			
R^1MgCl	R^2NCl_2, Et_2O, 5°, 1 h	$R^1R^2NH + (R^1)_2NR^2 + R^2NH_2$	68
C_{3-4}			
R^1M	$(R^2)_2NCl$	$(R^2)_2NH + R^1(R^2)_2N + (R^2)_2NN(R^2)_2$	70, 68
C_{3-8}			
RMgX	$Ph_2P(O)ONH_2$, THF	RNH_2	140, 140, 139
C_3			
i-PrMgCl	2,6-di-*t*-Bu-4-imino-cyclohexa-2,5-dienone (NH, *t*-Bu, *t*-Bu, O), Et_2O	*N*-(*i*-Pr) imine of 2,6-di-*t*-Bu-cyclohexa-2,5-dien-1,4-dione (NPr-*i*, *t*-Bu, *t*-Bu, O) (35)	169

R^1	R^2	R^1R^2NH	$(R^1)_2NR^2$	R^2NH_2
n-Pr	Et	(12)	(3)	(52)
n-Bu	Me	(22)	(5)	(43)
n-Bu	Et	(11)	(9)	(36)
n-C_5H_{11}	Et	(12)	(8)	(34)

R^1	M	R^2	Solvent	Temp	Time	$(R^2)_2NH$	$R^1(R^2)_2N$	$(R^2)_2NN(R^2)_2$	Refs.
i-Pr	K	*i*-Pr	petrol ether	rt, then reflux	—	(—)	(3)	(4.5)	70
n-Bu	MgCl	*n*-Bu	Et_2O	5°	1 h	(85)	(4)	(0)	68

R	X	Temp	RNH_2	Refs.
i-Pr	Cl or Br	–78°; to rt	(36)^m	140
n-Bu	Cl or Br	–78°; to rt	(50)^m	140
$Ph(CH_2)_2$	Br	–20°; rt, 12 h	(40)	139

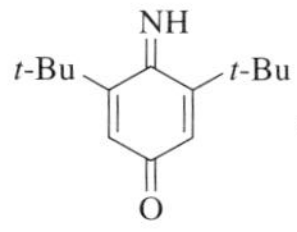

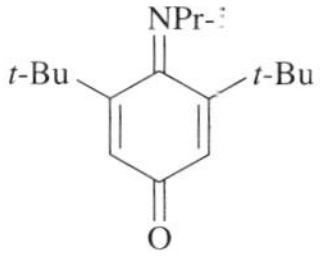

C_{3-8}

RMgX

1. $(4\text{-}CF_3C_6H_4)_2C{=}NOMs$ + CuCN•2 LiCl (ia), THF, HMPA, 0°, 30 min
2. H_3O^+

RNH_2 (179, 726)

R	X	
i-Pr	Br	(93)[o]
n-Bu	Cl	(96)[o]
t-Bu	Cl	(61)[o]
3,4-$(TBSO)_2C_6H_3(CH_2)_2$	Br	(87)[p]

C_{3-10}

RZnX

1. ArN=NTs, THF, –20°
2. Raney Ni, EtOH, reflux

ArNHR (356)

R	X	Ar	
$EtO_2C(CH_2)_2$	$EtO_2C(CH_2)_2$	$4\text{-}MeOC_6H_4$	(45)
n-C_5H_{11}	I	$4\text{-}EtO_2CC_6H_4$	(55)
n-C_8H_{17}	I	$4\text{-}EtO_2CC_6H_4$	(52)
$PhC(Me)_2CH_2$	Br	$4\text{-}EtO_2CC_6H_4$	(79)
$PhC(Me)_2CH_2$	Br	$3,5\text{-}(CF_3)_2C_6H_3$	(0)[q]

TABLE 1A. ACYCLIC ALIPHATIC CARBANIONS (*Continued*)

Substrate	Conditions	Product(s) and Yield(s) (%)	Refs.
C_{3-10}	Conditions 1: 1. Catalyst A (see Chart 1; 5 mol%), EtOH, rt 2. Substrate, then $PhSiH_3$ 3. t-$BuO_2CN{=}NCO_2Bu$-t, rt, time Conditions 2: 1. Catalyst B (see Chart 1; 2 mol%), i-PrOH, 0° 2. Substrate, then $PhSiH_3$, 0° 3. t-$BuO_2CN{=}NCO_2Bu$-t, 0°, time	I + II E = $N(CO_2Bu$-$t)NHCO_2Bu$-t	

R^1	R^2	R^3	R^4	R^5	Conditions	Time	I	II	Refs.
H	H	H	OH	H	1	3 h	(78)	(—)	215
H	H	H	OBn	H	1	3 h	(76)	(—)	215
H	H	H	OMe	OMe	1	7 h	(70)	(—)	215
H	H	H	CH_2OH	H	1	12 h	(22)	(—)	796
H	H	H	CH_2OH	H	2	2.5 h	(72)	(—)	215
Me	H	H	OH	H	2	2 h	(32)	(58)	215
H	H	H	OH	Me	1	5 h	(73) dr 1:1	(—)	215
H	H	H	CH_2Br	H	1	5 h	(90)	(—)	215
Me	Me	H	OH	H	1	8 h	(—)	(70)	215
Et	H	H	Me	H	1	10 h	(16)	(—)	796
Et	H	H	Me	H	2	3 h	(66)	(—)	215
H	H	Me	Et	H	1	5 h	(88)	(—)	215
Me	Me	Me	H	H	1	10 h	(14)	(—)	215
Me	Me	Me	H	H	2	3 h	(78)	(—)	215
H	H	H	CH_2COMe	H	1	3 h	(76)	(—)	215
H	H	H	CH_2Ph	H	1	4 h	(85)	(—)	215
H	H	H	CH_2Ph	H	2	2.5	(76)	(18)	215

C_{3-14}

1. $Co(BF_4)_3$•$6H_2O$ (6 mol%), ligand (6 mol%), EtOH, rt, 10 min
2. Substrate
3. $R^5SO_2N_3$, t-BuO_2H, rt, 5 min
4. Silane, rt, time

Ligand:

Product I + Product II — 215

R^1	R^2	R^3	R^4	R^5	Silane	Time	I	I:II
H	H	H	CH_2OBn	4-MeC_6H_4	$PhSiH_3$	48 h	(35)	—
H	H	H	CH_2OBn	4-MeC_6H_4	$(Me_2SiH)_2O$	48 h	(39)	—
H	H	H	CH_2OBn	2-MeO-5-MeC_6H_3	$(Me_2SiH)_2O$	—	(<20)	—
H	H	H	CH_2OBn	2-$MeO_2CC_6H_4$	$(Me_2SiH)_2O$	48 h	(28)	—
H	H	H	$CH_2OTBDPS$	4-MeC_6H_4	$PhSiH_3$	2 h	(55)	—
H	H	H	$CH_2OTBDPS$	4-MeC_6H_4	$(Me_2SiH)_2O$	18 h	(67)	—
H	H	H	$CH_2OTBDPS$	2-MeO-5-MeC_6H_3	$(Me_2SiH)_2O$	18 h	(19)	—
H	H	H	$CH_2OTBDPS$	2-$MeO_2CC_6H_4$	$(Me_2SiH)_2O$	18 h	(44)	—
H	H	H	$(CH_2)_2OTBDPS$	4-MeC_6H_4	$PhSiH_3$	3 h	(73)	—
H	H	H	$(CH_2)_2OTBDPS$	4-MeC_6H_4	$(Me_2SiH)_2O$	3 h	(85)	—
H	H	Me	$CH_2OTBDPS$	4-MeC_6H_4	$PhSiH_3$	—	(73)	—
H	H	Me	$CH_2OTBDPS$	4-MeC_6H_4	$(Me_2SiH)_2O$	—	(58)	—
H	H	Me	$CH_2OTBDPS$	2-MeO-5-MeC_6H_3	$(Me_2SiH)_2O$	—	(89)	—
H	H	Me	$CH_2OTBDPS$	2-$MeO_2CC_6H_4$	$(Me_2SiH)_2O$	—	(91)	—
H	H	Me	CH_2OBn	4-MeC_6H_4	$(Me_2SiH)_2O$	—	(40)	—
H	H	Me	CH_2OBn	2-MeO-5-MeC_6H_3	$(Me_2SiH)_2O$	—	(64)	—
H	H	Me	CH_2OBn	2-$MeO_2CC_6H_4$	$(Me_2SiH)_2O$	—	(76)	—
$CH_2OTBDPS$	H	Me	Me	4-MeC_6H_4	$PhSiH_3$	—	(63)	—
$CH_2OTBDPS$	H	Me	Me	4-MeC_6H_4	$(Me_2SiH)_2O$	—	(48)	—
$CH_2OTBDPS$	H	Me	Me	2-MeO-5-MeC_6H_3	$(Me_2SiH)_2O$	—	(83)	—
$CH_2OTBDPS$	H	Me	Me	2-$MeO_2CC_6H_4$	$(Me_2SiH)_2O$	—	(79)	89:11

Substrate	Conditions	Product(s) and Yield(s) (%)	Refs.
C_{3-14} $R^1R^2C{=}CR^3R^4$	(*Table continued from previous page.*)	$R^2CH(R^1)–C(R^3)(R^4)N_3$ (I) + $R^2CH(R^1)–CH(R^4)R^3$ (II)	215

R^1	R^2	R^3	R^4	R^5	Silane	Time	I	I:II
H	H	H	$(CH_2)_2CO_2Bn$	$4\text{-}MeC_6H_4$	$PhSiH_3$	10 h	(75)	96:4
H	H	H	$(CH_2)_2CO_2Bn$	$4\text{-}MeC_6H_4$	$(Me_2SiH)_2O$	10 h	(77)	—
H	H	H	$CH_2C_6H_3\text{-}3\text{-}CH_2OCH_2\text{-}4$	$4\text{-}MeC_6H_4$	$PhSiH_3$	20 h	(65)	—
H	H	H	$CH_2C_6H_3\text{-}3\text{-}CH_2OCH_2\text{-}4$	$4\text{-}MeC_6H_4$	$(Me_2SiH)_2O$	20 h	(62)	—
H	H	H	$(CH_2)_2Ph$	$4\text{-}MeC_6H_4$	$PhSiH_3$	2 h	(90)	89:11
H	H	H	$(CH_2)_2Ph$	$4\text{-}MeC_6H_4$	$(Me_2SiH)_2O$	3 h	(86)	—
H	H	H	$(CH_2)_2Ph$	$2\text{-}MeO\text{-}5\text{-}MeC_6H_3$	$(Me_2SiH)_2O$	5 h	(94)	—
H	H	H	$(CH_2)_2Ph$	$2\text{-}MeO_2CC_6H_4$	$(Me_2SiH)_2O$	4 h	(91)	—
H	H	H	$(CH_2)_2Ph$	Et	$PhSiH_3$	2 h	(>75)	77:23
H	H	H	$(CH_2)_2Ph$	Ph	$PhSiH_3$	10 h	(90)	90:10
H	H	Me	$(CH_2)_2Ph$	$4\text{-}MeC_6H_4$	$PhSiH_3$	3 h	(86)	—
H	H	Me	$(CH_2)_2Ph$	$4\text{-}MeC_6H_4$	$(Me_2SiH)_2O$	12 h	(90)	—
H	H	H	$(CH_2)_2COPh$	$4\text{-}MeC_6H_4$	$PhSiH_3$	24 h	(49)	—
H	H	H	$(CH_2)_2COPh$	$4\text{-}MeC_6H_4$	$(Me_2SiH)_2O$	24 h	(46)	—
Me	H	Me	$(CH_2)_2Ph$	$4\text{-}MeC_6H_4$	$PhSiH_3$	30 h	(66)	—
Me	H	Me	$(CH_2)_2Ph$	$4\text{-}MeC_6H_4$	$(Me_2SiH)_2O$	30 h	(48)	—
H	H	H	$(CH_2)_2$-2-naphthyl	$4\text{-}MeC_6H_4$	$PhSiH_3$	8 h	(72)	—
H	H	H	$(CH_2)_2$-2-naphthyl	$4\text{-}MeC_6H_4$	$(Me_2SiH)_2O$	12 h	(69)	—

Substrate	Conditions	Product(s) and Yield(s) (%)	Refs.
C_4			
n-BuMgCl	MeNHCl, Et_2O, 5°, 1 h	*n*-BuNHMe (14) + $MeNH_2$ (72)	68
$(t\text{-}Bu)_2Mg$	*t*-BuNHCl, dioxane, Et_2O, 5°, 2 h; rt overnight	$(t\text{-}Bu)_2NH$ (10)	67

C_4

Substrate	Conditions	Product(s) and Yield(s) (%)	Refs.
RLi	PhCH(Me)N(Li)OMe (ia), hexane, –78°; to rt, 2 h	PhCH(Me)NHR + PhC(Me)(R)NH₂	85

R	PhCH(Me)NHR	PhC(Me)(R)NH_2
n-Bu	(0)	(71)
s-Bu	(35.5)	(25)
t-Bu	(29.2)	(7.3)

Substrate	Conditions	Product(s) and Yield(s) (%)	Refs.
RLi R = *n*-Bu, *s*-Bu, *t*-Bu	BnN(Li)OMe + PhCHMeN(Li)OMe (1:19, ia), hexane, –78° to –10°	RNHBn (—)	85
R^1Li	1. $R^2N(Li)OR^3$ (ia), temp, time 2. ArCOCl	R^1R^2NCOAr	

R^1	R^2	R^3	Ar	Solvent(s)	Temp, Time	Yield	Refs.
n-Bu	Me	Me	Ph	Et_2O, hexane	–78°; to rt, 3 h	(63)	83, 82, 97
n-Bu	*n*-Pr	Me	Ph	Et_2O, hexane	–78°; to –15°, 3 h	(64)	83
n-Bu	*i*-Pr	Me	Ph	Et_2O, hexane	–78°; to rt, 18 h	(47)	83
n-Bu	*n*-C_5H_{11}	Bn	Ph[r]	THF	0° to 40°, 1-3 h	(70)	98
n-Bu	*i*-C_5H_{11}	Bn	4-$MeOC_6H_4$[r]	THF	0° to 40°, 1-3 h	(50)	98
n-Bu	Bn	Me	—	hexane	–78°; to –10°, 2 h	(68)[t]	85
n-Bu	PhCHMe[s]	Me	—	Et_2O, hexane	rt to 40°, 2 h	(68)[t]	82, 97
n-Bu	$Ph(CH_2)_2$	Me	Ph	Et_2O, hexane	–78°; to rt	(68)	83
s-Bu	Me	Me	Ph	—	—	(62)	97, 82
s-Bu	Bn	Me	Ph	hexane	–78°; to –10°, 2 h	(60)	85, 83
s-Bu	$Ph(CH_2)_2$	Me	—	hexane	–78°; to –10°, 2 h	(66)[t]	85
t-Bu	Me	Me	Ph	—	—	(30)	82
t-Bu	Bn	Me	—	hexane	–78°; to –10°, 2 h	(99)[t]	85
t-Bu	$Ph(CH_2)_2$	Me	—	hexane	–78°; to –10°, 2 h	(61)[t]	85

Substrate	Conditions	Product(s) and Yield(s) (%)	Refs.
R^1Li + R^2Li	BnN(Li)OMe (ia), hexane, –78° to –10°, 2 h	R^1NHBn + R^2NHBn	85

R^1	R^2	R^1Li:R^2Li	R^1NHBn	R^2NHBn
s-Bu	*n*-Bu	1:9.8	(27)	(47)
t-Bu	*n*-Bu	1:12.3	(29)	(50)
t-Bu	*s*-Bu	1:11.7	(19)	(60)

TABLE 1A. ACYCLIC ALIPHATIC CARBANIONS (*Continued*)

Substrate	Conditions	Product(s) and Yield(s) (%)	Refs.
C_4			
R^1Li	$R^2CH_2CH_2CH(R^3)N(OMe)CH(R^4)CH_2CH_2R^5$ (ia), hexane, –78°, 3 h; rt, 1-2 d	$R^2CH_2CH_2CH(R^3)N(R^1)CH(R^4)CH_2CH_2R^5$ **I** + $R^2CH_2CH_2CH(R^3)NHC(R^4)(R^1)CH_2CH_2R^5$ **II**	85
$R_2Cu(CN)Li_2$	1. TMSNHOTMS, THF, –50° to rt, 1 h 2. BzCl, pyridine	RNHBz **I** + ROH **II**	100
RMgX (x eq) X not specified	Me_2C=NOM, toluene	RNH_2	174a
C_{4-10}			
RMgBr	1,3-dimethylimidazolidin-2-one O-tosyloxime (NOTs) (ia), toluene, Et_2O, –78°, 15 min	1,3-dimethylimidazolidin-2-imine (NR)[v]	181

Products for ref. 85:

R^1	R^2	R^3	R^4	R^5	I	II
n-Bu	Ph	H	H	Ph	(5)	(5)
s-Bu	Ph	H	H	Ph	(47)	(5)
s-Bu	*c*-C_6H_{11}	H	H	*c*-C_6H_{11}	(9.5)	(—)
t-Bu	Ph	H	H	CH=CMe_2	(67)	(—)
t-Bu	Ph	H	H	Ph	(72)	(5)
t-Bu	*c*-C_6H_{11}	H	H	*c*-C_6H_{11}	(45)	(—)
t-Bu	Ph	Me	H	Ph	(trace)	(—)
t-Bu	Ph	Me	Me	Ph	(trace)	(—)

Products for ref. 100:

R	I	II
n-Bu	(48)	(18)
s-Bu	(60)	(5)
t-Bu	(80)	(10)

Products for ref. 174a:

R	x	M	
n-Bu	2	H	(15)
$Ph(CH_2)_2$	2	H	(25)
$Ph(CH_2)_2$	1	MgBr[u]	(48)

Products for ref. 181:

R	
t-Bu	(<8)
$Ph(CH_2)_2$	(94)
$Ph(CH_2)_2CHMe$	(96)

Reagent	Conditions	Product(s) and Yield(s) (%)	Refs.
"4-Cl$(CH_2)_4$Zn reagents"	$Me_2C=NOSO_2C_6H_3Me_3$-2,4,6, CuCN (cat), THF, rt, 3 h	4-Cl$(CH_2)_4NH_2$ (0)	177
RCu	1. $(4\text{-}CF_3C_6H_4)_2C=NOMs$, THF, HMPA, temp, time 2. BzCl, Et_3N	RNHBz R / Temp / Time: *n*-Bu, –23°, 30 min (92) *s*-Bu, –45°, 1 h (79) *t*-Bu, –23°, 1 h (60)	179
n-BuLi	1. Bn(RO_2C)C=N_2, THF, –78°, 20 min (forms **I**) 2. MeI, –78°, 3 h; to rt, 18 h (forms **II**)	Bn(RO_2C)C=NNHBu-*n* **I** + Bn(RO_2C)C(Me)N=NBu-*n* **II** R / I / II: Me (56) (—) Bu (—) (52)	205
	$CH_2=N_2$, Et_2O, rt	MeNHNHBu-*n* (53)	202
	1. $ZnCl_2$, THF, –73°, 10 min 2. CuCN, –30° 3. Ph_2NNHLi, 30 min 4. O_2, –78°	*n*-BuNHNPh$_2$ (18)	54
t-BuLi	2,3-diazabicyclo[2.2.1]hept-2-ene (ia), THF, –78°	2-*tert*-butyl-2,3-diazabicyclo[2.2.1]heptane (NBu-*t*, NH) (>90)	209
RLi	PhN=NPh, hexane or cyclohexane/THF, –78°, 2 h; rt, 10 h	PhN(R)NHPh R: *n*-Bu (88) *s*-Bu (73) *t*-Bu (47)	211, 797 211 211, 798

TABLE 1A. ACYCLIC ALIPHATIC CARBANIONS (*Continued*)

Substrate	Conditions	Product(s) and Yield(s) (%)	Refs.
C_{4-8}			
R^1M	$R^2O_2CN=NCO_2R^2$	$R^2O_2C-N(R^1)-NH-CO_2R^2$	

R^1	M	R^2	Solvent	Temp	Time	Yield (%)	Refs.
$CH\equiv C(CH)_2$	MgX^w	Me	Et_2O or THF	–78°	—	(>47)	799
$CH_2=CH(CH_2)_2$	MgX^w	Me	Et_2O or THF	–78°	—	(>42)	799
$CH_2=CH(CH_2)_3$	MgX^w	Me	Et_2O or THF	–78°	—	(>46)	799
t-C_5H_{11}	ZnBr	t-Bu	THF	rt	30 min	(75)	358
$MeCO_2(CH_2)_4$	ZnBr	t-Bu	THF	rt	3 h	(90)	358
(Z)-$TMSCH_2CH=CH(CH_2)_2$	MgX^w	Me	Et_2O or THF	–78°	—	(>23)	799
(Z)-$EtCH=CH(CH_2)_2$	MgX^w	Me	Et_2O or THF	–78°	—	(>35)	799
$Cl(CH_2)_5$	ZnBr	t-Bu	THF	0°	3 h	(81)	358
$NC(CH_2)_4$	ZnBr	t-Bu	THF	rt	1 h	(90)	358
$EtO_2C(CH_2)_5$	ZnBr	t-Bu	THF	rt	3 h	(90)	358
2-octyl	ZnBr	t-Bu	THF	rt	3 h	(94)	358

Substrate	Conditions	Product(s) and Yield(s) (%)	Refs.
C_4			
$R^1Cu(CN)M$	1. $(R^2)_2NNHLi$, THF, temp 1, time 1 2. Addend 3. O_2, temp 2, 30 min	$R^1NHN(R^2)_2$	

R^1	M	R^2	Temp 1	Time 1	Addend	Temp 2	Yield (%)	Refs.
n-Bu	ZnCl	Me	–78° to –40°	—	1,2-$(O_2N)_2C_6H_4$	—	(50)	55
n-Bu	ZnCl	Ph	–78° to –40°	—	1,2-$(O_2N)_2C_6H_4$	—	(34)	55
t-Bu	ZnCl	Ph	–78° to –40°	—	1,2-$(O_2N)_2C_6H_4$	—	(60)	55
t-Bu	Li	Ph	–40°	30 min	—	–78°	(30)	55, 54

Substrate	Conditions	Product(s) and Yield(s) (%)	Refs.
n-BuMgBr	1. 4-chloro-2-azidobenzothiazole (N_3), Et_2O, rt, 30 min 2. NH_4Cl, H_2O	2-amino-4-chlorobenzothiazole (NH_2, Cl) (95)	800

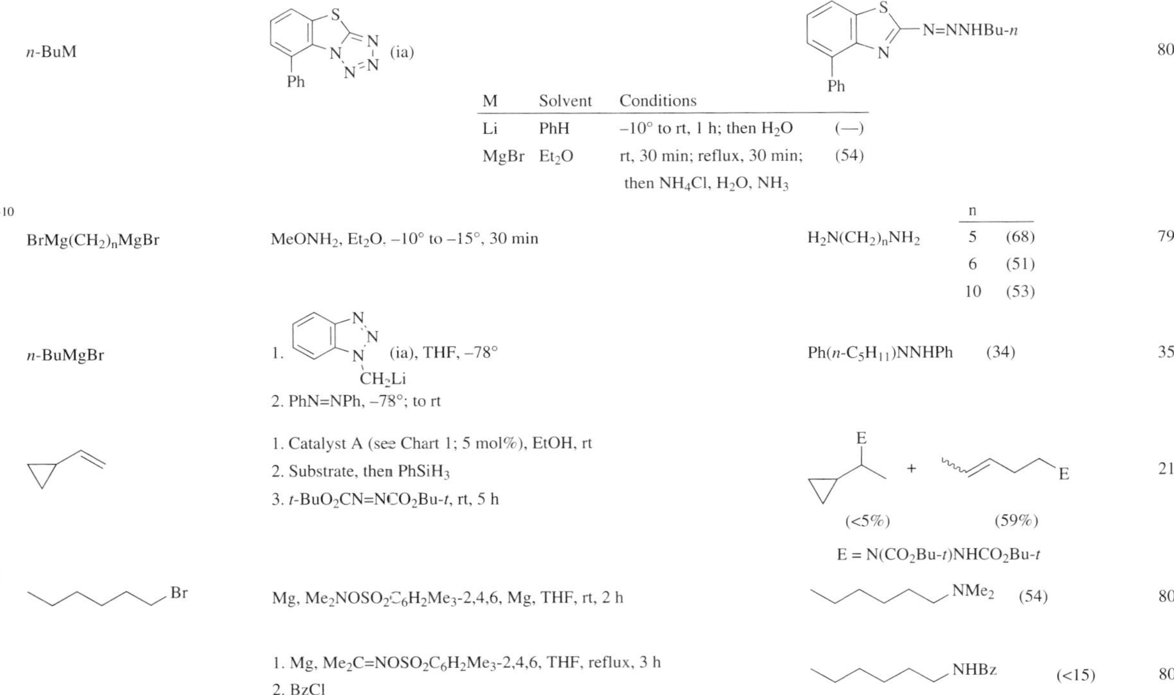

n-BuM	(ia)	N=NNHBu-n (Ph-substituted benzothiazole)	800
		M / Solvent / Conditions: Li, PhH, –10° to rt, 1 h; then H_2O (—); MgBr, Et_2O, rt, 30 min; reflux, 30 min; then NH_4Cl, H_2O, NH_3 (54)	
$C_{5\text{-}10}$			
$BrMg(CH_2)_nMgBr$	$MeONH_2$, Et_2O, –10° to –15°, 30 min	$H_2N(CH_2)_nNH_2$ n: 5 (68); 6 (51); 10 (53)	791
C_5			
n-BuMgBr	1. CH_2Li-benzotriazole (ia), THF, –78° 2. PhN=NPh, –78°; to rt	$Ph(n\text{-}C_5H_{11})NNHPh$ (34)	359
	1. Catalyst A (see Chart 1; 5 mol%), EtOH, rt 2. Substrate, then $PhSiH_3$ 3. $t\text{-}BuO_2CN{=}NCO_2Bu\text{-}t$, rt, 5 h	E + E (<5%) (59%) E = $N(CO_2Bu\text{-}t)NHCO_2Bu\text{-}t$	215
C_6			
Br	Mg, $Me_2NOSO_2C_6H_2Me_3$-2,4,6, Mg, THF, rt, 2 h	NMe_2 (54)	801
	1. Mg, $Me_2C{=}NOSO_2C_6H_2Me_3$-2,4,6, THF, reflux, 3 h 2. BzCl	NHBz (<15)	802

TABLE 1A. ACYCLIC ALIPHATIC CARBANIONS (*Continued*)

Substrate	Conditions	Product(s) and Yield(s) (%)	Refs.
C_{6-9}			
(diallyl-Y structure)	Conditions 1: 1. Catalyst A (see Chart 1; 5 mol%), EtOH, rt 2. Substrate, then $PhSiH_3$ 3. t-$BuO_2CN{=}NCO_2Bu$-t, rt, time Conditions 2: 1. Catalyst B (see Chart 1; 2 mol%), i-PrOH, 0° 2. Substrate, then $PhSiH_3$, 0° 3. t-$BuO_2CN{=}NCO_2Bu$-t, 0°, time	**I** + **II** + **III** E = $N(CO_2Bu$-$t)NHCO_2Bu$-t	215
C_8			
$CH_3(CH_2)_4CH_2CH_2ZnCl$ (structure)	O(morpholino)NOBz, $(Ph_3P)_2NiCl_2$ (cat), THF, rt, 10 min to 6 h	N-alkylmorpholine (structure) (58)	108
C_{8-9}			
RMgBr	1. $NOSO_2Ph$ (dioxolanone oxime structure), Et_2O, CH_2Cl_2, temp, time 2. HCl, EtOH, H_2O, reflux, 2-6 h	$RNH_3^+ Cl^-$	182

Y	Conditions	Time	**I**	**II**	**III**	**III** dr
O	1	15 h	(<5)	(—)	(68)	1.6:1
O	2	24 h	(<5)	(—)	(88)	2.5:1
CH_2	1	20 h	(40)	(24)	(8)	1.4:1
CH_2	2	15 h	(34)	(30)	(6)	5:1
$C(CO_2Et)_2$	1	15 h	(<5)	(—)	(62)	7:1
$C(CO_2Et)_2$	2	4 h	(<5)	(—)	(93)	9:1

R	Temp	Time	
$Ph(CH_2)_2$	rt	15 min	(90)
$PhCH_2CHMe$	0°	1 h	(89)
$PhCHMeCH_2$	0°	30 min	(92)

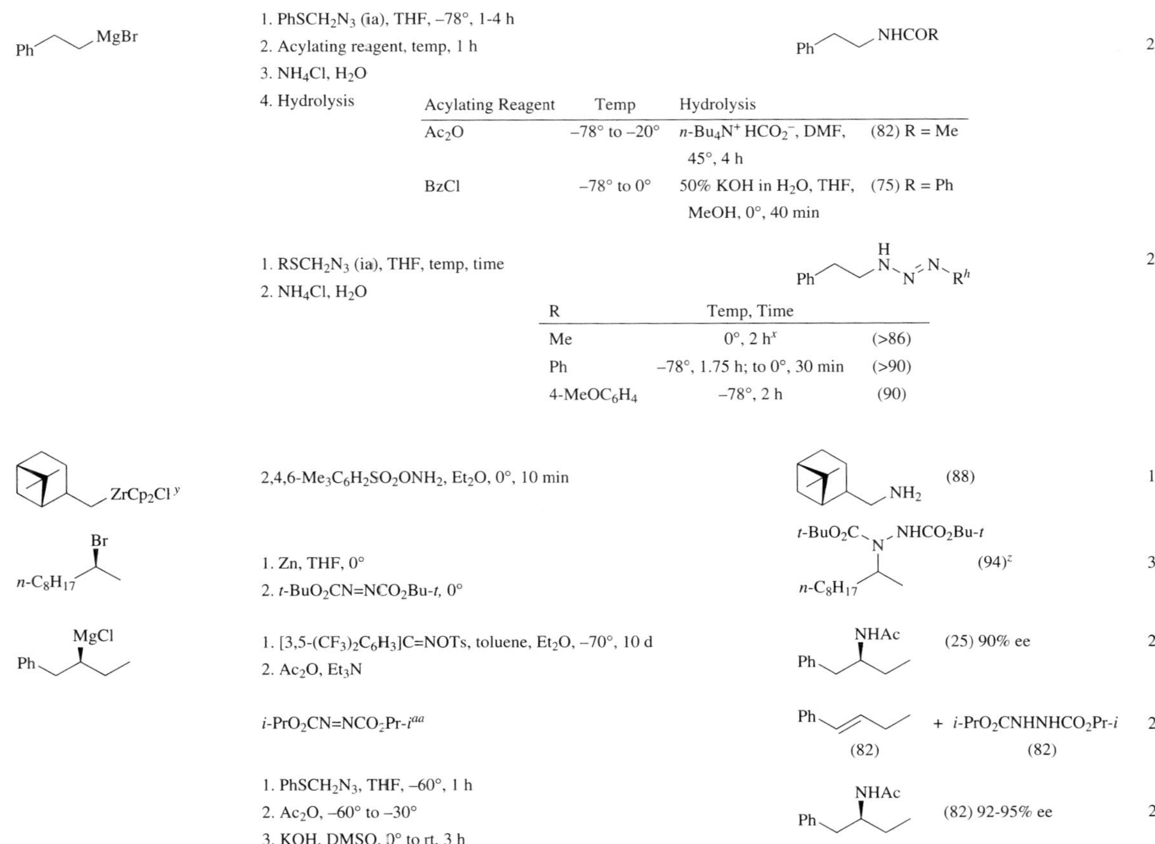

	Reactant	Conditions	Product(s) and Yield(s) (%)	Refs.
C_8	Ph(CH$_2$)$_2$MgBr	1. $PhSCH_2N_3$ (ia), THF, –78°, 1-4 h 2. Acylating reagent, temp, 1 h 3. NH_4Cl, H_2O 4. Hydrolysis	Ph(CH$_2$)$_2$NHCOR	274
		1. $RSCH_2N_3$ (ia), THF, temp, time 2. NH_4Cl, H_2O	Ph(CH$_2$)$_2$NHN=NR[h]	274
C_{10}	ZrCp$_2$Cl[y]	2,4,6-$Me_3C_6H_2SO_2ONH_2$, Et_2O, 0°, 10 min	NH_2 (88)	116
	Br, n-C_8H_{17}	1. Zn, THF, 0° 2. t-BuO_2CN=NCO_2Bu-t, 0°	t-BuO_2C-N(n-C_8H_{17})-$NHCO_2Bu$-t (94)[z]	357
	MgCl, Ph	1. [3,5-$(CF_3)_2C_6H_3$]C=NOTs, toluene, Et_2O, –70°, 10 d 2. Ac_2O, Et_3N	Ph, NHAc (25) 90% ee	220
		i-PrO_2CN=NCO_2Pr-i[aa]	Ph (82) + i-$PrO_2CNHNHCO_2Pr$-i (82)	220
		1. $PhSCH_2N_3$, THF, –60°, 1 h 2. Ac_2O, –60° to –30° 3. KOH, DMSO, 0° to rt, 3 h	Ph, NHAc (82) 92-95% ee	220

Acylating Reagent	Temp	Hydrolysis	
Ac_2O	–78° to –20°	n-Bu_4N^+ HCO_2^-, DMF, 45°, 4 h	(82) R = Me
BzCl	–78° to 0°	50% KOH in H_2O, THF, MeOH, 0°, 40 min	(75) R = Ph

R	Temp, Time	
Me	0°, 2 h[x]	(>86)
Ph	–78°, 1.75 h; to 0°, 30 min	(>90)
4-$MeOC_6H_4$	–78°, 2 h	(90)

TABLE 1A. ACYCLIC ALIPHATIC CARBANIONS (*Continued*)

Substrate	Conditions	Product(s) and Yield(s) (%)	Refs.
C_{10}			
Ph–CH₂–C*H(MgCl)–CH₂CH₃ (ca. 84% ee)	1. $ZnCl_2$, THF, –78°, 80 min 2. Bn_2NOBz, $CuCl_2$ (cat), –78° to rt, 2 h 3. H_2, Pd 3. Ac_2O, Et_3N	Ph–CH₂–C*H(NHAc)–CH₂CH₃ (18) 75% ee	113

[a] After 30 minutes at –15° the mixture was refluxed for one hour.

[b] The reaction was carried out with both R and S enantiomers (CMePh); the optical yields were 5-46%.

[c] The product is unstable; the yield was determined by NMR spectroscopy.

[d] Treatment of products **II** with BnOCOCl followed by cerium ammonium nitrate gave $RNHCO_2Bn$.

[e] The yield was estimated by reduction with Zn/HCl and titration with $NaNO_2$.

[f] X^2 was [benzo-fused cyclic N–S(O₂)/S(O₂) imide: $C_6H_4(SO_2)_2N$].

[g] Heating to 70° converted any Z-azo compounds into the E isomers.

[h] Some of the triazenes are isolated as mixtures of double-bond isomers.

[i] The yield is that of the amine hydrochloride after reduction of the triazene with RaNi.

[j] The product was converted in situ into the (*N*-4-methoxyphenylethyl)piperonylcarboxamide.

[k] The substrate was prepared in situ by reaction of the alkene $R^1R^2C{=}CH_2$ with $HZrCp_2Cl$ in THF at room temperature.

[l] $PhCHMeNH_2$ (21%) was also formed by addition of the zirconium reagent to C_1 of styrene.

[m] The yield is that of the hydrochloride.

[n] The substrate was *t*-BuMgCl.

[o] The yield is that of the *N*-benzoyl derivative.

[p] The yield is that of the fumarate.

[q] The initial uncleaved adduct was isolated in 41% yield.

[r] The reagent $R^2CH_2N(Li)OBn$ was prepared in situ by addition of R^2Li to $CH_2{=}NOBn$.

[s] The reagent was prepared either by reaction of PhCHMeNHOMe with MeLi or of $PhCHMeN(OMe)CO_2(CH_2)_2Br$ with two equivalents of *t*-BuLi. The yield in the latter reaction was 64%.

[t] The yield is that of the unbenzoylated amine.

[u] The reagent was prepared in situ by reaction of the oxime with one equivalent of EtMgBr.

[v] The products were converted into RNH_2 with CsOH in ethylene glycol at 150° or into RNHMe with $LiAlH_4$.

[w] X was not specified.

[x] No reaction occurred at −78°.

[y] The substrate was prepared in situ by reaction of (1*S*)-(−)-β-pinene with $HZrCp_2Cl$ in THF at room temperature.

[z] Treatment with TFA followed by reduction with RaNi/H_2 gave the racemic amine. Racemization most likely occurred during formation of the organozinc reagent.

[aa] The substrate was racemic.

TABLE 1B. CYCLIC ALIPHATIC CARBANIONS

Substrate	Conditions	Product(s) and Yield(s) (%)	Refs.
$C_{3\text{-}6}$			
R^1MgBr	$(R^2)_2NOBz + CuCl_2$ (x mol%, ia, slow addition), THF, rt, 15 min	$R^1N(R^2)_2$ R^1 / R^2 / x $c\text{-}C_3H_5$ / Bn / 15 (59) $c\text{-}C_6H_{11}$ / $CH_2CH{=}CH_2$ / 2.5 (84) $c\text{-}C_6H_{11}$ / Bn / 2.5 (88)	113
C_3			
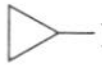 cyclopropyl–MgBr	1. $4\text{-}RC_6H_4N{=}NTs$, THF, –28°, 1 h 2. $CH_2{=}CHCH_2I$, *N*-methylpyrrolidinone, rt, 3 h 3. Zn, AcOH, TFA, 75°, 10 min	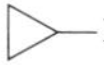 cyclopropyl–$NHC_6H_4R\text{-}4$ R Br (67) CO_2Et (62)	255
$C_{5\text{-}10}$			
RZnX	1. ArN=NTs, THF, –20°, 30 min 2. Raney Ni, EtOH, reflux, 1.5 h	ArNHR	356

R	X	Ar	
$c\text{-}C_5H_9$	I	$4\text{-}EtO_2CC_6H_4$	(71)
$c\text{-}C_5H_9$	I	$3,5\text{-}Me_2C_6H_3$	(76)
$c\text{-}C_6H_{11}$	I	$4\text{-}FC_6H_4$	(75)
$c\text{-}C_6H_{11}$	$c\text{-}C_6H_{11}$	$4\text{-}EtO_2CC_6H_4$	(62)
exo-1-norbornyl	*exo*-1-norbornyl	$4\text{-}MeOC_6H_4$	(71) exo:endo = 4:1
exo-1-norbornyl	*exo*-1-norbornyl	3-quinolyl	(62) exo:endo = 4:1
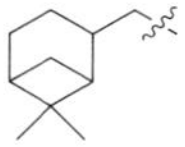	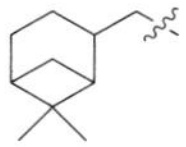	$4\text{-}MeOC_6H_4$	(67)
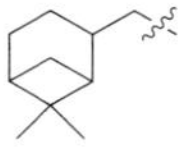	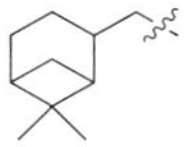	$4\text{-}EtO_2CC_6H_4$	(50)

Substrate	Conditions	Product(s) and Yield(s) (%)	Refs.
$C_{5\text{-}12}$			
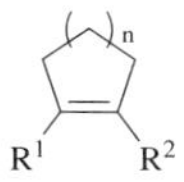 cycloalkene (R^1, R^2, n)	Conditions 1: 1. Catalyst A (see Chart 1; 5 mol%), EtOH, rt 2. Substrate, then $PhSiH_3$ 3. $t\text{-}BuO_2CN{=}NCO_2Bu\text{-}t$, rt, time Conditions 2: 1. Catalyst B (see Chart 1; 2 mol%), *i*-PrOH, 0° 2. Substrate, then $PhSiH_3$, 0° 3. $t\text{-}BuO_2CN{=}NCO_2Bu\text{-}t$, 0°, time	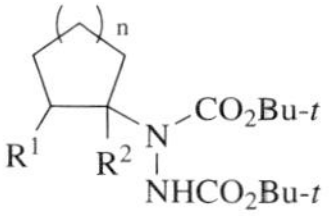 cycloalkane (R^1, n) bearing R^2 and $N(CO_2Bu\text{-}t)NHCO_2Bu\text{-}t$	215

n	R^1	R^2	Conditions	Time	
1	H	H	1	8 h	(74)
1	H	H	2	2 h	(94)
2	H	H	1	24 h	(24)
2	H	H	2	2 h	(90)
4	H	H	1	24 h	(62)
4	H	H	2	2.5 h	(95)
2	Me	Me	2	3 h	(79), dr = 1:1
2	—$(CH_2)_4$—		1	18 h	(78)
2	—$(CH_2)_4$—		2	18 h	(74)
2	H	Ph	1	8 h	(80)

C_6

Substrate	Conditions	Product(s) and Yield(s) (%)	Refs.
cyclohexyl-MgX	1. $BnONH_2$, Et_2O, –10° to –15° 2. HCl	cyclohexyl-$NH_3^+ Cl^-$ X: Cl (79); Br (62)	80
cyclohexyl-$ZrCp_2Cl$[a]	2,4,6-$Me_3C_6H_2SO_2ONH_2$, Et_2O, 0°, 10 min	cyclohexyl-NH_2 (75)	116
cyclohexyl-MgX excess X = Br	1. $TsONMe_2$, Et_2O, rt, 10 min 2. HCl	cyclohexyl-NMe_2 (13)	132
X = Cl or Br	1. $Ph_2P(O)ONH_2$, THF, –78°; to rt 2. HO_2CCO_2H	cyclohexyl-$NH_3^+ HO_2CCO_2^-$ (24)	140
cyclohexyl-MgX[b] 2 eq	Me_2C=NOH, toluene	cyclohexyl-NH_2 (12)	174a
cyclohexyl-Br	1. Mg, Me_2C=$NOSO_2C_6H_2Me_3$-2,4,6, THF, reflux, 3 h 2. BzCl	cyclohexyl-NHBz (<15)	802

TABLE 1B. CYCLIC ALIPHATIC CARBANIONS (*Continued*)

	Substrate	Conditions	Product(s) and Yield(s) (%)	Refs.
C_6	cyclohexyl-MgBr	1. $Me_2C=NOSO_2C_6H_2Me_3$-2,4,6, Et_2O, toluene, 0°, 2 h 2. Hydrolysis	cyclohexyl-NH_2 (40)[c]	803
	cyclohexyl-ZnCl excess	1. $Me_2C=NOSO_2C_6H_2Me_3$-2,4,6, CuCN (0.2 eq), THF, rt, 3 h 2. HCl	cyclohexyl-$NH_3^+Cl^-$ (20)	177
C_{6-10}	RMgBr	1. (4,4,5,5-tetramethyl-1,3-dioxolan-2-ylidene)-$NOSO_2Ph$, Et_2O, CH_2Cl_2, temp, time 1 2. HCl, EtOH, H_2O, reflux, time 2	$RNH_3^+ Cl^-$	182

R	Temp	Time 1	Time 2	
c-C_6H_{11}	0°	30 min	6 h	(92)[d]
1-norbornyl	rt	12 min	10 h	(64)
1-adamantyl	0°	30 min	10 h	(89)

R^1MgX — $[Ar_2C=NOR^2 + \text{addend}]$ (ia), 30 min — R^1NH_2

R^1	X	Ar	R^2	Addend	Solvent	Temp		Refs.
c-C_6H_{11}	Cl	4-$CF_3C_6H_4$	Ms	CuCN + 2 LiCl (0.2 eq)	THF, HMPA	0°	(80)[d]	179, 726
c-C_6H_{11}	Cl	3,5-$(CF_3)_2C_6H_3$	Ts	—	Et_2O, toluene	rt	(87)[d]	179
1-norbornyl	Cl	4-$CF_3C_6H_4$	Ms	CuCN + 2 LiCl (0.2 eq)	THF, HMPA	0°	(96)[e]	179
1-adamantyl	Br	4-$CF_3C_6H_4$	Ms	CuCN + 2 LiCl (0.2 eq)	THF, HMPA	0°	(82)[e]	179, 726

	Substrate	Conditions	Product(s) and Yield(s) (%)	Refs.
C_6	cyclohexyl-MgBr	$(EtO_2C)_2C=NC_6H_4OMe$-4 (ia), THF, –95°, 30 min	$(EtO_2C)_2CH$–N(C_6H_{11}-c)–C_6H_4-OMe-4 (48)	167
	(cyclohexyl)$_3$ZnMgBr	$PhN_2^+ BF_4^-$, THF, 0°, 1 h	Ph–N=N–cyclohexyl (0)	190

Carbon no.	Organometallic	Conditions	Product(s) and Yield(s) (%)	Refs.
C_{6-7}	cyclohexyl-MgBr	$R^1R^2C(N{=}N)$ (diazirine), Et_2O, 0° to rt, 1 h	$R^1R^2C(NC_6H_{11}\text{-}c)(NH)$ (diaziridine): R^1 = Me, R^2 = Me (62); R^1 = *i*-Pr, R^2 = *i*-Pr (40)[f]; R^1R^2 = —$(CH_2)_5$— (86)	208
	RMgX	1. $PhSCH_2N_3$, THF, Et_2O, –78° 2. Ac_2O, –78° 3. *n*-$Bu_4N^+ HCO_2^-$, DMF, rt, 1.5 h; 45°, 2 h	RNHAc: R = *c*-C_6H_{11} (93); R = 2-norbornyl[g] (70)	273, 274, 275
C_6	cyclohexyl-MgCl	$PhSO_2N_3$ (ia), Et_2O	$c\text{-}C_6H_{11}NHN{=}NSO_2Ph$ (—)	306
	cyclohexyl-MgX[b]	4-$FC_6H_4N_3$ (ia), Et_2O, rt, 1 h	$c\text{-}C_6H_{11}NH\text{-}C_6H_4\text{-}4\text{-}F$ (85)	279
	cyclohexyl-MgBr	$N_3\text{-}C_6H_4\text{-}C_6H_4\text{-}N_3$ (4,4′-diazidobiphenyl), Et_2O	$(c\text{-}C_6H_{11}NHN{=}N\text{-}C_6H_4\text{-})_2$ (79)	272
C_7	2-norbornyl-MgBr, exo:endo = 60:40	1. $ZnCl_2$ (50 mol%), Et_2O, THF, rt, 1 h 2. Bn_2NOBz, $CuCl_2$ (cat), THF, rt, 2 h	2-norbornyl-NBn_2 (57), endo:exo = 65:35	113
	endo-2-norbornyl-MgBr	1. $ZnCl_2$ (50 mol%), Et_2O, THF, –78°, 1 h 2. Bn_2NOBz, $CuCl_2$ (cat), THF, –78° to rt, 2 h	*endo*-2-norbornyl-NBn_2 (56), endo:exo > 95:5	113
	bicyclic bridgehead-Li	2,4,6-$Me_3C_6H_2SO_2ONMe_2$, Et_2O or Et_2O/THF, –10° to –15°; to rt, 15 h	bicyclic bridgehead-NMe_2 (54)	133

TABLE 1B. CYCLIC ALIPHATIC CARBANIONS (*Continued*)

Substrate	Conditions	Product(s) and Yield(s) (%)	Refs.
C_7			
Li	$ClNH_2$, Et_2O, sonication	NH_2 (39)	355
	Conditions 1: 1. Catalyst A (see Chart 1; 5 mol%), EtOH, rt 2. Substrate, then $PhSiH_3$ 3. t-$BuO_2CN{=}NCO_2Bu$-t, rt, 7 h Conditions 2: 1. Catalyst B (see Chart 1; 2 mol%), i-PrOH, 0° 2. Substrate, then $PhSiH_3$, 0° 3. t-$BuO_2CN{=}NCO_2Bu$-t, 0°, 7 h	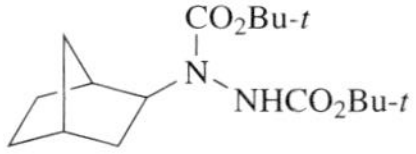 CO_2Bu-t, N, $NHCO_2Bu$-t Conditions: 1 (66); 2 (98)	215
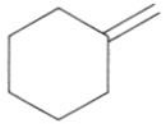	1. Catalyst A (see Chart 1; 5 mol%), EtOH, rt 2. Substrate, then $PhSiH_3$ 3. t-$BuO_2CN{=}NCO_2Bu$-t, rt, 5 h	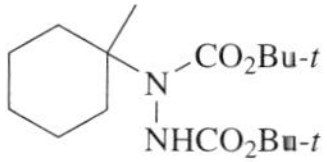 N, CO_2Bu-t, $NHCO_2Bu$-t (90)	215
 Cl	1. Li, cyclohexane, 90° 2. TsN_3, Et_2O, pentane, –15°, 30 min 3. Isolate triazene salt and heat to 75°/0.1 mm for 5 h[h]	 H_2N (27)	307
C_8			
	1. $Co(BF_4)_3{\bullet}6H_2O$, ligand, EtOH, rt, 10 min 2. Substrate 3. 4-$MeC_6H_4SO_2N_3$, t-BuO_2H, rt, 5 min 4. $PhSiH_3$, rt, 12 h Ligand: Ph, Ph, O, N, OK, t-Bu, OH, t-Bu	N_3 (56)	215

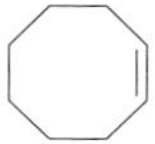

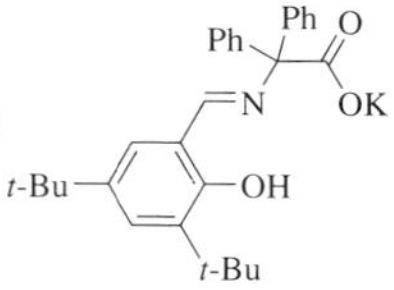

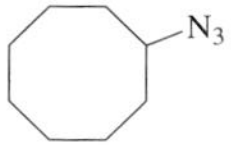

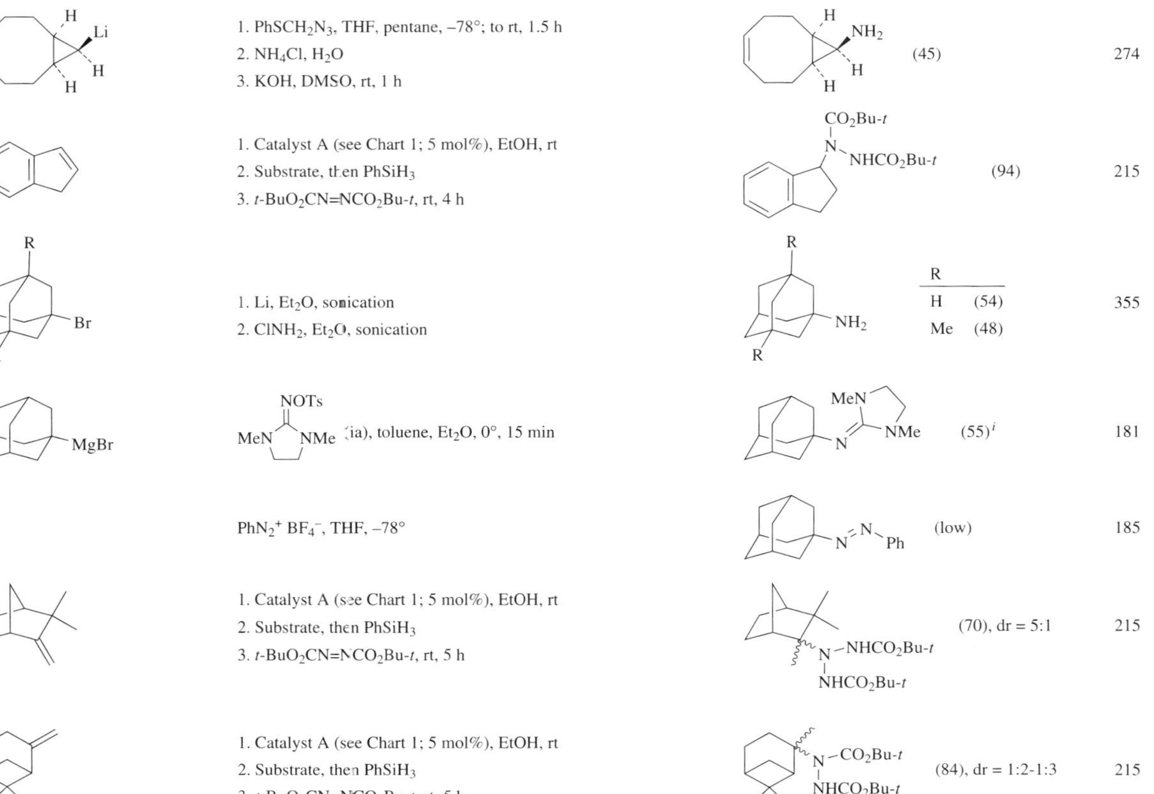

	Substrate	Conditions	Product(s)	Yield(s) (%)	Refs.
C_9		1. $PhSCH_2N_3$, THF, pentane, –78°; to rt, 1.5 h 2. NH_4Cl, H_2O 3. KOH, DMSO, rt, 1 h		(45)	274
		1. Catalyst A (see Chart 1; 5 mol%), EtOH, rt 2. Substrate, then $PhSiH_3$ 3. t-$BuO_2CN{=}NCO_2Bu$-t, rt, 4 h		(94)	215
$C_{10\text{-}12}$		1. Li, Et_2O, sonication 2. $ClNH_2$, Et_2O, sonication		R: H (54); Me (48)	355
C_{10}		(ia), toluene, Et_2O, 0°, 15 min		(55)[i]	181
		$PhN_2^+ BF_4^-$, THF, –78°		(low)	185
		1. Catalyst A (see Chart 1; 5 mol%), EtOH, rt 2. Substrate, then $PhSiH_3$ 3. t-$BuO_2CN{=}NCO_2Bu$-t, rt, 5 h		(70), dr = 5:1	215
		1. Catalyst A (see Chart 1; 5 mol%), EtOH, rt 2. Substrate, then $PhSiH_3$ 3. t-$BuO_2CN{=}NCO_2Bu$-t, rt, 5 h		(84), dr = 1:2-1:3	215

TABLE 1B. CYCLIC ALIPHATIC CARBANIONS (*Continued*)

Substrate	Conditions	Product(s) and Yield(s) (%)	Refs.
C_{10}	1. $Co(BF_4)_3$•$6H_2O$, ligand, EtOH, rt, 10 min 2. Substrate 3. 4-$MeC_6H_4SO_2N_3$, *t*-BuO_2H, rt, 5 min 4. Silane, rt, 12 h Ligand:	N_3 Silane / dr $PhSiH_3$ (89) 4:1 $(Me_2SiH)_2O$ (76) 4:1	215
C_{11} OMe, ZnPr-*i*, chiral non-racemic	1. 4-$EtO_2CC_6H_4N$=NTs, THF, –20° 2. Raney Ni, EtOH, reflux, 12 h	OMe, $NHC_6H_4CO_2Et$-4 (40) 88% ee; trans:cis = 98:2	356
C_{15} MgBr, Ph, Ph	2,4,6-$Me_3C_6H_2SO_2ONMe_2$, Et_2O or Et_2O/THF, –10° to –15°; to rt, 15 h	NMe_2, Ph, Ph (47)	133, 134
C_{20} Li	$ClNH_2$, Et_2O, sonication	NH_2 (41)	355

t-$BuO_2CN{=}NCO_2Bu$-t, Et_2O, sonication	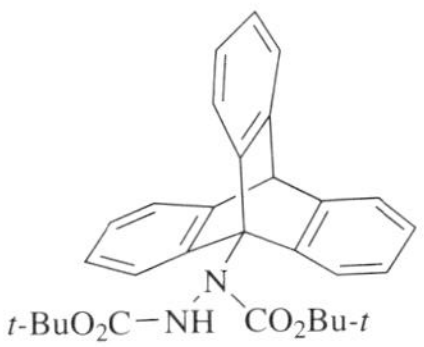 t-BuO_2C–NH–N(CO_2Bu-t) adduct	(48)	355

[a] The substrate was prepared in situ by addition of $HZrCp_2Cl$ to cyclohexene.

[b] X was not specified.

[c] The yield was determined by gas chromatography.

[d] The product was isolated as the *N*-benzoyl derivative.

[e] The product was isolated as the hydrochloride.

[f] The product was isolated as the oxalate.

[g] The substrate was a mixture of exo and endo isomers.

[h] *Caution!* Explosions and spontaneous ignition were encountered in this step.

[i] The product was converted into aminoadamantane with CsOH in ethylene glycol at 150° (71% yield) or into *N*-methylaminoadamantane with $LiAlH_4$ (55% yield).

TABLE 1C. ALLYLIC AND PROPARGYLIC CARBANIONS

	Substrate	Conditions	Product(s) and Yield(s) (%)	Refs.
C_3	MgBr, 2 eq	$MeONH_2$, Et_2O, –10° to –15°, 30 min	NH_2 (40)	791
	Li	1. MeONHLi (2 eq, ia), Et_2O, –78° to rt, 2 h 2. BzCl	NHBz (78)	82
	MgX, X not specified	(ia), Et_2O	(0) + (44-55)	161
$C_{3\text{-}4}$	MgX, R, X not specified	$MeO_2CN{=}NCO_2Me$, Et_2O or THF, –78°	N(CO_2Me)$NHCO_2Me$, R: H (>23); Me (>60)	799
C_3	MgX	RN_3, Et_2O	NHN=NR	
			X: Br, R: Ph, Temp: —, Time: — (48)	282
			X: Cl, R: $PhSO_2$, Temp: —, Time: — (—)[a]	306
			X: Br, R: 1-naphthyl, Temp: reflux, Time: 25 min (60)	281
$C_{3\text{-}4}$	R^1, In[b]	R^2N_3, DMF, rt	NHR^2 **I** + $(\;)_2NR^2$ **II**	269

R^1	R^2	Time	**I**	**II**
H	Ph	2.0 h	(90)	(5-8)
H	c-C_6H_{11}	2.5 h	(73)	(5-8)
H	n-C_6H_{11}	3.0 h	(70)	(5-8)
H	4-ClC_6H_4	3.0 h	(85)	(5-8)
H	4-MeC_6H_4	1.5 h	(88)	(5-8)
H	Bn	2.0 h	(80)	(—)
CO_2Me	Ph	1.5 h	(80)	(—)
CO_2Me	3-ClC_6H_4	2.5 h	(81)	(—)
CO_2Me	3,4-$Cl_2C_6H_3$	2.5 h	(80)	(—)
CO_2Me	4-$MeOC_6H_4$	1.0 h	(87)	(—)
CO_2Me	4-Cl-2-$MeOC_6H_3$	2.5 h	(80)	(—)
CO_2Me	3,4-$(MeO)_2C_6H_3$	3.0 h	(89)	(—)
CO_2Me	2-MeC_6H_4	3.0 h	(84)	(—)
CO_2Me	2-Br-4-MeC_6H_3	2.0 h	(85)	(—)
CO_2Me	Bn	2.5 h	(84)	(—)
CO_2Me	2-naphthyl	3.5 h	(75)	(—)

C_4

Substrate	Conditions	Product(s) and Yield(s) (%)	Refs.
$(CH_2{=}C(CH_3)CH_2)_2CuMgCl$	t-$BuO_2CN(M)OTs$, THF, –78°, 1.5 h	$CH_2{=}C(CH_3)CH_2NHCO_2Bu$-$t$ M: Li (66); MgCl (72)	127
$CH_2{=}CHCH_2CN$	1. LiHMDS (ia). THF, –78° 2. (+)-camphor oxaziridine (N-H), –78°; to rt, 26.5 h	camphor imine of $CH_3CH{=}C(CONH_2)N{=}$ (45)	151
$CH_3CH{=}CHCH_2SnBu_3$	t-$BuO_2CN{=}NCON(CH_2)_5$, $ZnCl_2$•OEt_2, CH_2Cl_2, –78° to rt, 2 h	t-BuO_2CHN-$N(CH(CH_3)CH{=}CH_2)CON(CH_2)_5$ (73) + t-BuO_2CNH-$N(CH_2CH{=}CHCH_3)CON(CH_2)_5$ (7)	410

Substrate	Conditions	Product(s) and Yield(s) (%)	Refs.
C_{4-10}			
R^1–C≡C–C(R^2)=C(R^3)R^4	1. Catalyst A (1-5 mol%), EtOH, rt 2. Substrate, then silane 3. $R^5O_2CN{=}NCO_2R^5$, rt, time	R^1–C≡C–C(R^2)(N(CO_2R^5)–$NHCO_2R^5$)–CH(R^3)R^4	216

R^1	R^2	R^3	R^4	R^5	Silane	Time	Yield (%)
TMS	H	H	H	Et	$PhSiH_3$	1 h	(73)
TMS	H	H	H	t-Bu	$PhSiH_3$	2 h	(83)
TMS	H	Me (H)	H (Me)	Et	$PhSiH_3$	1.5 h	(45)
TMS	H	Me (H)	H (Me)	t-Bu	$PhSiH_3$	3 h	(55)
TMS	Me	H	H	Et	$PhSiH_3$	2 h	(67)
TMS	Me	H	H	t-Bu	$PhSiH_3$	4 h	(42)
TMS	H	CH_2OH	H	t-Bu	$PhSiH_3$	2 h	(77)
TMS	H	$CH_2OTBDMS$	H	t-Bu	$PhSiH_3$	2 h	(63)
TMS	H	CO_2Me	H	t-Bu	$PhSiH_3$	4 h	(47)
TMS	Me	H	CH_2OH	Et	$PhSiH_3$	3 h	(66)
TMS	Me	H	CH_2OH	t-Bu	$(Me_2SiH)_2O$	4 h	(27)
n-Bu	H	H	H	Et	$PhSiH_3$	1.5 h	(72)
n-Bu	H	H	H	t-Bu	$PhSiH_3$	45 min	(78)
Ph	H	H	H	Et	$PhSiH_3$	1.5	(46)
Ph	H	H	H	t-Bu	$(Me_2SiH)_2O$	2 h	(56)
$(4\text{-}MeOC_6H_4CO_2)CMe_2$	H	H	H	Et	$PhSiH_3$	1 h	(75)
$(4\text{-}MeOC_6H_4CO_2)CMe_2$	H	H	H	t-Bu	$PhSiH_3$	1.5 h	(61)

Substrate	Conditions	Product(s) and Yield(s) (%)	Refs.
C_5			
cyclopentadienide Li^+	Me_2NOMs, THF, –20°, 30 min	1-(Me_2N)cyclopentadiene (47)	804
$Me_2C{=}CHCH_2ZrCp_2Cl$ [c]	2,4,6-$Me_3C_6H_2SO_2ONH_2$, Et_2O, 0°, 10 min	$Me_2C{=}CHCH_2NH_2$ (62)[d]	116

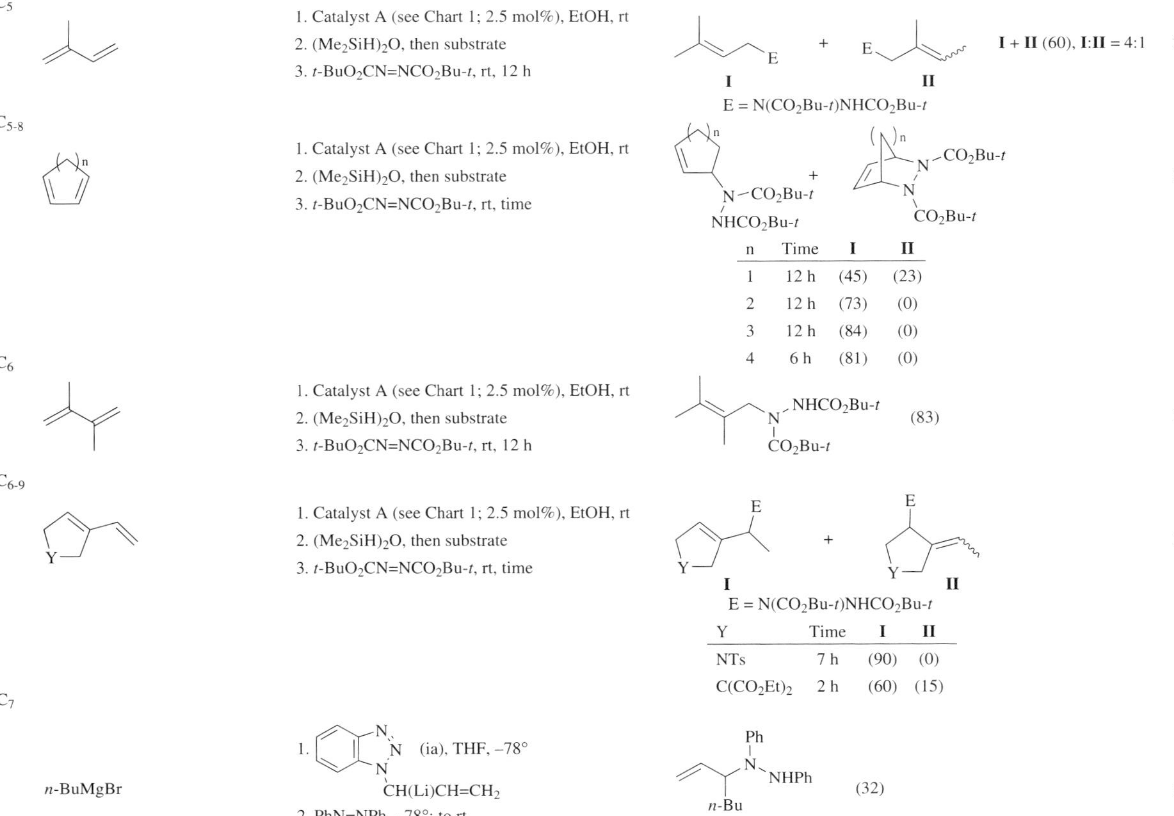

Substrate	Conditions	Product(s) and Yield(s) (%)	Refs.
C_5	1. Catalyst A (see Chart 1; 2.5 mol%), EtOH, rt 2. $(Me_2SiH)_2O$, then substrate 3. t-$BuO_2CN{=}NCO_2Bu$-t, rt, 12 h	**I** + **II** (60), **I**:**II** = 4:1 E = $N(CO_2Bu$-$t)NHCO_2Bu$-t	216
$C_{5\text{-}8}$	1. Catalyst A (see Chart 1; 2.5 mol%), EtOH, rt 2. $(Me_2SiH)_2O$, then substrate 3. t-$BuO_2CN{=}NCO_2Bu$-t, rt, time	**I** + **II**	216
C_6	1. Catalyst A (see Chart 1; 2.5 mol%), EtOH, rt 2. $(Me_2SiH)_2O$, then substrate 3. t-$BuO_2CN{=}NCO_2Bu$-t, rt, 12 h	(83)	216
$C_{6\text{-}9}$	1. Catalyst A (see Chart 1; 2.5 mol%), EtOH, rt 2. $(Me_2SiH)_2O$, then substrate 3. t-$BuO_2CN{=}NCO_2Bu$-t, rt, time	**I** + **II** E = $N(CO_2Bu$-$t)NHCO_2Bu$-t	216
C_7 n-BuMgBr	1. (ia), THF, –78° 2. PhN=NPh, –78°; to rt	(32)	359

n	Time	**I**	**II**
1	12 h	(45)	(23)
2	12 h	(73)	(0)
3	12 h	(84)	(0)
4	6 h	(81)	(0)

Y	Time	**I**	**II**
NTs	7 h	(90)	(0)
$C(CO_2Et)_2$	2 h	(60)	(15)

TABLE 1C. ALLYLIC AND PROPARGYLIC CARBANIONS (*Continued*)

Substrate	Conditions	Product(s) and Yield(s) (%)	Refs.
C_8			
	1. Catalyst A (see Chart 1; 2.5 mol%), EtOH, rt 2. $(Me_2SiH)_2O$, then substrate 3. t-$BuO_2CN{=}NCO_2Bu$-t, rt, 12 h	E **I** + E **II** E = $N(CO_2Bu$-$t)NHCO_2Bu$-t **I** + **II** (71), **I**:**II** = 7:1	216
C_9			
Ph MgCl	$ClNH_2$, Et_2O, –20°	Ph NH_2 (14)	805
 Li⁺	$(PhO)_2P(O)ONMe_2$, THF, –30° to 0°	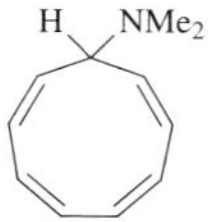 H NMe_2 (—)	146
	2,4,6-$Me_3C_6H_2SO_2ONMe_2$, Et_2O or Et_2O/THF, –10° to –15°; to rt, 15 h	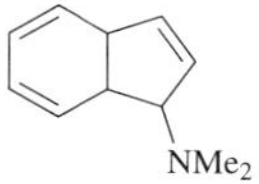 NMe_2 (69)	133
 Li⁺	$(PhO)_2P(O)ONMe_2$, THF, –30° to 0°	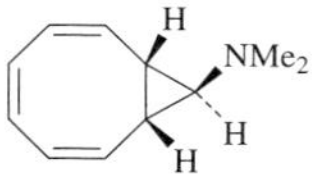 H NMe_2 H H (40)	146
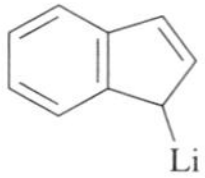 Li	2,4,6-$Me_3C_6H_2SO_2ONMe_2$, Et_2O or Et_2O/THF, –10° to –15°; to rt, 15 h	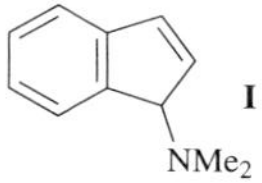 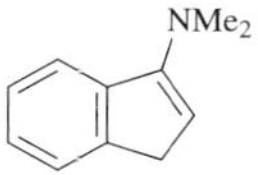 NMe_2 **I** + NMe_2 **II** **I** + **II** (57), **I**:**II** —	133
C_{10}			
Li⁺	$Ph_2P(O)ONH_2$, THF, –20°; rt, 12 h	NH_2 (37)	139

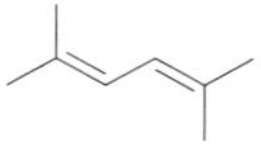

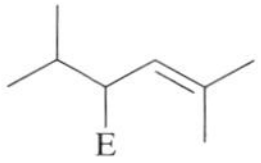

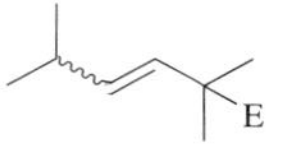

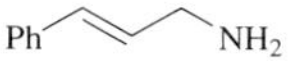

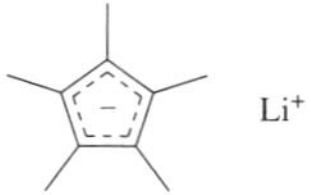

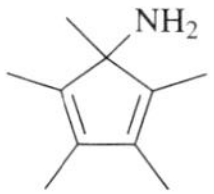

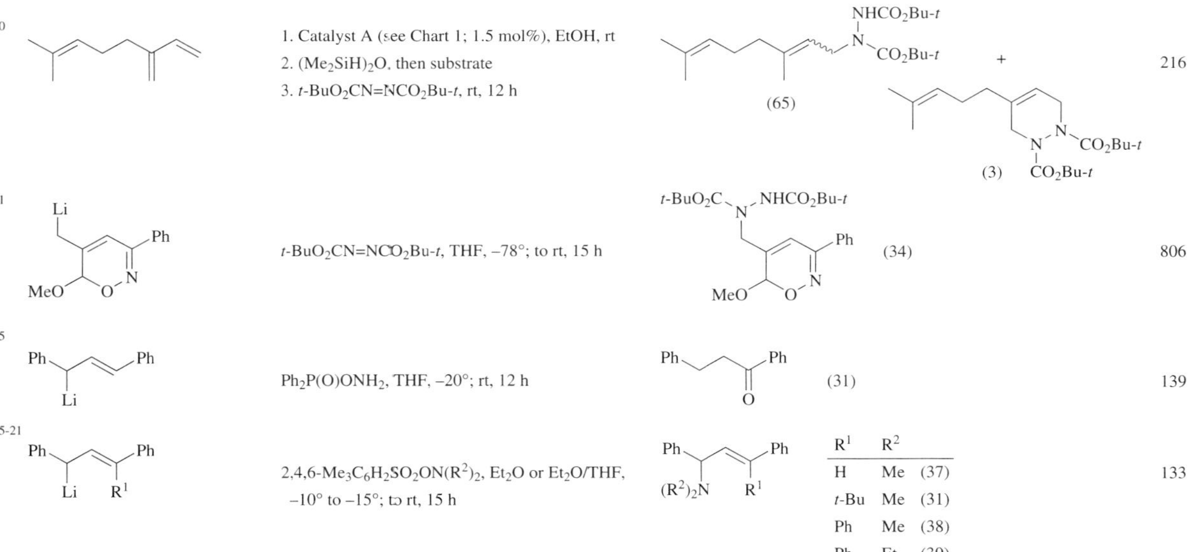

Substrate	Conditions	Product(s) and Yield(s) (%)	Refs.
C_{10}	1. Catalyst A (see Chart 1; 1.5 mol%), EtOH, rt 2. $(Me_2SiH)_2O$, then substrate 3. t-$BuO_2CN{=}NCO_2Bu$-t, rt, 12 h	(65) + (3)	216
C_{11}	t-$BuO_2CN{=}NCO_2Bu$-t, THF, –78°; to rt, 15 h	(34)	806
C_{15}	$Ph_2P(O)ONH_2$, THF, –20°; rt, 12 h	(31)	139
$C_{15\text{-}21}$	2,4,6-$Me_3C_6H_2SO_2ON(R^2)_2$, Et_2O or Et_2O/THF, –10° to –15°; to rt, 15 h	R^1 R^2 H Me (37) t-Bu Me (31) Ph Me (38) Ph Et (39)	133

[a] Pyrolysis of the triazene salt gave allyl azide in 13% yield.

[b] The substrates were prepared concurrently from the bromide with indium metal in the presence of one equivalent of NaI.

[c] The substrate was prepared in situ by addition of $HZrCp_2Cl$ to 3-methyl-1,2-butadiene.

[d] The product was characterized as the N-benzoyl derivative.

TABLE 1D. ARYLMETHYL AND HETEROARYLMETHYL CARBANIONS

Substrate	Conditions	Product(s) and Yield(s) (%)	Refs.
C_5	1. KNH_2, NH_3 (liquid), Et_2O 2. PhN=NPh	NHFh (39)	212
C_{5-6}	Conditions 1: 1. Catalyst A (see Chart 1; 2.5 mol%), EtOH, rt 2. $PhSiH_3$, t-BuO_2CN=NCO_2Bu-t, rt 3. Substrate, rt, time Conditions 2: 1. Catalyst B (see Chart 1; 1 mol %), i-PrOH, rt to 0° 2. $PhSiH_3$, t-BuO_2CN=NCO_2Bu-t, 0° 3. Substrate, 0°, time	t-BuO_2C N $NHCO_2Bu$-t	215
C_{6-8}	1. Base (x eq), THF, hexane, conditions 2. ArN=NAr, –78°, 10 min	NHAr	212

Y	R	Conditions	Time	
N	Me	1	8 h	(60)
N	Me	2	5 h	(88)
CH	H	1	—	(<5)
CH	Boc	1	11 h	(67)
CH	Ts	1	11 h	(74)

R	Base	x	Conditions	Ar	
H	LDA	2	rt, 1 h	Ph	(69)
H	LDA	1	rt, 1 h	4-ClC_6H_4	(36)
4-Me	n-BuLi	1	—	Ph	(47)
6-Me	LDA	2	rt, 1 h	Ph	(83)
4,6-Me_2	n-BuLi	1	—	Ph	(63)

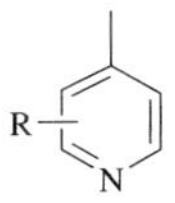

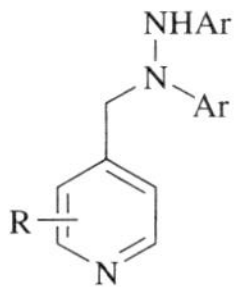

R	Ar		
H	Ph	(97)	212
H	2-ClC_6H_4	(53)	212
H	4-ClC_6H_4	(44)	212
2-Me	Ph	(82)	212
2,6-Me_2	Ph	(77)	212, 807

1. LDA (2 eq), THF, hexane, 1 h
2. ArN=NAr, –78°, 10 min

C_{6-7}

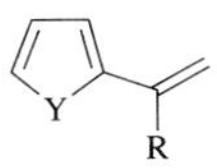

1. Catalyst A (see Chart 1; 2.5 mol%), EtOH, rt
2. $PhSiH_3$, t-BuO_2CN=NCO_2Bu-t, rt
3. Substrate, rt, time

t-BuO_2C–N(–$NHCO_2Bu$-t)

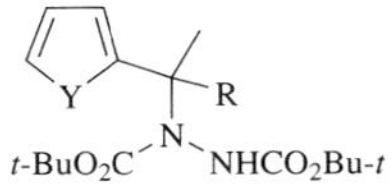

Y	R	Time		
O	H	12 h	(68)	215
S	H	18 h	(84)	
S	Me	7 h	(58)	

C_6

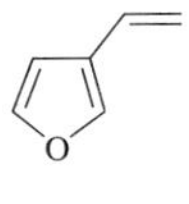

1. Catalyst A (see Chart 1; 2.5 mol%), EtOH, rt
2. $PhSiH_3$, t-BuO_2CN=NCO_2Bu-t, rt
3. Substrate, rt, 18 h

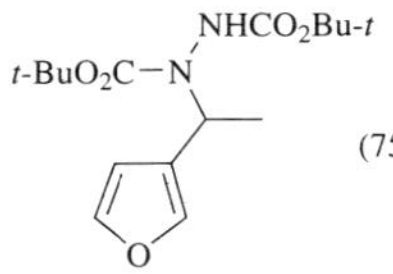

t-BuO_2C–N–$NHCO_2Bu$-t (75) 215

C_{6-8}

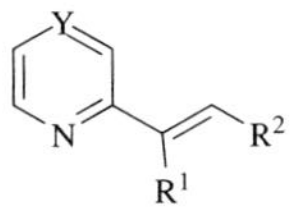

1. Catalyst A (see Chart 1; 2.5 mol%), EtOH, rt
2. $PhSiH_3$, t-BuO_2CN=NCO_2Bu-t, rt
3. Substrate, rt, time

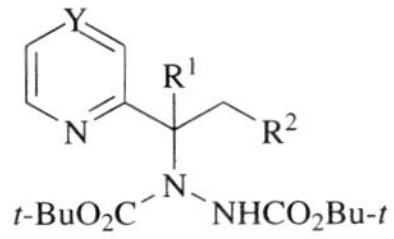

t-BuO_2C–N–$NHCO_2Bu$-t 215

Y	R^1	R^2	Time	
N	H	H	24 h	(63)
CH	H	H	4 h	(77)
CH	Me	H	5 h	(60)
CH	H	Me	5 h	(54)

TABLE 1D. ARYLMETHYL AND HETEROARYLMETHYL CARBANIONS (*Continued*)

Substrate	Conditions	Product(s) and Yield(s) (%)	Refs.
C_7			
$PhCH_2MgCl$ excess	NH_2Cl, Et_2O, 0°	$PhCH_2NH_2$ **I** (92)	805, 56
4 eq	NCl_3, Et_2O	**I** (32) + $PhCH_2NHCH_2Ph$ **II** (7)	77
excess	Br_2NH, Et_2O, 2-3°	**I** (34) + **II** (6)	75
	RNHCl, Et_2O, 5°, 1 h	$PhCH_2NHR$ **I** + RNH_2 **II** R / **I** / **II** Me (14) (70) Et (12) (75)	68
	$RNCl_2$, Et_2O, 5°, 1 h	**I** + **II** + $PhCH_2N(R)CH_2Ph$ **III** R / **I** / **II** / **III** Me (25) (43) (3) Et (19) (28) (6)	68
	R_2NCl, Et_2O, 5°, 1 h	$PhCH_2NR_2$ **I** + R_2NH **II** R / **I** / **II** Me (5) (95) Et (5) (89) *n*-Pr (5) (78)	68
2 eq	1. $BnONH_2$, Et_2O, –10° to –15° 2. HCl	$PhCH_2NH_3^+ Cl^-$ (79)	80
$PhCH_2Li$	1. MeONHLi (2 eq, ia), Et_2O 2. BzCl	$PhCH_2NHC(O)Ph$ (97)	82, 786

Reactant	Conditions	Product(s) and Yield(s) (%)	Refs.
$PhCH_2MgCl$	Me_2NOMs, Et_2O or THF, –30° to 0°	$PhCH_2NMe_2$ **I** (43)	134
$PhCH_2Br$	Mg, $Me_2NOSO_2C_6H_2Me_3$-2,4,6, Mg, THF, rt, 2 h	**I** (10)	801
$PhCH_2ZnCl$	R^1R^2NOBz, $(Ph_3P)_2NiCl_2$ (cat), THF, rt, 10 min to 6 h	$PhCH_2NR^1R^2$ **I** R^1 R^2 —$(CH_2)_5$— (68) —$(CH_2)_2O(CH_2)_2$— (85)	108
$(PhCH_2)_2Zn$	R^1R^2NOBz, $(CuOTf)_2$•PhH (cat), THF, rt, 1 h	**I** R^1 R^2 Bn Bn (91) —$(CH_2)_2O(CH_2)_2$— (80)	109 112, 109
$PhCH_2M$	$Ph_2P(O)ONH_2$, THF, –20°; rt, 12 h	$PhCH_2NH_2$ M Li (30) MgCl (70) MgBr (51)	139
	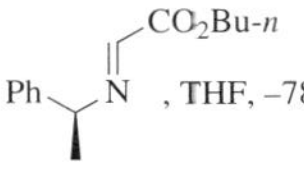 Ph(Me)CH–N=CH–CO_2Bu-*n*, THF, –78°; to rt, overnight	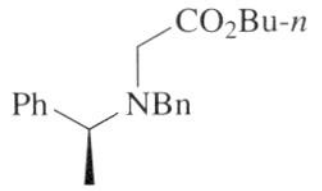 Ph(Me)CH–N(Bn)–CH_2CO_2Bu-*n* M MgCl (95) MgCl•Et_3Al (78) Cu (22) Cu•BF_3 (50) Ti(OPr-*i*)$_3$ (55)	164
$PhCH_2MX$ X not specified	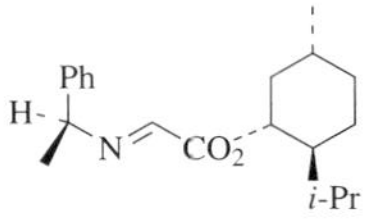 H-C(Ph)(Me)–N=CH–CO_2-(menthyl, *i*-Pr) (ia), Et_2O	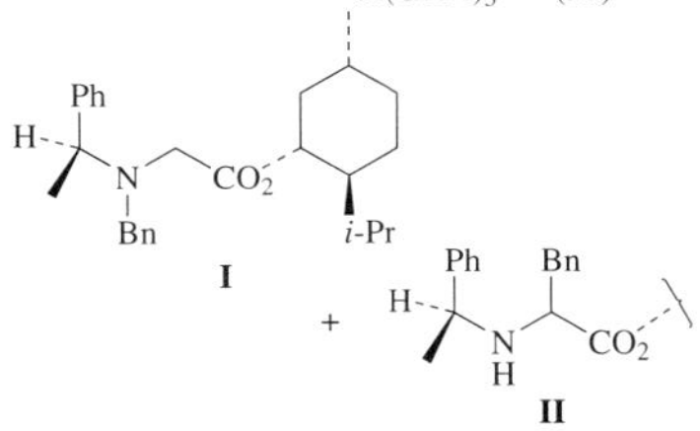 H-C(Ph)(Me)–N(Bn)–CH_2CO_2-(menthyl, *i*-Pr) **I** + H-C(Ph)(Me)–NH–CH(Bn)–CO_2-(menthyl) **II** M I + II I:II Mg (45-55) 100:0 Cd (55-70) 0:100[a]	161

TABLE 1D. ARYLMETHYL AND HETEROARYLMETHYL CARBANIONS (*Continued*)

Substrate	Conditions	Product(s) and Yield(s) (%)	Refs.
C_7			
$PhCH_2MgBr$	$4\text{-}MeOC_6H_4N{=}C(CO_2Et)_2$ (ia), THF, –95°, 30 min	$4\text{-}MeOC_6H_4N(Bn)CH(CO_2Et)_2$ (80)	167, 166
$(PhCH_2)_2Zn$ excess	$Me_2C{=}NOSO_2C_6H_2Me_3$-2,4,6, CuCN (cat), THF, rt, 3 h	$PhCH_2NH_2$ (56)	177
$PhCH_2MgCl$	1. $CH_2{=}N_2$, Et_2O, rt 2. HCl	$MeNHNHBn$ •HCl (41)	202, 199
$PhCH_2MgBr$	$Ph_2C{=}N_2$, Et_2O, rt	$Ph_2C{=}NNHBn$ (73)	202
	$PhN_2^+BF_4^-$, THF, 0°, 1 h	$PhCH_2N{=}NPh$ (0)	190
$PhCH_2M$	RCON=NCOR	$RCON(Bn)NHCOR$	

M	R	Solvent	Temp	Time	Yield (%)	Refs.
MgX [b]	MeO	Et_2O or THF	–78°	—	(>70)	799
ZnBr	*t*-BuO	THF	rt	30 min	(90)	358
MgBr	Ph	Et_2O	rt	—	(30)	794

Substrate	Conditions	Product(s) and Yield(s) (%)	Refs.
$PhCH_2MgCl$	RN_3, Et_2O	$RN{=}NNHNHBn$ [c]	

R	Temp	Time	Yield (%)	Refs.
Ph	reflux	30 min	(good)	270
$PhSO_2$ (ia)	—	—	(98)	306

	Substrate	Conditions	Product(s) and Yield(s) (%)	Refs.
	Ph⌒Li	1. N_3(=CH$_2$)Ph, THF, –78°; to rt, 2 h 2. HCl, then base	Ph⌒NH_2 (60)	278
	Ph⌒MgX[b]	6-N_3-2-methylbenzothiazole, Et_2O, 10-12 h	BnHNN=N-(2-methylbenzothiazol-6-yl) (43)	790
C_8	PhCH(Me)MgX	Ph, O, P(=O)ONMe$_2$, N-Me (oxazaphospholidine), THF, –15°	PhCH(Me)NMe$_2$ X: Cl (63) 30% ee; Br (40) 44% ee	147
	4-NC-C_6H_4-Me	1. LDA, THF, hexane, 1 h 2. ArN=NAr, –78°, 5 min	4-NC-C_6H_4-CH_2N(Ar)NHAr Ar: Ph (73); 2-ClC_6H_4 (68)	212
C_{8-13}	R^1MgBr	1. benzotriazol-1-yl-CH(R^2)Li (ia), THF, –78° 2. ArN=NAr, –78°; to rt	R^1CH(R^2)N(Ar)NHAr	359

R^1	R^2	Ar	
Me	Ph	Ph	(40)
n-Bu	Ph	Ph	(54)
n-Bu	Ph	4-ClC_6H_4	(51)
n-Bu	Ph	4-MeC_6H_4	(48)
n-Bu	4-MeC_6H_4	Ph	(57)
Ph	Ph	Ph	(52)

TABLE 1D. ARYLMETHYL AND HETEROARYLMETHYL CARBANIONS (*Continued*)

Substrate	Conditions	Product(s) and Yield(s) (%)	Refs.
C_{8-15} 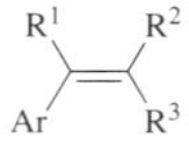 (Ar)(R¹)C=C(R²)(R³)	Conditions 1: 1. Catalyst A (see Chart 1; 5 mol%), EtOH, rt 2. Substrate and $PhSiH_3$ 3. t-$BuO_2CN{=}NCO_2Bu$-t, rt, time Conditions 2: 1. Catalyst A (see Chart 1; 5 mol %), EtOH, rt 2. $PhSiH_3$, t-$BuO_2CN{=}NCO_2Bu$-t, rt 3. Substrate, rt, time Conditions 3: 1. Catalyst B (see Chart 1; 5 mol%), *i*-PrOH, rt to 0° 2. Substrate, $PhSiH_3$ 3. t-$BuO_2CN{=}NCO_2Bu$-t, 0°, time Conditions 4: 1. Catalyst B (see Chart 1; 1 mol%), *i*-PrOH, rt to 0° 2. $PhSiH_3$, t-$BuO_2CN{=}NCO_2Bu$-t, 0° 3. Substrate, 0°, time	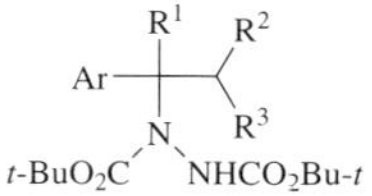 Ar(R¹)C(N(CO₂Bu-*t*)NHCO₂Bu-*t*)–CH(R²)(R³)	215

Ar	R¹	R²	R¹	Conditions	Time	
Ph	H	H	H	2	5 h	(86)
4-$H_2NC_6H_4$	H	H	H	2	20 h	(20-40)
4-$H_2NC_6H_4$	H	H	H	4	4 h	(66)
4-$FmocNHC_6H_4$	H	H	H	2	4 h	(98)
Ph	Me	H	H	2	2 h	(88)
Ph	H	Me	H	1	3 h	(88)
Ph	H	CH_2OH	H	1	1 h	(91)
Ph	Me	Me	H	1	20 h	(13)
Ph	Me	Me	Me	3	4 h	(51)

Substrate	Conditions	Product(s) and Yield(s) (%)	Refs.
C_8 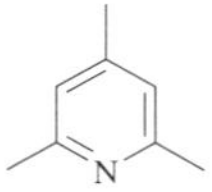 2,4,6-trimethylpyridine	1. BuLi, hexane, THF, –78° 2. PhN=NPh	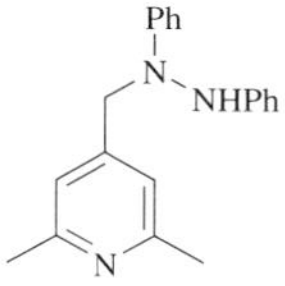 2,6-dimethyl-4-pyridyl-CH₂–N(Ph)–NHPh (65-77)	212

	Conditions	Product(s) and Yield(s) (%)	Refs.
C_9 Li	$2,4,6\text{-}Me_3C_6H_2SO_2ONMe_2$, Et_2O or Et_2O/THF, −10° to −15°; to rt, 15 h	NMe_2 + NMe_2 (57)	133
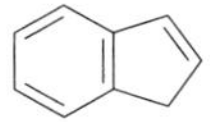	1. Catalyst A (see Chart 1; 5 mol%), EtOH, rt 2. Substrate, then $PhSiH_3$ 3. $t\text{-}BuO_2CN{=}NCO_2Bu\text{-}t$, rt, 4 h	$CO_2Bu\text{-}t$, N, $NHCO_2Bu\text{-}t$ (94)	215
C_{10} N	1. Base, solvent, temp, time 2. ArN=NAr	N, Ar, N, NHAr	212

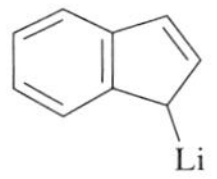

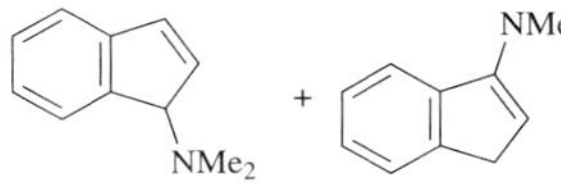

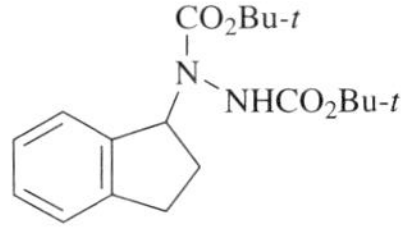

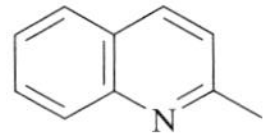

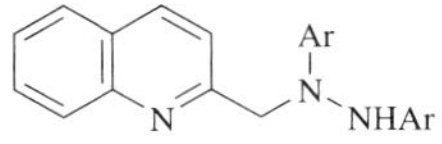

Base	Solvent	Temp, Time	Ar	
KNH_2, NH_3 (liq)	Et_2O	30 min	Ph	(50)
LDA	THF, hexane	rt, 1 h, then −78°	$2\text{-}ClC_6H_4$	(7)

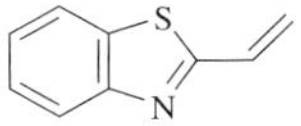

	Conditions	Product(s) and Yield(s) (%)	Refs.
S, N	1. Catalyst A (see Chart 1; 5 mol%), EtOH, rt 2. Substrate, then $PhSiH_3$ 3. $RO_2CN{=}NCO_2R$, rt, 4 h	S, N, N, $NHCO_2R$, CO_2R R: Et (54); t-Bu (<14)	215
N H	1. Catalyst A (see Chart 1; 2.5 mol%), EtOH, rt 2. $PhSiH_3$, $t\text{-}BuO_2CN{=}NCO_2Bu\text{-}t$ 3. Substrate, rt, 5 h	N H, N, $CO_2Bu\text{-}t$, $NHCO_2Bu\text{-}t$ (82)	215
N R	1. Catalyst A (see Chart 1; 2.5 mol%), EtOH, rt 2. $PhSiH_3$, $t\text{-}BuO_2CN{=}NCO_2Bu\text{-}t$ 3. Substrate, rt, 11 h	$NHCO_2Bu\text{-}t$, $t\text{-}BuO_2C{-}N$, N R R: Boc (76); Ts (85)	215

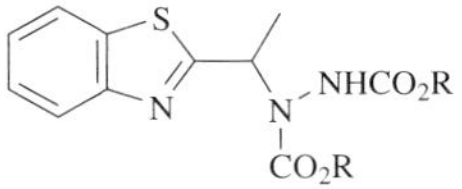

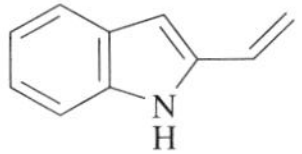

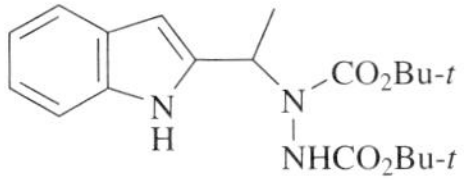

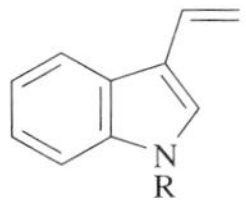

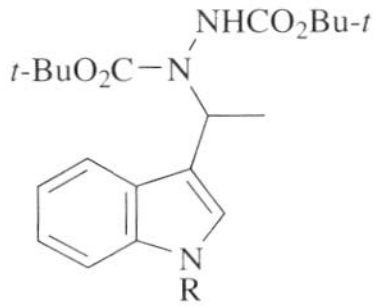

TABLE 1D. ARYLMETHYL AND HETEROARYLMETHYL CARBANIONS (*Continued*)

Substrate	Conditions	Product(s) and Yield(s) (%)	Refs.
C_{11}			
1-Naphthylmethyl-MgCl	$ClNH_2$	1-Naphthylmethyl-NH_2 (47)	805
Ph(C=CH$_2$)-cyclopropyl	Conditions 1: 1. Catalyst A (see Chart 1; 5 mol%), EtOH, rt 2. Substrate, then $PhSiH_3$ 3. t-BuO_2CN=NCO_2Bu-t, rt, 5 h Conditions 2: 1. Catalyst B (see Chart 1; 2 mol%), i-PrOH, 0° 2. Substrate, then $PhSiH_3$, 0° 3. t-BuO_2CN=NCO_2Bu-t, 0°, 4 h	Ph(Me)(cyclopropyl)C–N(CO_2Bu-t)–$NHCO_2Bu$-t Conditions 1 (48) Conditions 2 (60)	215
C_{12}			
1-Phenylcyclohexene	1. Catalyst A (see Chart 1; 5 mol%), EtOH, rt 2. Substrate, then $PhSiH_3$ 3. t-BuO_2CN=NCO_2Bu-t, rt, 4 h	1-Phenylcyclohexyl–N(CO_2Bu-t)–$NHCO_2Bu$-t (80)	215
C_{13}			
Ph_2CHLi	$MsONMe_2$, Et_2O or THF, –30° to 0°	Ph_2CHNMe_2 (84)	134
9-Lithiofluorene	RSO_2ONMe_2	9-(Dimethylamino)fluorene	

R	Solvent(s)	Temp, Time	Yield (%)	Refs.
Me	Et_2O or THF	–30° to 0°	(61)	134
2,4,6-$Me_3C_6H_2$	Et_2O or Et_2O/THF	–10° to –20° to rt; rt, 15 h	(95)	133

Reactant	Conditions	Product(s) and Yield(s) (%)	Refs.
	$Ph_2P(O)ONH_2$, THF, –20°; rt, 12 h	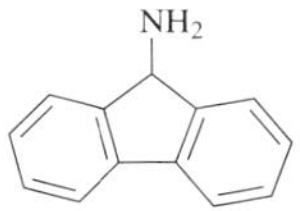 (30)	139
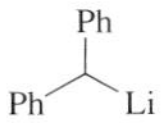 Ph(Ph)CHLi	$Ph_2P(O)ONH_2$, THF, –20°; rt, 12 h	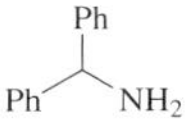 Ph_2CHNH_2 (41)	139
	1. LDA, THF, hexane, 1 h 2. ArN=NAr, –78°, 10 min	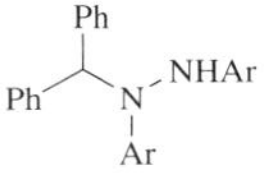 $Ph_2CHN(Ar)NHAr$ Ar: Ph (92); 2-ClC_6H_4 (20)	212
C_{14}	1. KNH_2, NH_3 (liq), Et_2O 2. PhN=NPh	(30)	212
C_{15}	1. *n*-BuLi (2.1 eq), THF, hexane, 0° 2. 2,4,6-$Me_3C_6H_2SO_2ONEt_2$, –78° to rt; rt, overnight	(43)	136
C_{15-21}	2,4,6-$Me_3C_6H_2SO_2ON(R^2)_2$, Et_2O or Et_2O/THF, –10° to –15°; to rt, 15 h	R^1 = H, R^2 = Me (37) R^1 = *t*-Bu, R^2 = Me (31) R^1 = Ph, R^2 = Me (38) R^1 = Ph, R^2 = Et (39)	133

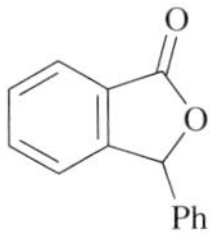

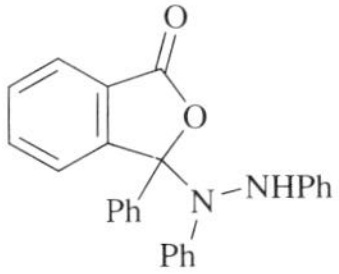

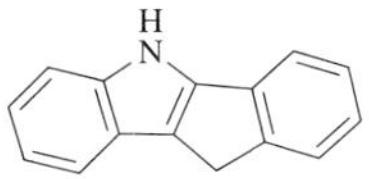

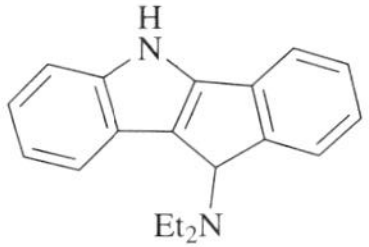

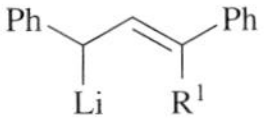

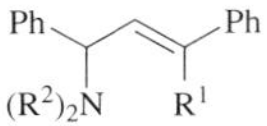

TABLE 1D. ARYLMETHYL AND HETEROARYLMETHYL CARBANIONS (*Continued*)

Substrate	Conditions	Product(s) and Yield(s) (%)	Refs.
C_{19}			
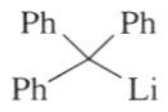 Ph_3C–Li	$ClNH_2$, Et_2O, sonication	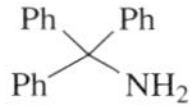 Ph_3C–NH_2 (67)	355
	RSO_2ONMe_2	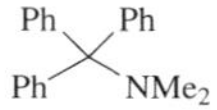 Ph_3C–NMe_2	
	R / Solvent(s) / Temp, Time		
	Me / Et_2O or THF / –30° to 0°	(60)	134
	2,4,6-$Me_3C_6H_2$ / Et_2O or Et_2O/THF / –10° to –20°; rt, 15 h	(30)	133
C_n			
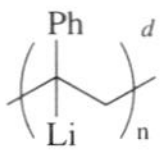 [poly(styryl)lithium, (CH(Ph)CH$_2$)$_n$ chain ending in Li][d]	1. MeONHLi, THF, Et_2O, hexane, –78° to –15°, 2 h 2. MeOH	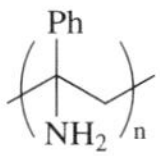 [same polymer chain ending in NH_2] (92)	808

[a] The reaction was carried out with both R and S enantiomers (CMePh); the optical yields were 5-46%.

[b] X was not specified.

[c] Some of the triazenes are isolated as mixtures of double-bond isomers.

[d] The substrate was prepared by addition of *sec*-butyllithium to styrene in benzene/THF at room temperature.

TABLE 2. VINYL AND ALLENYL CARBANIONS

Substrate	Conditions	Product(s) and Yield(s) (%)	Refs.
C_{2-3} MgBr, R = H, Me	$PhN_2^+BF_4^-$, THF	N=NPh, R (0)	185
C_3 Li	2,4,6-$Me_3C_6H_2SO_2ONEt_2$, Et_2O or Et_2O/THF, −10° to −20°; to rt, 14 h	NEt_2 (28)	133
C_{3-8} R^1, R^2, MgBr	1. $NOSO_2Ph$ dioxolane, Et_2O, PhCl, 0°; rt 2 h 2. pH 9 buffer	R^1 = H, R^2 = Me (93)[a] E:Z — R^1 = Ph, R^2 = H (100) E:Z 1:1	182
C_{4-14} R^1, R^2, MgBr	ArN_3, THF, 2 h	R^1, R^2, N=N–NHAr	286
C_6 MgBr (cyclohexenyl)	$PhN_2^+BF_4^-$, THF	N=NPh (0)	185

R^1	R^2	Ar	Yield (%)
Me	Me	Ph	(—)
H	Ph	Ph	(55)
H	Ph	4-MeC_6H_4	(72)
H	Ph	2-naphthyl	(48)
Ph	Ph	Ph	(63)
Ph	Ph	4-BrC_6H_4	(31)
Ph	Ph	4-MeC_6H_4	(56)
Ph	Ph	4-$MeOC_6H_4$	(55)

TABLE 2. VINYL AND ALLENYL CARBANIONS (*Continued*)

Substrate	Conditions	Product(s) and Yield(s) (%)	Refs.
$C_{7\text{-}12}$ 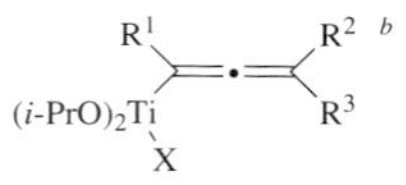 $R^1(i\text{-PrO})_2Ti(X)C{=}C{=}CR^2R^3$ [b] X = OCO_2Et or $OP(O)(OEt)_2$	$R^4O_2CN{=}NCO_2R^4$, Et_2O, –78° to rt, 1 h	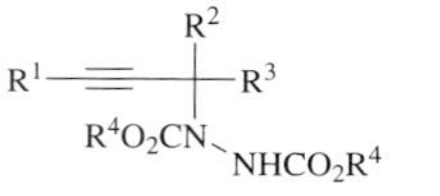 $R^1C{\equiv}C\text{–}C(R^2)(R^3)\text{–}N(CO_2R^4)NHCO_2R^4$	360

R¹	R²	R³	R⁴	
TMS	H	H	*t*-Bu	(49)
n-C_5H_{11}	H	H	*t*-Bu	(51)
TMS	*i*-Bu	H	*t*-Bu	(75)
TMS	*i*-Bu	H	Et	(52)
Me	*i*-Bu	H	*t*-Bu	(62)
n-Bu	Me	H	*t*-Bu	(61)
TMS	Me	$(CH_2)_3$Pr-*i*	*t*-Bu	(61)
TMS	Me	$(CH_2)_3$Pr-*i*	Et	(50)
TMS	Me	H	*t*-Bu (S) 94% ee	(77) 81% ee (S)
TMS	*n*-Bu	H	*t*-Bu[c] 96% ee	(73) 53% ee[a]
TMS	*i*-Bu	H	*t*-Bu[c] 96% ee	(74) 27% ee[a]

Substrate	Conditions	Product(s) and Yield(s) (%)	Refs.
C_8			
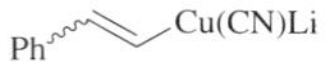 Ph~CH=CH–Cu(CN)Li	1. R^1R^2NLi, THF, –78° to –40°, 40 min 2. 1,2-$(O_2N)_2C_6H_4$, THF, –78° 3. O_2, –78°, 30 min; to rt	 Ph~CH=CH–NR^1R^2 R¹ R² *i*-Pr Me (60) *i*-Pr Bn (56)	55, 54
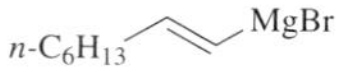 *n*-C_6H_{13}CH=CH–MgBr	1. $PhSCH_2N_3$ (ia), pentane, benzene, THF, Et_2O, –75° to 0°, 1 h; 0°, 1 h 2. Ac_2O, 30 min	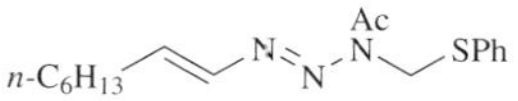 *n*-C_6H_{13}CH=CH–N=N–N(Ac)CH$_2$SPh (41)	274

C_{20}

Ar(Ar)C=C(Ar)MgBr

PhN_3

Ar(Ar)C=C(Ar)N=NNHPh

Ar	Solvent	Temp	Time		Refs.
Ph	Et_2O	rt	3 h	(59) (50)[d]	287, 288
4-$MeOC_6H_4$	THF	0°	1 h	(35)[d]	288

[a] The configuration was not reported.

[b] The substrate was prepared by reaction of $R^1C{\equiv}CCR^2R^3X$ [X = OCO_2Et or $OP(O)(OEt)_2$] with $Ti(OPr\text{-}i)_4$ and two equivalents of *i*-PrMgBr.

[c] The substrate was a single enantiomer of unspecified configuration.

[d] The substrate had ^{14}C in the 2-position.

TABLE 3. ETHYNYL CARBANIONS

Substrate	Conditions	Product(s) and Yield(s) (%)	Refs.
$C_{2\text{-}8}$			
$(R{-}C{\equiv}C)_3CuLi_2$	Me_2NX, Et_2O	$R{-}C{\equiv}C{-}NMe_2$ [a] R / X TMS / $Ph_2P(O)O$ (67) n-Pr / $Ph_2P(O)O$ (87)[b] n-Bu / $Ph_2P(O)O$ (78) t-Bu / $Ph_2P(O)O$ (71) c-C_6H_{11} / $Ph_2P(O)O$ (75) Ph[c] / MsO (52) PhS / $Ph_2P(O)O$ (45)	135
C_2			
$BrMg{-}C{\equiv}C{-}MgBr$	PhN_3, Et_2O, 10 d	$PhN{=}NHN{-}C{\equiv}C{-}NHN{=}NPh$[d] (1-2)	289, 809
$C_{2\text{-}8}$			
$R{-}C{\equiv}C{-}Li$	1. TsN_3, Et_2O, 0° 2. 2-$C_{10}H_7OH$, rt 3. H^+	4-R-5-[(2-hydroxy-1-naphthyl)-N=N]-1H-1,2,3-triazole R H (trace) n-Bu (31) Ph (10)	810
C_3			
$CH_3{-}C{\equiv}C{-}MgBr$	$N_3{-}C_6H_4{-}C_6H_4{-}N_3$, Et_2O	$(CH_3{-}C{\equiv}C{-}NHN{=}N{-}C_6H_4)_2$ (90)	272
C_8			
$Ph{-}C{\equiv}C{-}M$	$ClNEt_2$	$Ph{-}C{\equiv}C{-}NEt_2$ M / Solvent / Temp / Time Li / dioxane / 50°; rt / —; 3 d (2.3)[e] MgBr / Et_2O / rt / 12 h (1.7)	71
$(Ph{-}C{\equiv}C)_2CuLi$	$Ph_2P(O)ONH_2$, THF, –20°; rt, 12 h	$PhCH_2CN$[f] (38)	139
$(Ph{-}C{\equiv}C)_3CuLi_2$	$Ph(R)C{=}NOSO_2Ar$, Et_2O, 20°, 20 h	$Ph(R)C{=}N{-}C{\equiv}C{-}Ph$ R / Ar Me / Ph (39) Ph / 4-MeC_6H_4 (45)	178

[a] The yields of this product are based on two of the three acetylenic groups reacting.
[b] The yield was determined by NMR spectroscopy.
[c] (Phenylethynyl)lithium under these conditions produced no ethynylamine; the corresponding Grignard reagent gave only traces.
[d] The product was a mixture of two isomers.
[e] The yield reported is that of the crude product.
[f] 2-Phenylethynamine was formed as an intermediate.

TABLE 4. ARYL CARBANIONS

Substrate	Conditions	Product(s) and Yield(s) (%)	Refs.
C_6			
$PhCu(CN)ZnCl$	1. R^1R^2NLi, THF, –78° to –90°, 40 min 2. 1,2-$(O_2N)_2C_6H_4$, THF, –78° 3. O_2, –78°, 30 min	$PhNR^1R^2$ R^1 / R^2: H / Ph (69) *i*-Pr / *i*-Pr (70) Ph / Ph (76) Ph / Bn (68)	55
ArCu(CN)Li	1. (R)-1-(1-naphthyl)ethyl-NHLi, THF, –40°, 15 min 2. O_2, –78°, 20 min; to rt	1-(1-naphthyl)ethyl-NHAr Ar: Ph (48) 2-$MeOC_6H_4$ (55) 2-$BnOC_6H_4$ (46)	54
Ph_2CuLi	1. R^1R^2NH, solvent, reflux, 6 h 2. O_2, –78°	$PhNR^1R^2$ Solvent / R^1 / R^2: THF / Me / Ph (72) Et_2O / *n*-Bu / *n*-C_7H_{15} (64) Et_2O / Ph / Ph (94)	52
2,6-diethoxyphenyllithium (OEt, Li, OEt)	1. piperidino-NCu (5 eq), THF, hexane, reflux, 2 h 2. O_2	1-(2,6-diethoxyphenyl)piperidine (51)	52
PhMgCl	$ClNH_2$, petrol ether, 0°; rt, overnight	$PhNH_2$ (27)	58, 56
ArLi (3 eq)	$ClNH_2$ Et_2O	$ArNH_2$ Ar / Temp: Ph / –50° (33) 4-MeC_6H_4 / 0° (16)	58

TABLE 4. ARYL CARBANIONS (*Continued*)

Substrate	Conditions	Product(s) and Yield(s) (%)	Refs.
C_{6-7}			
$Ar^1MgCl•LiCl$	$Cl-N(R^2)-CH(R^1)-Ar^2$, THF, –45°, 15 min	$Ar^1-N(R^2)-CH(R^1)-Ar^2$	73
C_6			
PhMgCl (4 eq)	NCl_3, Et_2O	$PhNH_2$ (4) + Ph_2NH (1)	77
PhMgBr	R^1R^2NCl, Et_2O, 0°; rt, 12 h R^1, R^2 = Me, Me; Et, Et; —$(CH_2)_5$—	$PhNR^1R^2$ (0)	69
PhMgBr	H_2NOH, Et_2O, 0°, 30 min; rt, 15 min	$PhNH_2$ (4)	811
ArM	H_2NOMe (2 eq), THF, –15°	$ArNH_2$ (—)[a]	91

Ar^1	Ar^2	R^1	R^2	
Ph	Ph	H	Me	(57)
$4-IC_6H_4$	Ph	H	Et	(33)
$3-NCC_6H_4$	$4-BrC_6H_4$	H	Me	(70)
$3-NCC_6H_4$	1-naphthyl	Me	Me	(65)
$4-NCC_6H_4$	Ph	Me (R)	Me	(70) 99% ee
$2-EtO_2CC_6H_4$	Ph	H	Et	(67)
$2-EtO_2CC_6H_4$	Ph	Me (R)	Me	(67) 99% ee
$4-MeO_2CC_6H_4$	Ph	H	Me	(64)
$4-MeO_2CC_6H_4$	$4-BrC_6H_4$	H	Me	(73)
$4-MeO_2CC_6H_4$	Ph	Me (S)	$n-C_6H_{13}$	(34) 98% ee
1-naphthyl	Ph	H	Me	(56)

M = MgBr, $(CuMgBr)_{0.5}$, $Zn_{0.5}CuCN$
Ar = Ph, $3-BrC_6H_4$, $4-BrC_6H_4$, $3-MeC_6H_4$, $4-MeC_6H_4$, $3-MeOC_6H_4$, $4-MeOC_6H_4$

C_{6-10}

ArM + $RONH_2$ → $ArNH_2$

Ar	M	R	Addend	Solvent	Temp	Time	Yield	Refs.
Ph	Li	Me	—	Et_2O	–10° to –15°	—	$(63)^b$	792
Ph	Li	Bn	—	Et_2O	–10° to –15°	—	$(72)^b$	80
Ph	MgBr	Me	—	Et_2O	–10° to –15°	—	$(67)^b$	792
Ph	MgI	Me	—	Et_2O	–10° to –15°	—	$(0.3)^b$	792
Ph	MgBr	Bn	—	Et_2O	–10° to –15°	—	$(57)^b$	80
Ph	MgI	Bn	—	Et_2O	–10° to –15°	—	$(7)^b$	80
Ph	$Zn_{0.5}$	Me	—	THF	rt	3 h	$(41)^c$	177
Ph	ZnCl	Me	CuCN	THF	rt	3 h	(70)	176
Ph	Ph_2ZnLi	Me	CuCN	THF	rt	3 h	(92)	176
$4\text{-}BrC_6H_4$	MgBr	Me	—	Et_2O	–10° to –15°	—	$(73)^b$	792
$4\text{-}BrC_6H_4$	MgBr	Bn	—	Et_2O	–10° to –15°	—	$(58)^b$	80
$4\text{-}MeC_6H_4$	ZnCl	Me	CuCN	THF	rt	3 h	$(65)^c$	176
$4\text{-}MeOC_6H_4$	$Zn(C_6H_4OMe\text{-}4)$	Me	CuCN	THF	rt	3 h	$(65)^c$	176
1-naphthyl	MgBr	Bn	—	Et_2O	–10° to –15°	—	$(25)^b$	80

C_{6-8}

ArLi → ArNHBz

1. MeONHLi (2 eq, ia), Et_2O, hexane, –78° to –15°, 2 h
2. BzCl

Ar	Yield	Refs.
Ph	$(90)^d$	82, 83, 177, 786
$3\text{-}ClC_6H_4$	(46)	83
$2\text{-}MeOC_6H_4$	(98)	83, 786
$3\text{-}MeOC_6H_4$	(73)	83
$4\text{-}MeOC_6H_4$	(28)	83
$2\text{-}MeC_6H_4$	(96)	82
$4\text{-}MeC_6H_4$	(93)	83
$2\text{-}(i\text{-}Pr)_2NCOC_6H_4$	$(14)^e$	82
$2\text{-}EtC_6H_4$	(78)	83

TABLE 4. ARYL CARBANIONS (*Continued*)

Substrate	Conditions	Product(s) and Yield(s) (%)	Refs.
C_6			
Ar^1Li	1. $R^1N(Li)OR^2$ (ia), solvent(s), temp, time 2. Ar^2COCl	$Ar^1NR^1COAr^2$	

Ar^1	R^1	R^2	Solvent(s)	Temp, Time	Ar^2		Refs.
Ph	Me	Me	Et_2O, hexane	–78° to –15°, 3 h	Ph	(67)	97, 82
Ph	n-C_5H_{11}[f]	Bn	THF	0° to 40°, 1-3 h	4-PhC_6H_4	(47)	98
Ph	Bn[f]	Bn	THF	0° to 40°, 1-3 h	4-$MeOC_6H_4$	(37)	98
Ph	PhCHMe	Me	Et_2O, hexane	78° to –15°, 3 h; to 40°	Ph	(44)	97
2-(i-Pr)$_2$NCOC$_6$H$_4$	Me	Me	Et_2O, hexane	78° to –15°, 3 h	Ph	(0)	97

Substrate	Conditions	Product(s) and Yield(s) (%)	Refs.
C_{6-7}			
$Ar_2Cu(CN)Li_2$	1. RNHOTMS, THF, temp, time 2. HCl, then base	ArNHR	

Ar	R	Temp	Time		Refs.
Ph	TMS	–50°	1 h	(90) R = H[b]	100
Ph	Me	–50° to rt	2 h	(58)	101
Ph	i-Pr	–50° to rt	2 h	(64)	101
Ph	t-Bu	–50° to rt	2 h	(53)	101
4-FC_6H_4	i-Pr	–50° to rt	2 h	(45)	101
4-FC_6H_4	t-Bu	–50° to rt	2 h	(45)	101
3-$MeOC_6H_4$	Me	–50° to rt	2 h	(59)	101
3-$MeOC_6H_4$	t-Bu	–50° to rt	2 h	(73)	101
4-$MeOC_6H_4$	TMS	–50°	1 h	(70) R = H[b]	100
4-$MeOC_6H_4$	Me	–50° to rt	2 h	(88)	101
4-$MeOC_6H_4$	i-Pr	–50° to rt	2 h	(73)	101
4-MeC_6H_4	Me	–50° to rt	2 h	(57)	101
4-MeC_6H_4	i-Pr	–50° to rt	2 h	(67)	101
4-MeC_6H_4	t-Bu	–50° to rt	2 h	(65)	101

Reactant	Conditions	Product(s) and Yield(s) (%)	Refs.
C_6			
PhMgBr	1. Me_2NOTs, Et_2O, rt, 10 min 2. HCl	$PhNHMe_2^+ Cl^-$ (50)	132
$C_{6\text{-}10}$			
ArCu	CH_2=$CHCH_2O_2CN(Li)OR$, THF, –78°	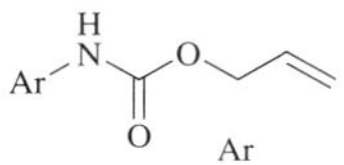 Ar–N(H)–C(=O)–O– (allyl)	130
C_6			
ArM	t-$BuO_2CN(Li)OTs$, THF	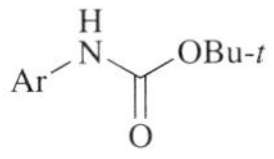 Ar–N(H)–C(=O)–OBu-t	
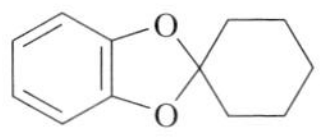 (benzodioxole spiro-cyclohexane)	1. n-BuLi, THF, Et_2O, hexane, 0°, 2 h; rt, 22 h 2. CuI, 0°, 15 min 3. t-$BuO_2CN(Li)OTs$, –78°, 30 min; 0°, 2 h	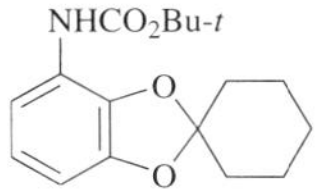 $NHCO_2Bu$-t (spiro compound) (45)	128
PhMgBr	1. Me_2NOTs, Et_2O, rt, 10 min 2. HCl	$PhNHMe_2^+ Cl^-$ (50)	132
$C_{6\text{-}10}$			
ArMgBr	2,4,6-$Me_3C_6H_2SO_2ONR_2$, Et_2O or Et_2O/THF, –10° to –15°; to rt, 15 h	$ArNR_2$	133

Product table for ArCu (ref. 130):

Ar	R	Time	
Ph	Ms	1 h	(51)
Ph	Ts	1 h	(51)
$3\text{-}FC_6H_4$	Ts	3 h	(44)
$4\text{-}MeOC_6H_4$	Ms	3 h	(52)
$4\text{-}MeOC_6H_4$	Ts	3 h	(68)
1-naphthyl	Ts	4 h	(57)

Product table for ArM:

Ar	M	Temp	Time		Refs.
Ph	Li	–78° to 0°	2 h	(10)	127
Ph	Cu	0°	2 h	(51)	127, 126
$4\text{-}FC_6H_4$	Cu	–78°	30 min	(50)	127
$2\text{-}MeOC_6H_4$	Cu	–78° to 0°	2 h	(73)	127

Product table for ArMgBr (ref. 133):

Ar	R	
Ph	Et	(42)
1-naphthyl	Me	(69)

TABLE 4. ARYL CARBANIONS (*Continued*)

Substrate	Conditions	Product(s) and Yield(s) (%)	Refs.
C_{6-7} ArBr	Mg, $Me_2NOSO_2C_6H_2Me_3$-2,4,6, Mg, THF, rt, 2 h	$ArNMe_2$	801

Ar	
Ph	(81)
2-MeC_6H_4	(60)
3-MeC_6H_4	(83)
4-MeC_6H_4	(79)
3-$MeOC_6H_4$	(78)
4-$MeOC_6H_4$	(80)

Substrate	Conditions	Product(s) and Yield(s) (%)	Refs.
C_{6-9} ArMgBr	$R^1R^2NOBz + CuCl_2$ (x mol%, ia, slow addition), THF, rt, 15 min	$ArNR^1R^2$	113

Ar	R^1	R^2	x	
Ph	*s*-Bu	H	5	(trace)
Ph	—$(CH_2)_2O(CH_2)_2$—		2.5	(37)
Ph	—$(CH_2)_2O(CH_2)_2$—		10	(68)
Ph	—$(CH_2)_5$—		2.5	(52)
Ph	—$(CH_2)_5$—		10	(64)
Ph	Bn	Bn	5	(20)
Ph	Bn	Bn	10	(92)
4-FC_6H_4	Bn	Bn	10	(58)
2-MeC_6H_4	Et	Et	2.5	(61)
2-MeC_6H_4	—$(CH_2)_2O(CH_2)_2$—		2.5	(7)
2-MeC_6H_4	—$(CH_2)_5$—		2.5	(75)
4-$MeOC_6H_4$	Et	Et	5	(65)
4-$MeOC_6H_4$	Et	Et	10	(75)
4-$MeOC_6H_4$	Et	Et	25	(26)
4-$MeOC_6H_4$	Et	Et	50	(8)
2,4,6-$Me_3C_6H_2$	Et	Et	2.5	(80)
2,4,6-$Me_3C_6H_2$	—$(CH_2)_5$—	Et	2.5	(61)

Reactant	Conditions	Product(s) and Yield(s) (%)	Refs.
PhMgBr x eq	Et_2NOCOR (slow ia during 7 min), $CuCl_2$ (3 mol%), $ZnCl_2$ (y eq), THF, rt; rt, 15 min	$PhNEt_2$	113

R	x	y	
EtO	2.2	0	(27)
t-Bu	2.2	0	(83)
Ph	1.1	0	(89)
Ph	2.2	0	(87)
Ph	2.2	0.1	(84)
4-$MeOC_6H_4$	2.2	0	(79)
4-$MeOC_6H_4$	2.2	0.1	(62)
4-$Me_2NC_6H_4$	2.2	0	(79)
4-$Me_2NC_6H_4$	2.2	0.1	(81)
2-MeC_6H_4	2.2	0	(89)
2-MeC_6H_4	2.2	0.1	(74)
2,4,6-$Me_3C_6H_2$	2.2	0	(82)
2,4,6-$Me_3C_6H_2$	2.2	0.1	(75)

Reactant	Conditions	Product(s) and Yield(s) (%)	Refs.
	1-(benzoyloxy)indole (N-OBz) (ia), $CuCl_2$ (2.5 mol%), THF, rt	1-phenylindole (N-Ph) (0)	113

$C_{6\text{-}7}$

Reactant	Conditions	Product(s) and Yield(s) (%)	Refs.
ArZnCl	R^1R^2NOBz, $(Ph_3P)_2NiCl_2$ (cat), THF, rt, 10 min to 6 h	$ArNR^1R^2$	108

Ar	R^1	R^2	
Ph	Et	Et	(71)
Ph	—$(CH_2)_5$—		(77)
Ph	—$(CH_2)_2O(CH_2)_2$—		(89)
4-ClC_6H_4	—$(CH_2)_2O(CH_2)_2$—		(84)
4-$MeOC_6H_4$	Et	Et	(92)
4-$MeOC_6H_4$	—$(CH_2)_5$—		(92)
4-$MeOC_6H_4$	—$(CH_2)_2O(CH_2)_2$—		(89)
2-MeC_6H_4	—$(CH_2)_2O(CH_2)_2$—		(73)
4-$CF_3C_6H_4$	—$(CH_2)_2O(CH_2)_2$—		(82)
4-NCC_6H_4	—$(CH_2)_2O(CH_2)_2$—		(56)
4-$EtO_2CC_6H_4$	—$(CH_2)_2O(CH_2)_2$—		(59)

TABLE 4. ARYL CARBANIONS (*Continued*)

Substrate	Conditions	Product(s) and Yield(s) (%)	Refs.
C_6			
Ph_2Zn	1. O(CH2CH2)2NOBz [morpholine-NOBz], $CuCl_2$ (2.5 mol%), THF, 0-5° 2. Substrate, THF, 0-5°, 1.5 h	N-phenylmorpholine (O, NPh) (67)	111
PhM x eq	piperidine-NOBz, Cu source (y eq), THF	N-phenylpiperidine (NPh)	113
	Et$_2$NOCOR, THF	$PhNEt_2$	113

M	x	Cu source	y	
Li	1	$CuBr•Me_2S$	1	(55)
Li	2	$CuBr•Me_2S$	1	(68)
Li	2	Li_2CuCl_3	1	(72)
MgBr	1	$CuBr•Me_2S$	1	(56)
MgBr	1	Li_2CuCl_3	1	(68)
ZnBr	2	$CuBr•Me_2S$	1	(68)
ZnBr	2	Li_2CuCl_3	1	(78)
ZnPh	1.1	CuCl	0.05	(88)

M	x	R	Temp	
MgBr	2.2	EtO	rt	(14)
MgBr	2.2	*t*-Bu	0°	(37)
MgBr	2.2	*t*-Bu	rt	(54)
MgBr	2.2	*t*-Bu	44°	(57)
MgBr	1.1	Ph	rt	(9)
MgBr	1.1	2-MeC_6H_4	rt	(18)
MgBr	1.1	4-$MeOC_6H_4$	rt	(17)
MgBr	1.1	4-$Me_2NC_6H_4$	rt	(21)
MgBr	1.1	2,4,6-$Me_3C_6H_2$	rt	(85)
Li	2.2	2,4,6-$Me_3C_6H_2$	–30°	(50)
Li	2.2	2,4,6-$Me_3C_6H_2$	0°	(46)
Li	2.2	2,4,6-$Me_3C_6H_2$	rt	(36)

C_{6-10}

Ar_2Zn 1.1 eq[g]

R^1R^2NOBz (ia), $(CuOTf)_2$•PhH (cat), THF, rt, 15 to 60 min

$ArNR^1R^2$

112, 109, 111

Ar	R^1	R^2	
Ph	*s*-Bu	H	(80)
Ph	*i*-Bu	H	(71)
Ph	Et	Et	(69)
Ph	—$(CH_2)_2O(CH_2)_2$—		(91)
Ph	—$(CH_2)_5$—		(91)
Ph	*i*-Pr	*i*-Pr	(72)
Ph	$CH_2CH{=}CH_2$	$CH_2CH{=}CH_2$	(96)
Ph	*t*-$BuCH_2CMe_2$	H	(74)
Ph	Bn	Bn	(94)
2-$MeOCH_2OC_6H_4$[h]	—$(CH_2)_2O(CH_2)_2$—		(95)
2-$Et_2NCOC_6H_4$[h]	—$(CH_2)_2O(CH_2)_2$—		(62)
3-FC_6H_4	—$(CH_2)_2O(CH_2)_2$—		(74)
4-FC_6H_4	—$(CH_2)_2O(CH_2)_2$—		(71) (74)[i]
4-ClC_6H_4	—$(CH_2)_2O(CH_2)_2$—		(93)
4-ClC_6H_4	Bn	Bn	(88)
4-$MeOC_6H_4$	—$(CH_2)_2O(CH_2)_2$—		(93) (71)[i]
4-$MeOC_6H_4$	—$(CH_2)_5$—		(95)
4-$MeOC_6H_4$	Bn	Bn	(87)
4-$AcOC_6H_4$	—$(CH_2)_2O(CH_2)_2$—		(76)
4-$TfOC_6H_4$	—$(CH_2)_2O(CH_2)_2$—		(95) (83)[i]
2-$O_2NC_6H_4$	—$(CH_2)_2O(CH_2)_2$—		(83)
2-$O_2NC_6H_4$	Bn	Bn	(97)
2,4-$(O_2N)_2C_6H_3$	—$(CH_2)_2O(CH_2)_2$—		(59)
2-MeC_6H_4	Et	Et	(70)
2-MeC_6H_4	—$(CH_2)_2O(CH_2)_2$—		(94)
2-MeC_6H_4	—$(CH_2)_5$—		(86)[i]
2-MeC_6H_4	*i*-Pr	*i*-Pr	(62)
3-$CF_3C_6H_4$	—$(CH_2)_2O(CH_2)_2$—		(74)

TABLE 4. ARYL CARBANIONS (*Continued*)

	Substrate	Conditions	Product(s) and Yield(s) (%)	Refs.
C_{6-10}	Ar_2Zn 1.1 eq[g] (*Table continued from previous page.*)		$ArNR^1R^2$	112, 109, 111
		Ar | R^1 | R^2		
		2-(4,4-dimethyl-4,5-dihydrooxazol-2-yl)]phenyl[h] | —$(CH_2)_2O(CH_2)_2$—	(55)	
		$4\text{-}NCC_6H_4$ | —$(CH_2)_2O(CH_2)_2$—	(76) (74)[i]	
		$4\text{-}NCC_6H_4$ | Bn | Bn	(95)	
		$4\text{-}EtO_2CC_6H_4$ | —$(CH_2)_2O(CH_2)_2$—	(77)	
		$4\text{-}EtO_2CC_6H_4$ | Bn | Bn	(99)	
		$4\text{-}(i\text{-}Pr)_2NCOC_6H_4$[h] | —$(CH_2)_2O(CH_2)_2$—	(88)	
		4-(2-methyl-1,3-dioxolan-2-yl)phenyl | —$(CH_2)_2O(CH_2)_2$—	(79)	
		$2,4,6\text{-}Me_3C_6H_2$ | *i*-Pr | *i*-Pr	(76)	
		1-naphthyl | —$(CH_2)_2O(CH_2)_2$—	(90)	
	ArMgX	$Ph_2P(O)ONH_2$, THF, –20°; rt, 12 h	$ArNH_2$ Ar | X Ph | Cl (35) Ph | Br (22) 1-naphthyl | Br (31)	139
C_6	PhMgBr 3 eq	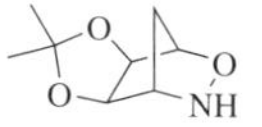 [structure: acetonide-protected bicyclic O,NH compound], –78° to 0°, 1 h	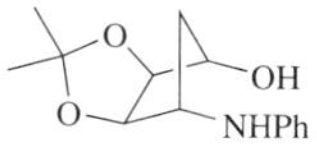 [structure: acetonide-protected cyclopentane with OH and NHPh] (~100)	104
	PhMgBr	$4\text{-}MeOC_6H_4N{=}C(CO_2Et)_2$ (ia), THF, –95°, 30 min	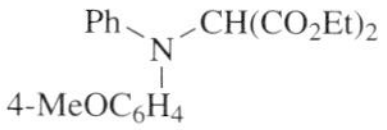 $Ph(4\text{-}MeOC_6H_4)N\text{-}CH(CO_2Et)_2$ (59)	167, 166
	PhMgBr 2 eq	1. PhCH=NOR, Et_2O, heat 2. HCl	$Ph_2CH\text{-}NHPh \cdot HCl$ R H (—) Me (—) Bn (68)	175

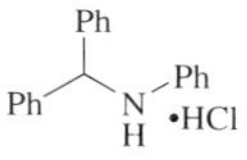

$C_{6\text{-}7}$				
	$ArMgX^{j}$ x eq	$Me_2C{=}NOR$, toluene	$ArNH_2$	174a

Ar	x	R	
Ph	1	MgBr	(62)
Ph	2	H	(35)
4-ClC_6H_4	2	H	(20)
2-$MeOC_6H_4$	2	H	(18)
4-$MeOC_6H_4$	2	H	(12)
4-MeC_6H_4	1	MgBr	(70)
4-MeC_6H_4	2	H	(31)

C_6				
	PhLi 3 eq	$Me_2C{=}NOH$	Ph, NHPh **I** + $PhNH_2$ **II**; **I** + **II** (—), **I**:**II** = 4:1	174a
	ArLi	1. $ZnCl_2$, THF, 0°; to rt 2. O, N, $Ni(acac)_2$ (cat), 2 h	CHO, NHAr	170

Ar	
Ph	(86)
2-$MeOC_6H_4$	(61)
3-$MeOC_6H_4$	(70)
4-$MeOC_6H_4$	(79)

$C_{6\text{-}10}$				
	ArMgBr	NOTs, MeN, NMe (ia), toluene, Et_2O	NAr, MeN, NMe	181

Ar	Temp	Time	
Ph	0°	15 min	$(98)^k$
2-$MeOC_6H_4$	0° to rt	30 min	$(99)^k$
2,4-$(MeO)_2C_6H_3$	0° to rt	30 min	$(96)^k$
2-MeC_6H_4	0°	15 min	$(96)^k$
4-$CF_3C_6H_4$	0°	15 min	$(88)^l$
2,6-$Me_2C_6H_3$	0° to rt	30 min	$(85)^m$
1-naphthyl	0°	30 min	$(>99)^k$

TABLE 4. ARYL CARBANIONS (*Continued*)

Substrate	Conditions	Product(s) and Yield(s) (%)	Refs.
C_6 PhMgBr	$R^1R^2C{=}N{-}OTs$ (ia)	$R^1R^2C{=}N{-}Ph$	181
C_{6-10} ArMgBr	$R_2C{=}NOSO_2C_6H_2Me_3$-2,4,6 (0.6-0.8eq), catalyst, Et_2O, toluene, 60°	$ArNH_2$[p]	803
	1. 2-(N-OSO$_2$R imino)-4,4,5,5-tetramethyl-1,3-dioxolane (ia), Et_2O, cosolvent, temp, time 1 2. Method A: HCl, MeOH, Et_2O, rt, time 2 or Method B: HCl, EtOH, H_2O, reflux, time 2	$ArNH_3^+ Cl^-$	182

R^1	R^2	Solvent	Temp	Time	
—$O(CH_2)_2O$—		toluene	0°	1 h	(66)
EtO	EtO	toluene	–30° to rt	12 h	(<10)
—$NMe(CH_2)_2O$—		toluene	rt	30 min	(97)
—$O(CH_2)_2NMe$—		CH_2Cl_2	0°	30 min	(0)[n]
—$NMe(CH_2)_2NMe$—		toluene	0°	15 min	(98)

Ar	R	Catalyst	Time	
Ph	Me	—	40 h	(58)
Ph	Me	CuI	11 h	(59)
Ph	Me	$MgCl_2$	22h	(56)
Ph	Ph	CuI	2.5 h	(59)
4-MeC_6H_4	Me	—	46 h	(36)
4-MeC_6H_4	Me	$MgCl_2$	20 h	(58)
1-naphthyl	Me	—	—	(0)

Ar	R	Cosolvent	Temp	Time 1	Method	Time 2	
Ph	Ph	PhCl	0°	0.5 h	A	1.5 h	(93)
4-FC_6H_4	Ph	CH_2Cl_2	rt	1 h	A	1.5 h	(90)
2-$MeOC_6H_4$	Ph	CH_2Cl_2	rt	0.5 h	A	2 h	(96)
3-$MeOC_6H_4$	Ph	CH_2Cl_2	rt	0.5 h	A	2.5 h	(90)
4-$MeOC_6H_4$	Ph	PhCl	0°	0.5 h	A	2.5 h	(96)
2,4-$(MeO)_2C_6H_3$	Ph	CH_2Cl_2	rt	1 h	A	6.5 h	(91)
4-MeC_6H_4	Me[a]	CH_2Cl_2	—	—	A	—	(86)
4-MeC_6H_4	Ph	PhCl	0°	0.5 h	A	1.5 h	(97)
4-MeC_6H_4	4-MeC_6H_4	CH_2Cl_2	—	—	A	—	(90)
4-MeC_6H_4	2,4,6-$Me_3C_6H_2$	CH_2Cl_2	—	—	A	—	(80)
4-$CF_3C_6H_4$	Ph	PhCl	0°	1 h	A	0.5 h	(94)
2,6-$Me_2C_6H_3$	Ph	PhCl	0°	1 h	B	3 h	(90)
1-naphthyl	Ph	CH_2Cl_2	rt	0.5 h	A	1 h	(93)

ArZnCl

1. $Me_2C{=}NOSO_2C_6H_2Me_3$-2,4,6 (2 eq), CuCN (10 mol%), DMPU (2 eq), rt, 3 h
2. conc. HCl, then base
3. BzCl

ArNHBz

Ar	
Ph	(78)
3-BrC_6H_4	(70)
4-$MeOC_6H_4$	(75)
4-MeC_6H_4	(72)
1-naphthyl	(79)

812

TABLE 4. ARYL CARBANIONS (*Continued*)

	Substrate	Conditions	Product(s) and Yield(s) (%)						Refs.
C_{6-7}	ArM	$Me_2C{=}NOSO_2C_6H_2Me_3$-2,4,6, CuCN (cat), THF	$ArNH_2$	R	M	Temp	Time		
				Ph	ZnCl	rt	3 h	(70)	176, 177
				Ph	Ph_2ZnLi	rt	3 h	(79)	176
				Ph	PhZn	rt	3 h	(44)	176
				Ph	Cu(CN)Li	rt	3 h	(76)[c]	177
				Ph	Cu(CN)ZnCl	0°	1 h	(57)	176
				Ph	$Ph_2ZnMgBr$	rt	3 h	(85)	177
				$4-BrC_6H_4$	ZnCl	rt	3 h	(33)	177
				$4-MeOC_6H_4$	ZnCl	rt	3 h	(54)[c]	176, 177
				$4-MeC_6H_4$	ZnCl	rt	3 h	(49)[c]	176, 177
				$4-EtO_2CC_6H_4$	ZnCl	rt	3 h	(51)	177
	ArM Ar = Ph, $3-BrC_6H_4$, $4-BrC_6H_4$, $4-MeOC_6H_4$, $3-MeC_6H_4$, $4-MeC_6H_4$; M = MgBr or Cu(CN)ZnCl	$Me_2C{=}NOSO_2C_6H_2Me_3$-2,4,6, THF, reflux, 1-2 h	$ArNH_2$ (—)[a]						183
	ArBr	1. Mg, $Me_2C{=}NOSO_2C_6H_4Me_3$-2,4,6, THF, reflux, 3 h 2. BzCl	ArNHBz	Ar					
				Ph	(52)				802
				$4-MeOC_6H_4$	(40)				
				$4-MeC_6H_4$	(53)				
				1-naphthyl	(40)				
C_{6-10}	ArMgBr	1. $[3,5-(CF_3)_2C_6H_3]_2C{=}NOTs$ (ia), Et_2O, toluene, rt, 30 min 2. BzCl, Et_3N	ArNHBz	Ar					
				Ph	(96)				179
				$4-FC_6H_4$	(86)				
				$2-MeOC_6H_4$	(72)				
				$3-MeOC_6H_4$	(94)				
				$4-MeOC_6H_4$	(98)				
				$4-CF_3C_6H_4$	(71)[b]				
				$2,4-Me_2C_6H_3$	(98)				
				1-naphthyl	(91)				

Reagent	Conditions	Product(s) and Yield(s) (%)	Refs.
C_{6-14}			
ArMgBr	Ph, Ph, Ph, Ph-cyclopentadienone NOTs (ia), THF, –78°, 45-90 min	Ph, Ph, Ph, Ph-cyclopentadienone NAr; Ar: Ph (95), 2,3,5,6-Me_4C_6H (65-68), 1-naphthyl (70), 2-naphthyl (78), 9-phenanthryl (83)	180
C_6			
PhCu(CN)Li	1. Me_2NNHLi, THF, –40°, 40 min 2. O_2, –73°, 30 min	PhNH–NMe_2 (40)	54
PhMgBr	$CH_2=N_2$, Et_2O, 0°	PhCH=N–NHPh (8)	200
	$CH_2=N_2$, Et_2O, rt	$PhCH_2$NH–NHPh (48)[b]	202
	$EtO_2CCH=N_2$, Et_2O, cooling	$Ph_2C(OH)CH=N$–NHPh (—)	199
	$(NC)_2C=N_2$, Et_2O, cooling	$(NC)_2C=N$–NHPh (66)	203
PhLi	diazocyclopentadiene (N_2), $FeCl_3$	Fe(C₅H₄N=NPh)₂ (—)	787
PhMgBr	diazo ketone (N_2, O), Et_2O, cooling	=NNHPh, O (—)	201
	$Ph_2C=N_2$, Et_2O, rt, 30 min	$Ph_2C=N$–NHPh (70)	202
	$(PhSO_2)_2C=N_2$ (ia), Et_2O, 30 min	$(PhSO_2)_2C=N$–NHPh (54)	204

TABLE 4. ARYL CARBANIONS (*Continued*)

Substrate	Conditions	Product(s) and Yield(s) (%)	Refs.
C_{6-7}			
$Ar_3ZnMgBr$	$PhN_2^+ BF_4^-$, THF, 0°, 1 h	Ar–N=N–Ph; Ar: Ph (87), $4\text{-}BrC_6H_4$ (74), $4\text{-}MeOC_6H_4$ (72), $4\text{-}MeC_6H_4$ (88)	190
ArM	1. $PhN_2^+ BF_4^-$, THF, addend, temp, time 2. HCl 3. $NaBH_4$, $NiCl_3$•6 H_2O, MeOH, 0°, 1 h	$ArNH_2$	190

Ar	M	Addend	Temp	Time	Yield
Ph	MgBr	—	–78°	3 h	(53)[p]
Ph	MgBr	TMEDA (0.3 eq)	–15°	3 h	(57)[p]
$(4\text{-}MeC_6H_4)_3$	ZnBr	—	–15°	1 h	(67)[p]

Substrate	Conditions	Product(s) and Yield(s) (%)	Refs.
C_{6-10}			
Ar^1MgBr	$Ar^2N_2^+ ZnCl_3^-$ (ia), Et_2O, reflux; rt, 60 min	Ar^1–N=N–Ar^2	

Ar^1	Ar^2	Yield	Refs.
Ph	$2\text{-}EtO_2CC_6H_4$	(poor)	187
Ph	$3\text{-}EtO_2CC_6H_4$	(43)	187
Ph	$4\text{-}EtO_2CC_6H_4$	(42)	187
Ph	1-naphthyl	(22)	188
Ph	2-naphthyl	(11)	188
$2\text{-}BrC_6H_4$	1-naphthyl	(3.5)	188
$2\text{-}BrC_6H_4$	2-naphthyl	(12)	188
$3\text{-}BrC_6H_4$	$3\text{-}MeOC_6H_4$	(17)	189
$3\text{-}BrC_6H_4$	1-naphthyl	(4)	188
$3\text{-}BrC_6H_4$	2-naphthyl	(12)	188
$4\text{-}BrC_6H_4$	1-naphthyl	(11)	188
$4\text{-}BrC_6H_4$	2-naphthyl	(13)	188

2-$MeOC_6H_4$	4-BrC_6H_4	(2)	189
2-$MeOC_6H_4$	2-$MeOC_6H_4$	(1.4)	189
2-$MeOC_6H_4$	3-BrC_6H_4	(9)	189
2-$MeOC_6H_4$	1-naphthyl	(12)	188
2-$MeOC_6H_4$	2-naphthyl	(11)	188
3-$MeOC_6H_4$	4-BrC_6H_4	(1)	189
3-$MeOC_6H_4$	2-$MeOC_6H_4$	(11)	189
3-$MeOC_6H_4$	3-$MeOC_6H_4$	(8)	189
3-$MeOC_6H_4$	4-$MeOC_6H_4$	(34)	189
3-$MeOC_6H_4$	1-naphthyl	(12)	188
3-$MeOC_6H_4$	2-naphthyl	(6)	188
4-$MeOC_6H_4$	3-BrC_6H_4	(10)	189
4-$MeOC_6H_4$	4-BrC_6H_4	(14)	189
4-$MeOC_6H_4$	2-$MeOC_6H_4$	(7)	189
4-$MeOC_6H_4$	4-$MeOC_6H_4$	(7)	189
4-$MeOC_6H_4$	1-naphthyl	(9)	188
4-$MeOC_6H_4$	2-naphthyl	(11)	188
2-MeC_6H_4	2-BrC_6H_4	(64)	186
2-MeC_6H_4	3-BrC_6H_4	(61)	186
2-MeC_6H_4	4-BrC_6H_4	(42)	186
2-MeC_6H_4	2-$MeOC_6H_4$	(18)	189
2-MeC_6H_4	3-$MeOC_6H_4$	(29)	189
2-MeC_6H_4	4-$MeOC_6H_4$	(9)	189
2-MeC_6H_4	2-$EtO_2CC_6H_4$	(36)	187
2-MeC_6H_4	3-$EtO_2CC_6H_4$	(56)	187
2-MeC_6H_4	4-$EtO_2CC_6H_4$	(63)	187
2-MeC_6H_4	1-naphthyl	(7)	188
2-MeC_6H_4	2-naphthyl	(12)	188
3-MeC_6H_4	2-BrC_6H_4	(40)	186
3-MeC_6H_4	3-BrC_6H_4	(52)	186
3-MeC_6H_4	4-BrC_6H_4	(68)	186
3-MeC_6H_4	2-$MeOC_6H_4$	(11)	189

TABLE 4. ARYL CARBANIONS (*Continued*)

Substrate	Conditions		Product(s) and Yield(s) (%)	Refs.
C_{6-10} Ar^1MgBr (*Table continued from previous page.*)			Ar^1–N=N–Ar^2	
	Ar^1	Ar^2		
	3-MeC_6H_4	3-$MeOC_6H_4$	(20)	189
	3-MeC_6H_4	4-$MeOC_6H_4$	(4)	189
	3-MeC_6H_4	2-$EtO_2CC_6H_4$	(45)	187
	3-MeC_6H_4	3-$EtO_2CC_6H_4$	(53)	187
	3-MeC_6H_4	4-$EtO_2CC_6H_4$	(42)	187
	3-MeC_6H_4	1-naphthyl	(23)	188
	3-MeC_6H_4	2-naphthyl	(6)	188
	4-MeC_6H_4	2-BrC_6H_4	(60)	186
	4-MeC_6H_4	3-BrC_6H_4	(68)	186
	4-MeC_6H_4	4-BrC_6H_4	(68)	186
	4-MeC_6H_4	2-$MeOC_6H_4$	(11)	189
	4-MeC_6H_4	3-$MeOC_6H_4$	(14)	189
	4-MeC_6H_4	4-$MeOC_6H_4$	(13)	189
	4-MeC_6H_4	2-$EtO_2CC_6H_4$	(49)	187
	4-MeC_6H_4	3-$EtO_2CC_6H_4$	(48)	187
	4-MeC_6H_4	4-$EtO_2CC_6H_4$	(45)	187
	4-MeC_6H_4	1-naphthyl	(20)	188
	4-MeC_6H_4	2-naphthyl	(7)	188
	1-naphthyl	2-$EtO_2CC_6H_4$	(trace)	188
	1-naphthyl	3-$EtO_2CC_6H_4$	(25)	188
	1-naphthyl	4-$EtO_2CC_6H_4$	(4)	188
	2-naphthyl	2-$EtO_2CC_6H_4$	(1)	188
	2-naphthyl	3-$EtO_2CC_6H_4$	(1)	188
	2-naphthyl	4-$EtO_2CC_6H_4$	(trace)	188

$C_{6\text{-}9}$

Ar^1ZnCl	$Ar^2N_2^+ BF_4^-$, Et_2O, $-10°$	$Ar^1{-}N{=}N{-}Ar^2$	192

Ar^1	Ar^2	Time	
Ph	4-ClC_6H_4	18 h	(10)
Ph	4-$MeOC_6H_4$	22 h	(2)
Ph	2,4,6-$Me_3C_6H_2$	24 h	(0)
4-ClC_6H_4	Ph	6 h	(35)
4-$MeOC_6H_4$	Ph	18 h	(20)
2,4,6-$Me_3C_6H_2$	Ph	22 h	(8)

TABLE 4. ARYL CARBANIONS (*Continued*)

Substrate	Conditions					Product(s) and Yield(s) (%)	Refs.
C_{6-7}							
Ar^1MgX^1	$Ar^2N_2^+ (X^2)^-$ (ia)					$Ar^1-N=N-Ar^2$	
Ar^1	X^1	Ar^2	X^2	Solvent	Conditions		
Ph	—	Ph	[q]	THF	–78°, 1 h; 70°, 1 h[r]	(71)	191
Ph	Br	Ph	BF_4	THF	–78°	(66)	185, 184
Ph	—	2-ClC_6H_4	[q]	THF	–78°, 1 h; 70°, 1 h[r]	(52)	191
Ph	—	4-ClC_6H_4	[q]	THF	–78°, 1 h; 70°, 1 h[r]	(80)	191
Ph	—	3-BrC_6H_4	[q]	THF	–78°, 1 h; 70°, 1 h[r]	(82)	191
Ph	—	4-BrC_6H_4	[q]	THF	–78°, 1 h; 70°, 1 h[r]	(85)	191
Ph	—	3-$MeOC_6H_4$	[q]	THF	–78°, 1 h; 70°, 1 h[r]	(81)	191
Ph	—	4-$MeOC_6H_4$	[q]	THF	–78°, 1 h; 70°, 1 h[r]	(91)	191
Ph	Br	1-naphthyl	Cl	Et_2O	$ZnCl_2$, reflux, 15 min	(5)	184
4-ClC_6H_4	—	Ph	[q]	THF	–78°, 1 h; 70°, 1 h[r]	(69)	191
4-ClC_6H_4	—	2-ClC_6H_4	[q]	THF	–78°, 1 h; 70°, 1 h[r]	(45)	191
4-ClC_6H_4	—	4-ClC_6H_4	[q]	THF	–78°, 1 h; 70°, 1 h[r]	(79)	191
4-ClC_6H_4	—	3-BrC_6H_4	[q]	THF	–78°, 1 h; 70°, 1 h[r]	(80)	191
4-ClC_6H_4	—	4-BrC_6H_4	[q]	THF	–78°, 1 h; 70°, 1 h[r]	(86)	191
4-ClC_6H_4	—	3-$MeOC_6H_4$	[q]	THF	–78°, 1 h; 70°, 1 h[r]	(86)	191
4-ClC_6H_4	—	4-$MeOC_6H_4$	[q]	THF	–78°, 1 h; 70°, 1 h[r]	(85)	191
4-$MeOC_6H_4$	Br	Ph	BF_4	THF	–78°	(74)	185
2-MeC_6H_4	Br	Ph	BF_4	THF	–78°	(87)	185
4-MeC_6H_4	—	Ph	[q]	THF	–78°, 1 h; 70°, 1 h[r]	(66)	191
4-MeC_6H_4	—	2-ClC_6H_4	[q]	THF	–78°, 1 h; 70°, 1 h[r]	(45)	191
4-MeC_6H_4	—	4-ClC_6H_4	[q]	THF	–78°, 1 h; 70°, 1 h[r]	(84)	191
4-MeC_6H_4	—	3-BrC_6H_4	[q]	THF	–78°, 1 h; 70°, 1 h[r]	(89)	191
4-MeC_6H_4	—	4-BrC_6H_4	[q]	THF	–78°, 1 h; 70°, 1 h[r]	(83)	191
4-MeC_6H_4	—	3-$MeOC_6H_4$	[q]	THF	–78°, 1 h; 70°, 1 h[r]	(84)	191
4-MeC_6H_4	—	4-$MeOC_6H_4$	[q]	THF	–78°, 1 h; 70°, 1 h[r]	(84)	191

C_6

Reactant	Conditions	Product(s) and Yield(s) (%)	Refs.
Ph_2Zn	$ArN_2^+ BF_4^-$, DMF, 0°	Ph–N=N–Ar	193

Ar	Time	
Ph	2.5 h	(95)
4-ClC_6H_4	15 min	(96)
4-$O_2NC_6H_4$	15 min	(57)
4-$MeOC_6H_4$	15 min	(72)

$C_{6\text{-}9}$

Reactant	Conditions	Product(s) and Yield(s) (%)	Refs.
$ArSn(R^1)_3$	4-O_2N-2-R^2-$C_6H_3N_2^+ BF_4^-$, (ia), MeCN, rt	4-O_2N-2-R^2-C_6H_3N=NAr	813

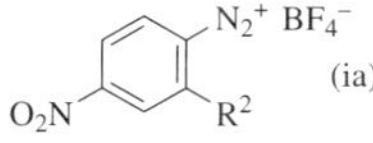

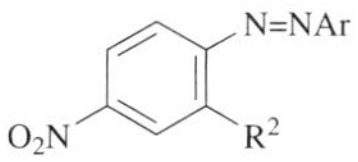

Ar	R^1	R^2	Time	
Ph	*n*-Bu	NO_2	18 h	(31)
4-ClC_6H_4	Me	NO_2	18 h	(14)
2-$MeOC_6H_4$	Me	H	27 h	(16)
2-$MeOC_6H_4$	*n*-Bu	NO_2	1.5 h	(66)
4-$MeOC_6H_4$	Me	H	7 h	(57)
4-$MeOC_6H_4$	Me	NO_2	2 min	(83)
3-CH_2OCH_2-4-C_6H_3	Me	NO_2	2 min	(78)
4-$TMSC_6H_4$	Me	NO_2	20 h	(36)
2-MeC_6H_4	Me	NO_2	16 h	(53)
3-MeC_6H_4	Me	NO_2	16 h	(67)
4-MeC_6H_4	Me	NO_2	16 h	(62)
4-EtC_6H_4	Me	NO_2	18 h	(54)
2,4,6-$Me_3C_6H_2$	*n*-Bu	NO_2	6 h	(65)

C_6

Reactant	Conditions	Product(s) and Yield(s) (%)	Refs.
PhMgBr	3,3-pentamethylenediazirine, Et_2O, 0°; to rt, 1 h	1-phenyl-3,3-pentamethylenediaziridine (PhN, HN) (62)[s]	208
PhLi	1. bicyclic diazene, DME, Et_2O, –35° to –20° 2. $FC_6H_4NO_2$-4, –20° to rt	PhN / 4-$O_2NC_6H_4$N bicyclic hydrazine (34)	210

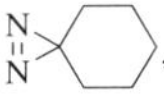

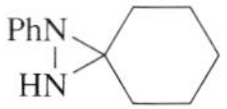

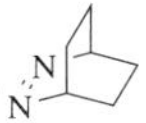

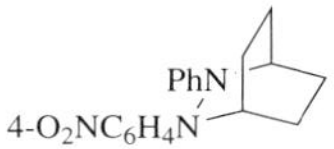

TABLE 4. ARYL CARBANIONS (*Continued*)

Substrate	Conditions	Product(s) and Yield(s) (%)	Refs.
C_6			
PhM	PhN=NPh	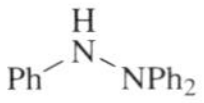 Ph–N(H)–NPh_2	
		M / Solvent(s) / Temp / Time	
		Li / hexane or cyclohexane, THF / –78°; rt / 2 h; 10 h (90)	211, 214, 506, 814
		K / PhH / 0° / 8 h (38)	815
		CaI / Et_2O / reflux / 12 h (18)	815
PhLi	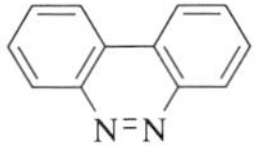 N=N (biaryl azo compound), Et_2O, 15 min	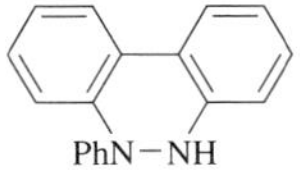 PhN–NH (biaryl hydrazine) (—)	214
C_{6-9}			
ArMgBr	1.  R-C_6H_4–N=NTs, THF, –20°, 1 h 2. CH_2=$CHCH_2I$, *N*-methylpyrrolidinone, rt, 3 h 3. Zn, AcOH, TFA, 75°, 2.5 h	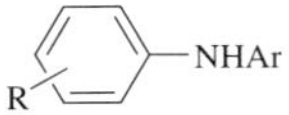 R-C_6H_4–NHAr	255

Ar	R	
4-IC_6H_4	4-EtO_2C^t	(63)
3-$TfOC_6H_4$	4-Br	(70)
3-$TfOC_6H_4$	4-MeO	(81)
3-$TfOC_6H_4$	3-TfO	(76)
3-$TfOC_6H_4$	4-EtO_2C^t	(80)
2-$EtO_2CC_6H_4$	4-Br	(80)
4-$EtO_2CC_6H_4$	2-Br	(65)
4-$EtO_2CC_6H_4$	4-Br	(83)
4-$EtO_2CC_6H_4$	4-I	(71)
4-$EtO_2CC_6H_4$	4-NC	(64)
2,4,6-$Me_2C_6H_2$	4-Br	(69)
2,4,6-$Me_2C_6H_2$	4-MeO	(83)

C_{6-10}

ArM + $R^1OCN=NR^2$ → $R^1C(=O)N(Ar)NHR^2$

Ar	M	R^1	R^2	Solvent	Temp	Time	Yield (%)	Refs.
Ph	Li	EtO	CO_2Et	THF	−78°	—	(100)	816
Ph	Li	*t*-BuO	CO_2Bu-*t*	THF	−78°	—	(60)	816
Ph	MgBr	MeO	CO_2Me	—	—	—	(poor)	794
Ph	MgBr	EtO	CO_2Et	THF	−78°	—	(100)	816
Ph	MgBr	Ph	Ph	Et_2O	rt	—	(50)	794
Ph	MgBr	Ph	COPh	Et_2O	rt	—	(30)	794
$4\text{-}MeOC_6H_4$	ZnI	*t*-BuO	CO_2Bu-*t*	THF	rt	3 h	(55)[a]	358
$4\text{-}MeSC_6H_4$	MgBr	*t*-BuO	CO_2Bu-*t*	THF	−78°	310 min	(>75)	816
$3,4\text{-}(MeO)_2C_6H_3$	MgBr	Et	CO_2Et	THF	−78°	—	(96)	816
$4\text{-}(n\text{-}C_5H_{11}O)C_6H_4$	MgBr	*t*-BuO	CO_2Bu-*t*	THF	−78°	—	(>60)	816
$4\text{-}(CF_3CH_2O)C_6H_4$	Li	*t*-BuO	CO_2Bu-*t*	THF	−78°	—	(61)	816
$2\text{-}CF_3C_6H_4$	MgBr	*t*-BuO	CO_2Bu-*t*	THF	−78°	—	(81)	816
$4\text{-}NCC_6H_4$	ZnBr	*t*-BuO	CO_2Bu-*t*	THF	70°	3 h	(66)	358
$4\text{-}EtO_2CC_6H_4$	ZnBr	*t*-BuO	CO_2Bu-*t*	THF	70°	3 h	(40)	358
6-MeO-2-naphthyl	MgBr	*t*-BuO	CO_2Bu-*t*	THF	−78°	—	(47)	816

C_{6-7}

ArM + $TMSCH_2N_3$ → $ArNH_2$

Ar	M	Solvent	Conditions	Yield (%)	Refs.
Ph	MgBr	Et_2O	rt, 3 h, then H_2O	(72)	264
$4\text{-}ClC_6H_4$	MgBr	Et_2O	rt, 3 h, then H_2O	(92)	264
$2\text{-}MeOC_6H_4$	Li	—	—	(35)	264
$2\text{-}MeOC_6H_4$	MgBr	Et_2O	rt, 3 h, then H_2O	(73)	264
$4\text{-}MeOC_6H_4$	MgBr	Et_2O	rt, 3 h, then H_2O	(69)	264
$2\text{-}Me_2NCH_2C_6H_4$	Li	—	—	(41)	264
$2,3\text{-}(MeO)_2C_6H_3$	Li	Et_2O	0° to rt, then HCl, NaOH	(78)	265
$2,6\text{-}Me_2C_6H_3$	MgBr	Et_2O	rt, 3 h, then H_2O	(79)	264

C_6

PhMgBr (2 eq) + $N_3(CH_2)_nN_3$, Et_2O → $PhN=N\text{-}NH(CH_2)_nNH\text{-}N=NPh$

n	Temp	Time	Yield (%)	Refs.
2	rt	10 min	(72)	271
5	—	—	(—)	290

TABLE 4. ARYL CARBANIONS (*Continued*)

Substrate	Conditions	Product(s) and Yield(s) (%)	Refs.
C_{6-10} ArM	CH_2=$CHCH_2N_3$, solvent, –78° to rt; then H_3O^+	$ArNH_2$	263

Ar	M	Solvent	Yield (%)
Ph	Li	*n*-hexane	(61)
Ph	MgBr	Et_2O	(76)
3-ClC_6H_4	MgBr	Et_2O	(68)
4-ClC_6H_4	MgBr	Et_2O	(52)
2-$MeOC_6H_4$	MgBr	Et_2O	(71)
3-$MeOC_6H_4$	MgBr	Et_2O	(55)
4-$MeOC_6H_4$	MgBr	Et_2O	(75)
2-Me-5-FC_6H_3	MgBr	Et_2O	(66)
2-naphthyl	MgBr	Et_2O	(77)

Substrate	Conditions	Product(s) and Yield(s) (%)	Refs.
C_{6-7} ArLi	1. CH_2=$C(N_3)R^1$, THF, –78°; to rt, 2 h 2. KOH	$ArNH_2$	278

Ar	R^1	Yield (%)
Ph	—[v]	(68)
2,3-$(MeO)_2C_6H_3$	Ph	(60)
2,6-$(MeO)_2C_6H_3$	—[v]	(70)
2-$R^2NHCOC_6H_4$[w]	—[v]	(52)

Substrate	Conditions	Product(s) and Yield(s) (%)	Refs.
C_{6-10} Ar^1MgBr	Ar^2N_3, Et_2O	Ar^1–NH–N=NAr^2[x]	

Ar^1	Ar^2	Temp	Time	Yield (%)	Refs.
Ph	Ph	0°	—	(71)	285, 284
Ph	4-$EtOC_6H_4$	reflux	30 min	(—)	280
Ph	4-MeC_6H_4	reflux	30 min	(—)	280
Ph	Bn	reflux	30 min	("good")	270
Ph	1-naphthyl	reflux	25 min	(64)	281, 280
4-$EtOC_6H_4$	Ph	reflux	30 min	(—)	280
4-MeC_6H_4	Ph	reflux	30 min	(—)	280
1-naphthyl	Ph	reflux	30 min	(—)	280

Carbon	Substrate	Conditions	Product(s) and Yield(s) (%)	Refs.
C_{6-12}	R-C_6H_4-MgBr	1. $PhSCH_2N_3$, THF, hexane, temp, time 1 2. NH_4Cl, H_2O 3. 50% KOH in H_2O, MeOH, THF, rt, time 2	R-C_6H_4-NH_2	275, 274

R	Temp, Time 1	Time 2	
H	–78°, 3 h; rt, 2 h	3 h	(66)
2-MeO	–78°, 1.5 h; 0°, 2 h	2 h	(78)
4-MeO	–78°, 2 h; 0°, 1 h	24 h	(50)
2,6-$(MeO)_2$	–78°, 45 min; to 0°, 3 h	3 h	(67)[y]
2-Me_2NCH_2	–78°, 15 min; 0°, 3 h	2 h	(85)
2-(t-BuO_2CNH)-5-Cl	–78°, 1.5 h; 0°, 2 h	19 min	(66)
2-(Et_2NCO)-5-(OCH_2)-6	–78° to 0°; 0°, 1 h	overnight	(71)
2-CH_2-(1,3-dimethylimidazolidin-2-yl)	–78°, 1.5 h; to 0°, 1.5 h	36 h	(60)[z]
2-MeO-5-Ph	0°, 2 h	24 h	(84)

Carbon	Substrate	Conditions	Product(s) and Yield(s) (%)	Refs.
C_6	PhMgBr	$RCOCH_2N_3$, Et_2O, cooling; rt, overnight	R(Ph)C(OH)CH_2NHN=NPh R: Me (40); Ph (50)	817
C_{6-7}	ArMgBr	$PhCOCH_2N_3$, Et_2O	Ar(Ph)C(OH)CH_2NHN=NAr Ar: 4-$MeOC_6H_4$ (71); 4-MeC_6H_4 (35)	789
C_6	PhMgBr	ArN=N$C_6H_4N_3$-4, Et_2O	ArN=N-C_6H_4-N=N-NHPh Ar: Ph (91); 4-MeC_6H_4 (—)	290
		4-$N_3C_6H_4COMe$, Et_2O	PhN=N-NH-C_6H_4-C(OH)(Me)Ph (84)	283

TABLE 4. ARYL CARBANIONS (*Continued*)

Substrate	Conditions	Product(s) and Yield(s) (%)	Refs.
C_6			
$PhMgX^{j}$ 2 eq	6-azido-2-methylbenzothiazole, Et_2O, 10-12 h	6-(PhNHN=N)-2-methylbenzothiazole (78)[x]	790
PhM	4-chloro-2-azidobenzothiazole (ia)	4-chloro-2-(NHN=NPh)benzothiazole[x]	800
		M: Li; Solvent: —; Temp: −10° to rt; Time: 1 h (63)	
		M: MgBr; Solvent: Et_2O; Temp: rt; Time: 10 h (100)	
PhMgX	R-substituted tetrazolo[5,1-b]benzothiazole (positions 4, 6) (ia)	R-2-(NHN=NPh)benzothiazole[x]	
	X: Br; R: H; Solvent: toluene; Temp: 100°; Time: 30 min	(80)	818
	X: Cl; R: 6-Cl; Solvent: toluene; Temp: 100°; Time: 30 min	(—)	819
	X: Br; R: 6-MeO; Solvent: toluene; Temp: 100°; Time: 30 min	(85)	818
	X: Br; R: 6-Me; Solvent: toluene; Temp: 100°; Time: 30 min	(—)	818
	X: Br; R: 4-Ph; Solvent: Et_2O; Temp: rt; reflux; Time: 30 min; 30 min	(65)	800
PhMgBr	naphtho-fused tetrazolobenzothiazole, toluene, 100°, 30 min	naphthothiazol-2-yl–N=NNHPh[x] (70)	818
	1,3-diazidobenzene, Et_2O, 0°	PhNHN=N–C_6H_4–N=NNHPh[x] (65)	289
	triazido-Y-triazine, Et_2O, 1 h	tris(N=NNHPh / PhNHN=N)-Y-triazine[x]; Y: CH (47), N (90)	788

	PhMgBr	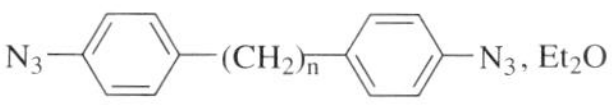 N_3–(C$_6$H$_4$)–$(CH_2)_n$–(C$_6$H$_4$)–N_3, Et_2O	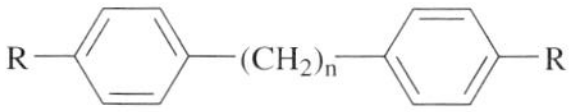 R–(C$_6$H$_4$)–$(CH_2)_n$–(C$_6$H$_4$)–R R = PhNHN=N— [x]	n 0 (79) 1 (—)	272 290
		$RCON_3$, Ǝt_2O, cooling; reflux, 15 min	RC(O)N=N–NHPh	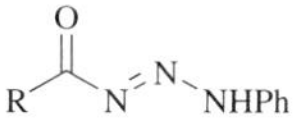 R H_2N (14-18) MeO (11-14) EtO (—) Ph (8) PhNH (0)	820 820, 284 820 820 820
		N_3CON_3, Et_2O, cooling	PhN=N–NH–C(O)NH_2	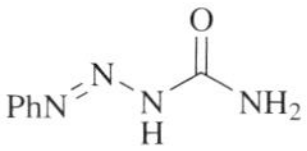 (18) [aa]	284
$C_{6\text{-}9}$	ArMgBr	1. Ph_3SiN_3, solvent, temp, time 2. HCl, reflux	$ArNH_2$		821

Ar	Solvent	Temp	Time	
Ph	Et_2O	100°	6 h	(56)
2,4,6-$Me_3C_6H_2$	toluene	120°	4 h	(26)

C_6	PhM	Ph_3SiN_3, Et_2O, 100°, 24 h	Ph_2(Ph)Si–N(M)–Ph	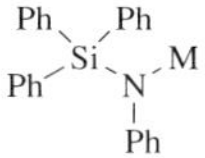 M MgBr (61) MgPh (—)	821
$C_{6\text{-}10}$	ArMgBr	1. $(PhO)_2F(O)N_3$ (ia), Et_2O, –73° to –69°, 2 h 2. 10% HCl, MeOH, rt, 200 min	$ArNH_3^+ Cl^-$	Ar Ph (51) 4-ClC_6H_4 (63) 2-MeC_6H_4 (33) 4-MeC_6H_4 (33) 1-naphthyl (28)[e]	334

TABLE 4. ARYL CARBANIONS (*Continued*)

Substrate	Conditions	Product(s) and Yield(s) (%)	Refs.
C_{6-10}			
ArM	1. $(PhO)_2P(O)N_3$ (ia), solvent, –73° to –69°, 2 h 2. $NaAlH_2(OCH_2CH_2OMe)_2$, toluene, –70°; to 0°, 1 h	$ArNH_2$	333, 334

Ar	M	Solvent	
Ph	MgBr	Et_2O	(73)[b]
4-ClC_6H_4	MgBr	Et_2O	(79)[b]
4-$MeOC_6H_4$	Li	THF	(84)
2,6-$(MeO)_2C_6H_3$	Li	THF	(72)
2,5-$(MeOCH_2O)_2C_6H_3$	Li	THF	(47)
4-MeC_6H_4	MgBr	Et_2O	(88)[b]
2,4,6-$Me_3C_6H_2$	MgBr	Et_2O	(67)
1-naphthyl	MgBr	Et_2O	(89)

Substrate	Conditions	Product(s) and Yield(s) (%)	Refs.
PhMgBr	1. TsN_3 (ia), Et_2O, –18° to –15°, 30 min 2. Isolate PhN=NN(MgBr)Ts 3. 120-130° (0.1-3.0 mmHg)	PhN_3 (82)	308
C_{6-13}			
ArMgBr	1. TsN_3, THF, 0° 2. Reagents	ArN_3	305

Ar	Reagents	
Ph	NaOH, H_2O	(50)
4-ClC_6H_4	$Na_4P_2O_7$, KOH, H_2O	(70)
4-$MeOC_6H_4$	$Na_4P_2O_7$, H_2O	(55)
4-MeC_6H_4	$Na_4P_2O_7$, KOH, H_2O	(73)
2,4,6-$Me_3C_6H_2$	$Na_4P_2O_7$, H_2O	(63)
2-*t*-BuC_6H_4	$Na_4P_2O_7$, H_2O	(42)
4-PhC_6H_4	NH_4Cl, H_2O	(68-79)
2-BnC_6H_4	$Na_4P_2O_7$, H_2O	(49)

Substrate	Conditions	Product(s) and Yield(s) (%)	Refs.
C_{6-13}			
ArMgBr	1. TsN_3, THF, 0° 2. RaNi, NaOH 3. HCl	$ArNH_3^+ Cl^-$	305

Ar	
3-ClC_6H_4	(41)
4-ClC_6H_4	(49)
2-$MeOC_6H_4$	(63)
4-$MeOC_6H_4$	(51)
2-MeC_6H_4	(82)
3-MeC_6H_4	(79)
4-MeC_6H_4	(66)
2,4-$Me_2C_6H_3$	(76)
2-*t*-BuC_6H_4	(19)
4-PhC_6H_4	(62)[e]
2-BnC_6H_4	(71)

Substrate	Conditions	Product(s) and Yield(s) (%)	Refs.
C_{6-7}			
2-lithio-benzene bearing R^1 and R^2 at the 1- and 3-positions	1. TsN_3, Et_2O, 0° 2. RaNi, 50% KOH, 0° to rt, 2 h	aniline (NH_2) bearing R^1 and R^2 at the 2- and 6-positions	310

R^1	R^2	
H	H	(80-85)
MeO	H	(75-80)
CH_2NMe_2	H	(52-55)
CONHMe	H	(37-40)
CONHMe	MeO	(34-38)

TABLE 4. ARYL CARBANIONS (*Continued*)

Substrate	Conditions			Product(s) and Yield(s) (%)	Refs.
C_{6-8}					
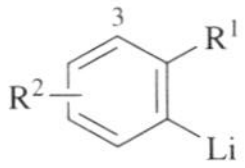 2-R¹-R²-phenyllithium (position 3 marked)	1. TsN_3 (ia), Et_2O, –70°; –70°, 5 h; to –10° 2. $NaBH_4$, $Bn_4N^+ HSO_4^-$, H_2O			2-R¹-R²-aniline (NH_2)	
	R^1	R^2			
	OCH_2OMe	H	(67)		311
	OCH_2OMe	3-Me	(72)[bb]		312
	$OCONEt_2$	H	(94)		311
	4,4-dimethyl-4,5-dihydrooxazol-2-yl	H	(50)		311
	$CONEt_2$	H	(40)		311
	$CONEt_2$	3-Cl	(31)		311
	$CONEt_2$	3-MeO	(55)		311
	$CONEt_2$	4-MeO	(34)		311
	$CONEt_2$	6-MeO	(66)		311
	$CONEt_2$	5-MeO-6-TMS	(69)		311
	$CONEt_2$	5-MeO-6-CH(TMS)$_2$	(47)		311
	$CONEt_2$	4-Me	(82)		311
	$CONEt_2$	3-MeO-4-Me	(69)		311
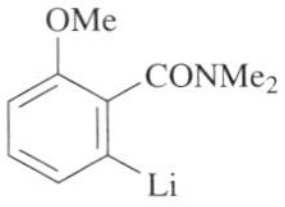 C_7 2-OMe-6-Li-C$_6$H$_3$CONMe$_2$ (OMe, $CONMe_2$, Li)	1.  (piperidino)NCu(CN)Li (5 eq), THF, 78°, 2 h 2. O_2, –78°			2-OMe-6-(piperidin-1-yl)-C$_6$H$_3$CONMe$_2$ (OMe, $CONMe_2$, N) (33)	53

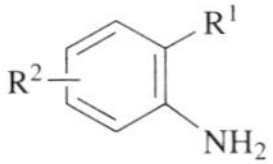

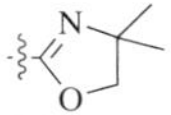

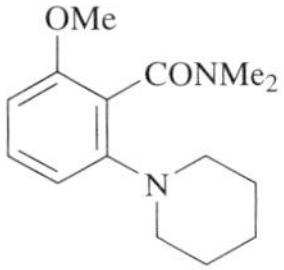

C_{7-11} R[1] (3) $CON(R^2)_2$, Li

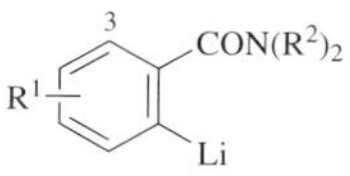

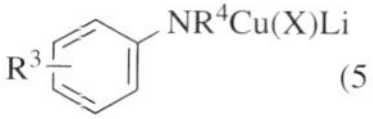

1. R^3-C_6H_4-$NR^4Cu(X)Li$ (5 eq), THF, –78°, 2 h
2. O_2, –78°

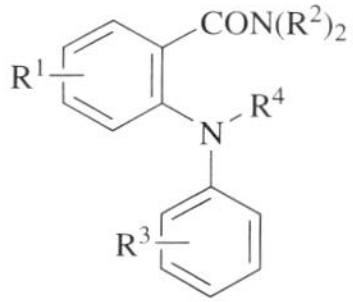

Product: R^1-C_6H_3($CON(R^2)_2$)-$N(R^4)$-C_6H_4-R^3 — 53

R^1	R^2	R^3	R^4	X	
H	Et	H	Me	Cl	(46)
H	Et	2-MeO	H	CN	(50)
3-MeO	Et	H	H	CN	(63)
3-MeO	Me	H	Me	CN	(33)
3-MeO	Et	H	TMS	CN	(54) (R^4 = H)
3-MeO	Me	3-MeO	H	CN	(36)
6-MeO	Me	H	Me	CN	(26)
3,5-$(MeO)_2$	Me	H	Me	Cl	(43)
3-MeO-4-OCH_2O-5	Me	H	Me	Cl	(48)
3-$(CH{=}CH)_2$-4	Et	H	Me	CN	(61)

C_{7-9} 2-Li-C_6H_4-$(CH_2)_n$-ONHMe

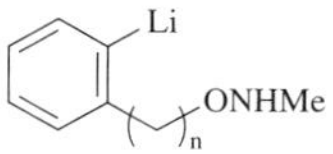

1. MeLi, *t*-BuLi
2. AcCl

Product: 2-(NMeAc)-C_6H_4-$(CH_2)_n$-OAc

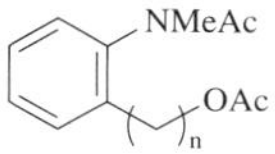

n		
1	$(13)^{cc}$	81
2	$(7)^{cc}$	82
3	$(12)^{cc}$	83

C_7 2-MgCl-C_6H_4-CO_2Me

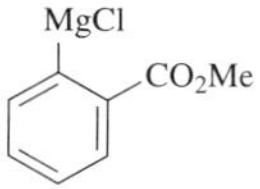

1. $ZnCl_2$, THF, –78°, 15 min
2. Bn_2NOBz, $CuCl_2$ (cat), THF, –35° to rt, 1 h

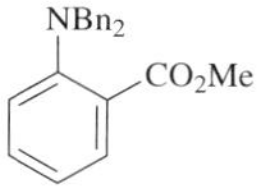

Product: 2-NBn_2-C_6H_4-CO_2Me (81) — 113

MgCl, $CO_2N(CH_2Ar)_2$, R, R, R, R
1:1 mixture of Ar = Ph, R = H and and Ar = C_6D_5, R = D

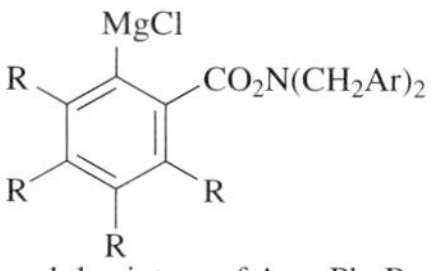

1. $ZnCl_2$ (1 eq), THF, –78°, –35°, 40 min
2. $CuCl_2$ (0.0015 mol%), –35° to rt
3. $TMSCHN_2$, MeOH, Et_2O

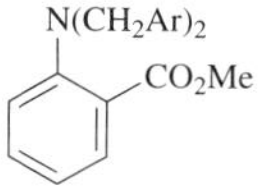

Product: 2-$N(CH_2Ar)_2$-C_6H_4-CO_2Me (50) d_0:d_4:d_{10}:d_{14} = 100:100:97:94[cc] — 113

TABLE 4. ARYL CARBANIONS (*Continued*)

Substrate	Conditions	Product(s) and Yield(s) (%)	Refs.
C_9 R = CH(TMS)$_2$	1. TMSCH$_2$N$_3$ 2. H_3O^+	(39)	267
C_{10}	RONH$_2$, Et$_2$O		
	R: Me; Conditions: –20° to rt, 4 h	(8)	822
	R: Bn; Conditions: –18° to –20°, 15 min; to rt, 30 min	(21)	823, 684, 822
C_{10-11} R[1], R[2]	1. H$_2$N–N(triazole), *t*-BuOK, DMSO, rt, 4 h 2. Ac$_2$O, pyridine	NHAc R[1] R[2] H H (38) H CN (64) Cl Cl (60)	824
C_{10} MgBr (2 eq)	PhCH=NOH or PhCH=NOMe, Et$_2$O, reflux	NH$_2$ (15)	175
ArMgBr Ar = 1-naphthyl	BzCH=N$_2$, Et$_2$O, rt, several h	Ar–NH–N=CH–C(O)Ph (7) + Ar–NH–N=CH–C(Ar)(OH)Ph (8-16)	825

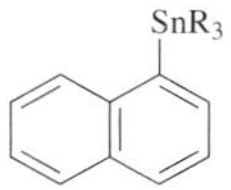	2,4-$(O_2N)_2C_6H_3N_2^+$ BF_4^-, MeCN, rt	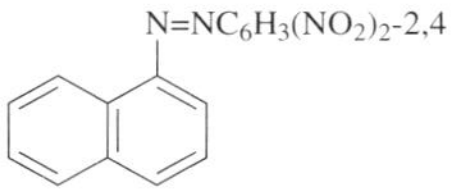	R / Time Me 6 h (55) *n*-Bu 1 h (71)	813
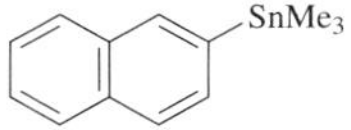	2,4-$(O_2N)_2C_6H_3N_2^+$ BF_4^-, MeCN, rt	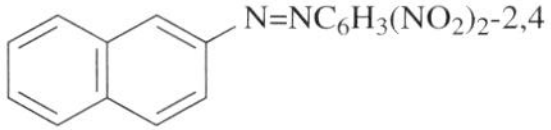	(76)	813
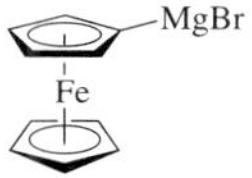	1. 4-BrC_6H_4N=NTs, THF, –20°, 1 h 2. CH_2=$CHCH_2I$, *N*-methylpyrrolidinone, rt, 3 h 3. Zn, AcOH, TFA, 75°, 15 min	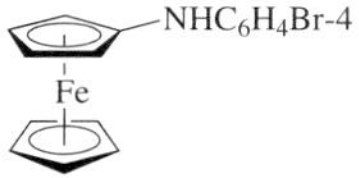	(58)	255
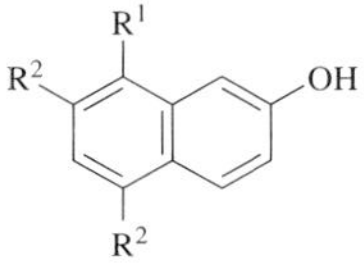	*t*-BuO_2CN=NCO_2Bu-*t*, catalyst (20 mol%)	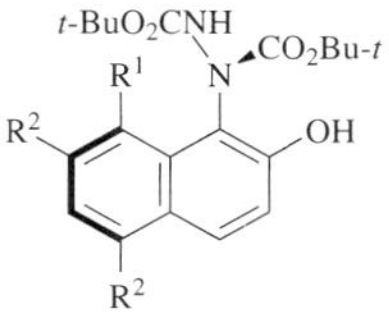		826

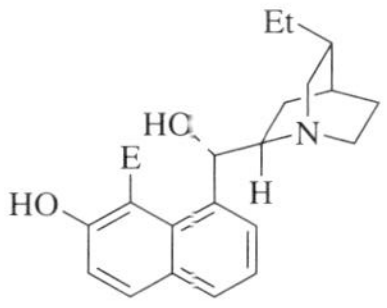

I

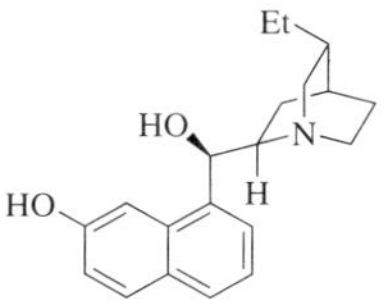

II

Catalyst	E
IA	H
IB	N(CO_2Bu-*t*)NH—CO_2Bu-*t*
IC	N(CO_2Bu-*t*)NH- - CO_2Bu-*t*

TABLE 4. ARYL CARBANIONS (*Continued*)

Substrate	Conditions	Product(s) and Yield(s) (%)	Refs.
C_{10} (Entry continued from previous page.)			

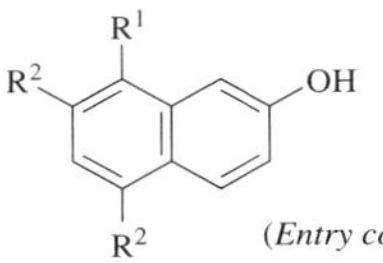

R^1	R^2	Catalyst	Solvent	Temp		% ee[dd]
NH_2	H	quinine	toluene	rt	(85)	16
NH_2	H	**IA**	$ClCH_2CH_2Cl$	–20°	(90)	88
NH_2	H	**II**	$ClCH_2CH_2Cl$	rt	(85)	–61
NH_2	H	**IB**	$ClCH_2CH_2Cl$	–20°	(87)	87
NH_2	H	**IC**	$ClCH_2CH_2Cl$	–20°	(91)	–96
NHMe	H	**IA**	$ClCH_2CH_2Cl$	–20°	(95)	33
NHMe	H	**IB**	$ClCH_2CH_2Cl$	–20°	(91)	93
NHMe	H	**IC**	$ClCH_2CH_2Cl$	–20°	(94)	–96
NHC_5H_{11}-*n*	H	**IA**	$ClCH_2CH_2Cl$	–20°	(98)	78
NHC_5H_{11}-*n*	H	**IB**	$ClCH_2CH_2Cl$	–20°	(95)	94
NHC_5H_{11}-*n*	H	**IC**	$ClCH_2CH_2Cl$	–20°	(98)	–94
NHBn	H	**IA**	$ClCH_2CH_2Cl$	–20°	(98)	48
NHBn	H	**IB**	$ClCH_2CH_2Cl$	–20°	(98)	92
NHBn	H	**IC**	$ClCH_2CH_2Cl$	–20°	(80)	–98
NH_2	Br	**IA**	$ClCH_2CH_2Cl$	–20°	(96)	80
NH_2	Br	**IB**	$ClCH_2CH_2Cl$	–20°	(85)	98
NH_2	Br	**IC**	$ClCH_2CH_2Cl$	–20°	(95)	–96

Substrate	Conditions	Product(s) and Yield(s) (%)	Refs.
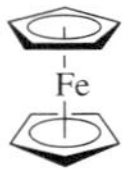 Fe	1. *n*-BuLi (5.4 eq), Et_2O, rt, 5 h 2. TsN_3 (ia), rt, 30 min; rt, overnight 3. 10% KOH in H_2O 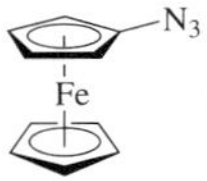	N_3 Fe (28) + N_3 Fe N_3 (6)	315
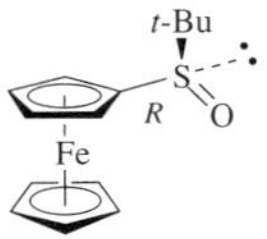 *t*-Bu S O *R* Fe	1. *n*-BuLi (1.3 eq), THF, pentane, rt, 2 h 2. TsN_3, 0°; rt, 4 h 3. $NaBH_4$, *n*-Bu_4^+ I^-, H_2O, rt, 48 h	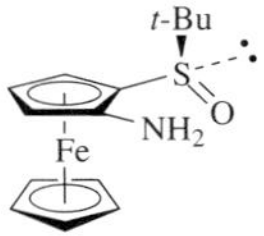 *t*-Bu S O NH_2 Fe (64), >98% de (R,R)	827

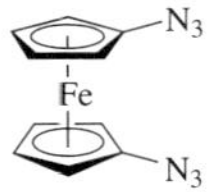

Substrate	Conditions	Product(s) and Yield(s) (%)	Refs.
C_{10}			
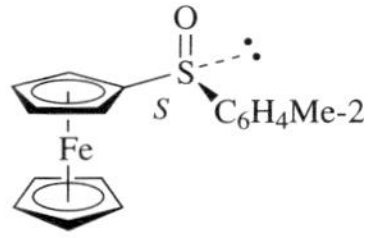 (S)-ferrocenyl sulfoxide, $S(O)C_6H_4Me$-2	1. LDA, THF, –78°, 40 min 2. TsN_3, –78°, 4 h; to rt 3. $NaBF_4$, $n\text{-}Bu_4^+\ I^-$, H_2O, rt, 48 h	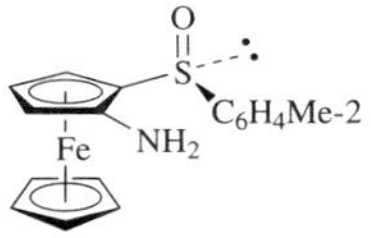 2-amino ferrocenyl sulfoxide, C_6H_4Me-2, NH_2 (67) >99% de (S,S)	827
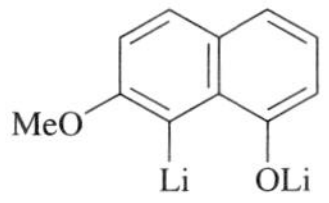 MeO, Li, OLi naphthalene	1. $PhSCH_2N_3$, THF, pentane, –78° to rt, 1.5 h 2. KOH, DMSO, rt, 1 h 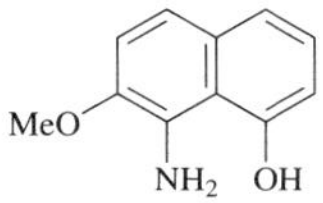	MeO, NH_2, OH naphthalene ("poor")	274
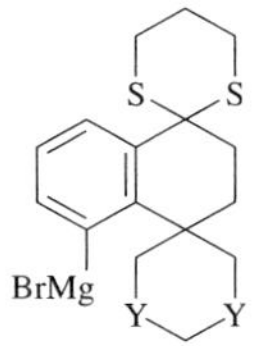 BrMg spiro dithiane, Y	1. $PhSCH_2N_3$, Et_2O, THF, hexane, –78° to 0°, 1 h 2. 50% KOH in H_2O, MeOH, THF, 30 min	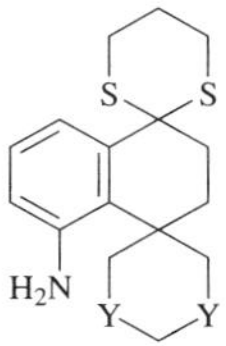 H_2N spiro dithiane, Y Y = O (88) Y = NMe (—)[ee]	274
C_{11}			
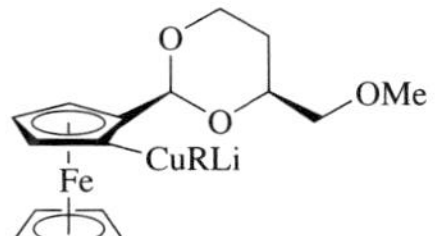 ferrocenyl dioxane, OMe, CuRLi R = 2-thienyl	TMSNHOTMS, Et_2O, THF, –50°, 1 h; to rt, overnight 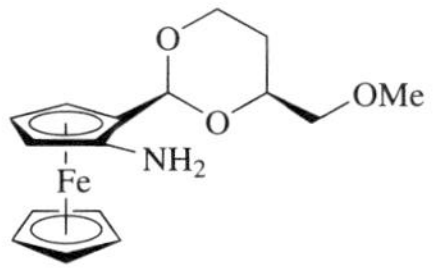	ferrocenyl dioxane, OMe, NH_2 (56)	102
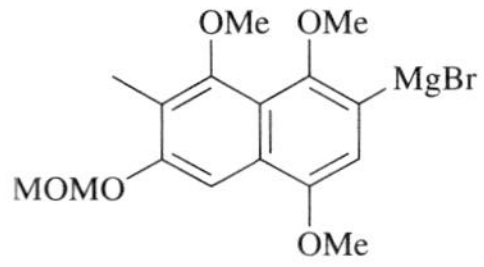 MOMO, OMe, OMe, MgBr, OMe naphthalene	1. $PhSCH_2N_3$ (ia), THF, hexane, –78°; to 0°; 0°, 1h; rt, 1 h 2. NH_4Cl, H_2O 3. 50% aq. KOH, MeOH, THF, rt, 2.5 h	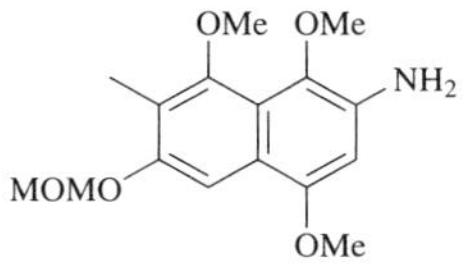 MOMO, OMe, OMe, NH_2, OMe naphthalene (71)	375
C_{12}			
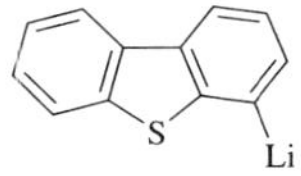 dibenzothiophene, Li	MeONHLi, Et_2O, hexane, –15°, 30 min; reflux 1 h 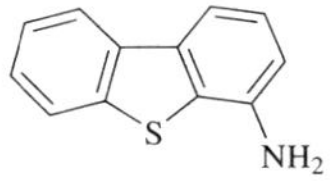	dibenzothiophene, NH_2 (55)	786

TABLE 4. ARYL CARBANIONS (*Continued*)

Substrate	Conditions	Product(s) and Yield(s) (%)	Refs.
C_{12} 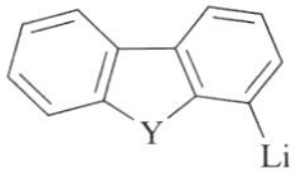 4-lithiodibenzo-Y (Y = O, S)	NH_2OMe, –20°, 2 h	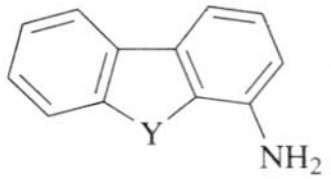 **I** (4-amino); Y = O (78); Y = S (28)	828, 829; 830
	1. $(PhO)_2P(O)N_3$, THF, –78°, 2 h; to –20°, 30 min 2. $NaAlH_2(OCH_2CH_2OMe)_2$, toluene, –78°; 0°, 1 h; rt, 30 min	**I**; Y = O (58); Y = S (62)	333, 334
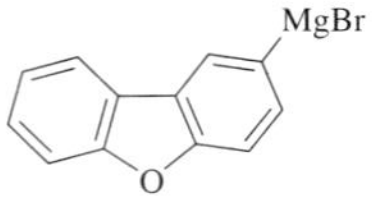 MgBr (dibenzofuran)	NH_2OMe	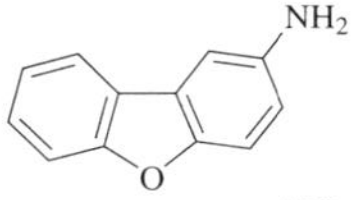 NH_2 (33)	828
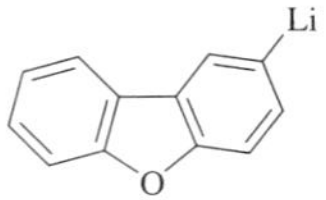 Li (dibenzofuran)	1. $(PhO)_2P(O)N_3$, THF, –78°, 2 h; to –20°, 30 min 2. $NaAlH_2(OCH_2CH_2OMe)_2$, toluene, –78°; 0°, 1 h; rt, 30 min	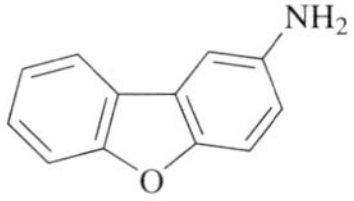 NH_2 (71)	333, 334
 BrMg, OMe (dibenzofuran)	NH_2OMe, Et_2O, 0°, then HCl	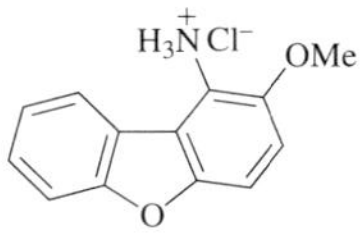 $H_3\overset{+}{N}Cl^-$, OMe (24)	831
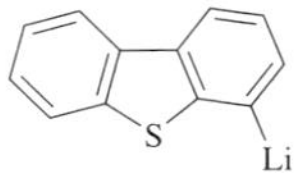 Li (dibenzothiophene)	MeONHLi (ia), Et_2O, –78° to –15°, 2 h	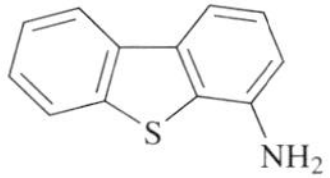 NH_2 (55)	82
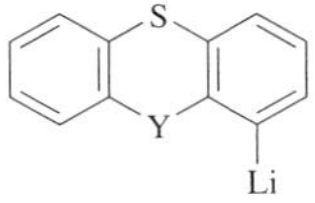 S, Y, Li	NH_2OMe, Et_2O	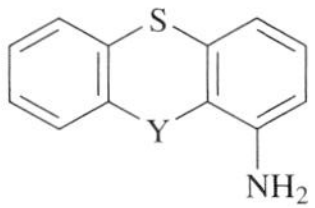 S, Y, NH_2 Y = O; Temp –20°; reflux; Time —; "several" h (59) Y = S; Temp 0-5°; rt; Time 15 min; 1 h (26)	832 833, 834

C_{24}

TsN_3

R^1	R^2	Solvent(s)	Temp	Time		
2,4,6-$Me_3C_6H_2$	H	Et_2O, hexanes	0°	2 h	(96)	314
2,4,6-$Me_3C_6H_2$	Me	THF	—	—	(95)	835

C_n

Polysulfone[ff] — 1. *n*-BuLi, THF, hexane, –70°; 2. $MeONH_2$ — Aminated polysulfone (—) — 836

1. *n*-BuLi (2.15 eq), THF, –78°, 15 min
2. TsN_3 (3 eq), to –50°, 1 h

(95)[gg] — 335

1. *n*-BuLi (1.2 eq), THF, –78°
2. MeONHLi, –78°

(17) — 837

1. *n*-BuLi (2.5 eq), THF, –78°
2. TsN_3 (3 eq), –78°, 15 min; to –50°, 90 min

(95) — 335

TABLE 4. ARYL CARBANIONS (*Continued*)

	Substrate	Conditions	Product(s) and Yield(s) (%)	Refs.
C_n 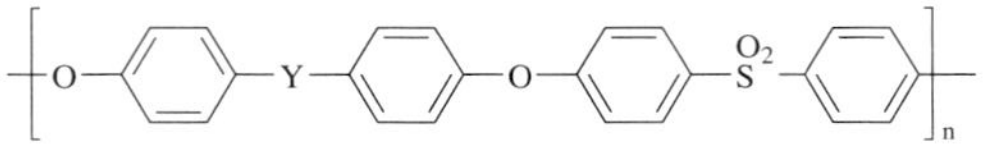	Y = bond or CMe_2	1. *n*-BuLi (2.1 eq), THF, –65° 2. $4\text{-AcNHC}_6H_4SO_2N_3$, to –50°, 15 min 3. H_2O, EtOH	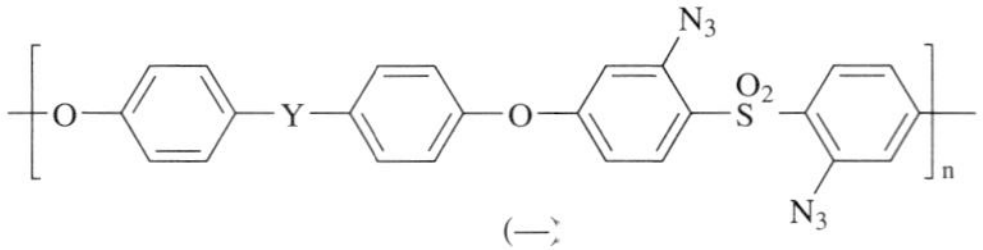 (—)	837

[a] This is a competitive kinetic study. No yields were reported.

[b] The product was isolated as the hydrochloride.

[c] The product was isolated as the *N*-benzoyl derivative.

[d] With PhMgBr at reflux the yield was 37% and with PhCuLi the yield was 83%.

[e] The yield is that of the amine.

[f] The reagent was prepared in situ by addition of *n*-BuLi or PhLi, respectively, to CH_2=NOBn.

[g] Unless otherwise noted, the substrates were prepared from the Grignard reagents and $ZnCl_2$.

[h] The substrate was prepared by ortholithiation followed by reaction with $ZnCl_2$. The catalyst in the subsequent amination was $CuCl_2$.

[i] This was the yield when 0.6 equivalents of Ar_2Zn were used.

[j] X was not specified.

[k] The product was converted into the arylamine with CsOH in ethylene glycol at 150° or into the *N*-methylarylamine with $LiAlH_4$.

[l] Hydrolysis with CsOH in ethylene glycol at 150° gave methyl 4-aminobenzoate; with $LiAlH_4$, partial loss of the fluorine and the methyl group was observed.

[m] No reaction occurred under the conditions of footnote k.

[n] PhC(=NTs)NMe$(CH_2)_2$OH was formed in 70% yield.

[o] This reagent is hygroscopic and reproducible results were obtained only with freshly prepared material.

[p] The yields were determined by gas chromatography.

[q] X^2 was 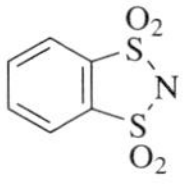 .

[r] Heating to 70° for one hour converted any Z-azo compound into the E isomer.

[s] The yield was determined by iodometry.

[t] These are corrected entries; Knochel, P.; Kofink, C. University of Munich, Germany. Personal communication, 2005.

[u] The yield was determined by NMR spectroscopy.

[v] R^1 was not specified but it was either Ph or *t*-Bu.

[w] R^2 was not specified.

[x] Some of the triazenes are isolated as mixtures of double-bond isomers.

[y] 2,6-Dimethoxy-4-(phenylthiomethyl)aniline was also formed in 16% yield.

[z] The product was indole after treatment of the amine with oxalic acid.

[aa] A later publication (ref. 820) reported a 0% yield for this reaction.

[bb] The product was isolated as the *N*-tert-butoxycarbonyl derivative.

[cc] Deuterium labeling indicates that the reaction is intermolecular.

[dd] Catalysts **IA**, **IB**, and **IC**, **II** gave atropisomers with the opposite absolute configurations.

[ee] A mixture of triazenes was obtained in low yield.

[ff] The type of polysulfone was not specified.

[gg] A mono-azide was obtained in 95% yield with 1.1 eq of *n*-BuLi and 1.5 eq of TsN_3. The corresponding reactions with $(PhO)_2P(O)N_3$ were not as clean.

TABLE 5. HETEROCYCLIC CARBANIONS

	Substrate	Conditions	Product(s) and Yield(s) (%)	Refs.
C_3	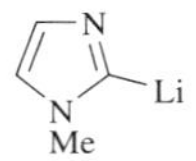 N, Li, N, Me	$ClNH_2$ or $MeONH_2$	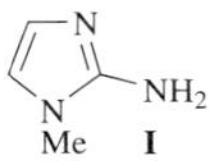 N, NH_2, N, Me **I** (0)	66
		1. 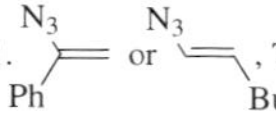 N_3, Ph or N_3, Bu-*t*, THF, –78° to rt, 2 h 2. HCl, then base	**I** (45)	278
	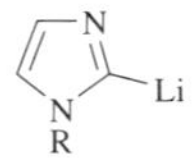 N, Li, N, R	1. PhN_3, rt, 1.5 h 2. HOAc, H_2O 3. HCl, 80-90°	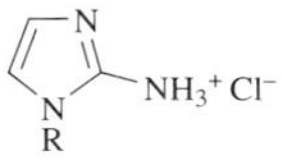 N, $NH_3^+ Cl^-$, N, R — R: Me (70); Ph (43); Bn (64)	66
C_4	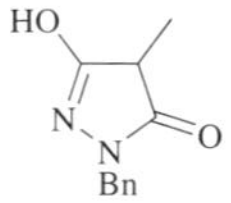 HO, N, N, O, Bn	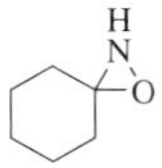 H, N, O, toluene, NaOH, H_2O, 0°, 10 min	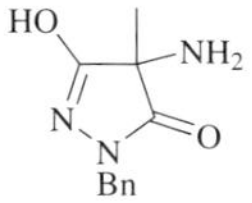 HO, NH_2, N, N, O, Bn (90) 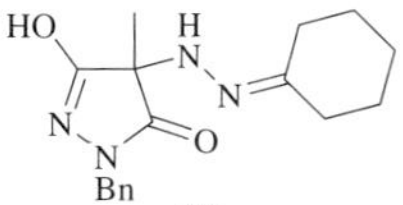 + HO, H, N, N, N, N, O, Bn (4)	149
	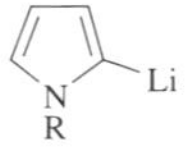 Li, N, R — R = Me or SO_2Ph	1. $PhSCH_2N_3$ (ia), THF, –75° to 0° 2. CuI, 0°, 1 h 3. KOH (50% aq.), MeOH, THF, rt, 3 h	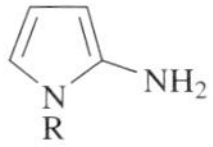 NH_2, N, R (0)	274
	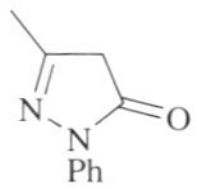 N, N, O, Ph	TsON=C(CN)$_2$, pyridine, 0°, 1 h	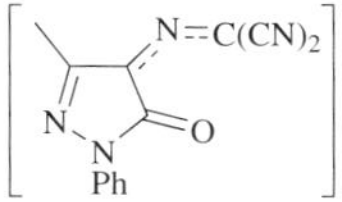 [N=C(CN)$_2$, N, N, O, Ph]$^-$ PyH$^+$ (50)	838
	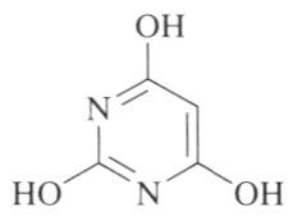 OH, N, HO, N, OH	TsON=C(CN)$_2$, pyridine, –30° to rt	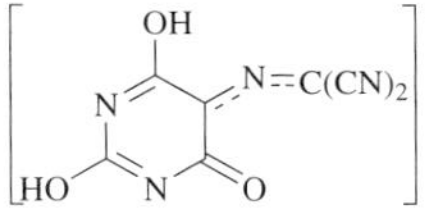 [OH, N=C(CN)$_2$, N, HO, N, O]$^-$ PyH$^+$ (50)	838

C_4

Substrate	Conditions	Product(s) and Yield(s) (%)	Refs.
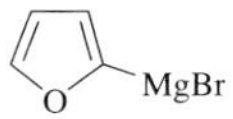 2-furyl-MgBr	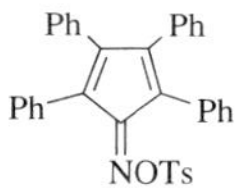 tetraphenylcyclopentadienone N-OTs (ia), THF, –78°, 10 min	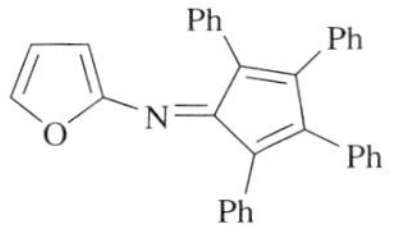 2-furyl-N=C(tetraphenylcyclopentadiene) (81)	180
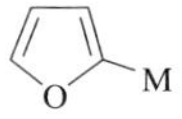 2-furyl-M, M = Li or MgBr	1. $PhSCH_2N_3$ (ia), THF, –75° to 0° 2. CuI, 0°, 1 h 3. KOH (50% aq.), MeOH, THF, rt, 3 h	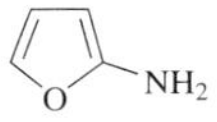 2-furyl-NH_2 (0)	274
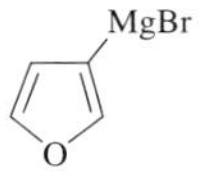 3-furyl-MgBr	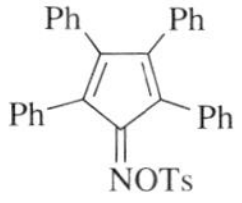 tetraphenylcyclopentadienone N-OTs (ia), THF, –78°, 10 min	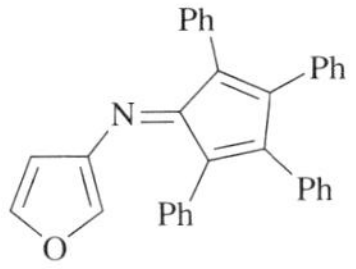 3-furyl-N=C(tetraphenylcyclopentadiene) (78)	180
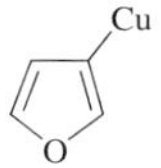 3-furyl-Cu	t-$BuO_2CN(Li)OTs$, THF, –78° to –40°, 2 h	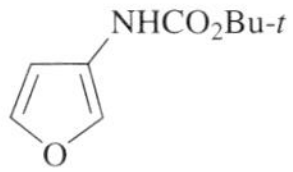 3-furyl-$NHCO_2Bu$-t (48)	127
 2,5-dihydrofuran	1. Catalyst B (see Chart 1; 2 mol%), i-PrOH, rt to 0° 2. Add substrate, then $PhSiH_3$, 0° 3. t-BuO_2CN=NCO_2Bu-t, 0°, 4 h	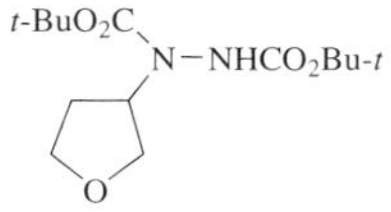 3-tetrahydrofuryl-$N(CO_2Bu$-$t)NHCO_2Bu$-t (81)	215
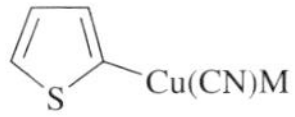 2-thienyl-Cu(CN)M	1. PhRNLi, THF, temp 1, time 1 2. Addend, THF, –78° 3. O_2, temp 2, time 2	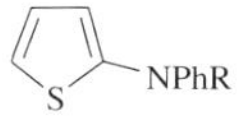 2-thienyl-NPhR	

M	R	Temp 1	Time 1	Addend	Temp 2	Time 2		Refs.
Li	Me	–40°	15 min	—	–78°; to rt	—	(52)	54
Li	Me	–40°	20 min	$Cu(NO_3)_2$	–78°	30 min	(70)	55
ZnCl	Bn	–78° to –40°	40 min	1,2-$(O_2N)_2C_6H_4$	–78°	30 min	(75)	55

Substrate	Conditions	Product(s) and Yield(s) (%)	Refs.
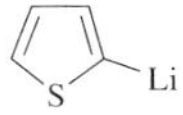 2-thienyl-Li	MeN(Li)OMe, Et_2O, hexane, –78° to –15°, 3 h	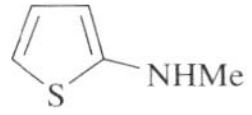 2-thienyl-NHMe (0)	97

TABLE 5. HETEROCYCLIC CARBANIONS (*Continued*)

Substrate	Conditions	Product(s) and Yield(s) (%)	Refs.
C_4			
2-thienyl-Cu	t-$BuO_2CN(Li)Ts$, THF, –78° to –40°, 2 h	2-thienyl-$NHCO_2Bu$-t (52)	127
(2-thienyl)$_2Cu(CN)Li_2$	RNHOTMS, THF	2-thienyl-NHR R / Temp / Time TMS / –50°; to rt / 1 h; — / (72) Me / –50° to rt / 2 h / (60) i-Pr / –50° to rt / 2 h / (65)	100 101 101
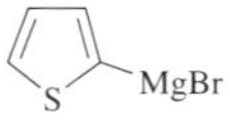 2-thienyl-MgBr	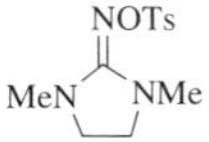 1,3-dimethyl-2-(tosyloxyimino)imidazolidine (MeN, NMe, NOTs) (ia), toluene, Et_2O, 0°, 15 min	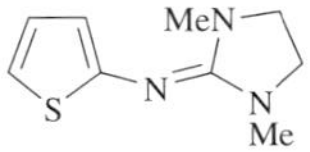 2-thienyl-N=(1,3-dimethylimidazolidin-2-ylidene) (MeN, N-Me) (84)[a]	181
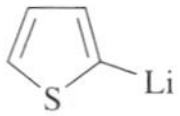 2-thienyl-Li	1. $ZnCl_2$, THF, 0°; to rt 2. 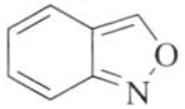 2,1-benzisoxazole (ia), $Ni(acac)_2$ (cat), THF, 2 h	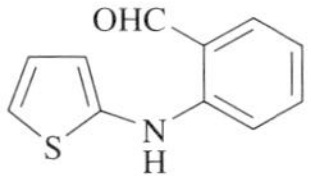 2-(2-thienyl-NH)benzaldehyde (OHC) (4-15)	170
2-thienyl-ZnBr	t-BuO_2CN=NCO_2Bu-t, THF, rt, 30 min	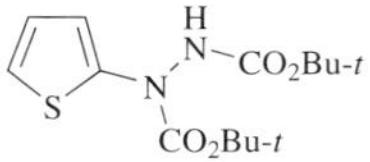 2-thienyl-N(CO_2Bu-t)-NH-CO_2Bu-t (80)	358
C_{4-5}			
3-R-2-thienyl-Li	1. TsN_3, Et_2O, –70° 2. $Na_4P_2O_7$, H_2O, rt, overnight	3-R-2-thienyl-N_3 R H (10) 1,3-dioxolan-2-yl (0)	316
R-3-thienyl-Li	1. TsN_3, Et_2O, –70° 2. $Na_4P_2O_7$, H_2O, rt, overnight	R-3-thienyl-N_3 R H (85) 2-Me (68) 4-Me (70) 2-(1,3-dioxolan-2-yl) (65) 4-(1,3-dioxolan-2-yl) (70)	316

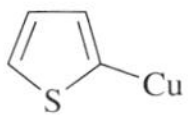

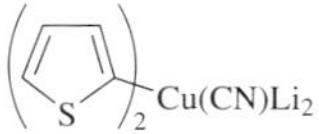

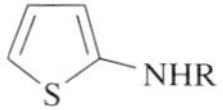

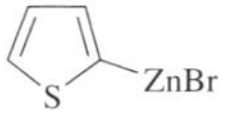

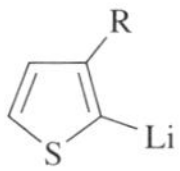

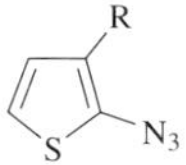

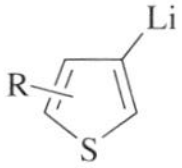

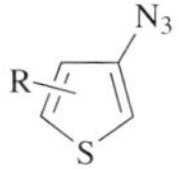

Substrate	Conditions	Product(s) and Yield(s) (%)	Refs.
C_4			
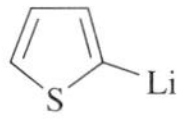 2-Thienyllithium	1. $PhSCH_2N_3$ (ia), THF, –75° to 0° 2. CuI, 0°, 1 h 3. KOH (50% aq.), MeOH, THF, rt, 3 h	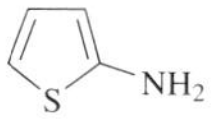 2-Aminothiophene (0)	274
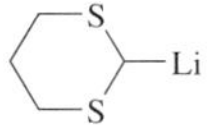 2-Lithio-1,3-dithiane	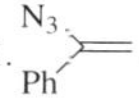 1. $N_3C(Ph)=CH_2$, THF, –78°; rt, 2 h 2. HCl	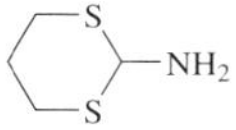 2-Amino-1,3-dithiane (64)	278
C_5			
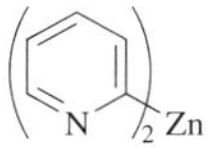 (2-Pyridyl)$_2$Zn	 Morpholino-NOBz (ia), $(CuOTf)_2$•PhH (cat), THF, rt, 15-60 min	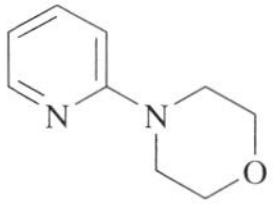 2-Morpholinopyridine (71)	112, 109
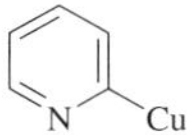 2-Pyridylcopper	t-$BuO_2CN(Li)OTs$, THF, 0°, 2.5 h	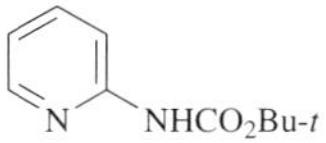 2-Pyridyl-$NHCO_2Bu$-t (53)	127
(2-Pyridyl)$_2Cu(CN)Li_2$	RNHOTMS, THF	2-Pyridyl-NHR	
		R: TMS, Temp: –60° to rt, Time: —, (60) R = H [b]	100
		R: Me, Temp: –50° to rt, Time: 2 h, (65)	101
		R: i-Pr, Temp: –50° to rt, Time: 2 h, (68)	101
		R: t-Bu, Temp: –50° to rt, Time: 2 h, (70)	101
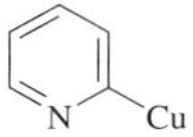 2-Pyridylcopper	$TsON(Li)CO_2CH_2CH=CH_2$, THF, –78°, 3 h	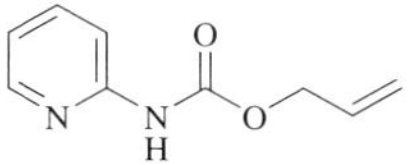 2-Pyridyl-$NHCO_2CH_2CH=CH_2$ (26)	130
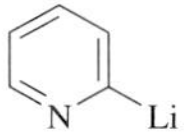 2-Pyridyllithium	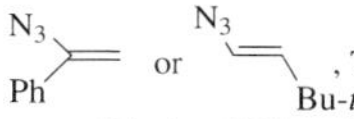 $N_3C(Ph)=CH_2$ or $N_3CH=CH$-Bu-t, THF, –78°; to rt, 2 h, then HCl, or KOH	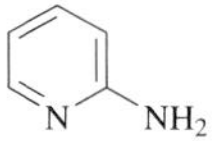 2-Aminopyridine (45)	278
	1. PhN_3, Et_2O 2. HOAc, H_2O 3. HCl, 80-90°	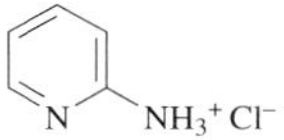 2-Pyridyl-$NH_3^+ Cl^-$ (38)	66

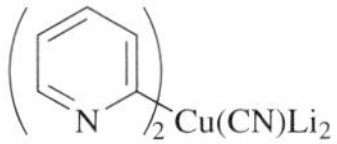

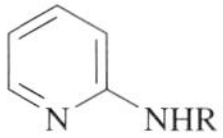

TABLE 5. HETEROCYCLIC CARBANIONS (*Continued*)

	Substrate	Conditions	Product(s) and Yield(s) (%)	Refs.
C_5	$(\text{3-pyridyl})_2Cu(CN)Li_2$	1. TMSNHOTMS, THF, –60° to rt 2. "Hydrolytic workup"[b]	3-aminopyridine (58)	100
	1-R-imidazo[4,5-b]pyrazine (N, N, N, R)	1. LDA, toluene, THF, –78°, 20 min 2. $ArSO_2N_3$, –78°; to rt, 1-2 h 3. $NaHCO_3$, H_2O	2-azido-1-R-imidazo[4,5-b]pyrazine (I) + (1-R-imidazo[4,5-b]pyrazin-2-yl)$_2$ (II) R = $MeOCH_2$, Ar = 4-MeC_6H_4: I (57), II (16)[c] R = $BnOCH_2$, Ar = 4-MeC_6H_4: I (69), II (11) R = —, Ar = 2,4,6-$Me_2C_6H_2$: I (0), II (—)	313
	5-methyl-2-thienyllithium	$N_3C(Ph)=CH_2$ or $N_3CH=CHBu$-*t*, THF, –78°; to rt, 2 h	2-amino-5-methylthiophene (58)	278
C_6	MeO_2C-substituted 3-Boc-2,2-dimethylthiazolidine (BocN, S)	$MeO_2CN=NCO_2Me$, 105°, 2 d	MeO_2C, $N(CO_2Me)NHCO_2Me$-substituted 3-Boc-2,2-dimethylthiazolidine (BocN, S) (—)	839
	1-methyl-5-[3-($NHCO_2Bu$-*t*)prop-1-ynyl]imidazole (N, N, Me)	1. *n*-BuLi (2.1 eq), THF, –75° 2. TsN_3, 10 min	2-azido-1-methyl-5-[3-($NHCO_2Bu$-*t*)prop-1-ynyl]imidazole (60)	840
C_7	2-benzothiazolyl-MgBr	1. $PhSCH_2N_3$ (ia), THF, –75° to 0° 2. CuI, 0°, 1 h 3. KOH (50% aq.), MeOH, THF, rt, 3 h	2-aminobenzothiazole (59)	274
	2-benzothiazolyl-Li	$N_3C(Ph)=CH_2$ or $N_3CH=CHBu$-*t*, THF, –78°; to rt, 2 h	2-aminobenzothiazole (53)	278
		1. PhN_3, Et_2O 2. HOAc, H_2O 3. HCl, 80-90°	2-benzothiazolyl-NH_3^+ Cl^- (52)	66

C$_8$

Reactant	Conditions	Product(s) and Yield(s) (%)	Refs.
3-Li-1-Et-indazole (Li, NEt, N)	PhN_3, Et_2O, –20°, 1 h	NHN=NPh, NEt, N (61)	841
indole (N, H)	4-$RC_6H_4SO_2N_3$	I (=NE, N, H) + II (—NHE, N, H); E = $SO_2C_6H_4R$-4	750
N-methylindole (N, Me)	$RC_6H_4SO_2N_3$, dioxane, 75-80°, 18-24 h	I (=NE, N, Me) + II (—NHE, N, Me); E = $SO_2C_6H_4R$	750

R	Solvent	Temp	Time	I + II [d]
O_2N	EtOH	reflux	6 h	(61)
MeO	dioxane	80°	26 h	(22)
Me	dioxane	75-80°	48 h	(46)

R	I	II
H	(54)	(22)
4-Cl	(60)	(16)
4-Br	(49)	(12)
3,4-Cl_2	(63)	(14)
4-AcNH	(67)	(5)
2-O_2N	(75)	(14)
3-O_2N	(74)	(6)
4-O_2N	(72)	(21)
3-O_2N,4-Cl	(82)	(8)
4-MeO	(44)	(22)
4-Me	(47)	(24)
2,4,6-Me_3	(34)	(15)
2,4,6-(i-Pr)$_3$	(32)	(24)

TABLE 5. HETEROCYCLIC CARBANIONS (*Continued*)

	Substrate	Conditions	Product(s) and Yield(s) (%)	Refs.
C_8 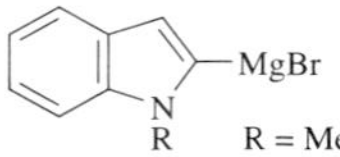	MgBr, N, R; R = Me, $PhSO_2$	1. $PhSCH_2N_3$ (ia), THF, –75° to 0° 2. CuI, 0°, 1 h 3. KOH (50% aq.), MeOH, THF, rt, 3 h	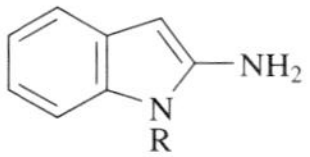 NH_2, N, R (0)	274
	Cu(CN)Li, O	1. PhMeNLi, THF, –40°, 20 min 2. $Cu(NO_3)_2$, THF, –78° 3. O_2, –78°, 30 min	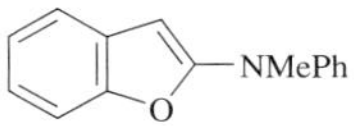 NMePh, O (76)	55, 54
	(O)$_2$ $Cu(CN)Li_2$	1. TMSNHOTMS, THF, –30° to rt, 18 h 2. TMSCl	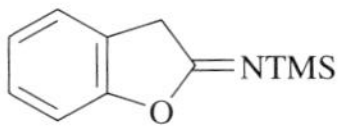 =NTMS, O (70)	100
	(S)$_2$ $Cu(CN)Li_2$	TMSNHOTMS, THF, –50°, 1 h; to rt	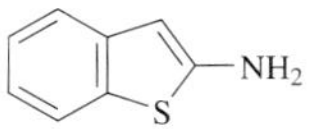 NH_2, S (58)	100
	Li, S	1. TsN_3, Et_2O, –70°, 5 h 2. $Na_4P_2O_7$, H_2O	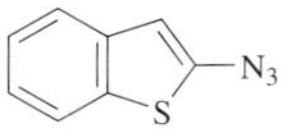 N_3, S (7)	316
	Li, S	1. TsN_3, Et_2O, –70°, 5 h 2. $Na_4P_2O_7$, H_2O	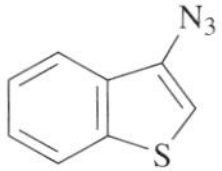 N_3, S (83)	316
	R, Li, S	TsN_3, Et_2O, hexane, –70°, 5 h; to –10°	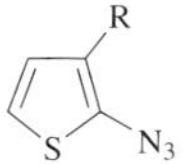 R, N_3, S R 2-thienyl (33) 3-thienyl (41)	779
	Li, R, S	TsN_3, Et_2O, hexane, –70°, 5 h; to –10°	N_3, R, S R 2-thienyl (72) 3-thienyl (73)	779
	R, Li, S	TsN_3, Et_2O, hexane, –70°, 5 h; to –10°	R, N_3, S R 2-thienyl (75) 3-thienyl (77)	779

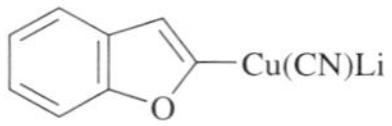

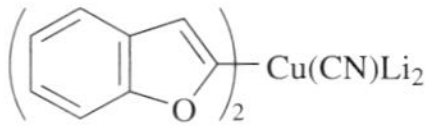

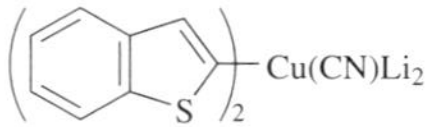

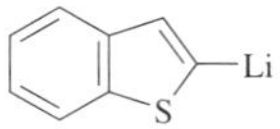

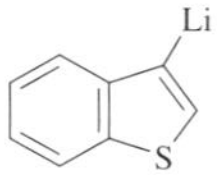

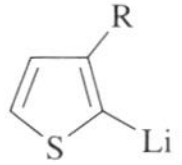

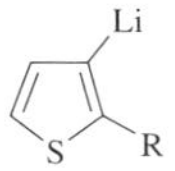

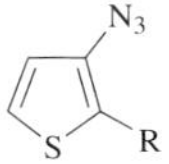

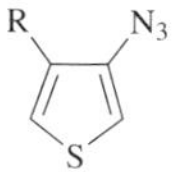

	Reactant	Conditions	Product(s) and Yield(s) (%)	Refs.
C_9	[indole structure: 3-MgCl, 2-CO_2Et, N-Bn]	1. 4-RC_6H_4N=NTs, THF, –20°, 1 h 2. CH_2=$CHCH_2I$, *N*-methylpyrroldinone, rt, 3 h 3. Zn, AcOH, TFA, 75°, 2.5 h	[indole structure: 3-NHC_6H_4R-4, 2-CO_2Et, N-Bn] R: Br (58); CO_2Et (71)	255
C_{14}	[3-phenylphthalide structure: Ph]	1. KNH_2, NH_3 (liq), Et_2O 2. PhN=NPh	[phthalide structure: 3-Ph, 3-N(Ph)NHPh] (30)	212

[a] The product was converted into *N*-methyl-2-thienylamine with $LiAlH_4$ but it could not be hydrolyzed to the amine.

[b] A hydrolytic workup was mentioned in the text but no details were given in the Experimental Section.

[c] With KHMDS, only the dimer **II** was obtained in 70% yield.

[d] Isomers **I** and **II** exist in equilibrium.

TABLE 6. ALDEHYDE ENOLATES

Substrate	Conditions	Product(s) and Yield(s) (%)	Refs.
C_2			
OEt	1. $ClNHCO_2Bn$, $CHCl_3$, MeOH, –78° 2. $CrCl_2$ 3. MeONa	BnO_2CNH, OMe, OEt (81)	343
	4-$O_2NC_6H_4N_2^+$ Cl^-, H_2O, 0-10°	O_2N, N, N, H, CHO (—)	842
OR R = Et, *i*-Bu, *n*-$C_{18}H_{37}$	1. MeO_2CN=NCO_2Me, rt 2. HCl (3% in MeOH), rt	MeO_2CNH, N, MeO_2C, OMe, OMe (—)	843
SR R = Et, Ph, 4-ClC_6H_4	1. EtO_2CN=NCO_2Et, PhH, rt 2. MeOH	EtO_2CNH, N, EtO_2C, SR, OMe (—)	844
OBu-*n*	1. 4-$O_2NC_6H_4N_3$, $CHCl_3$, 40°, 7 h 2. AcOH, PhH, 50°, 10 min	4-$O_2NC_6H_4NH$, OBu-*n*, OAc (84)	387
O, N	1. MeO_2CN=NCO_2Me, Et_2O, rt 2. HCl (3% in MeOH), rt	MeO_2CNH, N, MeO_2C, O, N, OMe (—)	843
OPh	1. RCON=NCOR, rt 2. HCl, MeOH R = MeO, EtO, Cl_3CCH_2O, Ph	RCONH, N, RCO, OPh, OMe (—)	240
OAr	1. MeO_2CN=NCO_2Me, rt 2. HCl, MeOH	MeO_2CNH, N, MeO_2C, OAr, OMe Ar 4-ClC_6H_4 (56) 4-$MeOC_6H_4$ (82) 4-MeC_6H_4 (85)	240
C_3			
OEt	4-$O_2NC_6H_4N_2^+$ Cl^-, H_2O, 0-10°	O_2N, N, N, H, CHO (—)	842

C_{3-9}

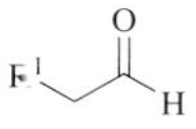

R^1CH_2CHO (1.5 eq)

1. $R^2O_2CN{=}NCO_2R^2$, L-proline (0.1 eq), MeCN
2. $NaBH_4$, EtOH

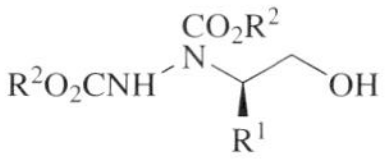

$R^2O_2CNH{-}N(CO_2R^2){-}CH(R^1){-}CH_2OH$ — 221

R^1	R^2	Temp	Time		% ee
Me	Bn	0° to rt	3 h	(97)	>95
n-Pr	Bn	0° to rt	3 h	(93)	>95
i-Pr	*t*-Bu	20°	—	(97)	92
i-Pr	Bn	0° to rt	3 h	(99)	96
n-Bu	Bn	0° to rt	3 h	(94)	97
Bn	Bn	0° to rt	3 h	(95)	>95

C_{3-7}

$R^2O_2CN{=}NCO_2R^2$, catalyst (x eq), rt

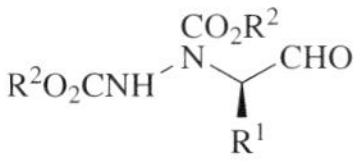

$R^2O_2CNH{-}N(CO_2R^2){-}CH(R^1){-}CHO$

R^1	R^2	Catalyst	x	Solvent	Time		% ee	Refs.
Me	Et	L-proline	0.5	CH_2Cl_2	45 min	(93)	92	222
Me	—	L-proline	—	neat	2 min	(100)	77	222
Me	Et	L-proline	0.02	CH_2Cl_2	5 h	(92)	84	222
Me	*i*-Pr	L-proline	0.1	CH_2Cl_2	105 min	(91)	88	222
Me	*t*-Bu	L-proline	0.1	CH_2Cl_2	205 min	(99)	89	222
Me	Bn	L-proline	0.2	CH_2Cl_2	3.5 h	(62)	54	229
Me	*t*-Bu	L-azetidinecarboxylic acid	0.2	CH_2Cl_2	22 h	(60)	74	229
Et	Et	L-proline	0,1	CH_2Cl_2	2 h	(77)	90	222
i-Pr	Bn	L-proline	0.1	CH_2Cl_2	4 h	(>90)	>90	229
n-C_5H_{11}	Bn	L-proline	0.2	CH_2Cl_2	1.25 h	(62)	91	229
n-C_5H_{11}	Bn	L-azetidinecarboxylic acid	0.2	CH_2Cl_2	15 h	(69)	72	229

C_3

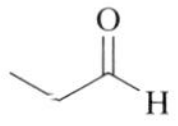

CH_3CH_2CHO

1. $EtO_2CN{=}NCO_2Et$, D-proline, 5°
2. $NaBH_4$

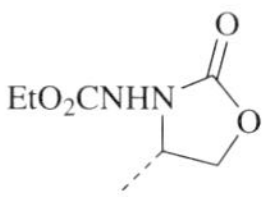

EtO_2CNHN-oxazolidinone (—) 92-93% ee — 224

Substrate	Conditions	Product(s) and Yield(s) (%)	Refs.
C_{3-9} R^1CH_2CHO	1. $R^2O_2CN{=}NCO_2R^2$, L-proline (0.1 eq), CH_2Cl_2, rt 2. $NaBH_4$, MeOH, then 0.5 N NaOH	3-(R^2O_2CNH)-4-R^1-oxazolidin-2-one	

R^1	R^2	Yield	% ee	Refs.
Me	Et	(67)	93	222, 224, 845
Et	Et	(77)	95	222
$CH_2{=}CHCH_2$	Et	(92)	93	222
i-Pr	Et	(83)	93	222
i-Pr	Bn	(70)	91	222
t-Bu	Et	(57)	91	222
Bn	Et	(68)	89	222

Substrate	Conditions	Product(s) and Yield(s) (%)	Refs.
C_4 $Me_2C{=}CHNR^1R^2$	R^3R^4NCl, O_2, dioxane, 0°, 2 h; rt, overnight; reflux, 5 h	$Me_2C(NR^3R^4)CHO$ **I** + $Me_2C(NR^1R^2)CHO$ **II**	74

R^1	R^2	R^3	R^4	**I**	**I + II**	**II**
Me	Me	Me	Me	(—)	(61)	(—)
Me	Me	Me	Me	(7)	(—)	(32)
Et	Et	Et	Et	(32)	(—)	(7)
—$(CH_2)_4$—		Me	Me	(53)	(—)	(36)
—$(CH_2)_2O(CH_2)_2$—		Me	Me	(42)	(—)	(15)
—$(CH_2)_5$—		Me	Me	(53)	(—)	(24)
—$(CH_2)_5$—		—$(CH_2)_5$—		(—)	(42)	(—)
t-Bu	MgBr	Me	Me	(88)	(—)	(0)

Substrate	Conditions	Product(s) and Yield(s) (%)	Refs.
C_{4-7} R^1R^2CHCHO	4-$MeC_6H_4SO_2N(Cl)Na$•x H_2O, L-proline (2 mol%), MeCN, rt	$R^1R^2C(NHTs)CHO$	78

R^1	R^2	Time	Yield	% ee
Me	Me	1 d	(83)	0
Me	Et	1 d	(81)	0
Et	Et	1 d	(78)	0
H	*i*-Pr	1 d	(86)	0
—$(CH_2)_5$—		2 d	(86)	0

Substrate	Conditions	Product(s) and Yield(s) (%)	Refs.
C_4			
butanal (CHO)	1. [$BnO_2CN{=}NCO_2Bn$, D-proline, MeCN] (ia), 0°, 2 h; to rt, 1 h; rt 2. $NaBH_4$, EtOH, 0°, 5 min	BnO_2CNH–N(CO_2Bn)–CH(Et)–CH_2OH (92), 96% ee	227
1-OTMS-butadiene (OTMS)	PhI=NTs, MeCN	$CH_2{=}CH$–CH(NHTs)–CHO (52)	172
2,3-dihydrofuran	4-$O_2NC_6H_4N_2^+$ Cl^-, H_2O, 0-10°	HO–CH_2CH_2–C(=N–NH–C_6H_4-4-NO_2)–CHO (—)	842
C_{4-8}			
R^1CH=CH–NR^2R^3	ArN_2^+ Cl^-, H_2O, NaOAc, pH 5-6	ArNHN=C(R^1)–CHO	195

R^1	R^2	R^3	Ar	
Et	—$(CH_2)_5$—		4-ClC_6H_4	(65)
Et	—$(CH_2)_5$—		4-$O_2NC_6H_4$	(41)
Et	—$(CH_2)_5$—		4-$HO_2CC_6H_4$	(53)
Ph	Et	Et	4-ClC_6H_4	(90)
Ph	Et	Et	4-$O_2NC_6H_4$	(94)
Ph	Et	Et	4-$MeOC_6H_4$	(76)
Ph	Et	Et	4-$HO_2CC_6H_4$	(89)

Substrate	Conditions	Product(s) and Yield(s) (%)	Refs.
C_4			
2-(i-Pr-methylidene)-1,3-dithiane (i-Pr)	4-$O_2NC_6H_4N_2^+$ BF_4^-, CH_2Cl_2, rt, 1 h	4-O_2N-C_6H_4–N=N–C(i-Pr)=(1,3-dithian-2-ylidene) (85)	846
C_{4-6}			
R^1CH_2CHO	1. (S)-2-[bis(Ar)(OTMS)methyl]pyrrolidine (10 mol%), $R^2O_2CN{=}NCO_2R^2$, CH_2Cl_2, rt, 15 min 2. $NaBH_4$, MeOH, 0° Ar = 3,5-$(CF_3)_2C_6H_3$	4-R^1-3-($NHCO_2R^2$)-oxazolidin-2-one	386

R^1	R^2		% ee
Et	Et	(79)	90
i-Pr	Et	(88)	97
i-Pr	i-Pr	(73)	92
allyl	Et	(81)	92
t-Bu	Et	(83)	97

TABLE 6. ALDEHYDE ENOLATES (*Continued*)

Substrate	Conditions	Product(s) and Yield(s) (%)	Refs.
C_{4-15} R[1](R[2])CH–CHO	For **I**: 1. Catalyst (0.5 eq), solvent, rt, 30 min 2. Substrate, 0°; rt, 1 h 3. $R^3O_2CN{=}NCO_2R^3$, rt, time For **II** from **I**: $NaBH_4$, CH_2Cl_2, EtOH, 0°, 30 min	**I**: R[1]R[2]C(CHO)E; **II**: oxazolidinone, N-$NHCO_2R^3$ E: $N(CO_2R^3)NHCO_2R^3$	

R^1	R^2	Catalyst	R^3	Time	I	% ee	II	% ee	Refs.
Me	Me	L-proline	Et	3 d	(83)	—	(—)	—	223
Me	Me	L-proline	Bn	3 d	(85)	—	(—)	—	223
Me	Et	L-proline	Et	—	(—)	—	(52)	28	847, 223
Me	Et	L-2-azetidinecarboxylic acid	Et	—	(—)	—	(—)	6	847
Me	*n*-Pr	L-proline	Et	3 d	(60)	—	(—)	—	223
Me	*n*-Pr	L-proline	Bn	3d	(60)	39	(—)	—	223
Et	Et	L-proline	Et	4 d	(55)	—	(—)	—	223
Et	Et	L-proline	Bn	4 d	(51)	—	(—)	—	223
—$(CH_2)_5$—		L-proline	Et	4 d	(—)	—	(26)	—	847, 223
Et	*n*-Bu	L-proline	Et	9 d	(—)	—	(35)	4	847, 223
Me	2-thienyl	L-proline	Bn	3 d	(60)	70	(—)	—	223
Me	Ph	L-proline	Et	3 d	(62)	80	(—)	—	223, 847
Me	Ph	L-proline	Et	3 d	(—)	—	(17)	81	847
Me	Ph	L-2-azetidinecarboxylic acid	Et	—	(—)	—	(—)	51	847
Me	Ph	L-proline	Bn	3 d	(—)	—	(83)	81	847, 223
Me	Ph	L-2-azetidinecarboxylic acid	Bn	—	(—)	—	(—)	52	847
Me	4-FC_6H_4	L-proline	Et	5 d	(26)	68	(—)	—	223
Me	4-FC_6H_4	L-proline	Bn	5 d	(29)	35	(—)	—	223
Me	4-ClC_6H_4	L-proline	Bn	3 d	(86)	61	(—)	—	223
Me	4-BrC_6H_4	L-proline	Bn	3 d	(70)	79	(—)	—	223
Me	4-$O_2NC_6H_4$	L-proline	Et	2 d	(85)	36	(—)	—	223

Me	4-$O_2NC_6H_4$	L-proline	Bn	2 d	(99)	56	(—)	—	223
Me	3-$MeOC_6H_4$	L-proline	Et	3 d	(62)	83	(—)	—	223
Me	4-$MeOC_6H_4$	L-proline	Et	5 d	(87)	76	(—)	—	223, 847
Me	3,5-$(MeO)_2C_6H_3$	L-proline	Et	6 d	(63)	85	(—)	—	223, 847
Me	3,4-$(BnO)_2C_6H_3$	L-proline	Bn	7 d	(58)	73	(—)	—	223
Me	4-NCC_6H_4	L-proline	Bn	3 d	(62)	53	(—)	—	223
Me	4-$CF_3C_6H_4$	L-proline	Et	5 d	(19)	—	(—)	—	223
Me	4-$CF_3C_6H_4$	L-proline	Bn	3 d	(40)	—	(—)	—	223
Me	4-$MeO_2CC_6H_4$	L-proline	Et	6 d	(50)	82	(—)	—	223, 847
Me	4-PhC_6H_4	L-proline	Bn	3 d	(53)	84	(—)	—	223
Et	Ph	L-proline	Et	3 d	(59)	80	(—)	—	223, 847
Et	Ph	L-2-azetidinecarboxylic acid	Et	—	(—)	49	(—)	—	847
Me	2-naphthyl	L-proline	Et	2.5 d	(54)	86	(—)	—	223, 847
Me	2-naphthyl	L-2-azetidinecarboxylic acid	Et	—	(—)	56	(—)	—	847

$C_{4\text{-}5}$

R^1–CH=CH–CHO (1.5 eq)

1. (S)-2-[$Ar_2C(OTMS)$]pyrrolidine (cat), R^2SH (1 eq), toluene
2. $R^3O_2CN{=}NCO_2R^3$ (1.3 eq), time
3. $NaBH_4$
4. NaOH

Ar = 3,5-$(CF_3)_2C_6H_3$

Product: oxazolidinone bearing R^3O_2CNHN and R^2S–CH(R^1) substituents — 225

R^1	R^2	R^3	Time		dr	ee %
Me	Et	Bn	16 h	(51)	88:12	>99
Me	Bn	Et	3.5 h	(57)	95:5	>99[a]
Me	Bn	Bn	16 h	(44)	89:11	>99
Et	Bn	Et	16 h	(42)	96:4	>99
Et	Et	Bn	16 h	(38)	95:5	97

TABLE 6. ALDEHYDE ENOLATES (*Continued*)

Substrate	Conditions	Product(s) and Yield(s) (%)	Refs.
C_{4-14} R^1R^2CH–CHO	$R^3SO_2N_3$, 2-R^4-pyrrolidine (1 eq), rt	$R^1R^2C(CHO)NHSO_2R^3$	386a

R^1	R^2	R^3	R^4	Solvent	Time	Yield	% ee
H	Ph	4-MeC_6H_4	CO_2H	EtOH	1 d	(26)	—
Me	Me	4-MeC_6H_4	CO_2H	EtOH	1 d	(42)	—
Me	Et	4-MeC_6H_4	CO_2H	EtOH	1 d	(49)	5
Me	*n*-Pr	4-MeC_6H_4	CO_2H	EtOH	1 d	(51)	12
Et	Et	4-MeC_6H_4	CO_2H	EtOH	1 d	(47)	—
—$(CH_2)_5$—		4-MeC_6H_4	CO_2H	EtOH	1 d	(52)	—
Et	*n*-Bu	2-$O_2NC_6H_4$	CO_2H	EtOH	1 d	(54)	28
Me	Ph	Me	CO_2H	EtOH	1 d	(33)	71
Me	Ph	*n*-C_4F_9	CO_2H	EtOH	1 d	(24)	29
Me	Ph	5-chloro-3-thienyl	CO_2H	EtOH	1 d	(36)	54
Me	Ph	2,5-dichloro-3-thienyl	CO_2H	EtOH	1 d	(27)	47
Me	Ph	6-chloro-5-bromo-2-pyridyl	CO_2H	EtOH	1 d	(24)	46
Me	Ph	2-$O_2NC_6H_4$	CO_2H	EtOH	1 d	(44)	56
Me	Ph	4-$O_2NC_6H_4$	CO_2H	EtOH	1 d	(52)	82
Me	Ph	2,4-$(O_2N)_2C_6H_3$	CO_2H	EtOH	1 d	(27)	45
Me	Ph	3,4-$(MeO)_2C_6H_3$	CO_2H	EtOH	1 d	(43)	67
Me	Ph	4-MeC_6H_4	CO_2H	EtOH	1 d	(35)	59
Me	Ph	2-$MeO_2CC_6H_4$	CO_2H	EtOH	1 d	(39)	8
Me	Ph	1-naphthyl	CO_2H	EtOH	1 d	(36)	65
Me	Ph	2-naphthyl	CO_2H	EtOH	1 d	(42)	63
Me	Ph	2,4,6-$(i\text{-}Pr)_3C_6H_2$	CO_2H	EtOH	1 d	(33)	50
Me	Ph	4-MeC_6H_4	H	EtOH	1 d	(36)	—
Me	Ph	4-MeC_6H_4	tetrazolyl	EtOH	1 d	(24)	66
Me	Ph	4-MeC_6H_4	CONHTs	DMSO	1 d	(25)	66

Me	Ph	4-MeC_6H_4	(1-pyrrolidinyl)methyl•CF_3CO_2H	DMSO	70 min	(23)	45
Me	Ph	2-$O_2NC_6H_4$	(1-pyrrolidinyl)methyl•CF_3CO_2H	DMSO	1 d	(26)	57
Me	Ph	4-MeC_6H_4	CH_2OH	EtOH	4 d	(<10)	—
Me	Ph	4-MeC_6H_4	$C(Ph)_2OH$	EtOH	1 d	(0)	—
Me	Ph	2-$O_2NC_6H_4$	$C(Ph)_2OTMS$	EtOH	1 d	(26)[b]	—
Me	Ph	2-$O_2NC_6H_4$	$C(C_{10}H_7)_2OTMS$	EtOH	1 d	(38)[b]	64
Me	Ph	2-$O_2NC_6H_4$	$C(Ph)_2CMe$	EtOH	1 d	(52)	55
Me	Ph	4-MeC_6H_4	CO_2H[c]	THF	7d	(21)	53
Me	Ph	4-MeC_6H_4	CO_2H[c]	MeCN	4d	(14)	54
Me	Ph	4-MeC_6H_4	CO_2H[c]	CH_2Cl_2	4d	(0)	—
Me	Ph	4-MeC_6H_4	CO_2H	*t*-BuOH	1 d	(25)	54
Me	Ph	4-MeC_6H_4	CO_2H	DMSO	2 h	(28)	60
Me	Ph	4-MeC_6H_4	CO_2H	[bmim][BF_4]	1 d	(38)	72
Me	Ph	2-$O_2NC_6H_4$	CO_2H	[bmim][BF_4]	1 d	(55)	66
Me	Ph	4-MeC_6H_4	CO_2H	[capemim][BF_4]	1 d	(53)	20
Me	2-$MeOC_6H_4$	2-$O_2NC_6H_4$	CO_2H	EtOH	1 d	(21)	72
Me	2-$MeOC_6H_4$	4-$O_2NC_6H_4$	CO_2H	EtOH	1 d	(21)	59
Me	3-$MeOC_6H_4$	2-$O_2NC_6H_4$	CO_2H	EtOH	1 d	(47)	84
Me	3-$MeOC_6H_4$	4-$O_2NC_6H_4$	CO_2H	EtOH	1 d	(49)	69
Me	4-$MeOC_6H_4$	2-$O_2NC_6H_4$	CO_2H	EtOH	1 d	(53)	86
Me	4-$MeOC_6H_4$	4-$O_2NC_6H_4$	CO_2H	EtOH	1 d	(44)	76
Me	2,4-$(MeO)_2C_6H_3$	2-$O_2NC_6H_4$	CO_2H	EtOH	1 d	(34)	45
Me	2,5-$(MeO)_2)C_6H_3$	2-$O_2NC_6H_4$	CO_2H	EtOH	1 d	(32)	54
Me	3,5-$(BnO)_2C_6H_3$	2-$O_2NC_6H_4$	CO_2H	EtOH	1 d	(31)	72
Me	4-(*t*-Bu)C_6H_4	2-$O_2NC_6H_4$	CO_2H	EtOH	1 d	(55)	61
Ph	Ph	4-MeC_6H_4	CO_2H	EtOH	1 d	(36)	—

C_5

O

1. $ClNHCO_2Et$, $CHCl_3$, MeOH, –78°
2. $CrCl_3$
3. H_2SO_4

$NHCO_2Et$, O, OMe (77)

343

TABLE 6. ALDEHYDE ENOLATES (*Continued*)

Substrate	Conditions	Product(s) and Yield(s) (%)	Refs.
C_{5-6} 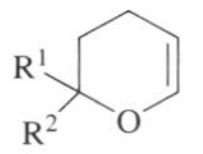	EtO_2CN_3, R^3OH, *hν*	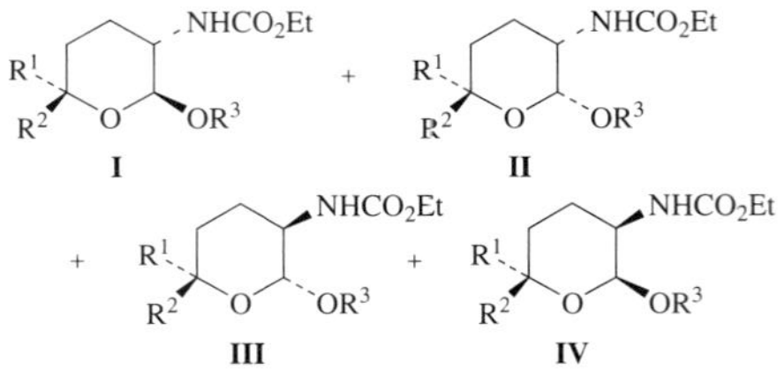 I + II + III + IV ($NHCO_2Et$, OR^3)	295
C_5 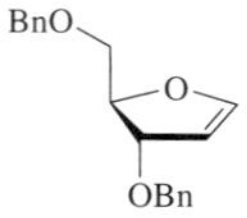	1. (Saltmen)Mn(N) (2 eq), 2,6-(*t*-Bu)$_2$,4-Me-pyridine, CH_2Cl_2 2. TFFA, –78°; to rt, 5-6 h	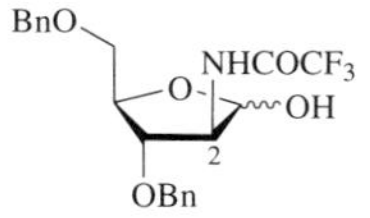 ($NHCOCF_3$, OH, OBn) (80), C2 de 86%	354
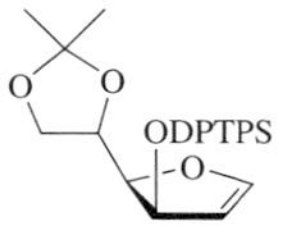 (ODPTPS)	1. (Saltmen)Mn(N) (2 eq), 2,6-(*t*-Bu)$_2$,4-Me-pyridine, CH_2Cl_2 2. TFFA, –78°; to rt, 5-6 h	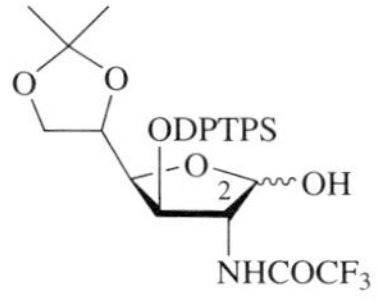 (ODPTPS, OH, $NHCOCF_3$) (80), C2 de 82%	354

R^1	R^2	R^3	I	II	III	IV
H	H	Me	(0)	(53)	(0)	(10)
H	H	*t*-Bu	(0)	(69)	(0)	(7)
$AcOCH_2$	H	Me	(31)	(25)	(0)	(11)
$AcOCH_2$	H	*t*-Bu	(36)	(20)	(0)	(trace)
H	MeO	*t*-Bu	(51)	(0)	(18)	(11)

C_6

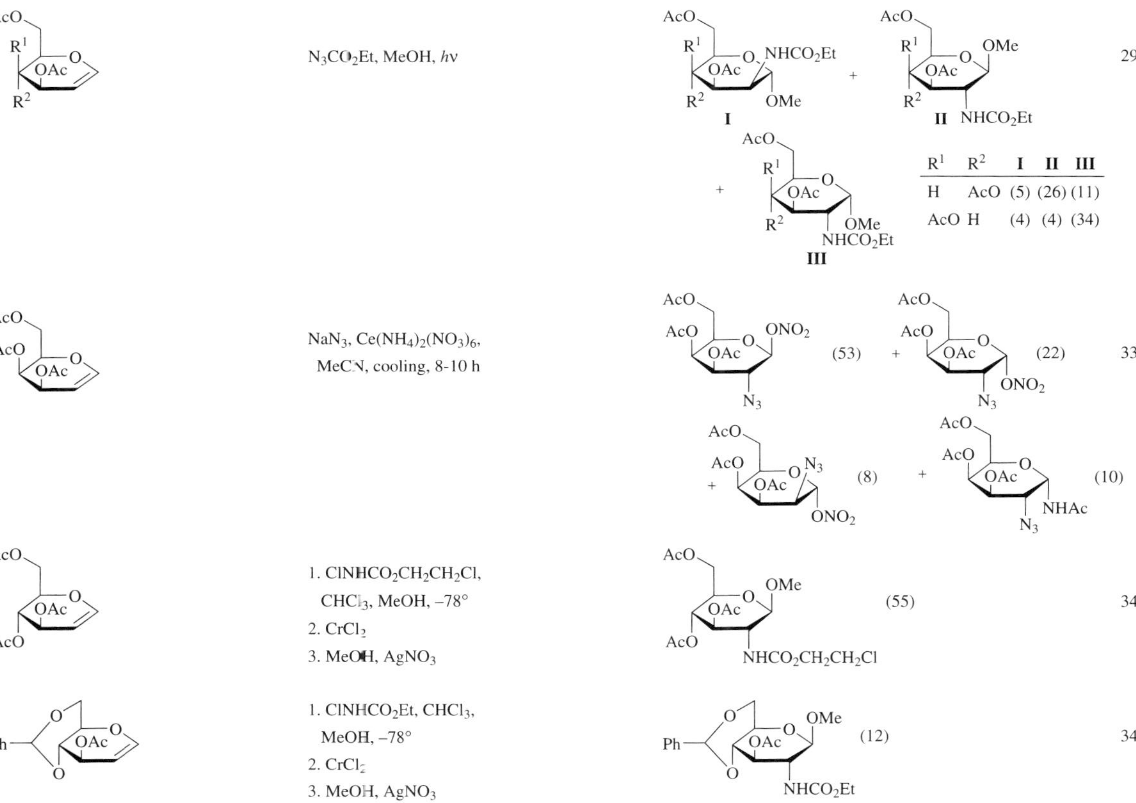

Conditions	Product(s) and Yield(s) (%)	Refs.
N_3CO_2Et, MeOH, $h\nu$	I + II + III	295
NaN_3, $Ce(NH_4)_2(NO_3)_6$, MeCN, cooling, 8-10 h	(53) + (22) + (8) + (10)	332
1. $ClNHCO_2CH_2CH_2Cl$, $CHCl_3$, MeOH, –78° 2. $CrCl_2$ 3. MeOH, $AgNO_3$	(55)	343
1. $ClNHCO_2Et$, $CHCl_3$, MeOH, –78° 2. $CrCl_2$ 3. MeOH, $AgNO_3$	(12)	343

R^1	R^2	I	II	III
H	AcO	(5)	(26)	(11)
AcO	H	(4)	(4)	(34)

TABLE 6. ALDEHYDE ENOLATES (*Continued*)

Substrate	Conditions	Product(s) and Yield(s) (%)	Refs.
C_6			
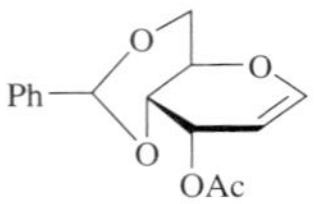 Ph, O, O, O, OAc	1. $ClNHCO_2CH_2CCl_3$, $CHCl_3$, MeOH, –78° 2. $CrCl_2$ 3. MeOH, $AgNO_3$ 4. AcOH, then Ac_2O	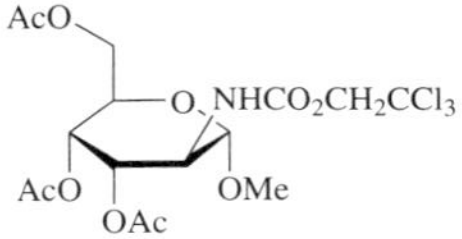 AcO, O, $NHCO_2CH_2CCl_3$, AcO, OAc, OMe (65)	343
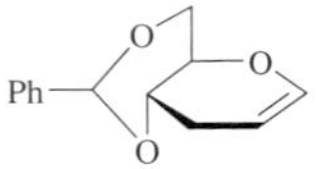 Ph, O, O, O	1. TFAA, CH_2Cl_2 2. (Saltmen)Mn(N) (1 eq), CH_2Cl_2; addition over 7 h	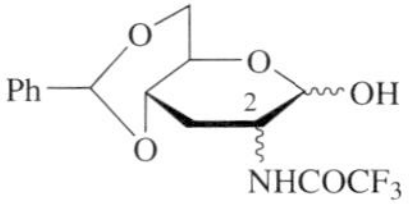 Ph, O, O, O, 2, OH, $NHCOCF_3$ (70), C2 de 0%	354
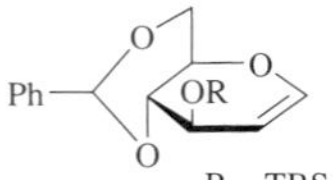 Ph, O, O, O, OR R = TBS	1. TFAA, CH_2Cl_2 2. (Saltmen)Mn(N) (1 eq), CH_2Cl_2; addition over 7 h	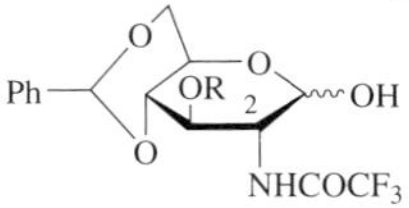 Ph, O, O, O, OR, 2, OH, $NHCOCF_3$ (75), C2 de 75%	354
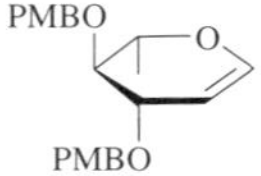 PMBO, O, PMBO	1. TFAA, CH_2Cl_2 2. (Saltmen)Mn(N) (1 eq), CH_2Cl_2; addition over 7 h	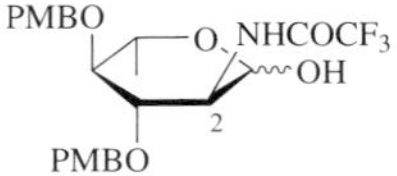 PMBO, O, $NHCOCF_3$, OH, 2, PMBO (60), C2 de 75%	354
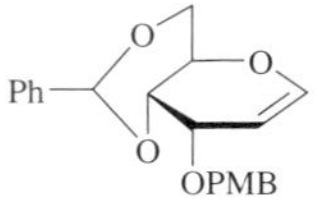 Ph, O, O, O, OPMB	1. TFAA, CH_2Cl_2 2. (Saltmen)Mn(N) (1 eq), CH_2Cl_2; addition over 7 h	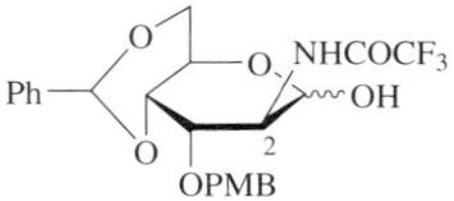 Ph, O, O, O, $NHCOCF_3$, OH, 2, OPMB R / (yield) / C2 % de: PMB (66) 71 TBS (68) 75	354
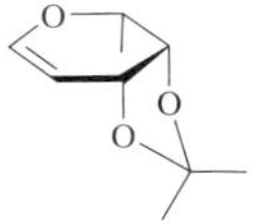 O, O, O	(Saltmen)Mn(N), TFAA, 2,6-(t-Bu)$_2$,4-Me-pyridine, –78° to rt	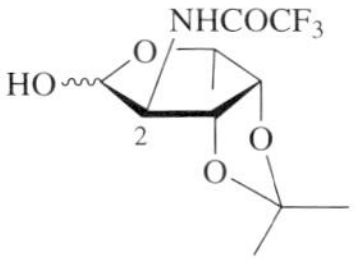 $NHCOCF_3$, HO, O, 2, O, O (69)	351
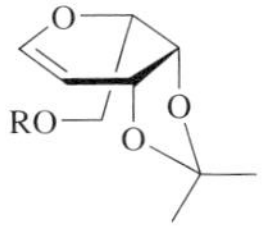 O, RO, O, O	1. TFAA, CH_2Cl_2 2. (Saltmen)Mn(N) (1 eq), CH_2Cl_2; addition over 7 h	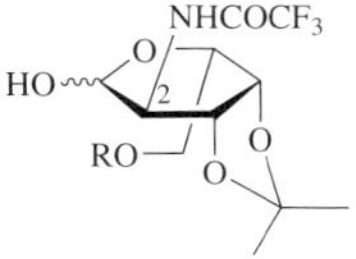 $NHCOCF_3$, HO, O, 2, RO, O, O R / (yield) / C2 de %: Bn (62) 87.5 TBDPS (64) 87.5	354

Substrate	Conditions	Product(s) and Yield(s) (%)	Refs.
C_8			
(hexahydro-1,3-benzodithiol-2-ylidene)(Ph)methane, cis-H,H	4-$O_2NC_6H_4N_2^+$ BF_4^-, CH_2Cl_2, rt	2-[Ph(N=$NC_6H_4NO_2$-4)C=] hexahydro-1,3-benzodithiole (58)	846
	EtO_2CN=NCO_2Et, CH_2Cl_2, rt, 6 h	2-[Ph(CO_2EtN–NHCO_2Et)C=] hexahydro-1,3-benzodithiole (79)	846
Ph–CH=CH–N(piperidine)	1. MeO_2CN=NCO_2Me 2. H_3O^+, H_2O	MeO_2CN(NHCO_2Me)CH(Ph)CHO (—)	247
C_9			
PhCH(Me)CHO	4-$MeC_6H_4SO_2N(Cl)Na$•x H_2O, pyrrolidine (10 mol%), $PhNMe_3^+$ Br_3^- (10 mol%), MeCN, rt, 1 d	PhC(Me)(NHTs)CHO (70)	78
C_{9-13}			
R^1CH(R^2)CHO	4-$MeC_6H_4SO_2N(Cl)Na$•x H_2O, L-proline (2 mol%), MeCN, microwave irradiation (y W)	R^1R^2C(NHTs)CHO 0% ee	78

R^1	R^2	y	Temp	Time	
Me	Ph	100	90°	30 min	(50)
Me	Ph	150	50°	30 min	(66)
Me	Ph	200	50°	40 min	(90)
Me	Ph	200	60°	30 min	(90)
Me	4-FC_6H_4	200	60°	30 min	(85)
Me	4-$O_2NC_6H_4$	200	60°	30 min	(89)
Me	4-MeC_6H_4	200	60°	30 min	(73)
Me	4-$CF_3C_6H_4$	200	60°	30 min	(79)
Me	4-NCC_6H_4	200	60°	30 min	(86)
Ph	Ph	200	60°	30 min	(88)
Me	naphthyl	200	60°	30 min	(91)
Me	4-(*i*-Pr)$C_6H_4CH_2$	200	rt	1 d	(83)

TABLE 6. ALDEHYDE ENOLATES (*Continued*)

Substrate	Conditions	Product(s) and Yield(s) (%)	Refs.
C_{10} 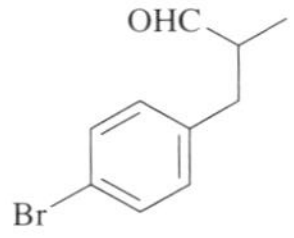 OHC, Br	1. 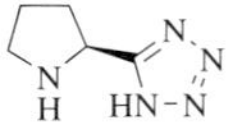 (pyrrolidinyl tetrazole, N, H, HN–N) (15 mol%), MeCN 2. Add $BnO_2CN{=}NCO_2Bn$, then substrate, rt, 3 h	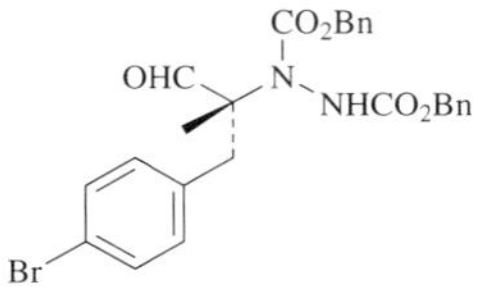 OHC, CO_2Bn, N, $NHCO_2Bn$, Br (95), 80% ee	385
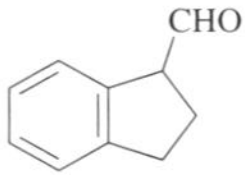 CHO	$BnO_2CN{=}NCO_2Bn$, L-proline (30 mol%), MeCN, rt, 1 h	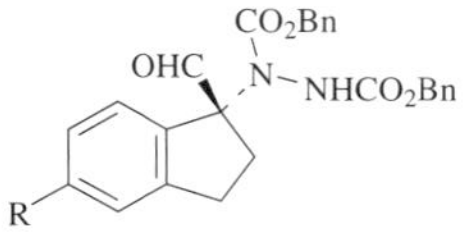 OHC, CO_2Bn, N, $NHCO_2Bn$, R (99%), >99% ee	226
$C_{10\text{-}11}$ 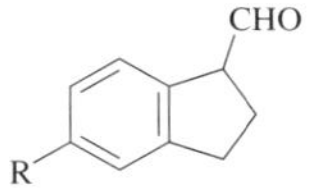 CHO, R	1. D-Proline (15 mol%), MeCN, $BnO_2CN{=}NCO_2Bn$ 2. Add substrate, rt, 4 h	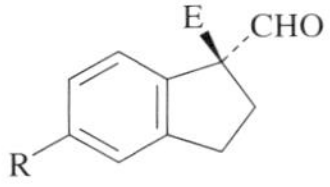 E, CHO, R R / % ee Br (75) >99 CO_2Me (96) >99 $E = N(CO_2Bn)NHCO_2Bn$	226

[a] With L-proline as the catalyst, both dr and ee values were considerably lower.

[b] The product was reduced with $NaBH_4$ prior to isolation.

[c] The catalyst loading was 40 mol%.

TABLE 7A. ACYCLIC KETONE ENOLATES

Substrate	Conditions	Product(s) and Yield(s) (%)	Refs.
C_3 	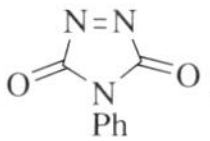 N=N, O, N, O, Ph, reflux	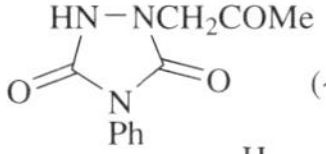 HN−NCH_2COMe, O, N, O, Ph (~100)	254
 OEt	$4\text{-}O_2NC_6H_4N_2^+\ Cl^-$, H_2O, 0–10°	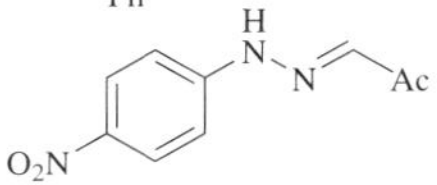 H, N, N, Ac, O_2N (—)	842
$C_{4\text{-}14}$ 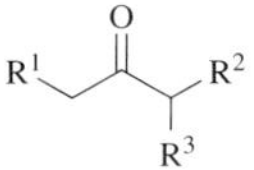 R[1], O, R[2], R[3]	1. $PhR^4NMnMe{\bullet}4\ LiBr$ (ia), THF, rt, 1 h 2. $R^5O_2CN{=}NCO_2R^5$, −30°; rt, 2.5 h	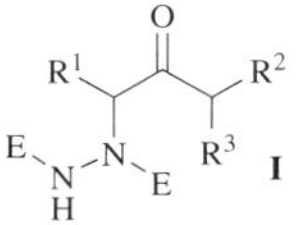 **I** 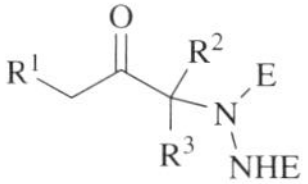 + **II**, $E = CO_2R^5$	388

R^1	R^2	R^3	R^4	R^5	I + II	I:II	I dr
H	H	Me	*n*-Bu	Et	(50)	50:50	—
Me	Me	Me	*n*-Bu	Et	(72)	90:10	—
Et	H	Et	Me	*t*-Bu	(60)	—	—
n-Pr	H	*n*-Pr	Me	Et	(90)	—	—
n-Pr	H	*n*-Pr	Me	*t*-Bu	(60)	—	—
n-C_5H_{11}	Me	Me	*n*-Bu	*t*-Bu	(60)	98:2	—
Et	Et	Bn	*n*-Bu	Et	(93)	98:2	3:1
Et	Et	Bn	*n*-Bu	*t*-Bu	(75)	90:10	3:1

Substrate	Conditions	Product(s) and Yield(s) (%)	Refs.
$C_{4\text{-}11}$ 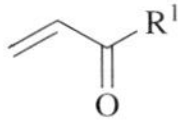 R[1], O	$R^2O_2CN{=}NCO_2R^2$, DABCO (cat), THF	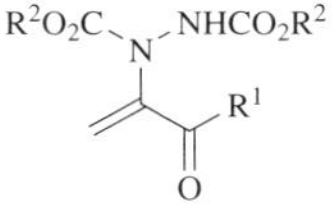 R^2O_2C, N, $NHCO_2R^2$, R[1], O	403

R^1	R^2	Temp	Time	
Me	Et	rt	8 h	(83)
Me	*t*-Bu	40°	24 h	(63)
Et	Et	rt	24 h	(78)
Et	*t*-Bu	40°	24 h	(34)
n-C_6H_{13}	Et	rt	24 h	(79)
n-C_7H_{15}	Et	rt	8 h	(61)
PhCH=CH	Et	rt	8 h	(90)
PhCH=CH	*t*-Bu	40°	24 h	(52)

TABLE 7A. ACYCLIC KETONE ENOLATES (*Continued*)

Substrate	Conditions	Product(s) and Yield(s) (%)	Refs.
C_{4-10} $R^1CH_2COCH_2R^2$	$R^3O_2CN{=}NCO_2R^3$, catalyst (x eq), rt	I + II, $E = CO_2R^3$	

R^1	R^2	R^3	Catalyst	x	Solvent	Time	I	I % ee[a]	II	Refs.
H	Me	Et	L-proline	0.2	MeCN	52 h	(—)	96	(<10)	228
H	Me	Et	L-proline	0.05	neat	65 h	(—)	93	(<10)	228
H	Me	Et	L-proline	0.1	MeCN	10 h	(73)	95	(7)	228
H	Me	*t*-Bu	L-proline	0.2	CH_2Cl_2	114 h	(49)	94	—	229
H	Me	*t*-Bu	L-azetidinecarboxylic acid	0.1	CH_2Cl_2	114 h	(54)	90	—	229
Me	Me	Et	L-proline	0.1	MeCN	60 h	(79)	94	—	228
H	Et	Et	L-proline	0.1	MeCN	20 h	(62)	98	(15)	228
H	*i*-Pr	Et	L-proline	0.1	MeCN	96 h	(52)	99	(17)	228
H	Bn	Et	L-proline	0.1	MeCN	24 h	(75)	98	(17)	228

Substrate	Conditions	Product(s) and Yield(s) (%)	Refs.
$R^1CH_2COCO_2Et$	1. bis(oxazoline)-$Cu(OTf)_2$ (R^3, R^4, R^5) (10 mol%), solvent 2. Add substrate, then $R^2O_2CN{=}NCO_2R^2$, rt, 16 h 3. L-Selectride, THF, –78°, 1 h; to rt 4. NaOH, H_2O, rt, 2 h 5. $TMSCHN_2$, MeOH, toluene, hexane, 15 min	oxazolidinone with R^1, $NHCO_2R^2$, MeO_2C	404

R^1	R^2	R^3	R^4	R^5	Solvent		% ee
Me	Bn	H	Ph	Ph	CH_2Cl_2	(45)	90
Me	Bn	Me	Ph	Ph	THF	(44)	92
i-Pr	Bn	H	Ph	Ph	CH_2Cl_2	(78)	95
i-Pr	Bn	H	Ph	Ph	THF	(60)	95
$CH_2CH{=}CH_2$	Bn	H	Ph	Ph	CH_2Cl_2	(62)	93
$CH_2CH{=}CH_2$	Bn	H	Ph	Ph	THF	(38)	90
i-Bu	Bn	H	Ph	Ph	CH_2Cl_2	(53)	96
i-Bu	Bn	H	Ph	Ph	THF	(51)	95
$(CH_2)_2CH{=}CH_2$	Bn	H	Ph	Ph	CH_2Cl_2	(52)	92
$(CH_2)_2CH{=}CH_2$	Bn	H	Ph	Ph	THF	(36)	94

$n\text{-}C_5H_{11}$	Bn	H	Ph	Ph	CH_2Cl_2	(63)	93
$n\text{-}C_5H_{11}$	Bn	H	Ph	Ph	THF	(48)	97
$c\text{-}C_6H_{11}CH_2$	Bn	H	Ph	Ph	CH_2Cl_2	(54)	96
$c\text{-}C_6H_{11}CH_2$	Bn	H	Ph	Ph	THF	(72)	96
Bn	Et	Me	H	Ph	CH_2Cl_2	(55)	68
Bn	Et	Me	H	Ph	THF	(47)	35
Bn	Bn	Me	H	Ph	CH_2Cl_2	(39)	90
Bn	Bn	Me	H	Ph	THF	(60)	82
Bn	Bn	Me	H	t-Bu	THF	(31)	7
Bn	Bn	H	Ph	Ph	THF	(57)	89
Bn	Bn	Me	Ph	Ph	THF	(58)	88

C_5

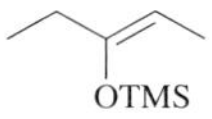

OTMS

PhI=NTs, MeCN

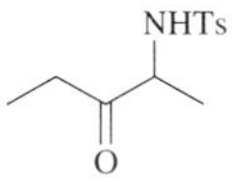

NHTs

O

(53) 172

$C_{5\text{-}11}$

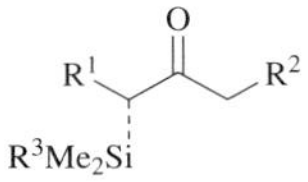

O

R^1 R^2

R^3Me_2Si

1. LDA, THF, 0°

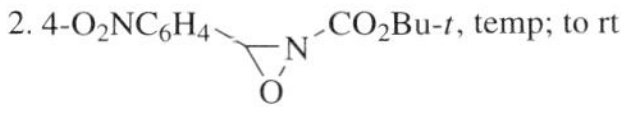

2. $4\text{-}O_2NC_6H_4$ N $CO_2Bu\text{-}t$, temp; to rt

O

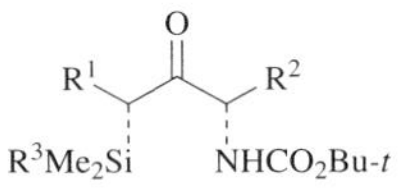

O

R^1 R^2

R^3Me_2Si $NHCO_2Bu\text{-}t$

156

R^1	R^2	R^3	Temp		% de
Me	Me	t-Bu	–100°	(27)	80
Et	Et	t-Bu	–100°	(37)	87
n-Pr	n-Pr	t-Bu	–100°	(29)	88
Me	Bn	t-Bu	–78°	(19)	41
Bn	Me	$t\text{-}C_6H_{13}$	–78°	(20)	83

TABLE 7A. ACYCLIC KETONE ENOLATES (*Continued*)

Substrate	Conditions	Product(s) and Yield(s) (%)	Refs.
C_{5-12} 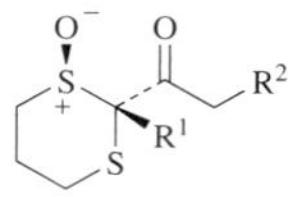	1. LiHMDS, THF, –78° 2. t-$BuO_2CN{=}NCO_2Bu$-t, –78°, 15 min 3. HOAc, –78° [b]	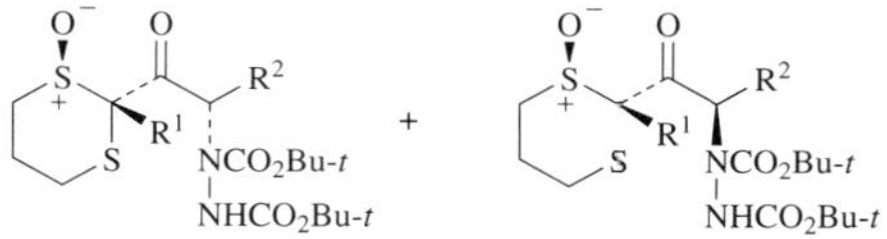 I + II (NCO_2Bu-t / $NHCO_2Bu$-t, R^1, R^2)	848

R^1	R^2	Enantiomer	I + II	I:II	% de
Me	Me	+/–	(69)	2:1	—
Et	H	+/–	(72)	—	—
Et	Me	+/–	(48)	>99:1	—
Et	*i*-Pr	+	(89)	—	72[c]
Et	*t*-Bu	+	(91)	—	—
Ph	Me	+/–	(37)	2:1	—
Et	Ph	+	(85)	—	—
Et	Bn	+	(93)	—	69[c]

Substrate	Conditions	Product(s) and Yield(s) (%)	Refs.
C_{5-6} 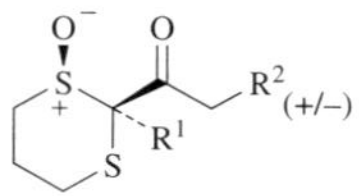 (+/–)	1. LiHMDS, THF, –78° 2. t-$BuO_2CN{=}NCO_2Bu$-t, –78°, 15 min 3. HOAc, –78° [b]	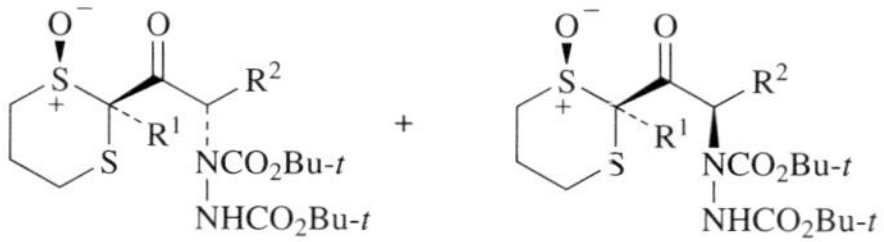 I + II (NCO_2Bu-t / $NHCO_2Bu$-t, R^1, R^2)	848

R^1	R^2	I + II	I:II
Me	Me	(76)	3:1
Et	H	(89)	—
Et	Me	(42)	12:1

Substrate	Conditions	Product(s) and Yield(s) (%)	Refs.
C_5 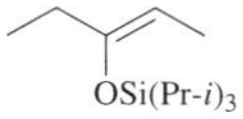 $OSi(Pr$-$i)_3$	NaN_3, $Ce(NH_4)_2(NO_3)_6$, MeCN, –20°	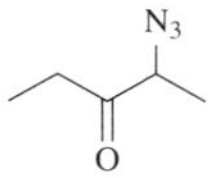 N_3 ketone (30)	331

Carbons	Substrate	Conditions	Product(s) and Yield(s) (%)	Refs.
C_{6-9}	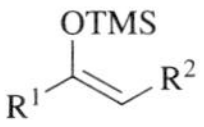 OTMS, R^1, R^2	PhI=NTs (0.67 eq), $CuClO_4$ (3-6 mol %), MeCN	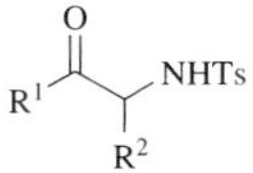 O, R^1, NHTs, R^2 R^1 / R^2 / Temp / Time: n-Bu / H / 0° / 1.5 h / (53) Ph / H / –20° / 3 h / (76) Ph / Me / 0° / 15 min / (58)	173
C_6	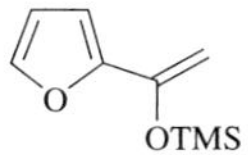 O, OTMS	EtO_2CN=NCO_2Et, PhH, 80°, 8 h	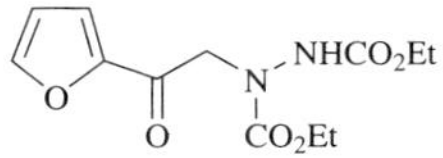 O, O, N, $NHCO_2Et$, CO_2Et (73)	243, 242
C_{6-8}	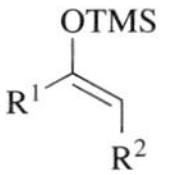 OTMS, R^1, R^2	1. EtO_2N_3, 100° 2. SiO_2	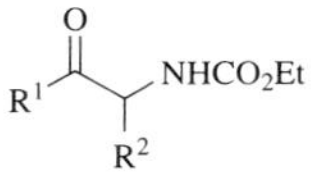 O, R^1, $NHCO_2Et$, R^2 R^1 / R^2: t-Bu / H / (65) n-Pr / Et / (56) Ph / H / (35)	296
C_7	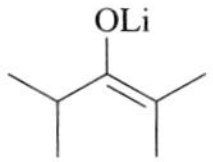 OLi	$PhN_2^+ BF_4^-$ (ia), THF, –78°	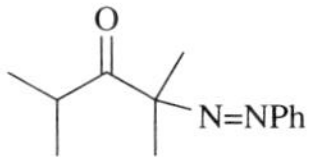 O, N=NPh (72)	185
C_8	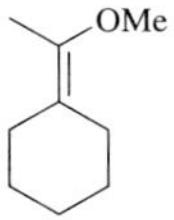 OMe	1. $ClNHCO_2Et$, $CHCl_3$, MeOH, –78° 2. $CrCl_2$ 3. H_2SO_4	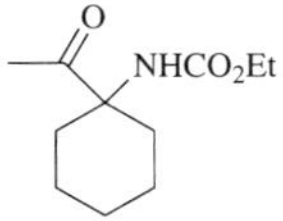 O, $NHCO_2Et$ (59)	343
C_{8-14}	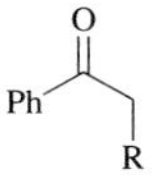 O, Ph, R	1. LiHMDS (ia), THF, –78°, 30 min 2. 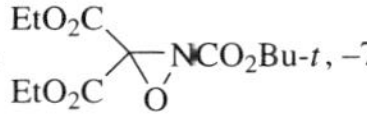 EtO_2C, EtO_2C, O, NCO_2Bu-t, –78°, 20 h; to rt	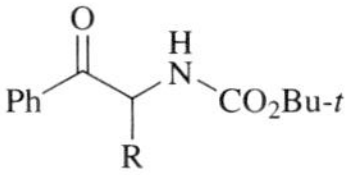 O, Ph, H, N, CO_2Bu-t, R R: H (21) Me (31) Ph (15)	155
C_{8-9}	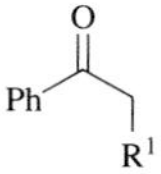 O, Ph, R^1	1. LDA (ia), THF, –78°, 1 h 2. 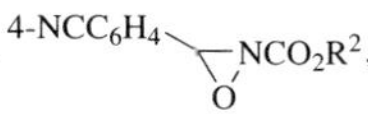 4-NCC_6H_4, O, NCO_2R^2, –78°; to rt, 2-3 h 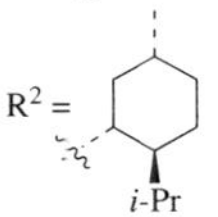 R^2 = i-Pr	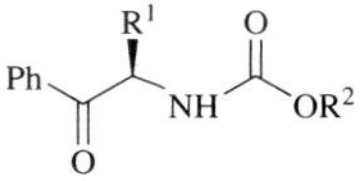 R^1, O, Ph, NH, OR^2, O R^1 / % de: H (59) — Me (62) 5	154

TABLE 7A. ACYCLIC KETONE ENOLATES (*Continued*)

	Substrate	Conditions	Product(s) and Yield(s) (%)	Refs.
C_8	OTMS, Ph	4-ClC_6H_4–oxaziridine–$NCONEt_2$, EtOH, H_2O (3:1), rt, 44 h	$Et_2NC(O)NHCH_2C(O)Ph$ (I) + $HOCH_2C(O)Ph$ (II); I + II (—), I:II = 1:5	155
C_{8-9}	OTMS, Ar, R	PhI=NTs, MeCN, rt	TsNH, R, O, Ar (see below)	172
C_8	OTMS, Ph	1. $EtO_2CN{=}NCO_2Et$, CH_2Cl_2, 0-5°, 3 h; rt, 2 h 2. H_2O, rt, 2 h	EtO_2C–NH–N(CO_2Et)–$CH_2C(O)Ph$ (56)	245
		1. 4-phenyl-1,2,4-triazoline-3,5-dione, CH_2Cl_2, 0-5°, 3 h; rt, 2 h 2. H_2O, rt, 2 h	1-(2-oxo-2-phenylethyl)-4-phenyl-1,2,4-triazolidine-3,5-dione (50)	245
C_{8-9}	OTMS, R, Ph	1. $BnO_2CN{=}NCO_2Bn$ + AgOTf (ia), CH_2Cl_2, –45°, 30 min 2. HF, THF	$RCH[N(CO_2Bn)NHCO_2Bn]C(O)Ph$ (see below)	244
		1. $BnO_2CN{=}NCO_2Bn$ + AgOTf + (*R*)-BINAP (ia), THF, cosolvent, –45°, time 2. HF, THF	BnO_2CHN–N(CO_2Bn)–CH(R)C(O)Ph (see below)	244

Ref. 172:

R	Ar	Time	
H	Ph	4.5 h	(95)
H	4-ClC_6H_4	>24 h	(99)
H	4-$O_2NC_6H_4$	6 h	(85)
H	4-$MeOC_6H_4$	3.5 h	(97)
H	2-MeC_6H_4	18 h	(51)
H	3-MeC_6H_4	11 h	(96)
H	4-MeC_6H_4	1 h	(97)
Me	Ph	—	(94)

Ref. 244 (AgOTf):

R	
H	(91)
Me	(84)

Ref. 244 ((*R*)-BINAP):

R	Cosolvent	Time		% ee
Me	2,4,6-$Me_3C_6H_3$	18 h	(95)	86
Et	—	3 h	(93)	59

C_8

OTMS / Ph

(Saltmen)Mn(N), CH_2Cl_2, pyridine, TFFA, –78° to rt, 3-4 h

CF_3CONH / O / Ph (69)

471

C_9

Ar / O

4-$MeC_6H_4SO_2N(Cl)Na$•x H_2O, L-proline (2 mol%), MeCN, rt

Ar / O / NHTs 0% ee

Ar	Time	
Ph	1 d	(83)
4-FC_6H_4	2 d	(82)
3-$MeOC_6H_4$	1 d	(74)
4-$MeOC_6H_4$	1 d	(71)
3-$CF_3C_6H_4$	2 d	(83)

78

Ph / O

1. Li base, Et_2O or THF
2. Me_2NOMs, –30° to 0°

Ph / O / NMe_2 (52)

134

O / Ph

1. LiHMDS, THF, –78°, 30 min
2. 4-NCC_6H_4–(oxaziridine)–NCO_2Bu-*t*, –78°, 30 min

O / Ph / $NHCO_2Bu$-*t* (36) + O / OH / Ph / C_6H_4CN-4 (25)

153, 157

1. LDA (ia), THF, –78°, 1 h
2. 2-NCC_6H_4–(oxaziridine)–$NCONEt_2$, –78°, 3 h; to rt, 1.5 h

O / Ph / $NHCONEt_2$ (60)

158

OR / Ph

PhI=NTs, $CuPF_6$, ligand (5.5-6 mol%), CH_2Cl_2, –40°

2,6-$Cl_2C_6H_4$ / N / N / $C_6H_3Cl_2$-2,6 **1**

O / N / N / O / Ph / Ph **2**

O / Ph / NHTs

R	Ligand	Time		% ee
Me	**1**	3 h	(87)	9 (R)
Me	**2**	3 h	(76)	10 (S)
Ac	**1**	5 h	(>95)	28 (R)
Ac	**2**	5 h	(61)	52 (R)
TMS	**1**	5 h	(>95)	18 (S)
TMS	**2**	4 h	(90)	16 (S)
TBDMS	**1**	3 h	(>95)	12 (S)
TBDMS	**2**	16 h	(92)	13 (S)

174

TABLE 7A. ACYCLIC KETONE ENOLATES (*Continued*)

Substrate	Conditions	Product(s) and Yield(s) (%)	Refs.
C_9 Ph–CH=C(OR)–CH$_3$	PhI=NTs, $CuPF_6$, **1** or **2** (structures on previous page, 6 mol%), CH_2Cl_2, –40°, 4 h	PhCH(NHTs)C(O)CH$_3$	174
		R / Ligand / % ee: TMS, **1**, (89), 20 (R); TBDMS, **1**, (91), 27 (R); TMS, **2**, (92), 19 (S); TBDMS, **2**, (>95), 21 (S)	
PhC(O)CH$_2$CH$_3$	$EtO_2CN{=}NCO_2Et$, $AlCl_3$, dioxane, rt, 48 h	PhC(O)CH(CH$_3$)N(CO$_2$Et)NHCO$_2$Et (43)	849
$C_{9\text{-}13}$ Ar–C(OTMS)=CH–R	$Cl_3CCH_2O_2CN{=}N$–C(O)–N(oxazolidinone), CF_3CH_2OH (1 eq), **3** (5 mol%), THF; **3** = t-Bu-bis(oxazoline)·Cu(OTf)$_2$	ArC(O)CH(R)N(CO$_2$CH$_2$CCl$_3$)NHC(O)N(oxazolidinone)	252
C_9 Ph–C(OTMS)=CH–CH$_3$	TsN(Cl)Na + OsO_4 (0.004 eq) + $(DHQD)_2PYR$ (0.008 eq) (ia), t-BuOH/H_2O (1:1), rt, 2 h	PhC(O)CH(NHTs)CH$_3$ (45), 85% ee	342

Ar	R	Temp	Time		% ee
Ph	Me	–20°	2 min	(95)	99
4-MeOC$_6$H$_4$	Me	–20°	<1 min	(96)	99
Ph	Et	–20°	30 min	(93)	98
Ph	CH$_2$CH=CH$_2$	–20°	2 h	(92)	97
Ph	*i*-Pr	–20°	3 h	(86)	99
Ph	*i*-Bu	–20°	2 h	(92)	98
4-MeOC$_6$H$_4$	*t*-Bu	–20°	6 h	(84)	98
6-MeO-2-naphthyl	Me	–20°	1 min	(96)	99
4-MeOC$_6$H$_4$	Ph	–50°	13 h	(94)	97
4-MeOC$_6$H$_4$	Bn	–78°	12 h	(94)	99

C_{15}	OTMS	PhI=NTs, MeCN	NHTs (70)	172

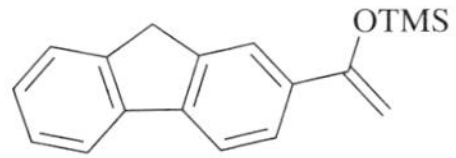

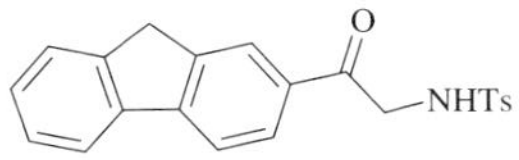

[a] The ee values are those of the crude products; some racemization occurred on silica chromatography.

[b] Some equilibration appears to occur even at –78°.

[c] The ee value was determined in a degradation product.

TABLE 7B. CYCLIC KETONE ENOLATES

Substrate	Conditions	Product(s) and Yield(s) (%)	Refs.
C_5 [structure]	4-$O_2NC_6H_4SO_2ONHCO_2Et$, Et_3N, CH_2Cl_2, rt, 2 h	[structure] $NHCO_2Et$ (12) + [structure] N^+–N^-–CO_2Et (10) + [structure] N–$NHCO_2Et$ (8)	399
$C_{5\text{-}10}$ [structure] $OSi(Pr\text{-}i)_3$, R_1, R^2, n	PhIO (1.5 eq), $TMSN_3$ (3 eq), TEMPO[a] (0.1 eq), solvent, –45°, 16 h	[structure] N_3, $OSi(Pr\text{-}i)_3$, N_3, R_1, R^2, n	850

R^1	R^2	n	Solvent	
H	H	0	toluene	(60) single isomer
H	H	1	CH_2Cl_2	(91) single isomer
—OCH_2O—		1	CH_2Cl_2	(82) 3:1 mixture of isomers
t-Bu	H	1	CH_2Cl_2	(71) single isomer
—O(EtO$_2$C)CH–CH(CO$_2$Et)O—		1	CH_2Cl_2	(67) 4:2:1 mixture of isomers

Substrate	Conditions	Product(s) and Yield(s) (%)	Refs.
C_5 [structure]	1. $MeO_2CCH{=}N_2$, rt, 7 d 2. SiO_2, H_2O	[structure] =$NNHCH_2CO_2Me$ (—)	207
	$(MeO_2C)_2C{=}N_2$, Et_2O, –30°	[structure] =$NNHCH(CO_2Me)_2$ (65)	207

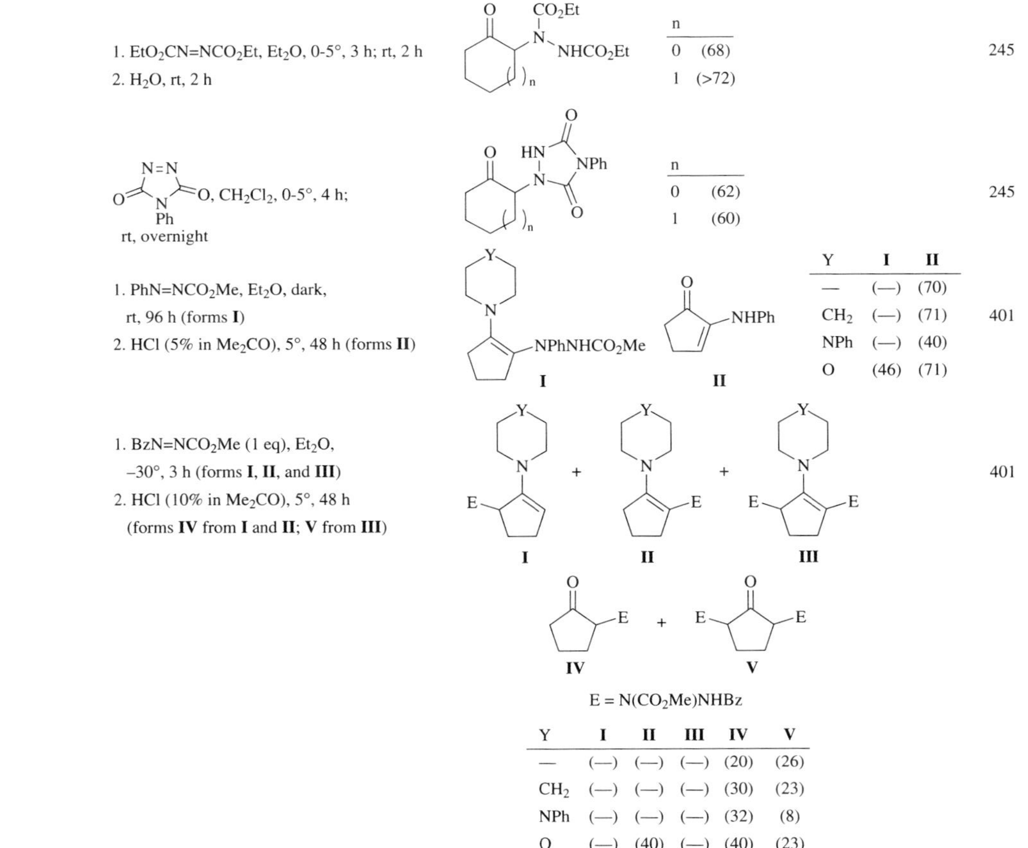

n	
0	(68)
1	(>72)

n	
0	(62)
1	(60)

Y	**I**	**II**
—	(—)	(70)
CH_2	(—)	(71)
NPh	(—)	(40)
O	(46)	(71)

Y	**I**	**II**	**III**	**IV**	**V**
—	(—)	(—)	(—)	(20)	(26)
CH_2	(—)	(—)	(—)	(30)	(23)
NPh	(—)	(—)	(—)	(32)	(8)
O	(—)	(40)	(—)	(40)	(23)

TABLE 7B. CYCLIC KETONE ENOLATES (*Continued*)

Substrate	Conditions	Product(s) and Yield(s) (%)	Refs.
C_5 1-morpholinocyclopentene	PhN=NCONH$_2$, MeOH, 0°; 5°, 2 h	2-[N-Ph-N-(NHCONH$_2$)]-1-morpholinocyclopentene (25) + spiro morpholino triazinone (HN, NH, NPh, O) (25-35)	851
	1. Ar^{1}N=NCOAr^2, solvent, 0°, time 1 2. 10% HCl, Me$_2$CO, temp, time 2	2-[N(Ar^1)NHCOAr^2]cyclopentanone	400

Ar^1	Ar^2	Solvent	Time 1	Temp	Time 2	
Ph	Ph	Et$_2$O	10-15 min	0°	few min	(55)
4-O$_2$NC$_6$H$_4$	Ph	PhH	2 d	rt	24-48 h	(64)
Ph	C$_6$H$_4$NO$_2$-4	PhH	2 d	rt	24-48 h	(63)

Substrate	Conditions	Product(s) and Yield(s) (%)	Refs.
morpholino-dihydrothiopyran (O$_2$S)	1. EtO$_2$CN=NCO$_2$Et, PhH, rt, 72 h 2. HCl, EtOH, H$_2$O, rt, 24h	O$_2$S ketone with N(CO$_2$Et)NHCO$_2$Et (86)	248
NHBu-*n* dihydrothiopyran (O$_2$S)	EtO$_2$CN=NCOR, MeCN, reflux, 3 h	*n*-BuHN, N(CO$_2$Et)NHCOR, O$_2$S	250

R	
OEt	(65)
Ph	(65)

Substrate	Conditions	Product(s) and Yield(s) (%)	Refs.
$C_{5\text{-}9}$ OTMS, ( )$_n$, Y	1. EtO$_2$CN$_3$, 100°, 15 h 2. SiO$_2$	O, ( )$_n$, Y, NHCO$_2$Et	296

Y	n	
CH$_2$	0	(40)
CH$_2$	1	(49)
CHBu-*t*	1	(36)

Substrate	Conditions	Product(s) and Yield(s) (%)	Refs.
C_5 4-OTMS-cyclohexene / dihydropyran (Y)	NaN_3, $Ce(NH_4)_2(NO_3)_6$, MeCN, –20°	3-azido-4-one; Y: — (65); O (47)	331
1-(cyclopentenyl)-2-(methoxymethyl)pyrrolidine	1. EtO_2CN_3, CH_2Cl_2, reflux 2. *h*ν, rt, 4 h	2-($NHCO_2Et$)cyclopentanone (18), 24% ee	303
	MsN=C(OEt)–N_3, CH_2Cl_2, rt, 30 min	2-[NHC(=NMs)OEt]cyclopentanone (25), 18% ee	303
1-(cyclopentenyl)-2,5-bis(methoxymethyl)pyrrolidine	1. MsN=C(OEt)–N_3, CH_2Cl_2, rt, 30 min 2. rt ; **I** converts slowly into **II**	**I** (18), >95% de; **II** 20% ee	303
1-(cyclopentenyl)pyrrolidine	$RCON_3$, CH_2Cl_2, 30 min; R = (camphorsultam N-yl, SO_2)	2-(NHCOR)cyclopentanone (5), 45% de + N-acyl imine of 2-(pyrrolidin-1-yl)cyclopentanone (NCOR) (22), >95% de	304
C_6 1-methoxycyclohexene	1. $ClNHCO_2CH_2CCl_3$, $CHCl_3$, MeOH, –78° 2. $CrCl_2$ 3. MeONa	2-($NHCO_2CH_2CCl_3$)cyclohexanone (86)	343
1-(OR^1)cyclohexene	4-$R^2C_6H_4SO_2ONHCO_2Et$, addend, CH_2Cl_2, rt	2-($NHCO_2Et$)cyclohexanone	

R^1	R^2	Addend	Time	Yield	Refs.
TMS	Me	Cs_2CO_3	24 h	(51)	119
TMS	O_2N	CaO	3.5 h	(67)	122
$O(CH_2)_2OTMS$	O_2N	CaO	2 h	(33)	122

TABLE 7B. CYCLIC KETONE ENOLATES (*Continued*)

Substrate	Conditions	Product(s) and Yield(s) (%)	Refs.
C_6 1-(2-R-pyrrolidin-1-yl)cyclohexene	4-$O_2NC_6H_4SO_2ONHCO_2Et$, Et_3N, CH_2Cl_2, rt, 3 h	2-($NHCO_2Et$)cyclohexanone **I** + 1-($NHCO_2Et$)-2-R-pyrrolidine **II** (—)	121
1-[(R,R)-2-(OTMS)-1,2-R-ethoxy]cyclohexene	4-$O_2NC_6H_4SO_2ONHCO_2Et$, Et_3N, CH_2Cl_2, rt, 3 h	2-($NHCHO_2Et$)cyclohexanone **I** + aziridine (NCC_2Et) **II**	397
1-(2,5-R^1-pyrrolidin-1-yl)-4-R^2-cyclohexene	$ArSO_2ONHCO_2Et$, addend, CH_2Cl_2, rt	2-($NHCO_2Et$)-4-R^2-cyclohexanone **I** + 1-($NHCO_2Et$)pyrrolidine **II** + 1-(cyclohexenyl)pyrrolidinium N^--NCO_2Et **III**	

R	I	% ee
CO_2Bu-*t* (S)	(14)	5
CH_2OTMS (S)	(12)	52
CH_2OMe (S)	(18)	77
CH_2OMe (R)	(21)	50

R	I	% ee	II	% de
Me (R,R)	(24)	68	(0)	—
$MeOCH_2$ (R,R)	(27)	63	(0)	—
Ph (S,S)	(36)	75	(19)	>95
$BnOCH_2$ (S,S)	(28)	60	(0)	—

R^1	R^2	Ar	Addend	Time	I	II	I:II	III	Refs.
H	H	4-$O_2NC_6H_4$	—	2 h	(—)	(—)	49:51	(0)	399
H	H	4-MeC_6H_4	Cs_2CO_3	24 h	(38)	(0)	—	(trace)	118
H	H	4-$O_2NC_6H_4$	Cs_2CO_3	24 h	(35)	(34)	—	(31)	118
Me	H	4-$O_2NC_6H_4$	Et_3N	3 h	(15)	(0)	—	(0)	120
H	Me	4-$O_2NC_6H_4$	Et_3N	3 h	(19)	(0)	—	(0)	120
H	*t*-Bu	4-$O_2NC_6H_4$	Et_3N	3 h	(24)	(0)	—	(0)	120

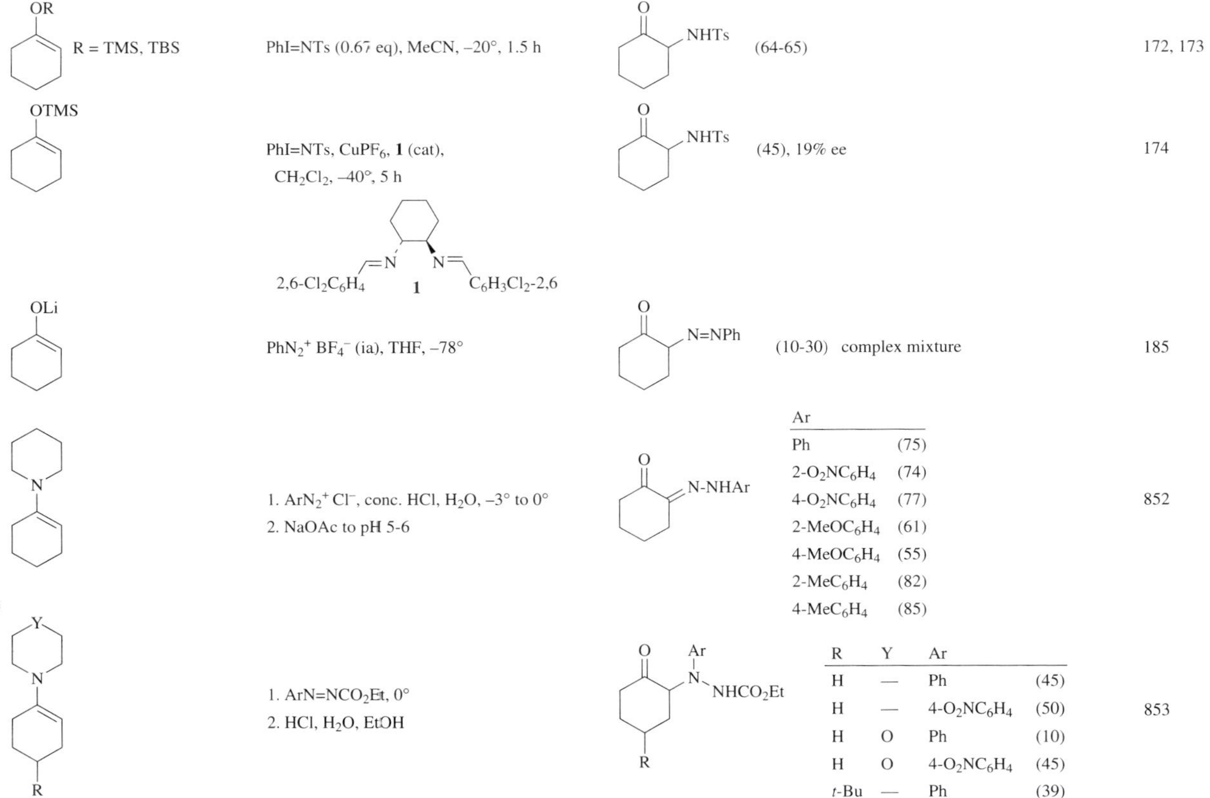

Reactant	Conditions	Product(s) and Yield(s) (%)	Refs.
R = TMS, TBS	PhI=NTs (0.67 eq), MeCN, –20°, 1.5 h	(64-65)	172, 173
OTMS	PhI=NTs, $CuPF_6$, **1** (cat), CH_2Cl_2, –40°, 5 h; 2,6-$Cl_2C_6H_4$ **1** $C_6H_3Cl_2$-2,6	(45), 19% ee	174
OLi	$PhN_2^+ BF_4^-$ (ia), THF, –78°	(10-30) complex mixture	185
	1. $ArN_2^+ Cl^-$, conc. HCl, H_2O, –3° to 0° 2. NaOAc to pH 5-6	N-NHAr: Ph (75); 2-$O_2NC_6H_4$ (74); 4-$O_2NC_6H_4$ (77); 2-$MeOC_6H_4$ (61); 4-$MeOC_6H_4$ (55); 2-MeC_6H_4 (82); 4-MeC_6H_4 (85)	852
C_{6-10}	1. $ArN=NCO_2Et$, 0° 2. HCl, H_2O, EtOH	see below	853

R	Y	Ar	
H	—	Ph	(45)
H	—	4-$O_2NC_6H_4$	(50)
H	O	Ph	(10)
H	O	4-$O_2NC_6H_4$	(45)
t-Bu	—	Ph	(39)

TABLE 7B. CYCLIC KETONE ENOLATES (*Continued*)

Substrate	Conditions	Product(s) and Yield(s) (%)	Refs.
C_{6-10}			
4-R-1-(Y-containing ring N)cyclohexene	1. $EtO_2CN{=}NCOAr$, Et_2O, 0°, 48 h 2. HCl, Me_2CO, H_2O, 48 h	2-[$N(CO_2Et)$-NHCOAr]-4-R-cyclohexanone R / Y / Ar: H / O / Ph $(84)^b$ H / O / 4-$O_2NC_6H_4$ $(36)^b$ *t*-Bu / O / Fh $(41)^b$ *t*-Bu / O / 4-$O_2NC_6H_4$ $(55)^b$ *t*-Bu / — / Fh $(50)^b$ *t*-Bu / — / 4-$O_2NC_6H_4$ $(37)^b$	853
4-R^1-1-morpholinocyclohexene	4-$(R^2)C_6H_4N{=}NCOC_6H_4(R^3)$-4, solvent, 0°, 48 h	2-[$N(C_6H_4(R^2)$-4)-$NCOC_6H_4(R^3)$-4]-4-R^1-1-morpholinocyclohexene R^1 / R^2 / R^3 / Solvent: H / H / H / Et_2O (23) H / H / O_2N / PhH (20) H / O_2N / H / PhH (80) *t*-Bu / H / H / Et_2O (41) *t*-Bu / H / O_2N / PhH (21) *t*-Bu / O_2N / H / PhH (74)	854
4-R-cyclohexanone, R = H, *t*-Bu	BzN=NBz, 100-110°, 20 h	2-[N(Bz)-NHBz]-4-R-cyclohexanone (20-25)	389
4-R-1-(Y-containing ring N)cyclohexene	ArCON=NCOAr, PhH, 7-8°; rt, 24 h	2-[N(COAr)-NHCOAr]-4-R-cyclohexanone Y / R / Ar: — / H / Ph (79) CH_2 / H / Ph (87) O / H / Ph (80) O / H / 4-MeC_6H_4 (81) O / *t*-Bu / Ph (74)	389, 855

C_6

Reactant	Conditions	Product(s) and Yield(s) (%)	Refs.
Cyclohexanone	$EtO_2CN{=}NCO_2Et$, K_2CO_3 or KOAc, 100°, 4 h	2-[N-(CO_2Et)-N'-($NHCO_2Et$)hydrazino]cyclohexanone (28-32)	390
	$RO_2CN{=}NCO_2R$, 100°, 24 h	2-[N-(CO_2R)-N'-($NHCO_2R$)hydrazino]cyclohexanone; R = Me (20)	246
		R = Et (48)	390, 246
1-(Trimethylsilyloxy)cyclohexene (OTMS)	1. $BnO_2CN{=}NCO_2Bn$ + AgOTf (ia), CH_2Cl_2, 0°, 30 min 2. HF, THF	2-[N-(CO_2Bn)-N'-($NHCO_2Bn$)hydrazino]cyclohexanone (91)	244
Cyclohexanone	$RO_2CN{=}NCO_2R$, catalyst (x eq)	2-E-cyclohexanone **I** + 2,6-di-E-cyclohexanone **II**; E = $N(CO_2R)NHCO_2R$	

R	Catalyst	x	Solvent	Temp	Time	I	% ee[c]	II	Refs.
Et	L-proline	0.1	$Cl(CH_2)_2Cl$	rt	23 h	(67)	84	(0)	228
Et	L-proline	0.2	$Cl(CH_2)_2Cl$	rt	44 h	(—)	84	(0)	228
Et	L-proline	0.2	CH_2Cl_2	rt	—	(46)	—	—	229
i-Pr	L-proline	0.2	MeCN	rt	6 h	(—)	59	(10)	228
t-Bu	L-proline	0.2	MeCN	rt	52 h	(—)	59	(10)	228
Bn	DL-proline	0.2	CH_2Cl_2	rt	24 h	(—)	—	(—)	229
Bn	L-proline	0.2	CH_2Cl_2	rt	24 h	(56)	61	(—)	229
Bn	L-azetidinecarboxylic acid	0.2	CH_2Cl_2	40°	—	(—)	6	(—)	229
Bn	L-azetidinecarboxylic acid	0.2	CH_2Cl_2	rt	24 h	60)	90	(—)	229

TABLE 7B. CYCLIC KETONE ENOLATES (*Continued*)

Substrate	Conditions	Product(s) and Yield(s) (%)	Refs.
C_{6-7} 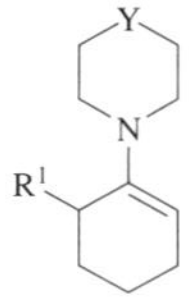	1. $R^2O_2CN{=}NCO_2R^2$ (1 eq), solvent, temp, time 2. HCl	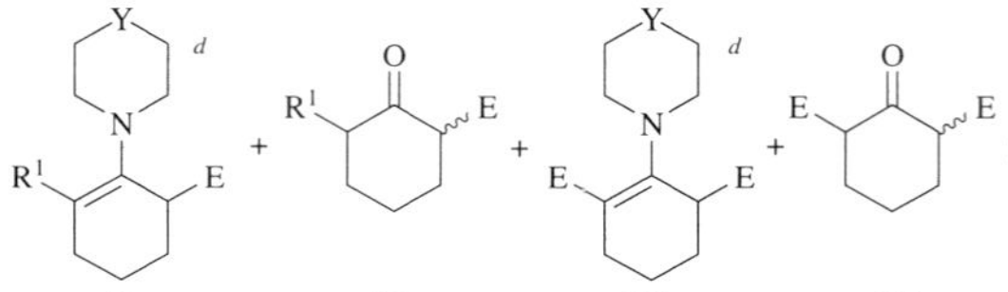 I + II + III + IV [d] $E = N(CO_2R^2)NHCO_2R^2$	246
	1. PhR^2NMnMe (ia), THF, rt, time 2. t-$BuO_2CN{=}NCO_2Bu$-t	I + II $E = N(CO_2Bu$-$t)NHCO_2Bu$-t	388
	t-$BuO_2CN{=}NCO_2Bu$-t, PhH, rt, 24 h	CO_2Bu-t, $NHCO_2Bu$-t (66), 38% ee (S)	

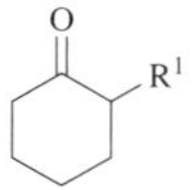

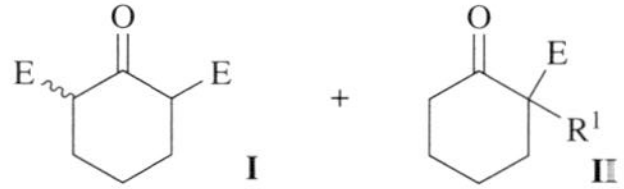

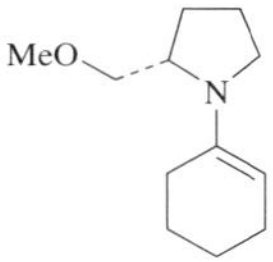

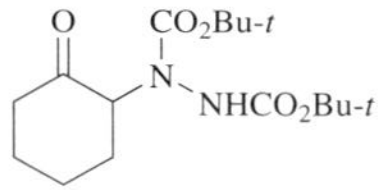

R^1	Y	R^2	Solvent	Temp, Time	I	II	III	IV
H	—	Me	Et_2O	rt, 20 h	(0)	(53)	(12)	(16)
H	—	Et	Et_2O	rt, 15 h	(0)	(38)	(48)	(0)
H	O	Me	Et_2O	rt, 48 h	(0)	(67)	(0)	(19)
H	O	Et	Et_2O	ice bath, then rt, 24 h	(45)	(18)	(0)	(28)
Me	O	Et	PhH	reflux, 4 h	(0)	(79)	(0)	(0)
H	CH_2	Me	Et_2O	ice bath, then rt, 15 h	(56)	(35)	(0)	(0)
H	CH_2	Et	Et_2O	ice bath, then rt, 15 h	(0)	(86)[e]	(0)	(0)

R^1	R^2	Time	I + II	I:II	I dr
H	Me	1 h	(25)	—	—
Me	Me	1 h	(52)	40:60	3:1
Me	n-Bu	0.5 h	(96)	88:12	3:1
Me	n-Bu	1 h	(84)	91:9	3:1
Me[e]	—	—	(52)	12:88	3:1

	NHR[1] cyclohexenone	$EtO_2CN{=}NCOR^2$, Et_2O, –20°					I + II + III	856

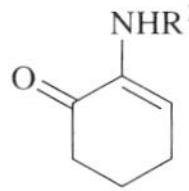

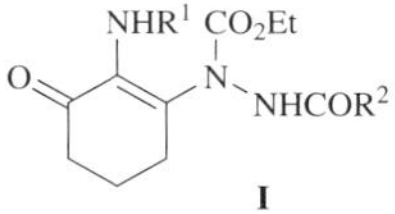

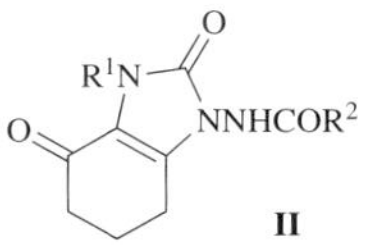

R^1	R^2	Time	**I**	**II**	**III**
n-Bu	EtO	2 h	(—)	(35)	(30)
n-Bu	Ph	2 h	(—)	(35)	(35)
t-Bu	EtO	2 h	(90)	(0)	(0)
t-Bu	Ph	48 h	(—)	(60)	(—)
Ph	EtO	2 h	(90)	(—)	(—)
Ph	Ph	15 min	(25)	(—)	(—)

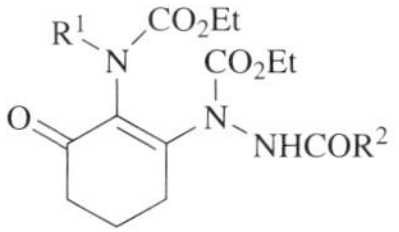

C_6

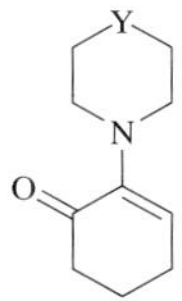

$R^1N{=}NCOR^2$, PhH

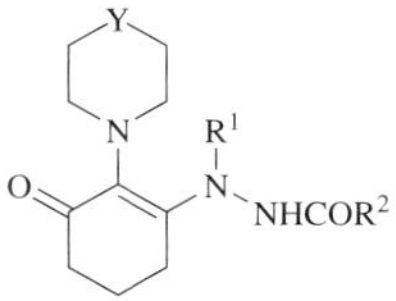

Y	R^1	R^2	Temp, Time		Refs.
—	EtO_2C	OEt	0°, 48-72 h; to rt	$(100)^f$	249
—	EtO_2C	Ph	0°, 48-72 h; to rt	$(58)^f$	249
O	EtO_2C	OEt	0°, 48-72 h; to rt	$(100)^f$	249
O	EtO_2C	Ph	0°, 48-72 h; to rt	$(100)^f$	249
O	EtO_2C	$C_6H_4NO_2$-4	0°, 48-72 h; to rt	$(100)^f$	249
O	4-$O_2NC_6H_4$	$C_6H_4NO_2$-4	reflux	(95)	857

TABLE 7B. CYCLIC KETONE ENOLATES (*Continued*)

Substrate	Conditions	Product(s) and Yield(s) (%)	Refs.
C_{6-10}			
OR[1], R[2], R[3], R[4], R[5] (cyclohexenyl ether)	NaN_3, $Ce(NH_4)_2(NO_3)_6$, MeCN, –20°	2-azido ketone (R[2], R[3], N$_3$, R[4], R[5])	331

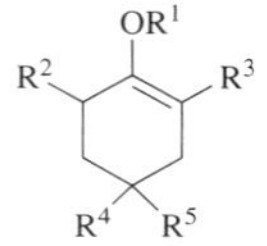

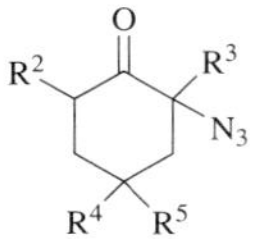

R[1]	R[2]	R[3]	R[4]	R[5]	Yield (%)
$SiMe_3$	H	H	H	H	(38)
$Si(Pr\text{-}i)_3$	H	H	H	H	(72)
$Si(Pr\text{-}i)_3$	Me	H	H	H	(56)[g]
$Si(Pr\text{-}i)_3$	H	Me	H	H	(49)
$Si(Pr\text{-}i)_3$	H	H	Me	H	(81)[g]
$Si(Pr\text{-}i)_3$	H	H	—$O(CH_2)_2O$—		(59)
$Si(Pr\text{-}i)_3$	H	*t*-Bu	H	H	(65)[g]

Substrate	Conditions	Product(s) and Yield(s) (%)	Refs.
C_6			
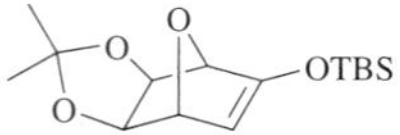 OTBS (bicyclic acetonide enol ether)	NaN_3, $Ce(NH_4)_2(NO_3)_6$, MeCN, –15°, 2 h; to rt	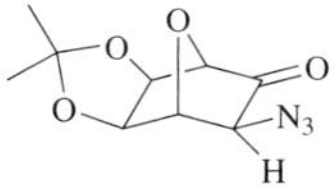 N_3, H (azido ketone) (70)	297
R-pyrrolidinyl cyclohexene enamine	1. EtO_2CN_3, CH_2Cl_2, reflux 2. *hν* 3. SiO_2	2-($NHCO_2Et$)cyclohexanone	
		R = $CO_2Bu\text{-}t$ (48) 3 (S) % ee	302
		R = CH_2OTMS (51) 35 (S) % ee	302
		R = CH_2OMe (40) 18 (S) % ee	302, 303
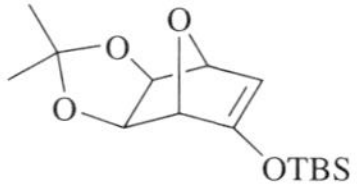 OTBS (bicyclic acetonide enol ether)	1. *t*-BuO_2CN_3, 60°, 36 h (forms **I**) For **II** from **I**: 1. *hν*, MeCN, 0°, 30 min 2. AcOH, then *n*-Bu_4N^+ F^-	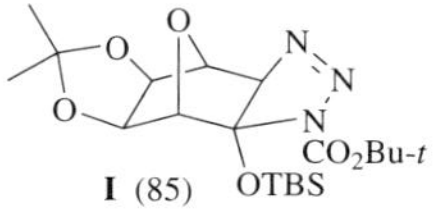 **I** (85) 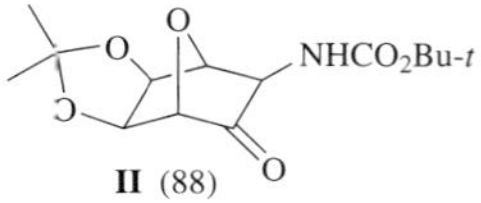 (triazoline, $CO_2Bu\text{-}t$, OTBS) + **II** (88) ($NHCO_2Bu\text{-}t$ ketone)	297
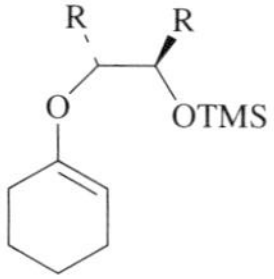 R, R, OTMS (cyclohexenyl ether)	EtO_2CN_3, CH_2Cl_2, 120°, 4 h	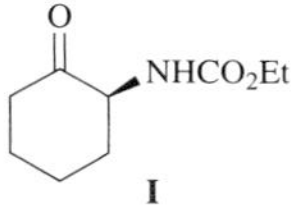 **I** (2-$NHCO_2Et$ cyclohexanone) + **II** ($NHCO_2Et$ dioxolane ketal)	397

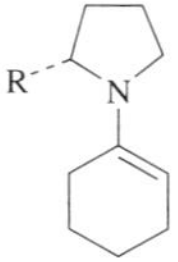

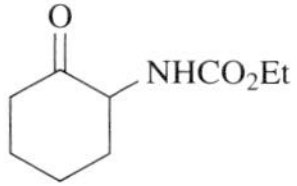

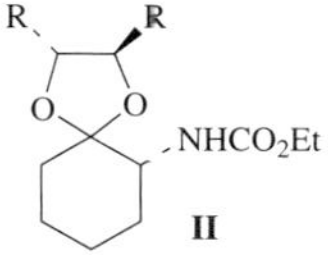

R	I	% ee	II	% de
Me (R,R)	(—)	—	(14)	69
CH_2OMe (R,R)	(—)	—	(14)	63
CH_2OBn (S,S)	(10)	76	(—)	—
Ph (S,S)	(—)	—	(13)	62

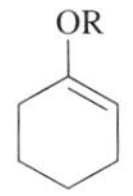

OR

(camphorsultam)-$NCON_3$, CH_2Cl_2, $h\nu$, 7 h

(camphorsultam)-NCONH-(2-oxocyclohexyl)

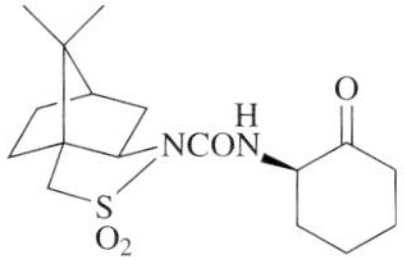

R		% de
TMS	(61)	60
Me	(53)	60

304

$C_{6\text{-}10}$

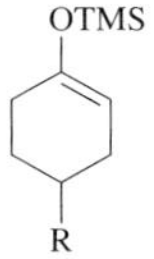

OTMS, R

[TsN(Cl)Na + OsO_4 (0.004 eq) + catalyst (0.008 eq)](ia), t-BuOH/H_2O (1:1), rt, 15 min

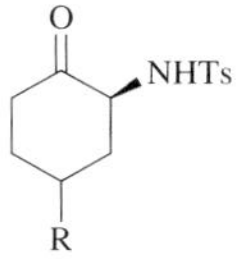

O, NHTs, R

R	Catalyst		% ee
H	$(DHQD)_2CLB$	(35)	92
Me	$(DHQD)_2PYR$	(40)	86
t-Bu	$(DHQD)_2CLB$	(34)	76

342

$C_{6\text{-}12}$

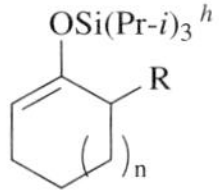

$OSi(Pr\text{-}i)_3$ [h], R, $( )_n$

"TsN=Se=NTs"(ia), CH_2Cl_2, 0° to rt, 3 h

O, R, NHTs, $( )_n$

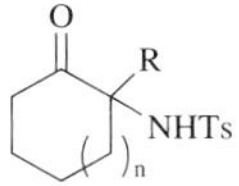

n	R	
0	Me	(23)
1	Me	(39)
1	Ph	(36)

348, 347

$C_{6\text{-}8}$

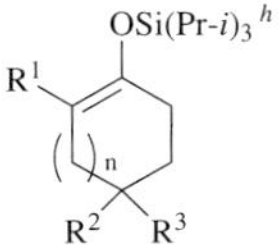

$OSi(Pr\text{-}i)_3$ [h], R^1, $( )_n$, R^2, R^3

"TsN=Se=NTs"(ia), CH_2Cl_2, 0° to rt, 3 h

$OSi(Pr\text{-}i)_3$, R^1, NHTs, $( )_n$, R^3, R^4 **I** + $OSi(Pr\text{-}i)_3$, TsNH **II**

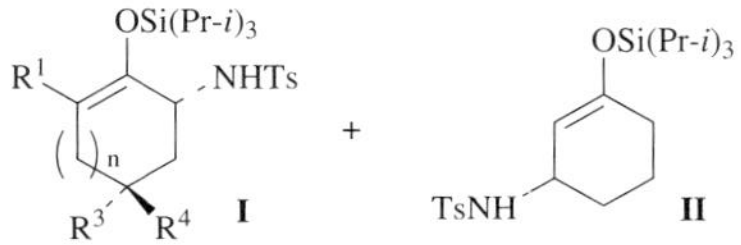

n	R^1	R^2	R^3	I	II
0	Me	H	H	(38)	(0)
1	H	H	H	(39)	(11)
1	Me	H	H	(49)	(0)
1	H	H	Me	(51)	(0)
1	H	Me	Me	(37)	(0)

348, 347

TABLE 7B. CYCLIC KETONE ENOLATES (*Continued*)

Substrate	Conditions	Product(s) and Yield(s) (%)	Refs.
C_7 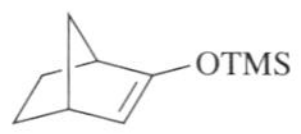 OTMS	1. $ClNHCO_2Bn$, MeOH, $CHCl_3$, –78° 2. $CrCl_2$ 3. MeOH	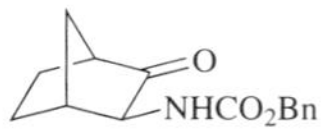 O, $NHCO_2Bn$ (19)	343
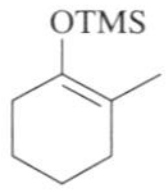 OTMS	$PhN_2^+BF_4^-$ (ia), THF, –78°	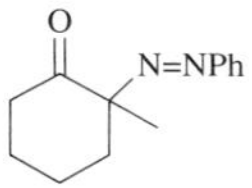 O, N=NPh (61)	185
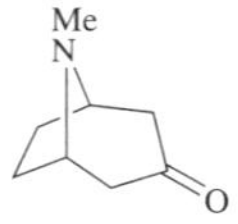 Me N, O	1. LDA, THF, –78°, 45 min 2. $EtO_2CN{=}NCO_2Et$, –78°, 30 min	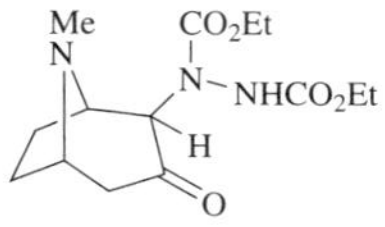 Me N, CO_2Et N $NHCO_2Et$, H, O (80)	391
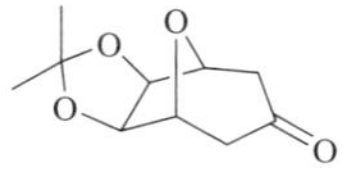 O, O, O, O	1. LDA, THF, hexane, –78°, 1 h 2. t-$BuO_2CN{=}NCO_2Bu$-t, CH_2Cl_2, –78°, 30 min	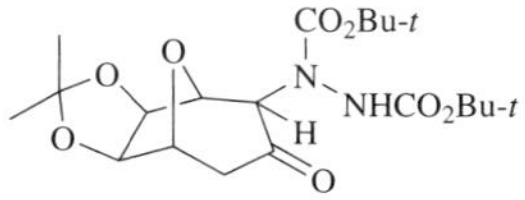 CO_2Bu-t N $NHCO_2Bu$-t, H, O (66)	349
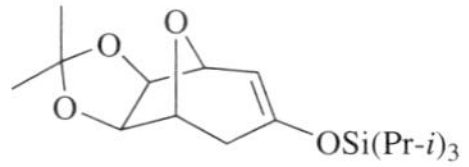 O, O, O, $OSi(Pr$-$i)_3$	"TsN=Se=NTs", CH_2Cl_2, rt, 2 h	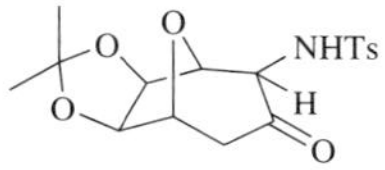 NHTs, H, O (81)	349
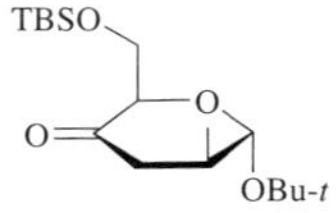 TBSO, O, O, OBu-t	1. LDA, THF, –78° 2. $BnO_2CN{=}NCO_2Bn$, –78°	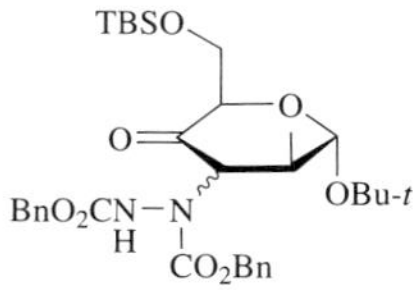 TBSO, O, O, $BnO_2CN(H)$–N(CO_2Bn), OBu-t (74)	217
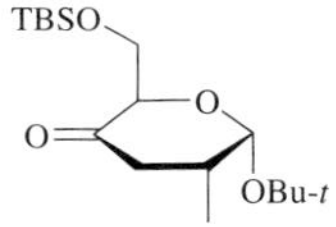 TBSO, O, O, OBu-t	1. LDA, THF, –78° 2. $BnO_2CN{=}NCO_2Bn$, –78°	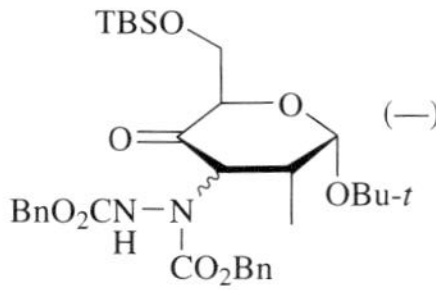 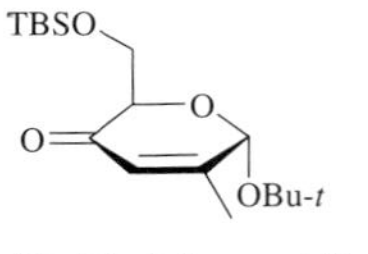 TBSO, O, O, $BnO_2CN(H)$–N(CO_2Bn), OBu-t (—) + TBSO, O, O, OBu-t ("substantial amounts")	217

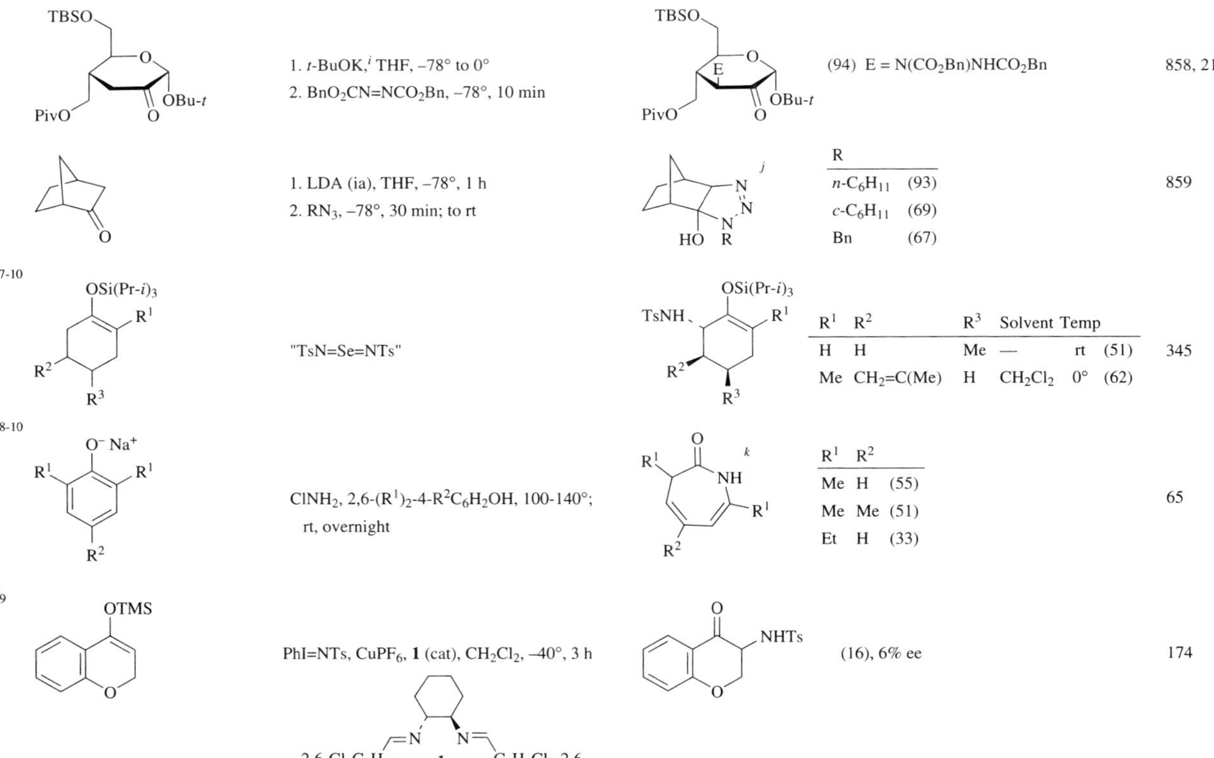

	Substrate	Conditions	Product(s) and Yield(s) (%)	Refs.
	TBSO, PivO, OBu-t	1. t-BuOK,[i] THF, –78° to 0° 2. $BnO_2CN{=}NCO_2Bn$, –78°, 10 min	TBSO, E, PivO, OBu-t (94) E = $N(CO_2Bn)NHCO_2Bn$	858, 217
		1. LDA (ia), THF, –78°, 1 h 2. RN_3, –78°, 30 min; to rt	HO, R [j] R: n-C_6H_{11} (93); c-C_6H_{11} (69); Bn (67)	859
C_{7-10}	$OSi(Pr$-$i)_3$, R^1, R^2, R^3	"TsN=Se=NTs"	TsNH, $OSi(Pr$-$i)_3$, R^1, R^2, R^3	345
C_{8-10}	$O^-\ Na^+$, R^1, R^1, R^2	$ClNH_2$, 2,6-$(R^1)_2$-4-$R^2C_6H_2OH$, 100-140°; rt, overnight	R^1, NH, R^1, R^2 [k]	65
C_9	OTMS	PhI=NTs, $CuPF_6$, **1** (cat), CH_2Cl_2, –40°, 3 h 2,6-$Cl_2C_6H_3$ **1** $C_6H_3Cl_2$-2,6	NHTs (16), 6% ee	174

R		
n-C_6H_{11}	(93)	
c-C_6H_{11}	(69)	
Bn	(67)	

R^1	R^2	R^3	Solvent	Temp	
H	H	Me	—	rt	(51)
Me	$CH_2{=}C(Me)$	H	CH_2Cl_2	0°	(62)

R^1	R^2	
Me	H	(55)
Me	Me	(51)
Et	H	(33)

TABLE 7B. CYCLIC KETONE ENOLATES (*Continued*)

Substrate	Conditions	Product(s) and Yield(s) (%)	Refs.
$C_{9\text{-}11}$ OTMS, R, n	$Cl_3CCH_2O_2CN{=}N$–C(O)–oxazolidinone, CF_3CH_2OH (1 eq), **2** (10 mol%), THF; **2**: t-Bu, Cu $(OTf)_2$, Bu-t	O, $CO_2CH_2CCl_3$, N, H, N, R, n, O, N, O	252
		R / n / Temp / Time / % ee: H 1 –78° 12 h (90) 21; Me 1 –78° 6 h (88) 96; H 2 –20° 0.5 h (51) 90; H 3 –78° overnight (94) 99	
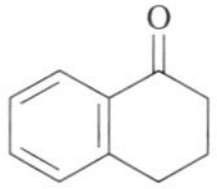 C_{10} O	1. *n*-BuLi, hexane, THF, 0°, 30 min 2. 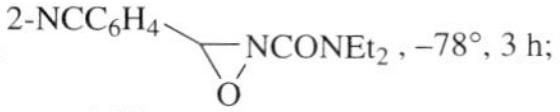 2-NCC_6H_4–oxaziridine–$NCONEt_2$, –78°, 3 h; to rt, 1.5 h	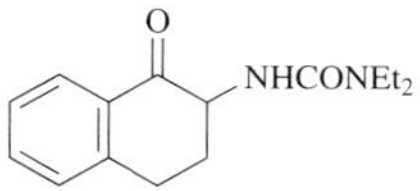 O, $NHCONEt_2$ (59)	158
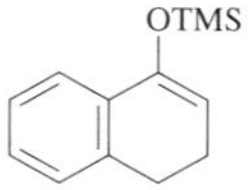 OTMS	PhI=NTs (0.67 eq), $CuClO_4$ (3-6 mol%), MeCN, 0°, 90 min	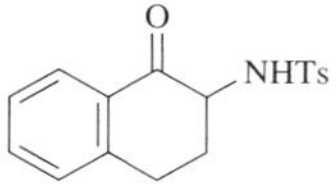 O, NHTs (53)	173
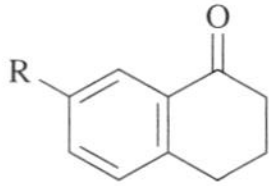 R, O	1. LDA, THF, hexane, –78° 2. $BnO_2CN{=}NCO_2Bn$, –78°, 3 min	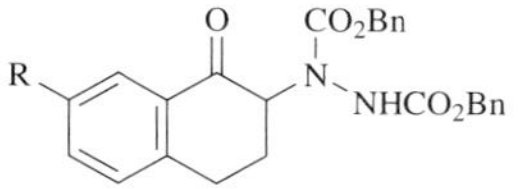 R, O, CO_2Bn, N, $NHCO_2Bn$; R: H (87), MeO (57)	393, 392
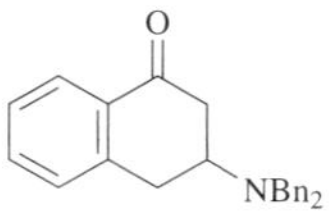 O, NBn_2	1. LDA, THF, HMPA, –78°, 30 min 2. $BnO_2CN{=}NCO_2Bn$, –78°, 1 h; rt, 24 h	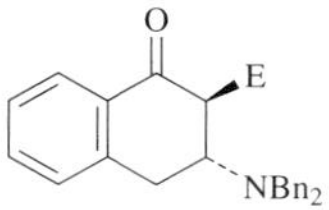 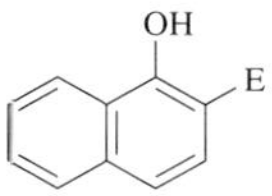 O, E, NBn_2 (45-60) + OH, E (11-28) E = $N(CO_2Bn)NHCO_2Bn$	860

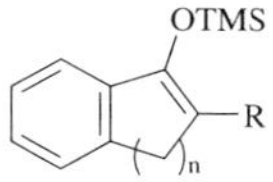

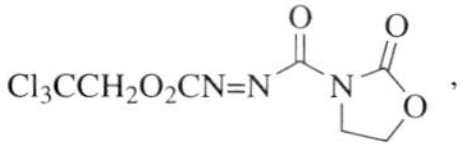

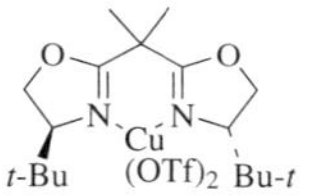

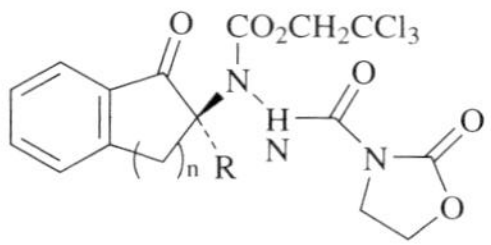

Substrate	Conditions	Product(s) and Yield(s) (%)	Refs.
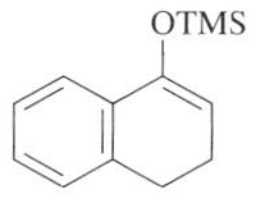 OTMS (1-(trimethylsilyloxy)-3,4-dihydronaphthalene)	1. $BnO_2CN{=}NCO_2Bn$ + $AgClO_4$ + (*R*)-BINAP + THF (ia), –45°, 5 h 2. HF, THF	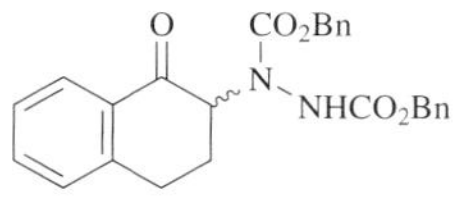 O, CO_2Bn, N, $NHCO_2Bn$ (82), 65% ee	244
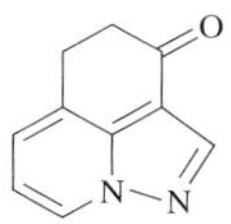 O, N–N	1. LiHMDS, THF, –78°, 30 min 2. $BnO_2CN{=}NCO_2Bn$, –78°, 3 min	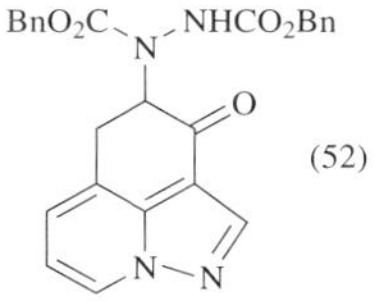 BnO_2C, N, $NHCO_2Bn$, O, N–N (52)	395, 396
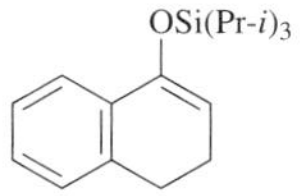 $OSi(Pr\text{-}i)_3$	PhIO (1.5 eq), $TMSN_3$ (3 eq), TEMPO[a], CH_2Cl_2, –45°, 16 h	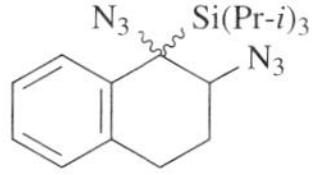 N_3, $Si(Pr\text{-}i)_3$, N_3 (41), 4:1 mixture of isomers	850
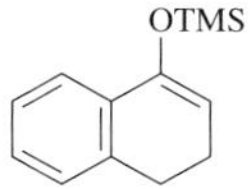 OTMS	TsN(Cl)Na, $(DHQD)_2CLB$ (0.008 eq), OsO_4 (0.004 eq), *t*-BuOH/H_2O (1:1), 10 min	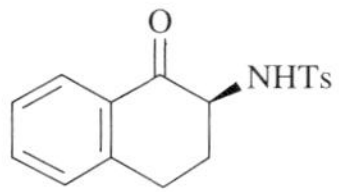 O, NHTs (28), 76% ee	342
$C_{10\text{-}13}$			
R, $OSi(Pr\text{-}i)_3$	TsN=Se=NTs, CH_2Cl_2, 0° to rt, 40 h	R, $OSi(Pr\text{-}i)_3$, NHTs R: H (71); $CH_2{=}CHCH_2$ (59)	348, 346
C_{10}			
OTMS, OMe	(Saltmen)Mn(N⁻, CH_2Cl_2, pyridine, TFFA, –78° to rt, 3-4 h	O, $NHCOCF_3$, OMe (78)	471

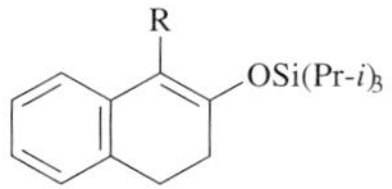

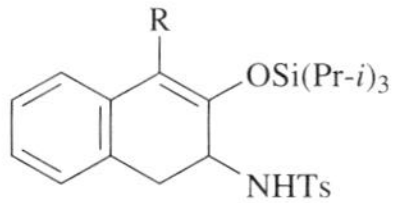

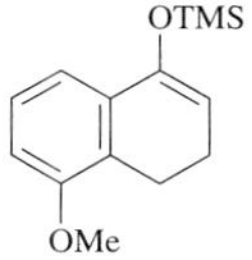

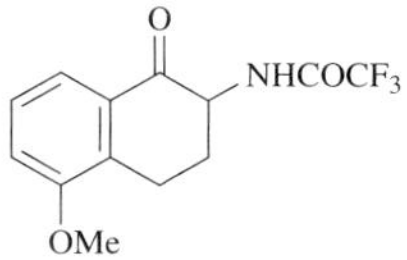

TABLE 7B. CYCLIC KETONE ENOLATES (*Continued*)

	Substrate	Conditions	Product(s) and Yield(s) (%)	Refs.
C_{10} 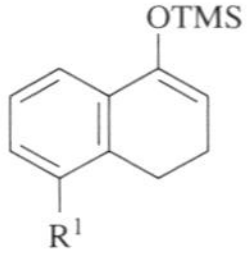	OTMS, R^1	1. **3** (2.2 eq), pyridine, CH_2Cl_2, –78° 2. TFFA, –78°, 6 h; to rt 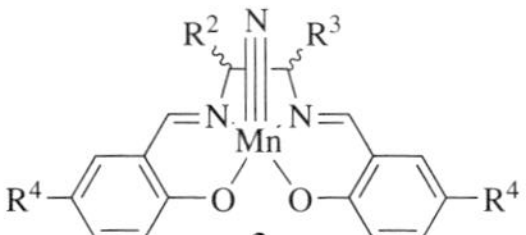 R^2, N, R^3, N, Mn, N, O, O, R^4, R^4, **3**	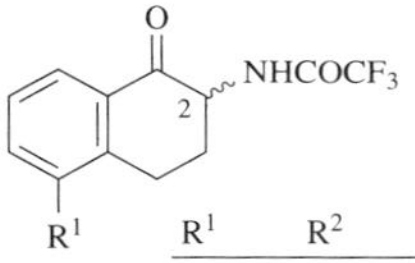 O, NHCOCF$_3$, 2, R^1	353
	OTMS x eq	**4** (y eq), R_2O, pyridine, CH_2Cl_2, addend N, H, H, N, Mn, N, O, O, **4**	O, NHR	352
	OTMS	1. **5** (2 eq) + pyridine (ia), CH_2Cl_2, –78° 2. TFFA, –78° to rt, 3-4 h N, N, Mn, N, O, O, **5**	NHCOCF$_3$, H, O (55)	471
C_{11}	OSi(Pr-*i*)$_3$, O, O, H	NaN_3, $Ce(NH_4)_2(NO_3)_6$, MeCN, –20°	O, N_3, O, O, H (60)	331

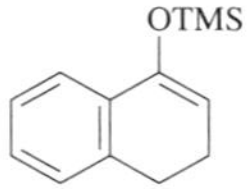

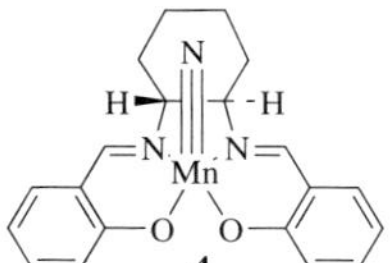

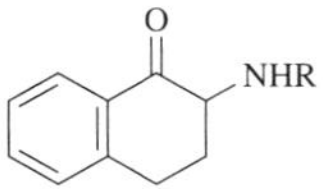

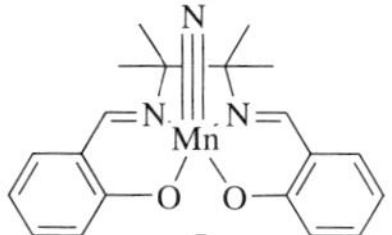

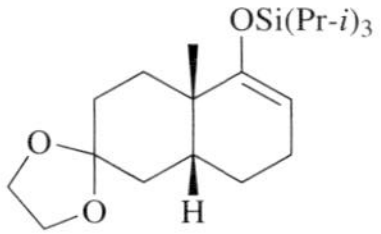

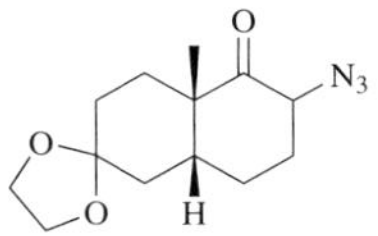

R^1	R^2	R^3		R^4		% ee C2
H	—$(CH_2)_4$—		(R,R)	H	(83)	75 (R)
H	Ph	Ph	(S,S)	*t*-Bu	(74)	65 (S)
MeO	—$(CH_2)_4$—		(R,R)	H	(58)	62 (R)
MeO	Ph	Ph	(S,S)	H	(55)	41 (S)

x	y	R	Addend	Temp	Time		% ee
10	1	Ts	pyridine *N*-oxide	0°	6 h	(76)	48
1	2	CF_3CO	—	–78° to rt	3 h	(58)	79

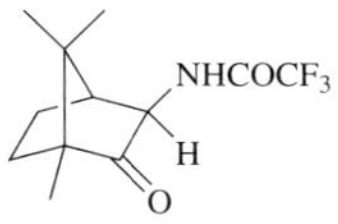

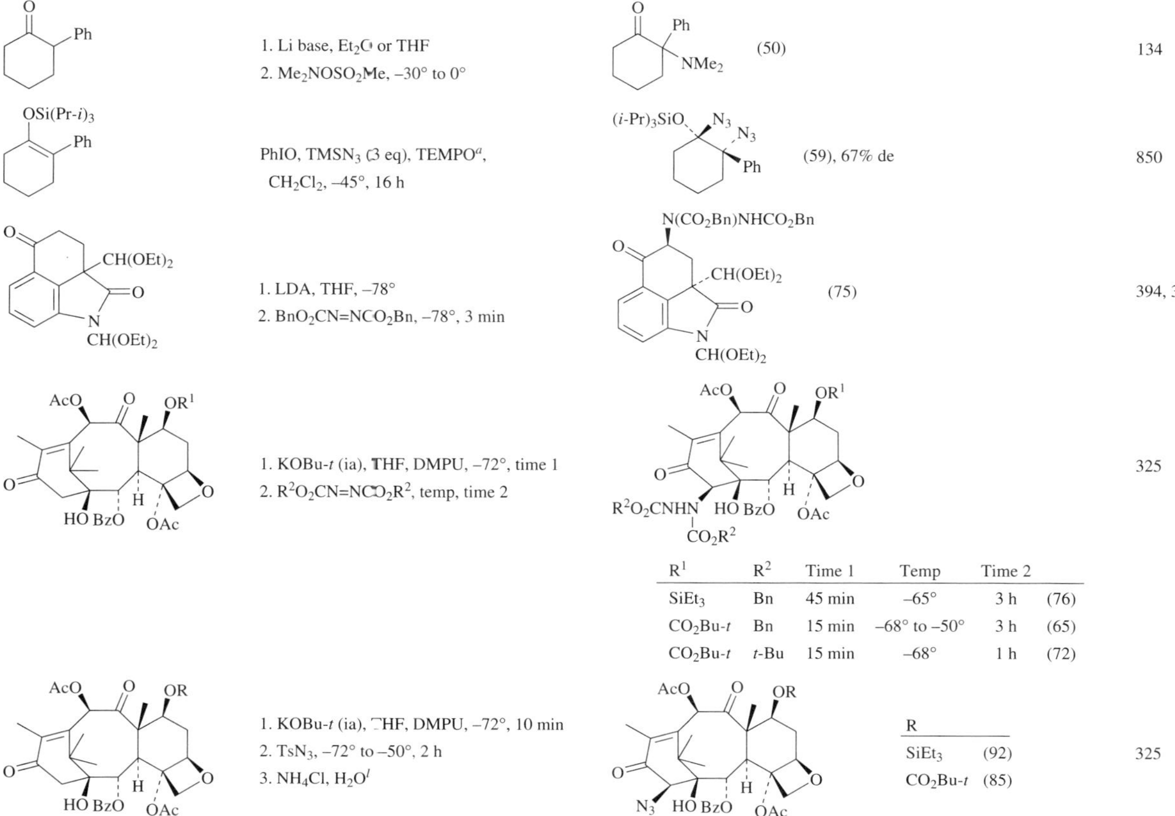

C$_{12}$

1. Li base, Et_2O or THF
2. Me_2NOSO_2Me, –30° to 0°

(50) 134

PhIO, $TMSN_3$ (3 eq), TEMPO[a], CH_2Cl_2, –45°, 16 h

(59), 67% de 850

1. LDA, THF, –78°
2. $BnO_2CN{=}NCO_2Bn$, –78°, 3 min

(75) 394, 392

C$_{20}$

1. KOBu-*t* (ia), THF, DMPU, –72°, time 1
2. $R^2O_2CN{=}NCO_2R^2$, temp, time 2

325

R^1	R^2	Time 1	Temp	Time 2	
$SiEt_3$	Bn	45 min	–65°	3 h	(76)
CO_2Bu-*t*	Bn	15 min	–68° to –50°	3 h	(65)
CO_2Bu-*t*	*t*-Bu	15 min	–68°	1 h	(72)

1. KOBu-*t* (ia), THF, DMPU, –72°, 10 min
2. TsN_3, –72° to –50°, 2 h
3. NH_4Cl, H_2O[l]

R	
$SiEt_3$	(92)
CO_2Bu-*t*	(85)

325

TABLE 7B. CYCLIC KETONE ENOLATES (*Continued*)

Substrate	Conditions	Product(s) and Yield(s) (%)	Refs.
C_{20}			
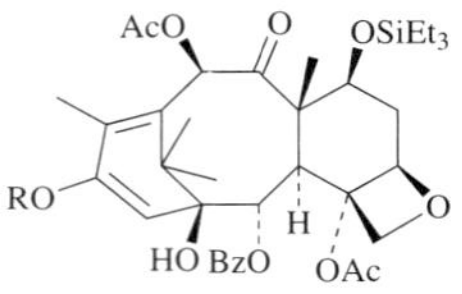 AcO, O, $OSiEt_3$, RO, HO BzO, H, O, OAc	NaN_3, $Ce(NH_4)_2(NO_3)_6$, MeCN, 0°	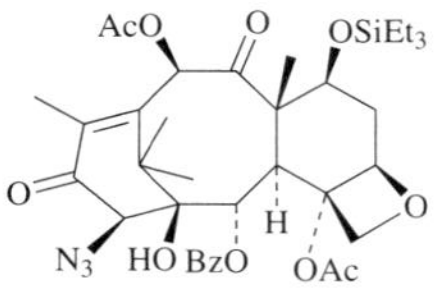 AcO, O, $OSiEt_3$, O, N_3, HO BzO, H, O, OAc R — Time Me_3Si — 1 h (95) $(i\text{-}Pr)_3Si$ — 1 d (55)	325

[a] TEMPO was added to suppress formation of the β-azido enol ether by a radical mechanism.

[b] The yield is that of the product before acid hydrolysis; the latter is reported to proceed in almost quantitative yield.

[c] The ee values are those of the crude products; some racemization occurred during chromatography on silica.

[d] This product was isolated before acid treatment.

[e] The substrate was the lithium enolate prepared with LDA at 0° to room temperature for one hour.

[f] The product is very resistant to acid hydrolysis.

[g] The product is a mixture of cis and trans isomers.

[h] TMS and TBDMS enol ethers gave negligible amounts of products.

[i] With LDA, lithium tetramethylpiperidide, or LiHMDS, reduction of the ketone occurred.

[j] With diethyl ketone, cyclohexanone, and various cyclohexenones, elimination of water occurred to give the triazole derivatives.

[k] The intermediate amination product ring expands to the tropolone product under the reaction conditions.

[l] The diazo compounds were formed when the reactions were quenched with acetic acid.

TABLE 8. IMINE AND HYDRAZONE ANIONS

Substrate	Conditions	Product(s) and Yield(s) (%)	Refs.
$C_{5\text{-}15}$ Me_2N–N=C(R^1)CH_2R^2	1. LDA, THF, 0°, 4-6 h 2. t-BuO_2CN=NCO_2Bu-t, –78°, 2-5 min	Me_2N–N=C(R^1)CH(R^2)N(CO_2Bu-t)$NHCO_2Bu$-t R^1 / R^2 Et / Me (85) n-Pr / Et (75) —$(CH_2)_4$— (78) Ph / Me (66) Bn / Ph (63)	327
	1. LDA, THF, 0°, 4-6 h 2. 2,4,6-(i-Pr)$_3C_6H_2SO_2N_3$, –78°, 2-5 min 3. NH_4Cl, H_2O	Me_2N–N=C(R^1)CH(R^2)$NHSO_2C_6H_2$(Pr-i)$_3$-2,4,6 R^1 / R^2 Et / Me (84) n-Pr / Et (78) —$(CH_2)_4$— (66) Ph / Me (69) Bn / Ph (65)	327
C_6 cyclohexylidene imine of R^1C(H)R^2–N[a]	1. Unspecified K base, DME, CH_2Cl_2, rt, 1.5 h 2. 4-$O_2NC_6H_4SO_2ONHCO_2Et$ (3 eq), LiOH (10 eq), –70°, 3 h; to rt 3. H_3O^+	2-($NHCO_2Et$)cyclohexanone **I** + 2-[N(CO_2Et)$NHCO_2Et$]cyclohexanone **II** R^1 / R^2 / I / % ee / II / % ee Ph / H / (32) / — / — / — Me / Ph / (25) / 34 / (10) / 8 $MeOCH_2$ / Bn / (10) / 36 / (5) / 9	123
	1. t-BuO_2CN=NCO_2Bu-t, toluene, reflux, 2-5 h 2. HO_2CCO_2H	2-[N(CO_2Bu-t)$NHCO_2Bu$-t]cyclohexanone R^1 / R^2 / / % ee Ph / H / (45) / — Me / Ph / (30) / — $MeOCH_2$ / Bn / (—) / 5	123

TABLE 8. IMINE AND HYDRAZONE ANIONS (*Continued*)

Substrate	Conditions	Product(s) and Yield(s) (%)	Refs.
C_6 R^1, H, R^2 [a] N=cyclohexylidene	1. LDA, HMPA, THF, 0°, 90 min 2. t-$BuO_2CN{=}NCO_2Bu$-t, 0°; to rt, time 3. HO_2CCO_2H	2-[N-(CO_2Bu-t)-N-$NHCO_2Bu$-t]cyclohexanone	123
$C_{6\text{-}13}$ t-$Bu\overset{*}{S}(O)$–N=C(CH_2R^1)$CH(R^2)R^3$	1. PhMeNMnMe•4 LiBr, THF, rt, 1 h 2. $R^4O_2CN{=}NCO_2R^4$, –30°; rt, 2.5 h 3. HCl	R^1, R^3, R^2, R^4O_2C–N–$NHCO_2R^4$ (**I**) + R^1, R^3, R^2, N–$NHCO_2R^4$, R^4O_2C (**II**)	388
$C_{9\text{-}10}$ Ph–C(=NR^2)–CH_2R^1	$EtO_2CN{=}NCO_2Et$, hexane, rt, 2 h	Ph–C(=NR^2)–CH(R^1)–N(CO_2Et)–$NHCO_2Et$	849

R^1	R^2	Time		% de
Ph	H	3 h	(23)	—
Me	Ph	4 h	(19)	29
$MeOCH_2$	Bn	8 h	(28)	33

R^1	R^2	R^3	*	R^4	I	I:II	I % ee
Me	Me	Me	R	Et	(50)	90:10	40
n-Pr	H	n-Pr	R	Et	(50)	—	65
n-C_5H_{11}	Me	Me	R,S	t-Bu	(65)	98:2	—
n-C_5H_{11}	Me	Me	R	t-Bu	(65)	98:2	68
Et	Et	Bn	R,S	t-Bu	(50)	99:1	—

R^1	R^2	
Me	Ph	(59)
Me	4-MeC_6H_4	(—)
Et	Ph	(—)
Et	2-MeC_5H_4	(—)
Et	4-MeC_5H_4	(—)

C_{10}

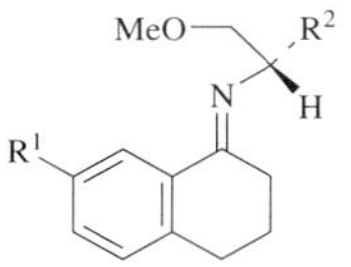

MeO, R^2, N, H, R^1

1. LDA, THF, hexane, –45°, 1.25 h
2. $R^3O_2CN{=}NCO_2R^3$, –78°, 3 min

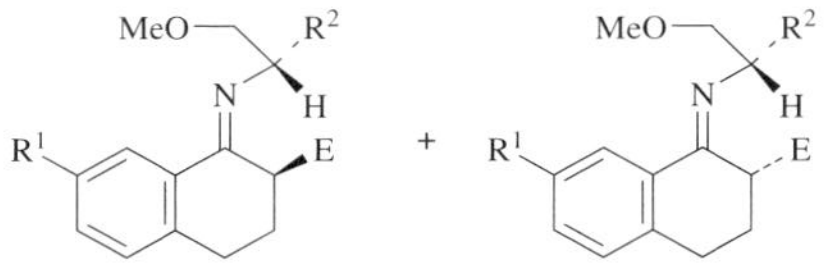

I + **II**

$E = N(CO_2R^3)NHCO_2R^3$

405

R^1	R^2	R^3	**I + II**	**I** % de
H	Bn	*t*-Bu	(86)	72
H	*i*-Bu	*t*-Bu	(82)	76
MeO	*i*-Pr	*t*-Bu	(—)	38
H	Bn	Bn	(65)	40

$R^3O_2CN{=}NCO_2R^3$, THF

I + II

405

R^1	R^2	R^3	Temp	Time	**I + II**	**I** % de
H	Bn	*t*-Bu	rt	24 h	(84)	74
H	*i*-Bu	*t*-Bu	rt	24 h	(85)	64
H	*i*-Pr	*t*-Bu	rt	24 h	(85)	58
H	*t*-Bu	*t*-Bu	rt	48 h	(85)	74
MeO	*i*-Pr	*t*-Bu	rt	20 h	(77)	—
H	Bn	Bn	0°	1 h	(72)	40

[a] This is a corrected structure; personal communication from L. Pellacani, Dipartimento di Chimica, Universitá "La Sapienza", Rome, Italy.

TABLE 9. CARBOXYLIC ACID DIANIONS

Substrate	Conditions	Product(s) and Yield(s) (%)	Refs.
C_{4-10} $R^1CH_2CO_2H$	1. LDA (2.2 eq), THF, 0° 2. R^2NH_2, –10° to rt, 1 h; rt to 30°, 2-5 h	$R^1CH(NH_2)CO_2H$	

R^1	R^2	Yield (%)	Refs.
$MeSCH_2CH_2$	MeO	(—)	861
i-Pr	Cl	(8)	79
i-Pr	$HOSO_2O$	(trace)	79
i-Pr	MeO	(34)	79
i-Pr	EtO	(22)	79
i-Pr	*i*-PrO	(25)	79
i-Pr	*t*-BuO	(13)	79
i-Pr	BnO	(trace)	79
i-Pr	$2,4,6\text{-}Me_3C_6H_2CO_2$	(4)	79
i-Pr	$2,4,6\text{-}Me_3\text{-}3,5\text{-}(O_2N)C_6CO_2$	(5)	79
i-Bu	MeO	(—)	861
Ph	MeO	(56)	79, 861
$Ph(CH_2)_2$	MeO	(24)	861

Substrate	Conditions	Product(s) and Yield(s) (%)	Refs.
C_{4-11}	1. LDA (2.2 eq), THF, HMPA, –15°, 15 min 2. $MeONH_2$[a] (3 eq), temp, time	$R^1CH(NH_2)CO_2H$	407

R^1	Temp	Time	Yield (%)
$MeSCH_2CH_2$	–15° to –10°; rt	2 h; overnight	(9)[b]
i-Pr	–15° to –10°; rt	2 h; overnight	(27)
$CH_2Pr\text{-}i$	–15° to –10°; rt	2 h; overnight	(11)[b]
Ph	–15° to –10°	2 h	(55)
Bn	–15° to –10°; rt	2 h; overnight	(70[b]

C_6

HO$_2$C, R, CO$_2$H

1. *n*-BuLi (3 eq), THF, hexane, temp 1, time 1
2. ClNH$_2$, Et$_2$O, temp 2, time 2

HO$_2$C, R, CO$_2$H, NH$_2$ — 668

R	Temp 1	Time 1	Temp 2	Time 2	
H	–50°; –40°; 0°	—; 0.5 h; 1.5 h	–50°; to rt	—; 1 h	(8)
PhS	–50° to 0°	1 h	–50° to –10°; 0°	—; 45 min	(3.7)

MeO, N, O, CO$_2$H

1. LDA (2.4 eq), THF, –10°, 30 min
2. HMPA, –5°, 30 min
3. 2,4,6-Me$_3$C$_6$H$_2$SO$_2$ONH$_2$, –5°; 0°, 3 h

MeO, N, O, CO$_2$H, H$_2$N (19) — 115

C_7

HO$_2$C, CO$_2$H

1. *n*-BuLi (3.3 eq), THF, hexane,
   –50° to 0°, 1 h
2. ClNH$_2$, Et$_2$O, –50° to –10°; 0°, 45 min

HO$_2$C, CO$_2$H, NH$_2$ (27) — 668

HO$_2$C, CO$_2$H

1. *n*-BuLi (3.3 eq), THF, hexane,
   –50° to 0°, 1 h
2. ClNH$_2$, Et$_2$O, –50° to –10°; 0°, 45 min

HO$_2$C, CO$_2$H, NH$_2$ **I** + HO$_2$C, CO$_2$H, NH$_2$ **II** — 668

I + **II** (19), **I**:**II** = 1:3

[a] Other aminating agents gave lower yields.

[b] The yield was determined by an amino acid analyzer.

TABLE 10A. ESTER ENOLATES

Substrate	Conditions	Product(s) and Yield(s) (%)	Refs.
C_2			
$BrZnCH_2CO_2Et$	2,4-$(O_2N)_2C_6H_3ONH_2$, THF	$H_2NCH_2CO_2Et$ (0)	93
$CH_3C(O)OBu$-*t*	1. LiHMDS, THF, –78°, 30 min 2. 4-NCC_6H_4-oxaziridine-NCO_2Bu-*t*, –78°, 30 min	*t*-$BuO_2CNHCH_2C(O)OBu$-*t* (35)	153, 157
	1. Base (ia), THF, –78°, 1 h 2. Ar-oxaziridine-$NCONEt_2$, –78°, 3 h; to rt, 1.5 h	$Et_2NCONHCH_2C(O)OBu$-*t* **I** + ArCH(OH)$CH_2C(O)OBu$-*t* **II**	

Base	Ar	I	II	Refs.
LDA	2-ClC_6H_4	(0)	(0)	158
LDA	4-ClC_6H_4	(31)	(10)	158
NaHMDS[a]	4-ClC_6H_4	(30)	(0)	155
LDA	2,6-$Cl_2C_6H_3$	(0)	(0)	158
LDA	2-NCC_6H_4	(55)	(7)	158
NaHMDS	2-NCC_6H_4	(33)	(23)	155
LDA	4-NCC_6H_4	(39)	(20)	158

Substrate	Conditions	Product(s) and Yield(s) (%)	Refs.
C_{2-5}			
$R^1CH_2CO_2R^2$	1. LDA (ia), THF, –78°, 1 h 2. **1**, –78° to rt, 2-3 h **1**: 4-NCC_6H_4-oxaziridine-N-C(O)O-menthyl (*i*-Pr)	$R^1CH(CO_2R^2)NHC(O)O$-menthyl (*i*-Pr)	154

R^1	R^2	Yield	% de
H	Bu-*t*	(60)	—
Me	Me	(57)	5
Me	Et	(52)	8
Me	*t*-Bu	(51)	7
i-Pr	Et	(49)	17

Substrate	Conditions	Product(s) and Yield(s) (%)	Refs.
C_{2-8}			
R⌒CO_2Et	1. LiHMDS, THF, –78° 2. (camphor-derived oxaziridine, NH), –78°, 3-7 h; to rt	I + II R: H, I (0), II (0); *n*-Pr, I (0), II (0); Ph, I (25), II (25)	151
C_2			
$CH_2=C(OPh)(OTMS)$	PhI=NTs, MeCN	$TsHNCH_2CO_2Ph$ (0)	172
$NaCH_2CO_2Et$	$PhN=NCO_2Et$	$EtO_2CNHN(Pn)CH_2CO_2Et$ (—)	159
$CH_2=C(OMe)_2$	$RO_2CN=NCO_2R$, PhH, rt	$RO_2CN(NHCO_2R)CH_2CO_2Me$ R: Me, Time 2 d (74); Et, short (52) + two 2:1 adducts (—)	251
C_{3-4}			
$R^1R^2C=C(OTMS)(OR^3)$	$EtO_2CN(OTMS)(TMS)$, 90°, 5 d	$R^1R^2C(NHCO_2Et)CO_2R^3$ R^1 Me, R^2 Me, R^3 Me (25); H, Me, Et (70); H, Et, Et (48)	105
C_3			
$CH_3CH_2CO_2Et$	1. LDA 2. Ar-oxaziridine-$NCONEt_2$ Ar = 2-NCC_6H_4 or 4-NCC_6H_4	$CH_3CH(NHCONEt_2)CO_2Et$ (<10)	155
C_{3-5}			
$R_2CHCH_2CO_2Et$	1. LDA (ia), THF, –78° 2. **2**, –78°, 4 h; to rt **2**: 4-ClC_6H_4-oxaziridine-N-C(O)-N(Me)-CH(Me)-CH(Ph)OTBDPS	$R_2CHCH(CO_2Et)NHC(O)N(Me)CH(Me)CH(Ph)OTBDPS$ R: H (low), dr —; Me (<10), dr 5:1 (S:R)	155

TABLE 10A. ESTER ENOLATES (*Continued*)

Substrate	Conditions	Product(s) and Yield(s) (%)	Refs.
C_3 OLi, OBu-*t*	$PhN_2^+BF_4^-$ (ia), THF, –78°	PhN=N, CO_2Bu-*t* (10-30) complex mixture	185
C_{3-9} R, OMe, OTMS	$PhN_2^+BF_4^-$, pyridine, 0°, 2 h	R, CO_2Me, N, PhNH **I** + R, CO_2Me, N, NHPh **II**	197, 196
C_3 CO_2Me	$EtO_2CN{=}NCO_2Et$, DABCO, THF, rt, 120 h	EtO_2CN, $NHCO_2Et$, CO_2Me (0)	403
C_{3-4} R^1, R^2, OR^3, OMe	1. $BnO_2CN{=}NCO_2Bn$ + AgOTf (ia), CH_2Cl_2, –45°, 0.5 h 2. HF, THF	R^1, R^2, CO_2Me, N, $NHCO_2Bn$, CO_2Bn	244
C_3 OTMS, OPh	1. $BnO_2CN{=}NCO_2Bn$ + $AgClO_4$ + (*R*)-BINAP (cat) (ia), toluene, THF, –45°, 3 h 2. HF, THF	BnO_2CHN, N, CO_2Bn, CO_2Ph (73), 5[illegible]% ee	244
C_{3-9} R^1, OR^2, OTMS R^2 = H, Ph, H, NMe_2	*t*-$BuO_2CN{=}NCO_2Bu$-*t*, $TiCl_4$, CH_2Cl_2, –80°	*t*-BuO_2CHN, N, H, R^1, CO_2R^2, CO_2Bu-*t*	411

R	I + II	I:II
Me	(59)	100:0
2-thienyl	(72)	90:10
Ph	(83)	88:12
Bn	(76)	100:0

R^1	R^2	R^3	
H	Me[b]	TMS	(91)
Me	Me	TMS	(97)
Me	Me	TBDMS	(95)

R^1		% de
Me	(70)	90
Et	(65)	84
i-Pr	(35)	—
n-Bu	(45)	78
i-Bu	(70)	81
Bn	(45)	91

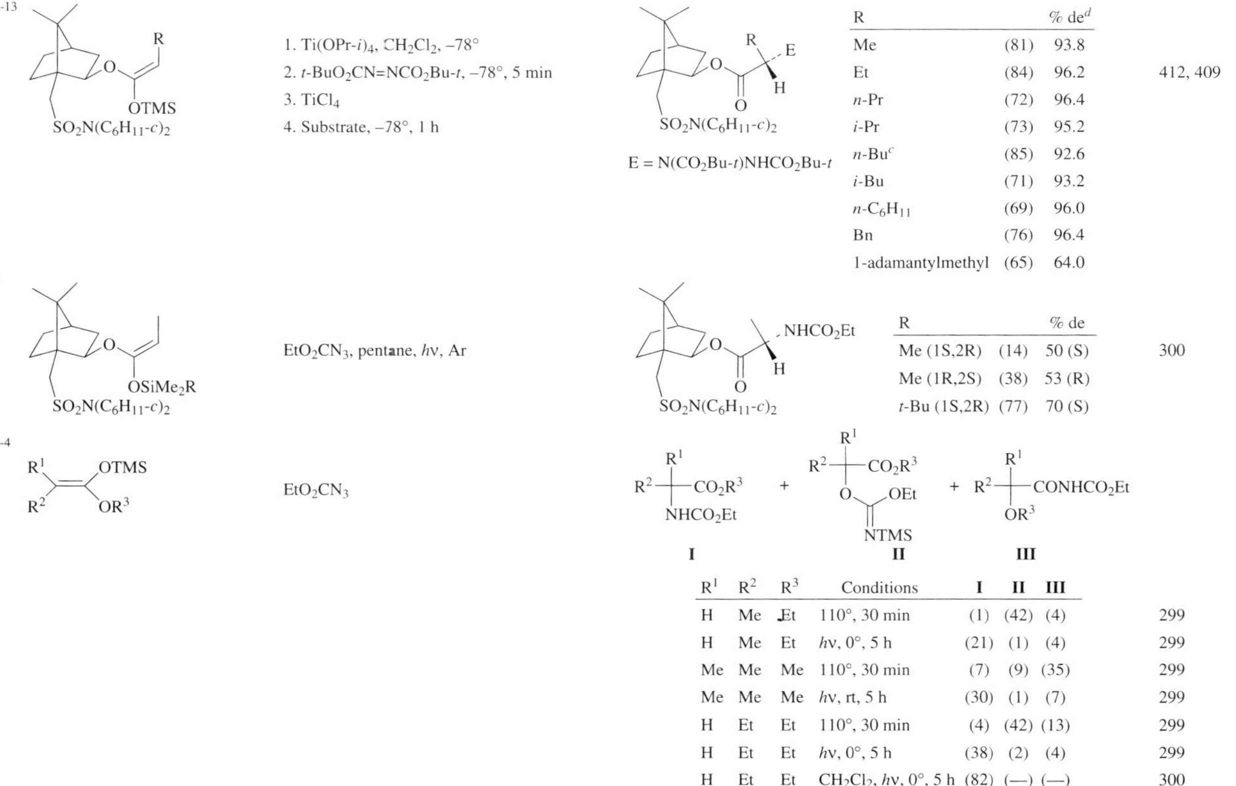

C$_{3-13}$

1. $Ti(OPr\text{-}i)_4$, CH_2Cl_2, –78°
2. $t\text{-}BuO_2CN{=}NCO_2Bu\text{-}t$, –78°, 5 min
3. $TiCl_4$
4. Substrate, –78°, 1 h

E = $N(CO_2Bu\text{-}t)NHCO_2Bu\text{-}t$

R		% de[d]
Me	(81)	93.8
Et	(84)	96.2
n-Pr	(72)	96.4
i-Pr	(73)	95.2
n-Bu[c]	(85)	92.6
i-Bu	(71)	93.2
n-C_6H_{11}	(69)	96.0
Bn	(76)	96.4
1-adamantylmethyl	(65)	64.0

412, 409

C$_3$

EtO_2CN_3, pentane, $h\nu$, Ar

R		% de
Me (1S,2R)	(14)	50 (S)
Me (1R,2S)	(38)	53 (R)
t-Bu (1S,2R)	(77)	70 (S)

300

C$_{3-4}$

EtO_2CN_3

R^1	R^2	R^3	Conditions	I	II	III	Refs.
H	Me	Et	110°, 30 min	(1)	(42)	(4)	299
H	Me	Et	$h\nu$, 0°, 5 h	(21)	(1)	(4)	299
Me	Me	Me	110°, 30 min	(7)	(9)	(35)	299
Me	Me	Me	$h\nu$, rt, 5 h	(30)	(1)	(7)	299
H	Et	Et	110°, 30 min	(4)	(42)	(13)	299
H	Et	Et	$h\nu$, 0°, 5 h	(38)	(2)	(4)	299
H	Et	Et	CH_2Cl_2, $h\nu$, 0°, 5 h	(82)	(—)	(—)	300

TABLE 10A. ESTER ENOLATES (*Continued*)

Substrate	Conditions	Product(s) and Yield(s) (%)	Refs.
C_4 TMSO, OTMS, OEt (silyl ketene acetal)	$PhN_2^+ BF_4^-$, pyridine, –35°, 18 h	TMSO, CO_2Et, N=NPh (22), dr >95:5	415
OTMS, OMe (dimethyl ketene acetal)	$PhN_2^+ BF_4^-$, pyridine, 0°, 2 h	CO_2Me, N=NPh (90)	196
$C_{4\text{-}10}$ R^1, R^2, OTMS, OR^3	$ArN_2^+ BF_4^-$, pyridine, 0°, 2 h	$R^1R^2C(CO_2R^3)(N{=}NAr)$	197
		R^1 Me, R^2 Me, R^3 Me, Ar Ph (90-92)	
		R^1 Me, R^2 Me, R^3 Me, Ar 4-ClC_6H_4 (72)	
		R^1 Me, R^2 Me, R^3 Me, Ar 4-$MeOC_6H_4$ (84)	
		R^1 Et, R^2 Ph, R^3 Ph, Ar Ph (90)	
		R^1 Ph, R^2 Et, R^3 Me, Ar Ph (90)	
$C_{4\text{-}9}$ $R^1CH{=}CHCO_2R^2$	Conditions 1: 1. Catalyst A (see Chart 1; 5 mol%), EtOH, rt 2. Substrate, then $PhSiH_3$ 3. t-$BuO_2CN{=}NCO_2Bu$-t, rt, time Conditions 2: 1. Catalyst B (see Chart 1; 2 mol%), i-PrOH, 0° 2. Substrate, then $PhSiH_3$, 0° 3. t-$BuO_2CN{=}NCO_2Bu$-t, 0°, time	$R^1CH_2CH(CO_2R^2)N(CO_2Bu$-$t)NHCO_2Bu$-$t$	
		R^1 Me, R^2 Et, Conditions 1, Time 12 h (66), dr —	215
		R^1 Me, R^2 Et, Conditions 2, Time 6 h (88), dr —	215
		R^1 Me, R^2 = 4,4-dimethyl-2-oxotetrahydrofuran-3-yl, Conditions 1, Time 20 h (75), dr 78:2	215
		R^1 Me, R^2 = 4,4-dimethyl-2-oxotetrahydrofuran-3-yl, Conditions 2, Time 3 h (74), dr 61:9	215
		R^1 Ph, R^2 Et, Conditions 1, Time — (87), dr 9:1	796
C_4 $EtO_2CCH_2CH_2CO_2Et$	1. LiHMDS, THF, –78°, 15 min 2. $RO_2CN{=}NCO_2R$, –78°, 7 min	$EtO_2CCH_2CH(CO_2Et)N(CO_2R)NHCO_2R$	408
		R (–)-menthyl (13), 2S:2R 1:1	
		R (–)-bornyl (49), 2S:2R 1:1	
		R (–)-isobornyl (41), 2S:2R 1:1	

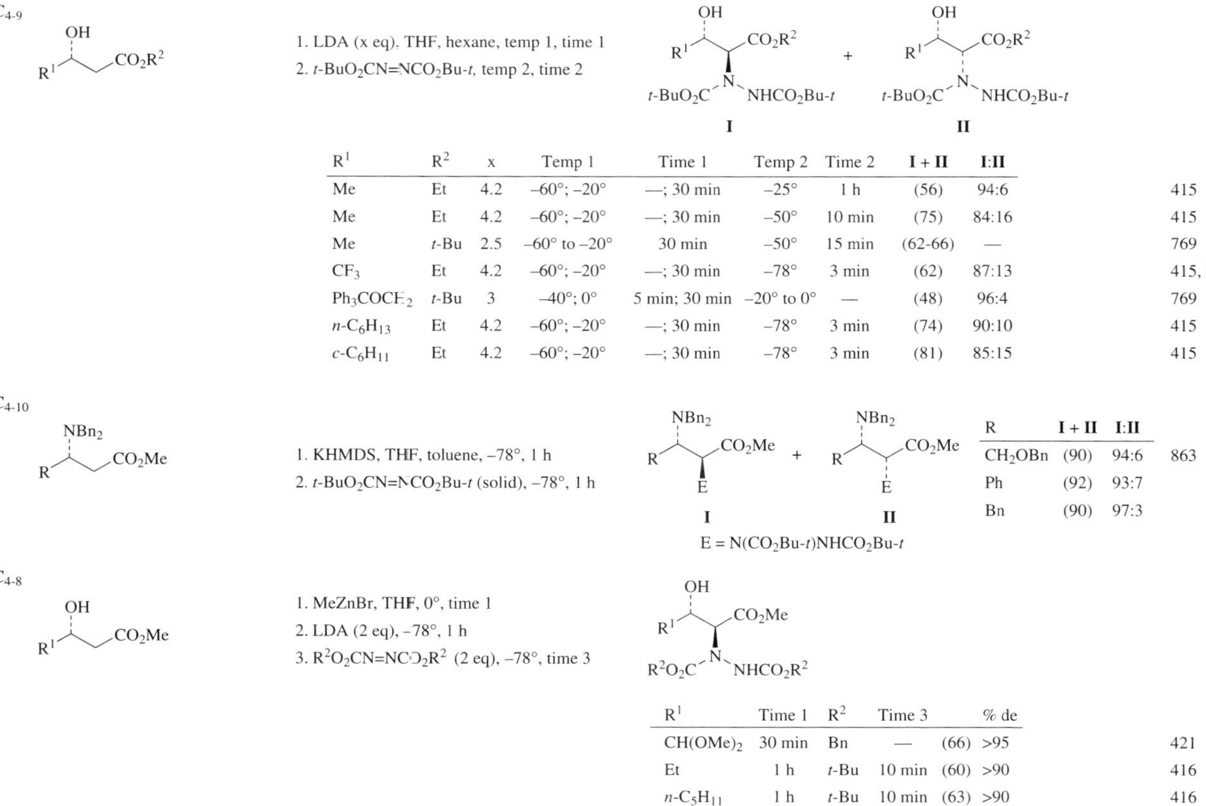

C_{4-9}

1. LDA (x eq), THF, hexane, temp 1, time 1
2. t-$BuO_2CN=NCO_2Bu$-t, temp 2, time 2

R^1	R^2	x	Temp 1	Time 1	Temp 2	Time 2	**I + II**	**I:II**	
Me	Et	4.2	–60°; –20°	—; 30 min	–25°	1 h	(56)	94:6	415
Me	Et	4.2	–60°; –20°	—; 30 min	–50°	10 min	(75)	84:16	415
Me	t-Bu	2.5	–60° to –20°	30 min	–50°	15 min	(62-66)	—	769
CF_3	Et	4.2	–60°; –20°	—; 30 min	–78°	3 min	(62)	87:13	415, 862
Ph_3COCH_2	t-Bu	3	–40°; 0°	5 min; 30 min	–20° to 0°	—	(48)	96:4	769
n-C_6H_{13}	Et	4.2	–60°; –20°	—; 30 min	–78°	3 min	(74)	90:10	415
c-C_6H_{11}	Et	4.2	–60°; –20°	—; 30 min	–78°	3 min	(81)	85:15	415

C_{4-10}

1. KHMDS, THF, toluene, –78°, 1 h
2. t-$BuO_2CN=NCO_2Bu$-t (solid), –78°, 1 h

E = $N(CO_2Bu$-$t)NHCO_2Bu$-t

R	**I + II**	**I:II**	
CH_2OBn	(90)	94:6	863
Ph	(92)	93:7	
Bn	(90)	97:3	

C_{4-8}

1. MeZnBr, THF, 0°, time 1
2. LDA (2 eq), –78°, 1 h
3. $R^2O_2CN=NCO_2R^2$ (2 eq), –78°, time 3

R^1	Time 1	R^2	Time 3		% de	
$CH(OMe)_2$	30 min	Bn	—	(66)	>95	421
Et	1 h	t-Bu	10 min	(60)	>90	416
n-C_5H_{11}	1 h	t-Bu	10 min	(63)	>90	416

TABLE 10A. ESTER ENOLATES (*Continued*)

Substrate	Conditions	Product(s) and Yield(s) (%)	Refs.
C_{4-9}			
OH, R^1, CO_2R^2	1. MeZnBr, THF, 0°, time 1 2. LDA (2 eq), –78°, 1 h 3. $R^3O_2CN{=}NCO_2R^3$ (2 eq), –78°, time 3	OH, R^1, CO_2R^2, R^3O_2C–N–$NHCO_2R^3$	

R^1	R^2	R^3	Time 1	Time 3		% de	Refs.
Me	Et	t-Bu	1 h	10 min	(63)	>90	416
$(MeO)_2CH$	Me	Bn	30 min	30 min	(66)	>95	422
$(i\text{-Pr})_3SiOCH_2$	Me	Bn	1 h	90 min	(52)	>95	423
Et	Me	t-Bu	1 h	10 min	(58)	>90	416
$n\text{-}C_5H_{11}$	Me	t-Bu	1 h	10 min	(66)	>90	416
Ph	Et	t-Bu	1 h	10 min	(69)	>90	416

Substrate	Conditions	Product(s) and Yield(s) (%)	Refs.
C_4			
OH, R^1, CO_2R^2	1. LDA (2 eq), THF 2. $t\text{-BuO}_2CN{=}NCO_2Bu\text{-}t$, –78°, 3 min	OH, R^1, CO_2R^2, $t\text{-BuO}_2C$–N–$NHCO_2Bu\text{-}t$	864

R^1	R^2		% de
H	Me	(58)	(64)
H	Et	(57)	(54)

Substrate	Conditions	Product(s) and Yield(s) (%)	Refs.
R, O, O, O (1,3-dioxan-4-one)	1. LDA 2. $t\text{-BuO}_2CN{=}NCO_2Bu\text{-}t$, –78°	R, O, O, O, $t\text{-BuO}_2C$–N–$NHCO_2Bu\text{-}t$	416

R		% de
Me	(74)	>90
$n\text{-}C_5H_{11}$	(66)	>90

Substrate	Conditions	Product(s) and Yield(s) (%)	Refs.
C_{4-14}			
R^3, R^2, R^1, O, O, O	1. Base, THF, temp 1 2. $t\text{-BuO}_2CN{=}NCO_2Bu\text{-}t$, temp 2, time 2	R^3, R^2, R^1, O, O, O, $t\text{-BuO}_2C$–N–$NHCO_2Bu\text{-}t$	

R^1	R^2	R^3	Base	Temp 1	Temp 2	Time 2		% de	Refs.
Me	H	H	LDA	—	–78°	—	(90)	90	416
Me	H	$(CH_2)_2Ph$	LDA	—	–78°	3 min	(95)	99	864
CF_3	H	*t*-Bu	*t*-BuLi	–75°	–75°	40 min	(97)	>96	865, 866
CF_3	Me	*t*-Bu	*t*-BuLi	–75°	–75°	40 min	(86)	>96	865
CF_3	*n*-Bu	*t*-Bu	*t*-BuLi	–75°	–75°	40 min	(80)	>96	865
CF_3	Ph	*t*-Bu	*t*-BuLi	–75°	–75°	40 min	(71)	>96	865

C_4

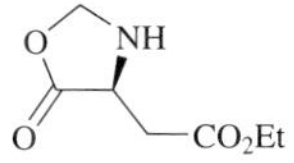

O NH, O, CO_2Et

1. Unspecified base
2. *i*-PrO_2CN=NCO_2Pr-*i*

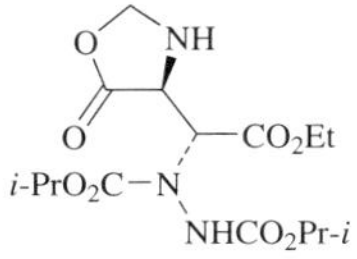

O NH, O, CO_2Et, *i*-PrO_2C–N, $NHCO_2Pr$-*i*

(52) single isomer

867

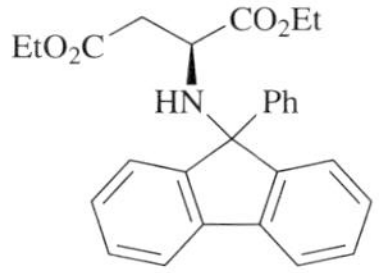

EtO_2C, CO_2Et, HN, Ph

1. LiHMDS (1.2 eq, ia), THF, HMPA, –78°; –55°, 1 h
2. RO_2CN=NCO_2R, –78°, 4.5 min

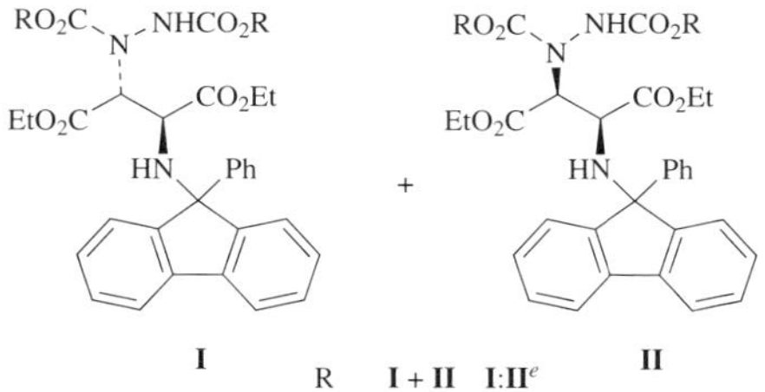

RO_2C, N, $NHCO_2R$, EtO_2C, CO_2Et, HN, Ph

I

\+

RO_2C, N, $NHCO_2R$, EtO_2C, CO_2Et, HN, Ph

II

R	**I + II**	**I:II**[e]
t-Bu	(80)	30:1
Bn	(75)	18:1

426

$C_{4\text{-}6}$

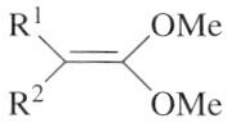

R^1, OMe, R^2, OMe

1. EtO_2CN_3 (0.5 eq), 35°, time
2. HCl, Me_2CO, 30 min

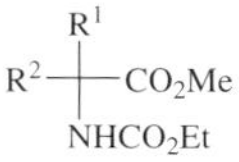

R^1, R^2, CO_2Me, $NHCO_2Et$

R^1	R^2	Time	
Me	Me	14 d	(10)
—$(CH_2)_4$—		4 d	(34)

298

TABLE 10A. ESTER ENOLATES (*Continued*)

Substrate	Conditions	Product(s) and Yield(s) (%)	Refs.
C_{4-7}			
$R^1CH(R^2)CO_2Me$	For **I**: 1. LDA, THF, –78° 2. R^3N_3 (ia), temp, time For **II** from **I**: 1. NH_4Cl, H_2O 2. NH_4OH, THF, rt, 12-24 h	**I** (triazolinone) + $R^1R^2C(NH_2)CONHR^3$ **II**	

R^1	R^2	R^3	Temp, Time	I	II	Refs.
Me	Me	$PhSCH_2$	–78°, 20 min; to –10°, 90 min	(85)	(78)	274, 275
Me	Me	1-adamantyl	–78°; rt, 3 h	(70)	(—)	867a
—$(CH_2)_5$—		$PhSCH_2$	–78° to –20°, 45 min	(—)	(79)	274
—CH_2CH=$CH(CH_2)_2$—		$PhSCH_2$	–78°; rt, 3 h	(—)	(83)	274, 275

Substrate	Conditions	Product(s) and Yield(s) (%)	Refs.
C_{4-6}			
R^1R^2C=$C(OMe)_2$	1. PhN_3 (0.6 eq), 70°, 4 h; 100°, 20 h (forms **I**) 2. HOAc, Me_2CO, rt, 24 h (forms **II**)	**I** (triazoline) + $R^1R^2C(NHPh)CO_2Me$ **II**	291

R^1	R^2	I	II
Me	Me	(44)	(11)[f]
—$(CH_2)_4$—		(38)	(7)[f]

Substrate	Conditions	Product(s) and Yield(s) (%)	Refs.
C_{4-9}			
$RCH(SiMe_2Ph)CH_2CO_2Et$	1. LDA, –78° 2. 2,4,6-(i-Pr)$C_6H_2SO_2N_3$, –78°, 2 h	$RCH(SiMe_2Ph)CH(N_3)CO_2Et$	868

R		% de
Me	(58)	>96
Ph	(64)	>96

Substrate	Conditions	Product(s) and Yield(s) (%)	Refs.
C_4			
$PhCH(Me)N(Bn)CH(Me)CH_2CO_2Bu$-$t$	1. LDA (ia), THF, –78°, 1 h 2. Reagent, –78°, 2 min	$PhCH(Me)N(Bn)CH(Me)CH(N_3)CO_2Bu$-$t$ **I** + $PhCH(Me)N(Bn)CH(Me)C(=N_2)CO_2Bu$-$t$ **II**	338

Reagent	I	% de	II
2,4,6-(i-Pr)$_3C_6H_2SO_2N_3$	(32)	>95	(9)
$Ph_2P(O)N_3$	(0)	—	(64)

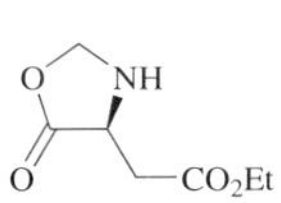

	Substrate	Conditions	Product(s) and Yield(s) (%)	Refs.
		1. MHMDS (2 eq), THF, –78° 2. $ArSO_2N_3$	I + II	867
C_5		1. KHMDS (1.3 eq, ia), THF, –78°; –55°, 1 h 2. 2,4,6-$(i\text{-Pr})_3C_6H_2SO_2N_3$, –78°, 6 min	(45) + (45)	426
		1. LDA (2.2 eq), THF, –78° to –20°, 10 min 2. HMPA 3. 2,4,6-$(i\text{-Pr})_3C_6H_2SO_2N_3$, –78°, 30 s	(77), 64% de	318
		1. NaHMDS, THF, –80°, 5 min 2. $t\text{-BuO}_2CN{=}NCO_2Bu\text{-}t$, –80°, 3 min	(87)	869
		1. LDA (4 eq), THF, –78° 2. $t\text{-BuO}_2CN{=}NCO_2Bu\text{-}t$	(55), 89% de	417

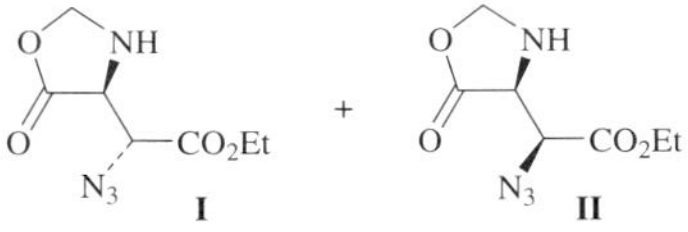

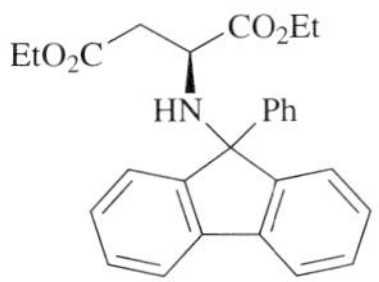

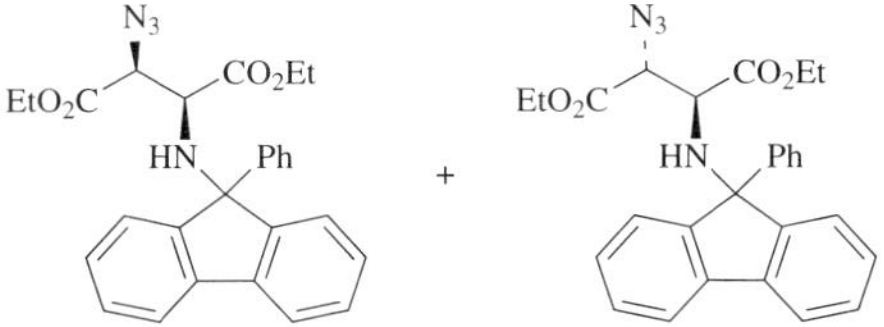

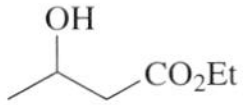

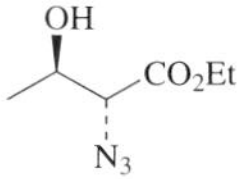

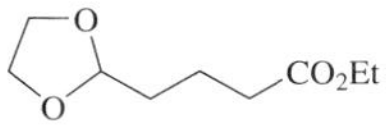

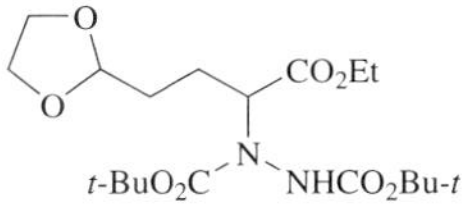

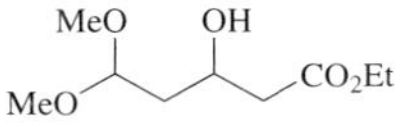

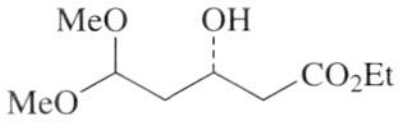

M	Ar	I + II	I:II
Li	4-MeC_6H_4	(90)	5:1
Li	2,4,6-$(i\text{-Pr})_3C_6H_2$	(80)	3:1
Na	4-MeC_6H_4	(86)	7:1
Na	2,4,6-$(i\text{-Pr})_3C_6H_2$	(89)	7:1
K	4-MeC_6H_4	(84)	5:1
K	2,4,6-$(i\text{-Pr})_3C_6H_2$	(84)	4:1

TABLE 10A. ESTER ENOLATES (*Continued*)

Substrate	Conditions	Product(s) and Yield(s) (%)	Refs.
C_5			
MeO, MeO, OH, CO_2Me	1. LDA, MeZnBr 2. $BnO_2CN{=}NCO_2Bn$, –78°	MeO, MeO, OH, CO_2Me, BnO_2C–N–$NHCO_2Bn$ (66), >98% de	424
dioxolane, OH, CO_2Bu-*i*	1. LDA, THF 2. $BnO_2CN{=}NCO_2Bn$, –78°	dioxolane, OH, CO_2Bu-*i*, BnO_2C–N–$NHCO_2Bn$ (61), 89% de	870
pyrrole-2-C(O)NH$(CH_2)_3$$CH_2CO_2Et$	1. LDA, THF, –30° 2. 2,4,6-(*i*-Pr)$_3C_6H_2SO_2N_3$, HMPA, –78°	pyrrole-2-C(O)NH$(CH_2)_3$CH(N_3)CO_2Et (70)	414
C_6			
n-Bu, OR^1, OR^2	PhI=NTs (0.8 eq), $Cu(MeCN)_4ClO_4$ (cat)	*n*-Bu, CO_2R^2, NHTs R^1 / R^2 / Temp / Time / (yield): TMS / Me / rt / — / (27) TMS / Ph / –20° / — / (50) TBDMS / Ph / –20° / 1 h / (50) $PhMe_2Si$ / Ph / rt / — / (0)	173
bornyl ester, Bu, O, $SO_2N(C_6H_{11}$-*c*$)_2$	1. LDA, THF, –78°, addend 2. *t*-$BuO_2CN{=}NCO_2Bu$-*t*	**I** (—) + **II** (—) [Bu, E, $SO_2N(C_6H_{11}$-*c*$)_2$] E = $N(CO_2Bu$-*t*$)NHCO_2Bu$-*t* Addend / **I**:**II**: — / 81:19 HMPA / 27:73	409
t-Bu, OTMS, OMe	piperidine-NCON=NCO_2Bu-*t*, $TiCl_4$, CH_2Cl_2, –78; to rt, 12.5 h	*t*-Bu, CO_2Me, piperidine-NCO, $NNHCO_2Bu$-*t* (15)	410

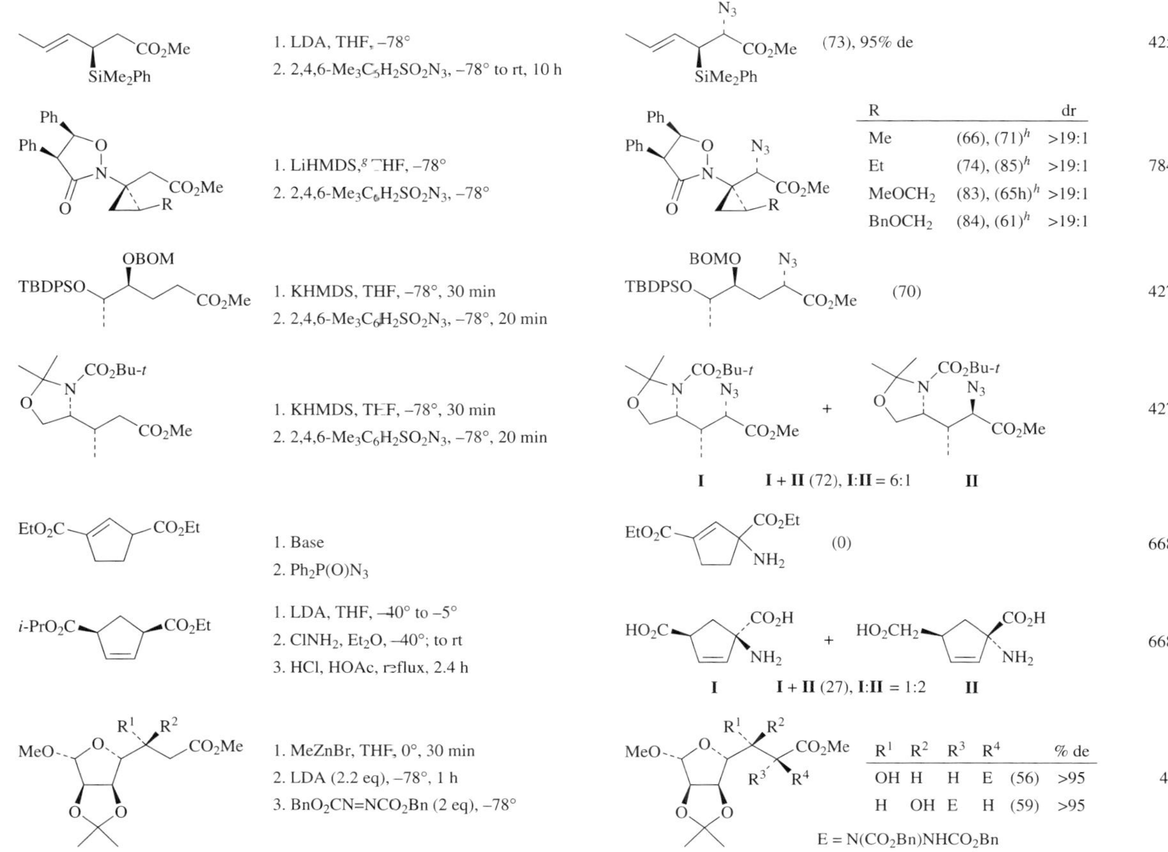

Carbon	Conditions	Product(s) and Yield(s) (%)	Refs.
	1. LDA, THF, –78° 2. 2,4,6-$Me_3C_5H_2SO_2N_3$, –78° to rt, 10 h	(73), 95% de	425
C_{6-7}	1. LiHMDS,[g] THF, –78° 2. 2,4,6-$Me_3C_6H_2SO_2N_3$, –78°	R = Me: (66), (71)[h], dr >19:1 R = Et: (74), (85)[h], dr >19:1 R = $MeOCH_2$: (83), (65h)[h], dr >19:1 R = $BnOCH_2$: (84), (61)[h], dr >19:1	784
C_6	1. KHMDS, THF, –78°, 30 min 2. 2,4,6-$Me_3C_6H_2SO_2N_3$, –78°, 20 min	(70)	427
	1. KHMDS, THF, –78°, 30 min 2. 2,4,6-$Me_3C_6H_2SO_2N_3$, –78°, 20 min	I + II (72), I:II = 6:1	427
C_7	1. Base 2. $Ph_2P(O)N_3$	(0)	668
	1. LDA, THF, –40° to –5° 2. $ClNH_2$, Et_2O, –40°; to rt 3. HCl, HOAc, reflux, 2.4 h	I + II (27), I:II = 1:2	668
	1. MeZnBr, THF, 0°, 30 min 2. LDA (2.2 eq), –78°, 1 h 3. $BnO_2CN{=}NCO_2Bn$ (2 eq), –78°	R^1 = OH, R^2 = H, R^3 = H, R^4 = E: (56), >95% de R^1 = H, R^2 = OH, R^3 = E, R^4 = H: (59), >95% de E = $N(CO_2Bn)NHCO_2Bn$	421

TABLE 10A. ESTER ENOLATES (*Continued*)

Substrate	Conditions	Product(s) and Yield(s) (%)	Refs.
2-(CH_2CO_2Me)-5-R^2-piperidine, N-R^1	1. KHMDS, toluene, THF, –78°, 30 min 2. 2,4,6-$Me_3C_6H_2SO_2N_3$, –78°, 3-5 min	I + II (α-azido esters, CO_2Me, N_3)	
		R^1 = H, R^2 = H: I + II (—), I:II 90:10	871
		R^1 = Cbz, R^2 = H: I + II (>80), I:II 80:20	872, 454
		R^1 = Cbz, R^2 = TBS: I + II (>80), I:II 95:5	872
TBSO-piperidine, N-R, CH_2CO_2Me	1. KHMDS, toluene, THF, –78°, 30 min 2. 2,4,6-$Me_3C_6H_2SO_2N_3$, –78°, 3 min	I + II (α-azido esters, CO_2Me, N_3)	
		R = Cbz: I + II (>80), I:II 90:10	872
		R = CO_2Me: I + II (91), I:II >95:5	873
C_8			
$PhCH_2CO_2Bu$-*t*	1. *n*-BuLi, THF, –70° 2. $MeONH_2$, –20° to –15°, 2 h; rt, overnight	$PhCH(NH_2)CO_2Bu$-*t* (0)	874
$PhCH(Li)CO_2Et$	Me_2NOSO_2Me, Et_2O or THF, –30° to 0°	$PhCH(NMe_2)CO_2Et$ (48)	134
C_{8-9}			
$PhCH(R)CO_2Et$	1. NaH, –70° to 0°, 25 min 2. 2,4-$(O_2N)_2C_6H_3ONH_2$, 0°, 35 min	$PhC(R)(NH_2)CO_2Et$ R = H (8) R = Me (31)	93
C_8			
$PhCH{=}C(OTMS)OEt$	2,4-$(O_2N)_2C_6H_3ONH_2$, THF, reflux, 4 h	$H_2NCH(Ph)CO_2Et$ (0)	93

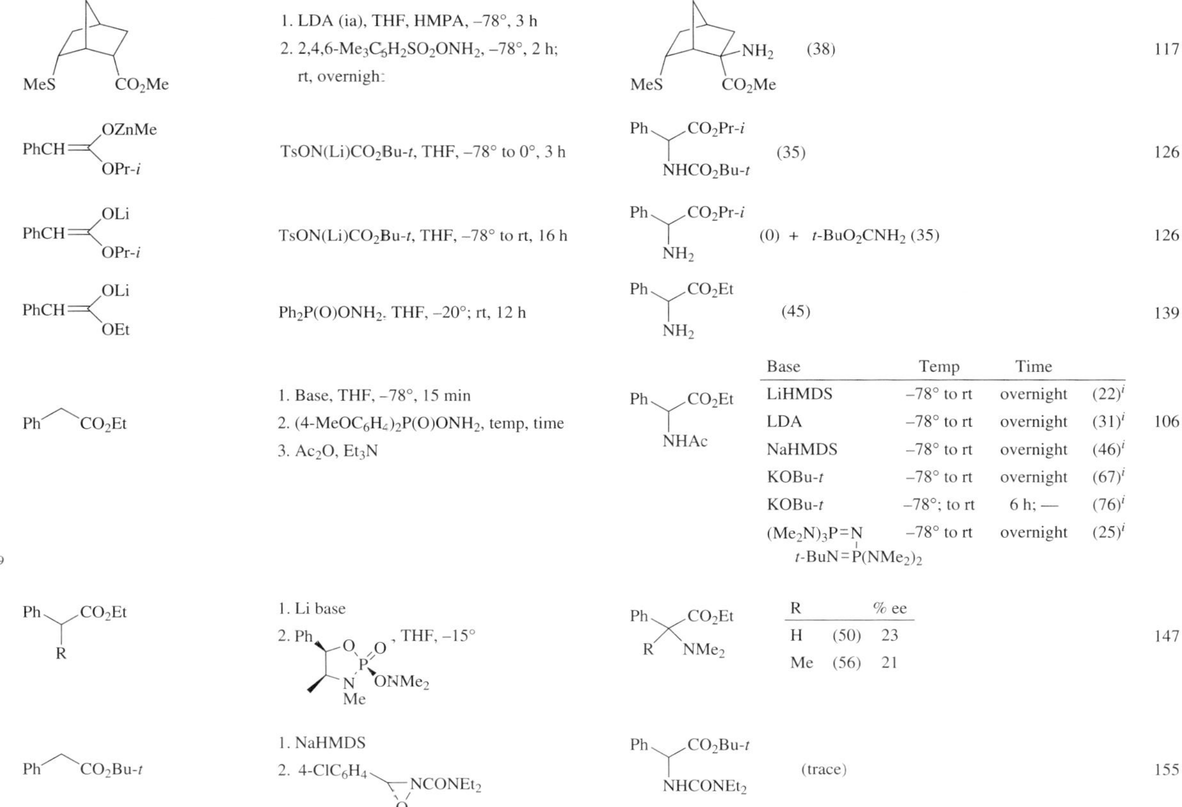

Substrate	Conditions	Product(s) and Yield(s) (%)	Refs.
C_8 MeS, CO_2Me (bicyclic)	1. LDA (ia), THF, HMPA, –78°, 3 h 2. 2,4,6-$Me_3C_5H_2SO_2ONH_2$, –78°, 2 h; rt, overnight	MeS, NH_2, CO_2Me (bicyclic) (38)	117
$PhCH{=}C(OZnMe)(OPr\text{-}i)$	$TsON(Li)CO_2Bu\text{-}t$, THF, –78° to 0°, 3 h	$PhCH(CO_2Pr\text{-}i)NHCO_2Bu\text{-}t$ (35)	126
$PhCH{=}C(OLi)(OPr\text{-}i)$	$TsON(Li)CO_2Bu\text{-}t$, THF, –78° to rt, 16 h	$PhCH(CO_2Pr\text{-}i)NH_2$ (0) + $t\text{-}BuO_2CNH_2$ (35)	126
$PhCH{=}C(OLi)(OEt)$	$Ph_2P(O)ONH_2$, THF, –20°; rt, 12 h	$PhCH(CO_2Et)NH_2$ (45)	139
$PhCH_2CO_2Et$	1. Base, THF, –78°, 15 min 2. $(4\text{-}MeOC_6H_4)_2P(O)ONH_2$, temp, time 3. Ac_2O, Et_3N	$PhCH(CO_2Et)NHAc$ Base / Temp / Time LiHMDS / –78° to rt / overnight / (22)[i] LDA / –78° to rt / overnight / (31)[i] NaHMDS / –78° to rt / overnight / (46)[i] KOBu-*t* / –78° to rt / overnight / (67)[i] KOBu-*t* / –78°; to rt / 6 h; — / (76)[i] $(Me_2N)_3P{=}N$–$P(NMe_2)_2{=}NBu\text{-}t$ / –78° to rt / overnight / (25)[i]	106
$C_{8\text{-}9}$ $PhCH(R)CO_2Et$	1. Li base 2. Ph, Me, N-Me oxazaphospholidine P(O)ONMe$_2$, THF, –15°	$PhC(R)(NMe_2)CO_2Et$ R / % ee H (50) 23 Me (56) 21	147
C_8 $PhCH_2CO_2Bu\text{-}t$	1. NaHMDS 2. 4-ClC_6H_4-oxaziridine-$NCONEt_2$	$PhCH(CO_2Bu\text{-}t)NHCONEt_2$ (trace)	155

TABLE 10A. ESTER ENOLATES (*Continued*)

Substrate	Conditions	Product(s) and Yield(s) (%)	Refs.
C_8			
Ph⌒CO_2Et	1. LiHMDS (ia), THF, –78°, 15 min 2. $RO_2CN{=}NCO_2R$, –78°, 7 min	Ph(2)CH(CO_2Et)N(CO_2R)$NHCO_2R$ R / 2S:2R (–)-menthyl (59) 2:1 (–)-bornyl (57) 1:1 (–)-isobornyl (42) 1:1	408
Et-CH=CH-CH₂-CH(OH)-CH₂-CO_2Me	1. MeZnBr, THF, 0°, 1 h 2. LDA, THF, –78°, 1 h 3. t-$BuO_2CN{=}NCO_2Bu$-t, –78°, 10 min	Et-CH=CH-CH₂-CH(OH)-CH(CO_2Me)-N(CO_2Bu-t)$NHCO_2Bu$-t (53), >98% de	418
Ph-CH=C(OR^1)(OR^2)	$R^3O_2CN{=}NCO_2R^3$, PhH, rt	$R^3O_2CNHN(CO_2R^3)$-C(Ph)=C(OR^1)(OR^2) R^1 / R^1 / R^3 / Time Me / Me / Me / 4 d (42) Me / Me / Et / 6 d (86) —$(CH_2)_2$— / Et / 2 d (42)	251
C_9			
Ph-CH₂-CH₂-CO_2Bn	1. LDA, THF, –78°, 30 min 2. t-$BuO_2CN{=}NCO_2Bu$-t, CH_2Cl_2, –70°	Ph-CH₂-CH(CO_2Bn)-N(CO_2Bu-t)$NHCO_2Bu$-t (60)	431, 848
3-Cl-4-BnO-C₆H₃-CH(OH)-CH₂-CO_2Me	1. MeZnBr, 0° 2. LDA (2 eq), –78° 3. t-$BuO_2CN{=}NCO_2Bu$-t (2 eq), –78°	3-Cl-4-BnO-C₆H₃-CH(OH)-CH(CO_2Me)-N(CO_2Bu-t)$NHCO_2Bu$-t (65), >98% de	420
3-BnO-4-MeO-C₆H₃-CH(OH)-CH₂-CO_2Bn	1. MeZnBr 2. LDA 3. t-$BuO_2CN{=}NCO_2Bu$-t	3-BnO-4-MeO-C₆H₃-CH(OH)-CH(CO_2Bn)-N(CO_2Bu-t)$NHCO_2Bu$-t (50), >95% de	875
Me₂C=CH-CH₂-CH₂-CH(OH)-CH₂-CO_2Me	1. MeZnBr, THF, 0°, 1 h 2. LDA (2.2 eq), –78°, 1 h 3. t-$BuO_2CN{=}NCO_2Bu$-t, –78°, 30 min	Me₂C=CH-CH₂-CH₂-CH(OH)-CH(CO_2Me)-N(CO_2Bu-t)$NHCO_2Bu$-t (66), >95% de	876, 419

1. MeZnBr, 0°
2. LDA (2 eq), –78°
3. t-BuO$_2$CN=NCO$_2$Bu-t, –78°

OH, CO$_2$Me, t-BuO$_2$C, N, NHCO$_2$Bu-t (55), >98% de 419

Ph, CO$_2$Bn

1. Base, THF, –78°, 30 min
2. ArSO$_2$N$_3$, THF, –78°, time

Ph, CO$_2$Bn, N$_3$ **I** + Ph, CO$_2$Bn, N$_3$, N$_3$ **II** + Ph, CO$_2$Bn, N$_2$ **III** 318

Base	Ar	Time	**I**	**II**	**III**
LDA	2,4,6-(i-Pr)$_3$C$_6$H$_2$	2 min	(73)	(—)	(1)
LDA	4-MeC$_6$H$_4$	0.5 min	(48)	(—)	(24)
LDA	4-O$_2$NC$_6$H$_4$	15 min	(5)	(—)	(68)
KHMDS	2,4,6-(i-Pr)$_3$C$_6$H$_2$	1 min	(48)	(20)	(—)
KHMDS	2,4,6-(i-Pr)$_3$C$_6$H$_2$	1 min[j]	(72)	(8)	(—)

R^1, R^2, CO$_2$Et

1. LDA (ia), THF, –78° to –30°, 1 h
2. HMPA, –78°
3. 2,4,6-(i-Pr)$_3$C$_6$H$_2$SO$_2$N$_3$, –78°, 1 h

R^1, R^2, CO$_2$Et, N$_3$ 413

R^1	R^2	
H	H	(60)
H	MeO	(58)
H	MOMO	(51)
—CH$_2$OCH$_2$—		(52)

Ph, N, Bn, Ph, CO$_2$Bu-t

1. Base (ia), THF, –78°, 1 h
2. 2,4,6-Me$_3$C$_6$H$_2$SO$_2$N$_3$, –78°, 2 min

Ph, N, Bn, Ph, CO$_2$Bu-t, N$_3$ **I** + Ph, N, Bn, Ph, CO$_2$Bu-t, N$_2$ **II** 338

Base	**I**	de	**II**
LDA	(57)	85	(17)
KN(Pr-i)$_2$	(64)	55	(<2)

TABLE 10A. ESTER ENOLATES (*Continued*)

Substrate	Conditions	Product(s) and Yield(s) (%)	Refs.
C_9			
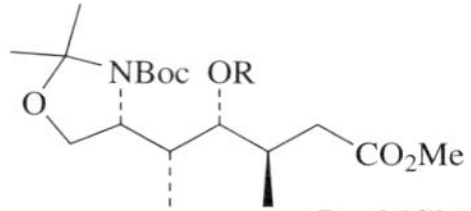 NBoc, OR, CO_2Me; R = MOM	1. KHMDS, THF, –78°, 30 min 2. 2,4,6-$Me_3C_6H_2SO_2N_3$, –78°, 20 min	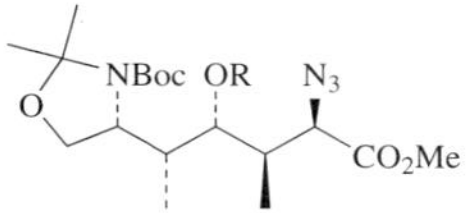 NBoc, OR, N_3, CO_2Me (65)	427
C_{10}			
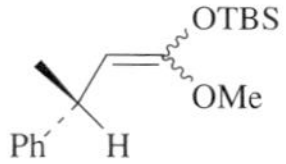 OTBS, OMe, Ph, H	EtO_2CN_3, pentane, *hν*, 0°, 8 h	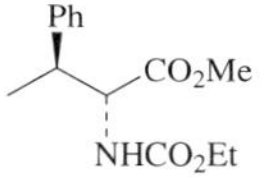 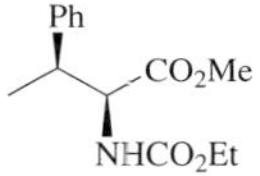 Ph, CO_2Me, $NHCO_2Et$ **I** + Ph, CO_2Me, $NHCO_2Et$ **II** **I** + **II** (61), **I**:**II** = 7:3	301
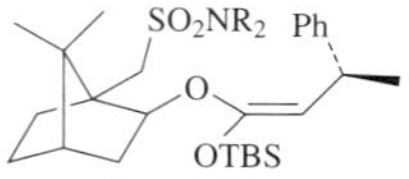 SO_2NR_2, Ph, O, OTBS; R = *c*-C_6H_{11}	EtO_2CN_3, pentane, *hν*, 0°, 8 h	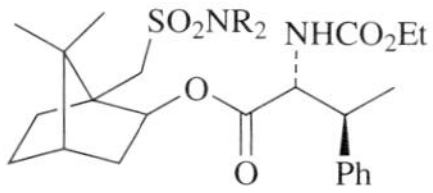 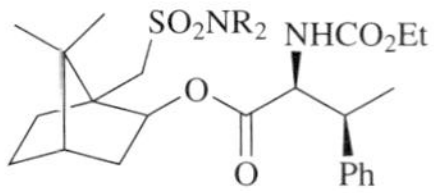 SO_2NR_2, $NHCO_2Et$, O, Ph **I** + SO_2NR_2, $NHCO_2Et$, O, Ph **II** **I** + **II** (84), **I**:**II** = 88:12	301
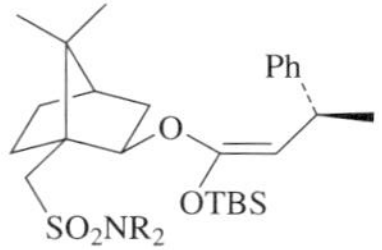 Ph, O, OTBS, SO_2NR_2; R = *c*-C_6H_{11}	EtO_2CN_3, pentane, *hν*, 0°, 8 h	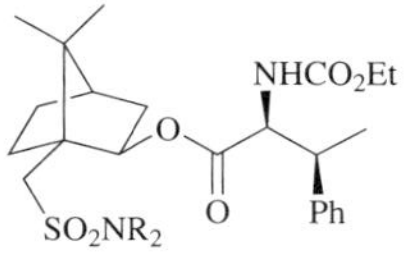 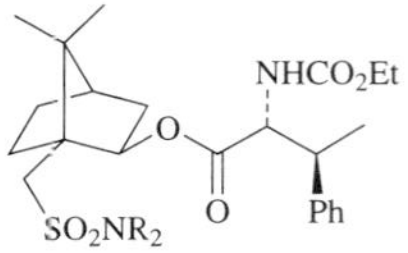 $NHCO_2Et$, O, Ph, SO_2NR_2 **I** + $NHCO_2Et$, O, Ph, SO_2NR_2 **II** **I** + **II** (89), **I**:**II** = 77:23	301
C_{12}			
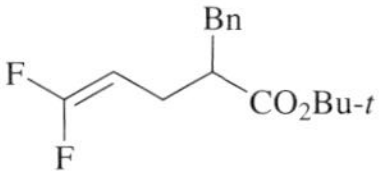 F, F, Bn, CO_2Bu-*t*	1. LDA, THF, –78°, 30 min 2. $Ph_2P(O)ONH_2$, –78°, 30 min	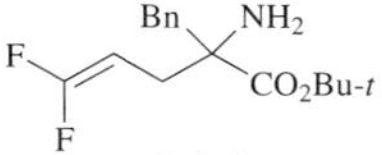 F, F, Bn, NH_2, CO_2Bu-*t* (20)	142
	1. LDA, THF, –78°, 30 min 2. EtO_2CN=NCO_2E, –78°, 3 min	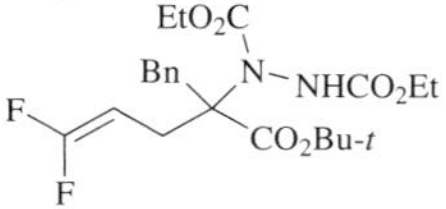 F, F, Bn, EtO_2C, N–$NHCO_2Et$, CO_2Bu-*t* (44)	142
C_{14}			
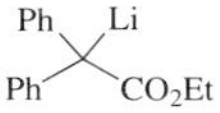 Ph, Ph, Li, CO_2Et	Me_2NOSO_2Me, Et_2O or THF, –30° to 0°	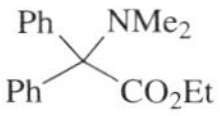 Ph, Ph, NMe_2, CO_2Et (34)	134

Carbon	Substrate	Conditions	Product(s) and Yield(s) (%)	Refs.
	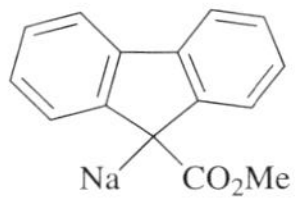 Na, CO_2Me (fluorenyl)	Me_2NOSO_2Me, Et_2O or THF, –30° to 0°	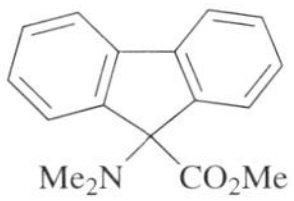 Me_2N, CO_2Me (fluorenyl) (54)	134
	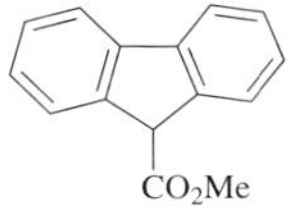 CO_2Me (fluorenyl)	1. KOMe, MeOH, PhH 2. 2,4-$(O_2N)_2C_6H_3ONH_2$, rt, overnight	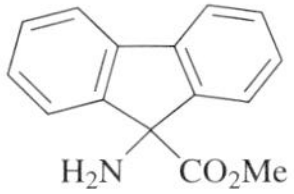 H_2N, CO_2Me (fluorenyl) (50)	94,877
	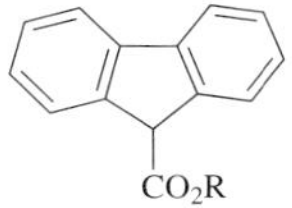 CO_2R (fluorenyl)	1. Li base 2. $Ph_2P(O)ONH_2$, THF, –20°; rt, overnight	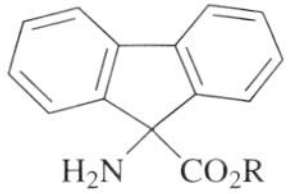 H_2N, CO_2R (fluorenyl) R: Me (47); Bu-*t* (78)	139
	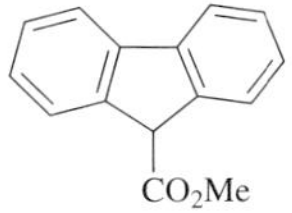 CO_2Me (fluorenyl)	1. NaH, glyme, rt 2. TsN_3, rt; 35-40°, 1 h	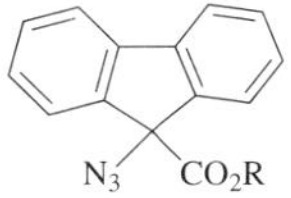 N_3, CO_2R (fluorenyl) (57)	483
C_{17}	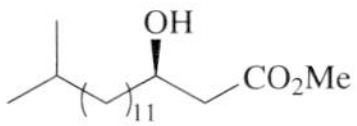 OH, $(\ )_{11}$, CO_2Me	1. MeZnBr, 0°, h 2. LDA (2 eq), –78°, 1 h 3. *t*-$BuO_2CN{=}NCO_2Bu$-*t*, –78°, 2 h	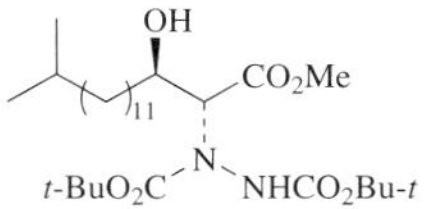 OH, $(\ )_{11}$, CO_2Me, N, *t*-BuO_2C, $NHCO_2Bu$-*t* (72), >95% de	878

[a] Among a range of bases (LiHMDS, NaHMDS, KHMDS, BuLi, NaH, and NaOBu-*t*), NaHMDS gave the highest yield.

[b] The substrate contained 6% of the Z isomer.

[c] Reaction of the kinetic lithium enolate with *t*-$BuO_2CN{=}NCO_2Bu$-*t* gave the product in 62% yield and 62% de.

[d] The values are for the crude products.

[e] Use of LiHMDS without HMPA, *n*-BuLi-LiHMDS, or KHMDS resulted in **I**:**II** ratios of 1:1 to 2.5:1.

[f] The number is the over-all yield from the substrate ketene acetal.

[g] With KHMDS as the base, the yields were higher but the diastereomeric ratios were lower.

[h] The yields are from the two diastereomers of the substrate, respectively.

[i] The number is the percent conversion.

[j] The enolate was added to the azide.

TABLE 10B. THIOESTER ENOLATES

Substrate	Conditions	Product(s) and Yield(s) (%)	Refs.
C_{2-8}			
$R^1CH_2C(=S)YR^2$	1. BuLi (ia), THF, –70°, 20 min 2. $R^3O_2CN{=}NCO_2R^3$, –70°, to rt	$R^1CH[N(CO_2R^3)NHCO_2R^3]C(=S)YR^2$ R[1] | R[2] | Y | R[3] | H | Me | S | Et | (54) H | Me | S | t-Bu | (57) H | Et | O | Et | (57) H | Et | S | Et | (57) Ph | Et | S | Et | (0)	477
$RCH{=}C(SEt)_2$	$MeO_2CN{=}NCO_2Me$, PhH, rt, time	$MeO_2C(NHCO_2Me)N$–$C(R){=}C(SEt)_2$ R | Time | H | 3 d | (56) Ph | 2 d | (90)	251
$R^1R^2C{=}C(STMS)SMe$	1. t-$BuO_2CN{=}NCO_2Bu$-t, CH_2Cl_2, rt, time 2. HOAc, H_2O	$R^1R^2C[N(CO_2Bu$-$t)NHCO_2Bu$-$t]C(=S)SMe$ R[1] | R[2] | Time | H | H | 3 h | (37) Me | H | 1 h | (72) Me | Me | 3 h | (90) n-C_5H_{11} | H | 1 h | (75) Ph | H | 2 h | (69) Bn | H | 1 h | (76) n-C_8H_{17} | H | 1 h | (80)	253
C_3			
$CH_3CH{=}C(OTMS)SEt$	1. t-$BuO_2CN{=}NCO_2Bu$-t + AgOTf (ia), CH_2Cl_2, 0°, 3 h 2. HF, THF	$CH_3CH(COSEt)N(CO_2Bu$-$t)NHCO_2Bu$-t (92)	244

Substrate	Conditions	Product(s) and Yield(s) (%)		Refs.
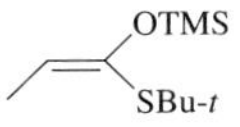 OTMS, SBu-*t*	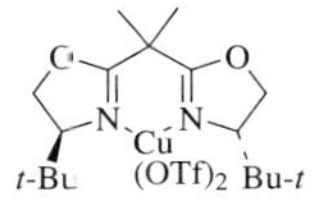 **1**, **2** (0.1 mol%), THF, CF_3CH_2OH, −78°, overnight 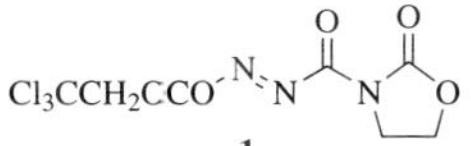 Cl_3CCH_2CCO–N=N–CO–N(oxazolidinone) **1** Cu(OTf)$_2$ bis(oxazoline), *t*-Bu, Bu-*t* **2**	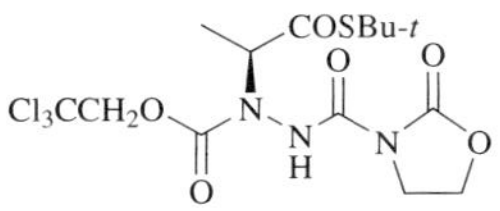 Cl_3CCH_2O, COSBu-*t*, N, N, H, O, O, O, N, O	(89), 84% ee	252
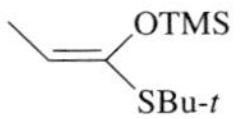 OTMS, SBu-*t*	**1**, **2** (0.1 mol%), THF, CF_3CH_2OH, −20°, 20 h	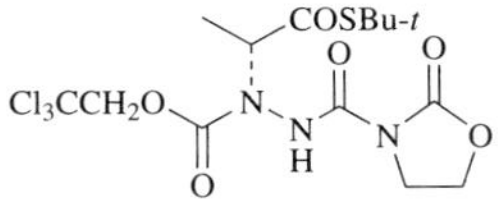 Cl_3CCH_2O, COSBu-*t*, N, N, H, O, O, O, N, O	(85), 96% ee	252

TABLE 11. LACTONE ENOLATES

Substrate	Conditions	Product(s) and Yield(s) (%)	Refs.
C_4			
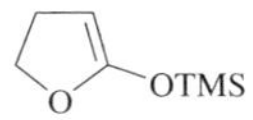 OTMS	$EtO_2CN(TMS)OTMS$, 90°, 5 d	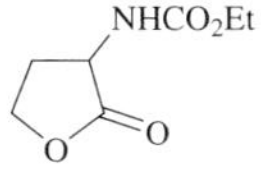 $NHCO_2Et$ (44)	105
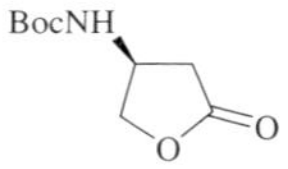 BocNH	1. LDA (2.1 eq), THF, −78° to −20°, 1 h 2. TsN_3, −78°, 1 h 3. TMSCl, −78° to 0°	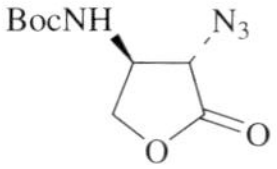 BocNH, N_3 (58), 100% de	879
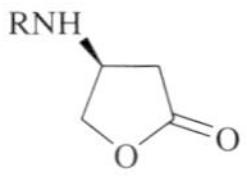 RNH, R = Boc or Cbz	1. LiHMDS (2 eq, ia), THF, −78°, 30 min 2. TsN_3, THF, −78°, 5 min	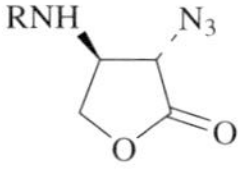 RNH, N_3 (55)[a]	880
C_5			
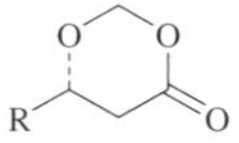 R	1. LDA 2. t-$BuO_2CN{=}NCO_2Bu$-t, −78°	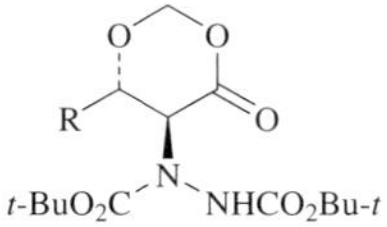 R, t-BuO_2C–N–$NHCO_2Bu$-t	416
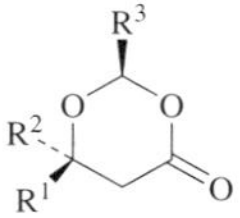 R^1, R^2, R^3	1. Base, THF, temp 1 2. t-$BuO_2CN{=}NCO_2Bu$-t, temp 2, time 2	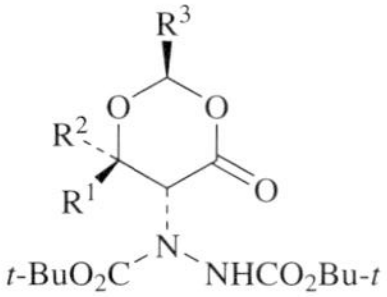 R^1, R^2, R^3, t-BuO_2C–N–$NHCO_2Bu$-t	

R		% de	Refs.
Me	(74)	>90	416
n-C_5H_{11}	(66)	>90	

R^1	R^2	R^3	Base	Temp 1	Temp 2	Time 2		% de	Refs.
Me	H	H	LDA	—	−78°	—	(90)	90	416
Me	H	$(CH_2)_2Ph$	LDA	—	−78°	3 min	(95)	99	864
CF_3	H	t-Bu	t-BuLi	−75°	−75°	40 min	(97)	>96	865, 866
CF_3	Me	t-Bu	t-BuLi	−75°	−75°	40 min	(86)	>96	865
CF_3	n-Bu	t-Bu	t-BuLi	−75°	−75°	40 min	(80)	>96	865
CF_3	Ph	t-Bu	t-BuLi	−75°	−75°	40 min	(71)	>96	865

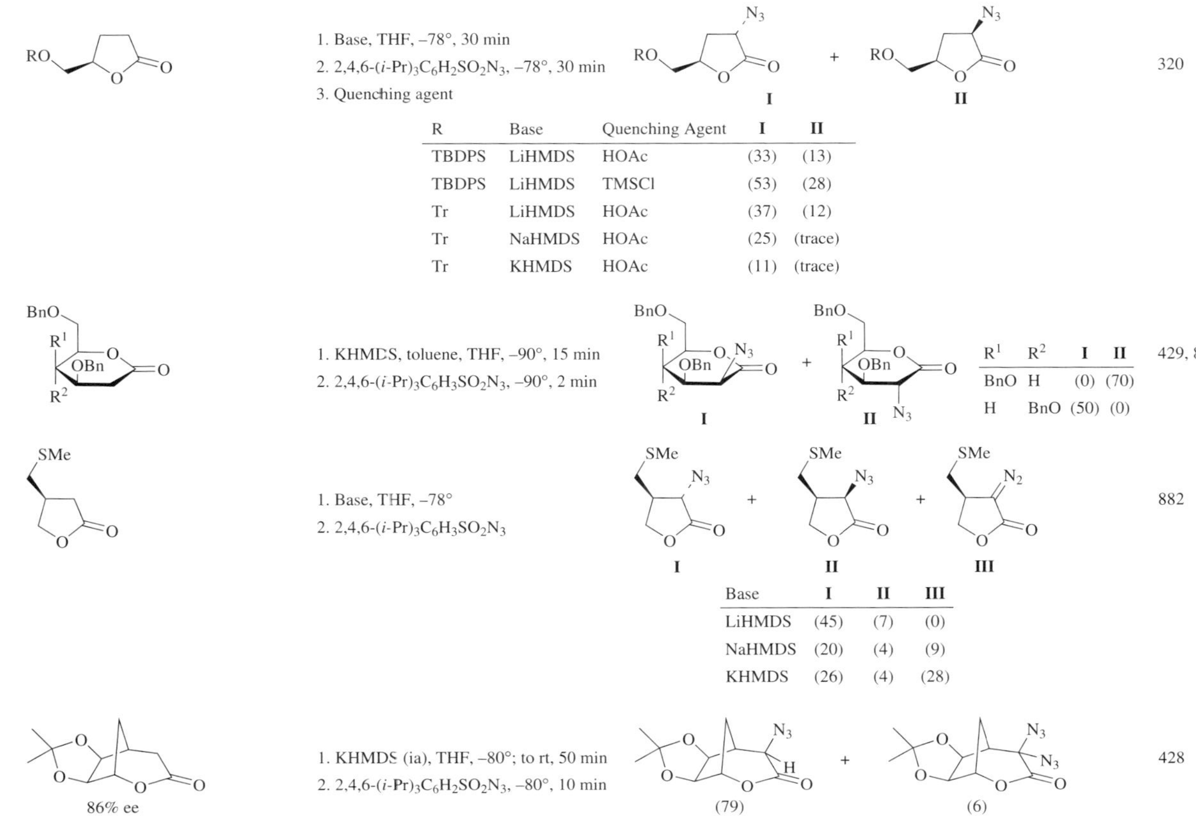

R	Base	Quenching Agent	I	II
TBDPS	LiHMDS	HOAc	(33)	(13)
TBDPS	LiHMDS	TMSCl	(53)	(28)
Tr	LiHMDS	HOAc	(37)	(12)
Tr	NaHMDS	HOAc	(25)	(trace)
Tr	KHMDS	HOAc	(11)	(trace)

R^1	R^2	I	II
BnO	H	(0)	(70)
H	BnO	(50)	(0)

Base	I	II	III
LiHMDS	(45)	(7)	(0)
NaHMDS	(20)	(4)	(9)
KHMDS	(26)	(4)	(28)

TABLE 11. LACTONE ENOLATES (*Continued*)

	Substrate	Conditions	Product(s) and Yield(s) (%)	Refs.
	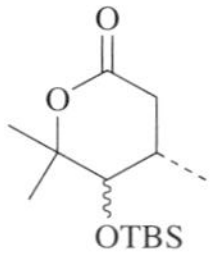 OTBS	1. KHMDS (ia), toluene, THF, –78°, 30 min 2. 2,4,6-$(i\text{-Pr})_3C_6H_2SO_2N_3$, –78°, 1-2 min	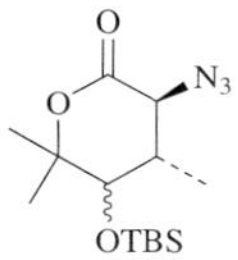 N_3, OTBS (63)	883
C_9	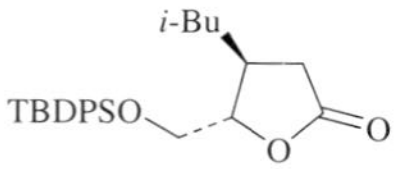 *i*-Bu, TBDPSO	1. LiHMDS (ia), THF, –78°, 60 min 2. 2,4,6-$(i\text{-Pr})_3C_6H_2SO_2N_3$, –78°, 30 min	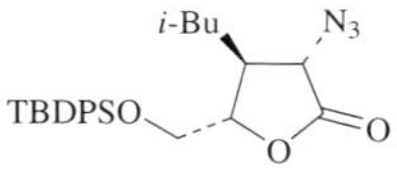 *i*-Bu, N_3, TBDPSO (69) 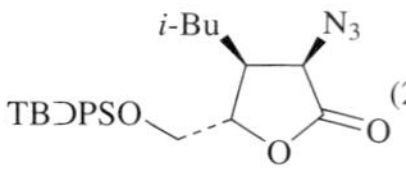 + *i*-Bu, N_3, TBDPSO (26)	884
	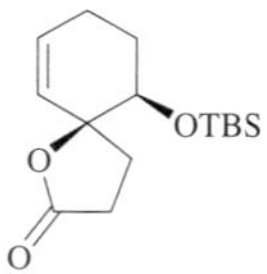 OTBS	1. LDA (ia), THF, –78°, 30 min 2. 2,4,6-$(i\text{-Pr})_3C_6H_2SO_2N_3$, –78°, 15 min	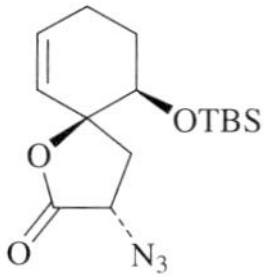 OTBS, N_3 (89)	885
C_{11}	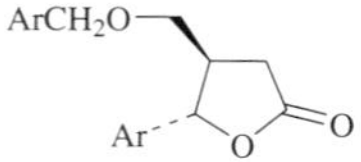 $ArCH_2O$, Ar	1. LiHMDS, THF, –78°, 45 min 2. $4\text{-}O_2NC_6H_4SO_2N_3$, –78°, 10 min 3. Quench conditions	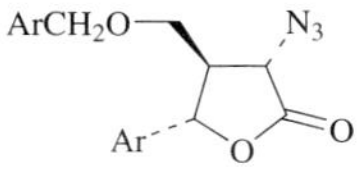 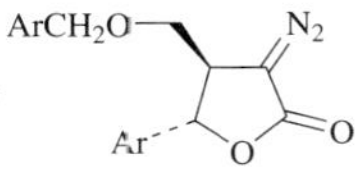 $ArCH_2O$, N_3, Ar **I** + $ArCH_2O$, N_2, Ar **II**	

Ar	Quench conditions	I	II	Refs.
Ph	AcOH, to rt	(58)	(0)	326, 886
3,4-$(OCH_2O)C_6H_3$	AcCl, to rt; then DMAP, THF, rt, 16 h	(27)	(40)	326

Substrate	Conditions	Product(s) and Yield(s) (%)	Refs.
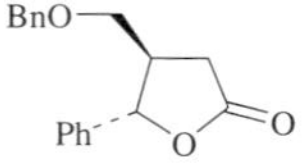 BnO, Ph	1. LDA 2. $4\text{-}O_2NC_6H_4SO_2N_3$ 3. pH 7 buffer	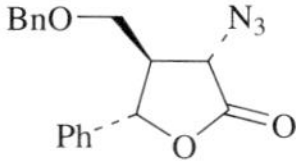 BnO, N_3, Ph (45) 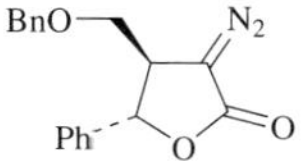 + BnO, N_2, Ph (6)	326

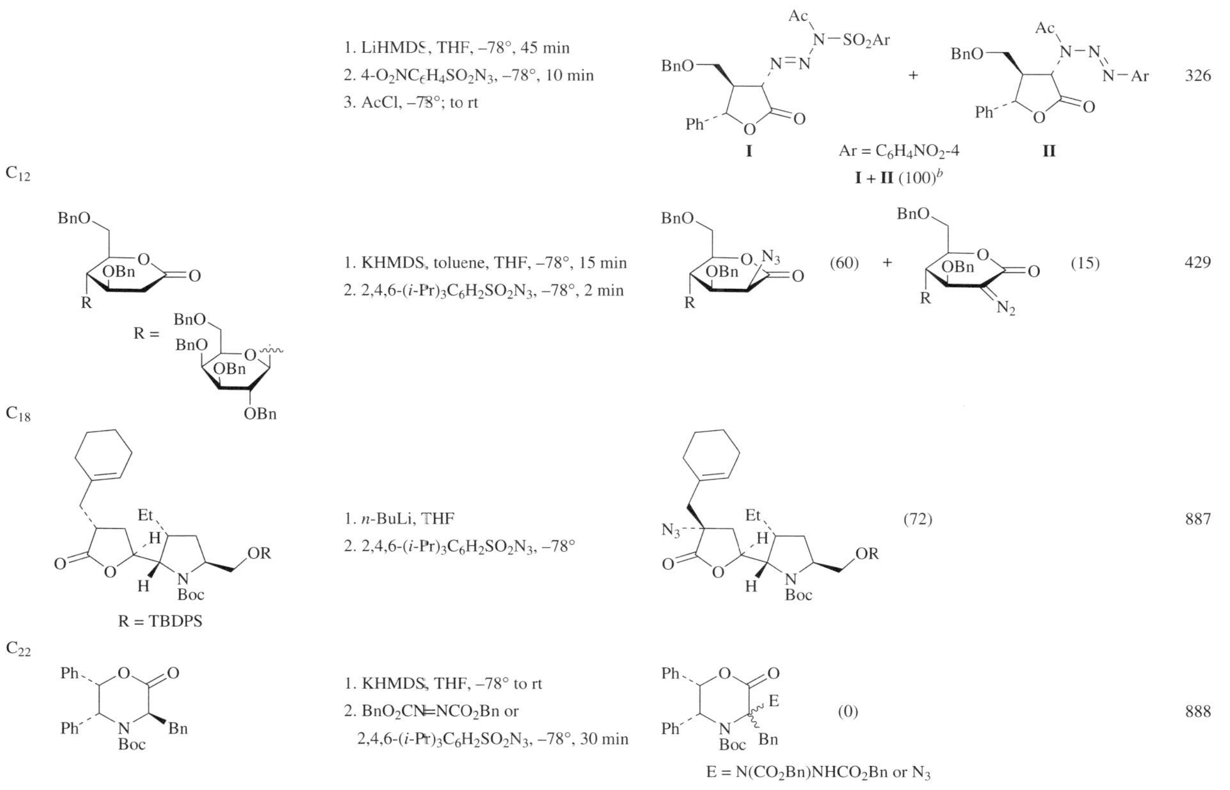

Conditions	Product(s) and Yield(s) (%)	Refs.
1. LiHMDS, THF, −78°, 45 min 2. 4-$O_2NC_6H_4SO_2N_3$, −78°, 10 min 3. AcCl, −78°; to rt	I + II (100)[b]	326
1. KHMDS, toluene, THF, −78°, 15 min 2. 2,4,6-(i-Pr)$_3C_6H_2SO_2N_3$, −78°, 2 min	(60) + (15)	429
1. n-BuLi, THF 2. 2,4,6-(i-Pr)$_3C_6H_2SO_2N_3$, −78°	(72)	887
1. KHMDS, THF, −78° to rt 2. $BnO_2CN{=}NCO_2Bn$ or 2,4,6-(i-Pr)$_3C_6H_2SO_2N_3$, −78°, 30 min	(0)	888

[a] The use of KHMDS or 2,4,6-(i-Pr)$_3C_6H_2SO_2N_3$ gave lower yields.

[b] The ratio **I**:**II** was not reported. On treatment with DMAP in THF, the 3-R azide and the diazo compound were obtained in 48% and 51% yields, respectively.

TABLE 12. AMIDE ENOLATES

Substrate	Conditions	Product(s) and Yield(s) (%)	Refs.
C_2			
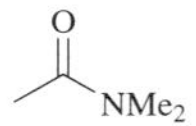 [structure] NMe_2	1. LDA (ia), THF, –78°, 1 h 2. **1**, –78°; to rt, 2-3 h 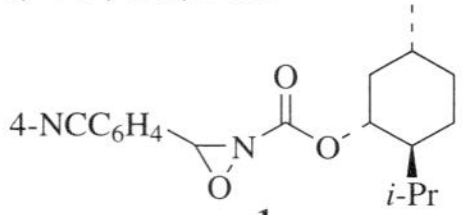 4-NCC_6H_4 ... *i*-Pr **1**	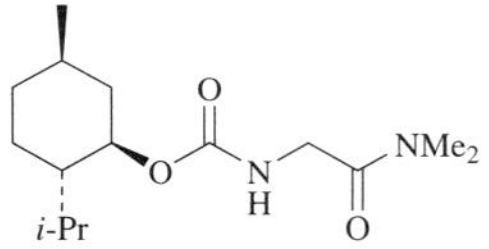 [structure] *i*-Pr ... NMe_2 (56)	154, 158
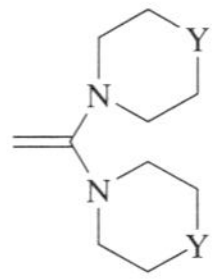 [structure] Y	$RO_2CN{=}NCO_2R$, PhH, rt, 3 d	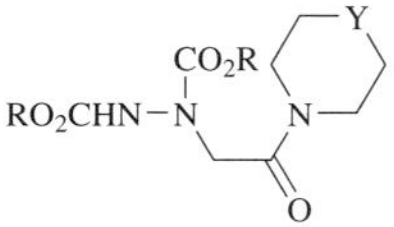 $RO_2CHN-N(CO_2R)$ ... Y Y / R: CH_2 Me (41); O Et (18)	251
C_{3-9}			
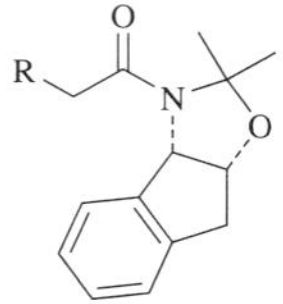 [structure] R	1. *n*-BuLi, THF, –78° 2. CuCN, –78° to –5° 3. $TsON(Li)CO_2Bu$-*t*, –78°, 30 min	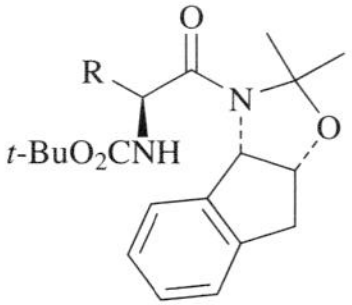 [structure] R, *t*-BuO_2CNH R (yield) % de: Me (63) 99; *i*-Pr (57) 99; *n*-Bu (58) 96; *i*-Bu (72) 99; *t*-Bu (52) 99; Ph (51) 99; Bn (77) 99	889
C_{3-5}			
 $R^1CH_2CONMe_2$	1. LDA, THF, hexane, –78°, 40 min 2. $R^2O_2CN{=}NCO_2R^2$, –78°, 4 min R^2 = isobornyl	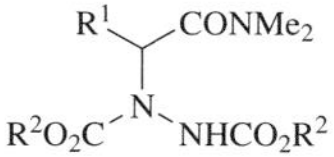 $R^1CH(CONMe_2)N(CO_2R^2)NHCO_2R^2$ R^1 (yield) % de: Me (72) (0); *i*-Pr (87) (0)	408
C_{3-8}			
 $RCH_2CONMePh$	1. LDA, THF, –78°, 60 min 2. $(PhO)_2P(O)N_3$, –78°, 5 min 3. $(t\text{-}BuO_2C)_2O$, –78° to rt, 6 h	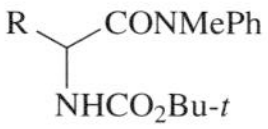 $RCH(CONMePh)NHCO_2Bu$-*t* R: Me (74); Et (76); 2-thienyl (70); Ph (80)	336

C_{3-11}

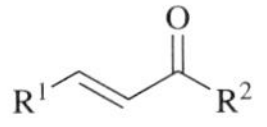

I R^2 =

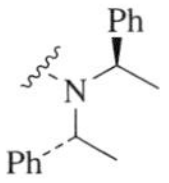

II R^2 =

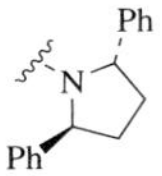

III R^2 =

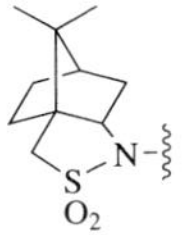

1. Tris(dipivaloylmethanato)manganese(III) (5 mol%) + *i*-PrOH (ia), 0°
2. $R^3O_2CN{=}NCO_2R^3$, *i*-PrOH
3. Ph_3SiH, 0°, time

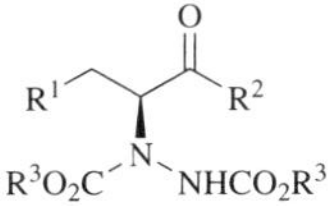

R^3O_2C N $NHCO_2R^3$

R^1	R^2	R^3	Time		dr
H	**III**	*t*-Bu	1 h	(56)	91:1
Me	**III**	*t*-Bu	1 h	(84)	98:2
$MeS(CH_2)_2$	**III**	*t*-Bu	1 h	(72)	96:4
n-Pr	**I**	*t*-Bu	1.5 h	(—)[a]	—
n-Pr	**II**	*t*-Bu	1.5 h	(78)	89:11
n-Pr	**III**	Et	1.5 h	(—)[a]	—
n-Pr	**III**	*i*-Pr	1.5 h	(47)	90:10
n-Pr	**III**	*t*-Bu	1.5 h	(81)	96:4
i-Pr	**III**	*t*-Bu	1.5 h	(78)	99:1
n-C_7H_{15}	**III**	*t*-Bu	1.5 h	(80)	96:4
$Ph(CH_2)_2$	**III**	*t*-Bu	1.5 h	(68)	97:3

890

C_{3-6}

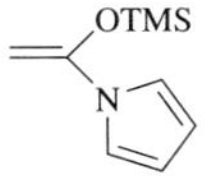

OTMS

1, **2** (5 mol%), THF, CF_3CH_2OH

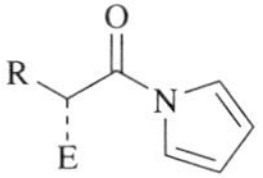

I + 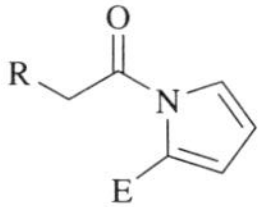 **II**

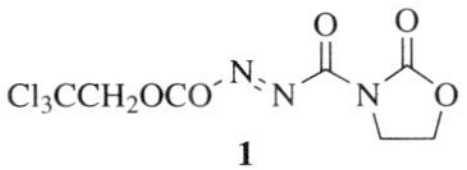

Cl_3CCH_2OCO N N

1

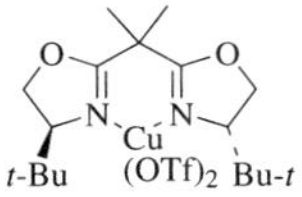

t-Bu Cu $(OTf)_2$ Bu-*t*

2

E = Cl_3CCH_2OCO N N H

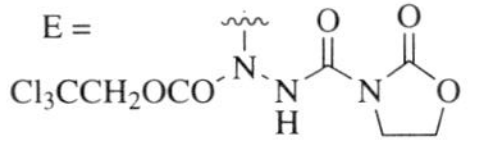

R	Temp	Time	I	% ee	II
Me	–78°	30 min	(96)	(99)	(0)
$CH_2{=}CH\text{-}CH_2$	–20°	5 min	(73)	(98)	(18)
i-Pr	–20°	5 min	(65)	(99)	(23)
t-Bu	–20°	5 min	(0)	(—)	(80)

252

C_{5-9}

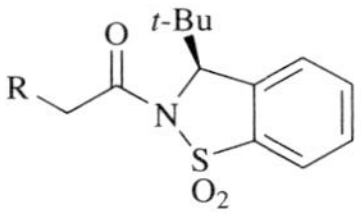

t-Bu O_2

1. NaHMDS, THF, –78°, 30 min
2. 2,4,6-$(i\text{-}Pr)_3C_6H_2SO_2N_3$, –78°, 2 min

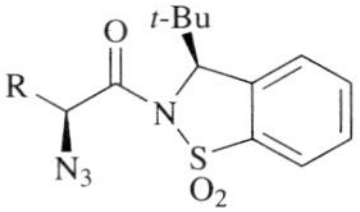

t-Bu N_3 O_2

R		% de
$CH_2CH{=}CH_2$	(96)	>96
Bn	(85)	98

891

TABLE 12. AMIDE ENOLATES (*Continued*)

Substrate	Conditions	Product(s) and Yield(s) (%)	Refs.
C_7 Bn$_2$N–CO–CH(Me)–CH(Me)–CH$_2$–CO–R; R = (camphorsultam)	1. KHMDS, THF, –78°, 30 min 2. 2,4,6-(i-Pr)$_3C_6H_2SO_2N_3$, –78°, 1-2 min	Bn$_2$N–CO–CH(Me)–CH(Me)–CH(N$_3$)–CO–R (78)	776
C_8 PhCH$_2$CONR^{1}R^2	1. Base (x eq), THF, –70° 2. MeONH$_2$, temp, 2 h; rt, overnight	PhCH(NH$_2$)CONR^{1}R^2 R^1 / R^2 / Base / x / Temp H / H / n-BuLi / 2 / –20° to –15° / (0) H / t-Bu / LDA / 2 / –20° to –15° / (49) Et / Et / LDA / 1 / –20° / (15)	874
PhCH$_2$CONHR	1. Base (2 eq), THF, 1 h 2. ArN=NAr, –78°, 5 min	PhCH(N(Ar)NHAr)CONHR R / Ar / Base H / Ph / NaNH$_2$ / (29) Ph / Ph / NaNH$_2$ / (49) Ph / 2-ClC$_6$H$_4$ / LDA / (6)	212
(R^1,R^2,R^3-C$_6$H$_2$)CH$_2$CON(Me)CH(Me)CH(OH)Ph	1. LDA (2 eq), THF, 78°, 1 h; 0°, 15 min; rt, 5 min 2. t-BuO$_2$CN=NCO$_2$Bu-t, –105°, 1 h; to rt	(R^1,R^2,R^3-C$_6$H$_2$)CH(N(CO$_2$Bu-t)NHCO$_2$Bu-t)CON(Me)CH(Me)CH(OH)Ph R^1 / R^2 / R^3 / % de MeO / MeO / H / (89) / >90 —OCH$_2$O— / H / (91) / >90 MeO / MeO / MeO / (90) / >90 BnO / MeO / BnO / (86) / >90	764, 892, 765
C_{8-12} (t-Bu benzisothiazoline dioxide)–N–COCH$_2$R	1. NaHMDS (ia), THF, –78°, 30 min 2. 2,4,6-(i-Pr)$_3C_6H_2N_3$, –78°, 2 min	(t-Bu benzisothiazoline dioxide)–N–COCH(N$_3$)R R 3-TMSOC$_6$H$_4$ (92) 1-naphthyl (96)	893

[a] The reaction products were a complex mixture.

TABLE 13. *N*-ACYLOXAZOLIDINONE ENOLATES

Substrate	Conditions	Product(s) and Yield(s) (%)	Refs.
C_{3-9}			
R-CH₂C(O)-oxazolidinone	1. LDA, THF, –78°, 20 min 2. **1**, CH_2Cl_2, –78°, 7 min (**1**)	R: Me (85)[a]; Bn (92)[a]	219
R¹-CH₂C(O)-(4-*i*-Pr)oxazolidinone	1. LDA, THF, –78°, 15-40 min 2. $R^2O_2CN{=}NCO_2R^2$, –78° 3. HCl, time (quench; for R^2 = isobornyl: HOAc)	E = $N(CO_2R^2)NHCO_2R^2$	

R^1	R^2	Time		% de	Refs.
Me	*t*-Bu	0 min	(92)	—	432
Me	Bn	0 min	(91)	80[b]	432, 408
Me	(–)-isobornyl	4 min	(56)	100	408
t-Bu	Bn	0 min	(85)	94[b]	432
Bn	Bn	0 min	(90)	88[b]	432

Substrate	Conditions	Product(s) and Yield(s) (%)	Refs.
C_3			
propionyl-(4-*i*-Pr)oxazolidinone	1. LDA, THF, –78°, 40 min 2. $RO_2CN{=}NCO_2R$, –78°, 4 min	E = $N(CO_2R)NHCO_2R$	408

R		% de
Bn	(—)	80
(–)-isobornyl	(88)	100

TABLE 13. *N*-ACYLOXAZOLIDINONE ENOLATES (*Continued*)

Substrate	Conditions	Product(s) and Yield(s) (%)	Refs.
C_{3-8} (R-CH$_2$CO-oxazolidinone, 4-Bn)	1. LDA, THF, –78°, 30 min 2. *t*-$BuO_2CN{=}NCO_2Bu$-*t*, CH_2Cl_2, 30-180 s	E = N(CO_2Bu-*t*)NHCO_2Bu-*t*	431
C_{3-9}	1. Base, THF, –78°, 30 min 2. $ArSO_2N_3$, –78°, time 2 3. Acid, temp 3, time 3 (quench)	I (azide) + II (diazo)	318, 433

R		2S:2R
Me	(92)	98:2
i-Pr	(95)	98:2
$CH_2CH{=}CH_2$	(94)	98:2
t-Bu	(96)	>99:1
$MeC_2C(CH_2)_2$[c]	(51)	>95:5
Ph	(96)	97:3
Bn	(91)	97:3

R	Base	Ar	Time 2	Acid	Temp 3	Time 3	I	I % de	II
Me	KHMDS	2,4,6-(*i*-Pr)$_3C_6H_2$	1-2 min	HOAc	—	—	(74)[d]	94	(0)
i-Pr	KHMDS	2,4,6-(*i*-Pr)$_3C_6H_2$	1-2 min	HOAc	—	—	(77)[d]	96	(0)
$CH_2CH{=}CH_2$	KHMDS	2,4,6-(*i*-Pr)$_3C_6H_2$	1-2 min	HOAc	—	—	(78)[d]	94	(0)
t-Bu	KHMDS	2,4,6-(*i*-Pr)$_3C_6H_2$	1-2 min	HOAc	—	—	(90)[d]	>98	(0)
Ph	KHMDS	2,4,6-(*i*-Pr)$_3C_6H_2$	1-2 min	HOAc	—	—	(82)[d]	82	(0)
Bn	NaHMDS	4-$O_2NC_6H_4$	1 h	TMSCl	–78°	1 h	(0)	—	(85)
Bn	KHMDS	4-$O_2NC_6H_4$	—	HOAc	—	—	(15)	—	(70)
Bn	KHMDS	4-MeC_6H_4	—	HOAc	—	—	(51)	—	(26)
Bn	KHMDS	4-MeC_6H_4	—	TFA	—	—	(0)	—	(57)
Bn	LDA	2,4,6-(*i*-Pr)$_3C_6H_2$	1 min	HOAc	–78° to rt	12 h	(74)	—	(0)
Bn	NaHMDS	2,4,6-(*i*-Pr)$_3C_6H_2$	30 s	HOAc	–78° to rt	3 h	(59)	—	(0)
Bn	KHMDS	2,4,6-(*i*-Pr)$_3C_6H_2$	1-2 min	HOAc	–78° to rt	3 h	(92)	94	(0)

	Reactant	Conditions	Product(s) and Yield(s) (%)	Refs.
C_{3-4}	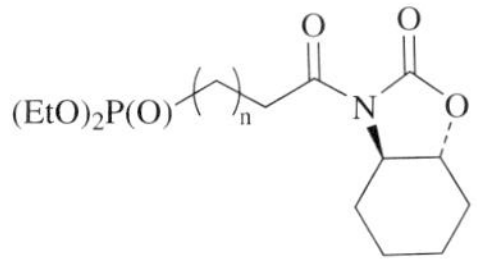 $(EtO)_2P(O)$	1. NaHMDS, THF, –78°, 30 min 2. 2,4,6-$(i\text{-}Pr)_3C_6H_2SO_2N_3$, THF (precooled), –78°, 30 min	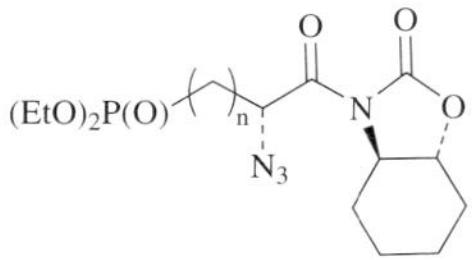 $(EtO)_2P(O)$ N_3 n = 1 (66) de >98% n = 2 (64) de —	894
C_3	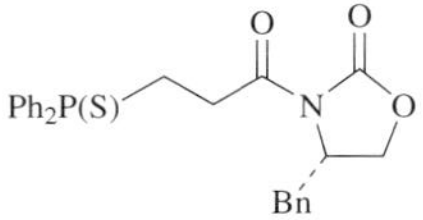 $Ph_2P(S)$ Bn	1. KHMDS (ia), THF, –78°, 30 min 2. 2,4,6-$(i\text{-}Pr)_3C_6H_2SO_2N_3$, –78°, 2 min	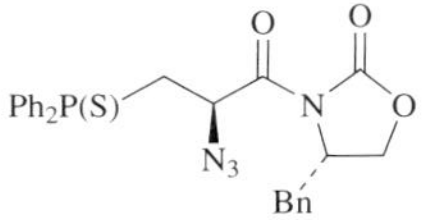 $Ph_2P(S)$ N_3 Bn (85), 94% de	458
C_4	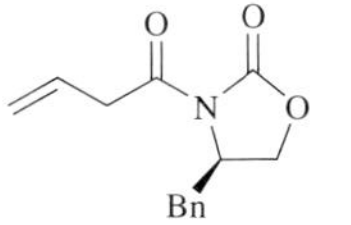 Bn	1. LDA, THF, –78°, 30 min 2. HMPA, –78°, 15 min 3. $t\text{-}BuO_2CN{=}NCO_2Bu\text{-}t$, –78°, 1 h	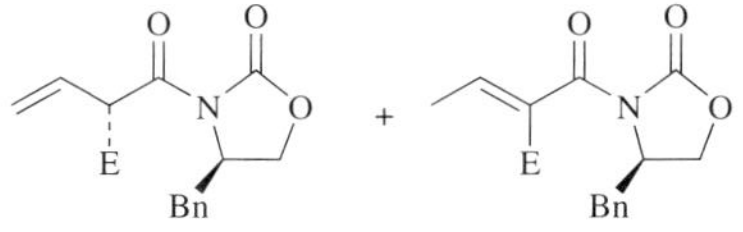 **I** (44) + **II** (18) 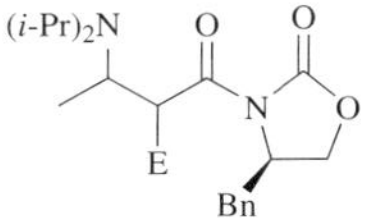 + $(i\text{-}Pr)_2N$ (7) E = $N(CO_2Bu\text{-}t)NHCO_2Bu\text{-}t$	895
		1. NaHMDS, THF, –78°, 30 min 2. HMPA, –78°, 15 min 3. $t\text{-}BuO_2CN{=}NCO_2Bu\text{-}t$, –78°, 1 h	**I** (61) + **II** (21)	895

TABLE 13. *N*-ACYLOXAZOLIDINONE ENOLATES (*Continued*)

Substrate	Conditions	Product(s) and Yield(s) (%)	Refs.
C_{4-5} (R[1], Bn)	$R^2O_2CN{=}NCO_2R^2$, catalyst (10 mol%), $ClCH_2CH_2Cl$; Catalyst A; Catalyst B	I + II; E = $N(CO_2R^2)NHCO_2R^2$	896
C_4 RO, Bn; R = TBDMS	1. NaHMDS, THF. –78°, 30 min 2. t-$BuO_2CN{=}NCO_2Bu$-t, –78°, 1 h	(71), 80% de; E = $N(CO_2Bu$-$t)NHCO_2Bu$-t	895
Bn	1. LDA, THF, –78°, 30 min 2. Addend 3. t-$BuO_2CN{=}NCO_2Bu$-t, –78°, 1 h	I + II; E = $N(CO_2Bu$-$t)NHCO_2Bu$-t	895

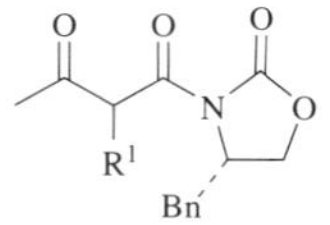

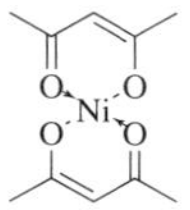

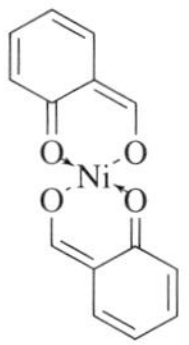

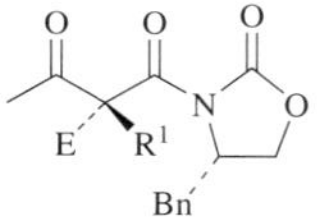

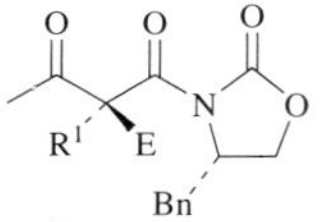

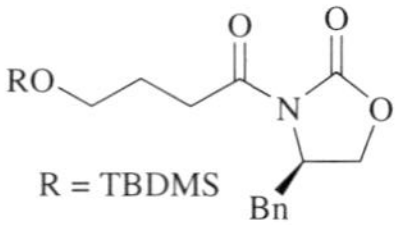

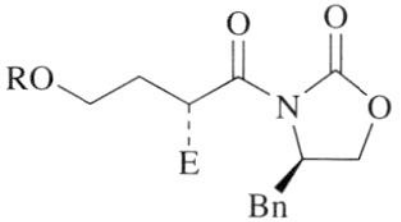

R^1	R^2	Catalyst	Temp	Time	I + II	I:II
H	Et	A	rt	18 h	(94)	73:27
H	Et	B	rt	12 h	(97)	74:26
H	Bn	A	0°	4 d	(100)	89:11
H	Bn	B	rt	17 h	(100)	70:30
H	Bn	B	0°	72 h	(100)	69:31
H	Bn	B	50°	17 h	(100)	63:27
Me	Et	B	50°	—	(0)	—

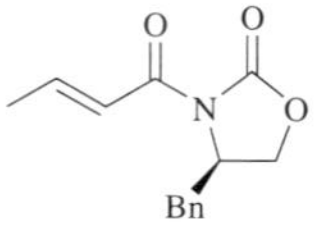

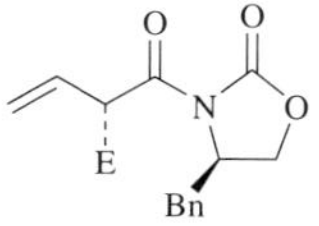

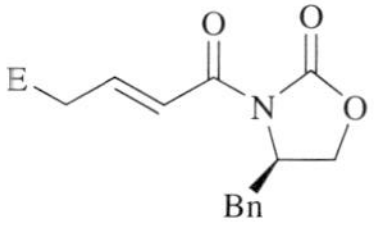

Addend	I	II
—	(51)	(33)
HMPA (1 eq)	(61)	(21)
HMPA (5 eq)	(52)	(11)

	Substrate	Conditions	Product(s) and Yield(s) (%)	Refs.
C_5	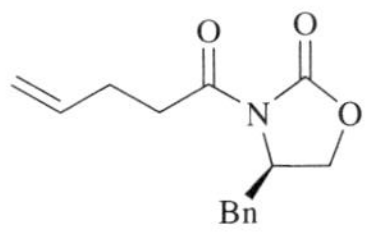 (S)-4-benzyl-3-(pent-4-enoyl)oxazolidin-2-one (Bn)	1. LDA, THF, –78°, 30 min 2. t-$BuO_2CN=NCO_2Bu$-t, –78°, 3 min	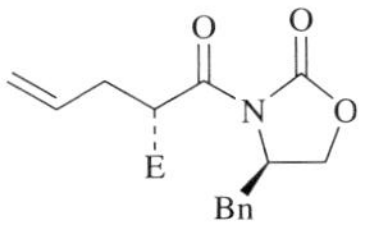 (86), 98% de E = $N(CO_2Bu$-t$)NHCO_2Bu$-t	895
	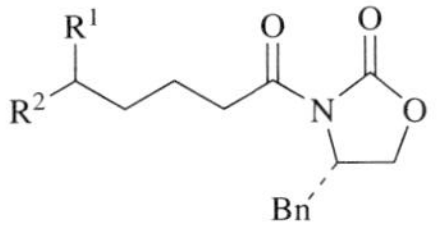 $R^1R^2CH(CH_2)_3CO$-oxazolidinone (Bn)	1. Base, THF, temp 1, time 1 2. t-$BuO_2CN=NCO_2Bu$-t, temp 2, time 2	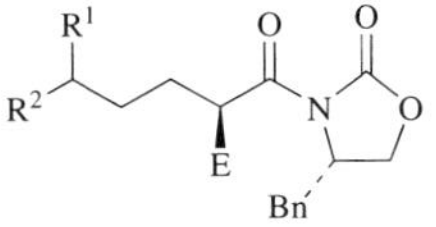 E = $N(CO_2Bu$-t$)NHCO_2Bu$-t	

R^1	R^2	Base	Temp 1	Time 1	Temp 2	Time 2		Refs.
MeO	MeO	LDA	–78°	—	–78°	—	(93)	897
—O(CH$_2$)$_2$O—		NaHMDS	–80°	30 min	–80°	3 min	(67)	896, 898

	Substrate	Conditions	Product(s) and Yield(s) (%)	Refs.
	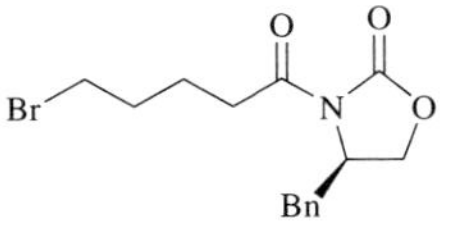 $Br(CH_2)_4CO$-oxazolidinone (Bn)	1. LDA (ia), THF, –78°, 45 min 2. t-$BuO_2CN=NCO_2Bu$-t, CH_2Cl_2, –78°, 15 min	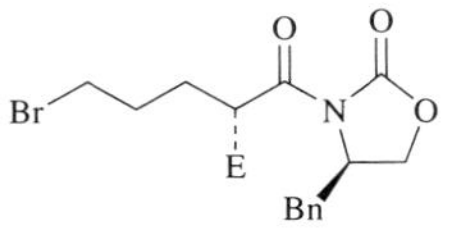 (84) E = $N(CO_2Bu$-t$)NHCO_2Bu$-t	441
	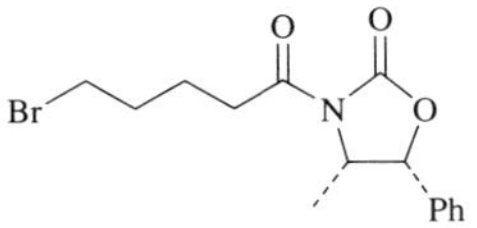 $Br(CH_2)_4CO$-oxazolidinone (Me, Ph)	1. LDA (ia), THF, hexane, –78°, 80 min 2. t-$BuO_2CN=NCO_2Bu$-t, –78°, 1 h 3. DMPU, –78° to rt	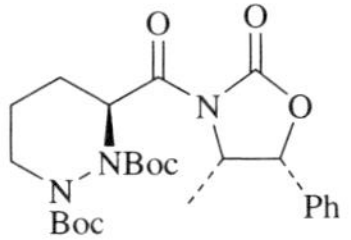 NBoc, N-Boc piperidine product (70)	441
	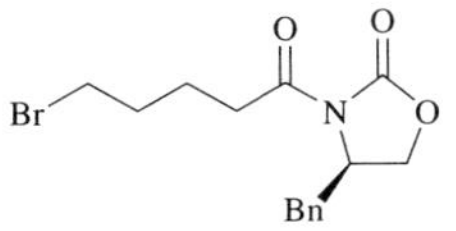 $Br(CH_2)_4CO$-oxazolidinone (Bn)	1. LDA (ia), THF, –78°, 2 h 2. t-$BuO_2CN=NCO_2Bu$-t, CH_2Cl_2, < –70°, 15 min 3. Bu_4NI (0.15 eq),[e] –78°, 15 min; –20°, 18 h	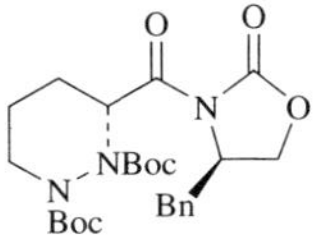 NBoc, N-Boc piperidine product (91), 94% de	442, 441
C_{5-7}	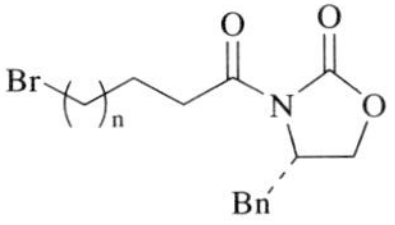 $Br(CH_2)_n(CH_2)_2CO$-oxazolidinone (Bn)	1. KHMDS, THF, –78°, 30 min 2. 2,4,6-(i-Pr)$_3C_6H_2SO_2N_3$	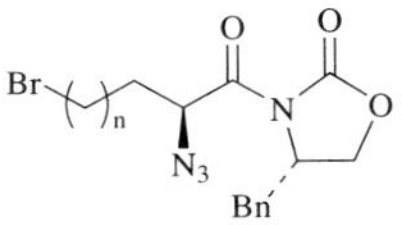 N_3 product (see below)	443

n		% de
2	(60-70)	95
3	(40)[f]	95
4	(60-70)	95

TABLE 13. *N*-ACYLOXAZOLIDINONE ENOLATES (*Continued*)

	Substrate	Conditions	Product(s) and Yield(s) (%)	Refs.
C_5	R[1], R[2], N_3, O, O, N, O, Bn mixture of R^1 = T; R^2 = H and R^1 = H and R^2 = T	1. KHMDS (ia), THF, –78°, 30 min 2. 2,4,6-$(i\text{-Pr})_3C_6H_2SO_2N_3$, THF, –78°, 1-2 min	R[1], R[2], N_3, O, O, N, O, N_3, Bn (46), 82.5% de	899
	PhSe, O, O, N, O, Bn	1. KHMDS (ia), toluene, THF, –78°, 40 min 2. 2,4,6-$(i\text{-Pr})_3C_6H_2SO_2N_3$, THF, –78°, 4 min	PhSe, O, O, N, O, N_3, Bn (80)	456
C_6	O, O, N, O, Ph, R, R	1. Tris(dipivaloylmethanato)manganese(III) (5 mol%), *i*-PrOH (ia), 0° 2. *t*-BuO_2CN=NCO_2Bu-*t*, *i*-PrOH 3. Ph_3SiH, 0°, 90 min	O, O, N, O, E, Ph, R, R E = $N(CO_2Bu\text{-}t)NHCO_2Bu\text{-}t$ R: H (75) dr 68:32; Ph (51) dr 65:35	890, 900
	OTBDPS, O, O, N, O, MeN, CO_2Bu-*t*, Bn	1. KHMDS, THF, –78°, 30 min 2. 2,4,6-$(i\text{-Pr})_3C_6H_2SO_2N_3$, –78°, 1 min	OTBDPS, O, O, N, O, MeN, CO_2Bu-*t*, N_3, Bn (>40), 93% de	447
$C_{6\text{-}7}$	R[1], O, O, N, O, R[2]	1. Base (x eq), THF, –78° 2. 2,4,6-$(i\text{-Pr})_3C_6H_2SO_2N_3$, –78°, time	R[1], O, O, N, O, N_3, R[2] **I** + R[1], O, O, N, O, N_2 **II**	

R^1	R^2	Base	x	Time	**I**	**II**	Refs.
TBDPSO	Ph	KHMDS	—	2 min	(82)	(—)	901, 902
BzO	Bn	NaHMDS	1.2	3 min	(73)	(—)	903
Me	Bn	KHMDS	1.2	—	(10)	(20)	437
Me	Bn	KHMDS	1.5	—	(76)	(—)	437

C_6

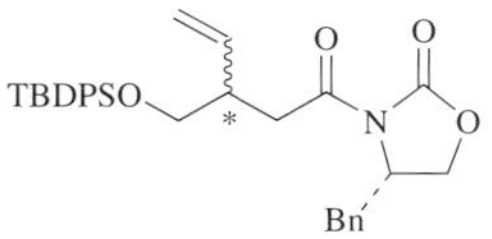

1. KHMDS, THF, –78°, 30 min
2. 2,4,6-(i-Pr)$_3C_6H_2SO_2N_3$, –78°, 2 min

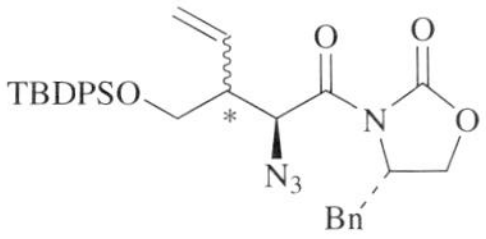

* Config.	
R	(86)
S	(80)

904

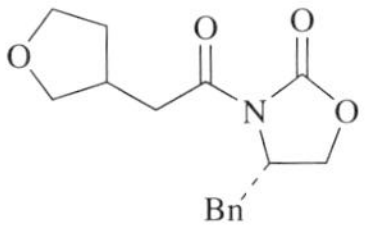

1. KHMDS, THF, –78°, 30 min
2. 2,4,6-(i-Pr)$_3C_6H_2SO_2N_3$, –78°, 2 min

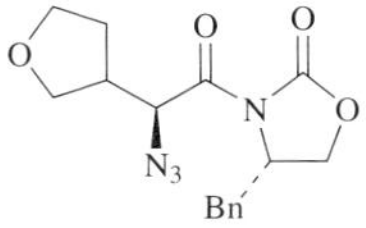

(84) 905

$C_{6\text{-}12}$

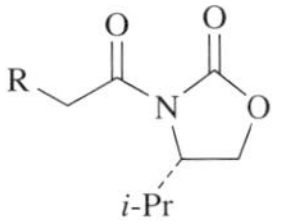

1. KHMDS (ia), THF, –78°, 45 min
2. 2,4,6-(i-Pr)$_3C_6H_2SO_2N_3$, –78°, 5 min

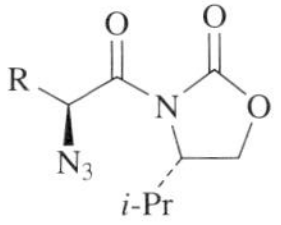

R	
CMe_2CO_2Bn	(55)
1-adamantyl	(27)

906

C_7

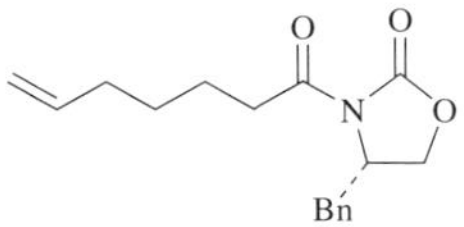

1. KHMDS, THF, –78°
2. 2,4,6-(i-Pr)$_3C_6H_2SO_2N_3$, –78°

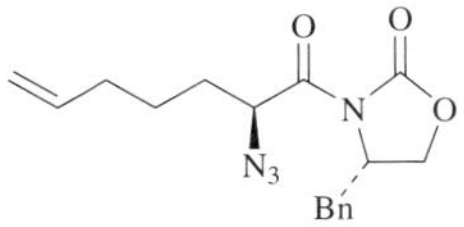

(67) 907

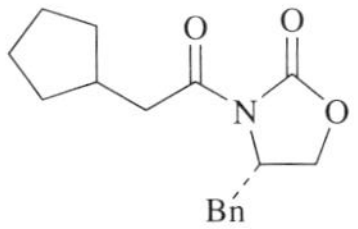

1. KHMDS, THF, –78°
2. 2,4,6-(i-Pr)$_3C_6H_2SO_2N_3$, –78°

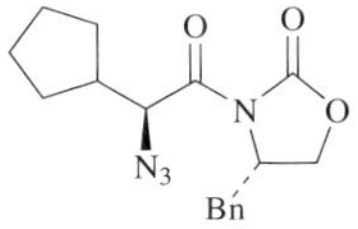

(—) 908

TABLE 13. *N*-ACYLOXAZOLIDINONE ENOLATES (*Continued*)

	Substrate	Conditions	Product(s) and Yield(s) (%)	Refs.
C_{7-8}	NR^2CO_2Bu-*t*, R^1, O, O, N, O, Bn R^1, R^2 = H, H; H, Me; Me, H; Me,Me	1. KHMDS, THF, –78° 2. 2,4,6-(*i*-Pr)$_3C_6H_2SO_2N_3$, –78°	NR^2CO_2Bu-*t*, R^1, O, O, N, O, N_3, Bn (40-50), de >95%	451
C_7	N, N, Bn, O, O, N, O, Ph	1. KHMDS (ia), THF, –50°, 45 min 2. 2,4,6-(*i*-Pr)$_3C_6H_2SO_2N_3$, –78° 2 min	N, N, Bn, O, O, N, O, N_3, Ph (72)	909
	N, Cbz, H, O, O, N, O, R^1, R^2	1. KHMDS, THF, –78°, 30 min 2. 2,4,6-(*i*-Pr)$_3C_6H_2SO_2N_3$, –78°, 5 min	N, Cbz, H, O, O, N, O, N_3, R^1, R^2 R^1 R^2 threo:erythro H H (—) 1:2 Me Ph (68) 100:0	871 454
	O, O, N, O, Bn, O, O, N, O, Bn	1. KHMDS (2 eq, ia), THF, –78°, 30 min 2. 2,4,6-(*i*-Pr)$_3C_6H_2SO_2N_3$ (2 eq), –78°, 30 min	O, O, N, O, N_3, Bn, N_3, O, O, N, O, Bn (35) + O, N_3, O, N_3 (8)	910
	O, O, N, O, Bn, O, O, N, O, Bn	1. KHMDS (2 eq, ia), THF, –78°, 30 min 2. 2,4,6-(*i*-Pr)$_3C_6H_2SO_2N_3$ (2 eq), –78°, 30 min	H, 2,4,6-(*i*-Pr)$_3C_6H_2SO_2NH$, O, O, O, N, O, Bn (—)	910

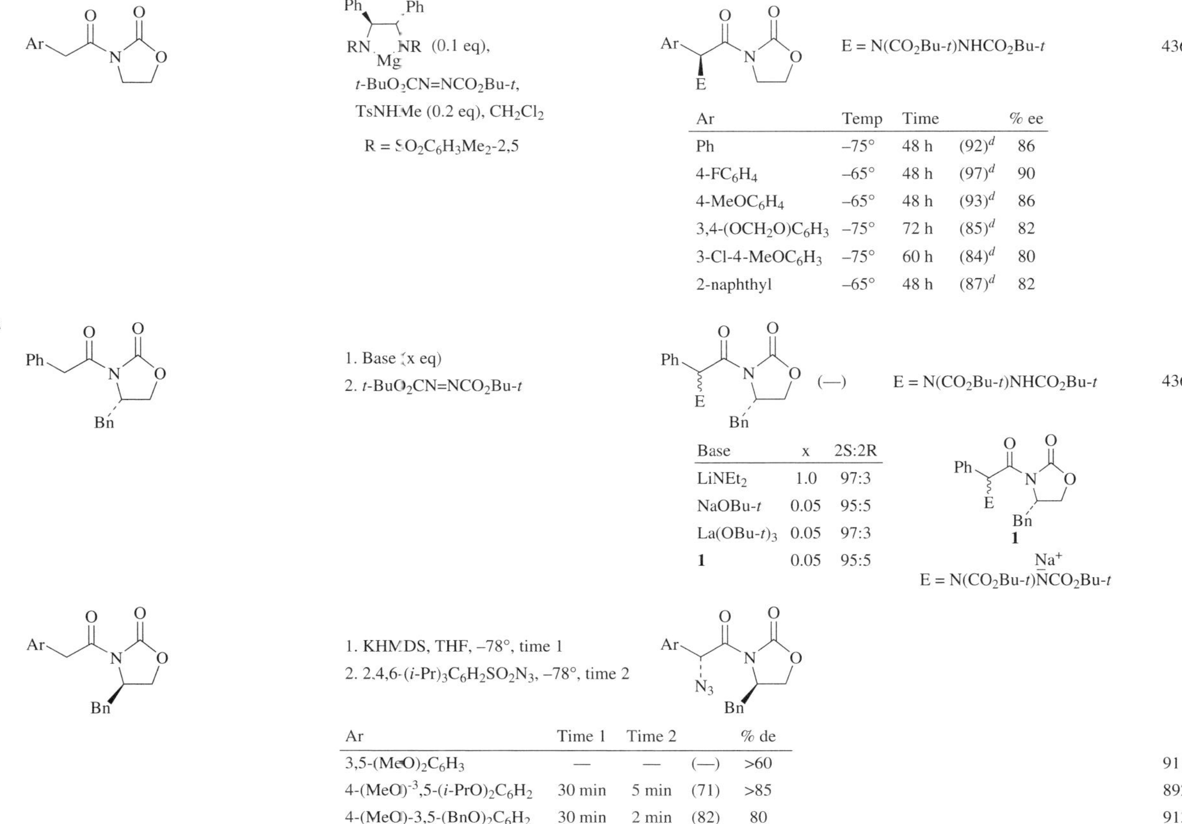

Ar	Temp	Time		% ee
Ph	–75°	48 h	(92)[d]	86
4-FC$_6$H$_4$	–65°	48 h	(97)[d]	90
4-MeOC$_6$H$_4$	–65°	48 h	(93)[d]	86
3,4-(OCH$_2$O)C$_6$H$_3$	–75°	72 h	(85)[d]	82
3-Cl-4-MeOC$_6$H$_3$	–75°	60 h	(84)[d]	80
2-naphthyl	–65°	48 h	(87)[d]	82

Base	x	2S:2R
LiNEt$_2$	1.0	97:3
NaOBu-t	0.05	95:5
La(OBu-t)$_3$	0.05	97:3
1	0.05	95:5

Ar	Time 1	Time 2		% de	
3,5-(MeO)$_2$C$_6$H$_3$	—	—	(—)	>60	911
4-(MeO)-3,5-(i-PrO)$_2$C$_6$H$_2$	30 min	5 min	(71)	>85	892, 912
4-(MeO)-3,5-(BnO)$_2$C$_6$H$_2$	30 min	2 min	(82)	80	913

TABLE 13. *N*-ACYLOXAZOLIDINONE ENOLATES (*Continued*)

Substrate	Conditions	Product(s) and Yield(s) (%)	Refs.
C_{8-9}	1. KHMDS, THF, –78°, time 2. 2,4,6-(i-Pr)$_3C_6H_2SO_2N_3$, –78°, 2 min		
		Ar / Time / % de	
		3-$BnOC_6H_4$ / 30 min / (30-50) / —	914
		3-BnO-4-$MeOC_6H_3$ / 23 min / (90) / 92	915
		4-$ClC_6H_4CH_2$ / 30 min / (68) / >98	914
C_8	1. KHMDS, THF, –78°, 30 min 2. 2,4,6-(i-Pr)$_3C_6H_2SO_2N_3$, –78°, 2 min	(26) + (26)	916
C_{8-22}	1. KHMDS, THF, –78°, time 1 2. 2,4,6-(i-Pr)$_3C_6H_2SO_2N_3$, –78°, time 2		

Ar	Time 1	Time 2		% de	Refs.
Ph	15-45 min	1-2 min	(82)	82	450
4-FC_6H_4	30 min	2 min	(67)	—	917
4-ClC_6H_4	20 min	5 min	(—)	—	918
3,5-$(MeO)_2C_6H_3$	15-45 min	1-2 min	(78)	80	450
3-TBSO-4-$MeOC_6H_3$	20 min	5 min	(75)	—	918
3,5-$(BnO)_2C_6H_3$	—	—	(81)	—	919
3,5-$(BnO)_2$-4-MeC_6H_2	15-45 min	1-2 min	(81)	76	450
3-(CH_2=$CHCH_2O$)-4-Me-5-$BnOC_6H_2$	15-45 min	1-2 min	(75)	80	450
3,5-(3,4-$Cl_2C_6H_3CH_2)_2C_6H_3$	15-45 min	1-2 min	(76)	76	450

NRCO$_2$Bu-*t*

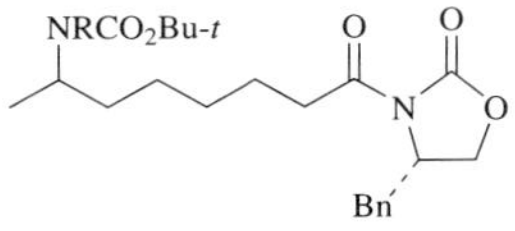

1. KHMDS (2 eq), THF, –78°, 20 min
2. 2,4,6-(*i*-Pr)$_3$C$_6$H$_2$SO$_2$N$_3$, –78°, 1 min

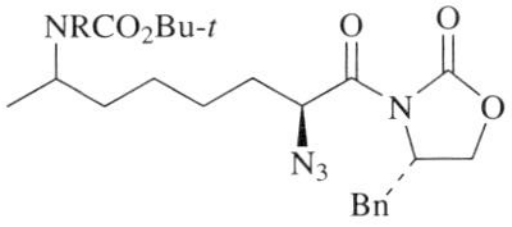

R	
H	(54)
Me	(51)

452

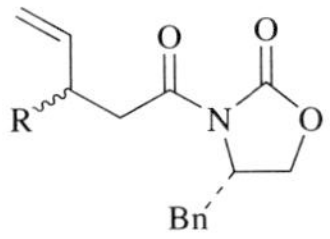

C$_{8-11}$

1. KHMDS (ia), THF, –78°, 30 min
2. 2,4,6-(*i*-Pr)$_3$C$_6$H$_2$SO$_2$N$_3$, –78°, 1-2 min

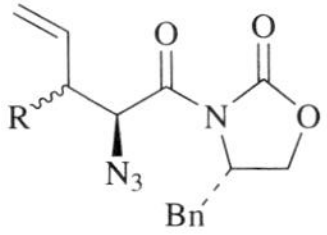

R	
i-Pr (R,S)	(63)
Ph (R)	(68)
Ph (S)	(68)

920

C$_{8-13}$

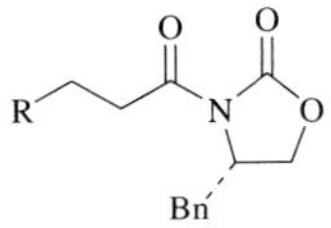

1. KHMDS (ia), THF, –78°, 60 min
2. 2,4,5-(*i*-Pr)$_3$C$_6$H$_2$SO$_2$N$_3$, –78°, 60 min

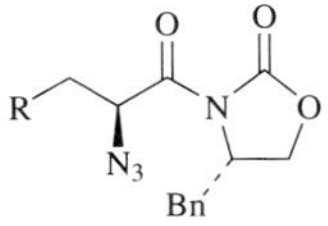

R	
c-C$_5$H$_9$	(—)
c-C$_6$H$_{11}$CH$_2$	(61)
c-C$_8$H$_{15}$	(—)
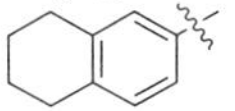	(—)
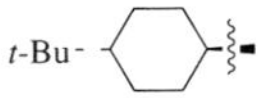 *t*-Bu-	(—)

438

C$_8$

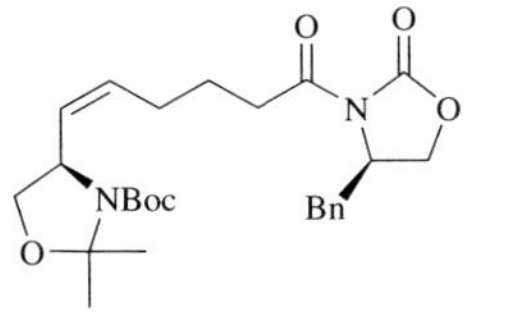

1. KHMDS (ia), THF, –78°, 80 min
2. 2,4,6-(*i*-Pr)$_3$C$_6$H$_2$SO$_2$N$_3$
(moce of addition), temp, time

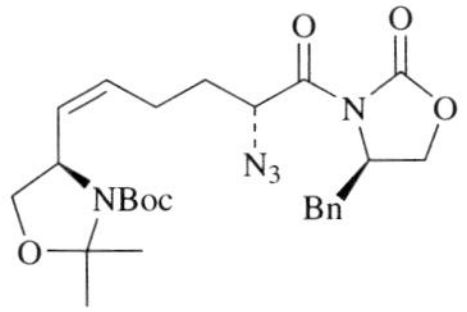

Moce of Addition	Temp	Time	
azide solution precooled to –78°	–78°	220 sec	(20-40)
azide solution precooled to –95°, insulated cannula	–95° to –100°	220 sec	(45-82)
azide added as a solid	–78°	200 sec	(75-95)

440

TABLE 13. *N*-ACYLOXAZOLIDINONE ENOLATES (*Continued*)

Substrate	Conditions	Product(s) and Yield(s) (%)	Refs.
C_9 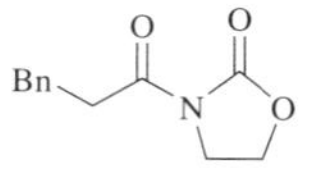	1. Li base, CuI, THF 2. TosON(Li)CO_2Bu-*t*, –50°, 1 h	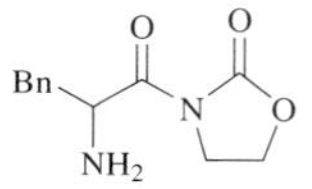 (55)	126
	1. LiHMDS (ia), THF, –78°, 60 min 2. 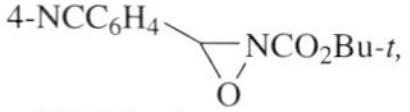 4-NCC_6H_4–(oxaziridine)–NCO_2Bu-*t*, –78°, 30 min	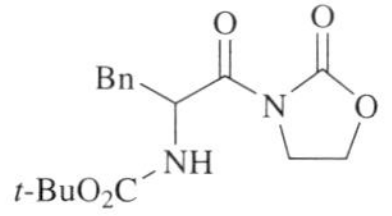 (33)	153, 157
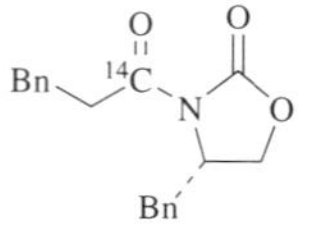	1. LDA (ia), THF, –78°, 1.5 h 2. *t*-$BuO_2CN{=}NCO_2$Bu-*t*, THF, –78°, 30 min	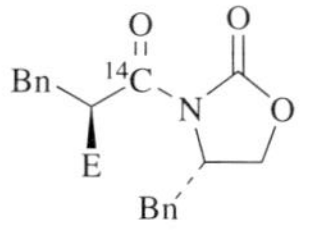 (46) E = N(CO_2Bu-*t*)$NHCO_2$Bu-*t*	921
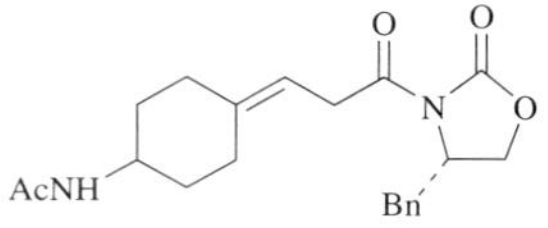	1. KHMDS (1.9 eq), THF, –78°, 30 min 2. *t*-$BuO_2CN{=}NCO_2$Bu-*t* (ia), CH_2Cl_2, –78°, 3 min	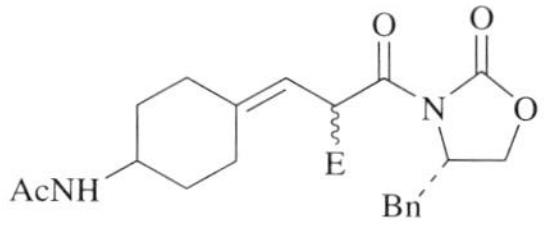 (53)[h] E = N(CO_2Bu-*t*)$NHCO_2$Bu-*t*	453
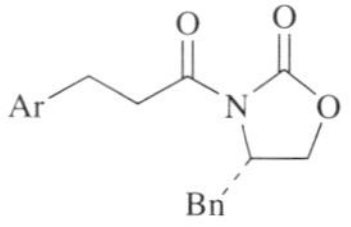	1. KHMDS, THF, –78°, time 1 2. 2,4,6-(*i*-Pr)$_3C_6H_2SO_2N_3$, –78°, time 2	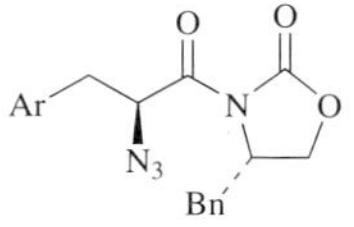	

Ar	Time 1	Time 2		dr	Refs.
Ph	15 min	2 min	(—)	—	444
3-$BnOC_6H_4$	—	—	(59)	95:5	774, 922
3-(*i*-PrO)-4-$MeOC_6H_3$	30 min	2 min	(83)	—	923

	Reactant	Conditions	Product(s) and Yield(s) (%)	Refs.
	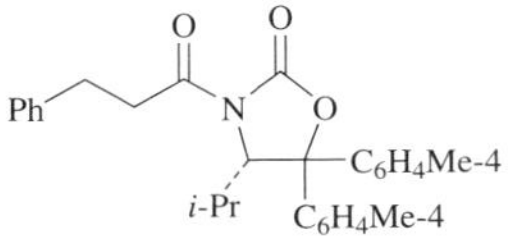 Ph–CH₂CH₂–C(O)–oxazolidinone (4-i-Pr, 5,5-bis(C_6H_4Me-4))	1. KHMDS (ia), THF, –78°, 30 min 2. 2,4,6-(i-Pr)$_3C_6H_2SO_2N_3$, –78°, 2 min	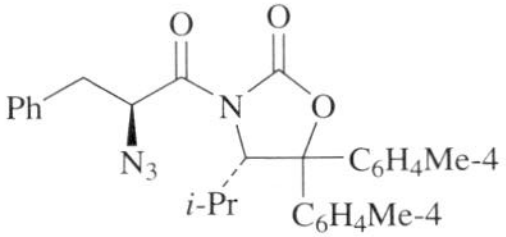 Ph–CH₂–CH(N_3)–C(O)–oxazolidinone (4-i-Pr, 5,5-bis(C_6H_4Me-4)) (65), 92% de	924
C_{10-11}	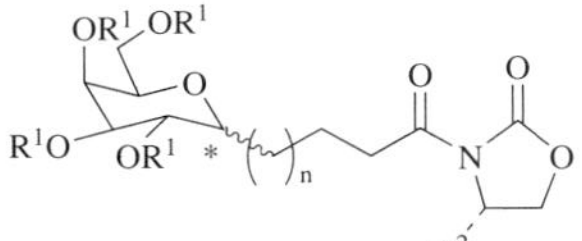 Glycoside (OR^1)–*–$(CH_2)_n$–CH₂CH₂–C(O)–oxazolidinone (R^2)	1. Base, THF, –78°, 30 min 2. t-BuO$_2$CN=NCO$_2$Bu-t (ia), –78°, 5 min	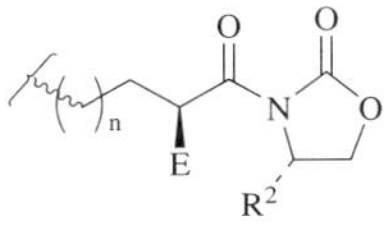 Glycoside–$(CH_2)_n$–CH₂–CH(E)–C(O)–oxazolidinone (R^2) E = N(CO$_2$Bu-t)NHCO$_2$Bu-t	

R^1	R^2	*	n	Base		% de	Refs.
Me	i-Pr	α	1	KHMDS	(61)	92	439
Bn	H	β	1	LDA	(70)	73[i]	459
Bn	i-Pr	α	1	KHMDS	(72)	72	439
Bn	i-Pr	β	1	KHMDS	(68)	98	439
Bn	i-Pr	β	2	LDA	(65)	98	459

	Reactant	Conditions	Product(s) and Yield(s) (%)	Refs.
C_{10}	Benzene-bis(CH₂C(O)–oxazolidinone (4-Bn))	1. KHMDS (2.1 eq) (ia), THF, –78°, 30 min 2. 2,4,6-(i-Pr)$_3C_6H_2SO_2N_3$ (2.2 eq), –78°, 10 min	Benzene-bis(CH(N_3)C(O)–oxazolidinone (4-Bn))	925

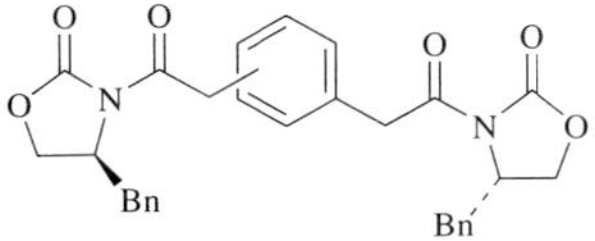

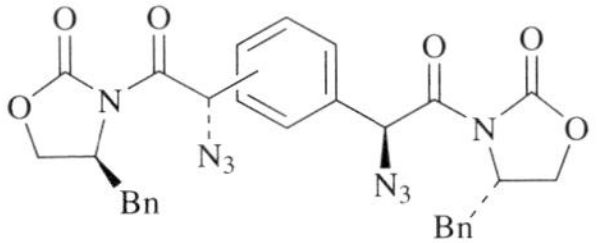

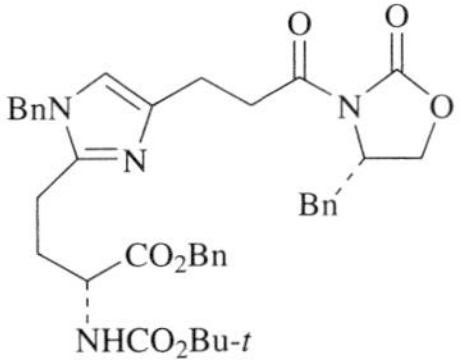

Connection		% de
1,3	(73)	>95
1,4	(65)	60

	Reactant	Conditions	Product(s) and Yield(s) (%)	Refs.
	1-Bn-imidazol-4-yl–CH₂CH₂–C(O)–oxazolidinone (4-Bn), 2-substituent CH₂CH₂CH(CO$_2$Bn)NHCO$_2$Bu-t	1. KHMDS 2. 2,4,6-(i-Pr)$_3C_6H_2SO_2N_3$	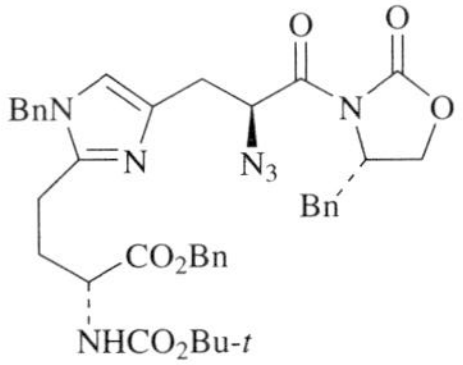 1-Bn-imidazol-4-yl–CH₂–CH(N_3)–C(O)–oxazolidinone (4-Bn), 2-substituent CH₂CH₂CH(CO$_2$Bn)NHCO$_2$Bu-t (>62), 52% de	449

TABLE 13. *N*-ACYLOXAZOLIDINONE ENOLATES (*Continued*)

Substrate	Conditions	Product(s) and Yield(s) (%)	Refs.
C_{10} 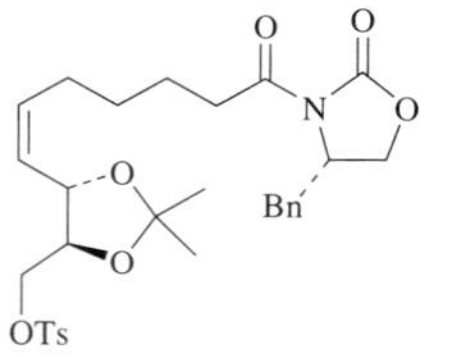 (structure: O, O, N, O, Bn, O, O, OTs)	1. KHMDS, THF, –78°, 30 min 2. 2,4,6-$(i\text{-}Pr)_3C_6H_2SO_2N_3$, –78°, 3 min	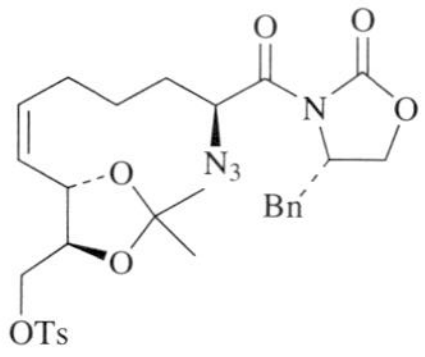 (structure: O, O, N, O, N_3, Bn, O, O, OTs) (73)	455
C_{11} 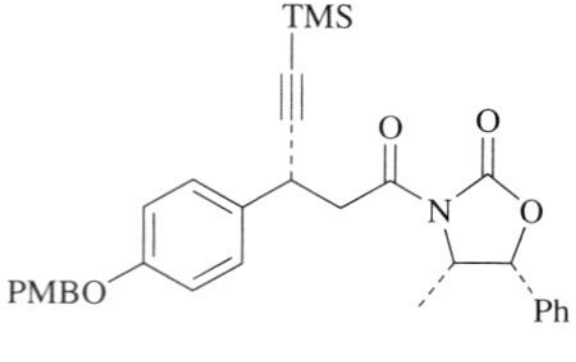 (structure: TMS, O, O, N, O, PMBO, Ph)	1. KHMDS, THF, –100°, 30 min 2. 2,4,6-$(i\text{-}Pr)_3C_6H_2SO_2N_3$ (3 eq), –78°, 5 min	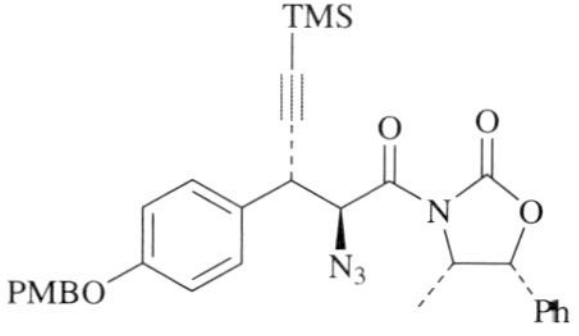 (structure: TMS, O, O, N, O, N_3, PMBO, Ph) (50), 80% de	782
$C_{11\text{-}13}$ (structure: R, O, O, Ar, N, O, Ph)	1. KHMDS, THF 2. 2,4,6-$(i\text{-}Pr)_3C_6H_2SO_2N_3$, –78°	(structure: R, O, O, Ar, N, O, N_3, Ph)	
		Ar / R / % de	
		2,6-Me_2-4-$MeOC_6H_2$ / *i*-Pr / (80) >95	926
		2-naphthyl / Me / (94) >95	927
C_{12} (structure: O, O, N, O, N, R, Ph)	1. KHMDS, THF, –78°, 30 min 2. 2,4,6-$(i\text{-}Pr)_3C_6H_2SO_2N_3$, –78°, 5 min	(structure: O, O, N, O, N_3, N, R, Ph)	
		R	
		CO_2Bu-*t* (40)	928
		2,4,6-$Me_3C_6H_2SO_2$ (59)	929, 930
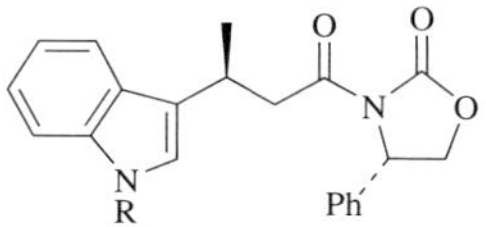 (structure: O, O, N, O, N, R, Ph) R = 2,4,6-$Me_3C_6H_2SO_2$	1. KHMDS (ia), THF, –78°, 30 min 2. 2,4,6-$(i\text{-}Pr)_3C_6H_2SO_2N_3$ (ia), –78°, 5 min	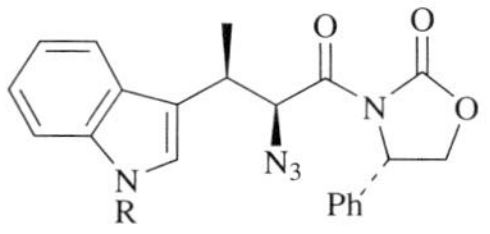 (structure: O, O, N, O, N_3, N, R, Ph) (62)	929, 930

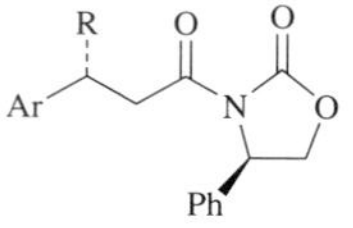

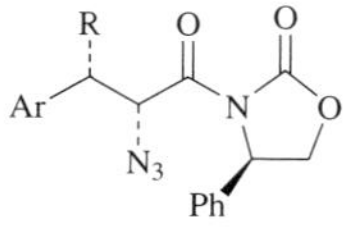

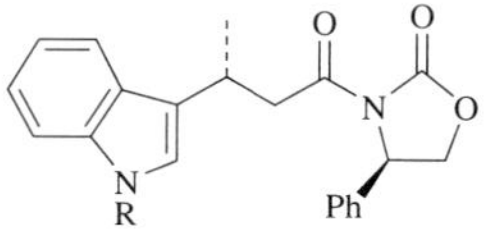

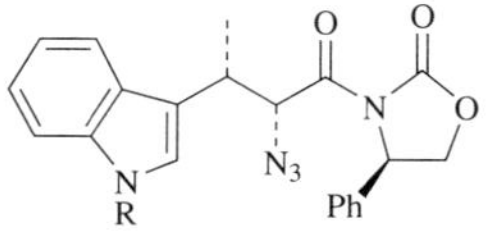

Reactant	Conditions	Product(s) and Yield(s) (%)	Refs.
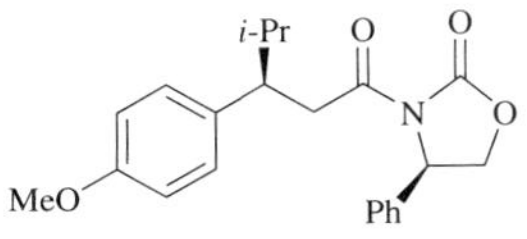 MeO, i-Pr, O, O, N, O, Ph	1. KHMDS (ia), THF, –78°, 30 min 2. 2,4,6-$(i\text{-Pr})_3C_6H_2SO_2N_3$, –78°, 15 min	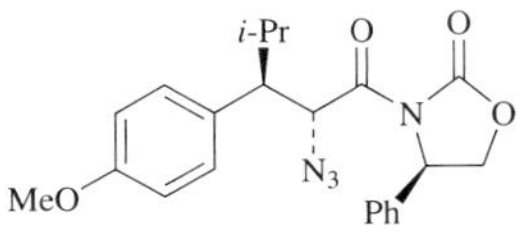 i-Pr, MeO, N_3, Ph (85), >90 de	434
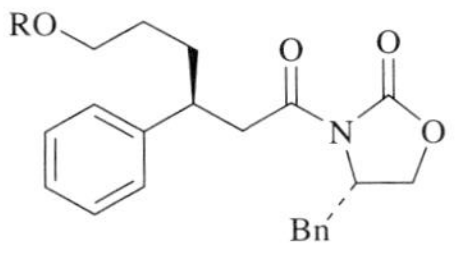 RO, O, O, N, O, Bn R = $(t\text{-BuO})_2P(O)$	1. KHMDS (ia), THF, –78°, 40 min 2. 2,4,6-$(i\text{-Pr})_3C_6H_2SO_2N_3$, –78°, 2 min	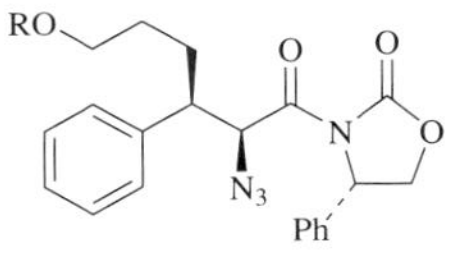 RO, N_3, Ph R: TBS (95); Bn (73)	931
C_{13} O, O, N, O, Bn	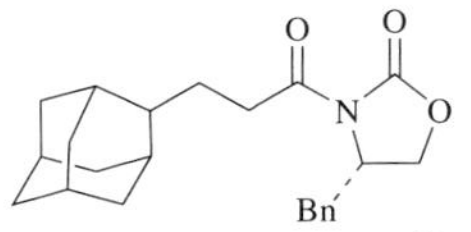 1. LDA, THF, –78°, 30 min 2. $t\text{-BuO}_2CN{=}NCO_2Bu\text{-}t$, –78°, 1 h	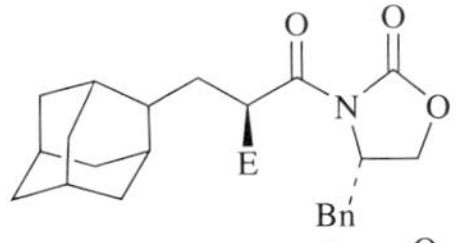 E, Bn (82) E = $N(CO_2Bu\text{-}t)NHCO_2Bu\text{-}t$	438
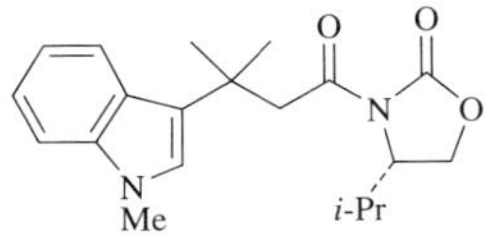 N, Me, O, O, N, O, i-Pr	1. KHMDS (ia), THF, –78° 2. 2,4,6-$(i\text{-Pr})_3C_6H_2SO_2N_3$, –78°, 1 min	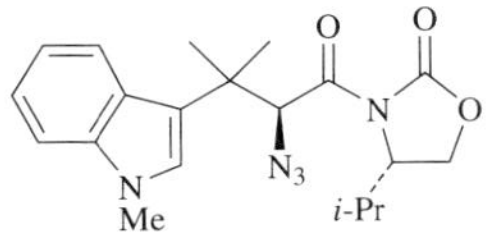 N, Me, N_3, i-Pr (70), >98% de	932
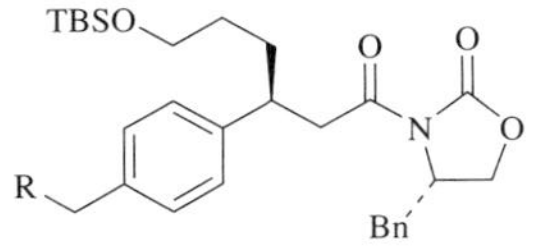 TBSO, R, O, O, N, O, Bn	1. KHMDS (1.2 eq, ia), THF, –78°, 40 min 2. 2,4,6-$(i\text{-Pr})_3C_6H_2SO_2N_3$, –78°, 2 min	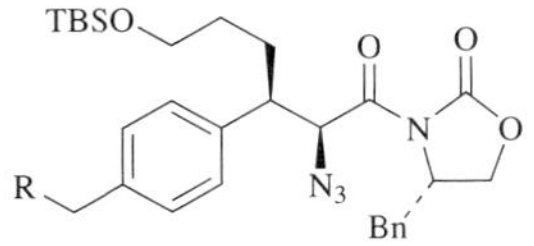 TBSO, R, N_3, Bn (35) R = $P(O)(OBu\text{-}t)_2$	457
C_{15} Ph, Ph, O, O, N, O, Ph	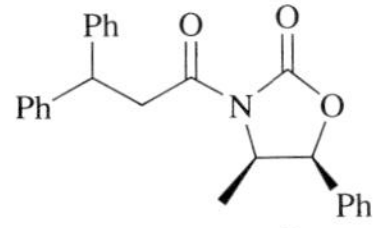 1. KHMDS, THF, –78° 2. 2,4,6-$(i\text{-Pr})_3C_6H_2SO_2N_3$, –78°	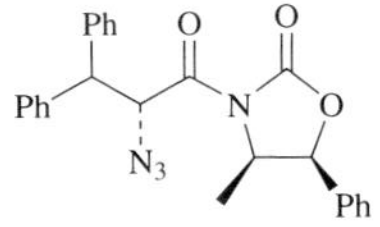 Ph, Ph, N_3, Ph (94)	933
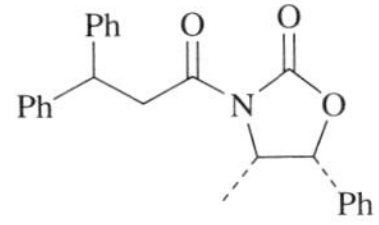 Ph, Ph, O, O, N, O, Ph	1. KHMDS, THF, –78° 2. 2,4,6-$(i\text{-Pr})_3C_6H_2SO_2N_3$, –78°	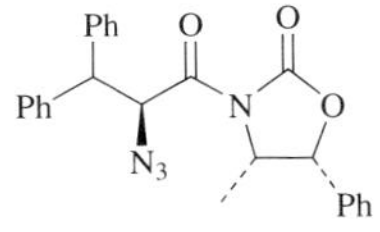 Ph, Ph, N_3, Ph (—)	933

TABLE 13. *N*-ACYLOXAZOLIDINONE ENOLATES (*Continued*)

Substrate	Conditions	Product(s) and Yield(s) (%)	Refs.
C_{15} OEt, *i*-Pr, N, N, OEt, O, O, N, O, 5, *i*-Pr	1. KHMDS (ia), THF, –78°, 30 min 2. 2,4,6-(*i*-Pr)$_3$C$_6$H$_2$SO$_2$N$_3$, –78°, 2 min	OEt, *i*-Pr, N, N, OEt, 5, N$_3$, O, O, N, O, *i*-Pr (84)	934
C_{17} O, O, N, O, Ph	1. KHMDS, THF, –78° 2. 2,4,6-(*i*-Pr)$_3$C$_6$H$_2$SO$_2$N$_3$, –78°	O, O, N, O, N$_3$, Ph (85)	935
O, O, N, O, Ph	1. KHMDS, THF, –78° 2. 2,4,6-(*i*-Pr)$_3$C$_6$H$_2$SO$_2$N$_3$, –78°	O, O, N, O, N$_3$, Ph (—)	935
H, N, N, 5, O, O, N, O, Bn	1. KHMDS (2.3 eq), THF, –78°, 30 min 2. 2,4,6-(*i*-Pr)$_3$C$_6$H$_2$SO$_2$N$_3$, –78°, 3 min 3. HOAc, –78° to rt, overnight 4. NaHCO$_3$, H$_2$O	H, N, N, 5, N$_3$, O, O, N, O, Bn (34) + 5, N$_3$, O, NH$_2$ (24)	445, 446
R, R, R, Cl, O, O, N, O, Bn R = BnO	1. KHMDS, THF, –78°, 15-45 min 2. 2,4,6-(*i*-Pr)$_3$C$_6$H$_2$SO$_2$N$_3$, –78°, 1-2 min	R, R, R, Cl, N$_3$, O, O, N, O, Bn (77), >90% de	450

C_{17}	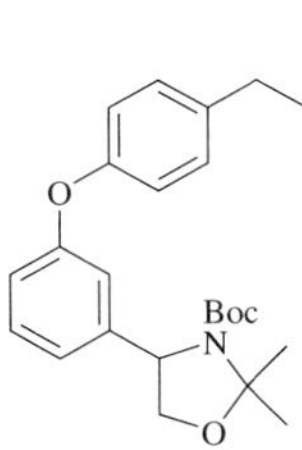 1. KHMDS, THF, –78°, 30 min 2. 2,4,6-$(i\text{-Pr})_3C_6H_2SO_2N_3$, –78°, 2 min	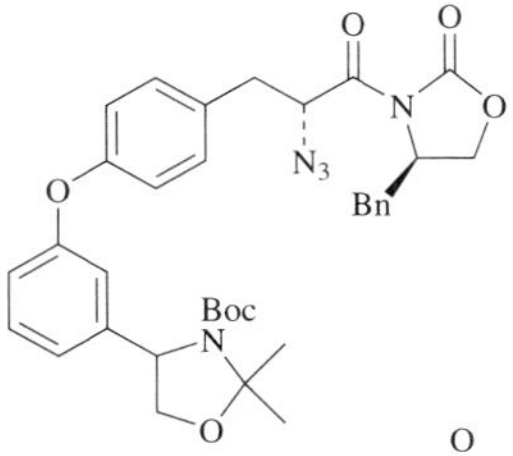 (85)	448
C_{18}	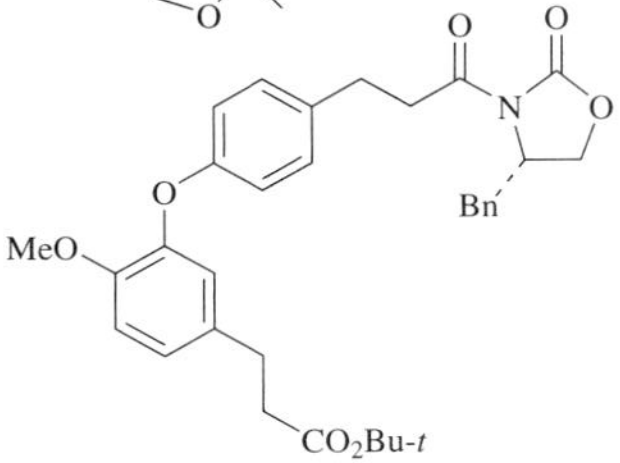 1. KHMDS, THF, –78°, 15 min 2. 2,4,6-$(i\text{-Pr})_3C_6H_2SO_2N_3$, –78°, 2 min	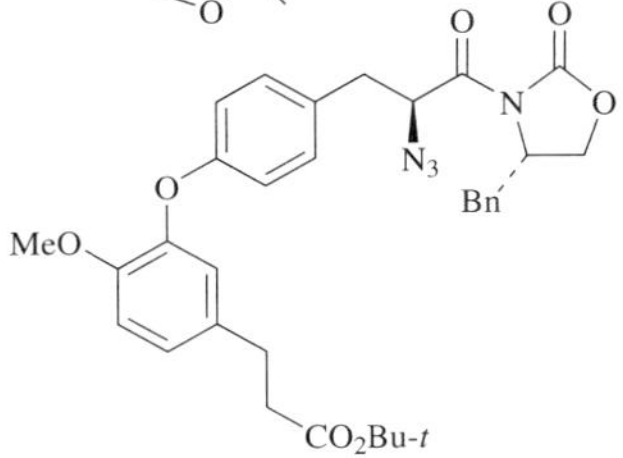 (83-86),[d] 95% de[g]	444
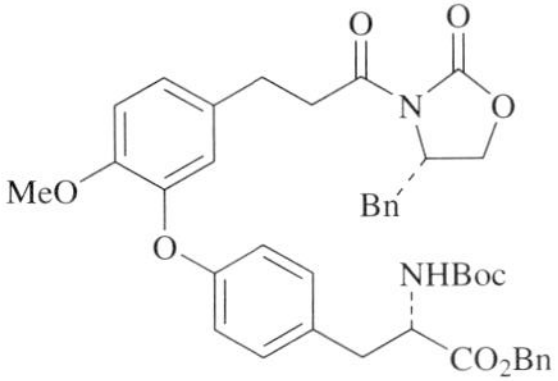	1. KHMDS (2.2 eq), THF, –78°, 15 min 2. 2,4,5-$(i\text{-Pr})_3C_6H_2SO_2N_3$, –78°, 1-2 min	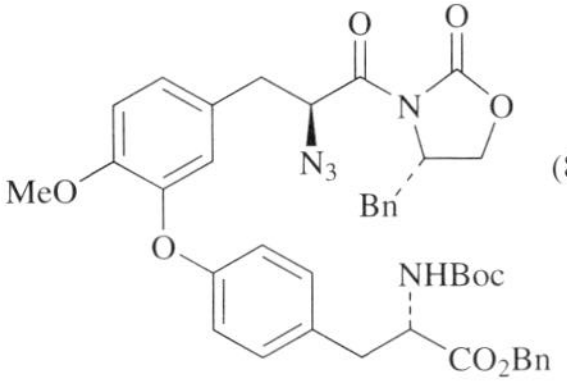 (85)	433
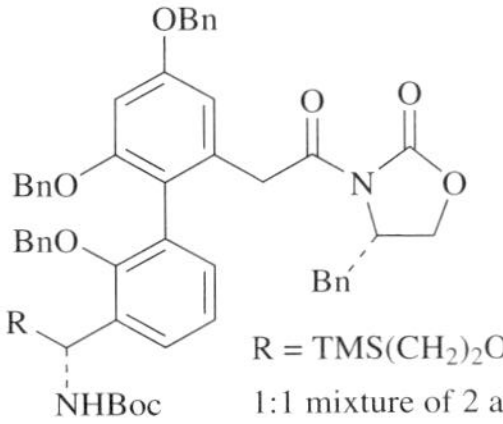 R = $TMS(CH_2)_2O_2C$ 1:1 mixture of 2 atropisomers	1. KHMDS (2.2 eq), THF, –78°, 15 min 2. 2,4,6-$(i\text{-Pr})_3C_6H_2SO_2N_3$, –78°, 1 min	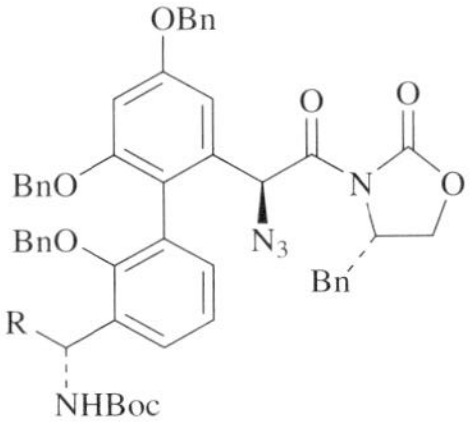 (60), 86% de[j]	450

TABLE 13. *N*-ACYLOXAZOLIDINONE ENOLATES (*Continued*)

Substrate	Conditions	Product(s) and Yield(s) (%)	Refs.
C_{19-21}			
MEMO, R, O, O, N, O, Ph	1. KHMDS (1 eq), NaH (1 eq), THF, –78°, 30 min 2. 2,4,6-(*i*-Pr)$_3$C$_6$H$_2$SO$_2$N$_3$, –78°, 15 min	MEMO, R, O, O, N, O, N_3, Ph R: *t*-Bu (54); Ph (73)	435, 936
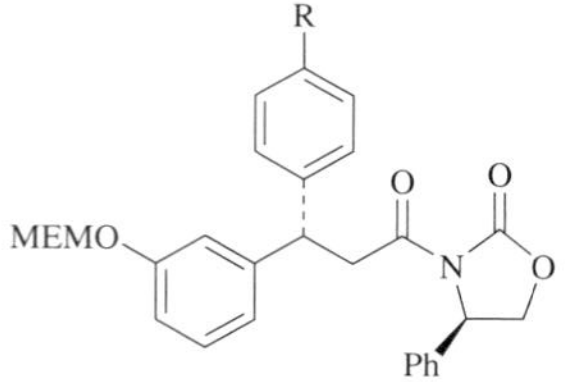 MEMO, R, O, O, N, O, Ph	1. KHMDS (1 eq), NaH (1 eq), THF, –78°, 30 min 2. 2,4,6-(*i*-Pr)$_3$C$_6$H$_2$SO$_2$N$_3$, –78°, 15 min	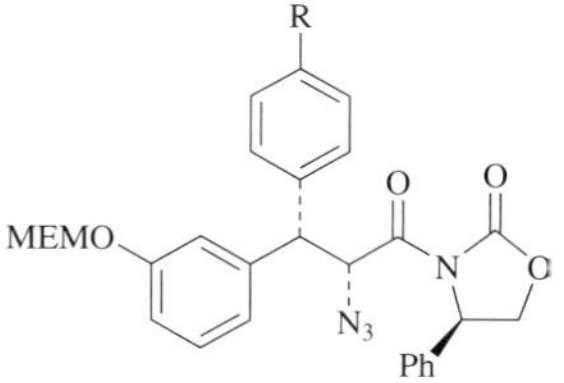 MEMO, R, O, O, N, O, N_3, Ph R: *t*-Bu (51); Ph (60)	435, 936
C_{24}			
BnO, O, O, CN, O, O, N, O, Bn	1. KHMDS (2.2 eq), THF, –78°, 15-45 min 2. 2,4,6-(*i*-Pr)$_3$C$_6$H$_2$SO$_2$N$_3$, –78°, 1-2 min	BnO, O, O, CN, O, O, N, O, N_3, Bn (61), 84% de	450

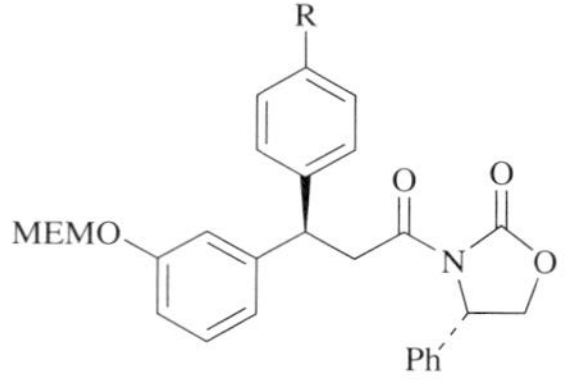

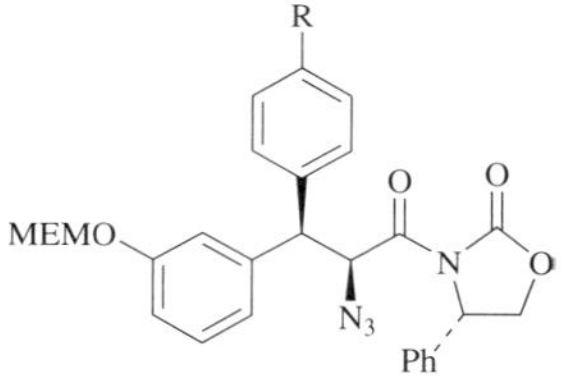

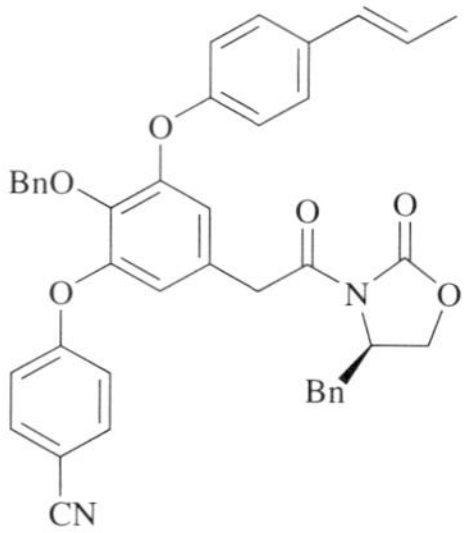

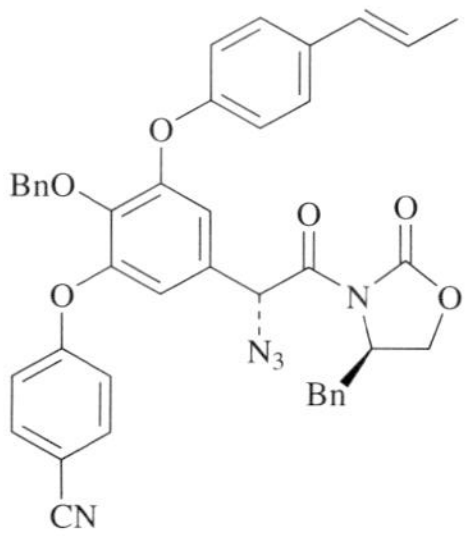

[a] The chiral auxiliary could not be cleaved to give the hydrazine derivative.

[b] Reagents where R^1 = Me or Et gave products with poorer dr values.

[c] The treatment time with LDA was 5 minutes.

[d] The number is the yield of pure major product.

[e] Addition of Bu_4NI prevents the reverse reaction of the initially formed adduct; occurrence of a reverse reaction has been disputed (Ref. 441).

[f] Excess 2,4,6-(i-Pr)$_3C_6H_2SO_2N_3$ was added early in the enolization step to partially counteract cyclization of the enolate.

[g] The values are those of the crude product.

[h] The product epimerized on attempted removal of the chiral auxiliary.

[i] No configurations were assigned to the two diastereomers.

[j] The numbers are the for the two atropisomers.

TABLE 14. LACTAM ENOLATES

Substrate	Conditions	Product(s) and Yield(s) (%)	Refs.
$C_{3\text{-}4}$			
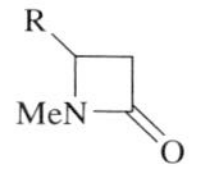 R, MeN (β-lactam)	1. LDA, Et_2O, –70° 2. TsN_3, –78° to –50°, 1 h 3. TMSCl, reflux, 6 h	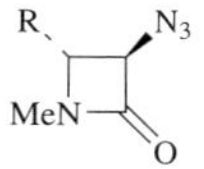 R, N_3, MeN R: EtS (52); Me (39)	339
C_3			
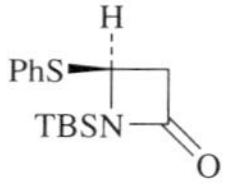 H, PhS, TBSN	1. LDA 2. 2-$C_{10}H_7SO_2N_3$	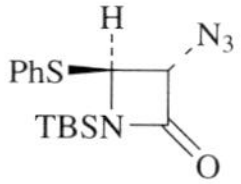 H, PhS, N_3, TBSN (66)	780
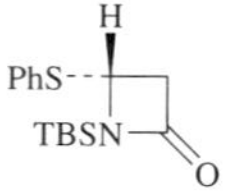 H, PhS, TBSN	1. LDA 2. 2-$C_{10}H_7SO_2N_3$ 3. TMSCl	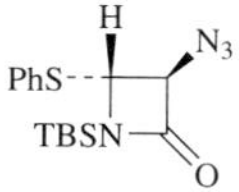 H, PhS, N_3, TBSN (56)	780
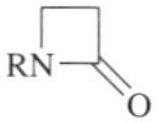 RN R = $C(CO_2Me)_2C_6H_4OBn$-4	1. LDA, THF, –78°, 2 h 2. TsN_3, –78°, 1 h 3. TMSCl, rt. 1 h	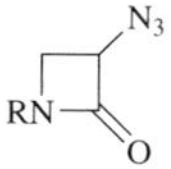 N_3, RN (76)	777
C_4			
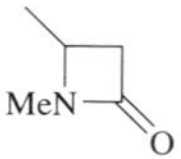 MeN	1. LDA, Et_2O, –70° 2. $TMSN_3$, –78° to –50°, 1 h 3. TMSCl, reflux, 6 h	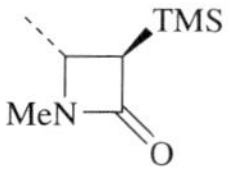 TMS, MeN (64)	339
$C_{4\text{-}10}$			
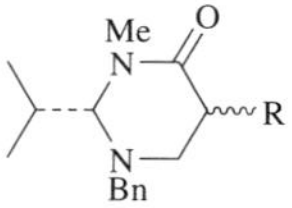 Me, N, O, R, N, Bn	1. LDA, THF, hexane, –78°, 2 h 2. t-$BuO_2CN{=}NCO_2Bu$-t, –78°, 6 h; rt, 3-6 h	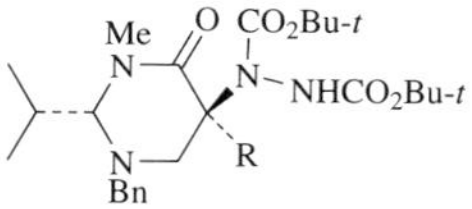 Me, N, O, CO_2Bu-t, N, $NHCO_2Bu$-t, R, N, Bn R: Me (45) >98 % de; Et (67) >98 % de; Bn (73) >98 % de	460
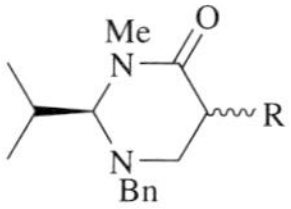 Me, N, O, R, N, Bn	1. LDA, THF, hexane, –78°, 2 h 2. t-$BuO_2CN{=}NCO_2Bu$-t, –78°, 6 h; rt, 3-6 h	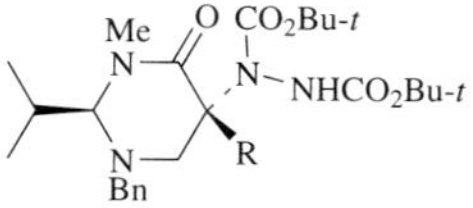 Me, N, O, CO_2Bu-t, N, $NHCO_2Bu$-t, R, N, Bn R: Me (41) >98 % de; Et (37) >98 % de; Bn (25) >98 % de	460

	Substrate	Conditions	Product(s) and Yield(s) (%)	Refs.
C_5	t-BuO_2C, CbzN, O	1. LiHMDS, –73° 2. t-$BuO_2CN{=}NCO_2Bu$-t	t-BuO_2C, CbzN, O, E **I** + t-BuO_2C, CbzN, O, E **II** **I + II** (76), **I**:**II** = 10:1 E = $N(CO_2Bu\text{-}t)NHCO_2Bu\text{-}t$	937
C_7	t-BuO_2C, N, O, H	1. LiHMDS, THF, –78° 2. $Ph_2P(O)ONH_2$	t-BuO_2C, N, O, NH_2, H (47)	143
C_8	N, Boc, O, BocO	1. KHMDS (ia), THF, –78°, 80 min 2. 2,4,6-(i-Pr)$_3C_6H_2SO_2N_3$ (precooled to –95°, insulated cannula)	N_3, N, Boc, O, BocO (36), 17% de	440
C_9	S, N, O, MeO_2C, CO_2Me	1. LiHMDS, THF, –70°, 4 h 2. TsN_3, –70° 3. TMSCl, to rt	H, S, N_3, N, O, MeO_2C, CO_2Me (68)	339
	H, N, O, TBDPSO	1. LDA (ia), THF, –78°, 20 min 2. 2,4,6-(i-Pr)$_3C_5H_2SO_2N_3$, –78°, 3 h	H, N, O, N_3, TBDPSO (65), 72% de	462
	H, N, O, TBDPSO	1. LDA (ia), THF, –78°, 20 min 2. 2,4,6-(i-Pr)$_3C_5H_2SO_2N_3$,[a] –78°, 10 min	H, N, O, N_3, TBDPSO (20) + H, N, O, N_2, TBDPSO (64)	461
	R, N, O, N, Cl	1. KOBu-t, THF, –78°, 30 min 2. 2,4,6-(i-Pr)$_3C_6H_2SO_2N_3$, THF, –78°, 30 min	R, N, O, N_3, N, Cl R: Me (57); PMB (92)	938

TABLE 14. LACTAM ENOLATES (*Continued*)

Substrate	Conditions	Product(s) and Yield(s) (%)	Refs.
C_9 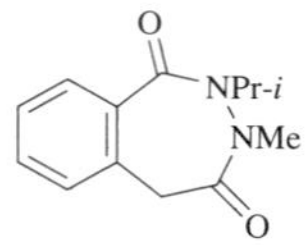	1. KHMDS, THF, –78° 2. 2,4,6-$(i\text{-Pr})_3C_6H_2SO_2N_3$, –78°	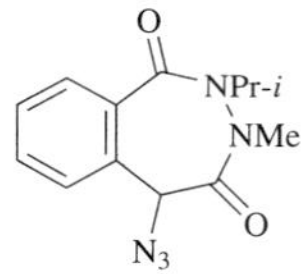 NPr-*i*, NMe, O, O, N_3 (—)	939
C_{10} 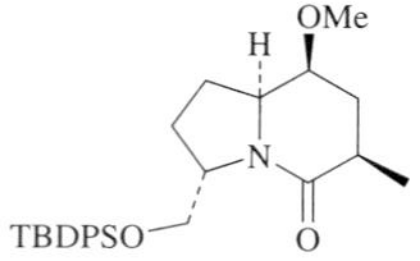 OMe, H, N, O, TBDPSO	1. *t*-BuLi, THF, –78°, 1 h 2. 2,4,6-$(i\text{-Pr})_3C_6H_2SO_2N_3$, –78°, 3 h	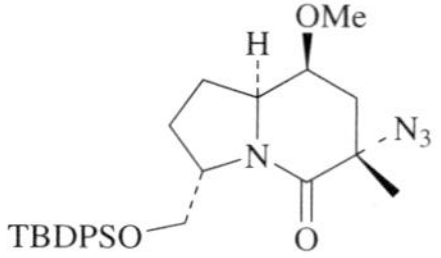 OMe, H, N, N_3, O, TBDPSO (63)	940
C_{10} 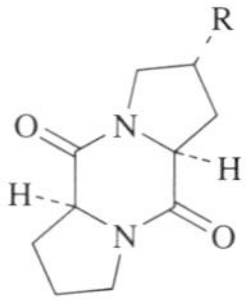 R, O, N, H, H, N, O	1. KHMDS, THF, –78° 2. 2,4,6-$(i\text{-Pr})_3C_6H_2SO_2N_3$, –78°	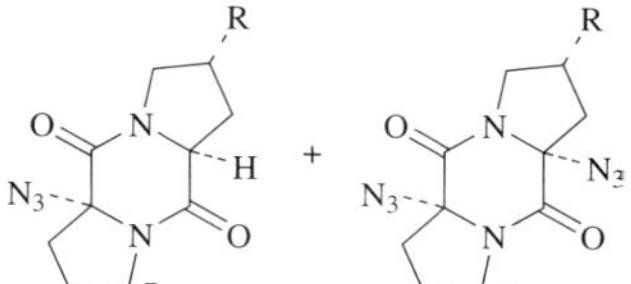 I + II R: H, I (42), II (0) R: OTBDMS, I (22), II (22)	941
$C_{11\text{-}21}$ 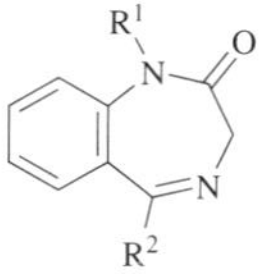 R^1, N, O, N, R^2	1. KHMDS, THF, –78°, 5 min 2. 2,4,6-$(i\text{-Pr})_3C_6H_2SO_2N_3$, –78°, 5 min	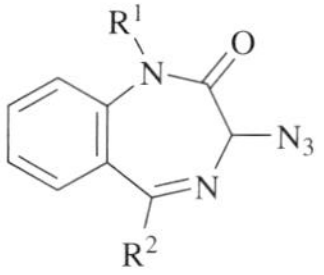 R^1, N, O, N_3, N, R^2	942

R^1	R^2	
Me	Et	(83)
Me	*n*-Pr	(76)
Me	*i*-Pr	(86)
Me	$MeO(CH_2)_2OCMe_2$	(81)
Me	Ph	(88)
CH_2CF_3	*i*-Pr	(50)
CH_2CF_3	*i*-Pr	(94)[b]
CH_2CF_3	Ph	(77)
i-Pr	Ph	(89)
$4\text{-}MeOC_6H_4CH_2$	*i*-Pr	(84)
$4\text{-}MeOC_6H_4CH_2$	Ph	(89)

Substrate	Conditions	Product(s) and Yield(s) (%)	Refs.
C_{11}	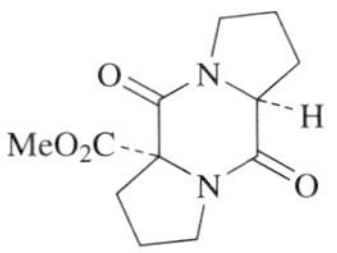 1. KHMDS, THF, –78° 2. 2,4,6-$(i\text{-Pr})_3C_6H_2SO_2N_3$, –78°	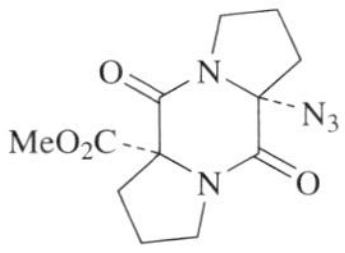 (79)	941
C_{12}	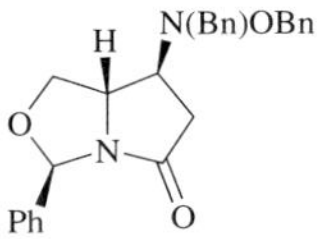 1. LDA (ia), THF, –78°, 30 min 2. $RO_2CN{=}NCO_2R$, –78°, 8 h	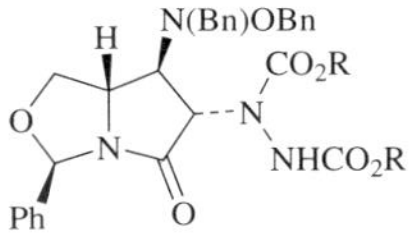 R: *t*-Bu (70); Bn (85)	943, 944
$C_{15\text{-}18}$ (R = CMe=CH₂, 1-cyclohexenyl, Ph)	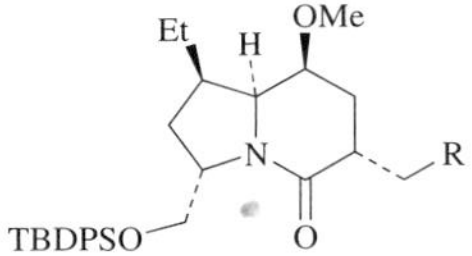 1. *t*-BuLi, THF 2. 2,4,6-$(i\text{-Pr})_3C_6H_2SO_2N_3$, –78°	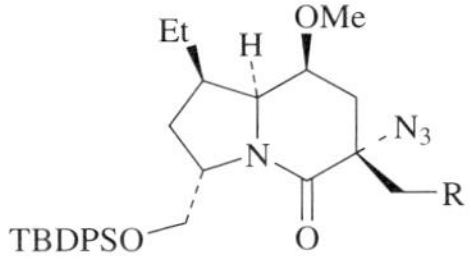 R: $CMe{=}CH_2$ (72); 1-cyclohexenyl (78); Ph (60)	887
C_{16}	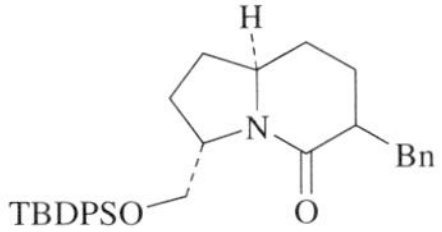 1. *t*-BuLi, THF, pentane, –78°, 50 min 2. 2,4,6-$(i\text{-Pr})_3C_6H_2SO_2N_3$, –78°, 6 h	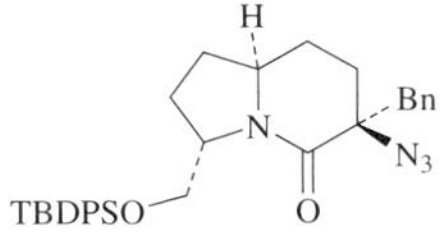 (27) 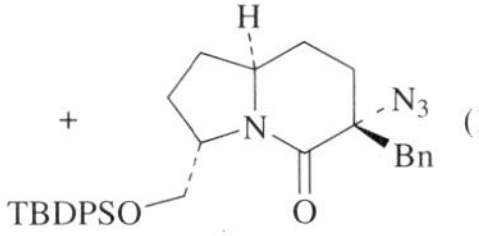 + (15)	462
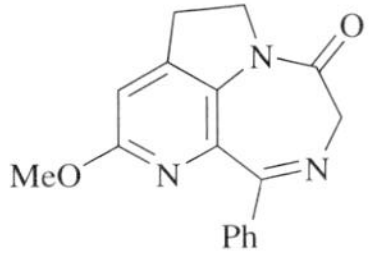	1. *t*-BuOK, THF 2. 2,4,6-$(i\text{-Pr})_3C_6H_2SO_2N_3$, –78°	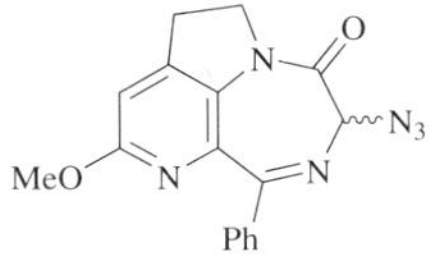 (26)	945
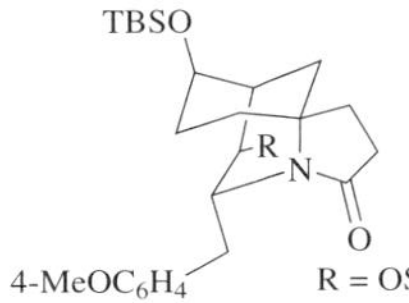 C_{18} (R = $OSiEt_3$)	1. LDA (ia), THF, –78°; to 0°, 30 min 2. 2,4,6-$(i\text{-Pr})_3C_6H_2SO_2N_3$, –78°, 5 min	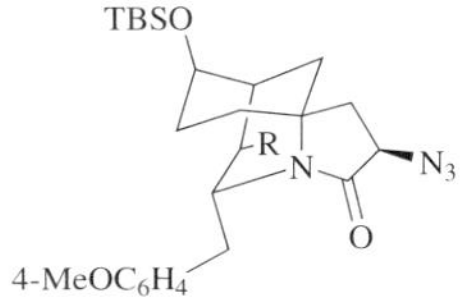 (59) 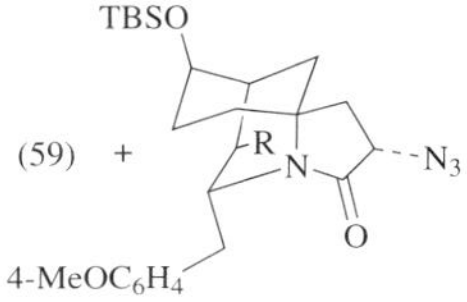 + (24)	946

[a] No reaction occurred with $t\text{-BuO}_2CN{=}NCO_2Bu\text{-}t$.

[b] The base was KOBu-*t* (2 eq).

TABLE 15. CYANO-STABILIZED CARBANIONS

Substrate	Conditions	Product(s) and Yield(s) (%)	Refs.
C_3 CH$_3$CH$_2$CN	1. *n*-BuLi, hexane, THF, 0°, 30 min 2. 2-NCC$_6$H$_4$-oxaziridine-NCONEt$_2$, –78°, 3 h; to rt, 1.5 h	CH$_3$CH(CN)NHCONEt$_2$ (56)	158
C_{3-4} RCH$_2$CN	1. LiHMDS (ia), THF, –78°, 1 h 2. (+)-oxaziridine, –78°, time; to rt	I + II R / Time / I / II Me / 4.5 h / (36) / (0) Et / 6.5 h / (83) / (17)	151
C_3 (NC)$_2$CH$^-$ Na$^+$	1. 2,4,6-Me$_3$C$_6$H$_2$SO$_2$ONH$_2$, THF, 0°, 2.5 h 2. TsOH	(NC)$_2$CH–NH$_3^+$ TsO$^-$ (55)	463
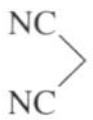 NCCH$_2$CN	1. LiHMDS (ia), THF, –78°, 1 h 2.  (–)-oxaziridine, –78°, 7 h; to rt	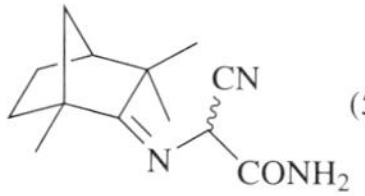 imine-CH(CN)CONH$_2$ (57), 23% de	151
	1. LiHMDS (ia), THF, –78°, 1 h 2. 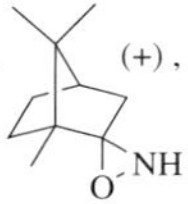 (+)-oxaziridine, –78°, 6 h; to rt	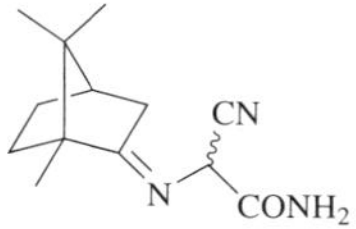 imine-CH(CN)CONH$_2$ (82), dr 1:1	151
	(NC)$_2$C=NOTs, pyridine, Et$_2$O, 0°	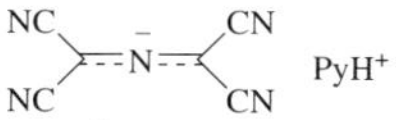 (NC)$_2$C=N$^-$=C(CN)$_2$ PyH$^+$ (55)	838
C_4 CH$_2$=CHCH$_2$CN	1. LiHMDS (ia), THF, –78°, 1 h 2. (+)-oxaziridine, –78°, 26.5 h; to rt	imine-C(=CHCH$_3$)CONH$_2$ (45)	151

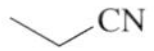

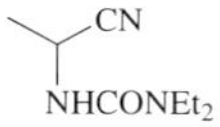

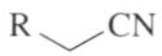

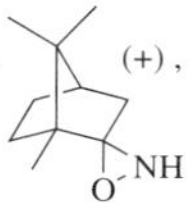

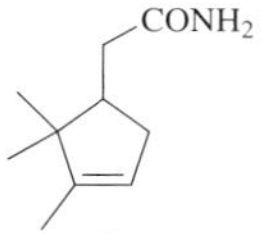

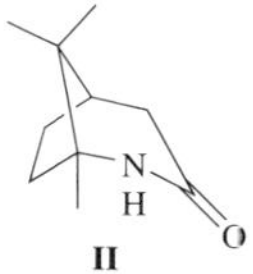

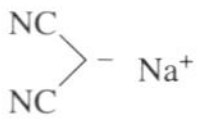

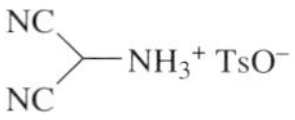

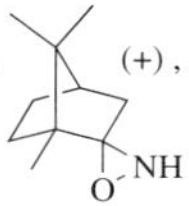

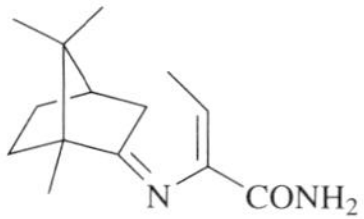

Substrate	Conditions	Product(s) and Yield(s) (%)	Refs.
(CH3CH=CHCN)	Conditions 1: 1. Catalyst A (see Chart 1; 5 mol%), EtOH, rt 2. Substrate, then $PhSiH_3$ 3. t-$BuO_2CN{=}NCO_2Bu$-t, rt, time Conditions 2: 1. Catalyst B (see Chart 1; 2 mol%), i-PrOH, 0° 2. Substrate, then $PhSiH_3$, 0° 3. t-$BuO_2CN{=}NCO_2Bu$-t, 0°, time	$CH_3CH_2CH(CN)N(CO_2Bu$-$t)NHCO_2Bu$-t Conditions 1, Time 18 h (46) Conditions 2, Time 2.5 h (45)	215
$C_{6\text{-}15}$			
TMSO(NC)CH–R	1. LDA 2. $Ph_2P(O)ONMe_2$, –78° to 20°, 5 h	$RC(O)NMe_2$ R: 2-furyl (77) 2-thienyl (80) 2-(1-methylpyrrolyl) (68) 2-pyridinyl (96) Ph (76) 2-ClC_6H_4 (92) 4-ClC_6H_4 (95) 4-BrC_6H_4 (91) 4-$MeOC_6H_4$ (75) 4-$Me_2NC_6H_4$ (98) $2,4$-$(HO)_2C_6H_3$ (67) E-PhCH=CH (35) 1-naphthyl (80) 2-naphthyl (90) 4-BzC_6H_4 (51)	947
C_8			
$PhCH_2CN$	1. n-BuLi 2. $2,4$-$(O_2N)_2C_6H_4ONH_2$	$PhCH(CN)NH_2$ (7)	93
	1. Base, THF, –78°, 15 min 2. $(4$-$MeOC_6H_4)_2P(O)ONH_2$, –78° to rt; rt, overnight 3. Ac_2O, Et_3N	$PhCH(CN)NHAc$ Base: LiHMDS (59) NaHMDS (64) KOBu-t (67)	106

TABLE 15. CYANO-STABILIZED CARBANIONS (*Continued*)

Substrate	Conditions	Product(s) and Yield(s) (%)	Refs.
C_8			
Ph, NC	1. Li base, THF or ether 2. $MeSO_2ONMe_2$, –30° to 0°	Ph, NC, NMe_2 **I** (69)	134
	LDA, $Ph_2P(O^{18})ONMe_2$	**I** (65)	84
	1. Li base, THF 2. Ph, O, P, O, $ONMe_2$, N, Me, –15°	**I** (62), 8% ee[a]	147
C_{8-12}			
Ar, NC	1. LiHMDS (ia), THF, –78°, 1 h 2. NH, O (+), –78°, time; to rt	Ar, N, $CONH_2$ **I** + Ar, N, $CONH_2$ **II** + $CONH_2$ **III**	151
	1. LiHMDS (ia), THF, –78°, 1 h 2. HN, O (–), –78°; to rt	Ar, N, $CONH_2$	151

Ar	Time	I + II	% de	III
Ph	4.5 h	(78)[b]	25	(0)
4-ClC_6H_4	5 h	(80)	16	(0)
4-$O_2NC_6H_4$	6 h	(0)	—	(21)
4-$MeOC_6H_4$	9 h	(75)	5	(0)
1-naphthyl	4.5 h	(80)	33	(0)
2-naphthyl	4.5 h	(73)	33	(0)

Ar		% de
Ph	(55)	50
1-naphthyl	(48)	33
2-naphthyl	(31)	52

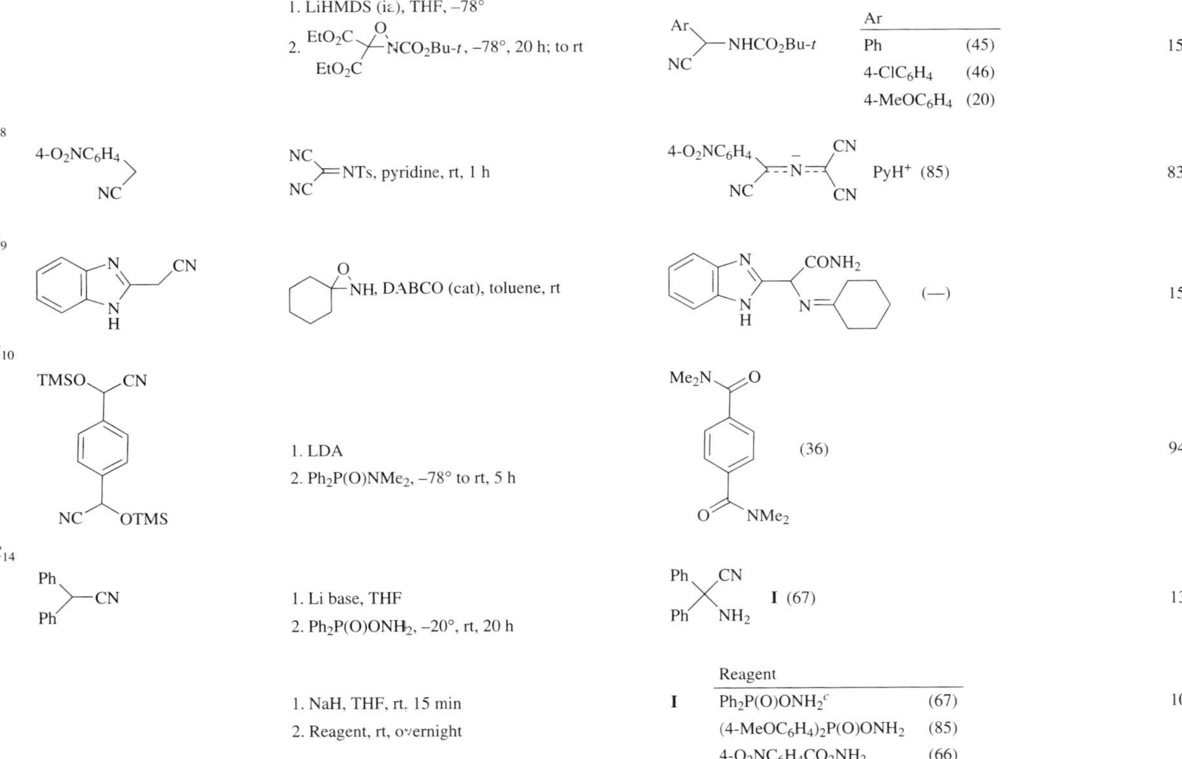

Reactant	Conditions	Product(s) and Yield(s) (%)	Refs.
	1. LiHMDS (ia), THF, −78° 2. EtO_2C, EtO_2C oxaziridine NCO_2Bu-*t*, −78°, 20 h; to rt	Ar(NC)CH–$NHCO_2Bu$-*t* Ar: Ph (45); 4-ClC_6H_4 (46); 4-$MeOC_6H_4$ (20)	155
C_8 4-$O_2NC_6H_4CH_2CN$	$(NC)_2C$=NTs, pyridine, rt, 1 h	4-$O_2NC_6H_4$(NC)C=$\bar{N}$=C(CN)$_2$ PyH^+ (85)	838
C_9 2-(cyanomethyl)benzimidazole	cyclohexane spiro-oxaziridine NH, DABCO (cat), toluene, rt	benzimidazolyl–CH($CONH_2$)–N=cyclohexylidene (—)	150
C_{10} TMSO(CN)CH–C_6H_4–CH(CN)OTMS	1. LDA 2. $Ph_2P(O)NMe_2$, −78° to rt, 5 h	$Me_2NC(O)C_6H_4C(O)NMe_2$ (36)	947
C_{14} Ph_2CHCN	1. Li base, THF 2. $Ph_2P(O)ONH_2$, −20°, rt, 20 h	$Ph_2C(CN)NH_2$ **I** (67)	139
	1. NaH, THF, rt, 15 min 2. Reagent, rt, overnight	**I** Reagent: $Ph_2P(O)ONH_2$[c] (67); $(4\text{-}MeOC_6H_4)_2P(O)ONH_2$ (85); 4-$O_2NC_6H_4CO_2NH_2$ (66)	106

TABLE 15. CYANO-STABILIZED CARBANIONS (*Continued*)

Substrate	Conditions	Product(s) and Yield(s) (%)	Refs.
C_{14} Ph, Ph–CH–CN	(cyclohexane spiro-oxaziridine) NH, DABCO (cat), toluene, rt	(spiro imidazolidinone: HN, O, Ph, Ph, N H) (82)	150
	1. Li base, Et_2O or THF 2. $MeSO_2ONMe_2$, –30° to 0°	Ph, Ph, CN, NMe_2 (67)	134
	1. NaH, glyme, rt 2. TsN_3, rt; 35-40°, 1 h	Ph, Ph, CN, N_3 (18)[d]	483

[a] Racemization probably occurred during isolation by treatment with acid (pH 4.5).

[b] **I** is the major isomer.

[c] LiHMDS was used as the base.

[d] The reported yield is that of the crude product.

TABLE 16. NITRONATES

Substrate	Conditions	Product(s) and Yield(s) (%)	Refs.
C_{1-6}			
$R^1R^2CH–NO_2$	1. KH, THF, rt; 40°, 15 min 2. TsN_3, –10° to 0°; 0°, 1 h	$R^1R^2C(N_3)Ts$ R^1 / R^2: H / H (0) Me / H (37) Me / Me (49) —$(CH_2)_5$— (56) Me / 2-methyl-1,3-dioxolan-2-yl-$(CH_2)_2$ (35)	464

TABLE 17. SULFONE-STABILIZED CARBANIONS

Substrate	Conditions	Product(s) and Yield(s) (%)	Refs.
C_5			
3-(NHBu-*n*)-dihydrothiopyran SO_2 (structure)	$EtO_2CN{=}NCOR$, MeCN, reflux, 3 h	$EtO_2CN(NHCOR)$-substituted NHBu-*n* dihydrothiopyran SO_2 (structure); R = OEt (60), Ph (60)	250
C_7			
$PhSO_2CH_2CuLi$	Me_2NOSO_2R, Et_2O or THF, 0° R = Me, Ph, 4-MeC_6H_4, or 2,4,6-$Me_3C_6H_2$	$PhSO_2CH_2NMe_2$ (22)	134
$PhSO_2Me$	1. *n*-BuLi, THF, hexane, 0°, 30 min 2. 2-NCC_6H_4-oxaziridine-$NCONEt_2$, –78°, 3 h; to rt, 90 min	$PhSO_2CH_2NHCONEt_2$ (43)	158
C_{13}			
$PhO_2SCH(Ph)CuLi$	1. Me_2NOSO_2R, Et_2O or THF, 0° (forms **I** as interm.) 2. $PhO_2SCH(Ph)CuLi$ (forms **II**) R = Me, Ph, 4-MeC_6H_4, or 2,4,6-$Me_3C_6H_2$	$PhO_2SCH(Ph)NMe_2$ **I** (—) $PhC_2SCH(Ph)CH(Ph)NMe_2$ **II** (28)	134
C_{15}			
Bn-substituted bicyclic sultam with $CH_2OTBDPS$ (structure)	1. *t*-BuLi, THF, pentane, –78°, 55 min 2. 2,4,6-(*i*-Pr)$_3C_6H_2SO_2N_3$, –78°, 6 h	Bn/N_3 sultam diastereomer (40) + N_3/Bn sultam diastereomer (33) (structures, OTBDPS)	465

TABLE 18. PHOSPHORUS-STABILIZED CARBANIONS

Substrate	Conditions	Product(s) and Yield(s) (%)	Refs.
C_1			
$(EtO)_2P(O)CH_2P(O)(OEt)_2$	1. NaH, THF, rt, 15 min 2. See table. Reagent / Temp / Time: $2,4\text{-}(O_2N)_2C_6H_3ONH_2$, rt, 12 h (15) $Ph_2P(O)ONH_2$, –70°; rt, 2 h; 10 h (50)	$(EtO)_2P(O)CH(NH_2)P(O)(OEt)_2$	95
$(EtO)_2P(O)Me$	1. *n*-BuLi, THF, hexane, 0°, 30 min 2. 2-NCC_6H_4-oxaziridine-$NCONEt_2$, –78°, 3 h; to rt, 1.5 h	$(EtO)_2P(O)CH_2NHCONEt_2$ (51)	158
$C_{1\text{-}8}$			
$(EtO)_2P(O)CH_2R$	1. *n*-BuLi, THF, –78° 2. *t*-$BuO_2CN{=}NCO_2Bu$-*t*, –78°, few min	$(EtO)_2P(O)CH(R)N(CO_2Bu\text{-}t)NHCO_2Bu\text{-}t$ R: H (53); Me (65); *i*-Pr (65); *t*-Bu (71); Ph (75); Bn (50)	770
C_2			
$(EtO)_2P(O)CH_2CO_2Me$	1. NaH, DME 2. $2,4,6\text{-Me}_3C_6H_2SO_2ONH_2$, 30°, 30 min	$(EtO)_2P(O)CH(NH_2)CO_2Me$ (47)	704
$(EtO)_2P(O)CH_2CO_2Bn$	1. NaH, THF, rt, 1 h 2. $Ph_2P(O)ONH_2$, THF, –78°, 2 h 3. HO_2CCO_2H	$(EtO)_2P(O)CH(NH_3^+)CO_2Bn$ $HO_2CCO_2^-$ (60)	141
$(EtO)_2P(O)CH^-CH_3$ Cu^+	$MsON(Li)CO_2CH_2CH{=}CH_2$, THF, –78° to –60°, 1.5 h	$(EtO)_2P(O)CH(CH_3)NHCO_2CH_2CH{=}CH_2$ (58)	130

TABLE 18. PHOSPHORUS-STABILIZED CARBANIONS (*Continued*)

Substrate	Conditions	Product(s) and Yield(s) (%)	Refs.
C_{2-7}			
	$TsON(Li)CO_2Bu$-*t*	R / Me (80) / Ph (50)	126
C_{2-8}			
	1. *n*-BuLi, THF, –78° 2. $CuBr \cdot Me_2S$ 3. $TsON(Li)CO_2Bu$-*t*, THF, –78°	R, % de / Me (59) 30 / Ph (49) 35	129
C_2	1. *n*-BuLi, THF, –78° 2. *t*-$BuO_2CN{=}NCO_2Bu$-*t*, –78°, few min	(33) mixture of two diastereomers	770
	1. LDA, THF, –30° 2. *t*-$BuO_2CN{=}NCO_2Bu$-*t*, –30°, 5 min	(41), 52% de	948
	1. LDA, THF, –30° 2. *t*-$BuO_2CN{=}NCO_2Bu$-*t*, –30°, 5 min	(46), 83% de	948
	1. LDA, THF, –78° 2. *t*-$BuO_2CN{=}NCO_2Bu$-*t*, –78°, 30 min	R^1 R^2 R^3 dr / H Ph Me (55) 2:1 / Ph H Me (58) 1.5:1 / Ph H *i*-Pr (65) 1:4	949

C_{2-9}

1. LDA, THF, –78°
2. t-BuO$_2$CN=NCO$_2$Bu-t, –78°, 30 min

R^1	R^2	R^3		dr
H	Ph	Me	(50)	1:1
Ph	H	Me	(54)	1.5:1
Ph	H	i-Pr	(62)	1:3

949

1. n-BuLi, hexane, cosolvent, –78°, 30 min
2. 2,4,6-(i-Pr)$_3$C$_6$H$_2$SO$_2$N$_3$, –78°, 5 h
3. Ac$_2$O, –78° to rt

N–SO$_2$C$_6$H$_2$(Pr-i)$_3$-2,4,6

R^1	R^2	R^3	Cosolvent		dr[a]
H	Bu-t	Me	THF	(72)	4:1
H	Bu-t	Ph	THF	(56-79)	11:1
H	Bu-t	Ph	Et$_2$O	(75)	13:1[b]
Me	Bu-t	Ph	THF	(70-93)	>20:1
H	Bu-t	Bn	THF	(63)	2:1
H	CEt$_3$	Ph	THF	(79)	3:1
Me	PhCHMe (S)	Ph[c]	THF	(52)	>20:1
Me	isobornyl	Ph[d]	THF	(74)	5.3:1

317

C_3

1. LDA, THF, –78°
2. 2,4,6-(i-Pr)$_3$C$_6$H$_4$SO$_2$N$_3$, –78°

Complex mixture

895

C_{4-13}

1. **1** (10 mol%), CH$_2$Cl$_2$
2. Substrate, then BnO$_2$CN=NCO$_2$Bn, rt, time

1

R^1	R^2	R^3	Time		% ee
Me	Me	Et	48 h	(75)	85
—(CH$_2$)$_3$—		Et	48 h	(98)	95
—(CH$_2$)$_4$—		Et	48 h	(98)	94
Ph	Me	Me	48 h	(97)	94
Ph	Me	Et	48 h	(85)	92
Bn	Me	Et	48 h	(60)	95
Ph	CH$_2$CH=CH$_2$	Et	140 h	(85)	98
2-naphthyl	Me	Et	48 h	(93)	92

468

TABLE 18. PHOSPHORUS-STABILIZED CARBANIONS (*Continued*)

Substrate	Conditions	Product(s) and Yield(s) (%)	Refs.
C_4			
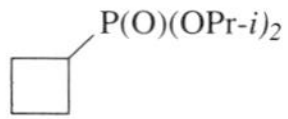 $P(O)(OPr\text{-}i)_2$	1. *n*-BuLi or LDA, THF, –78° 2. $EtO_2CN{=}NCO_2Et$	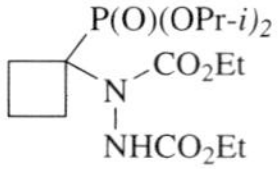 $P(O)(OPr\text{-}i)_2$, CO_2Et, $NHCO_2Et$ (35)	773
$C_{4\text{-}6}$			
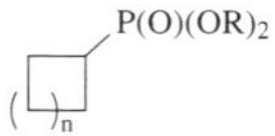 $P(O)(OR)_2$	1. *n*-BuLi, THF, –78°, 30 min 2. 2,4,6-(*i*-Pr)$_3C_6H_2SO_2N_3$, –78°, 1 h	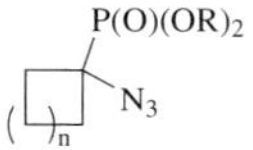 $P(O)(OR)_2$, N_3 n R 1 Et (74) 1 *i*-Pr (77) 2 Et (76) 2 *i*-Pr (69) 3 Et (69) 3 *i*-Pr (72)	773
C_7			
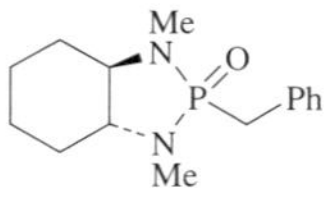 Me, N, O, P, Ph, N, Me	1. *n*-BuLi, THF, –78° 2. CuBr•Me_2S 3. TsNO(Li)CO_2Bu-*t*, THF, –78°	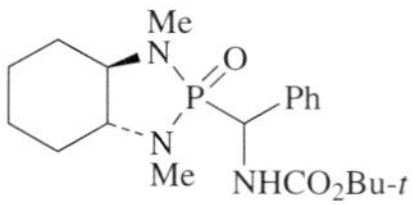 Me, N, O, P, Ph, N, Me, $NHCO_2Bu$-*t* (49), 52% de	129
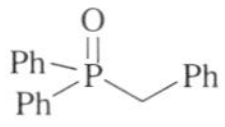 O, Ph, P, Ph, Ph	1. $LiN(Et)_2$, PhH, rt, 1 h 2. PhN_3, rt, 18 h	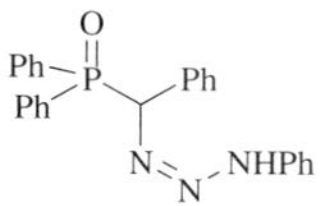 O, Ph, P, Ph, Ph, N=N, NHPh (26)	467
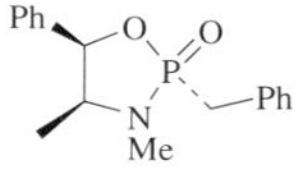 Ph, O, O, P, Ph, N, Me	1. LDA, THF, –78°, 30 min 2. 4-$O_2NC_6H_4SO_2N_3$, –78°; to rt	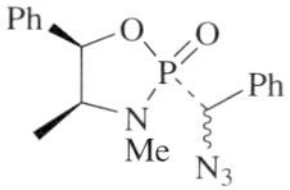 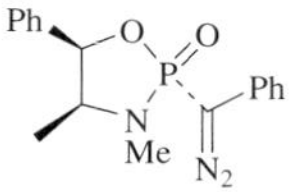 Ph, O, O, P, Ph, N, Me, N_3 (—) + Ph, O, O, P, Ph, N, Me, N_2 (19)	632
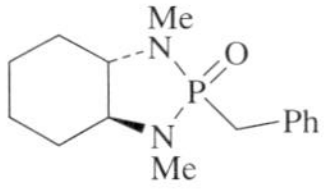 Me, N, O, P, Ph, N, Me	1. *n*-BuLi, THF, –78°, 30 min 2. 4-$O_2NC_6H_4SO_2N_3$, –78°, 4 h	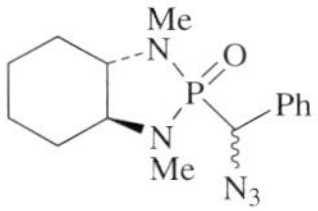 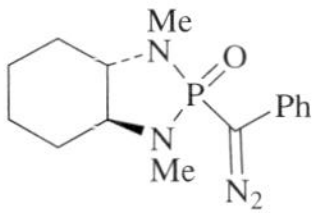 Me, N, O, P, Ph, N, Me, N_3 (—) + Me, N, O, P, Ph, N, Me, N_2 (33)	632

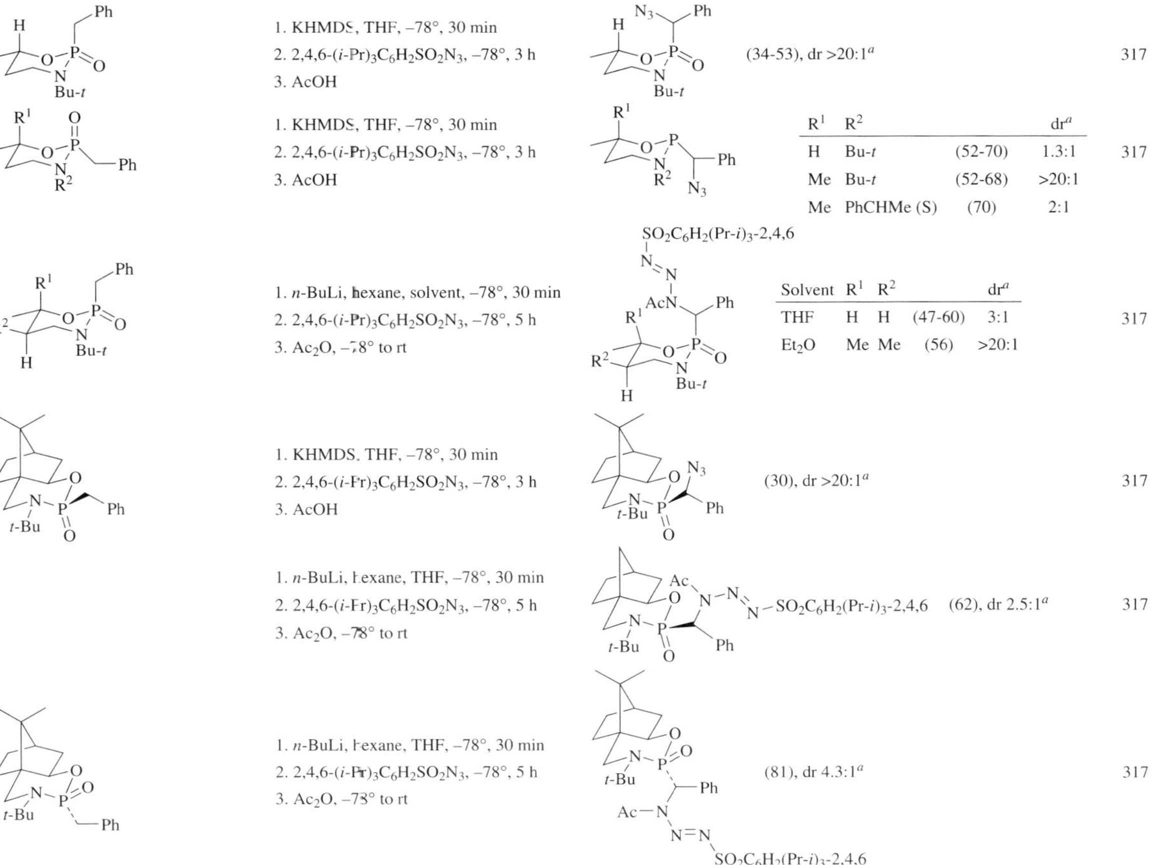

	1. KHMDS, THF, –78°, 30 min 2. 2,4,6-(i-Pr)$_3$C$_6$H$_2$SO$_2$N$_3$, –78°, 3 h 3. AcOH	(34-53), dr >20:1[a]	317
	1. KHMDS, THF, –78°, 30 min 2. 2,4,6-(i-Pr)$_3$C$_6$H$_2$SO$_2$N$_3$, –78°, 3 h 3. AcOH	R[1] / R[2] / / dr[a] H / Bu-t / (52-70) / 1.3:1 Me / Bu-t / (52-68) / >20:1 Me / PhCHMe (S) / (70) / 2:1	317
	1. n-BuLi, hexane, solvent, –78°, 30 min 2. 2,4,6-(i-Pr)$_3$C$_6$H$_2$SO$_2$N$_3$, –78°, 5 h 3. Ac$_2$O, –78° to rt	Solvent / R[1] / R[2] / / dr[a] THF / H / H / (47-60) / 3:1 Et$_2$O / Me / Me / (56) / >20:1	317
	1. KHMDS, THF, –78°, 30 min 2. 2,4,6-(i-Pr)$_3$C$_6$H$_2$SO$_2$N$_3$, –78°, 3 h 3. AcOH	(30), dr >20:1[a]	317
	1. n-BuLi, hexane, THF, –78°, 30 min 2. 2,4,6-(i-Pr)$_3$C$_6$H$_2$SO$_2$N$_3$, –78°, 5 h 3. Ac$_2$O, –78° to rt	(62), dr 2.5:1[a]	317
	1. n-BuLi, hexane, THF, –78°, 30 min 2. 2,4,6-(i-Pr)$_3$C$_6$H$_2$SO$_2$N$_3$, –78°, 5 h 3. Ac$_2$O, –78° to rt	(81), dr 4.3:1[a]	317

TABLE 18. PHOSPHORUS-STABILIZED CARBANIONS (*Continued*)

Substrate	Conditions	Product(s) and Yield(s) (%)	Refs.
C_7			
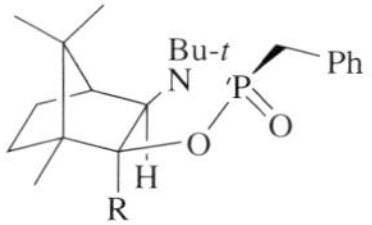 Bu-*t*, N, P, O, Ph, H, R	1. *n*-BuLi, hexane, THF, –78°, 30 min 2. 2,4,6-(*i*-Pr)$_3C_6H_2SO_2N_3$, –78°, 5 h 3. Ac_2O, –78° to rt	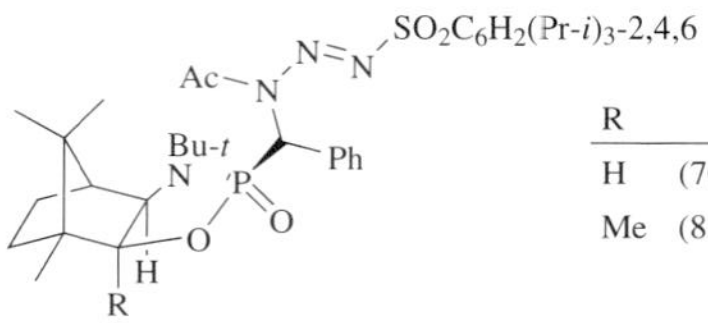 Ac, N, N=N, $SO_2C_6H_2$(Pr-*i*)$_3$-2,4,6, Bu-*t*, N, P, O, Ph, H, R R / dr[a]: H (70) 6:1; Me (85) 9:1	317
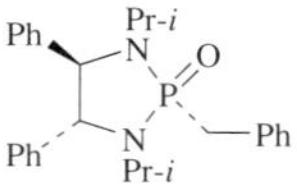 Pr-*i*, N, P, O, Ph, Ph, Ph, N, Pr-*i*	1. *n*-BuLi, hexane, THF, –78°, 30 min 2. 2,4,6-(*i*-Pr)$_3C_6H_2SO_2N_3$, –78°, 3 h 3. H_2O	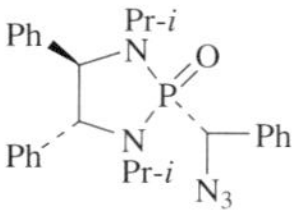 Pr-*i*, N, P, O, Ph, Ph, Ph, N, Pr-*i*, N_3 (100)[e], dr 2:1[a]	317
C_9			
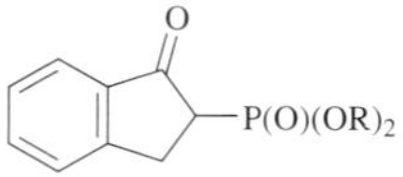 O, P(O)(OR)$_2$	$EtO_2CN{=}NCO_2Et$, **2** (2.5 mol%), Me_2CO, rt 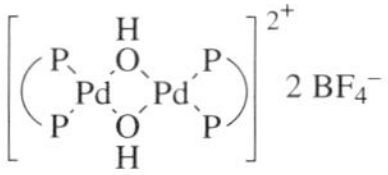 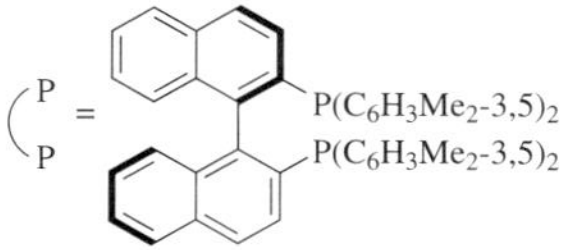	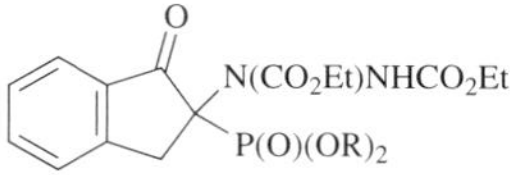 O, $N(CO_2Et)NHCO_2Et$, P(O)(OR)$_2$ R / Time / % ee: Me 35 h (81) 99; Et 20 h (92) 99; *i*-Pr 60 h (68) 99	238

[(P P)Pd(μ-OH)$_2$Pd(P P)]$^{2+}$ 2 BF_4^- **2**; (P P) = binaphthyl bearing P($C_6H_3Me_2$-3,5)$_2$ and P($C_6H_3Me_2$-3,5)$_2$

[a] The values are for the diastereomeric ratios in the crude products.

[b] Cleavage of the product followed by hydrolysis gave (*S*)-phosphono glycine.

[c] The configuration on phosphorus was not established.

[d] The substrate was a mixture of cis and trans isomers.

[e] The reported yield is that of crude product.

TABLE 19. ENOLATES OF α,β-UNSATURATED CARBONYL COMPOUNDS

Substrate	Conditions	Product(s) and Yield(s) (%)	Refs.
C_{4-10} 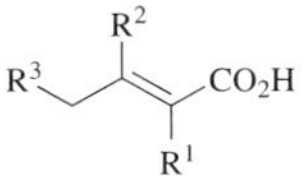 (structure with R[1], R[2], R[3], CO_2H)	1. $LiNEt_2$ (2.2 eq,[a] ia), THF, –70°; 0°, 15 min 2. $Ph_2P(O)ONH_2$, –70°, 25 min; rt, 2 h	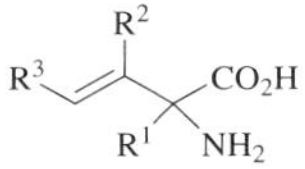 (structure with R[1], R[2], R[3], CO_2H, NH_2)	144
	1. $LiNEt_2$ (2.2 eq,[a] ia), THF, –70°, 30 min 2. $EtO_2CN{=}NCO_2Et$, –70°, 15 min	(structure with R[1], R[2], R[3], CO_2H, $N(CO_2Et)NHCO_2Et$) **I** + (structure with R[1], R[2], R[3], CO_2H, $N(CO_2Et)NHCO_2Et$) **II**	144
C_{4-11} (structure with R[1], R[2], CO_2R^3)	1. LDA, THF, addend, –78°, 70 min 2. $EtO_2CN{=}NCO_2Et$, –78°, 3 min 3. MeOH, –78°; to rt	(structure with R[1], R[2], CO_2R^3, $N(CO_2Et)NHCO_2Et$) **I** + (structure with R[1], R[2], CO_2R^3, $N(CO_2Et)NHCO_2Et$) **II**	469, 950

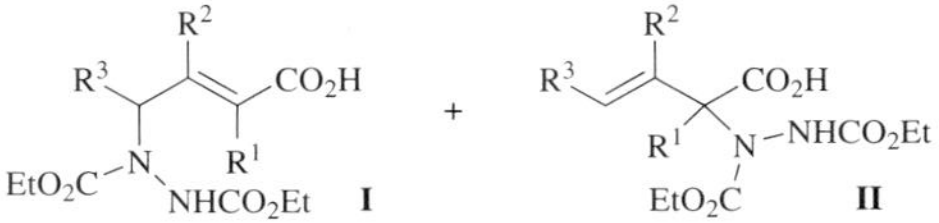

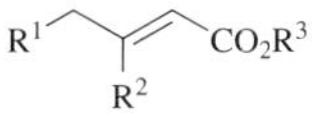

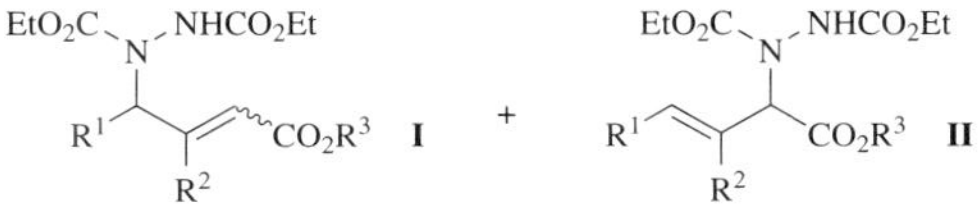

Products of the first entry (ref. 144):

R[1]	R[2]	R[3]	Yield (%)
H	H	H	(14)
Me	H	H	(45)
Me	H	Me	(64)
H	t-Bu	H	(28)
H	Ph	H	(45)

Products of the second entry (ref. 144):

R[1]	R[2]	R[3]	**I**	**II**
H	H	H	(69)	(0)
Me	H	H	(65)	(3)
H	Me	H	(68)	(0)
H	H	Me	(51)	(0)
Me	H	Me	(50)	(0)
H	H	Et	(62)	(0)
H	Ph	H	(74)	(0)

Products of the C_{4-11} entry (refs. 469, 950):

R[1]	R[2]	R[3]	Addend	**I**	**I** E:Z	**II**
H	H	Me	HMPA	(64)	>99:1	(0)
H	Me	Et	HMPA	(63)	2:1	(12)
H	Me	Et	—	(71)	1:1	(7)
H	Me	Et	$ZnCl_2$[b]	(61)	1:1	(25)
Bn	H	Me	HMPA	(71)	>99:1	(0)

TABLE 19. ENOLATES OF α,β-UNSATURATED CARBONYL COMPOUNDS (*Continued*)

Substrate	Conditions	Product(s) and Yield(s) (%)	Refs.
C_4			
[structure: N-crotonyl oxazolidinone, Bn]	1. LDA, THF, –78°, 30 min 2. Addend 3. t-$BuO_2CN{=}NCO_2Bu$-t, –78°, 1 h	I + II E = $N(CO_2Bu$-$t)NHCO_2Bu$-t Addend / I / II — / (51) / (33) HMPA (1 eq) / (61) / (21) HMPA (5 eq) / (52) / (11)	895
[structure: OMe, OTMS diene]	1. $EtO_2CN{=}NCO_2Et$, $TiCl_4$, CH_2Cl_2, –78° 2. Substrate, –78°, 30 min	EtO_2C–N(NHCO$_2$Et)–CH$_2$CH=CHCO$_2$Me (68) + [structure: CO_2Me, EtO_2C–N–$NHCO_2Et$] (17)	469, 950
$C_{4\text{-}5}$			
[structure: R, CO_2Et[c], Cl_3Sn]	1. $EtO_2CN{=}NCO_2Et$, THF, –10°, time; to –78° 2. MeOH, –78° to rt	[structure: R, CO_2Et, E] I + [structure: R, CO_2Et, E] II E = $N(CO_2Et)NHCO_2Et$ R / Time / I / II H / 30 min / (53) / (5) Me / 100 min / (26) / (3)	469
C_5			
[structure: CO_2Et, Bu_3Sn]	1. $EtO_2CN{=}NCO_2Et$, $ZnCl_2$, CH_2Cl_2, –78° 2. Substrate, –78° to 0°, 40 min	[structure: CO_2Et, EtO_2C–N–$NHCO_2Et$] (75) + [structure: CO_2Et, N(CO_2Et)$NHCO_2Et$] (5)	469, 950
[structure: $GeMe_3$[d], R^1, R^2, CO_2R^3]	1. $EtO_2CN{=}NCO_2Et$, $ZnCl_2$, CH_2Cl_2, –78° 2. Substrate, –78° to temp, time	[structure: EtO_2C–N–$NHCO_2Et$, R^1, R^2, CO_2R^3] R^1 / R^2 / R^3 / Temp / Time / / E:Z H / Me / Et / rt / 2 h / (71) / 3:1 —$(CH_2)_3$— / / Et / 5° / 40 min / (88) / 8:1 —$(CH_2)_4$— / / Me / 0° / 30 min / (55) / 6:1	469

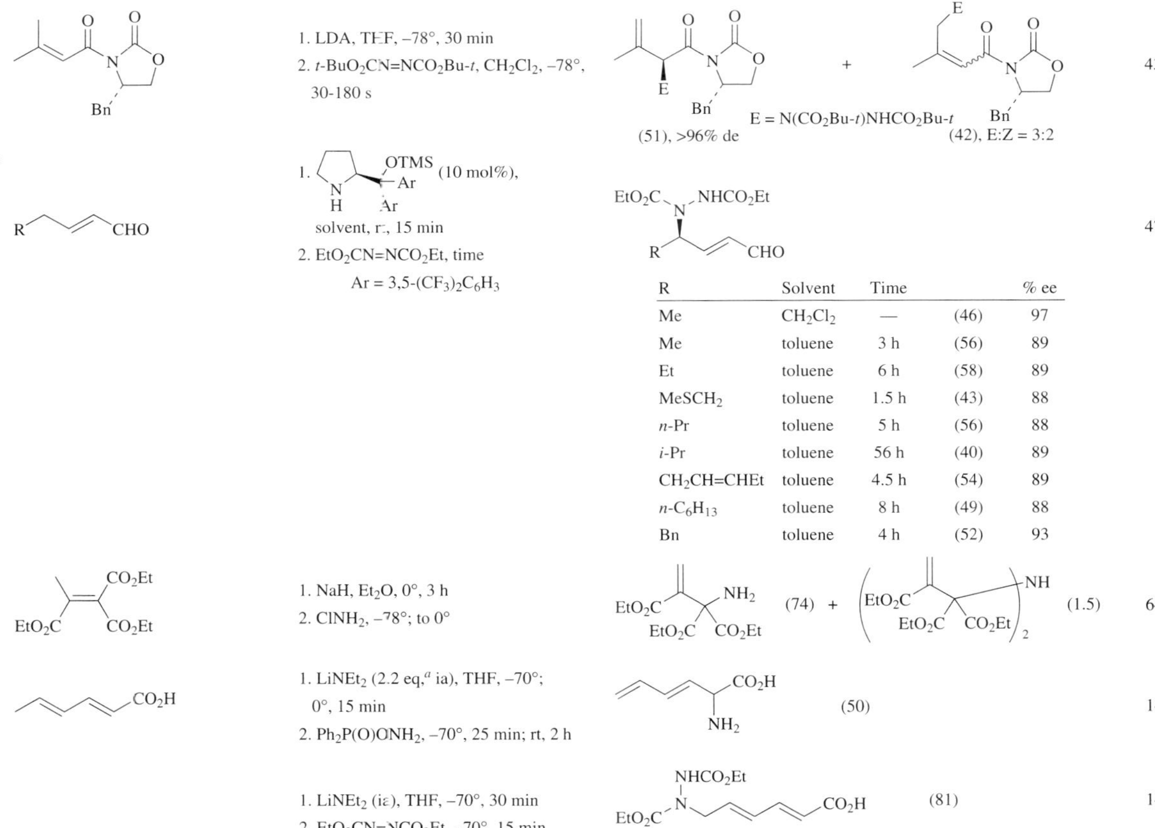

Reactant	Conditions	Product(s) and Yield(s) (%)	Refs.
(oxazolidinone, Bn)	1. LDA, THF, –78°, 30 min 2. t-$BuO_2CN{=}NCO_2Bu$-t, CH_2Cl_2, –78°, 30-180 s	(51), >96% de + (42), E:Z = 3:2 E = $N(CO_2Bu$-$t)NHCO_2Bu$-t	431
C_{5-11}			
R–CH₂CH=CHCHO	1. (pyrrolidine OTMS, Ar, Ar) (10 mol%), solvent, rt, 15 min 2. $EtO_2CN{=}NCO_2Et$, time Ar = 3,5-$(CF_3)_2C_6H_3$	EtO_2C–N(NHCO₂Et)–CH(R)CH=CHCHO	470
C_6			
EtO_2C, CO_2Et, CO_2Et	1. NaH, Et_2O, 0°, 3 h 2. $ClNH_2$, –78°; to 0°	(74) + (1.5)	64
CO_2H	1. $LiNEt_2$ (2.2 eq,[a] ia), THF, –70°; 0°, 15 min 2. $Ph_2P(O)ONH_2$, –70°, 25 min; rt, 2 h	(50)	144
	1. $LiNEt_2$ (ia), THF, –70°, 30 min 2. $EtO_2CN{=}NCO_2Et$, –70°, 15 min	(81)	144

R	Solvent	Time		% ee
Me	CH_2Cl_2	—	(46)	97
Me	toluene	3 h	(56)	89
Et	toluene	6 h	(58)	89
$MeSCH_2$	toluene	1.5 h	(43)	88
n-Pr	toluene	5 h	(56)	88
i-Pr	toluene	56 h	(40)	89
$CH_2CH{=}CHEt$	toluene	4.5 h	(54)	89
n-C_6H_{13}	toluene	8 h	(49)	88
Bn	toluene	4 h	(52)	93

TABLE 19. ENOLATES OF α,β-UNSATURATED CARBONYL COMPOUNDS (*Continued*)

Substrate	Conditions	Product(s) and Yield(s) (%)	Refs.
C_{7-8} 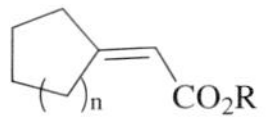 2-cycloalkylidene acetate (CO_2R, ring size n)	1. LDA, HMPA, THF, –78°, 70 min 2. $EtO_2CN{=}NCO_2Et$, –78°, 3 min 3. MeOH, –78°	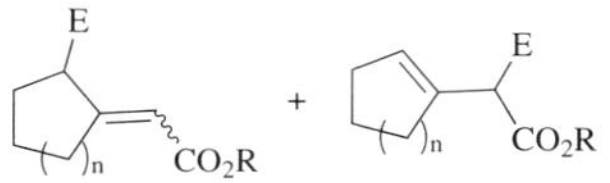 **I** + **II** E = $N(CO_2Et)NHCO_2Et$ R, n, **I**, E:Z, **II**: Et, 1, (65), 1:2, (22) Me, 2, (14), 1:1.5, (55) Me, 2, (13), 10:1, (38)[e]	469
C_8 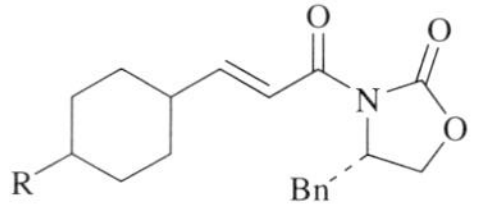 R = NHAc (oxazolidinone substrate, Bn)	1. LiHMDS, THF, 78° 2. t-$BuO_2CN{=}NCO_2Bu$-t	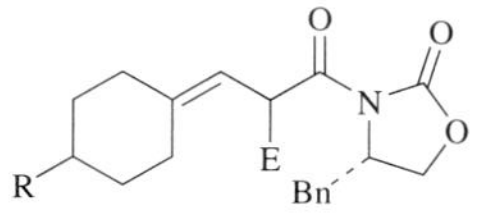 E = $N(CO_2Bu$-$t)NHCO_2Bu$-t (0)	453
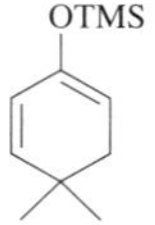 OTMS (dienol silyl ether)	(Saltmen)Mn(N), CH_2Cl_2, pyridine, TFFA, –78° to rt, 3-4 h	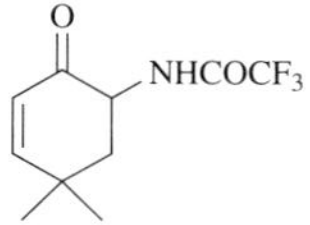 $NHCOCF_3$ (50)	471
C_{8-10} 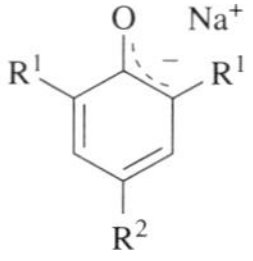 O Na+, R^1, R^1, R^2	$ClNH_2$, 2,6-$(R^1)_2$-4-$R^2C_6H_2OH$, 100-140°; rt, overnight	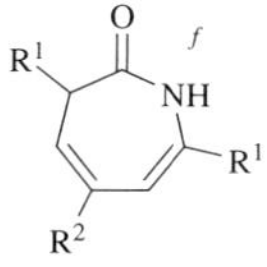 [f] R^1, R^2: Me, H, (55) Me, Me, (51) Et, H, (33)	65

[a] This is a corrected value; personal communication from R. Mestres, Department of Chemistry, University of Valencia, Spain, 2006.

[b] The $ZnCl_2$ was added after the formation of the lithium dienolate.

[c] The substrate was generated in situ from the lithium dienolate and $SnCl_4$.

[d] The substrate was generated in situ from the lithium dienolate and $ClGeMe_3$.

[e] Protonation was carried out at 0°.

[f] The intermediate formed upon amination of the enolate undergoes subsequent ring expansion to give the product shown.

TABLE 20. ENOLATES OF α–CYANO CARBONYL AND β–DICARBONYL COMPOUNDS

	Substrate	Conditions	Product(s) and Yield(s) (%)	Refs.
C_{3-10}	$RC^{-}(CO_2Et)_2$ Na^+	$ClNH_2$ (ia), Et_2O, morpholine, 0°, 2 h; rt, overnight; reflux, 5 h	$(EtO_2C)_2C(R)NH_2$ R: H (92), Me (85), Et (89), *i*-Pr (71), *s*-Bu (83), Ph (70), Bn (72)	62
C_{3-9}	$RCH(CO_2Et)_2$	1. NaH, THF, rt, 25 min 2. 2,4-$(O_2N)_2C_6H_3ONH_2$, rt, overnight	$(EtO_2C)_2C(R)NH_2$ R: H (55)[a], Me (31)[b], Ph (65)	93
C_{3-10}	$RCH(CO_2Et)_2$	1. NaH, THF, rt, 25 min 2. 2,4-$(O_2N)_2C_6H_3ONH_2$, rt, overnight 3. 6 N HCl, reflux 4. Et_3N	$RCH(NH_2)CO_2H$ R: Me (98), Et (74), Et_2OCCH_2 (61), *n*-Bu (46-57), Bn (73)	93
C_{3-9}	$RCH(CO_2Et)_2$	1. NaH, THF, rt, 15 min 2. Reagent, rt, overnight	$(EtO_2C)_2C(R)NH_2$ R, Reagent: H, $Ph_2P(O)ONH_2$ (57)[b] H, $(4\text{-}MeOC_6H_4)_2P(O)ONH_2$ (41) H, 4-$O_2NC_6H_4CO_2NH_2$ (52) Ph, $Ph_2P(O)ONH_2$ (31)[b] Ph, $(4\text{-}MeOC_6H_4)_2P(O)ONH_2$ (92) Ph, 4-$O_2NC_6H_4CO_2NH_2$ (99)	106

TABLE 20. ENOLATES OF α–CYANO CARBONYL AND β–DICARBONYL COMPOUNDS (*Continued*)

Substrate	Conditions	Product(s) and Yield(s) (%)	Refs.
C_3			
HN, NH, Y, O, O	O–NH, , NaOH, toluene	HN, NH, Y, O, O, NH_2; Y: O (78), S (82)	150
O, HN, NH, O, O	O–NH, , toluene, H_2O, NH_4OH, rt, 30 min	O, HN, HN, O, O, =N–, O, NH, NH, O, $O^- NH_4^+$ (38)	149
HN, NH, O, O	O–NH, , toluene, DABCO, rt, 12 h	HN, HN, O, O, =N– (79)	149
NC, O, NR^1R^2	O–NH, , toluene, NaOH, H_2O, 0°, 10 min	H_2N, O, O, NR^1R^2, NH_2 **I** + H_2N, O, O, NR^1R^2, H_2N, NH_2 **II**	149
RHN, O, O, NHR	1. O–NH, (2 eq), toluene, DABCO, rt, 12 h (forms **I**) 2. EtOH, reflux, 15 min; rt, 12 h (forms **II**)	RHN, O, O, NHR, H_2N, NH_2 **I**; RHN, O, O, NHR, NH **II**	149

R^1	R^2	**I**	**II**
H	Me	(62)	(0)
Me	Me	(81)	(0)
H	*n*-Pr	(13)	(0)
—$(CH_2)_4$—		(59)	(0)
—$(CH_2)_2O(CH_2)_2$—		(55)	(0)
—$(CH_2)_5$—		(61)	(0)
H	Bn	(17)	(74)

C_{3-4}

EtO_2C–CH(R)–CO_2Et

1. NaOEt (2.9 eq), [camphor-derived oxaziridine, NH], EtOH, rt
2. Substrate, rt, time

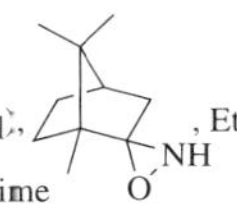

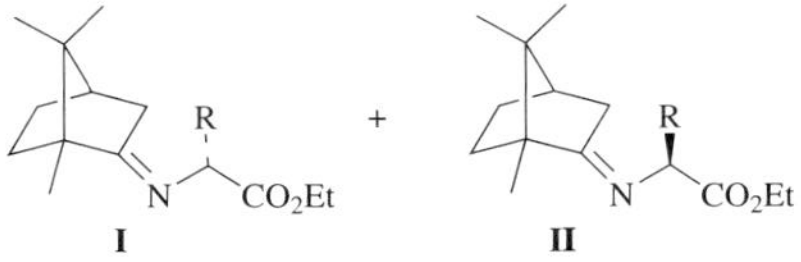

I + II (imines with N–CH(R)–CO_2Et)

R	I	II
n-Pr	(47)	(—)
Ph	(91)	(96)
3-ClC_6H_4	(73)	(—)
4-BrC_6H_4	(52)	(73)
2-$EtO_2CC_6H_4$	(93)	(90)
Bn	(83)	(—)
n-C_8H_{17}	(39)	(—)

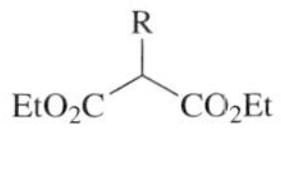

151

R	Time	I	II	I + II
H	3 h	(—)	(—)	(68)
Me	4 h	(45)	(45)	(—)

C_{3-9}

R^1C(O)–CH(R^2)–CO_2Et

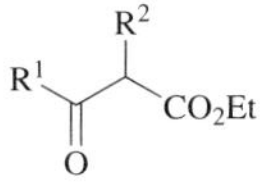

1. LiHMDS (1.1 eq, ia), THF, –78°, 1 h
2. 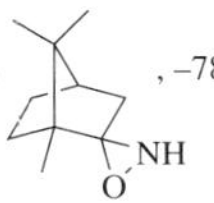 [camphor-derived oxaziridine, NH], –78° to rt, 2 h; rt, time

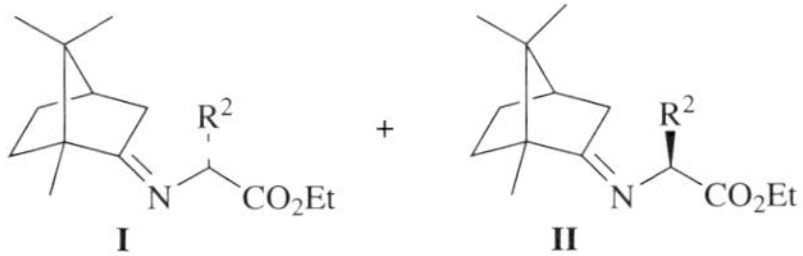

I + II (imines with N–CH(R^2)–CO_2Et)

151

R^1	R^2	Time	I + II
EtO	H	6 h	(50)
Me	H	48 h	(43)
EtO	Me	56 h	(20)
EtO	Bu	49 h	(14)
EtO	Ph	60 h	(0)

NC–CH(R^1)–CO_2R^2

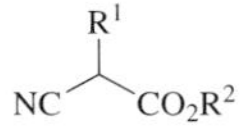

1. LiHMDS (ia), THF, –78°, 30 min
2. 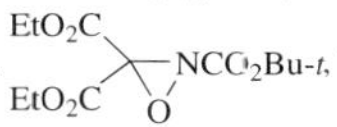 $(EtO_2C)_2C$–O–NCO_2Bu-*t* (oxaziridine), 78°, 20 h; to rt

NC–C(R^1)(CO_2R^2)–$NHCO_2Bu$-*t*

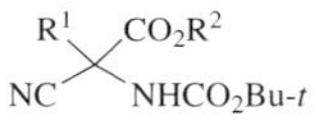

R^1	R^2	
H	Et	(33)
H	Bu-*t*	(20)
Ph	Et	(50)

155

TABLE 20. ENOLATES OF α–CYANO CARBONYL AND β–DICARBONYL COMPOUNDS (*Continued*)

Substrate	Conditions	Product(s) and Yield(s) (%)	Refs.
C_3			
NC–CH$_2$–COR	(NC)$_2$C=NOTs, pyridine, Et_2O	(NC)$_2$C=N$^-$=C(CN)COR PyH$^+$ R / Temp / Time NH_2 / reflux / 10 h (45) OMe / — / — (51) OEt / — / — (46)	838
$C_{3\text{-}4}$			
R–CH$_2$–CO_2Et R = CN, CO_2Et, Ac	1. Na base 2. PhNHN=C(CO_2Et)$_2$	PhNH–N(CH(CO_2Et)$_2$)–CH(R)–CO_2Et (—)	159
$C_{3\text{-}5}$			
R^1–CH$_2$–COR^2	1. Na base 2. R^3N=NCO$_2$Et	R^1CH(COR^2)–N(R^3)–NHCO$_2$Et ("very poor") R^1 / R^2 / R^3 NC / EtO / EtO_2C NC / EtO / Ph Ac / EtO / Ph Ac / Me / Ph	159
C_3			
R^1–CH$_2$–CO_2Et	R^2O_2CN=NCO$_2R^2$, KOAc	R^1C(CO_2Et)[N(CO_2R^2)–NHCO_2R^2]$_2$ R^1 / R^2 / Temp NC / Et / 60° (25) EtO_2C / Me / rt (>90) EtO_2C / Et / rt (80-90)	478
R^1C(=Y)CH$_2$C(=Y)R^1	1. *n*-BuLi (ia), THF, –70°, 20 min 2. R^2CON=NCOR^2, –78°; to rt	[R^1C(=Y)]$_2$CH–N(COR^2)–NHCOR^2 R^1 / Y / R^2 MeO / O / EtO (88) MeO / O / *i*-PrO (83) MeO / O / *t*-BuO (80) MeO / S / EtO (62) MeO / S / *t*-BuO (54) Me_2N / O / EtO (85) Me_2N / O / *i*-PrO (76) Me_2N / O / *t*-BuO (82) Me_2N / O / *N*-morpholinyl (76) Me_2N / S / *t*-BuO (57)[b]	477

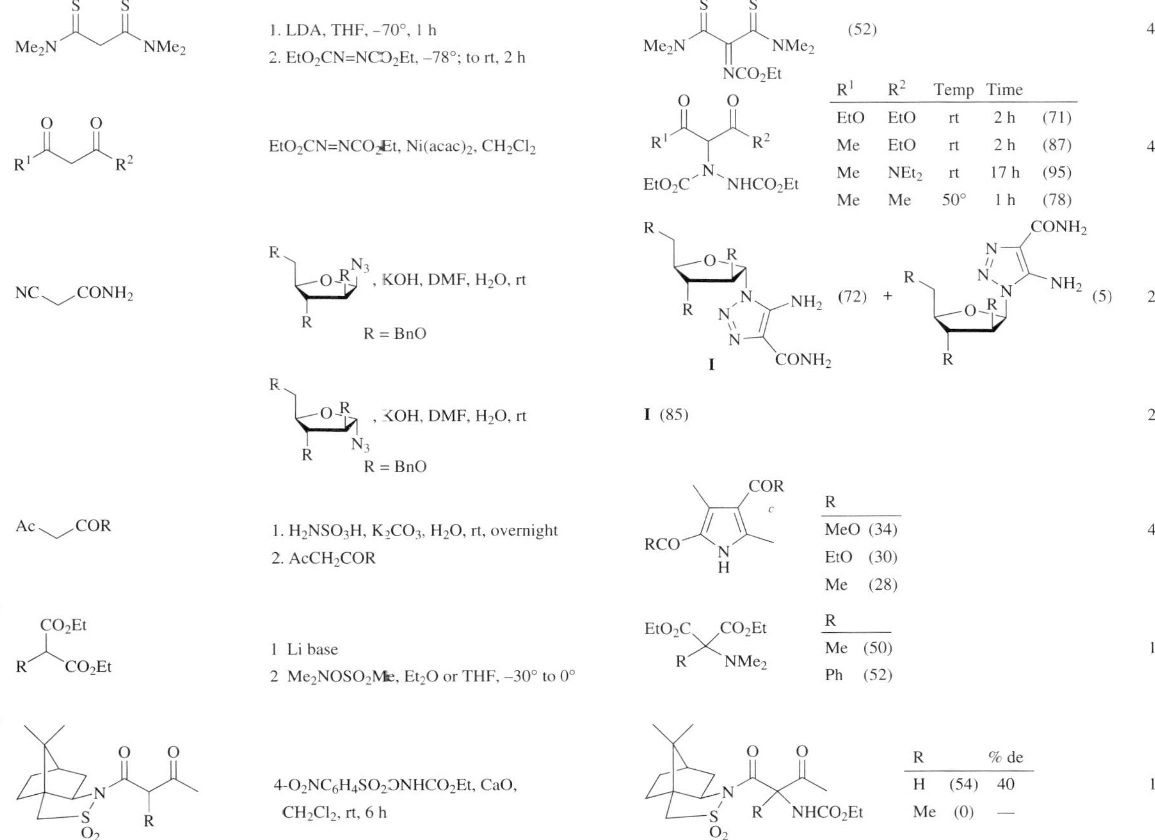

C_{3-5}	1. LDA, THF, –70°, 1 h 2. $EtO_2CN{=}NCO_2Et$, –78°; to rt, 2 h	(52)	477
	$EtO_2CN{=}NCO_2Et$, $Ni(acac)_2$, CH_2Cl_2	see table below	479
C_3	, KOH, DMF, H_2O, rt R = BnO	**I** (72) + (5)	276
	, KOH, DMF, H_2O, rt R = BnO	**I** (85)	276
C_{4-5}	1. H_2NSO_3H, K_2CO_3, H_2O, rt, overnight 2. $AcCH_2COR$	see table below [c]	473
C_{4-9}	1 Li base 2 Me_2NOSO_2Me, Et_2O or THF, –30° to 0°	see table below	134
C_{4-5}	4-$O_2NC_6H_4SO_2ONHCO_2Et$, CaO, CH_2Cl_2, rt, 6 h	see table below	124

R^1	R^2	Temp	Time	
EtO	EtO	rt	2 h	(71)
Me	EtO	rt	2 h	(87)
Me	NEt_2	rt	17 h	(95)
Me	Me	50°	1 h	(78)

R	
MeO	(34)
EtO	(30)
Me	(28)

R	
Me	(50)
Ph	(52)

R		% de
H	(54)	40
Me	(0)	—

TABLE 20. ENOLATES OF α–CYANO CARBONYL AND β–DICARBONYL COMPOUNDS (*Continued*)

	Substrate	Conditions	Product(s) and Yield(s) (%)	Refs.
C_{4-7}	Ac–CH(R)–CO_2Et (x eq)	4-$O_2NC_6H_4SO_2ONHCO_2Et$ (y eq), CaO (z eq), CH_2Cl_2, rt	Ac–C(R)(CO_2Et)–$NHCO_2Et$ **I** + Ac–C(CO_2Et)($NHCO_2Et$)–EtO_2CHN **II** + Ac–C(R)(CO_2Et)–NCO_2Et(–$NHCO_2Et$) **III** R, x, y, z, Time, I, II, III H, —, —, 3, 3 h, (2), (41), (0) H, 5, 1, 2, —, (58), (8), (0) Me, 5, 1, 1, 1 h, (58), (0), (4) *i*-Pr, 1, 3, 6[d], 5 h, (28), (0), (5)	124
C_{4-6} 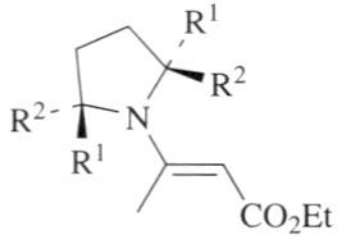	1-(1-methyl-2-ethoxycarbonylvinyl)-2,5-R^1,R^2-pyrrolidine (CO_2Et)	4-$O_2NC_6H_4SO_2ONHCO_2Et$, CH_2Cl_2, rt	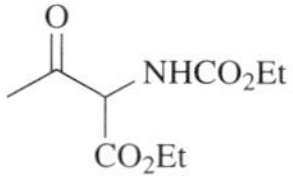 CH_3CO–CH($NHCO_2Et$)–CO_2Et R^1, R^2, Time, % de H, H, 45 h, (40), — H, Me, 2 h, (33), — (R) CH_2OMe, H, 2 h, (36), 80 (S)	125
C_{4-10} 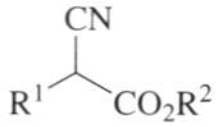	R^1–CH(CN)–CO_2R^2 R^2 = (camphor-derived O–, $CH_2SO_2N(C_6H_{11}\text{-}c)_2$) 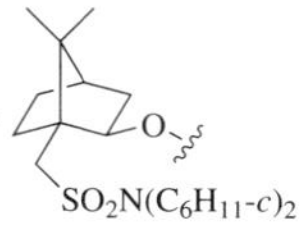	1. LiHMDS, THF, hexane, –78°, 1 h 2. $Ph_2P(O)ONH_2$, –78°; rt, 12 h	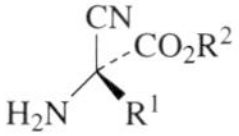 H_2N–C(R^1)(CN)–CO_2R^2 R^1, % de Me, (87), 56 *n*-Pr, (84), 52 *i*-Pr, (78), 40 *i*-Bu, (84), 44 Bn, (91), 60	475
C_4 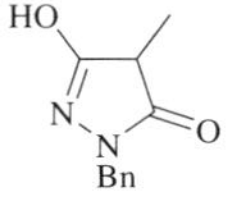	1-Bn-3-hydroxy-4-methylpyrazol-5-one (HO, N–N, Bn, O)	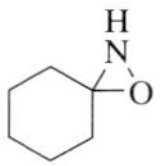 3-spirocyclohexyl oxaziridine (H, N, O), toluene, NaOH, H_2O, 0°, 10 min	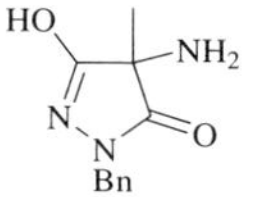 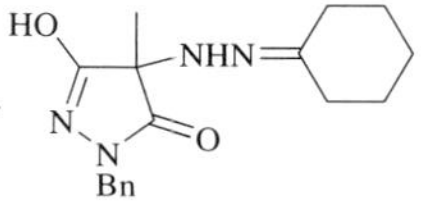 4-NH_2-4-methyl-1-Bn-3-hydroxypyrazol-5-one (90) + 4-(cyclohexylidene-NHN)-4-methyl-1-Bn-3-hydroxypyrazol-5-one (4)	149
	RO_2C–CH(CH_3)–CO_2R	1. LiHMDS (ia), THF, –78°, 30 min 2.  $(EtO_2C)_2C$–O–$NCO_2Bu\text{-}t$ (oxaziridine), –78°, 20 h; to rt	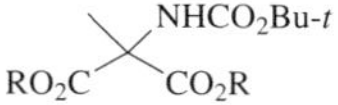 RO_2C–C(CH_3)($NHCO_2Bu\text{-}t$)–CO_2R R Me (30) Et (22)	155

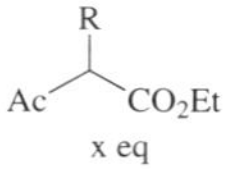

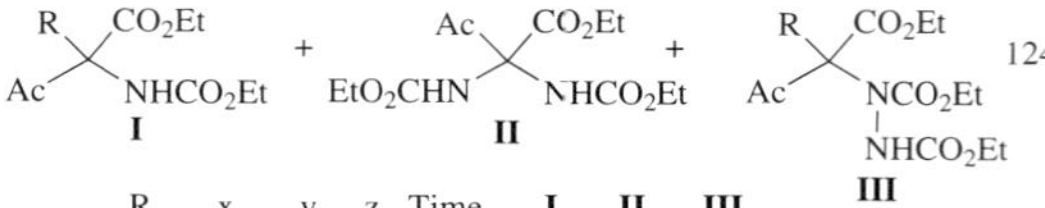

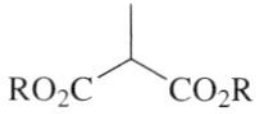

	Substrate	Conditions	Product(s) and Yield(s) (%)		Refs.
C_4	EtO_2C–CH(Me)–CO_2Ph	t-BuO_2CN=NCO_2Bu-t, **1** (5 mol%), toluene, rt, 16 h	EtO_2C–C(Me)(CO_2Ph)–$N(CO_2Bu$-$t)NHCO_2Bu$-t	(99), 90% ee (R)	481
$C_{4\text{-}6}$	R^1CO–CH(R^2)–COR^3	t-BuO_2CN=NCO_2Bu-t, **2** (2 mol%), THF, –60°	R^1CO–C(R^2)(NCO_2Bu-t–$NHCO_2Bu$-t)–COR^3		951

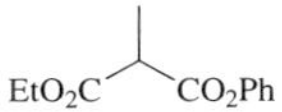

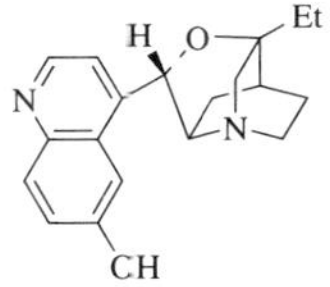

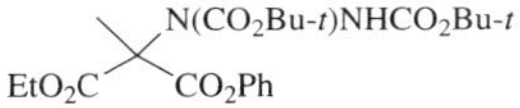

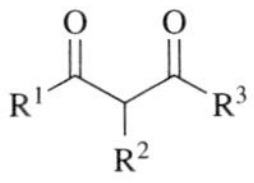

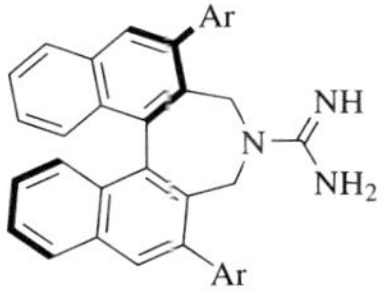

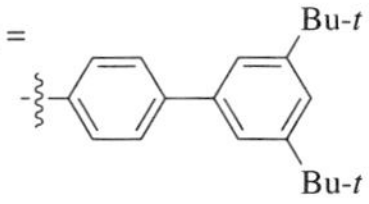

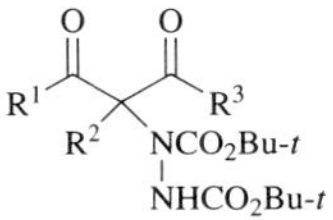

1 (catalyst structure; labels: H, O, Et, N, N, CH)

2 (catalyst structure; labels: Ar, Ar, N, NH, NH_2), Ar = 3,5-di-t-butylbiphenyl-4'-yl (labels: Bu-t, Bu-t)

R^1	R^2	R^3	Time		% ee
H	Me	OBu-t	0.5 h	(99)	83
Me	Me	OEt	0.5 h	(99)	85 (S)
Me	Me	OBu-t	3 h	(>99)	88
Me	Et	OBu-t	24 h	(54)	62
Et	Me	OBu-t	24 h	(90)	86

TABLE 20. ENOLATES OF α–CYANO CARBONYL AND β–DICARBONYL COMPOUNDS (*Continued*)

Substrate	Conditions	Product(s) and Yield(s) (%)	Refs.
C_{4-6} 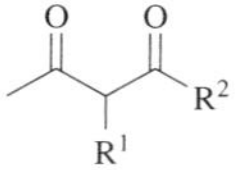 (R[1], R[2] β-diketone)	$R^3O_2CN{=}NCO_2R^3$, catalyst (10 mol%), $ClCH_2CH_2Cl$ Catalyst A, Catalyst B Y = (4-benzyl-2-oxo-oxazolidin-3-yl; Bn)	 I + II $E = N(CO_2R^3)NHCO_2R^3$	896
C_{4-13} NC–CH(R[1])–CO_2Bu-*t*	$R^2O_2CN{=}NCO_2R^2$, **1** (5 mol%), toluene **1**	R^2O_2C–N(NHCO$_2R^2$)–C(R[1])(NC)(CO_2Bu-*t*)	481

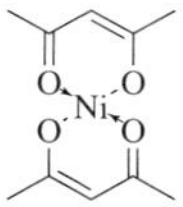

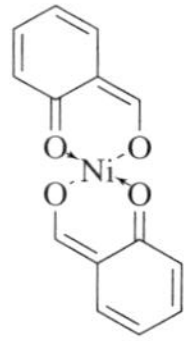

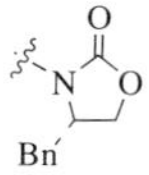

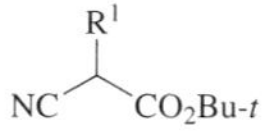

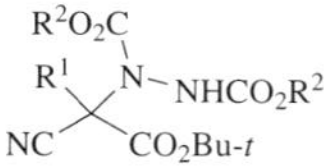

R[1]	R[2]	R[3]	Catalyst	Temp	Time	I + II	I:II
H	Y	Et	A	rt	18 h	(94)	73:27
H	Y	Et	B	rt	12 h	(97)	74:26
H	Y	Bn	A	0°	4 d	(100)	89:11
H	Y	Bn	B	rt	17 h	(100)	70:30
H	Y	Bn	B	0°	72 h	(100)	69:31
H	Y	Bn	B	50°	17 h	(100)	63:27
Me	Me	Et	B	50°	20 h	(96)	—
Me	OEt	Et	B	50°	23 h	(73)	—
Me	Y	Et	B	50°	—	(0)	—

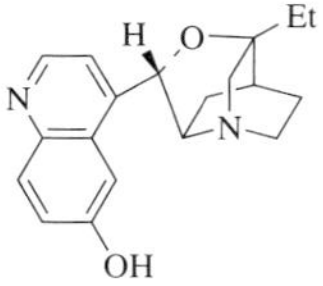

R^1	R^2	Temp	Time		% ee
Me	*t*-Bu	—	—	("excellent")	("lower")
i-Bu	*t*-Bu	—	—	("excellent")	("lower")
2-thienyl	*t*-Bu	–78°	16-20 h	(99)	97
Ph	Et	–78°	30 s	(>95)	84
Ph	Cl_3CCH_2	–78°	30 s	(>95)	7
Ph	*t*-Bu	–78°	4 h	(>95)	>98
Ph	*t*-Bu	rt	45 min	(>95)	90
Ph	Bn	–78°	30 s	(>95)	64
2-FC_6H_4	*t*-Bu	–50°	16-20 h	(99)	98
4-ClC_6H_4	*t*-Bu	–78°	16-20 h	(99)	98 (S)
4-$O_2NC_6H_4$	*t*-Bu	–50°	16-20 h	(99)	91
4-$MeOC_6H_4$	*t*-Bu	–78°	16-20 h	(95)	89
3-MeC_6H_4	*t*-Bu	–78°	16-20 h	(99)	97
2-naphthyl	*t*-Bu	–78°	16-20 h	(99)	98

C_4

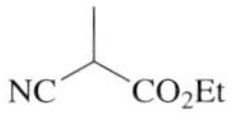

NC–CH(CH_3)–CO_2Et

$BnO_2CN{=}NCO_2Bn$, catalyst (10 mol%), toluene, –78°, 30 min

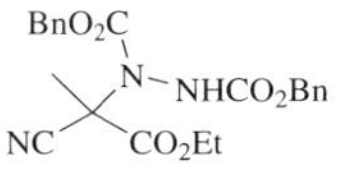

BnO_2C–N(–$NHCO_2Bn$)–C(CH_3)(NC)CO_2Et

Catalyst		% ee
A	(75)	35
B	(74)	23

232

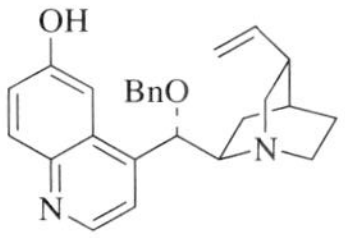

Catalyst A

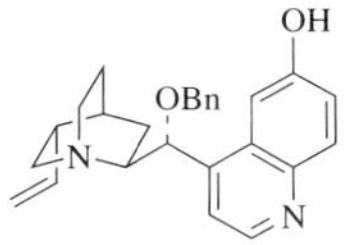

Catalyst B

TABLE 20. ENOLATES OF α–CYANO CARBONYL AND β–DICARBONYL COMPOUNDS (*Continued*)

Substrate	Conditions	Product(s) and Yield(s) (%)	Refs.
C_{4-9}			
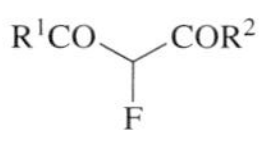 R^1CO–CH(F)–COR^2	1. 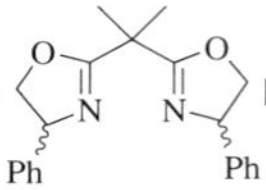 bis(oxazoline) [*S,S* (A) or *R,R* (B), 5 mol%], $Cu(OTf)_2$ (5 mol%), solvent, rt, 3 h 2. Substrate and $R^3O_2CN{=}NCO_2R^3$, rt, 2 d	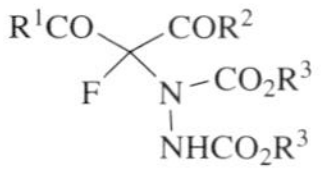 R^1CO–C(F)(COR^2)–$N(CO_2R^3)NHCO_2R^3$	237

R^1	R^2	R^3	Catalyst or Addend	Solvent		% ee or dr
Me	OMe	Et	A	CH_2Cl_2	(94)	94
Me	OMe	Et	A	hexane	(90)	86
Me	OMe	Bn	A	CH_2Cl_2	(95)	92
Me	OEt	Et	A	CH_2Cl_2	(90)	93
Me	OEt	Et	A	hexane	(84)	90
Me	OEt	Et	A	toluene	(95)	85
Me	OEt	Et	A	MeCN	(82)	20
Me	OEt	Bn	A	CH_2Cl_2	(73)	91
Me	OEt	Bn	A	hexane	(—)	88
Me	(2*S*,3*R*)-menthyloxy	Bn	TMEDA	CH_2Cl_2	(31)	43:57
Me	(2*S*,3*R*)-menthyloxy	Bn	A	CH_2Cl_2	(75)	87.5:12.5
Me	(2*S*,3*R*)-menthyloxy	Bn	B	CH_2Cl_2	(95)	1.5:98.5
Me	NPh_2	Bn	A	CH_2Cl_2	(88)	92
t-Bu	OEt	Et	A	CH_2Cl_2	(84)	93
t-Bu	OEt	Bn	A	CH_2Cl_2	(84)	93
Ph	OEt	Et	A	CH_2Cl_2	(85)	87
Ph	OEt	Bn	A	CH_2Cl_2	(78)	81

Substrate	Conditions	Product(s) and Yield(s) (%)	Refs.
C_4			
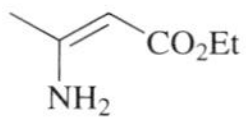 $CH_3C(NH_2){=}CHCO_2Et$	$RO_2CN{=}NCO_2R$, solvent	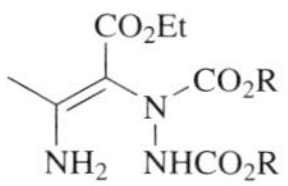 $CH_3C(NH_2){=}C(CO_2Et)N(CO_2R)NHCO_2R$	230

R	Solvent	
Me	Et_2O	(—)
Et	MeOH	(96)[e]

C_4

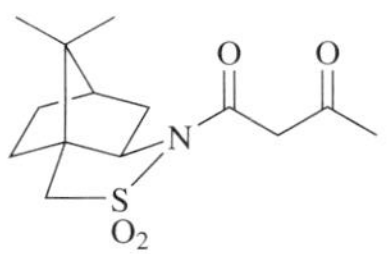

$BnO_2CN{=}NCO_2Bn$, catalyst (0.06 eq), MeCN, reflux, 24 h

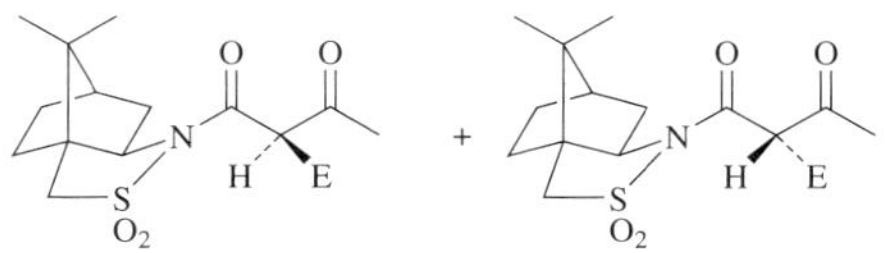

I major product + **II**

$E = N(CO_2Bn)NHCO_2Bn$

Catalyst	I + II	% de
$RuH_2(PPh_3)_4$	(100)	81
$RuCl_2(PPh_3)_3$	(100)	79
PPh_3	(86)	22

476

$C_{5\text{-}7}$

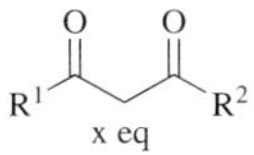

x eq

1. H_2NSO_3H, NaOH, H_2O, rt, few minutes (forms **I**)
2. $R^1C(O)CH_2C(O)R^2$ (forms **II**)

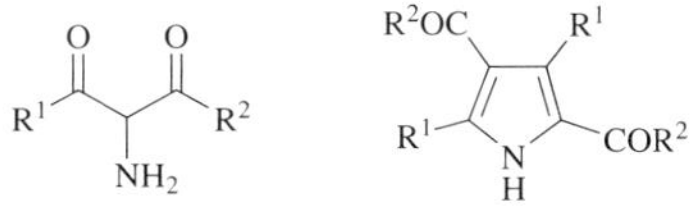

R^1 R^2 NH_2 **I**

R^2OC R^1 R^1 COR^2 N H **II**

R^1	R^2	x	I	II
Me	Me	1	(100)	(0)
Me	Me	2	(—)	(—)
Et	Et	2	(—)	(—)
Me	OEt	2	(—)	(30)
Me	NHPh	2	(—)	(47)

472

$C_{5\text{-}6}$

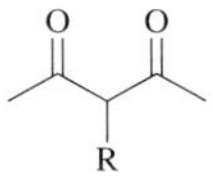

1. 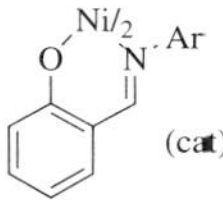 (cat), toluene, BrC_8F_{17}, 60°
2. $EtO_2CN{=}NCO_2Et$, reflux, 3 d

$Ar = 4\text{-}(n\text{-}C_{10}F_{21})C_6H_4$

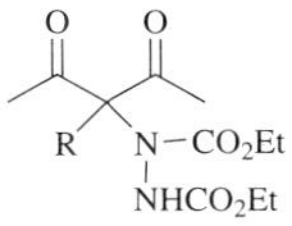

R $N-CO_2Et$ $NHCO_2Et$

R	
H	(96)
Me	(40)[f]

480

TABLE 20. ENOLATES OF α–CYANO CARBONYL AND β–DICARBONYL COMPOUNDS (*Continued*)

Substrate	Conditions	Product(s) and Yield(s) (%)	Refs.
$C_{5\text{-}15}$ $R^1COCH_2COR^2$	$R^3NHCON{=}NCO_2R^4$, $ZnCl_2$, CH_2Cl_2 (forms **I**, then **II**)	$R^1COCH(COR^2)N(CONHR^3)NHCO_2R^4$ **I**; imidazolone (R^1, NR^3, R^2CO, $NHCO_2R^4$) **II**	952
C_5 ethyl 2-methylacetoacetate	$BnO_2CN{=}NCO_2Bn$, **3** (5 mol%), MeOH, rt, 62 h	$CH_3COC(CH_3)(CO_2Et)N(CO_2Bn)NHCO_2Bn$ (57), 95% ee (R)	239

R^1	R^2	R^3	R^4	Temp	Time	I	II
Me	Me	$Cl(CH_2)_2$	Me	rt	4 h	>73	(—)
Me	Me	$c\text{-}C_6H_{11}$	Me	rt	4 h	>86	(—)
Me	Me	$4\text{-}MeOC_6H_4$	Me	0°	5 h	(—)	(47)
Me	OEt	$Cl(CH_2)_2$	Me	rt	4 h	(—)	(69)
Me	Ph	$Cl(CH_2)_2$	Me	rt	23 h	(—)	(68)
Me	Ph	Ph	Et	0°	2.5 h	(—)	(77)
Me	Ph	$2,4\text{-}F_2C_6H_3$	Et	0°	5 h	(—)	(90)
Me	Ph	$4\text{-}FC_6H_4$	Et	0°	2 h	(—)	(86)
Ph	OEt	$Cl(CH_2)_2$	Me	rt	3 h	(—)	(65)
Ph	OEt	$4\text{-}FC_6H_4$	Et	0°	2 h	(—)	(88)
Ph	OEt	$c\text{-}C_6H_{11}$	Me	rt	2 h	(>92)	(—)
Ph	Ph	$Cl(CH_2)_2$	Me	rt	10 h	(—)	(93)
Ph	Ph	$c\text{-}C_6H_{11}$	Me	rt	4 h	(>90)	(—)
Ph	Ph	$2,4\text{-}F_2C_6H_3$	Et	0°	5 h	(—)	(57)
Ph	Ph	$3\text{-}ClC_6H_4$	Et	0°	2 h	(—)	(72)

3: $[(P\text{–}P)Pd(OH_2)(NCMe)]^{2+}$ 2 PF_6^-; (P–P) = BINAP (PPh_2, PPh_2)

C_5

CO_2Et

t-$BuO_2CN{=}NCO_2Bu$-t,

Ar, N, H, N, H, S, NMe_2 (10 mol%), toluene, rt, "slow"

Ar = 3,5-$(CF_3)_2C_6H_3$

O, CO_2Et, N−$NHCO_2Bu$-t, t-BuO_2C (76) 15% ee

233

$C_{5\text{-}11}$

R^1, O, CO_2R^3, R^2

1. O, N, N, O, Ph, Ph (S,S) (x mol%),

$Cu(OTf)_2$, CH_2Cl_2

2. Substrate, then $BnO_2CN{=}NCO_2Bn$, rt, 16 h

R^1, O, CO_2Bn, N, $NHCO_2Bn$, R^2, CO_2R^3

235

R^1	R^2	R^3	x		% ee
Me	Me	Et	0.2	(91)	96
Me	Me	Et	10	(98)	98
Me	Me	t-Bu	10	(86)	98
—$(CH_2)_3$—		Et	0.5	(96)	99
—$(CH_2)_3$—		Et	10	(98)	99
Et	Me	Et	0.5	(98)	98
—$(CH_2)_4$—		Et	0.5	(96)	99
Me	$CH_2CH{=}CH_2$	t-Bu	0.5	(80)	98
i-Pr	Me	t-Bu	0.5	(89)	98
i-Pr	Me	t-Bu	10	(96)	98
—$(CH_2)_5$—		Et	0.5	(70)	99
Ph	Me	Et	0.5	(81)	87
Ph	Me	Et	10	(85)	95
Bn	Me	t-Bu	0.5	(79)	98
Bn	Me	t-Bu	10	(84)	98

TABLE 20. ENOLATES OF α–CYANO CARBONYL AND β–DICARBONYL COMPOUNDS (*Continued*)

	Substrate	Conditions	Product(s) and Yield(s) (%)	Refs.
C_5 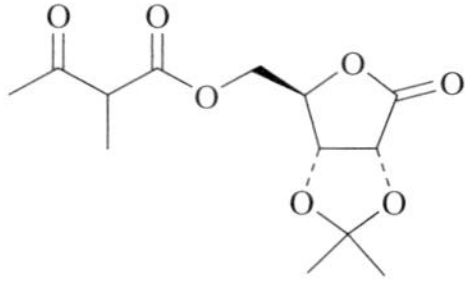	[structure]	$RO_2CN{=}NCO_2R$, catalyst (x eq), MeCN, rt, 24 h	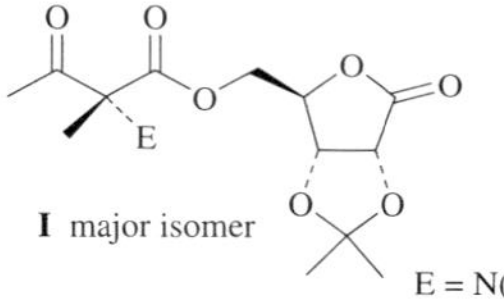 **I** major isomer 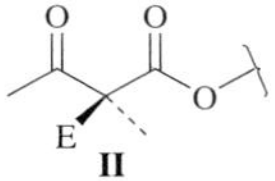 + **II**; E = $N(CO_2R)NHCO_2R$ (see table below)	476
C_6 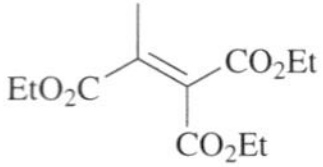	EtO_2C, CO_2Et, CO_2Et [structure]	1. NaH, Et_2O, 0°, 3 h 2. NH_2Cl, –78°; to 0°	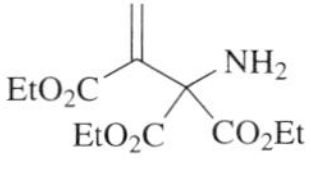 EtO_2C, NH_2, EtO_2C, CO_2Et **I** (74) 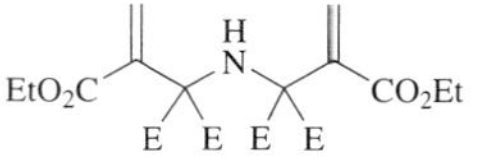 + EtO_2C, H N, CO_2Et, E E E E (1.5) E = CO_2Et	64
	EtO_2C, CO_2Et, CO_2Et [structure]	1. NaH, PhH, 0°; rt, 1 h 2. Add suspension of Na salt in Et_2O to $ClNH_2$ at 0°; rt, 2 h	**I** (65)	63
	EtO_2C, CO_2Et, CO_2Et [structure]	1. NaH, Et_2O, 0°, 3 h 2. NH_2Cl, –78°; to 0°	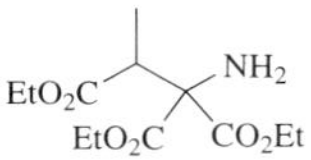 EtO_2C, NH_2, EtO_2C, CO_2Et (—)	64

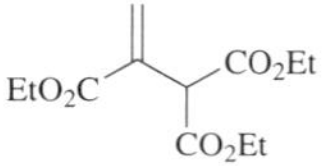

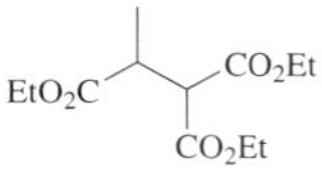

R	Catalyst	x	**I** + **II**	% de
Et	$RuCl_2(PPh_3)_3$	0.03	(92)	13
Et	$RuH_2(PPh_3)_4$	0.03	(72)	33
Et	PPh_3	0.09	(83)	48
i-Pr	$RuH_2(PPh_3)_4$	0.03	(69)	40
t-Bu	$RuCl_3(PPh_3)_3$	0.09	(75)	0
t-Bu	$RuH_2(PPh_3)_4$	0.09	(55)	19
Bn	$RuCl_2(PPh_3)_3$	0.03	(98)	5
Bn	$RuH_2(PPh_3)_4$	0.03	(64)	34
Bn	PPh_3	0.09	(65)	0

C_{6-7} — Substrate: 2-(ethoxycarbonyl)cycloalkanone, 5 eq

Conditions: 4-$O_2NC_6H_4SO_2ONHCO_2Et$ (1 eq), CaO (2 eq), CH_2Cl_2, rt

Products: 2-(ethoxycarbonylamino)-2-(ethoxycarbonyl)cycloalkanone (I) + aziridine NCO_2Et (II)

n	Time	I	II
1	1 h	(49)	(4)
2	3 h	(40)	(10)

Ref. 124

C_{6-10} — Substrate: $NC-CH(R^1)-CONHR^2$

Conditions: cyclohexane spiro-oxaziridine (NH), toluene, NaOH, H_2O, rt, 12 h

Products: spiro imidazolidinone I ($CONHR^2$) + spiro imidazolidinone II (H_2NCH_2O, NR^2)

R^1	R^2	I	II
i-Pr	2,6-$Me_2C_6H_3$	(—)	(81)
c-C_5H_{11}	$PhCH_2CH_2$	(62)	(—)
4-$ClC_6H_4CH_2$	Me	**I** + **II** (54)	
4-$O_2NC_6H_4CH_2$	3,4-$(MeO)_2C_6H_3(CH_2)_2$	(47)	(—)
2-$MeOC_6H_4CH_2$	Me	(35)	(—)
4-$MeOC_6H_4CH_2$	*i*-Pr	(81)	(—)
4-$MeOC_6H_4CH_2$	Ph	(—)	(58)
4-$MeOC_6H_4CH_2$	2,6-$Me_2C_6H_3$	(—)	(98)
4-$Me_2NC_6H_4CH_2$	*i*-Pr	(52)	(—)

Ref. 149

C_6 — Substrate: ethyl 2-ethyl-3-oxobutanoate

Conditions: $BnO_2CN{=}NCO_2Bn$, catalyst, CH_2Cl_2, –25°, 7 d

Product: $MeCOC(Et)(CO_2Et)N(CO_2Bn)NHCO_2Bn$

Catalyst		% ee
cinchonine	(72)	47[g]
cinchonidine	(72)	27[g]

Ref. 231

Substrate: ethyl 2-oxocyclopentanecarboxylate

Conditions: $RO_2CN{=}NCO_2R$, **3** (5 mol%), MeOH, rt

Product: 2-[N-(CO_2R)-N'-($NHCO_2R$)hydrazino]-2-(CO_2Et)cyclopentanone

R	Time		% ee
i-Pr	31 h	(89)	97
Bn	1 h	(73)	93 (R)

3 = $[(P\text{–}P)Pd(OH_2)(NCMe)]^{2+}$ 2 PF_6^-, (P–P) = BINAP (PPh_2, PPh_2)

Ref. 239

TABLE 20. ENOLATES OF α–CYANO CARBONYL AND β–DICARBONYL COMPOUNDS (*Continued*)

Substrate	Conditions	Product(s) and Yield(s) (%)	Refs.
C_6			
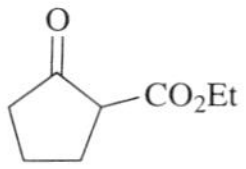 O, CO_2Et	$EtO_2CN{=}NCO_2Et$, **4** (cat), toluene, rt 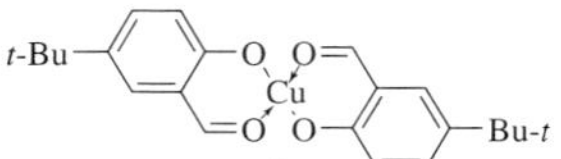 *t*-Bu, O, O, Cu, O, O, Bu-*t* **4**	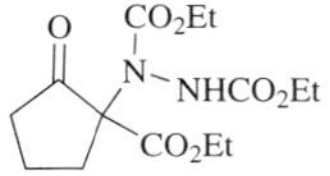 O, CO_2Et, N, $NHCO_2Et$, CO_2Et (95)	953
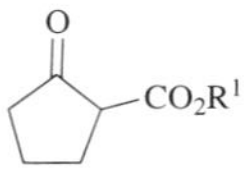 O, CO_2R^1	1. **5** (12.5 mol%), $M(OTf)_3$ (9 mol%), 4 Å MS, solvent, rt, overnight 2. Substrate 3. *t*-$BuO_2CN{=}NCO_2Bu$-*t*, temp 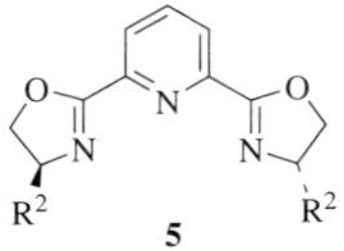 O, N, N, O, N, R^2, N, R^2 **5**	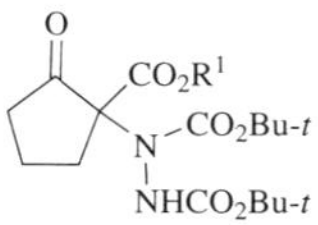 O, CO_2R^1, N, CO_2Bu-*t*, $NHCO_2Bu$-*t*	954

R^1	R^2	M	Solvent	Temp		% ee	Config.
Et	*i*-Pr	Yb	CH_2Cl_2	rt	(85)	68	R
Et	1-adamantyl	Yb	CH_2Cl_2	rt	(74)	0	—
Et	*i*-Pr	Eu	CH_2Cl_2	0°	(82)	62	R
Et	1-adamantyl	Eu	CH_2Cl_2	0°	(88)	12	S
Et	*i*-Pr	La	CH_2Cl_2	0°	(84)	52	R
Et	1-adamantyl	La	CH_2Cl_2	0°	(79)	18	R
t-Bu	*i*-Pr	Sc	CH_2Cl_2	rt	(0)	—	—
t-Bu	*i*-Pr	Yb	CH_2Cl_2	–41° to 0°	(70)	70	R
t-Bu	1-adamantyl	Yb	CH_2Cl_2	rt	(31)	66	S
t-Bu	*i*-Pr	Eu	CH_2Cl_2	rt	(55)	67	R
t-Bu	*i*-Pr	Eu	CH_2Cl_2	–41° to 0°	(81)	>95	R
t-Bu	1-adamantyl	Eu	CH_2Cl_2	–41°	(66)	86	S
t-Bu	*i*-Pr	La	CH_2Cl_2	0°	(78)	22	R
t-Bu	1-adamantyl	La	CH_2Cl_2	–41°	(71)	84	S
1-adamantyl	*i*-Pr	Yb	CH_2Cl_2	0°	(62)	55	R
1-adamantyl	*i*-Pr	La	MeCN	rt	(—)	89	R
1-adamantyl	*i*-Pr	Eu	MeCN	rt	(—)	93	R
1-adamantyl	*i*-Pr	Eu	MeCN	0°	(—)	95	R
1-adamantyl	*i*-Pr	Eu	MeCN	–41°	(—)	100	R
1-adamantyl	*i*-Pr	Eu	$ClCH_2CH_2Cl$	0°	(—)	73	R
1-adamantyl	*i*-Pr	Eu	CH_2Cl_2	0°	(—)	71	R
1-adamantyl	*i*-Pr	Eu	DME	0°	(—)	30	R
1-adamantyl	*i*-Pr	Eu	THF	0°	(—)	50	R
1-adamantyl	*i*-Pr	Eu	cyclohexane	0°	(—)	0	—

TABLE 20. ENOLATES OF α–CYANO CARBONYL AND β–DICARBONYL COMPOUNDS (*Continued*)

Substrate	Conditions	Product(s) and Yield(s) (%)	Refs.
C_{6-7} 2-(CO_2R)-cycloalkanone (ring size n)	$BnO_2CN{=}NCO_2Bn$, CH_2Cl_2, catalyst, –25°	2-COR-2-[$N(CO_2Bn)NHCO_2Bn$]-cycloalkanone	231

n	R	Catalyst	Time		% ee	Config.
1	Et	KOAc	15 min	(>80)	—	—
1	Et	quinine	2 min	(—)	26[h]	—
1	Et	quinidine	1 min	(—)	38[h]	—
1	Et	cinchonine	5 min	(95)	88 (R)	R
1	Et	cinchonidine	5 min	(95)	87 (S)	S
1	Bn	cinchonine	2 min	(99)	54[i]	—
1	Bn	cinchonidine	10 min	(99)	76[i]	—
2	Et	cinchonine	48 h	(92)	84 (R)	R
2	Et	cinchonidine	24 h	(81)	77 (S)	S

Substrate	Conditions	Product(s) and Yield(s) (%)	Refs.
2-(CO_2Et)-cycloalkanone (ring size n)	t-$BuO_2CN{=}NCO_2Bu$-t, **2** (x mol%), THF, –60°	2-CO_2Et-2-[$N(CO_2Bu$-$t)NHCO_2Bu$-t]-cycloalkanone	951

2: binaphthyl-based azepine guanidine (N–C(=NH)NH_2) bearing Ar groups; Ar = 4-(3,5-di-t-Bu-phenyl)phenyl (Bu-t, Bu-t)

n	x	Time		% ee	Config.
1	0.05	4 h	(100)	97 (R)	R
1	2	5 min	(100)	97 (R)	R
2	2	24 h	(>99)	98	—

C_{6-11}

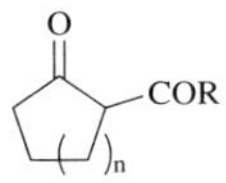

COR

t-BuO$_2$CN=NCO$_2$Bu-t, **1** (5 mol%), toluene

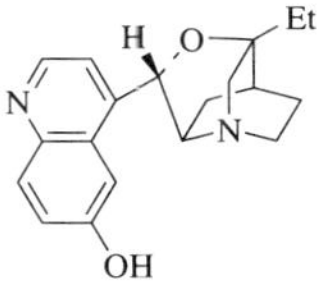

1

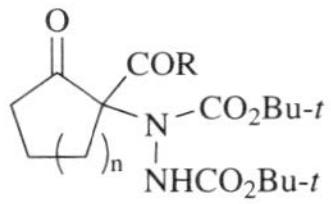

R	n	Temp	Time		% ee
OBu-t	1	–52°	66 h	(99)	89
Et	2	rt	143 h	(86)	83
CH_2Bu-t	1	–50°	91 h	(90)	83

481

C_6

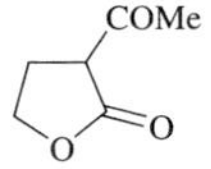

COMe

i-PrO$_2$CN=NCO$_2$Pr-i, **3** (5 mol%), MeOH, rt, 9 h

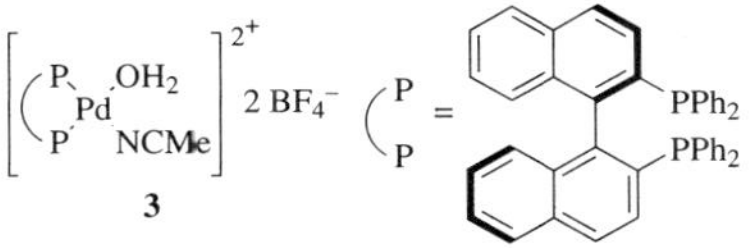

3

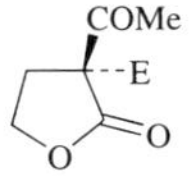

(93), 93% ee E = N(CO$_2$Pr-i)NHCO$_2$Pr-i

239

C_{6-7}

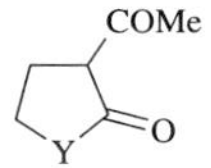

COMe

t-BuO$_2$CN=NCO$_2$Bu-t, **2** (cat, 2 mol%), THF, –60°

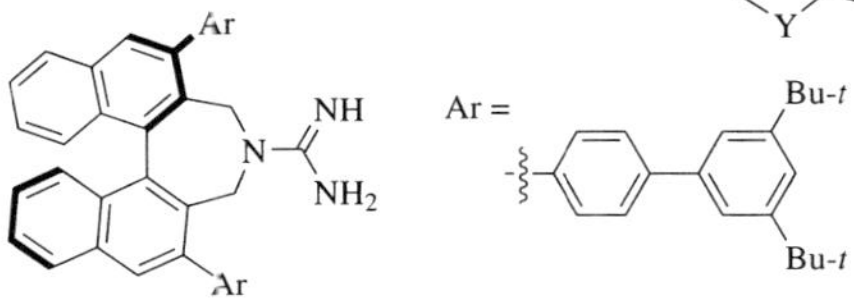

2

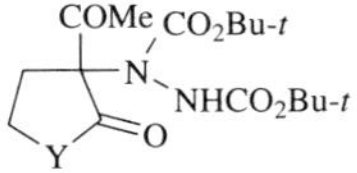

Y	Time		% ee
O	24 h	(>99)	15
CH_2	5 h	(99)	91

951

TABLE 20. ENOLATES OF α–CYANO CARBONYL AND β–DICARBONYL COMPOUNDS (*Continued*)

Substrate	Conditions	Product(s) and Yield(s) (%)	Refs.
C_{6-9} COR (lactone)	$BnO_2CN{=}NCO_2Bn$, CH_2Cl_2, catalyst, –25°	COR, N(CO_2Bn)$NHCO_2Bn$ (lactone)	231
		R / Catalyst / Time / / % ee[i]	
		Me / cinchonine / 2 min / (91) / 49	
		Me / cinchonidine / 2 min / (96) / 42	
		i-Pr / cinchonine / 25 min / (91) / 60	
		i-Pr / cinchonidine / 40 min / (91) / 64	
		t-Bu / cinchonine / 4 d / (68) / 51	
		t-Bu / cinchonidine / 4 d / (51) / 57	
C_{6-8} Y, N, R, R, O	$EtO_2CN{=}NCO_2Et$	Y, N, CO_2Et, N, $NHCO_2Et$, R, R, O	955
		R / Y	
		H / — / (100)	
		H / O / (100)	
		Me / — / (—)	
		Me / O / (—)	
C_6 NHR^1, O	$R^2CON{=}NCOR^2$, MeCN, reflux, 3 h	NHR^1, COR^2, N, $NHCOR^2$, O	250
		R^1 / R^2	
		n-Bu / EtO / (46)	
		n-Bu / Ph / (67)	
		t-Bu / EtO / (50)	
		t-Bu / Ph / (40)	
		Ph / EtO / (65)	
		Ph / Ph / (65)	
Y, N, O	1. $EtO_2CN{=}NCOAr$, PhH, 0°, 2-3 d (forms **I**) 2. H_3O^+, rt, 30 d (forms **II**)	**I** (Y, N, E, O) + **II** (OH, E, O)	249
		Y / Ar / **I** / **II**	
		— / Ph / (59) / (40)	
		— / 4-$O_2NC_6H_4$ / (50) / (45)	
		O / Ph / (67) / (—)	
		O / 4-$O_2NC_6H_4$ / (65) / (—)	
		E = $N(CO_2Et)NHCOAr$	

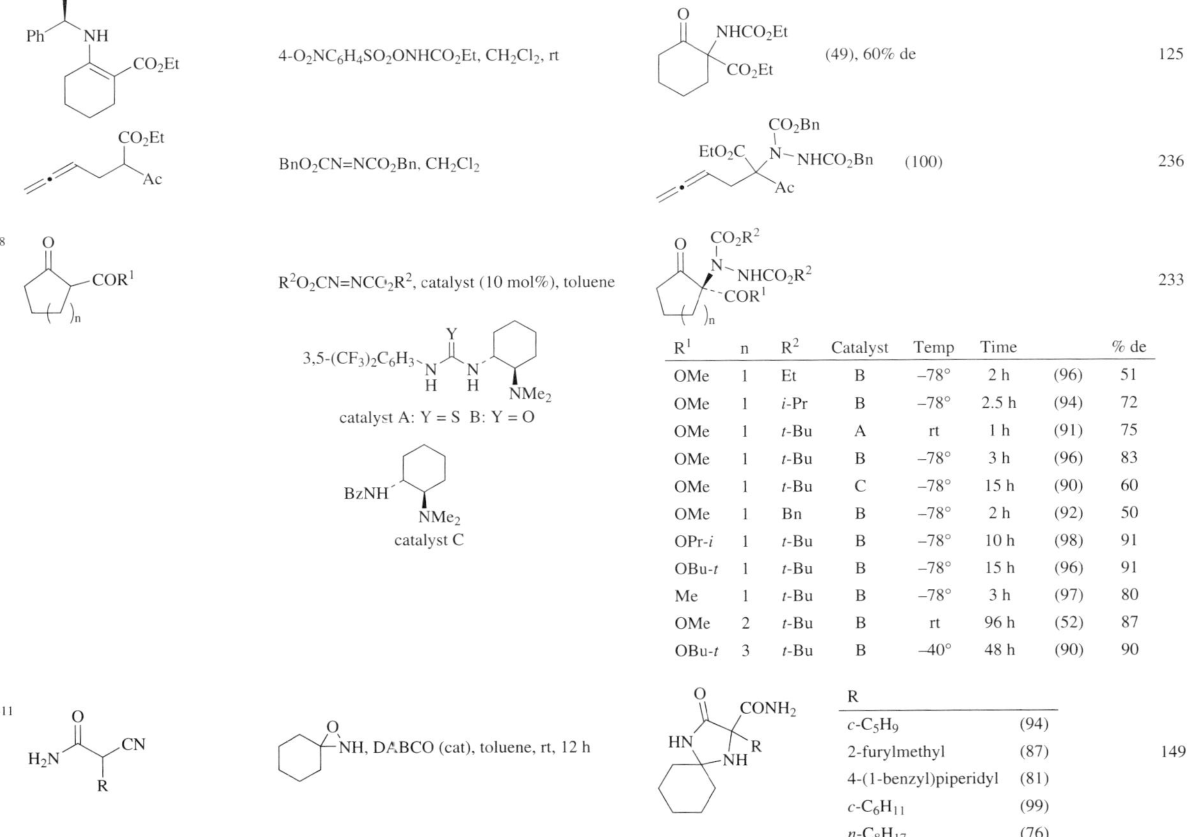

R^1	n	R^2	Catalyst	Temp	Time		% de
OMe	1	Et	B	–78°	2 h	(96)	51
OMe	1	*i*-Pr	B	–78°	2.5 h	(94)	72
OMe	1	*t*-Bu	A	rt	1 h	(91)	75
OMe	1	*t*-Bu	B	–78°	3 h	(96)	83
OMe	1	*t*-Bu	C	–78°	15 h	(90)	60
OMe	1	Bn	B	–78°	2 h	(92)	50
OPr-*i*	1	*t*-Bu	B	–78°	10 h	(98)	91
OBu-*t*	1	*t*-Bu	B	–78°	15 h	(96)	91
Me	1	*t*-Bu	B	–78°	3 h	(97)	80
OMe	2	*t*-Bu	B	rt	96 h	(52)	87
OBu-*t*	3	*t*-Bu	B	–40°	48 h	(90)	90

R	
c-C_5H_9	(94)
2-furylmethyl	(87)
4-(1-benzyl)piperidyl	(81)
c-C_6H_{11}	(99)
n-C_8H_{17}	(76)

TABLE 20. ENOLATES OF α–CYANO CARBONYL AND β–DICARBONYL COMPOUNDS (*Continued*)

Substrate	Conditions	Product(s) and Yield(s) (%)	Refs.
C_{8-13} 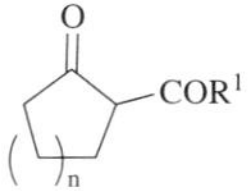 2-COR^1-cycloalkanone (ring size n)	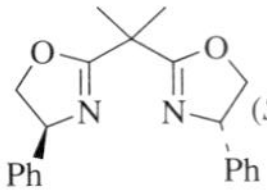 1. bis(oxazoline) ligand, Ph, Ph (*S,S*) (x mol%), $Cu(OTf)_2$ (x mol%), CH_2Cl_2, rt, 2 h 2. Substrate, then $R^2O_2CN{=}NCO_2R^2$, temp, time	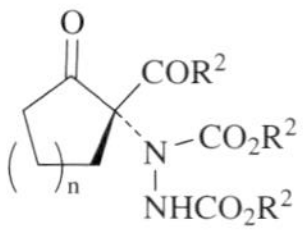 2-COR^2-2-[$N(CO_2R^2)NHCO_2R^2$]-cycloalkanone	956

n	R^1	R^2	x	Temp	Time	Yield	% ee
1	*s*-Bu	Bn	10	rt	18 h	(60)	89
1	*s*-Bu	Bn	10	–24°	18 h	(60)	94
2	Me	Bn	1	rt	18 h	(94)	80
2	Me	Bn	10	rt	18 h	(82)	84
2	Et	Bn	10	rt	18 h	(83)	94
2	*i*-Pr	Bn	10	rt	18 h	(86)	94
2	*s*-Bu	Bn	1	rt	18 h	(76)	94
2	*s*-Bu	Bn	10	rt	18 h	(76)	95
2	Ph	Et	10	rt	40 h	(74)	77
2	Ph	Bn	10	rt	18 h	(87)	91
3	*s*-Bu	Bn	10	rt	18 h	(90)	83
3	*s*-Bu	Bn	10	–24°	18 h	(89)	85

Substrate	Conditions	Product(s) and Yield(s) (%)	Refs.
C_8 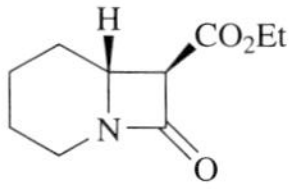 bicyclic β-lactam (H, CO_2Et, N, O)	1. LDA, THF 2. TsN_3 3. TMSCl	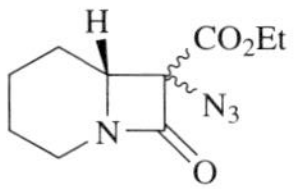 bicyclic β-lactam (H, CO_2Et, N_3, N, O) (—)	482
C_9 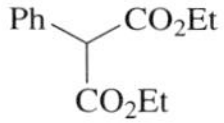 $PhCH(CO_2Et)_2$	1. NaH, DMF 2. 2,4-$(O_2N)_2C_6H_3ONH_2$	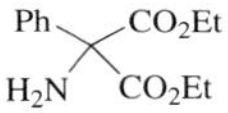 $PhC(NH_2)(CO_2Et)_2$ **I** (53)	877
	1. Li base, THF 2. $Ph_2P(O)ONH_2$, –20°; rt, 12 h	**I** (31)	139

Substrate	Conditions	Product(s) and Yield(s) (%)	Refs.
PhCH(CN)CO_2Et	1. NaH, THF 2. 2,4-$(O_2N)_2C_6H_3ONH_2$	PhC(H_2N)(CN)CO_2Et (54)	93
	1. Li base 2. 2,4,6-$Me_3C_6H_2SO_2ONMe_2$, Et_2O or THF, −10° to −20° ; rt, 15 h	PhC(Me_2N)(CN)CO_2Et (95)	133
	1. NaH, THF, rt, 15 min 2. Reagent, rt, overnight	PhC(H_2N)(CN)CO_2Et **I** Reagent: $Ph_2P(O)ONH_2$ (96)[b]; $(4\text{-}MeOC_6H_4)_2P(O)ONH_2$ (75); $4\text{-}O_2NC_6H_4CO_2NH_2$ (85)	106
	1. Li base, THF 2. $Ph_2P(O)ONH_2$, −20°; rt, 12 h	**I** (>96)	139
	1. LiHMDS or NaHMDS 2. 4-ClC_6H_4-oxaziridine-$NCONEt_2$	PhC(NC)($NHCONEt_2$)CO_2Et (0)	155
PhCH(CONHPh)$_2$	cyclohexane spiro-oxaziridine (NH), DABCO, toluene, rt, 12 h	PhC(NH_2)(CONHPh)$_2$ (72)	149
PhCH(CN)CO_2Et	1. LiHMDS (ia), THF, −78°, 1 h 2. camphor-derived spiro-oxaziridine (NH) (+), −78°, 31 h; to rt	N-imine (Ph, H, CN) (34) + N-imine diastereomer (Ph, H, CN) (34) + N-imine (Ph, $CONH_2$) (9)	151
PhCH(CN)CO_2Et	$t\text{-}BuO_2CN{=}NCO_2Bu\text{-}t$, Ar-NH-C(S)-NH-cyclohexyl-NMe_2 thiourea (10 mol%), toluene, −78° Ar = 3,5-$(CF_3)_2C_6H_3$	PhC(NC)[N($CO_2Bu\text{-}t$)$NHCO_2Bu\text{-}t$]CO_2Et (93), 73% ee	233

TABLE 20. ENOLATES OF α–CYANO CARBONYL AND β–DICARBONYL COMPOUNDS (*Continued*)

Substrate	Conditions	Product(s) and Yield(s) (%)	Refs.
C_{9-13} 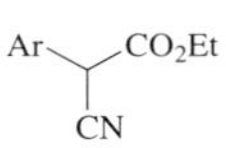 Ar–CH(CN)–CO_2Et	$RO_2CN{=}NCO_2R$, catalyst (x mol%), solvent, –78° 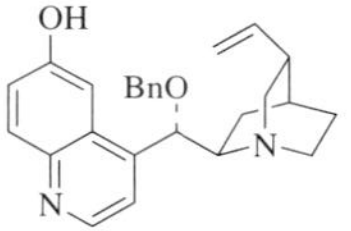 catalyst A 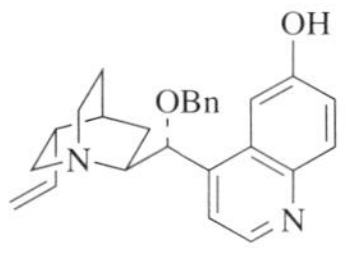 catalyst B	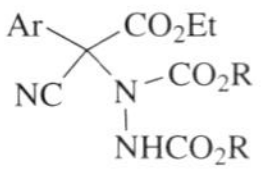 Ar–C(CN)(CO_2Et)–N(CO_2R)–$NHCO_2R$	232

Ar	R	Catalyst	x	Solvent	Time		% ee
Ph	Et	A	10	toluene	1 min	(—)	85
Ph	*i*-Pr	A	10	toluene	45 min	(—)	44
Ph	*t*-Bu	A	10	Et_2O	3 h	(—)	70
Ph	*t*-Bu	A	10	THF	12 h	(—)	10
Ph	*t*-Bu	A	10	CH_2Cl_2	11 h	(—)	77
Ph	*t*-Bu	A	5	toluene	2 h	(92)	95
Ph	*t*-Bu	B	5	toluene	4 h	(92)	97
Ph	Bn	A	10	toluene	1 min	(—)	89
4-FC_6H_4	*t*-Bu	A	5	toluene	1 h	(97)	94
4-FC_6H_4	*t*-Bu	B	5	toluene	2 h	(95)	96
4-ClC_6H_4	*t*-Bu	A	5	toluene	30 min	(96)	93 (S)
4-ClC_6H_4	*t*-Bu	B	5	toluene	30 min	(94)	97
4-BrC_6H_4	*t*-Bu	A	5	toluene	1 h	(99)	93
4-BrC_6H_4	*t*-Bu	B	5	toluene	2 h	(97)	96
4-BrC_6H_4	Bn	A	5	toluene	1 min	(86)	91
4-BrC_6H_4	Bn	B	5	toluene	1 min	(83)	92
4-$MeOC_6H_4$	*t*-Bu	A	10	toluene	5 min	(96)	94
4-$MeOC_6H_4$	*t*-Bu	B	10	toluene	10 min	(96)	97
2-MeC_6H_4	Bn	A	5	toluene	1 h	(71)	82
2-MeC_6H_4	Bn	B	5	toluene	3.5 h	(72)	87
4-MeC_6H_4	*t*-Bu	A	5	toluene	8 h	(96)	94
4-MeC_6H_4	*t*-Bu	B	5	toluene	8 h	(96)	96
1-naphthyl	*t*-Bu	A	10	toluene	8 h	(99)	93
1-naphthyl	*t*-Bu	B	10	toluene	12 h	(98)	99

Carbons	Substrate	Conditions	Product(s) and Yield(s) (%)	Refs.
C_{9-10}	$RCH(CO_2Et)_2$	1. NaH, glyme, rt 2. TsN_3, rt; 35° to 40°, 1 h	$RC(N_3)(CO_2Et)_2$ R: Ph (71)[j]; 7-cycloheptatrienyl (65)[j]	483
C_9	$4\text{-}BnOC_6H_4CH(CO_2Et)_2$	1. NaH, THF, HMPA, rt, 2 h 2. TsN_3, 50°, 2.5 h	$4\text{-}BnOC_6H_4C(N_3)(CO_2Et)_2$ (76)	777, 957
	ethyl 2-oxocyclooctanecarboxylate	TsN_3, Et_3N, Et_2O, rt, 140 h	ethyl 1-azido-2-oxocyclooctanecarboxylate (74)[k] + $TsNHCO(CH_2)_6C(=N_2)CO_2Et$ (20)	319
C_{10}	$NCCH(R^1)CONR^2R^3$	1. 1-oxa-2-azaspiro[2.5]octane (cyclohexane spiro-oxaziridine, NH), NaOH, toluene, H_2O, rt, 12 h (forms **I** + **II**) 2. **I** or **II**, EtOH, addend, reflux, 5 min (forms **III**)	**I**: spiro-imidazolidinone (cyclohexane) bearing R^1 and $CONR^2R^3$ + **II**: $H_2NCH_2OC(R^1)(N{=}C_6H_{10})CONR^2R^3$ **III**: $H_2NOCC(NH_2)(R^1)CONR^2R^3$	149

R^1	R^2	R^3	**I**	**II**	Addend	**III**
$4\text{-}MeOC_6H_4CH_2$	—$(CH_2)_2O(CH_2)_2$—		(48) + 39% **III**	(—)	—	(95)
$4\text{-}ClC_6H_4CH_2$	—$(CH_2)_5$—		(62) **I + III**	(—)	—	(82)
$2\text{-}MeOC_6H_4CH_2$	—$(CH_2)_5$—		(—)	(32)	HCl	(90)[a]
$4\text{-}MeOC_6H_4CH_2$	Ph	Me	(—)	(43)	HCl	(98)[a]

Carbons	Substrate	Conditions	Product(s) and Yield(s) (%)	Refs.
C_{10-11}	1-oxo-indan/tetralin-2-CO_2R ($(CH_2)_n$)	$t\text{-}BuO_2CN{=}NCO_2Bu\text{-}t$, Ar-NH-C(=S)-NH-(cyclohexyl-2-NMe_2) thiourea (10 mol%), toluene Ar = $3,5\text{-}(CF_3)_2C_6H_3$	2-E-2-CO_2R ketone E = $N(CO_2Bu\text{-}t)NHCO_2Bu\text{-}t$	233

n	R	Temp	Time		% ee
1	t-Bu	–78°	4 h	(93)	90
2	Me	–40°	5 h	(99)	97

Substrate	Conditions	Product(s) and Yield(s) (%)	Refs.
C_{10}	$RO_2CN{=}NCO_2R$, **3** (5 mol%), ionic liquid [bmim]Y, rt	E = $N(CO_2R)NHCO_2R$	239
C_{10-11}	$R^4O_2CN{=}NCO_2R^4$, **3** (as above, 5 mol%), solvent, rt	E = $N(CO_2R^4)NHCO_2R^4$	239

3: $[(P\text{-}P)Pd(OH_2)(NCMe)]^{2+}$ 2 X^-; P–P = BINAP (PPh_2, PPh_2)

R	X	Y	Time		% ee
Et	BF_4	BF_4	30 min	(88)	0
Et	PF_6	BF_4	1 h	(90)	0
Et	SbF_6	SbF_6	30 min	(91)	87
Et	PF_6	SbF_6	1 h	(89)	79
Et	SbF_6	PF_6	1 h	(96)	97
Et	PF_6	PF_6	1 h	(93)	85
i-Pr	SbF_6	PF_6	12 h	(96)	91
i-Pr	PF_6	PF_6	18 h	(81)	71
i-Pr	SbF_6	SbF_6	12 h	(92)	89

n	R^1	R^2	R^3	R^4	X	Solvent	Time		% ee
1	H	H	Et	Et	PF_6	MeOH	30 min	(94)	94
1	H	H	Et	*i*-Pr	PF_6	MeOH	30 min	(98)	97
1	H	H	Bn	*i*-Pr	PF_6	MeOH	30 min	(99)	94
1	MeO	MeO	Et	*i*-Pr	PF_6	Me_2CO	170 h	(71)	95
2	H	H	Me	*t*-Bu	BF_4	MeOH	106 h	(56)	95
2	H	H	Et	Et	PF_6	MeOH	1 h	(75)	91
2	H	MeO	Et	*t*-Bu	SbF_6	MeOH	144 h	(66)	91

C_{10}

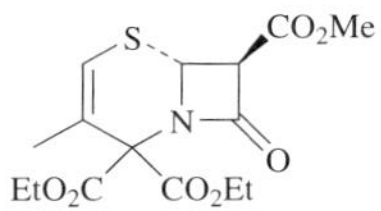

1. LiHMDS, THF, –70°, 5 min
2. TsN_3, –70°
3. TMSCl, –70°; to rt, 2 h

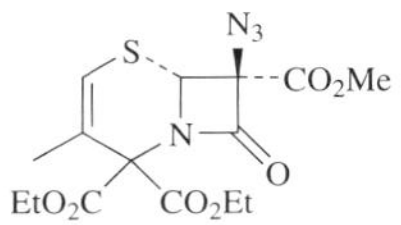

(43) 339

C_{10-12}

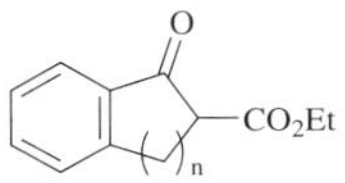

RSO_2N_3, Et_3N solvent

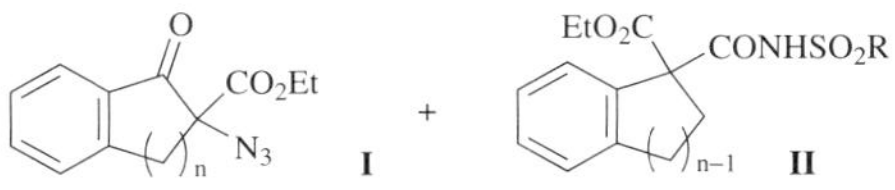

I + II + products of diazo transfer **III**

n	R	Solvent	Temp	Time	I	II	III	Refs.
1	4-$O_2NC_6H_4$	THF	0°; rt	30 min; 2 h	(0)	(0)	(94)	321
1	4-MeC_6H_4	THF	rt	6 d	(16)	(0)	(80)	321
2	4-$O_2NC_6H_4$	THF	rt	20 h	(21)	(47)	(0)	321
2	4-MeC_6H_4	THF	rt	4 d	(75)	(0)	(trace)	321
3	4-$O_2NC_6H_4$	THF	rt	6 d	(31)	(34)	(14)	319
3	4-MeC_6H_4	THF	rt	5 d	(31)	(0)	(66)	321
3	4-MeC_6H_4	MeCN	rt	1 d	(20)	(56)	(18)	319
3	4-MeC_6H_4	DMF	rt	10 h	(32)	(18)	(10)	319
3	Me	THF	rt	3 d	(24)	(30)	(28)	319
3	Me	MeCN	rt	15 h	(19)	(60)	(8)	319
3	2,4,6-$(i\text{-}Pr)_3C_6H_2$	THF	rt	10 d	(72)	(0)	(0)	319
3	2,4,6-$(i\text{-}Pr)_3C_6H_2$	MeCN	rt	4 d	(80)	(0)	(0)	319
3	2,4,6-$(i\text{-}Pr)_3C_6H_2$	CH_2Cl_2	0°	1 h	(0)	(71)	(15)	319

C_{11}

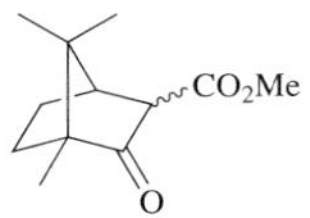

TsN_3, Et_3N, Et_2O, rt, 190 h

(20) + (56) 319

TABLE 20. ENOLATES OF α–CYANO CARBONYL AND β–DICARBONYL COMPOUNDS (*Continued*)

Substrate	Conditions	Product(s) and Yield(s) (%)	Refs.
C_{11}			
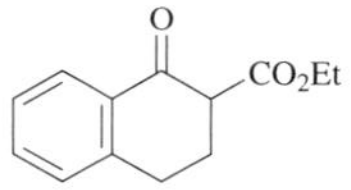 CO_2Et	*t*-$BuO_2CN{=}NCO_2Bu$-*t*, **2** (2 mol%), THF, –60°, 1 h; 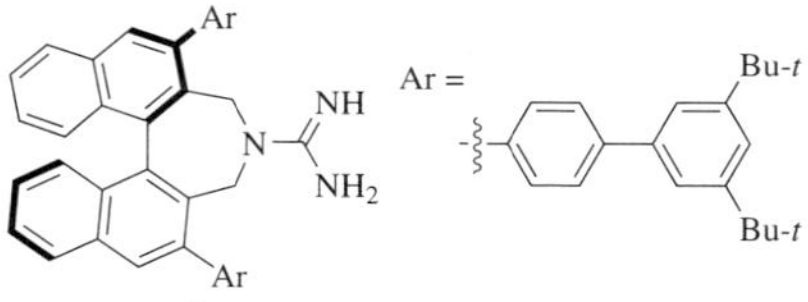 Ar = ; Bu-*t*; NH; NH_2; Ar; **2**	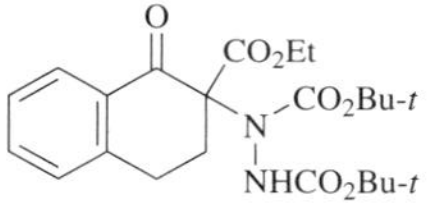 CO_2Et, CO_2Bu-*t*, $NHCO_2Bu$-*t* (>99), 97% ee	951
C_{13}			
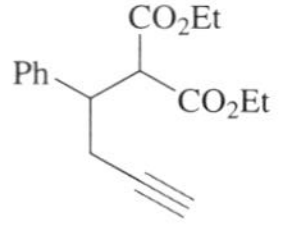 CO_2Et, Ph, CO_2Et	1. NaH, THF, rt, 25 min 2. 2,4-$(O_2N)_2C_6H_3ONH_2$, rt, 20 h	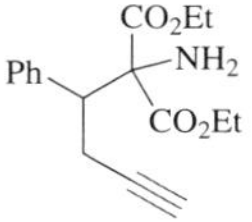 CO_2Et, NH_2, Ph, CO_2Et (62)	474
C_{17}			
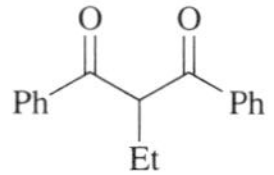 Ph, Ph, Et	$EtO_2CN{=}NCO_2Et$, catalyst (x eq), MeCN, rt, 24 h	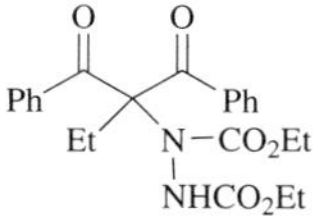 Ph, Ph, Et, $N-CO_2Et$, $NHCO_2Et$ Catalyst / x: Ph_3P 1 (100) $RuCl_2(PPh_3)_3$ 0.06 (83)	476
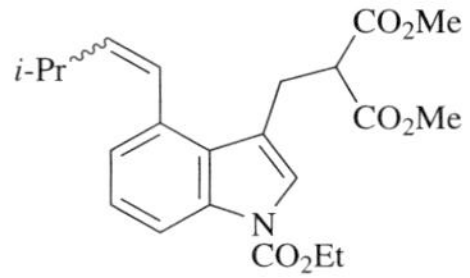 *i*-Pr, CO_2Me, CO_2Me, N, CO_2Et	1. NaH, THF, HMPA, rt, 2 h 2. TsN_3, rt; reflux, 2 h	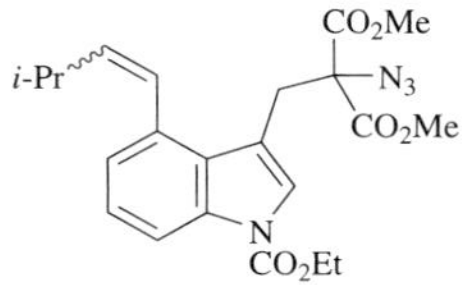 *i*-Pr, CO_2Me, N_3, CO_2Me, N, CO_2Et (62)	958

C_{19}	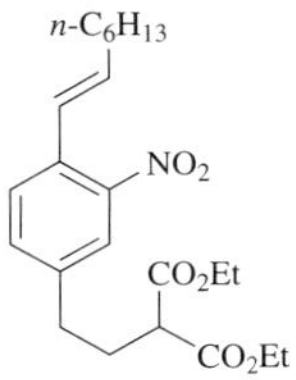 n-C_6H_{13} / NO_2 / CO_2Et / CO_2Et	$(4\text{-}MeOC_6H_4)_2P(O)ONH_2$, NaH, THF, rt	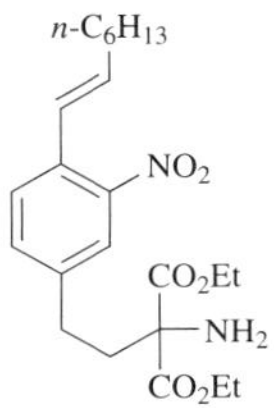 n-C_6H_{13} / NO_2 / CO_2Et / NH_2 / CO_2Et (31) 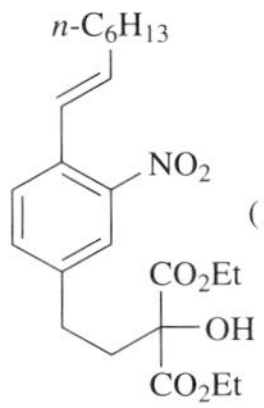 + n-C_6H_{13} / NO_2 / CO_2Et / OH / CO_2Et (15)	145

[a] The product was isolated as the hydrochloride.

[b] The base used was LDA.

[c] The aminated product is formed as an intermediate, which then reacts with another molecule of the starting diketone to yield the pyrrole derivative.

[d] Lithium hydroxide (6 eq) was also added.

[e] The number is the yield of crude product.

[f] With catalyst recovered from the BrC_8F_{17} phase, the yield was 92%.

[g] The absolute configuration was not determined but it was opposite to that obtained with the other catalyst.

[h] Quinine and quinidine gave the opposite enantiomers as the major products.

[i] In each example, cinchonine and cinchonidine gave the opposite enantiomers as the major products.

[j] The number is the yield of unpurified product.

[k] With the corresponding 5-, 6-, and 7-membered keto esters, only the ring-opened diazo esters were obtained.

[l] The isomeric 1-carbethoxy-2-keto analogs (n = 1, 2) only gave the products of diazo transfer and/or ring contraction.

TABLE 21. INTRAMOLECULAR AMINATIONS

	Substrate	Conditions	Product(s) and Yield(s) (%)	Refs.
C_3	Ph–N(OCOBu-*t*)–$CH_2CH_2CO_2Me$	NaH, KH, or LDA, THF, –80° to 15°	1-Ph-2-CO_2Me-aziridine (0)	490
	BnONH–CH_2CH_2–CO–(3-methyl-4-methyl-5-phenyl-imidazolidin-2-one, N-1)	1. $TiCl_4$, CH_2Cl_2, rt, 15 min 2. Add to Et_3N (2 eq), CH_2Cl_2, rt, 30 min	aziridin-2-yl–CO–N(imidazolidinone) (81) + epimeric aziridin-2-yl–CO–N(imidazolidinone) (14)	488
$C_{4\text{-}12}$	$R^1CH=C(NO_2)R^2$	4-$O_2NC_6H_4SO_2ONHCO_2Et$, CaO, neat, rt, 20 min	N-CO_2Et aziridines **I** + **II** (see below)	959

R^1	R^2	I	II
Me	Me	(38)	(5)
Et	Me	(58)	(6)
i-Pr	Me	(63)	(12)
—$(CH_2)_4$—		(89)	(0)
c-C_6H_{11}	Me	(70)	(8)
n-C_6H_{13}	$(CH_2)_2CH{=}CH_2$	(71)	(4)
Z-$EtCH{=}CH(CH_2)_2$	Me	(70)	(7)
$Ph(CH_2)_2$	Me	(70)	(16)
$Ph(CH_2)_2$	Et	(72)	(13)

	Substrate	Conditions	Product(s) and Yield(s) (%)	Refs.
$C_{4\text{-}6}$	R–CH(NHOBn)–CH_2–CO–N(imidazolidinone)	1. MX, CH_2Cl_2 2. Add to Et_3N (2 eq), CH_2Cl_2, rt, 30 min	3-R-aziridin-2-yl–CO–N(imidazolidinone) (see below)	487, 488

R	MX	
Me	$TiCl_4$	(97)
Me	$AlMe_2Cl$	(80)
Et	$AlMe_2Cl$	(67)
n-Pr	$AlMe_2Cl$	(70)

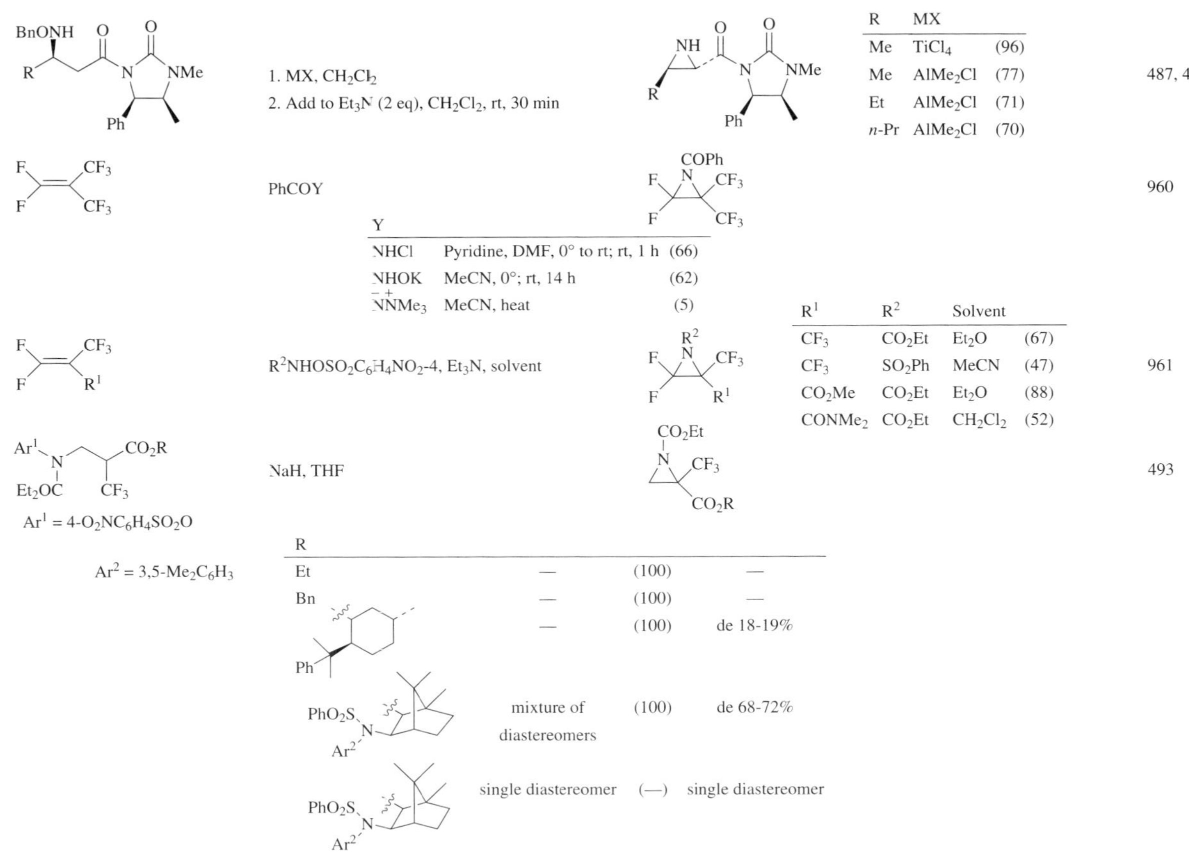

Substrate	Conditions	Product(s) and Yield(s) (%)	Refs.
C_{4-6} BnONH, R, imidazolidinone (O, NMe, Ph)	1. MX, CH_2Cl_2 2. Add to Et_3N (2 eq), CH_2Cl_2, rt, 30 min	NH aziridine, R, imidazolidinone (O, NMe, Ph) R / MX: Me / $TiCl_4$ (96); Me / $AlMe_2Cl$ (77); Et / $AlMe_2Cl$ (71); *n*-Pr / $AlMe_2Cl$ (70)	487, 488
C_4 $F_2C{=}C(CF_3)_2$	PhCOY	COPh aziridine (F, F, CF_3, CF_3) Y: $\bar{N}HCl$ Pyridine, DMF, 0° to rt; rt, 1 h (66); $\bar{N}HOK$ MeCN, 0°; rt, 14 h (62); $\bar{N}\overset{+}{N}Me_3$ MeCN, heat (5)	960
$F_2C{=}C(CF_3)R^1$	$R^2NHOSO_2C_6H_4NO_2$-4, Et_3N, solvent	R^2 aziridine (F, F, CF_3, R^1) R^1 / R^2 / Solvent: CF_3 / CO_2Et / Et_2O (67); CF_3 / SO_2Ph / MeCN (47); CO_2Me / CO_2Et / Et_2O (88); $CONMe_2$ / CO_2Et / CH_2Cl_2 (52)	961
$Ar^1N(COEt)CH_2CH(CF_3)CO_2R$ (Et_2OC) Ar^1 = 4-$O_2NC_6H_4SO_2O$ Ar^2 = 3,5-$Me_2C_6H_3$	NaH, THF	CO_2Et aziridine (CF_3, CO_2R) R: Et — (100) —; Bn — (100) —; 8-phenylmenthyl — (100) de 18-19%; PhO_2S-N(Ar^2)-bornyl, mixture of diastereomers (100) de 68-72%; PhO_2S-N(Ar^2)-bornyl, single diastereomer (—) single diastereomer	493

TABLE 21. INTRAMOLECULAR AMINATIONS (*Continued*)

Substrate	Conditions	Product(s) and Yield(s) (%)	Refs.
$C_{4\text{-}5}$			
(butenolide, R^1, R^2)	4-$O_2NC_6H_4SO_2ONHCO_2Et$ (5-6 eq), CaO, CH_2Cl_2	(N-CO_2Et aziridine-fused lactone, R^1, R^2)	495
$C_{5\text{-}10}$			
Ar−NH, R^1, R^2, R^3, R^4; Ar = 2,4,6-$Me_3C_6H_2SO_2O$	Et_3N, Et_2O, rt	(NH aziridine, R^1–R^4)	492
$C_{5\text{-}11}$			
(α-alkylidene γ-butyrolactone, R^1, R^2)	4-$O_2NC_6H_4SO_2ONHCO_2Et$ (x eq), CaO, CH_2Cl_2, water bath	I + II (spiro N-CO_2Et aziridines)	495

R^1	R^2	
H	H	(trace)
H	Me	(30)
MeO	H	(trace)

R^1	R^2	R^3	R^4	
H	CO_2Me	CO_2Me	CO_2Me	(100)[a]
H	CO_2Me	CN	CO_2Me	(80)[b]
H	CN	CO_2Et	CO_2Et	(90)[b]
Me	CO_2Me	CN	CO_2Me	(76)[b]
H	4-$O_2NC_6H_4$	CO_2Me	CN	(32)[b]

R^1	R^2	x	**I**	**II**
H	H	2	(39)	(0)
Me	H	3	(45)	(0)
H	Me	3	(42)	(39)
Me	Me	3	(52)	(0)
n-C_5H_{11}	H	5	(47)	(0)
H	n-C_5H_{11}	5	(28)	(24)
Ph	H	5	(60)	(0)

C_{6-13}

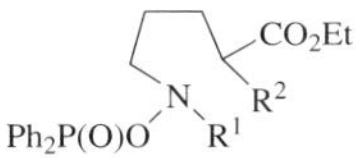

Ph$_2$P(O)O–N(R^1) pyrrolidine, CO$_2$Et, R^2 | Base | I (2-CO$_2$Et-2-R^2-pyrrolidine, NH) + II (Y=N–(CH$_2$)$_4$–CO$_2$Et) | 496

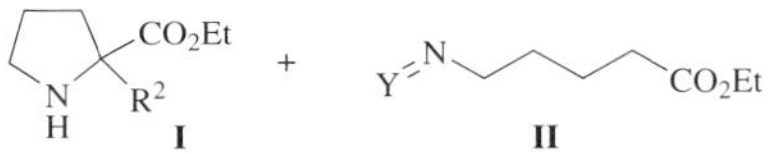

R^1	R^2	Base	Y	**I**	**II**
Me	H	LDA	—	(75)	(0)
Me	CO_2Et	*t*-BuOK	—	(95)	(0)
$CH_2CH=CH_2$	H	LDA	—	(20)	(0)
Bn	H	LDA	PhCH	(47)	(40)
Bn	CO_2Et	*t*-BuOK	—	(81)	(0)
CHMePh	H	LDA	PhMeC	(10)	("almost exclusively")
CMe_2Ph	H	LDA	—	(0)	(0)

C_7

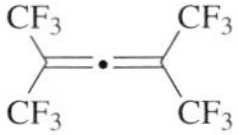

$(CF_3)_2C=C=C(CF_3)_2$ | $EtO_2CNHOSO_2C_6H_4NO_2$-4, Et_3N, CH_2Cl_2, Freon® 113, rt, 3 h | N-CO$_2$Et aziridine with CF$_3$, CF$_3$ and =C(CF$_3$)$_2$ (70) | 961

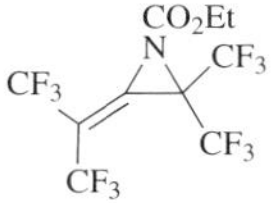

C_{7-10}

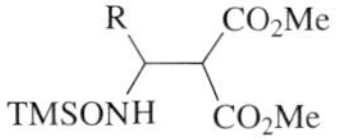

TMSONH–CH(R)–CH(CO$_2$Me)$_2$ | *t*-BuOK (0.25 eq), CH_2Cl_2, THF, rt, 4 h | 2,2-bis(CO$_2$Me)-3-R-aziridine (NH) | 491

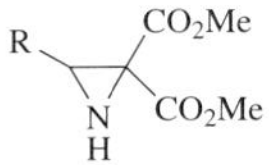

R	
n-Pr	(57)
i-Pr	(78)
i-Bu	(65)
Ph	(0)
n-C_7H_{15}	(54)

C_{7-11}

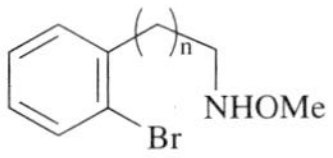

2-Br-C$_6$H$_4$–(CH$_2$)$_n$–CH$_2$NHOMe | 1. MeLi, addend, Et_2O, –78°, 30 min; to –15°, 3 h 2. AcCl, pyridine | benzo-fused N-COMe heterocycle

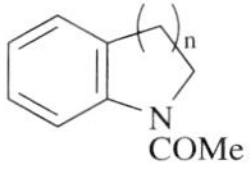

n	Addend		Refs.
0	*n*-BuLi	(21)	83
1	*n*-BuLi	(43)	82, 97
2	*t*-BuLi	(64)	83
3	*t*-BuLi	(24)	83
4	—	(0)	83

TABLE 21. INTRAMOLECULAR AMINATIONS (*Continued*)

	Substrate	Conditions	Product(s) and Yield(s) (%)	Refs.
$C_{7\text{-}11}$ 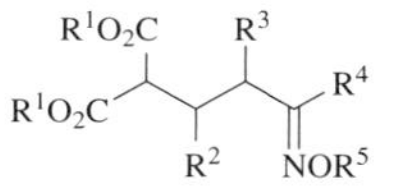	R^1O_2C, R^1O_2C, R^2, R^3, R^4, NOR^5	DBU, CH_2Cl_2, 0°, 30 min	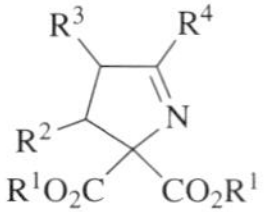 R^2, R^3, R^4, N, R^1O_2C, CO_2R^1	497
C_8	BnO_2C, CO_2Bn, NOMs	DBU, DMF, rt, 5 d	N, BnO_2C, CO_2Bn (83)	497
C_9	O, MeN, N, O, NHOBn, Ph, Ph	$AlMe_2Cl$, Et_3N, CH_2Cl_2, 0°	O, MeN, N, O, H N, Ph, Ph (75), 100% de	489
	O, MeN, N, O, NHOBn, Ph, Ph	$AlMe_2Cl$, Et_3N, CH_2Cl_2, 0°	O, MeN, N, O, H N, Ph, Ph (75), 100% de	489

R^1	R^2	R^3	R^4	R^5	
Bn	H	H	Me	Ms	(87) (96)[c]
Bn	H	Me	Me	Ms	(87) (79)[c]
Bn	*i*-Pr	H	Me	Ms	(96) (79)[c]
Bn	*n*-C_5H_{11}	H	Me	Ms	(83)
Me	Ph	H	H	Ms	(0)[d]
Me	Ph	H	Me	Ac	(0)
Me	Ph	H	Me	$COCF_3$	(57)
Me	Ph	H	Me	Ms	(99)
Me	Ph	H	CO_2Me	Ms	(83) (80)[c]

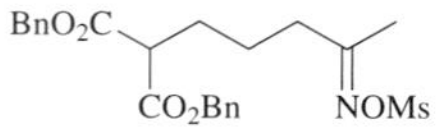

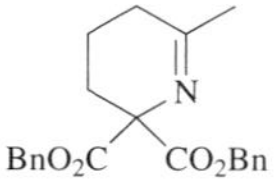

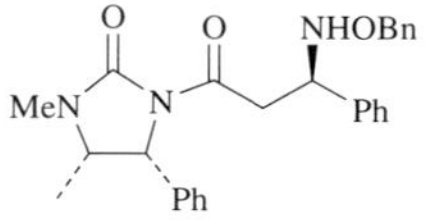

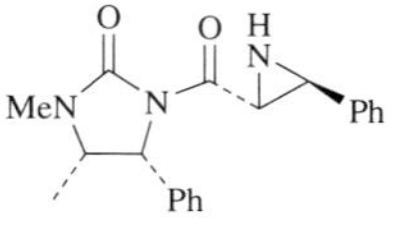

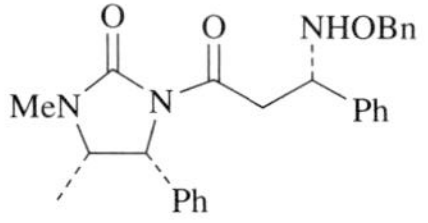

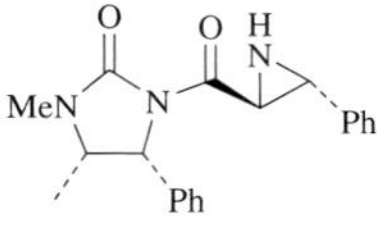

$C_{9\text{-}10}$ (Ph–C(=$\overset{+}{N}NMe_3$ I^-)–CH(R^1)–CH₃ structure) — NaOR[2], R[2]OH → 2-phenyl-2H-azirine product

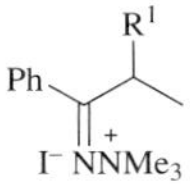

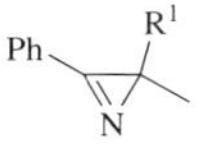

R^1	R^2	Temp	Time	Yield	Refs.
H	t-Bu	50°	1 h	(29%)	494
Me	i-Pr	—	—	(85%)	962

$C_{9\text{-}17}$ (oxime ether, $OC_6H_3(NO_2)_2$-2,4) — NaH + Na(CN)BH$_3$ (ia), dioxane, rt; 50°, 10 h → 8-hydroxy-1,2,3,4-tetrahydroquinoline

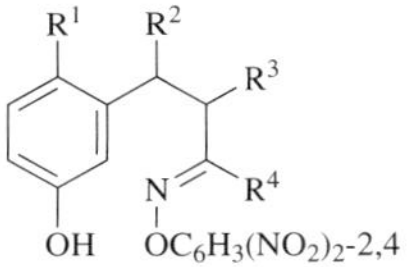

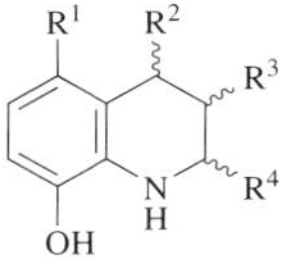

R^1	R^2	R^3	R^4	Yield	cis:trans	Refs.
H	H	H	H	(0)	—	498
Br	H	H	Me	(92)	—	499
H	H	H	Me	(78)	—	499
H	Me	H	Me	(83)	5:1	499
H	H	H	i-Pr	(83)	—	499
H	H	Me	Et	(70)	5:1	499
H	H	H	CH=CHPh	(64)	—	499

$C_{10\text{-}17}$ (oxime ether, $OC_6H_3(NO_2)_2$-2,4) — 1. NaH, dioxane, 50°, 20 h; 2. DDQ, HOAc, reflux, 2 h → 8-hydroxyquinoline (Ref. 499)

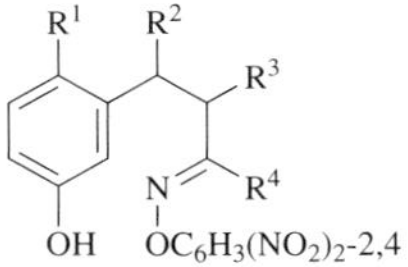

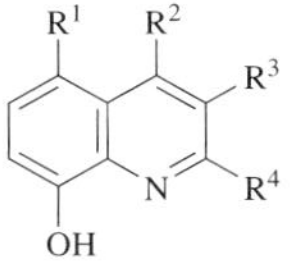

R^1	R^2	R^3	R^4	Yield
Br	H	H	Me	(90)
H	H	H	Me	(80)
H	H	H	Et	(75)
H	Me	H	Me	(74)
H	H	H	i-Pr	(84)
H	H	Me	Et	(72)
H	H	—OCMe$_2$O—		(62)
H	H	H	CH=CHPh	(60)

Substrate	Conditions	Product(s) and Yield(s) (%)	Refs.
C_{10}			
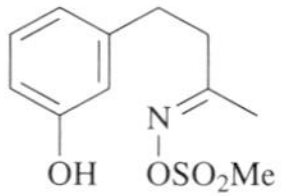 OH, N–OSO_2Me	NaH, dioxane, reflux	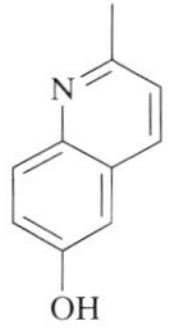 OH, N (20) + OH, N (6)	498
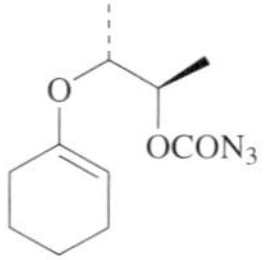 O, $OCON_3$	1. *hν*, CH_2Cl_2, rt, 4 h 2. BF_3•Et_2O, MeOH, reflux, 2 h	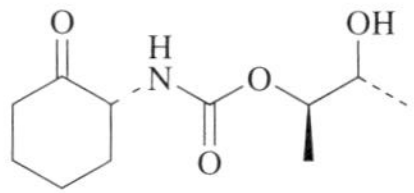 O, H, N, O, O, OH (20), 95% de	963
C_{12}			
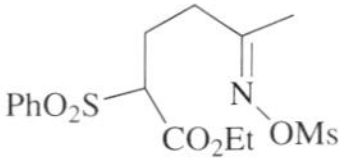 PhO_2S, CO_2Et, N, OMs	DBU, CH_2Cl_2, 0°, 30 min	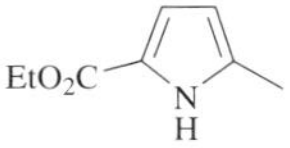 EtO_2C, N, H (48)	497
C_{12-14}			
R^2, R^1, R^3, N, OH, N, H	1. MsCl, Et_3N, CH_2Cl_2, 0°, 1 h 2. C_6F_5COCl, rt, 4 h 3. $NaHCO_3$, H_2O, Me_2CO, rt, 20 h	R^1, R^2, R^3, N, OH, N, COC_6F_5	964
		R^1 R^2 R^3	
		H H H (63) single diastereomer	
		Me H H (72) mixture of two diastereomers	
		H Me Me (74) single diastereomer	
C_{13}			
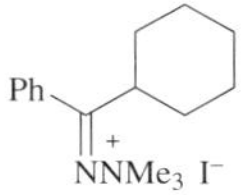 Ph, $\overset{+}{N}NMe_3$ I^-	NaOPr-*i*, *i*-PrOH, 40°, 1 h	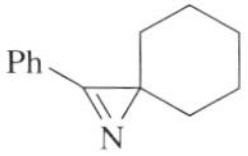 Ph, N (80)	494

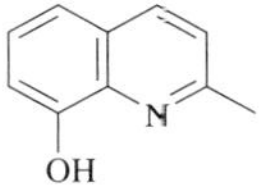

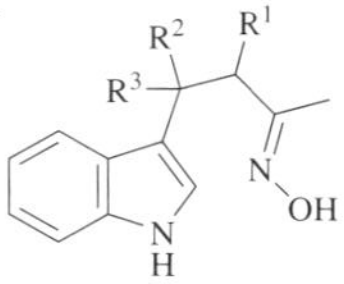

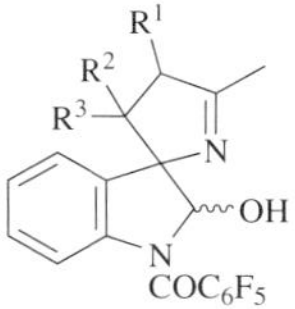

C$_{13}$

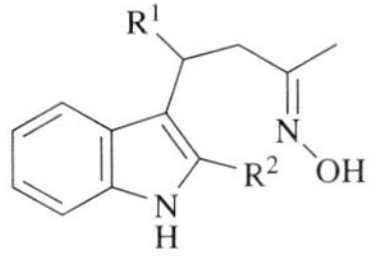

MsCl, Et_3N, CH_2Cl_2, 0°

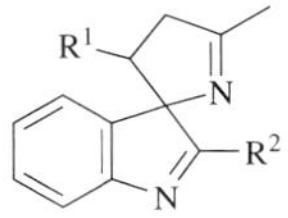

R^1	R^2	Time	
H	Me	1 h	(95)
CO_2Bu-*t*	H	6 h	(68) mixture of two isomers

964

C$_{13}$

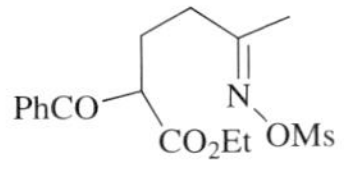

DBU, CH_2Cl_2, 0°, 30 min

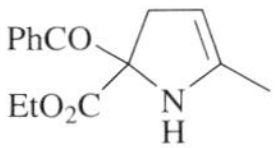 (80)

497

C$_{13\text{-}16}$

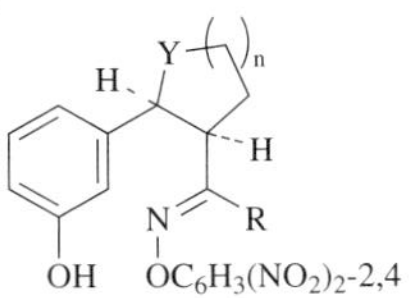

NaH, Na(CN)BH_3, dioxane, rt, 5 h

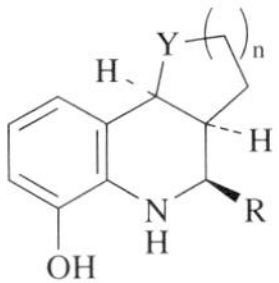

I

+

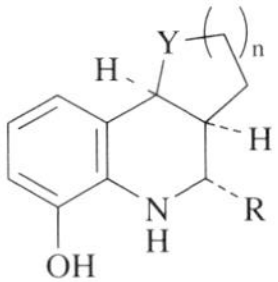

II

Y	R	n	**I**	**II**
CH_2	Me	1	(65)	(28)
CH_2	Me	2	(80)	(0)
NBoc	*n*-Bu	1	(93)	(0)

499

C$_{14}$

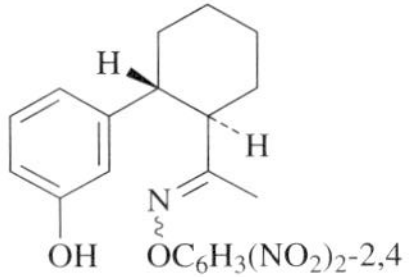

NaH, Na(CN)BH_3, dioxane, rt, 17 h

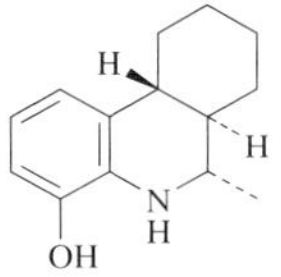 (52) + 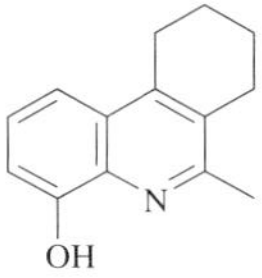 (12)

499

TABLE 21. INTRAMOLECULAR AMINATIONS (*Continued*)

Substrate	Conditions	Product(s) and Yield(s) (%)	Refs.
C_{14}			
OH, N, $OC_6H_3(NO_2)_2$-2,4	NaH, dioxane, rt, 4 h	OH, N (98)	499, 498
C_{15-16}			
O, NHOMe, 4-$R^1C_6H_4$, $C_6H_4R^2$-4	NaOMe, MeOH, 60°, 10 min; rt, overnight	O, H, N, 4-$R^1C_6H_4$, $C_6H_4R^2$-4	
		R^1 R^2	
		H H (94)	485[e]
		H Cl (80)	485[e]
		Cl H (66)	485[e]
		Br H (83)	485[e]
		H Br (90)	485[e]
		Me H (64)	485[e], 486
C_{16}			
O, NHOMe, 4-$MeOC_6H_4$, Ph	MeOH, reflux, 15 min	O, H, N, 4-$MeOC_6H_4$, Ph (65)	965
C_{20}			
Ph, O, P, O, NMe_2, CN, O, O, $()_3$	1. LDA, THF, –10°, 1 h; rt, 36 h 2. CH_2N_2	Ph, O, P, OMe, Me_2N, CN, O, O, $()_3$ (7)[f]	85

[a] The reaction was carried out in CH_2Cl_2.

[b] The yield includes that of the preparation of the substrate.

[c] The yield is that of the one-pot reaction of the oxime with MsCl and Et_3N followed by addition of DBU.

[d] Beckmann fragmentation occurred exclusively.

[e] The products were initially considered to be the primary enamines. The correct assignment was made in a later publication.[486]

[f] The yield was 10% with the substrate labeled with ^{13}C at the cyano carbon and one deuterium in one of the methyl groups. The isotopic labeling confirmed the intramolecularity of the reaction.

REFERENCES

[1] Wöhler, F. *Ann. Physik* **1828**, *12*, 253.

[2] Erdik, E.; Ay, M. *Chem. Rev.* **1989**, *89*, 1947.

[3] Schmitz, E. *Uspekhi Khim.* **1976**, *45*, 54; *Russ. Chem. Rev.* **1976**, *45*, 16.

[4] Kron, K. *Nachr. Chem., Tech. Lab.* **1987**, *35*, 1047.

[5] Mulzer, J.; Altenbach, H. J.; Brown, M.; Krohn, K.; Reissig, H.-U. In *Organic Synthesis Highlights*; VCH: Weinheim, Germany, 1991; p 43.

[6] Boche, G. In *Methoden der organischen Chemie (Houben-Weyl)*; Georg Thieme Verlag: Stuttgart, 1995; Vol. E21e, p 5133.

[7] Greck, G.; Genêt, J.-P. *Synlett* **1997**, 741.

[8] *Modern Amination Methods*; Ricci, A., Ed.: Wiley-VCH: Weinheim 2000.

[9] Dembech, P.; Seconi, C.; Ricci, A. *Chem. Eur. J.* **2000**, *6*, 1281.

[10] Kovacic, P.; Lowery, M. K.; Field, K. W. *Chem. Rev.* **1970**, *70*, 639.

[11] Tamura, Y.; Minamikawa, J.; Ikeda, M. *Synthesis* **1977**, 1.

[12] Andreae, S.; Schmitz, E. *Synthesis* **1991**, 327.

[13] Narasaka, K.; Kitamura, M. *Eur. J. Org. Chem.* **2005**, 4505.

[14] Parmerter, S. M. *Org. React.* **1959**, *10*, 1.

[15] Lang-Fugmann, S.; Lang-Fugmann, S. In *Methoden der organischen Chemie (Houben-Weyl)*; Georg Thieme Verlag: Stuttgart, New York 1992; Vol. E16d/1 p 99.

[16] Enders, E. In *Methoden der organischen Chemie (Houben-Weyl)*; Georg Thieme Verlag: Stuttgart, 1967; Vol. 10/2, p 456.

[17] Fahr, E.; Lind, H. *Angew. Chem., Int. Ed. Engl.* **1966**, *5*, 372.

[18] Süling, C. In *Methoden der organischen Chemie (Houben-Weyl)*; Georg Thieme Verlag: Stuttgart, 1965; Vol. 10/3, p 721.

[19] L'Abbé, G. *Ind. Chim. Belge* **1989**, *34*, 519.

[20] Scriven, E. F. V.; Turnbull, K. *Chem. Rev.* **1988**, *88*, 298.

[21] Engel, A. In *Methoden der organischen Chemie (Houben-Weyl)*; Georg Thieme Verlag: Stuttgart, 1990; Vol. 16a/2, p 1182.

[22] Hassner, A. In *Methoden der organischen Chemie (Houben-Weyl)*; Georg Thieme Verlag: Stuttgart, New York, 1992; Vol. E16d/2, p 1283.

[23] Bräse, S.; Gil, C.; Knepper, K.; Zimmermann, V. *Angew. Chem., Int. Ed.* **2005**, *44*, 5188.

[24] Du Bois, J.; Tomooka, C. S.; Hong, J.; Carreira, E. M. *Acc. Chem. Res.* **1997**, *30*, 364.

[25] Tomooka, C. S.; Iikura, H.; Carreira, E. M. In *Modern Amination Methods*; Ricci, A., Ed.: Wiley-VCH: Weinheim 2000, chapter 5.

[26] List, B. *Tetrahedron* **2002**, *58*, 5573.

[27] Duthaler, R. O. *Angew. Chem., Int. Ed.* **2003**, *42*, 975.

[28] Erdik, E. *Tetrahedron* **2004**, *60*, 8747.

[29] Greck, C.; Drouillat, B.; Thomassigny, C. *Eur. J. Org. Chem.* **2004**, 1377.

[30] Janey, J. M. *Angew. Chem., Int. Ed.* **2005**, *44*, 4292.

[31] Duthaler, R. O. *Tetrahedron* **1994**, *50*, 1539.

[32] Calmes, M.; Daunis, J. *Amino Acids*, **1999**, *16*, 215.

[33] Vogt, H.; Bräse, S. *Org. Biomol. Chem.* **2007**, 406.

[34] Krüger, G. In *Methoden der organischen Chemie (Houben-Weyl)*; Georg Thieme Verlag: Stuttgart, 1992; Vol. 16d/1, p 618.

[35] Backes, J.; Braun, M.; Maercker, A.; von Rague-Schleyer, P.; Bransdma, L.; Lambert, C.; Saalfrank, R. W.; Subramanian, L. R. In *Methoden der organischen Chemie (Houben-Weyl)*; Georg Thieme Verlag: Stuttgart, New York, 1992; Vol. E19d, p 1.

[36] Leroux, F.; Schlosser, M.; Zohar, E.; Marek, I. In *The Chemistry of Organolithium Compounds*; Patai, Z.; Marek, I., Eds.; John Wiley & Sons: Chichester, 2004, p 435.

[37] Gschwend, H. W.; Rodriguez, H. R. *Org. React.* **1979**, *26*, 1.

[38] Nützel, K. In *Methoden der organischen Chemie (Houben-Weyl)*; Georg Thieme Verlag: Stuttgart, 1973; Vol. 13/2a, p 47.

[39] Wakefield, B. J. *Organomagnesium Methods in Organic Synthesis*; Academic Press: New York NY, 1995.

[40] Nützel, K. In *Methoden der organischen Chemie (Houben-Weyl)*; Georg Thieme Verlag: Stuttgart, 1973; Vol. 13/2a, p 553.
[41] Knochel, P.; Millot, N.; Rodriguez, A. L. *Org. React.* **2001**, *58*, 417.
[42] Arya, P.; Qin, H. *Tetrahedron* **2000**, *56*, 917.
[43] Klar, G.; Kramolowski, R. In *Methoden der organischen Chemie (Houben-Weyl)*; Georg Thieme Verlag: Stuttgart, 1993; Vol. E15a, p 463.
[44] Heathcock, C. H. *Modern Enolate Chemistry*; VCH: Weinheim, Germany, 1992.
[45] Evans, D. A. In *Asymmetric Synthesis*; Morrison, J. D., Ed.; Academic Press: New York, 1984; Vol. 3, p 1.
[46] Rasmussen, J. K. *Synthesis* **1977**, 91.
[47] Pawlenko, S. In *Methoden der organischen Chemie (Houben-Weyl)*; Georg Thieme Verlag: Stuttgart, 1980; Vol. E13/5, p 1.
[48] Pawlenko, S. In *Methoden der organischen Chemie (Houben-Weyl)*; Georg Thieme Verlag: Stuttgart, 1980; Vol. E15/1, p 404.
[49] Brownbridge, P. *Synthesis* **1983**, 1.
[50] Pawlenko, S. In *Methoden der organischen Chemie (Houben-Weyl)*; Georg Thieme Verlag: Stuttgart, 1993; Vol. E15/2, p 1742.
[51] Amado-Bedolla, C.; Salmón-Ferrer, R.; Lester, W. A., Jr.; Vázquez-Martínez, J. A.; Aspuru-Guzik, A. *J. Chem. Phys.* **2007**, *126*, 204308.
[52] Yamamoto, H.; Maruoka, K. *J. Org. Chem.* **1980**, *45*, 2739.
[53] Iwao, M.; Reed, J. N.; Snieckus, V. *J. Am. Chem. Soc.* **1982**, *104*, 5531.
[54] Alberti, A.; Canè, F.; Dembech, P.; Lazzari. D.; Ricci, A.; Seconi, G. *J. Org. Chem.* **1996**, *61*, 1677.
[55] Cané, F.; Brancaleoni, D.; Dembech, P.; Ricci, A.; Seconi, G. *Synthesis* **1997**, 545.
[56] Coleman, G. H.; Hauser, C. R. *J. Am. Chem. Soc.* **1928**, *50*, 1193.
[57] Coleman, G. H.; Yager, C. B. *J. Am. Chem. Soc.* **1929**, *51*, 567.
[58] Coleman, G. H.; Hermanson, J. L.; Johnson, H. L. *J. Am. Chem. Soc.* **1937**, *59*, 1896.
[59] Coleman, G. H.; Blomquist, R. F. *J. Am. Chem. Soc.* **1941**, *63*, 1692.
[60] Boche, G.; Lohrenz, J. C. W. *Chem. Rev.* **2001**, *101*, 697.
[61] Coleman, G. H.; Soroos, H.; Yager, C. B. *J. Am. Chem. Soc.* **1933**, *55*, 2075.
[62] Horiike, M.; Oda, J.; Inouye, Y.; Ohno, M. *Agr. Biol. Chem.* **1969**, *33*, 292.
[63] Dowd, P.; Kaufman, C. *J. Org. Chem.* **1979**, *44*, 3956.
[64] Hand, E. S.; Baker, D. C. *Int. J. Pept. Protein Res.* **1984**, *23*, 420.
[65] Paquette, L. A. *J. Am. Chem. Soc.* **1963**, *85*, 3288.
[66] Tortov, G. A.; Burykin, V. V.; Koblik, A. V. *Khim. Geterotsikl. Soedin.* **1972**, *11*, 1552; *Engl. Transl.* p 1403.
[67] Klages, F.; Nober, G.; Kircher, F.; Bock, M. *Liebigs Ann. Chem.* **1941**, *547*, 1.
[68] Coleman, G. H. *J. Am. Chem. Soc.* **1933**, *55*, 3001.
[69] Le Fèvre, R. J. W. *J. Chem. Soc.* **1932**, 1745.
[70] Kuffner, F.; Seifried, W. *Monatsh. Chem.* **1952**, *83*, 748.
[71] Wolf, V.; Kowitz, F. *Liebigs Ann. Chem.* **1960**, *638*, 33.
[72] Coleman, G. H.; Andersen, H. P.; Hermanson, J. L. *J. Am. Chem. Soc.* **1934**, *56*, 1381.
[73] Sinha, P.; Knochel, P. *Synlett* **2006**, 3304.
[74] Wada, T.; Oda, J.-i.; Inouye, Y. *Agr. Biol. Chem.* **1972**, *36*, 799.
[75] Coleman, G. H.; Yager, C. B.; Soroos, H. *J. Am. Chem. Soc.* **1934**, *56*, 965.
[76] Tcherniak, M. J. *Bull. Soc. Chim. Fr.* **1876**, *25*, 160.
[77] Coleman, G. H.; Buchanan, M. A.; Paxson, W. L. *J. Am. Chem. Soc.* **1933**, *55*, 3669.
[78] Baumann, T.; Bächle, M.; Bräse, S. *Org. Lett.* **2006**, *8*, 3797.
[79] Yamada, S.-i.; Oguri, T.; Shioiri, T. *J. Chem. Soc., Chem. Commun.* **1972**, 623.
[80] Shverdina, N. I.; Kocheshkov, K. A. *Izv. Akad. Nauk SSSR, Ser. Khim.* **1941**, 75; *Chem. Abstr.* **1943**, *37*, 3006.
[81] Beak, P,; Basha, A.; Kokko, B. *J. Am. Chem. Soc.* **1984**, *106*, 1511.
[82] Beak, P.; Basha, A.; Kokko, B.: Loo, D. K. *J. Am. Chem. Soc.* **1986**, *108*, 6016.
[83] Beak, P.; Selling, G. W. *J. Org. Chem.* **1989**, *54*, 5574.

[84] Beak, P.; Li, J. *J. Am. Chem. Soc.* **1991**, *113*, 2796.
[85] Beak, P.; Conser Basu, K.; Li, J. J. *J. Org. Chem.* **1999**, *64*, 5218.
[86] Boche, G.; Wagner, H.-U. *J. Chem. Soc., Chem. Commun.* **1984**, 1591.
[87] Armstrong, D. R.; Snaith, R.; Walker, G. T. *J. Chem. Soc., Chem. Commun.* **1985**, 789.
[88] McKee, M. L. *J. Am. Chem. Soc.* **1985**, *107*, 859.
[89] Bühl, M.; Schaefer, H. F., III *J. Am. Chem. Soc.* **1993**, *115*, 364.
[90] Glukhovtsev, M. N.; Pross, A.; Radom, L. *J. Am. Chem. Soc.* **1995**, *117*, 9012.
[91] Erdik, E.; Eroglu, F.; Kâhya, D. *J. Phys. Org. Chem.* **2005**, *18*, 950.
[92] Kokko, B. PhD. Dissertation, University of Illinois at Champaign-Urbana, 1983.
[93] Radhakrishna, A. S.; Loudon, G. M.; Miller, M. J. *J. Org. Chem.* **1979**, *44*, 4836.
[94] Sheradsky, T.; Salemnick, G.; Nir, Z. *Tetrahedron* **1972**, *28*, 3833.
[95] Sturtz, G.; Couthon, H. *C. R. Hebd. Scéances Acad. Sci.* **1993**, *316/II*, 181.
[96] Boyles, D. C.; Curran, T. T.; Partlett, R. V., IV *Org. Proc. Res. Dev.* **2002**, *6*, 230.
[97] Kokko, B. J.; Beak, P. *Tetrahedron Lett.* **1983**, *24*, 561.
[98] Basha, A.; Brooks, D. W. *J. Chem. Soc., Chem. Commun.* **1987**, 305.
[99] Buck, P.; Köbrich, G. *Tetrahedron Lett.* **1967**, 1563.
[100] Casarini, A.; Dembech, P.; Lazzari, D.; Marini, E.; Reginato, G.; Ricci, A.; Seconi, G. *J. Org. Chem.* **1993**, *58*, 5620.
[101] Bernardi, P.; Dembech, P.; Fabbri, G.; Ricci, A.; Seconi, G. *J. Org. Chem.* **1999**, *64*, 641.
[102] Riant, O.; Samuel, O.; Flessner, T.; Taudien, S.; Kagan, H. B. *J. Org. Chem.* **1997**, *62*, 6733.
[103] West, R.; Boudjouk, P. *J. Am. Chem. Soc.* **1973**, *95*, 3987.
[104] Lin, C.-C.; Wang, Y.-C.; Hsu, J.-L.; Chiang, C.-C.; Su, D.-W.; Yan, T.-H. *J. Org. Chem.* **1997**, *62*, 3806.
[105] Loreto, M. A.; Pellacani, L.; Tardella, P. A. *J. Chem. Res. (S)* **1988**, 304.
[106] Smulik, J. A.; Vedejs, E. *Org. Lett.* **2003**, *5*, 4187.
[107] Shen, Y.; Friestad, G. K. *J. Org. Chem.* **2002**, *67*, 6236.
[108] Berman, A. M.; Johnson, J. S. *Synlett* **2005**, 1799.
[109] Berman, A. M.; Johnson, J. S. *J. Am. Chem. Soc.* **2004**, *126*, 5680.
[110] Berman, A. M.; Johnson, J. S. *J. Org. Chem.* **2005**, *70*, 364.
[111] Berman, A. M.; Johnson, J. S. *Org. Synth.* **2006**, *83*, 31.
[112] Berman, A. M.; Johnson, J. S. *J. Org. Chem.* **2006**, *71*, 219.
[113] Campbell, M. J.; Johnson, J. S. *Org. Lett.* **2007**, *9*, 1521.
[114] Ning, R. Y. *Chem. Eng. News* **1973**, *51*, 36.
[115] Hansen, J. J.; Krogsgaard-Larsen, P. *J. Chem. Soc., Perkin Trans. 1* **1980**, 1826.
[116] Zheng, B.; Srebnik, M. *J. Org. Chem.* **1995**, *60*, 1912.
[117] Glass, R. S.; Hojjatie, M.; Sabahi, M.; Steffen, L. K.; Wilson, G. S. *J. Org. Chem.* **1990**, *55*, 3797.
[118] Fioravanti, S.; Pellacani, L.; Tardella, P. A. *Gazz. Chim. Ital.* **1997**, *127*, 41.
[119] Barani, M,; Fioravanti, S.; Peliciani, L.; Tardella, P. A. *Tetrahedron* **1994**, *50*, 11235.
[120] Fioravanti, S.; Loreto, M. A.; Pellacani, L.; Tardella, P. A. *J. Org. Chem.* **1985**, *50*, 5365.
[121] Fioravanti, S.; Loreto, M. A.; Pellacani, L.; Tardella, P. A. *J. Chem. Res. (S)* **1987**, 310.
[122] Barani, M.; Fioravanti, S.; Loreto, A. M.; Pellacani, L.; Tardella, P. A. *Tetrahedron* **1994**, *50*, 3829.
[123] Fioravanti, S.; Olivieri, L.; Pellacani, L.; Tardella, P. A. *J. Chem. Res. (S)* **1998**, 338.
[124] Fioravanti, S.; Morreale, A.; Pellacani, L.; Tardella, P. A. *Tetrahedron Lett.* **2001**, *42*, 1171.
[125] Felice, E.; Fioravanti, S.; Pellacani, L.; Tardella, P. A. *Tetrahedron Lett.* **1999**, *40*, 4413.
[126] Genêt, J.-P.; Mallart, S.; Greck, C.; Piveteau, E. *Tetrahedron Lett.* **1991**, *32*, 2359.
[127] Greck, C.; Bischoff, L.; Girard, A.; Hajicek, J.; Genêt, J.-P. *Bull. Soc. Chim. Fr.* **1994**, *131*, 429.
[128] Shopova, M.; Vassileva, E.; Fugier, C.; Henry-Basch, E. *Synth. Commun.* **1997**, *27*, 1661.
[129] Ferreira, F. Ph. D. Dissertation, Université P. & M. Curie, Paris, France, 1996; Lavergne, D., École Nationale Superieure de Chimie, Paris, France, unpublished work; both cited (references 11a and 11b on p 101) by Genêt, J. P.; Greck, C.; Lavergne, D, in *Modern Amination Methods*; Ricci, A., Ed.: Wiley-VCH: Weinheim 2000.
[130] Greck, C.; Bischoff, L.; Ferreira, F.; Genêt, J.-P. *J. Org. Chem.* **1995**, *60*, 7010.

[131] Boche, G.; Boie, C.; Bosold, F.; Harms, K.; Marsch, M. *Angew. Chem., Int. Ed. Engl.* **1994**, *33*, 115.

[132] Barton, D. H. R.; Bould, L.; Clive, D. L. J.; Magnus, P. D.; Hase, T. *J. Chem. Soc. C* **1971**, 2204.

[133] Boche, G.; Mayer, N.; Bernheim, M.; Wagner, K. *Angew. Chem., Int. Ed. Engl.* **1978**, *17*, 687.

[134] Bernheim, M. Ph. D. Dissertation, University of Munich, Germany, 1981; quoted in ref. 137, ref. 2b.

[135] Boche, G.; Bernheim, M.; Niessner, M. *Angew. Chem., Int. Ed. Engl.* **1983**, *22*, 53.

[136] Abraham, T.; Curran, T. *Tetrahedron* **1982**, *38*, 1019.

[137] Boche, G. In *Encyclopedia of Reagents for Organic Synthesis*; L. A. Paquette, Ed.; Wiley: New York, 1995; p 2100.

[138] Klötzer, W.; Stadlwieser, J.; Raneburger, J. *Org. Synth. Coll. Vol. 7* **1986**, 8.

[139] Boche, G.; Bernheim, M.; Schrott, W. *Tetrahedron Lett.* **1982**, *23*, 5399.

[140] Sosnovsky, G.; Purgstaller, K. *Z. Naturforsch.* **1989**, 44*b*, 582.

[141] Colvin, E. W.; Kirby, G. W.; Wilson, A. C. *Tetrahedron Lett.* **1982**, *23*, 3835.

[142] Kendrick, D. A.; Kolb, M. *J. Fluorine Chem.* **1989**, *45*, 265.

[143] Baldwin, J. E.; Adlington, R. M.; Jones. R. H.; Schofield, C. J.; Zarocostas, C.; Greengrass, C. W. *J. Chem. Soc., Chem. Commun.* **1985**, 194.

[144] Aurell, M. J.; Gil, S.; Martínez, P. V.; Parra, M.; Tortajada, A.; Mestres, R. *Synth. Commun.* **1991**, *21*, 1833.

[145] Sun, C.; Bittman, R. *J. Org. Chem.* **2006**, *71*, 2200.

[146] Boche, G.; Bernheim, M.; Lawaldt, D.; Ruisinger, B. *Tetrahedron Lett.* **1979**, 4285.

[147] Boche, G.; Schrott, W. *Tetrahedron Lett.* **1982**, *23*, 5403.

[148] Andreae, S.; Schmitz, E. In *Encyclopedia of Reagents for Organic Synthesis*; L. A. Paquette, Ed.; Wiley: New York, 1995; p 3810.

[149] Andreae, S.; Schmitz, E.; Wulf, J.-P.; Schulz, B. *Liebigs Ann. Chem.* **1992**, 239.

[150] Wulff, J.-P.; Andreae, S. unpublished results quoted in Andreae, S.; Schmitz, E. *Synthesis* **1991**, 327 (ref. 58).

[151] Page, P. C. B.; Limousin, C.; Murrell, V. L. *J. Org. Chem.* **2002**, *67*, 7787.

[152] Chen, B.-C.; Zhou, P.; Davis, F. A.; Ciganek, E. *Org. React.* **2003**, *62*, 1.

[153] Vidal, J.; Damestoy, S.; Guy, L.; Hannachi, J.-C.; Aubry, A.; Collet, A. *Chem. Eur. J.* **1997**, *3*, 1691.

[154] Armstrong, A.; Atkin, M. A.; Swallow, S. *Tetrahedron: Asymmetry* **2001**, *12*, 535.

[155] Armstrong, A.; Edmonds, I. D.; Swarbrick, M. E.; Treweeke, N. R. *Tetrahedron* **2005**, *61*, 8423.

[156] Enders, D.; Poiesz, C.; Joseph, R. *Tetrahedron: Asymmetry* **1998**, *9*, 3709.

[157] Vidal, J.; Guy, L.; Stérin, S.; Collet, A. *J. Org. Chem.* **1993**, *58*, 4791.

[158] Armstrong, A.; Atkin, M. A.; Swallow, S. *Tetrahedron Lett.* **2000**, *41*, 2247.

[159] Ghosh, T. N.; Guha, P. C. *J. Indian Inst. Sci.* **1933**, 16*A*, 103; *Chem. Abstr.* **1934**, *28*, 2691.

[160] Fiaud, J.-C.; Kagan, H. B. *Tetrahedron Lett.* **1970**, 1813.

[161] Fiaud, J.-C.; Kagan, H. B. *Tetrahedron Lett.* **1971**, 1019.

[162] van Vliet, M. R. P.; Jastrzebski, J. T. B. H.; van Koten, G.; Vrieze, K. *J. Organomet. Chem.* **1983**, *251*, C17.

[163] van Vliet, M. R. P.; Jastrzebski, J. T. B. H.; Klaver, W. J.; Goubitz, K.; van Koten, G. *Recl. Trav. Chim. Pays-Bas* **1987**, *106*, 132.

[164] Yamamoto, Y.; Ito, W. *Tetrahedron* **1988**, *44*, 5415.

[165] Uneyama, K.; Yan, F.; Hirama, S.; Katagiri, T. *Tetrahedron Lett.* **1996**, *37*, 2045.

[166] Niwa, Y.; Takayama, K.; Shimizu, M. *Tetrahedron Lett.* **2001**, *42*, 5473.

[167] Niwa, Y.; Takayama, K.; Shimizu, M. *Bull. Chem. Soc. Jpn.* **2002**, *75*, 1819.

[168] Dai, W.; Srinivasan, R.; Katzenellenbogen, J. A. *J. Org. Chem.* **1989**, *54*, 2204.

[169] Bracht, J.; Rieker, A. *Synthesis* **1977**, 708.

[170] Baum, J. S.; Condon, M. E.; Shook, D. A. *J. Org. Chem.* **1987**, *52*, 2983.

[171] Klerks, J. M.; Jastrzebski, J. T. B. H.; van Koten, G.; Vrieze, K. *J. Organomet. Chem.* **1982**, *224*, 107.

[172] Lim, B.-W.; Ahn, K.-H. *Synth. Commun.* **1996**, *26*, 3407.

[173] Evans, D. A.; Faul, M. M.; Bilodeau, M. T. *J. Am. Chem. Soc.* **1994**, *116*, 2742.

174 Adam, W.; Roschmann, K. J.; Saha-Möller, C. R. *Eur. J. Org. Chem.* **2000**, 557.
174 a. Alvernhe, G.; Laurent, A. *Tetrahedron Lett.* **1972**, 1007.
175 Busch, M.; Hobein, R. *Chem. Ber.* **1907**, *40*, 2096.
176 Erdik, E.; Daskapan, T. *Synth. Commun.* **1999**, *29*, 3989.
177 Erdik, E.; Daskapan, T. *J. Chem. Soc., Perkin Trans. 1* **1999**, 3139.
178 Würthwein, E.-U.; Weigmann, R. *Angew. Chem., Int. Ed. Engl.* **1987**, *26*, 923.
179 Tsutsui, H.; Ichikawa, T.; Narasaka, K. *Bull. Chem. Soc. Jpn.* **1999**, *72*, 1869.
180 Hagopian, R. A.; Therien, M. J.; Murdoch, J. R. *J. Am. Chem. Soc.* **1984**, *106*, 5753.
181 Kitamura, M.; Chiba, S.; Narasaka, K. *Bull. Chem. Soc. Jpn.* **2003**, *76*, 1063.
182 Kitamura, M.; Suga, T.; Chiba, S.; Narasaka, K. *Org. Lett.* **2004**, *6*, 4619.
183 Erdik, E.; Ömür, Ö. *Appl. Organomet. Chem.* **2005**, *19*, 887.
184 Hodgson, H. H.; Marsden, E. *J. Chem. Soc.* **1945**, 274.
185 Garst, M. E.; Lukton, D. *Synth. Commun.* **1980**, *10*, 155.
186 Nomura, Y. *Bull. Chem. Soc. Jpn.* **1961**, *34*, 1648.
187 Nomura, Y.; Anzai, H. *Bull. Chem. Soc. Jpn.* **1962**, *35*, 111.
188 Nomura, Y.; Anzai, H. *Bull. Chem. Soc. Jpn.* **1964**, *37*, 970.
189 Nomura, Y.; Anzai, H.; Tarao, R.; Shiomi, K. *Bull. Chem. Soc. Jpn.* **1964**, *37*, 967.
190 Erdik, E.; Kocuglu, M. *Main Group Metal Chem.* **2002**, *25*, 621.
191 Barbero, M.; Degani, I.; Dughera, S.; Fochi, R.; Perracino, P. *Synthesis* **1998**, 1235.
192 Curtin, D. Y.; Ursprung, J. A. *J. Org. Chem.* **1956**, *21*, 1221.
193 Curtin, D. Y.; Tveten, J. L. *J. Org. Chem.* **1961**, *26*, 1764.
194 Neumann, W. P.; Wicenec, C. *Chem. Ber.* **1991**, *124*, 2297.
195 Crary, J. W.; Quayle, O. R.; Lester, C. T. *J. Am. Chem. Soc.* **1956**, *78*, 5584.
196 Sakakura, T.; Tanaka, M. *J. Chem. Soc., Chem. Commun.* **1985**, 1309.
197 Sakakura, T.; Hara, M.; Tanaka, M. *J. Chem. Soc., Perkin Trans. 1* **1994**, 289.
198 Eistert, B.; Regitz, M.; Heck, G.; Schwall, H. In *Methoden der organischen Chemie (Houben-Weyl)*; Georg Thieme Verlag: Stuttgart, 1968; Vol. 10/4, p 709.
199 Zerner, E. *Monatsh. Chem.* **1913**, *34*, 1609.
200 Zerner, E. *Monatsh. Chem.* **1913**, *34*, 1631.
201 Forster, M. O.; Cardwell, D. *J. Chem. Soc.* **1913**, 861.
202 Coleman, G. H.; Gilman, H.; Adams, C. E.; Pratt, P. E. *J. Org. Chem.* **1938**, *3*, 99.
203 Ciganek, E. *J. Org. Chem.* **1965**. *30*, 4198.
204 Diekmann, J. *J. Org. Chem.* **1965**, *30*, 2272.
205 Takamura, N.; Yamada, S.-i. *Chem. Pharm. Bull.* **1976**, *24*, 800.
206 Balli, H.; Gipp, R. *Liebigs Ann. Chem.* **1966**, *699*, 133.
207 Huisgen, R.; Bihlmeier, W.; Reissig, H.-U. *Angew. Chem., Int. Ed. Engl.* **1979**, *18*, 331.
208 Schmitz, E.; Ohme, R. *Chem. Ber.* **1961**, *94*, 2166.
209 Nelsen, S. F.; Landis, R. T., II *J. Am. Chem. Soc.* **1973**, *95*, 2719.
210 Neugebauer, F. A.; Weger, H. *Chem. Ber.* **1979**, *112*, 1076.
211 Katritzky, A. R.; Wu, J.; Verin, S. V. *Synthesis* **1995**, 651.
212 Kaiser, E. M.; Bartling, G. J. *J. Org. Chem.* **1972**, *37*, 490.
213 Bozzini, S.; Stener, A. *Ann. Chim. (Rome)* **1968**, *58*, 169.
214 Wittig, G.; Schuhmacher, A. *Chem. Ber.* **1955**, *88*, 234.
215 Waser, J.; Gaspar, B.; Nambu, H.; Carreira, E. M. *J. Am. Chem. Soc.* **2006**, *128*, 11693.
216 Waser, J.; Gonzáles-Gómez, J. C.; Nambu, H.; Huber, P.; Carreira, E. M. *Org. Lett.* **2005**, *7*, 4249.
217 Udodong, U. E.; Fraser-Reid, B. *J. Org. Chem.* **1988**, *53*, 2132.
218 Brimble, M. A.; Heathcock, C. H.; Nobin, G. N. *Tetrahedron: Asymmetry* **1996**, *7*, 2007.
219 Harris, J. M.; McDonald, R.; Vederas, J. C. *J. Chem. Soc., Perkin Trans. 1* **1996**, 2669.
220 Hoffmann, R. W.; Hölzer, B.; Knopff, O. *Org. Lett.* **2001**, *3*, 1945.
221 List, B. *J. Am. Chem. Soc.* **2002**, *124*, 5656.
222 Bøgevig, A.; Juhl, K.; Kumaragurubaran, N.; Zhuang, W.; Jørgensen, K. A. *Angew. Chem., Int. Ed.* **2002**, *41*, 1790.
223 Baumann, T.; Vogt, H.; Bräse, S. *Eur. J. Org. Chem.* **2007**, 266.
224 Iwamura, H.; Mathew, S. P.; Blackmond, D. G. *J. Am. Chem. Soc.* **2004**, *126*, 11770.

[225] Marigo, M.; Schulte, T.; Franzén, J.; Jørgensen, K. A. *J. Am. Chem. Soc.* **2005**, *127*, 15710.
[226] Suri, J. T.; Steiner, D. D.; Barbas, C. F., III *Org. Lett.* **2005**, *7*, 3885.
[227] Kotkar, S. P.; Sudalai, A. *Tetrahedron: Asymmetry* **2006**, *17*, 1738.
[228] Kumaragurubaran, N.; Juhl, K.; Zhuang, W.; Bøgevig, A.; Jørgensen, K. A. *J. Am. Chem. Soc.* **2002**, *124*, 6254.
[229] Thomassigny, C.; Prim, D.; Greck, C. *Tetrahedron Lett.* **2006**, *47*, 1117.
[230] Diels, O. *Liebigs Ann. Chem.* **1922**, *429*, 1.
[231] Pihko, P. M.; Pohjakallio, A. *Synlett* **2004**, 2115.
[232] Liu, X.; Li, H.; Deng, L. *Org. Lett.* **2005**, *7*, 167.
[233] Xu, X.; Yabuta, T.; Yuan, P.; Takemoto, Y. *Synlett* **2006**, 137.
[234] Ghosh, A. K.; Mathivanan, P.; Capiello, J. *Tetrahedron: Asymmetry* **1998**, *9*, 1.
[235] Marigo, M.; Juhl, K.; Jørgensen, K. A. *Angew. Chem., Int. Ed.* **2003**, *42*, 1367.
[236] Ma, S.; Jiao, N.; Zheng, Z.; Ma, Z.; Lu, Z.; Ye, L.; Deng, Y.; Chen, G. *Org. Lett.* **2004**, *6*, 2193.
[237] Huber, D. P.; Stanek, K.; Togni, A. *Tetrahedron: Asymmetry* **2006**, *17*, 658.
[238] Kim, S. M.; Kim, H. R.; Kim, D. Y. *Org. Lett.* **2005**, *7*, 2309.
[239] Kang, Y. K.; Kim, D. Y. *Tetrahedron Lett.* **2006**, *47*, 4565.
[240] Firl, J.; Sommer, S. *Tetrahedron Lett.* **1970**, 1925.
[241] Firl, J.; Sommer, S. *Tetrahedron Lett.* **1970**, 1929.
[242] Sasaki, T.; Ishibashi, Y.; Ohno, M. *Heterocycles* **1983**, *20*, 1933.
[243] Sasaki, T.; Ishibashi, Y.; Ohno, M. *J. Chem. Res. (M)* **1984**, 1972.
[244] Yamashita, Y.; Ishitani, H.; Kobayashi, S. *Can. J. Chem.* **2000**, *78*, 666; Kobayashi, S.; Yamashita, Y.; Ishitani, H. *Chem. Lett.* **1999**, 307.
[245] Moriarty, R. M.; Prakash, I. *Synth. Commun.* **1985**, *15*, 649.
[246] Risaliti, A.; Marchetti, L. *Ann. Chim. (Rome)* **1963**, *53*, 718.
[247] Firl, J.; Sommer, S. *Tetrahedron Lett.* **1969**, 1137.
[248] Fatutta, S.; Pitacco, G.; Russo, C.; Valentin, E. *J. Chem. Soc., Perkin Trans. 1* **1992**, 2045.
[249] Forchiassin, M.; Pitacco, G.; Risalti, A.; Russo, C.; Valentin, E. *J. Heterocycl. Chem.* **1983**, *20*, 305.
[250] Benedetti, F.; Bozzini, S.; Fatutta, S.; Forchiassin, M.; Nitti, P.; Pitacco, G.; Russo, C. *Gazz. Chim. Ital.* **1991**, *121*, 401.
[251] Hall, J. H.; Woiciechowska, M. *J. Org. Chem.* **1978**, *43*, 3348.
[252] Evans, D. A.; Johnson, D. S. *Org. Lett.* **1999**, *1*, 595.
[253] Beslin, P.; Marion, P. *Tetrahedron Lett.* **1992**, *33*, 935.
[254] Cookson, R. C.; Gilani, S. S. H.; Stevens, I. D. R. *J. Chem. Soc. C.* **1967**, 1905.
[255] Sapountzis, I.; Knochel, P. *Angew. Chem., Int. Ed.* **2004**, *43*, 897.
[256] Dimroth, O. *Chem. Ber.* **1906**, *39*, 3905.
[257] Brinckman, F. E.; Haiss, H. S.; Robb, R. A. *Inorg. Chem.* **1965**, *4*, 936.
[258] Smith, R. H., Jr.; Michejda, C. J. *Synthesis* **1983**, 476.
[259] Smith, R. H., Jr.; Denlinger, C. L.; Kupper, R.; Mehl, A. F.; Michejda, C. J. *J. Am. Chem. Soc.* **1986**, *108*, 3726.
[260] Danilov, S. N.; Yastrebov, L. N.; Burova, L. N. *Zh. Obshch. Khim.* **1970**, *40*, 2248; *Engl. Transl.* p 2235.
[261] Sieh, D. H.; Wilbur, D. J.; Michejda, C. J. *J. Am. Chem. Soc.* **1980**, *102*, 3883.
[262] Sieh, D. H.; Michejda, C. J. *J. Am. Chem. Soc.* **1981**, *103*, 442.
[263] Kabalka, G. W.; Li, G. *Tetrahedron Lett.* **1997**, *38*, 5777.
[264] Nishiyama, K.; Tanaka, N. *J. Chem. Soc., Chem. Commun.* **1983**, 1322.
[265] Kelly, T. R.; Maguire, M. P. *Tetrahedron* **1985**, *41*, 3033.
[266] Perrin, P.; Aubert, F.; Lellouche, J. P.; Beaucourt, J. P. *Tetrahedron Lett.* **1986**, *27*, 6193.
[267] Okazaki, R.; Unno, M.; Inamoto, N. *Chem. Lett.* **1987**, 2293.
[268] Smith, R. H., Jr.; Mehl, A. F.; Shantz, D. L., Jr.; Chmurny, G. N.; Michejda, C. J. *J. Org. Chem.* **1988**, *53*, 1467.
[269] Yadav, J. S.; Madhuri, C.; Reddy, B. V. S.; Reddy, G. S. K. K.; Sabitha, G. *Synth. Commun.* **2002**, *32*, 2771.
[270] Dimroth, O. *Chem. Ber.* **1905**, *38*, 670.

[271] Forster, M. O.; Fierz, H. E.; Joshua, W. P. *J. Chem. Soc.* **1908**, 1070.
[272] Pochinok, V. Ya.; Shrobovich, V. A.; Portnyagina, V. A.; Polyanskaya, A.L. *Ukr. Khim. Zh.* **1959**, *25*, 774; *Chem. Abstr.* **1960**, *54*. 13034i.
[273] Trost, B. M.; Pearson, W. H. *J. Am. Chem. Soc.* **1983**, *105*, 1054.
[274] Pearson, W. H. Ph. D. Dissertation, University of Wisconsin, Madison, 1982; Univ. Mictofilms Int., Order No. 8301883T.
[275] Trost, B. M.; Pearson, W. H. *J. Am. Chem. Soc.* **1981**, *103*, 2483.
[276] Tolman, R. L.; Smith, C. W.; Robins, R. K. *J. Am. Chem. Soc.* **1972**, *94*, 2530.
[277] Hassner, A.; Belinka, B. A., Jr. *J. Am. Chem. Soc.* **1980**, *102*, 6185.
[278] Hassner, A.; Munger, P.; Belinka, B. A., Jr. *Tetrahedron Lett.* **1982**, *23*, 699.
[279] Kumar, S. H. M.; Reddy, S. B. V.; Anjaneyulu, S.; Yadav, J. S. *Tetrahedron Lett.* **1999**, *40*, 8305.
[280] Dimroth, O.; Eble, M.; Gruhl. W. *Chem. Ber.* **1907**, *40*, 2390.
[281] Pochinok, V. Ya.; El'gort, R. G. *Ukr. Khim. Zh.* **1949**, *15*, 311; *Chem. Abstr.* **1951**, *48*, 3320c.
[282] Pochinok, V. Ya. *Zh. Obshch. Khim.* **1946**, *16*, 1303; *Chem. Abstr.* **1947**, *41*, 3066h. 283. Pochinok, V. Ya.; Kalashnikova, E. S. *Ukr. Khim. Zh.* **1951**, *17*, 517; *Chem. Abstr.* **1951**, *48*, 10640d.
[284] Bertho, A. *J. Prakt. Chem.* **1901**, *63*, 101.
[285] Dimroth, O. *Chem. Ber.* **1903**, *36*, 909.
[286] Akimova, G. S.; Kolokol'tseva, I. G.; Chistokletov, V. N.; Petrov, A. A. *Zh. Org. Khim.* **1968**, *4*, 954; *Engl. Transl.* p 927.
[287] Jones, W. M.; Maness, D. D. *J. Am. Chem. Soc.* **1970**, *92*, 5457.
[288] Lee, C. C.; Ko, E. C. F. *Can. J. Chem.* **1976**, *54*, 3041.
[289] Kleinfeller, H. *J. Prakt. Chem.* **1928**, *119*, 61; Kleinfeller, H.; Bönig, G. *J. Prakt. Chem.* **1931**, *132*, 175.
[290] Pochinok. A. V.; Pochinok, V. Ya.; Kondratenko, P. A. *Ukr. Khim. Zh.* **1984**, *50*, 884; *Chem. Abstr.* **1985**, *102*, 61876m.
[291] Scarpati, R.; Sica, D. *Gazz. Chim. Ital.* **1963**, *93*, 942.
[292] Dimroth, O. *Liebigs Ann. Chem.* **1910**, *373*, 336.
[293] Babudri, F.; Di Nunno, L.; Florio, S.; Valzano, S. *Tetrahedron* **1984**, *40*, 1731.
[294] Pochinok, V. Ya. *Ukr. Khim. Zh.* **1949**, *15*, 302; *Chem. Abstr.* **1954**, *48*, 3285d.
[295] Kozlowska-Gramsz, E.; Descotes, G. *Tetrahedron Lett.* **1981**, *22*, 563.
[296] Lociuro, S.; Pellacani, L.; Tardella, P. A. *Tetrahedron Lett.* **1983**, *24*, 593.
[297] Auberson, Y.; Vogel, P. *Tetrahedron* **1990**, *46*, 7019.
[298] Scarpati, R.; Graziano, M. L.; Nicolaus, R. A. *Gazz. Chim. Ital.* **1970**, *100*, 665.
[299] Cipollone, A.; Loreto, M. A.; Pellacani, L.; Tardella, P. A. *J. Org. Chem.* **1987**, *52*, 2584.
[300] Loreto, M. A.; Pellacani, L.; Tardella, P. A. *Tetrahedron Lett.* **1989**, *30*, 2975.
[301] Fioravanti, S.; Loreto, M. A.; Pellacani, L.; Sabbatini, F.; Tardella, P. A. *Tetrahedron: Asymmetry* **1994**, *5*, 473.
[302] Fioravanti, S.; Loreto, M. A.; Pellacani, L.; Tardella, P. A. *Tetrahedron: Asymmetry* **1990**, *1*, 931.
[303] Fioravanti, S.; Pellacani, L.; Ricci, D.; Tardella, P. A. *Tetrahedron: Asymmetry* **1997**, *8*, 2261.
[304] Del Signore, G.; Fioravanti, S.; Pellacani, L.; Tardella, P. A. *Tetrahedron* **2001**, *57*, 4623.
[305] Smith, P. A. S.; Rowe, C. D.; Bruner, L. B. *J. Org. Chem.* **1969**, *34*, 3430.
[306] Ito, S.; Hirabayashi, T.; Matsumoto, K. *Bull. Chem. Soc. Jpn.* **1970**, *43*, 2254.
[307] Reed, J. O.; Lwowski, W. *J. Org. Chem.* **1971**, *36*, 2864.
[308] Ito, S. *Bull. Chem. Soc. Jpn.* **1966**, *39*, 635.
[309] Hakimelahi, G. H.; Just, G. *Synth. Commun.* **1980**, *10*, 429.
[310] Narasimhan, N. S.; Ammanamanchi, R. *Tetrahedron Lett.* **1983**, *23*, 4733.
[311] Reed, J. N.; Snieckus, V. *Tetrahedron Lett.* **1983**, *24*, 3795.
[312] Reed, J. N.; Snieckus, V. *Tetrahedron Lett.* **1984**, *25*, 5505.
[313] Wan, Z.-K.; Woo, G. H. C.; Snyder, J. K. *Tetrahedron* **2001**, *57*, 5497.
[314] Gavenonis, J.; Tilley, T. D. *Organometallics* **2002**, *21*, 5549.
[315] Nesmeyanov, A. N.; Drozd, V. N.; Sazonova, V. A. *Dokl. Akad. Nauk SSSR* **1963**, *150*, 321; *Engl. Transl.* p 416.
[316] Spagnolo, P.; Zanirato, P. *J. Org. Chem.* **1978**, *43*, 3539.
[317] Denmark, S.; Chatani, N.; Pansare, S. V. *Tetrahedron* **1992**, *48*, 2191.

[318] Evans, D. A.; Britton, T. C.; Ellman, J. A. Dorow, R. L. *J. Am. Chem. Soc.* **1990**, *112*, 4011.
[319] Benati, L.; Nanni, D.; Spagnolo, P. *J. Org. Chem.* **1999**, *64*, 5132.
[320] Shiro, Y.; Kato, K.; Fujii, M.; Ida, Y.; Akita, H. *Tetrahedron* **2006**, *62*, 8687.
[321] Benati, L.; Calestani, G.; Nanni, D.; Spagnolo, P. *J. Org. Chem.* **1998**, *63*, 4679.
[322] Lombardo, L.; Mander, L. N. *Synthesis* **1980**, 368.
[323] Coates, R. M.; Kang, H.-Y. *J. Org. Chem.* **1987**, *52*, 2065.
[324] Uyehara, T.; Takehara, N.; Ueno, M.; Sato, T. *Bull. Chem. Soc. Jpn.* **1995**, *68*, 2687.
[325] Battaglia, A.; Baldelli, E.; Bombardelli, E.; Carenzi, G.; Fontana, G.; Gelmi, M. L.; Guerrini, A.; Pocar, D. *Tetrahedron* **2005**, *61*, 7737.
[326] Brown, R. C. D.; Bataille, C. J. R.; Bruton, G.; Hinks, J. D.; Swain, N. A. *J. Org. Chem.* **1994**, *66*, 6719.
[327] Enders, D.; Joseph, R.; Poiesz, C. *Tetrahedron* **1998**, *54*, 10069.
[328] Charette, A. B.; Wurz, R. P.; Ollevier, T. *Helv. Chim. Acta* **2002**, *85*, 4468.
[329] Wurz, R. P.; Lin, W.; Charette, A. B. *Tetrahedron Lett.* **2003**, *44*, 8845.
[330] A repetition of the reaction of triethyl phosphonoacetate with trifluoromethanesulfonyl azide gave exclusively the diazo transfer product rather than the azide as reported in reference 309. Charette, A. B.; Marcoux, D. Départment de Chimie, Université de Montréal, Montréal, Canada. Personal communication, 2007.
[331] Magnus, P.; Barth, L. *Tetrahedron* **1995**, *51*, 11075.
[332] Lemieux, R. U.; Ratcliffe, R. M. *Can. J. Chem.* **1979**, *57*, 1244.
[333] Mori, S.; Aoyama, T.; Shioiri, T. *Tetrahedron Lett.* **1984**, *25*, 429.
[334] Mori, S.; Aoyama, T.; Shioiri, T. *Chem. Pharm. Bull.* **1986**, *34*, 1524.
[335] Guiver, M. D.; Robertson, G. P. *Macromolecules* **1995**, *28*, 294.
[336] Villalgordo, J. M.; Linden. A.; Heimgartner, H. *Helv. Chim. Acta* **1996**, *79*, 213.
[337] Villalgordo, J. M; Enderli, A.; Linden, A.; Heimgartner, H. *Helv. Chim. Acta* **1995**, *78*, 1983.
[338] Bunnage, M. E.; Burke, A. J.; Davies, S. G.; Millican, N. L.; Nicholson, R. L.; Roberts, P. M.; Smith, A. D. *Org. Biomol. Chem.* **2003**, 3708.
[339] Kühlein, K.; Jensen, H. *Liebigs Ann. Chem.* **1974**, 369.
[340] Sharpless, K. B. *Angew. Chem., Int. Ed. Engl.* **1996**, *35*, 451.
[341] Reddy, K. L.; Sharpless, K. B. *J. Am. Chem. Soc.* **1998**, *120*, 1207.
[342] Phukan, P.; Sudelai, A. *Tetrahedron: Asymmetry* **1998**, *9*, 1001.
[343] Driguez, H.; Vermes, J.-P.; Lessard, J. *Can. J. Chem.* **1978**, *56*, 119.
[344] Driguez, H.; Lessard, J. *Can. J. Chem.* **1977**, *55*, 720.
[345] Magnus, P.; Mugrage, B. *J. Am. Chem. Soc.* **1990**, *112*, 462.
[346] Magnus, P.; Coldham, I. *J. Am. Chem. Soc.* **1991**, *113*, 672.
[347] Magnus, P.; Lacour, J.; Bauta, W.; Mugrage, B.; Lynch, V. *J. Chem. Soc., Chem. Commun.* **1991**, 1362.
[348] Magnus, P.; Lacour. J.; Coldham, I.; Mugrage, B.; Bauta, W. B. *Tetrahedron* **1995**, *51*, 11087.
[349] Gethin, D. M.; Simpkins, N. S. *Tetrahedron* **1997**, *53*, 14417.
[350] Du Bois, J.; Hong, J.; Carreira, E. M. *J. Am. Chem. Soc.* **1996**, *118*, 915.
[351] Carreira, E. M.; Hong, J.; Du Bois, J.; Tomooka, C. S. *Pure Appl. Chem.* **1998**, *70*, 1097.
[352] Minakata, S.; Ando, T.; Nishimura, M.; Ryu, I.; Komatsu, M. *Angew. Chem., Int. Ed. Engl.* **1998**, *37*, 3392.
[353] Svenstrup, N.; Bøgevig, A.; Hazell, R. G.; Jørgensen, K. A. *J. Chem. Soc., Perkin Trans. 1* **1999**, 1559.
[354] Du Bois, J.; Tomooka, C. S.; Hong, J.; Carreira, E. M. *J. Am. Chem. Soc.* **1997**, *119*, 3179.
[355] Kraus, G. A. U. S. Patent 5,599,998 (1997).
[356] Sinha, P.; Kofink, C. C.; Knochel, P. *Org. Lett.* **2006**. *8*, 3741.
[357] Guijarro, A.; Rieke, R. D. *Angew. Chem., Int. Ed. Engl.* **1998**, *37*, 1679.
[358] Velarde-Ortiz, R.; Guijarro, A.; Rieke, R. D. *Tetrahedron Lett.* **1998**, *39*, 9157.
[359] Katritzky, A. R.; Verin, S. V.; Yang, B. *Org. Prep. Proc. Int.* **1996**, *28*, 97.
[360] An, D. K.; Hirakawa, K.; Okamoto, S.; Sato, F. *Tetrahedron Lett.* **1999**, *40*, 3737.
[361] Boyer, J. H.; Mack, C. H.; Goebel, N.; Morgan, L. R., Jr. *J. Org. Chem.* **1958**, *23*, 1051.
[362] Hartwig, J. F. *Acc. Chem. Res.* **1998**, *31*, 852.

[363] Hartwig, J. F. *Angew. Chem., Int. Ed. Engl.* **1998**, *37*, 2046.
[364] Wolfe, J. P.; Wagah, S.; Marcoux, J.-F.; Buchwald, S. L. *Acc. Chem. Res.* **1998**, *31*, 805.
[365] Hartwig, J. F. In *Modern Amination Methods*; Ricci, A., Ed.: Wiley-VCH: Weinheim, 2000, Chapter 7.
[366] Hartwig, J. F. In *Handbook of Organopalladium Chemistry for Organic Synthesis*; Negishi, E.-I., Ed.; Wiley: New York, 2002, Vol. 1, p1051.
[367] Lipshutz, B. H.; Ueda, H. *Angew. Chem., Int. Ed.* **2000**, *39*, 4492.
[368] Muci, A. R.; Buchwald, S. L. *Top. Curr. Chem.* **2002**, *219*, 131.
[369] Wolfe, J. P.; Buchwald, S. L. *Org. Synth.* **2002**, *78*, 23.
[370] Okano, K.; Tokuyama, H.; Fukuyama, T. *Org. Lett.* **2003**, *5*, 2987.
[371] Tasler, S.; Lipshutz, B. H. *J. Org. Chem.* **2003**, *68*, 1190.
[372] Schlummer, B.; Scholz, U. *Adv. Synth. Catal.* **2004**, *346*, 1599.
[373] Christensen, H.; Kiil, S.; Dam-Johansen, K.; Nielsen, O.; Sommer, M. B. *Org. Process Res. Dev.* **2006**, *10*, 762.
[374] Antilla, J. C.; Buchwald, S. L. *Org. Lett.* **2001**, *3*, 2077.
[375] Trost, B. M.; Pearson, W. H. *Tetrahedron Lett.* **1983**, *24*, 263.
[376] Minisci, F. *Top. Curr. Chem.* **1976**, *62*, 1.
[377] Stollé, R.; Adam, G. *J. Prakt. Chem.* **1925**, *111*, 167.
[378] Stollé, R.; Reichert, W. *J. Prakt. Chem.* **1929**, *123*, 75.
[379] Zaltsgendler, I.; Leblanc, Y.; Bernstein, M. A. *Tetrahedron Lett.* **1993**, *34*, 2441.
[380] Mitchell, H.; Leblanc, Y. *J. Org. Chem.* **1994**, *59*, 682.
[381] Leblanc, Y.; Boudreault, N. *J. Org. Chem.* **1995**, *60*, 4268.
[382] Dufresne, C.; Leblanc, Y.; Berthelette, C.; McCooeye, C. *Synth. Commun.* **1997**, *27*, 3613.
[383] Lenarsic, R.; Kocevar, M.; Polanc, S. *J. Org. Chem.* **1999**, *64*, 2558.
[384] Bombek, S.; Pozgan, F.; Kocevar, M.; Polanc, S. *J. Org. Chem.* **2004**, *69*, 2224.
[385] Chowdari, N. S.; Barbas, C. F., III *Org Lett* **2005**, *7*, 867.
[386] Franzén, J.; Marigo, M.; Fielenbach, D.; Wabnitz, T.C.; Kjaersgaard, A.; Jørgensen, K. A. *J. Am. Chem. Soc.* **2005**, *127*, 18296.
[386] a. Vogt, H.; Baumann, T.; Nieger, M.; Bräse, S. *Eur. J. Org. Chem.* **2006**, 5315.
[387] Huisgen, R.; Möbius, L.; Szeimies, G. *Chem. Ber.* **1965**, *98*, 1138.
[388] Dessole, G.; Bernardi, L.; Bonini, B. F.; Capitò, E.; Fochi, M.; Herrera, R. P.; Ricci, A.; Cahiez, G. *J. Org. Chem.* **2004**, *69*, 8525.
[389] Marchetti, L. *J. Chem. Soc., Perkin Trans. 2* **1978**, 382.
[390] Huisgen, R.; Jakob, F. *Liebigs Ann. Chem.* **1954**, *590*, 37.
[391] Majewski, M.; Zheng, G.-Z. *Can. J. Chem.* **1992**, *70*, 2618.
[392] Gmeiner, P.; Bollinger, B. *Tetrahedron Lett.* **1991**, *32*, 5927.
[393] Gmeiner, P.; Bollinger, B. *Liebigs Ann. Chem.* **1992**, 273.
[394] Gmeiner, P.; Bollinger, B.; Mierau, J.; Höfner, G. *Arch. Pharm. (Weinheim, Ger.)* **1995**, *328*, 609.
[395] Gmeiner, P.; Sommer, J. *Arch. Pharm. (Weinheim, Ger.)* **1994**, *327*, 435.
[396] Gmeiner, P.; Sommer, J.; Mierau, J.; Höfner, G. *Bioorg. Med. Chem. Lett.* **1993**, *3*, 1477.
[397] Fioravanti, S.; Loreto, M. A.; Pellacani, L.; Tardella, P. A. *Tetrahedron* **1991**, *47*, 5877.
[398] Du Bois, J.; Hong, J.; Carreira, E. M. *J. Am. Chem. Soc.* **1996**, *118*, 915.
[399] Pellacani, L.; Pulcini, P.; Tardella, P. A. *J. Org. Chem.* **1982**, *47*, 5023.
[400] Benedetti, F.; Forchiassin, M.; Russo, C. Risaliti, A. *Gazz. Chim. Ital.* **1985**, *115*, 663.
[401] Ballaben, E.; Forchiassin, M.; Nitti, P.; Russo, C. *Gazz. Chim. Ital.* **1993**, *123*, 387.
[402] Ciganek, E. *Org. React.* **1997**, *51*, 201.
[403] Kamimura, A.; Gunjigake, Y.; Mitsudera, H.; Yokoyama, S. *Tetrahedron Lett.* **1998**, *39*, 7323.
[404] Juhl, K.; Jørgensen, K. A. *J. Am. Chem. Soc.* **2002**, *124*, 2420.
[405] Gmeiner, P.; Bollinger, B. *Tetrahedron* **1994**, *50*, 10909.
[406] Green, J. R. *Science in Synthesis* **2006**, 8*a*, 427.
[407] Oguri. T.; Shioiri, T.; Yamada, S.-i. *Chem. Pharm. Bull.* **1975**, *23*, 167.
[408] Harris, J. M.; Bolessa, E. A.; Mendonca, A. J.; Feng, S.-C.; Vederas, J. C. *J. Chem. Soc., Perkin Trans. 1* **1995**, 1945.
[409] Oppolzer, W.; Moretti, R. *Helv. Chim. Acta* **1986**, *69*, 1923.

[410] Yamamoto, Y.; Yumoto, M.; Yamada, J.-i. *Tetrahedron Lett.* **1991**, *32*, 3079.
[411] Gennari, C.; Colombo, L.; Bertolini, G. *J. Am. Chem. Soc.* **1986**, *108*, 6394.
[412] Oppolzer, W.; Moretti, R. *Tetrahedron* **1988**, *44*, 5541.
[413] Molina, P.; Fresneda, P. M.; Sanz, M. *J. Org. Chem.* **1999**, *64*, 2540.
[414] Fresneda, P. M.; Molina, P.; Sanz, M. A. *Tetrahedron Lett.* **2001**, *42*, 851.
[415] Guenti, G.; Banfi, L.; Narisano, E. *Tetrahedron* **1988**, *44*, 5553.
[416] Greck, C.; Bischoff, L.; Ferreira, F.; Pinel, C.; Piveteau, E.; Genêt, J.-P. *Synlett*, **1993**, 475.
[417] Ciufolini, M. A.; Xi, N. *J. Chem. Soc., Chem. Commun.* **1994**, 1867.
[418] Greck, C.; Bischoff, L.; Genêt, J.-P. *Tetrahedron: Asymmetry* **1995**, *6*, 1989.
[419] Greck, C.; Ferreira, F.; Genêt, J.-P. *Tetrahedron Lett.* **1996**, *37*, 2031.
[420] Girard, A.; Greck, C.; Ferroud, D.; Genêt, J.-P. *Tetrahedron Lett.* **1996**, *37*, 7967.
[421] Drouillat, B.; Poupardin, O.; Bourdreux, Y.; Greck, C. *Tetrahedron Lett.* **2003**, *44*, 2781.
[422] Poupardin, O.; Greck, C.; Genêt, J.-P. *Tetrahedron Lett.* **2000**, *41*, 8795.
[423] Bourdreux, Y.; Drouillat, B.; Greck, C. *Synlett* **2005**, 2086.
[424] Poupardin, O.; Greck, C.; Genêt, J.-P. *Synlett* **1998**, 1279.
[425] Panek, J. S.; Beresis, R.; Xu, F.; Yang, M. *J. Org. Chem.* **1991**, *56*, 7341.
[426] Fernández-Megía, E.; Paz, M. M.; Sardina, F. J. *J. Org. Chem.* **1994**, *59*, 7643.
[427] Hanessian, S.; Wang, W.; Gai, Y. *Tetrahedron Lett.* **1996**, *37*, 7477.
[428] Kapeller, H.; Griengl, H. *Tetrahedron* **1997**, *53*, 14635.
[429] Kumar, J. S. D.; Dupradeau, F.-Y.; Strouse, M. J.; Phelps, M. E.; Toyokuni, T. *J. Org. Chem.* **2001**, *66*, 3220.
[430] Ager, D. J. *Chem. Rev.* **1996**, *96*, 835.
[431] Evans, D. A.; Britton, T. C.; Dorow, R. L.; Dellaria, J. F., Jr. *Tetrahedron* **1988**, *44*, 5525.
[432] Trimble, L. A.; Vederas, J. C. *J. Am. Chem. Soc.* **1986**, *108*, 6397.
[433] Evans, D. A.; Britton, T. C. *J. Am. Chem. Soc.* **1987**, *109*, 6881.
[434] Lin, J.; Liao, S.; Han, Y.; Qiu, W.; Hruby, V. *Tetrahedron: Asymmetry* **1997**, *8*, 3213.
[435] Lin, J.; Liao, S.; Hruby, V. J. *J. Peptide Res.* **2005**, *65*, 105.
[436] Evans, D. A.; Nelson, S. G. *J. Am. Chem. Soc.* **1997**, *119*, 6452.
[437] Nishida, A.; Fuwa, M.; Fujikawa, Y.; Nakahata, E.; Furuno, A.; Nakagawa, M. *Tetrahedron Lett.* **1998**, *39*, 5983.
[438] Tilley, J. W.; Danho, W.; Shiuey, S.-J.; Kulesha, I.; Sarabu, R.; Swistok, J.; Makofske, R.; Olsen, G. L.; Chiang, E.; Rusiecki, V. K.; Wagner, R.; Michalewski, J.; Triscari, J.; Nelsen, D.; Chiruzzo, F. Y.; Weatherford. S. *Int. J. Pept. Protein Res.* **1992**, *39*, 322.
[439] Ben, R. N.; Orellana, A.; Arya, P. *J. Org. Chem.* **1998**, *63*, 4817.
[440] Derrer, S.; Davies, J. E.; Holmes, A. B. *J. Chem. Soc., Perkin Trans. 1* **2000**, 2943.
[441] Hale, K. J.; Cai, J.; Delisser, V.; Manaviazar, S.; Peak, S. A.; Bhatia, G. S.; Collins, T. C.; Jogiya, N. *Tetrahedron* **1996**, *52*, 1047.
[442] Decicco, C. P.; Leathers, T. *Synlett* **1995**, 615.
[443] Lundquist, J. T.; Dix, T. A. *Tetrahedron Lett.* **1998**, *39*, 775.
[444] Evans, D. A.; Ellman, J. A. *J. Chem. Soc., Perkin Trans. 1* **1989**, *111*, 1063.
[445] Whitman, D. B.; Askew, B. C.; Duong, L. T.; Fernandez-Metzler, C.; Halczenko, W.; Hartman, G. D.; Hutchinson, J. H.; Leu, C.-T.; Prueksaritanont, T.; Rodan, G. A.; Rodan, S. B.; Duggan, M. E. *Bioorg. Med. Chem. Lett.* **2004**, *14*, 4411.
[446] Whitman, D. B., Merck Research Laboratories, West Point, PA. Personal communication, 2005.
[447] Broka, C. A.; Ehrler, J. *Tetrahedron Lett.* **1991**, *32*, 5907.
[448] Stone, M. J.; van Dyk, M. S.; Booth, P. M.; Williams, D. H. *J. Chem. Soc., Perkin Trans. 1* **1991**, 1629.
[449] Evans, D. A.; Lundy, K. M. *J. Am. Chem. Soc.* **1992**, *114*, 1495.
[450] Evans, D. A.; Evrard, D. A.; Rychnovsky, S. D.; Früh, T.; Whittingham, W. G.; DeVries, K. M. *Tetrahedron Lett.* **1992**, *33*, 1189.
[451] Kennedy, K. J.; Lundquist, J. T., IV; Simandan, T. L.; Beeson, C. C.; Dix, T. A. *Bioorg. Med. Chem. Lett.* **1997**, *7*, 1937.
[452] Kennedy, K. J.; Lundquist, J. T., IV; Simandan, T. L.; Kokko, K. P.; Beeson, C. C.; Dix, T. A. *J. Peptide Res.* **2000**, *55*, 348.

453 Noguchi, H.; Aoyama, T.; Shioiri, T. *Heterocycles* **2002**, *58*, 471.
454 Chung, H.-K.; Kim, H.-W.; Chung, K.-H. *Heterocycles* **1999**, *51*, 2983.
455 Taunton, J.; Collins, J. L.; Schreiber, S. L. *J. Am. Chem. Soc.* **1996**, *118*, 10412.
456 Woiwode, T. F.; Wandless, T. J. *J. Org, Chem.* **1999**, *64*, 7670.
457 Liu, D.-G.; Wang, X.-Z.; Gao, Y.; Li, B.; Yang, D.; Burke, T. R., Jr. *Tetrahedron* **2002**, *58*, 10423.
458 Gilbertson, S. R.; Chen, G.; McLoughlin, M. *J. Am. Chem. Soc.* **1994**, *116*, 4481.
459 Arya, P.; Ben, R. N.; Qin, H. *Tetrahedron Lett.* **1998**, *39*, 6131.
460 Castellanos, E.; Reyes-Rangel, G.; Juaristi, E. *Helv. Chim. Acta* **2004**, *87*, 1016.
461 Shimizu, M.; Nemoto, H.; Kakuda, H.; Takahata, H. *Heterocycles* **2003**, *59*, 245.
462 Hanessian, S.; Sailes, H.; Munro, A.; Therrien, E. *J. Org. Chem.* **2003**, *68*, 7219.
463 Taylor, E. C.; Sun, J.-H. *Synthesis* **1980**, 801.
464 Koft, E. R. *J. Org. Chem.* **1987**, *52*, 3466.
465 Hanessian, S.; Sailes, H.; Therrien, E. *Tetrahedron* **2003**, *59*, 7047.
466 Wulf, J.-P,; Sienkewicz, K.; Makosza, M.; Schmitz, E. *Liebigs Ann. Chem.* **1991**, 537.
467 Hoffmann, H. *Chem. Ber.* **1962**, *95*, 2563.
468 Bernardi, L.; Zhuang, W.; Jørgensen, K. A. *J. Am. Chem. Soc.* **2005**, *127*, 5772.
469 Yamamoto, Y.; Hatsuya, S.; Yamada, J.-i. *J. Org. Chem.* **1990**, *55*, 3113.
470 Bertelsen, S.; Marigo, M.; Brandes, S.; Dinér, P.; Jørgensen, K. A. *J. Am. Chem. Soc.* **2006**, *128*, 12973.
471 Du Bois, J.; Hong, J.; Carreira, E. M. *J. Am. Chem. Soc.* **1996**, *118*, 915.
472 Schmitz, E.; Jänisch, K. *Z. Chem.* **1971**, *11*, 458.
473 Tamura, Y.; Kato, S.; Ikeda, M. *Chem. Ind. (London)* **1971**, 767.
474 Sofia, M. J.; Katzenellenbogen, J. A. *J. Org. Chem.* **1985**, *50*, 2331.
475 Badorrey, R.; Cativiela, C.; Diaz-de-Villegas, M. D.; Gálvez, J. A. *Tetrahedron: Asymmetry* **1995**, *6*, 2787.
476 Lumbierres, M.; Marchi, C.; Moreno-Mañas, M.; Sebastián, R. M.; Vallribera, A.; Lago, E.; Molins, E. *Eur. J. Org. Chem.* **2001**, 2321.
477 Hartke, K.; Brutsche, A.; Gerber, H.-D. *Liebigs Ann. Chem.* **1992**, 927.
478 Diels, O.; Behncke, H. *Chem. Ber.* **1924**, *57*, 653.
479 Nelson, J. H.; Howells, P. N.; DeLullo, G. C.; Landen, G. L.; Henry, R. A. *J. Org. Chem.* **1980**, *45*, 1246.
480 Meseguer, M.; Mareno-Mañas, M.; Vallribera, A. *Tetrahedron Lett.* **2000**, *41*, 4093.
481 Saaby, S.; Bella, M.; Jørgensen, K. A. *J. Am. Chem. Soc.* **2004**, *126*, 8120.
482 Golding, B. T.; Smith, A. J. *J. Chem. Soc., Chem. Commun.* **1980**, 702.
483 Weininger, S. J.; Kohen, S.; Mataka, S.; Koga, G.; Anselme, J.-P. *J. Org. Chem.* **1974**, *39*, 1591.
484 Benati, L.; Nanni, D.; Spagnolo, P. *J. Chem. Soc., Perkin Trans. 1* **1997**, 457.
485 Blatt, A. H. *J. Am. Chem. Soc.* **1939**, *61*, 3494.
486 Cromwell, N. H.; Barker, N. G.; Wankel, R. A.; Vanderhorst, P. J.; Olson, F. W.; Anglin, J. H., Jr. *J. Am. Chem. Soc.* **1951**, *73*, 1044.
487 Cardillo, G.; Casolari, S.; Gentilucci, L.; Tomasini, C. *Angew. Chem., Int. Ed. Engl.* **1996**, *35*, 1848.
488 Bongini, A.; Cardillo, G.; Gentilucci, L.; Tomasini, C. *J. Org. Chem.* **1997**, *62*, 9148.
489 Cardillo, G.; Gentilucci, L.; Tolomelli, A. *Tetrahedron Lett.* **1999**, *40*, 8261.
490 Pereira, M. M.; Santos, P. P. O.; Reis, L. V.; Lobo, A. M.; Prabhakar, S. *J. Chem. Soc., Chem. Commun.* **1993**, 38.
491 Cardillo, G.; Gentilucci, L.; Gianotti, M.; Perciaccante, R.; Tolomelli, A. *J. Org. Chem.* **2001**, *66*, 8657.
492 Métra, P.; Hamelin, J. *J. Chem. Soc., Chem. Commun.* **1980**, 1038.
493 Colantoni, D.; Fioravanti, S.; Pellacani, L.; Tardella, P. A. *Org. Lett.* **2004**, *6*, 197.
494 Sato, S. *Bull. Chem. Soc. Jpn.* **1968**, *41*, 1440.
495 Gasperi, T.; Loreto, M. A.; Tardella, P. A.; Veri, E. *Tetrahedron Lett.* **2003**, *44*, 4953.
496 Sheradsky, T.; Yusupova, L. *Tetrahedron Lett.* **1995**, *36*, 7701.
497 Yoshida, M.; Uchiyama, K.; Narasaka, K. *Heterocycles* **2000**, *52*, 681.
498 Uchiyama, K.; Hayashi, Y.; Narasaka, K. *Synlett* **1997**, 445.

[499] Uchiyama, K.; Ono, A.; Hayashi, Y.; Narasaka, K. *Bull. Chem. Soc. Jpn.* **1998**, *71*, 2945.
[500] Glaser, H.; Möller, F.; Pieper, G.; Schröter, R.; Spielberger, G.; Söll, H. In *Methoden der organischen Chemie (Houben-Weyl)*; Georg Thieme Verlag: Stuttgart, 1957; Vol. 11/1, p 3.
[501] *Chemistry of the Amino Group*; Patai, S. Ed.; Wiley: New York, 1968.
[502] Mitsunobu, O. In *Comprehensive Organic Synthesis;* Trost, B. M.; Fleming, I., Eds.; Pergamon Press: Oxford, 1991; Vol. 6, p 65.
[503] Hemmer, R.; Lürken, W. In *Methoden der organischen Chemie (Houben-Weyl)*; Georg Thieme Verlag: Stuttgart, New York, 1992; Vol. 16d/2, p 646.
[504] *Chemistry of the Amino, Nitroso, Nitro, and Related Groups, Part 2*; Patai, S. Ed.; Wiley: Chichester, 1996.
[505] Müller, T. E.; Beller, M. *Chem. Rev.* **1998**, *98*, 675.
[506] Meier, R. *Chem. Ber.* **1953**, *86*, 1483.
[507] Beringer, F. M.; Farr, J. A., Jr.; Sands, S. *J. Am. Chem. Soc.* **1953**, *75*, 3984.
[508] Sand, J.; Singer, F. *Liebigs Ann. Chem.* **1903**, *329*, 190.
[509] Müller, E.; Metzger, H. *Chem. Ber.* **1956**, *89*, 396.
[510] Kato, K.; Mukaiyama, T. *Chem. Lett.* **1990**, 1395.
[511] Wieland, H. *Chem. Ber.* **1903**, *36*, 2315.
[512] Baker, E. B.; Sisler, H. H. *J. Am. Chem. Soc.* **1953**, *75*, 5193.
[513] Oddo, B. *Gazz. Chim. Ital.* **1909**, *39*, 659.
[514] Waters, W. L.; Marsh, P. G. *J. Org. Chem.* **1975**, *40*, 3344.
[515] Waters, W. L.; Marsh, P. G. *J. Org. Chem.* **1975**, *40*, 3349.
[516] Klages, F.; Sitz, H.; Heinle, R. *Chem. Ber.* **1959**, *92*, 2606.
[517] Klages, F.; Heinle, R.; Sitz, H.; Specht, E. *Chem. Ber.* **1963**, *96*, 2387.
[518] Rasmussen, J. K.; Hassner, A. *J. Org. Chem.* **1974**, *39*, 2558.
[519] Bachman, G. B.; Hokama, T. *J. Org. Chem.* **1960**, *25*, 178.
[520] Olah, G. A.; Rochin, C. *J. Org. Chem.* **1987**, *52*, 701.
[521] Elfehail, F.; Dampawan, P.; Zajac, W., Jr. *Synth. Commun.* **1980**, *10*, 929.
[522] Dampawan, P.; Zajac, W. W., Jr. *J. Org. Chem.* **1982**, *47*, 1176.
[523] Detty, M. R.; Logan, M. E. In *Encyclopedia of Reagents for Organic Synthesis*; L. A. Paquette, Ed.; Wiley: New York, 1995; p 2908.
[524] Touster, O. *Org. React.* **1953**, *7*, 327.
[525] Metzger, H. In *Methoden der organischen Chemie (Houben-Weyl)*; Georg Thieme Verlag: Stuttgart, 1968; Vol. 10/4, p 17.
[526] Unterhalt, B. In *Methoden der organischen Chemie (Houben-Weyl)*; Georg Thieme Verlag: Stuttgart, New York 1990; Vol. E14b, p 295.
[527] Sénéchal-Tocquer, M.-C.; Sénéchal, D.; Le Bihan, J.-Y.; Gentric, D.; Caro, B.; Gruselle, M.; Jaouen, G. *J. Organomet. Chem.* **1992**, *433*, 261.
[528] Metzger, H. In *Methoden der organischen Chemie (Houben-Weyl)*; Georg Thieme Verlag: Stuttgart, 1968; Vol. 10/4, p 236.
[529] Hemmer, R.; Lürken, W. In *Methoden der organischen Chemie (Houben-Weyl)*; Georg Thieme Verlag: Stuttgart, New York, 1992; Vol. 16d/2, p 878.
[530] Bewad, I. *J. Prakt. Chem.* **1901**, *63*, 94.
[531] Berti, C. *Synthesis* **1983**, 793.
[532] Kato, K.; Mukaiyama, T. *Chem. Lett.* **1990**, 1917.
[533] Kato, K.; Mukaiyama, T. *Bull. Chem. Soc. Jpn.* **1991**, *64*, 2948.
[534] Kato, K.; Mukaiyama, T. *Chem. Lett.* **1990**, 1137.
[535] Boschan, R.; Merrow, R. T.; Van Dolah, R. W. *Chem. Rev.* **1955**, *55*, 485.
[536] Moureu, C. *C. R. Hebd. Scéances Acad. Sci.* **1901**, *132*, 837.
[537] Hepworth, H. *J. Chem. Soc.* **1921**, 251.
[538] Wislicenus, W.; Waldmüller, M. *Chem. Ber.* **1908**, *41*, 3334.
[539] Kornblum, N. *Org. React.* **1962**, *12*, 101.
[540] Feuer, H.; Blecker, L. R.; Jans, R. W., Jr.; Frost, J. W. *J. Heterocycl. Chem.* **1979**, *16*, 481 and earlier papers in this series.
[541] Elfehail, F. E.; Zajac, W. W., Jr. *J. Org. Chem.* **1981**, *46*, 5151.

[542] Hassner, A.; Larkin, J. M.; Dowd, J. E. *J. Org. Chem.* **1968**, *33*, 1733.
[543] Cushman, M.; Mathew, J. *Synthesis* **1982**, 397.
[544] Kobayashi, Y. *Bull. Chem. Soc. Jpn.* **1973**, *46*, 3462.
[545] Schaub, R. E.; Fulmor, W.; Weiss, M. J. *Tetrahedron* **1964**, *20*, 373.
[546] Chadwick, D. J.; Cottrell, W. R. T.; Meakins, G. D. *J. Chem. Soc., Perkin Trans. 1* **1972**, 655.
[547] Suginome, H.; Kurokawa, Y. *Bull. Chem. Soc. Jpn.* **1989**, *62*, 1343.
[548] Curran, T. T.; Flynn, G. A.; Rudisill, D. E.; Weintraub, P. M. *Tetrahedron Lett.* **1995**, *36*, 4761.
[549] Feuer, H.; Vincent, B. F., Jr. *J. Org. Chem.* **1964**, *29*, 939.
[550] Feuer, H.; Savides, C. *J. Am. Chem. Soc.* **1959**, *81*, 5826.
[551] Feuer, H.; Spinicelli, L. F. *J. Org. Chem.* **1976**, *41*, 2981.
[552] Fetell, A. I.; Feuer, H. *J. Org. Chem.* **1978**, *43*, 497.
[553] Griswold, A. A.; Starcher, P. S. *J. Org. Chem.* **1966**, *31*, 357.
[554] Sheehan, D.; Vellturo, A. F. South African Patent 67 05,789 (1968); *Chem. Abstr.* **1969**, *70*, 57261k.
[555] Özbal, H.; Zajac, W. W., Jr. *J. Org. Chem.* **1981**, *46*, 3082.
[556] Dampawan, P.; Zajac, W. W., Jr. *Synthesis* **1983**, 545.
[557] Rank, W. *Tetrahedron Lett.* **1991**, *32*, 5353.
[558] Evans, P. A.; Longmire, J. M. *Tetrahedron Lett.* **1994**, *35*, 8345.
[559] Zuman, P.; Shah, B. *Chem. Rev.* **1994**, *94*, 1621.
[560] Vogt, P. F.; Miller, M. J. *Tetrahedron* **1998**, *54*, 1317.
[561] Joghyuk, L.; Li, C.; Ann, H. W.; George, B. R. *Chem. Rev.* **2002**, *102*, 1019.
[562] Wieland, H.; Roseeu, A. *Chem. Ber.* **1912**, *45*, 494.
[563] Wieland, H.; Offenbächer, M. *Chem. Ber.* **1914**, *47*, 2111.
[564] Wieland, H.; Reverdy, A. *Chem. Ber.* **1915**, *48*, 1117.
[565] Wieland, H.; Reverdy, A. *Chem. Ber.* **1915**, *48*, 1112.
[566] Wieland, H.; Roth, K. *Chem. Ber.* **1920**, *53*, 210.
[567] Wieland, H.; Kögl, F. *Chem. Ber.* **1922**, *55*, 1798.
[568] Maruyama, K. *Bull. Chem. Soc. Jpn.* **1964**, *37*, 1013.
[569] Gilman, H.; McCracken, R. *J. Am. Chem. Soc.* **1927**, *49*, 1052.
[570] Kopp, F.; Sapountzis, I.; Knochel, P. *Synlett* **2003**, 885.
[571] Belousova, S. P.; Vasil'ev, N. V.; Kolomiets, A. F.; Nikolaev, K. M.; Sokol'skii, G. A.; Fokin, A. V. *Izv. Akad. Nauk, Ser. Khim.* **1984**, 1198; *Engl. Transl.* p 1103.
[572] Vasil'ev, N. V.; Kolomiets, A. F.; Sokol'skii, G. A. *Zh. Org. Khim.* **1981**, *17*, 1321; *Engl. Transl.* p 1171.
[573] Momiyama, N.; Yamamoto, H. *Org. Lett.* **2002**, *4*, 3579.
[574] Guo, H.-M.; Cheng, L.; Cun, L.-F.; Gong, L.-Z.; Mi, A.-Q.; Jiang, Y.-Z. *Chem. Commun.* **2006**, 429.
[575] Momiyama, N.; Yamamoto, H. *J. Am. Chem. Soc.* **2003**, *125*, 6038.
[576] Sasaki, T.; Ishibashi, Y.; Ohno. M. *Chem. Lett.* **1983**, 863.
[577] Sasaki, T.; Mori, K.; Ohno, M. *Synthesis* **1985**, 279.
[578] Sasaki, T.; Mori, K.; Ohno, M. *Synthesis* **1985**, 280.
[579] Lewis, J. W.; Myers, P. L.; Ormerod, J. A. *J. Chem. Soc., Perkin Trans. 1* **1972**, 2521.
[580] Abramovitch, R. A.; Challand, S. R.; Yamada, Y. *J. Org. Chem.* **1975**, *40*, 1541.
[581] Kresze, G.; Ascherl, B.; Braun, H. *Org. Prep. Proc. Int.* **1987**, *19*, 329.
[582] Schenk, C.; Beekes, M. L.; van der Drift, J. A. M.; de Boer, T. J. *Recl. Trav. Chim. Pays-Bas* **1980**, *99*, 278.
[583] Schlenk, C.; Beekes, M. L.; de Boer, T. J. *Recl. Trav. Chim. Pays-Bas* **1980**, *99*, 246.
[584] Lub, J.; Beekes, M. L.; de Boer, T. J. *Recl. Trav. Chim. Pays-Bas* **1986**, *105*, 22.
[585] Filip, S. V.; Seewald, N. *Synthesis* **2005**, 3565.
[586] Oppolzer, W.; Tamura, O. *Tetrahedron Lett.* **1990**. *31*, 991.
[587] Oppolzer, W.; Tamura, O.; Deerberg, J. *Helv. Chim. Acta* **1992**, *75*, 1965.
[588] Oppolzer, W.; Merifield, E. *Helv. Chim. Acta* **1993**, *76*, 957.
[589] Oppolzer, W.; Cintas-Moreno, P.; Tamura, O. *Helv. Chim. Acta* **1993**, *76*, 187.
[590] Oppolzer, W.; Bochet, C. G.; Merifield, E. *Tetrahedron Lett.* **1994**, *35*, 7015.

[591] Ludwig, S. N.; Unkefer, C. J. *J. Labeled Comp. Radiopharm.* **1996**, *38*, 239.
[592] Annunziata, R.; Benaglia, M.; Cinquini, M.; Cozzi, F.; Scolaro, A. *Gazz. Chim. Ital.* **1995**, *125*, 65.
[593] Otaka, A.; Mitsuyama, E.; Kinoshita, T.; Tamamura, H.; Fujii, N. *J. Org. Chem.* **2000**, *65*, 4888.
[594] Davison, E. C.; Fox, M. E.; Holmes, A. B.; Roughley, S. D.; Smith, C. J.; Williams, G. M.; Davies, J. E.; Raithby, P. R.; Adams, J. P.; Forbes, I. T.; Press, N. J.; Thompson, M. J. *J. Chem. Soc., Perkin Trans. 1* **2002**, 1494.
[595] Oppolzer, W.; Tamura, O.; Sundarababu, G.; Signer, M. *J. Am. Chem. Soc.* **1992**, *114*, 5900.
[596] Felber, H.; Kresze, G.; Braun, H.; Vasella, A. *Tetrahedron Lett.* **1984**, *25*, 5381.
[597] Bartoli, G. *Acc. Chem. Res.* **1984**, *17*, 109.
[598] Bartoli, G.; Marcantoni, E.; Petrini, M. *J. Chem. Soc., Chem. Commun.* **1993**, 1373.
[599] Barboni, L.; Bartoli, G.; Marcantoni, E.; Petrini, M.; Dalpozzo, R. *J. Chem. Soc., Perkin Trans. 1* **1990**, 2133.
[600] Bartoli, G.; Palmieri, G.; Petrini, M.; Bosco, M.; Dalpozzo, R. *Gazz. Chim. Ital.* **1990**, *120*, 247.
[601] Bartoli, G.; Marcantoni, E.; Petrini, M.; Dalpozzo, R. *J. Org. Chem.* **1990**, *55*, 4456.
[602] Bartoli, G.; Marcantoni, E.; Petrini, M. *J. Chem. Soc., Chem. Commun.* **1991**, 793.
[603] Bartoli, G.; Marcantoni, E.; Petrini, M. *J. Org. Chem.* **1992**, *57*, 5834.
[604] Yost, Y.; Gutmann, H. R.; Muscoplat, C. C. *J. Chem. Soc. (C)* **1971**, 2119.
[605] Gilman, H.; McCracken, R. *J. Am. Chem. Soc.* **1929**, *51*, 821.
[606] Sapountzis, I.; Knochel, P. *J. Am. Chem. Soc.* **2002**, *124*, 9390.
[607] Dalpozzo, R.; Bartoli, G. *Curr. Org. Chem.* **2005**, *9*, 163.
[608] Dobbs, A. *J. Org. Chem.* **2001**, *66*, 638.
[609] Sitzmann, M. E.; Kaplan, L. A.; Angres, I. *J. Org. Chem.* **1977**, *42*, 563.
[610] Rathore, R.; Lin, Z.; Kochi, J. K. *Tetrahedron Lett.* **1993**, *34*, 1859.
[611] Briere, R.; Rassat, A. *Bull. Soc. Chim. Fr.* **1965**, 378.
[612] Chapelet-Letourneux, G.; Lemaire, H.; Rassat, A. *Bull. Soc. Chim. Fr.* **1965**, 444.
[613] Lemaire, H.; Marechal, Y.; Ramasseul, R.; Rassat, A. *Bull. Soc. Chim. Fr.* **1965**, 372.
[614] Hoffmann, A. K.; Feldman, A. M.; Gelblum, E. *J. Am. Chem. Soc.* **1964**, *86*, 646.
[615] Enders, E. In *Methoden der organischen Chemie (Houben-Weyl)*; Georg Thieme Verlag: Stuttgart, 1965; Vol. 10/3, p 490.
[616] Dumic, M; Kuruncev, D.; Kovacevic, K.; Polak, L.; Kolbah, D. In *Methoden der organischen Chemie (Houben-Weyl)*; Georg Thieme Verlag: Stuttgart, New York, 1990; Vol. E14b/1, p 450.
[617] Phillips, R. R. *Org. React.* **1959**, *10*, 143.
[618] Enders, E. In *Methoden der organischen Chemie (Houben-Weyl)*; Georg Thieme Verlag: Stuttgart, 1965; Vol. 10/3, p 522.
[619] Schröter, R. In *Methoden der organischen Chemie (Houben-Weyl)*; Georg Thieme Verlag: Stuttgart, 1957; Vol. 11/1 p 531.
[620] Regitz, M. *Angew. Chem., Int. Ed. Engl.* **1967**, *6*, 733.
[621] Regitz, M. *Synthesis* **1972**, 351.
[622] Regitz, M.; Maas, G. *Diazo Compounds*; Academic Press: Orlando, 1986.
[623] Böhshar, M.; Fink, J.; Heydt, H.; Wagner, O.; Regitz, M. In *Methoden der organischen Chemie (Houben-Weyl)*; Georg Thieme Verlag: Stuttgart, 1990; Vol. E14b/2, p 961.
[624] Ye, T.; McKervey, M. A. *Chem. Rev.* **1994**, *94*, 1091.
[625] Doering, W. von E.; DePuy, C. H. *J. Am. Chem. Soc.* **1953**, *75*, 5955.
[626] Ando, W.; Tanikawa, H.; Sekiguchi, A. *Tetrahedron Lett.* **1983**, *39*, 4245.
[627] Hazen, G. G.; Weinstock, L. M.; Connell, R.; Bollinger, F. W. *Synth. Commun.* **1981**, *11*, 947.
[628] Hazen, G. G.; Bollinger, F. W.; Roberts, F. E.; Russ, W. K.; Seman, J. J.; Staskiewicz, S. *Org. Synth.* **1996**, *75*, 144.
[629] Tuma, L. D. *Thermochim. Acta* **1994**, *243*, 161.
[630] Gisin, B.; Brenner, M. *Helv. Chim. Acta* **1970**, *53*, 1030.
[631] Davis, F. A.; Yang, B.; Deng, J. *J. Org. Chem.* **2003**, *68*, 5147.
[632] Moody, C. J.; Morfitt, C. N.; Slawin, A. M. Z. *Tetrahedron: Asymmetry* **2001**, *12*. 1657.
[633] Brown, H. C.; Kramer, G. W.; Levy, A. B.; Midland, M. M. *Organic Synthesis via Boranes*; Wiley: New York, 1975.

[634] Carboni, B.; Vaultier, M. *Bull. Soc. Chim. Fr.* **1995**, *132*, 1003.
[635] Brown, H. C.; Kim, K.-W.; Srebnik, M.; Singaram, B. *Tetrahedron* **1987**, *43*, 4071.
[636] Mueller, R. H. *Tetrahedron Lett.* **1976**, 2925.
[637] Brown, H. C.; Heydkamp, W. R.; Breuer, E.; Murphy, W. S. *J. Am. Chem. Soc.* **1964**, *86*, 3565.
[638] Rathke, M. W.; Millard, A. A. *Org. Synth. Coll. Vol. 6*, **1988**, 943.
[639] Mikhailov, B. M.; Shagova, E. A.; Etinger, M. Yu. *J. Organomet. Chem.* **1981**, *220*, 1.
[640] Genêt, J.-P.; Hajicek, J.; Bischoff, L.; Greck, C. *Tetrahedron Lett.* **1992**, *33*, 2677.
[641] Jigajinni, V. B.; Pelter, A.; Smith, K. *Tetrahedron Lett.* **1978**, 181.
[642] Brown, H. C.; Midland, M. M.; Levy, A. B. *Tetrahedron* **1987**, *43*, 4079.
[643] Kabalka, G. W.; Goudgaon, N. M.; Liang, Y. *Synth. Commun.* **1988**, *18*, 1363.
[644] Carboni, B; Vaultier, M.; Courgeon, T.; Carrié, R. *Bull. Soc. Chim. Fr.* **1989**, 844.
[645] Brown, H. C.; Salunkhe, A. M.; Singaram, B. *J. Org. Chem.* **1991**, *56*, 1170.
[646] Fernandez, E.; Hooper, M. W.; Knight, F. I.; Brown, J. *J. Chem. Soc., Chem. Commun.* **1997**, 173.
[647] Brown, H. C.; Kim, K.-W.; Cole, T. E.; Singaram, B. *J. Am. Chem. Soc.* **1986**, *108*, 6761.
[648] Matheson, D. S. *Acc. Chem. Res.* **1988**, *21*, 294.
[649] Kabalka, G. W.; Ferrell, J. W. *Synth. Commun.* **1979**, *9*, 443.
[650] O'Brien, C. *Chem. Rev.* **1964**, *64*, 81.
[651] Mayer, D. In *Methoden der organischen Chemie (Houben-Weyl)*; Georg Thieme Verlag: Stuttgart, 1977; Vol. 7/2c, p 2272 and references cited therein.
[652] Fisher, L. E.; Muchowski, J. M. *Org. Prep. Proced. Int.* **1990**, *22*, 399.
[653] Maruoka, K.; Yamamoto, H. In *Comprehensive Organic Synthesis*; Trost, B. M.; Fleming, I., Eds., 1991, Vol. 6, p 763.
[654] Cram, D. J.; Hatch, M. J. *J. Am. Chem. Soc.* **1953**, *75*, 33.
[655] Egushi, S.; Ishii, Y. *Bull. Chem. Soc. Jpn.* **1963**, *36*, 1434.
[656] Alvernhe, G.; Laurent, A. *J. Chem. Res. (S)* **1978**, 28; *J. Chem. Res. (M)* **1978**. 501.
[657] Ricart, G.; Couturier, D. *C. R. Hebd. Scéances Acad. Sci.* **1977**, *284*, 191 and references cited therein.
[658] Hoch, J. *C. R. Hebd. Scéances Acad. Sci.* **1934**, *198*, 1865.
[659] Campbell, K. N.; Campbell, B. K.; McKenna, J. F.; Chaput, E. P. *J. Org. Chem.* **1943**, *8*, 103 and references cited therein.
[660] Rewicki, D.; Tuchscherer, C. *Angew. Chem., Int. Ed. Engl.* **1972**, *11*, 44.
[661] Spencer, H. *Chem. Britain* **1981**, 17.
[662] Müller, E. In *Methoden der organischen Chemie (Houben-Weyl)*; Georg Thieme Verlag: Stuttgart, 1967; Vol. 10/4, p 827.
[663] Andree, R.; Kluth, J. F.; Hanefeld, W. In *Methoden der organischen Chemie (Houben-Weyl)*; Georg Thieme Verlag: Stuttgart, New York, 1990; Vol. 16a/2, p 856.
[664] Goehring, R, R. In *Encyclopedia of Reagents for Organic Synthesis*; L. A. Paquette, Ed.; Wiley: New York, 1995; p 1052.
[665] Sisler, H. H.; Omietanski, G. *Inorg. Synth.* **1957**, *5*, 91.
[666] Coleman, G. H.; Johnson, H. L. *Inorg. Synth.* **1939**, *1*, 59.
[667] Schmitz, E.; Schramm, S.; Flamme, W.; Bricker, U. *Z. Anorg. Allgem. Chem.* **1973**, *396*, 178.
[668] Allan, R. D.; Duke, R. K.; Hambley, T. W.; Johnston, G. A. R.; Mewett, K. N.; Quickert, N.; Tran, W. *Aust. J. Chem.* **1997**, *49*, 785.
[669] Coleman, G. H.; Goheen, G. E. *Inorg. Synth.* **1939**, *1*, 62.
[670] Noyes, W. A. *Inorg. Synth.* **1939**, *1*, 65.
[671] Bartsch, R. A.; Cho, B. R.. *J. Am. Chem. Soc.* **1979**, *101*, 3587.
[672] Cho, B. R,; Namgoong, S. K.; Kim, T. R. *J. Chem. Soc., Perkin Trans. 2* **1987**, 853.
[673] Noack, M.; Göttlich, R. *Eur. J. Org. Chem.* **2002**, 3171.
[674] Bachand, C.; Driguez, H.; Paton, J. M.; Touchard, D.; Lessard, J. *J. Org. Chem.* **1974**, *39*, 3136.
[675] Andree, R.; Kluth, J. In *Methoden der organischen Chemie (Houben-Weyl)*; Georg Thieme Verlag: Stuttgart, 1990; Vol. 16a/1, p 214.
[676] Theilacker, W.; Ebke, K. *Angew. Chem.* **1956**, *68*, 303.
[677] Palazzo, G.; Rogers, E. F.; Marini-Bettòlo, G. B. *Gazz. Chim. Ital.* **1954**, *84*, 915.
[678] Choong, I. C.; Ellman, J. A. *J. Org. Chem.* **1999**, *64*, 6528.

[679] Foot, O. F.; Knight, D. W. *Chem. Commun.* **2000**, 975.
[680] Chimiak, A.; Kolasa, T. *Bull. Acad. Pol. Sci. Chim.* **1974**, *22*, 195; *Chem. Abstr.* **1974**, *80*, 132725b.
[681] Kokko, B. J. In *Encyclopedia of Reagents for Organic Synthesis*; L. A. Paquette, Ed.; Wiley: New York, 1995; p 3511.
[682] Hjeds, H. *Acta Chem. Scand.* **1965**, *19*, 1764.
[683] Traube, W.; Ohlendorf, H.; Zander, H. *Chem. Ber.* **1920**, *53*, 1477.
[684] Knox, G. R.; Pauson, P. L.; Willison, D.; Solcanova, E.; Toma, S. *Organometallics* **1990**, *9*, 301.
[685] Rees, D. C.; Hamilton, N. M. In *Encyclopedia of Reagents for Organic Synthesis*; L. A. Paquette, Ed.; Wiley: New York, 1995; p 332.
[686] Bumgardner, C. L.; Lilly, R. L. *Chem. Ind. (London)* **1962**, 559.
[687] Bellettini, J. R.; Olsen, E. R.; Teng, M.; Miller, M. J. In *Encyclopedia of Reagents for Organic Synthesis*; L. A. Paquette, Ed.; Wiley: New York, 1995; p 2189.
[688] Sheradsky, T. *J. Heterocycl. Chem.* **1967**, *4*, 413.
[689] Legault, C.; Charette, A. *J. Org. Chem.* **2003**, *68*, 7119.
[690] Marmer, W. N.; Maerker, G. *J. Org. Chem.* **1972**, *37*, 3520.
[691] Carpino, L. A.; Giza, C. A.; Carpino, B. A. *J. Am. Chem. Soc.* **1959**, *81*, 955.
[692] Boche, G. In *Encyclopedia of Reagents for Organic Synthesis*; L. A. Paquette, Ed.; Wiley: New York, 1995; p 3270.
[693] Carpino, L. A. *J. Am. Chem. Soc.* **1960**, *82*, 3133.
[694] Psiorz, M.; Zinner, G. *Synthesis* **1984**, 217.
[695] Biloski, A. J.; Ganem, B. *Synthesis* **1983**, 537.
[696] Erdik, E. In *Encyclopedia of Reagents for Organic Synthesis*; L. A. Paquette, Ed.; Wiley: New York, 1995; p 2764.
[697] Wallace, R. G. *Aldrichimica Acta* **1980**, *13*, 3.
[698] Wallace, R. G. *Org. Prep. Proced. Int.* **1982**, *14*, 265.
[699] King, F. D.; Walton, D. R. M. *Synthesis* **1975**, 788.
[700] Tamura, Y.; Ikeda, M. *Yuki Gosei Kagaku Kyokai Shi* **1974**, *32*, 136.
[701] Boche, G. In *Encyclopedia of Reagents for Organic Synthesis*; L. A. Paquette, Ed.; Wiley: New York, 1995; p 3277.
[702] Krause, J. G. *Synthesis* **1972**, 140.
[703] Johnson, C. R.; Kirchhoff, R. A.; Corkins, H. G. *J. Org. Chem.* **1974**, *39*, 3458.
[704] Scopes, D. I. C.; Kluge, A. F.; Edwards, J. A. *J. Org. Chem.* **1977**, *42*, 376.
[705] Tamura, Y.; Minamikawa, J.; Sumoto, K.; Fujii. S.; Ikeda, M. *J. Org. Chem.* **1973**, *38*, 1239.
[706] Koziara, A.; Novalinska, M.; Zwierzak, A. *Synth. Commun.* **1993**, *23*, 2127.
[707] Fioravanti, S.; Morreale, A.; Pellacani, L.; Tardella, P. A. *Tetrahedron Lett.* **2003**, *44*, 3031.
[708] Chapman, T. M.; Freedman, E. A. *Synthesis* **1971**, 591.
[709] Knight, F. I.; Brown, J. M.; Lazzari, D.; Ricci, A.; Blacker, A. J. *Tetrahedron* **1997**, *53*, 11411.
[710] Boche, G. In *Encyclopedia of Reagents for Organic Synthesis*; L. A. Paquette, Ed.; Wiley: New York, 1995; p 2240.
[711] Klötzer, W.; Baldinger, H.; Karpitschka, E. M.; Knoflach, J. *Synthesis* **1982**, 592.
[712] Harger, M. J. P. *J. Chem. Soc., Chem. Commun.* **1979**, 768.
[713] Harger, M. J. P. *J. Chem. Soc., Perkin Trans. 1* **1981**, 3284.
[714] Boche, G. In *Encyclopedia of Reagents for Organic Synthesis*; L. A. Paquette, Ed.; Wiley: New York, 1995; p 2066.
[715] Boche, G.; Sommerlade, R. H. *Tetrahedron* **1986**, *42*, 2703.
[716] Yaquanc, J. J.; Masse, G.; Sturtz, G. *Synthesis* **1985**, 807.
[717] Shustov, G. V.; Kadorkina, G. K.; Varlamov, S. V.; Kachanov, A. V.; Kostyanovsky, R. G.; Rauk, A. *J. Am. Chem. Soc.* **1992**, *114*, 1616.
[718] Page, P. C. B.; Murrell, V. L.; Limousin, C.; Laffan, D. D. P.; Bethell, D.; Slawin, A. M. Z.; Smith, T. A. D. *J. Org. Chem.* **2000**, *65*, 4204.
[719] Vidal, J.; Hannachi, J.-C.; Hourdin, G.; Mulatier, J.-C.; Collet, A. *Tetrahedron Lett.* **1998**, *39*, 8845.
[720] Vidal, J.; Damestoy, S.; Collet, A. *Tetrahedron Lett.* **1995**, *36*, 1439.

[721] Watanabe, H.; Hashizume, Y.; Uneyama, K. *Tetrahedron Lett.* **1992**, *33*, 4333.
[722] Evans, D. A.; Barnes, D. M. In *Encyclopedia of Reagents for Organic Synthesis*; L. A. Paquette, Ed.; Wiley: New York, 1995; p 4958.
[723] Hellmann, H.; Teichmann, K. *Chem. Ber.* **1956**, *89*, 1134.
[724] Erdik, E. In *Encyclopedia of Reagents for Organic Synthesis*; L. A. Paquette, Ed.; Wiley: New York, 1995; p 41.
[725] Oxley, P.; Short, W. F. *J. Chem. Soc.* **1948**, 1514.
[726] Tsutsui, H.; Hayashi, Y.; Narasaka, K. *Chem. Lett.* **1997**, 317.
[727] Erdik, E. In *Encyclopedia of Reagents for Organic Synthesis*; L. A. Paquette, Ed.; Wiley: New York, 1995; p 4826.
[728] Barbero, M.; Crisma, M.; Degani, I.; Fochi, R.; Perracino, P. *Synthesis* **1998**, 1171.
[729] Cohen, S. G.; Nicholson, J. *J. Org. Chem.* **1965**, *30*, 1162.
[730] Bock, H.; Baltin, E.; Kroner, J. *Chem. Ber.* **1966**, *99*, 3337.
[731] Knight, G. T.; Loadman, M. J. R.; Saville, B.; Wildgoose, J. *J. Chem. Soc., Chem. Commun.* **1974**, 193.
[732] Stoner, E. J. In *Encyclopedia of Reagents for Organic Synthesis*; L. A. Paquette, Ed.; Wiley: New York, 1995; p 1790.
[733] Mackay, D.; Pilger, C. W.; Wong, L. L. *J. Org. Chem.* **1973**, *38*, 2043.
[734] Little, R. D.; Bregant, T. M. In *Encyclopedia of Reagents for Organic Synthesis*; L. A. Paquette, Ed.; Wiley: New York, 1995; p 572.
[735] Rutjes, F. P. J. T.; Paz, M. M.; Hiemstra, H.; Speckamp, N. *Tetrahedron Lett.* **1991**, *32*, 6629.
[736] Klinge, M.; Vederas, J. C. In *Encyclopedia of Reagents for Organic Synthesis*; L. A. Paquette, Ed.; Wiley: New York, 1995; p 1586.
[737] Leblanc, Y. In *Encyclopedia of Reagents for Organic Synthesis*; L. A. Paquette, Ed.; Wiley: New York, 1995; p 1532.
[738] Harris, J. M.; Bolessa, E. A.; Vederas, J. C. *J. Chem. Soc., Perkin Trans. 1* **1995**, 1951.
[739] Vorbrüggen, H.; Krolikiewicz, K. *Synthesis* **1979**, 35.
[740] Pearson, W. J.; Ramamoorthy, P. S. In *Encyclopedia of Reagents for Organic Synthesis*; L. A. Paquette, Ed.; Wiley: New York, 1995; p 2393.
[741] Hassner, A. In *Encyclopedia of Reagents for Organic Synthesis*; L. A. Paquette, Ed.; Wiley: New York, 1995; p 219.
[742] Thomas, A. V. In *Encyclopedia of Reagents for Organic Synthesis*; L. A. Paquette, Ed.; Wiley: New York, 1995; p 2242.
[743] Nikolaev, V. A. In *Encyclopedia of Reagents for Organic Synthesis*; L. A. Paquette, Ed.; Wiley: New York, 1995; p 3306.
[744] Cavender, C. J.; Shiner, V. J., Jr. *J. Org. Chem.* **1972**, *37*, 3567.
[745] Fritschi, S.; Vasella, A. *Helv. Chim. Acta* **1991**, *74*, 2024.
[746] Heydt, H.; Regitz, M. In *Encyclopedia of Reagents for Organic Synthesis*; L. A. Paquette, Ed.; Wiley: New York, 1995; p 4943.
[747] Regitz, M.; Hocker, J.; Liedhegener, A. *Org. Synth., Coll. Vol. 5*, **1973**, 179.
[748] Leffler, J. E.; Tsuno, Y. *J. Org. Chem.* **1963**, *28*, 902.
[749] Mander, L. N. In *Encyclopedia of Reagents for Organic Synthesis*; L. A. Paquette, Ed.; Wiley: New York, 1995; p 5174.
[750] Harmon, R. E.; Wellman, G.; Gupta, S. K. *J. Org. Chem.* **1973**, *38*, 11.
[751] Davies, H. M. L.; Cantrell, W. R., Jr.; Romines, K. R.; Baum, J. S. *Org. Synth.* **1992**, *70*, 93.
[752] Dürr, H.; Hauck, G.; Brück, W.; Kober, H. *Z. Naturforsch.* **1981**, 86*b*, 1149.
[753] Roush, W. R.; Feitler, D.; Rebek, J. *Tetrahedron Lett.* **1974**, 1391.
[754] Weinreb, S. M.; Heintzelman, G. R. In *Encyclopedia of Reagents for Organic Synthesis*; L. A. Paquette, Ed.; Wiley: New York, 1995; p 562.
[755] Du Bois, J.; Tomooka, C. S.; Hong, J.; Carreira, E. M.; Day, M. W. *Angew. Chem., Int. Ed. Engl.* **1997**, *36*, 1645.
[756] Jepsen, A. S.; Roberson, M.; Hazell, R. G.; Jørgensen, K. A. *Chem. Commun.* **1998**, 1599.
[757] Theodoridis, G. *Tetrahedron* **2000**, *56*, 2339 and references cited therein.
[758] Kocienski, P. J. *Protecting Groups*; Thieme: Stuttgart, 2005.

759 Marshalkin, M. F.; Yakhontov, L. N. *Uspekhi Khim.* **1986**, *55*, 1785; *Russian Chem. Rev.* **1986**, *55*, 1016.
760 Ding, H.; Fristad, G. K. *Org. Lett.* **2004**, *6*, 637.
761 Fernández, R.; Ferrete, A.; Llera, J. M.; Magriz, A.; Martín-Zamora, E.; Díez, E.; Lassaletta, J. M. *Chem. Eur. J.* **2004**, *10*, 737.
762 Robinson, F. P.; Brown, R. K. *Can. J. Chem.* **1961**, *39*, 1171.
763 Alexakis, A.; Lensen, N.; Mangeney, P. *Synlett* **1991**, 625.
764 Anakabe, E.; Vicario, J. L.; Badía, D.; Carrillo, L.; Yoldi, V. *Eur. J. Org. Chem.* **2001**, 4343.
765 Vicario, J. L.; Badía, D.; Domínguez, E.; Crespo, A.; Carillo, L.; Anakabe, E. *Tetrahedron Lett.* **1999**, *40*, 7123.
766 Gilchrist, T. L.; Hughes, D.; Wasson, R. *Tetrahedron Lett.* **1987**, *28*, 1573.
767 LiBassi, G.; Ventura, P.; Monguzzi, R.; Pifferi, G. *Gazz. Chim. Ital.* **1977**, *107*, 253.
768 Milcent, R.; Guevrekian-Soghomoniantz, M.; Barbier, G. *J. Heterocycl. Chem.* **1986**, *23*, 1845.
769 Banfi, L.; Cascio, G.; Guanti, G.; Manghisi, E.; Narisano, E.; Riva, R. *Tetrahedron* **1994**, *50*, 11967.
770 Maffre, D.; Dumy, P.; Vidal, J.-P.; Escale, R.; Girard, J.-P. *J. Chem. Res. (S)* **1994**, 30.
771 Hassner, A. In *Methoden der organischen Chemie (Houben-Weyl)*; Georg Thieme Verlag: Stuttgart, 1990; Vol. 16a/2, p 1275.
772 Hemmer, R.; Lürken, W. In *Methoden der organischen Chemie (Houben-Weyl)*; Georg Thieme Verlag: Stuttgart, New York, 1992; Vol. 16d/2, p 956.
773 Guéguen, C.; About-Jaudet, E.; Collignon, N.; Savignac, P. *Synth. Commun.* **1996**, *26*, 4131.
774 Bänteli, R.; Brun, I.; Hall, P.; Metternich, R. *Tetrahedron Lett.* **1999**, *40*, 2109.
775 Maiti, S. N.; Singh, M. P.; Micetich, R. G. *Tetrahedron Lett.* **1986**, *27*, 1423.
776 Liang, B.; Carroll, P. J.; Joullié, M. M. *Org. Lett.* **2000**, *2*, 4157.
777 Wasserman, H. H.; Hlasta, D. J.; Tremper, A. W.; Wu, J. S. *J. Org. Chem.* **1981**, *46*, 2999.
778 Rolla, F. *J. Org. Chem.* **1982**, *47*, 4327.
779 Spagnolo, P.; Zanirato, P.; Gronowitz, S. *J. Org. Chem.* **1982**, *47*, 3177.
780 Nishida, A.; Shibasaki, M.; Ikegami, S. *Tetrahedron Lett.* **1984**, *25*, 765.
781 Vaultier, M.; Knouzi, N.; Carrié, R. *Tetrahedron Lett.* **1983**, *24*, 763.
782 Shaw, A. N.; Dolle, R. E.; Kruse, L. I. *Tetrahedron Lett.* **1990**, *31*, 5081.
783 Manis, P. A.; Rathke, M. W. *J. Org. Chem.* **1980**, *45*, 4952.
784 Es-Sayed, M.; Gratkowski, C.; Krass, N.; Meyers, A. I.; de Meijere, A. *Synlett* **1992**, 962.
785 Patonay, T.; Hoffman, R. V. *J. Org. Chem.* **1995**, *60*, 2368.
786 Beak, P.; Kokko, B. J. *J. Org. Chem.* **1982**, *47*, 2823.
787 Knox, G. R. *Proc. Chem. Soc.* **1959**, 56.
788 Pochinok, V. Ya.; Avramenko, L. F.; Grigorenko, T. F.; Pochinok, A. V.; Sidorenko, I, A.; Bovchaljuk, L. N. *Ukr. Khim. Zh.* **1979**, *45*, 975; *Chem. Abstr.* **1980**, *92*, 76464z.
789 Pochinok, V. Ya.; Mikhailyuchenko, N. K. *Ukr. Khim. Zh.* **1955**, *21*, 625; *Chem. Abstr.* **1955**, *50*, 14599i.
790 Skripnik, L. I.; Pochinok, V. Ya. *Khim. Geterotsikl. Soedin.* **1967**, *3*, 292; *Engl. Transl.* p 221.
791 Brown, R.; Jones, W. E. *J. Chem. Soc.* **1946**, 781.
792 Shverdina, N. I.; Kocheshkov, K. A. *Zh. Obshch, Khim.* **1938**, *8*, 1825; *Chem. Abstr.* **1939**, *33*, 5804.
793 Silver, M. S.; Shafer, P. R.; Nordlander, J. E.; Rüchardt, C.; Roberts, J. D. *J. Am. Chem. Soc.* **1960**, *82*, 2646.
794 Stollé, R.; Reichert, W. *J. Prakt. Chem.* **1929**, *122*, 344.
795 Carpino, L. A.; Terry, P. H.; Crowley, P. J. *J. Org. Chem.* **1961**, *26*, 4336.
796 Waser, J.; Carreira, E. M. *Angew. Chem., Int. Ed.* **2004**, *43*, 4099.
797 Kaiser, E. M.; Bartling, G. J. *Tetrahedron Lett.* **1969**, 4357.
798 Yeung, D. W. K.; Warkentin, J. *Can. J. Chem.* **1976**, *54*, 1345.
799 Rutjes, F. P. J. T.; Hiemstra, H.; Mooiweer, H. H.; Speckamp, W. N. *Tetrahedron Lett.* **1988**, *29*, 6975.
800 Skripnik, L. I.; Pochinok, V. Ya. *Khim. Geterotsikl. Soedin.* **1968**, *4*, 474; *Engl. Transl.* p 353.
801 Erdik, E.; Ates, S. *Synth. Commun.* **2006**, *36*, 2813.

[802] Erdik, E.; Daskapan, T. *Tetrahedron Lett.* **2002**, *43*, 6237.
[803] Erdik, E.; Ay, M. *Synth. React. Inorg. Met.-Org. Chem.* **1989**, *19*, 663.
[804] Bernheim, M.; Boche, G. *Angew. Chem., Int Ed. Engl.* **1980**, *19*, 1010.
[805] Coleman, G. H.; Forrester, R. A. *J. Am. Chem. Soc.* **1936**, *58*, 27.
[806] Unger, C.; Zimmer, R.; Reissig, H.-U.; Würthwein, E.-U. *Chem. Ber.* **1991**, *124*, 2279.
[807] Kaiser, E. M.; Bartling, G. J.; Foy, T. *Org. Syn.* **1973**, *53*, 1829 [sic]; *Chem. Abstr.* **1974**, *81*, 105213x.
[808] Quirk, R. P.; Cheng, P. L. *Macromolecules* **1986**, *19*, 1291.
[809] Kleinfeller, H.; Bönig, G. *J. Prakt. Chem.* **1931**, *132*, 175.
[810] Robson, E.; Tedder, J. M.; Webster, B. *J. Chem. Soc.* **1963**, 1863.
[811] Weißberger, A.; Fasold, K.; Bach, H. *J. Prakt. Chem.* **1930**, *124*, 29.
[812] Daskapan, T. *Tetrahedron Lett.* **2006**, *47*, 2879.
[813] Neumann W. P.; Wicenec, C. *Chem. Ber.* **1991**, *124*, 2297.
[814] Holt, P. F.; Hughes, B. P. *J. Chem. Soc.* **1954**, 764.
[815] Gilman, H.; Bailie, J. C. *J. Org. Chem.* **1937**, *2*, 84.
[816] Demers, J. P.; Klaubert, D. H. *Tetrahedron Lett.* **1987**, *28*, 4933.
[817] Pochinok, V. Ya. *Zh. Obshch. Khim.* **1946**, *16*, 1306; *Chem. Abstr.* **1947**, *41*, 3066f.
[818] Pochinok, V. Ya.; Zaitseva, S. D.; El'gort, R. G. *Ukr. Khim. Zh.* **1951**, *17*, 509; *Chem. Abstr.* **1954**, *48*, 11392i.
[819] Pochinok, V. Ya.; Zaitseva, S. D. *Ukr. Khim. Zh.* **1960**, *26*, 351; *Chem. Abstr.* **1961**, *55*, 4485i.
[820] Bertho A. *J. Prakt. Chem.* **1927**, *116*, 101.
[821] Wiberg, N.; Joo, W.-C. *J. Organomet. Chem.* **1970**, *22*, 333.
[822] Acton, E. M.; Silverstein, R. M. *J. Org. Chem.* **1959**, *24*, 1487.
[823] Nesmeyanov, A. N.; Perevalova, E. G.; Golovaya, R. V.; Shilovtseva, L. S. *Dokl. Akad. Nauk SSSR* **1955**, *102*, 535.
[824] Makosza, M.; Podraza, R. *Eur. J. Org. Chem.* **2000**, 193.
[825] Canonica, L.; Tedeschi, C. *Gazz. Chim. Ital.* **1951**, *84*, 175.
[826] Brandes, S.; Bella, M.; Kjoersgaard, A.; Jørgensen, K. A. *Angew. Chem., Int. Ed.* **2006**, *45*, 1147.
[827] Priego, J.; Mancheño, O. G.; Cabrera, S.; Carretero, J. C. *J. Org. Chem.* **2002**, *67*, 1346.
[828] Willis, H. B. *Iowa State Coll. J. Sci.* **1943**, *18*, 98; *Chem. Abstr.* **1944**, *38*, 739.
[829] Gilman, H.; Ingham, R. K. *J. Am. Chem. Soc.* **1953**, *75*, 4843.
[830] Gilman, H.; Avakian, S. *J. Am. Chem. Soc.* **1946**, *68*, 1514.
[831] Gilman, H.; Avakian, S. *J. Am. Chem. Soc.* **1946**, *68*, 580.
[832] Gilman, H.; Van Ess, M. W.; Willis, H. B.; Stuckwish, C. G. *J. Am. Chem. Soc.* **1940**, *62*, 2606.
[833] Gilman, H.; Swayampati, D. R. *J. Am. Chem. Soc.* **1957**, *79*, 208.
[834] Gilman, H.; Stuckwisch, C. G. *J. Am. Chem. Soc.* **1943**, *65*, 1461.
[835] Sasaki, S.; Hatsushiba, H.; Yoshifuji, M. *Chem. Commun.* **1998**, 2221.
[836] Hinke, E.; Staude, E. *J. Appl. Polym. Sci.* **1991**, *42*, 2951.
[837] Guiver, M. D.; Robertson, G. P.; Foley, S. *Macromolecules* **1995**, *28*, 7612.
[838] Perchais, J.; Fleury, J.-P. *Tetrahedron* **1974**, *30*, 999.
[839] Woodward. R. B.; Heusler, K.; Gosteli, J.; Naegeli, P.; Oppolzer, W.; Ramage, R.; Ranganathan, S.; Vorbrüggen, H. *J. Am. Chem. Soc.* **1966**, *68*, 852.
[840] Lindel, T.; Hochgürtel, M. *Tetrahedron Lett.* **1998**, *39*, 2541.
[841] Tertov, B. A.; Onishchenko, P. P. *Zh. Obshch. Khim.* **1971**, *41*, 1594; *Engl. Transl.* p 1601.
[842] Seefelder, M.; Eilingsfeld, H. *Angew. Chem., Int. Ed. Engl.* **1963**, *2*, 484.
[843] Firl, J.; Sommer, S. *Tetrahedron Lett.* **1969**, 1133.
[844] Firl, J.; Sommer, S. *Tetrahedron Lett.* **1972**, 4713.
[845] Iwamura, H.; Wells, D. H., Jr.; Mathew, S. P.; Klussmann, M.; Armstrong, A.; Blackmond, D. G. *J. Am. Chem. Soc.* **2004**, *126*, 16312.
[846] Carey, F. A.; Neergaard, J. R. *J. Org. Chem.* **1971**, *36*, 2731.
[847] Vogt, H.; Vanderheiden, S.; Bräse, S. *Chem. Commun.* **2003**, 2448.
[848] Page, P. C. B.; McKenzie, M. J.; Allin, S. M.; Buckle, D. R. *Tetrahedron* **2000**, *56*, 9683.
[848] a. Page, P. C. B.; Allin, S. M,; Collington, E. W.; Carr, R. A. E. *Tetrahedron Lett.* **1994**, *35*, 2427.
[849] Barluenga, J.; Gómez, N.; Palacios, F.; Gotor, V. *Synthesis* **1981**, 563.

[850] Magnus, P.; Ros, M. B.; Hulme, C. *J. Chem. Soc., Chem. Commun.* **1995**, 263.
[851] Benedetti, F.; Forchiassin, M.; Russo, C.; Nitti, P. *Gazz. Chim. Ital.* **1988**, *118*, 695.
[852] Shvedov, V. I.; Altukhova, L. B.; Grinev, A. N. *Zh. Org. Khim.* **1965**, *1*, 879; *Engl. Transl.* p 882.
[853] Forchiassin, M.; Risaliti, A.; Russo, C. *Tetrahedron* **1981**, *37*, 2921.
[854] Bigotto, A.; Forchiassin, M.; Risaliti, A.; Russo, C. *Tetrahedron Lett.* **1979**, 4761.
[855] Marchetti, L.; Tosi, G. *Tetrahedron Lett.* **1971**, 3071.
[856] Benedetti, F.; Forchiassin, M.; Pispisa, G.; Nitti, P.; Pitacco, G.; Russo, C.; Valentin, E. *Gazz. Chim. Ital.* **1990**, *120*, 327.
[857] Forchiassin, M.; Pitacco, G.; Russo, C.; Valentin, E. *Gazz. Chim. Ital.* **1982**, *112*, 335.
[858] Udodong, U. E.; Fraser-Reid, B. *J. Org. Chem.* **1989**, *54*, 2103.
[859] Yao, L.; Smith, B. T.; Aubé, J. *J. Org. Chem.* **2004**, *69*, 1720.
[860] Gmeiner, P.; Hummel, E. *Synthesis* **1994**, 1026.
[861] Yamada, S.; Oguri, T.; Shioiri, T. Japanese Patent 74 00217 (1974); *Chem. Abstr.* **1974**, *80*, 121321r.
[862] Shimizu, M.; Yokota, T.; Fujimori, K.; Fujisawa, T. *Tetrahedron. Asymmetry* **1993**, *4*, 835.
[863] Capone, S.; Guaragna, A.; Pulumbo, G.; Pedarella, S. *Tetrahedron* **2005**, *61*, 6575.
[864] Genêt, J.-P.; Juge, S.; Mallart, S. *Tetrahedron Lett.* **1985**, *29*, 6765.
[865] Sting, R. A.; Seebach, D. *Tetrahedron* **1996**, *52*, 279.
[866] Gautschi, M.; Seebach, D. *Angew. Chem., Int. Ed. Engl.* **1992**, *31*, 1083.
[867] Kim, K. H.; Kil, K.-e.; Ko, D. H. *Bull. Korean Chem. Soc.* **2002**, *23*, 655.
[867] a. Quast, H.; Seiferling, B. *Tetrahedron Lett.* **1982**, *23*, 4681.
[868] Chabaud, L.; Landais, Y. *Tetrahederon Lett.* **2003**, *44*. 6995.
[869] Schmidt, U.; Riedl, B. *Synthesis* **1993**, 809.
[870] Ciufolini, M. A.; Xi, N. *J. Org. Chem.* **1997**, *62*, 2320.
[871] Chung, K.-H.; An, S.-O. *J. Korean Chem. Soc.* **1995**, *39*, 431.
[872] Chung, H.-K.; Kim, H.-W.; Chung, K.-H. *Bull. Korean Chem. Soc.* **1999**, *20*, 325.
[873] Herdeis, C.; Held, W. A.; Kirfel, A.; Schwabenländer, F. *Tetrahedron* **1996**, *52*, 6409.
[874] Oguri. T.; Shioiri, T.; Yamada, S.-i. *Chem. Pharm. Bull.* **1975**, *23*, 173.
[875] Poupardin, O.; Ferreira, F.; Genêt, J.-P.; Greck, C. *Tetrahedron Lett.* **2001**, *42*, 1523.
[876] Ferreira, F.; Greck, C.; Genêt, J. P. *Bull. Soc. Chim. Fr.* **1997**, *134*, 615.
[877] Sheradsky, T.; Nir, Z. *Tetrahedron Lett.* **1969**, 77.
[878] Lebeeuw, O.; Phansavath, P.; Genêt, J.-P. *Tetrahedron Lett.* **2003**, *44*, 6383.
[879] Nitta, H.; Hatanaka, M.; Ishimaru, T. *J. Chem. Soc., Chem. Commun.* **1987**, 51.
[880] Hanessian, S.; Vanasse, B.; Yang, H.; Alpegiani, M. *Can. J. Chem.* **1993**, *71*, 1407.
[881] Dupradeau, F.-Y.; Hakomori, S.-I.; Toyokuni, T. *J. Chem. Soc., Chem. Commun.* **1995**, 221.
[882] Lee, C.-S.; Lee, K.-I; Hamilton, A. D. *Tetrahedron Lett.* **2001**, *42*, 211.
[883] Tarver, J. E., Jr.; Joullié, M. *J. Org. Chem.* **2004**, *69*, 815.
[884] Hanessian, S.; Moitessier, N.; Wilmouth, S. *Tetrahedron* **2000**, *56*, 7643.
[885] Eipert, M.; Maichle-Mössmer, C.; Maier, M. E. *Tetrahedron* **2003**, *59*, 7949.
[886] Brown, R. C. D.; Hinks, J. D. *Chem. Commun.* **1998**, 1895.
[887] Hanessian, S.; Therrien, E.; Granberg, K.; Nilsson, I. *Bioorg. Med. Chem. Lett.* **2002**, *12*, 2907.
[888] Clark, C. Ph. D. Dissertation, Louisiana State University, 2005.
[889] Zheng, N.; Armstrong, J. D., III; McWilliams, J. C.; Volante, R. P. *Tetrahedron Lett.* **1997**, *38*, 2817.
[890] Sato, M.; Gunji, Y.; Ikeno, T.; Yamada, T. *Chem. Lett.* **2005**, *34*, 316.
[891] Ahn, K. H.; Kim, S.-K.; Ham, C. *Tetrahedron Lett.* **1998**, *39*, 6321.
[892] Vergne, C.; Bouillon, J.-P.; Chastanet, J.; Bois-Choussy, M.; Zhu, J. *Tetrahedron: Asymmetry* **1998**, *9*, 3095
[893] Ku, H.-Y.; Jung, J.; Kim, S.-H.; Kim, H. Y; Ahn, K. H.; Kim, S.-G. *Tetrahedron: Asymmetry* **2006**, 17.1111.
[894] Reyes-Rangel, G.; Marañón, V.; Avila-Ortiz, G.; Anaya de Parrodi, C.; Quintero, L.; Juaristi, E. *Tetrahedron* **2006**, *62*, 8404.
[895] Zhang, P. J. Ph. D. Dissertation, Université Catholique de Louvain, Belgium, 2003.

[896] Clariana, J.; Gálvez, N.; Marchi, C.; Moreno-Mañas, M.; Vallribera, A.; Molins, E. *Tetrahedron* **1999**, *55*, 7331.

[897] Nakamura, Y.; Shin, C.-g. *Chem. Lett.* **1991**, 1953.

[898] Schmidt, U.; Riedl, B. *J. Chem. Soc., Chem. Commun.* **1992**, 1186.

[899] Parry, R. J.; Ju. S.; Baker, B. J. *J. Labeled Compd. Radiopharm.* **1991**, *29*, 633.

[900] The configuration of the oxazolidinones is R as shown in Table 1 of ref. 890 and not S as specified in the text. Tohru Yamada, Department of Chemistry, Keio University, Yokohama, Japan. Personal communication, 2007.

[901] Sugiyama, H.; Shioiri, T.; Yokokawa, F. *Tetrahedron Lett.* **2002**, *43*, 3489.

[902] Ardá, A.; Jiménez, C.; Rodríguez, J. *Eur. J. Org. Chem.* **2006**, 3645.

[903] Wen, S.-J.; Yao, Z.-J. *Org. Lett.* **2004**, *6*, 2721.

[904] Sabol, J. S.; Flynn, G. A.; Friedrich, D.; Huber, E. W. *Tetrahedron Lett.* **1997**, *38*, 3687.

[905] Thompson, W. J.; Ghosh, A. K.; Holloway, M. K.; Lee, H. Y.; Munson, P. M.; Schwering, J. E.; Wai, J.; Darke, P. L.; Zugay, J.; Emini, E. A.; Schleif, W. A.; Huff, J. R.; Anderson, P. S. *J. Am. Chem. Soc.* **1993**, *115*, 801.

[906] Ogilvie, W.; Bailey, M.; Poupart, M.-A.; Abraham, A.; Bhavsar, A.; Bonneau, P.; Bordeleau, J.; Bousquet, Y.; Chabot, C.; Duceppe, J.-S.; Fazal, G.; Goulet, S.; Grand Maitre, C.; Guse, I.; Halmos, T.; Lavallée, P.; Leach, M.; Malenfant, E.; O'Meara, J.; Plante, R.; Plouffe, C.; Poirier, M.; Soucy, F.; Yoakim, C.; Déziel, R. *J. Med. Chem.* **1997**, *40*, 4113.

[907] Ripka, A. S.; Bohacek, R. S.; Rich, D. H. *Bioorg. Med. Chem. Lett.* **1998**, *8*, 357.

[908] Belshaw, P. J.; Schreiber, S. L. *J. Am. Chem. Soc.* **1997**, *119*, 1805.

[909] Guerin, D. J.; Miller, S. J. *J. Am. Chem. Soc.* **2002**, *124*, 2134.

[910] Neset, S.; Hope, H.; Undheim, K. *Tetrahedron* **1997**, *53*, 10459.

[911] Stone, M. J.; Maplestone, R. A.; Rahman, S. K.; Williams, D. H. *Tetrahedron Lett.* **1991**, *32*, 2663.

[912] Beugelmans, R.; Bois-Choussy, M.; Vergne, C.; Bouillon, J.-P.; Zhu, J. *J. Chem. Soc., Chem. Commun.* **1996**, 1029.

[913] Pearson, A. J.; Chelliah, M. V.; Bignan, G. C. *Synthesis* **1997**, 536.

[914] Pearson, A. J.; Park, J. G. *J. Org. Chem.* **1992**, *57*, 1744.

[915] Pearson, A. J.; Shin, H. *Tetrahedron* **1992**, *48*, 7527.

[916] Ami, E.; Rajesh, S.; Wang, J.; Kimura, T.; Hayashi, Y.; Kiso, Y.; Ishida, T. *Tetrahedron Lett.* **2002**, *43*, 2931.

[917] Hale, J. J.; Mills, S. G.; MacCoss, M.; Finke, P. E.; Cascieri, M. A.; Sadowski, S.; Ber, E.; Chicchi, G. G.; Kurtz, M.; Metzger, J.; Eiermann, G.; Tsou, N. N.; Tattersall, F. D.; Rupniak, N. M. J.; Williams, A. R.; Rycroft, W.; Hargreaves, R.; MacIntyre, D. E. *J. Med. Chem.* **1998**, *41*, 4607.

[918] Pearson, A. J.; Zhang, P.; Lee, K. *J. Org. Chem.* **1996**, *61*, 6581.

[919] Freund, E.; Vitali, F.; Linden, A.; Robinson, J. A. *Helv. Chim. Acta* **2000**, *83*, 2572.

[920] Evans, M. C.; Johnson, R. L. *Tetrahedron* **2000**, *56*, 9801.

[921] Lee, H. T.; Hicks, J. L.; Johnson, D. R. *J. Labeled Compd. Radiopharm.* **1991**, *29*, 1065.

[922] Lee, H.-Y.; Sohn, J.-H.; Kwon, B.-M. *Bioorg. Med. Chem. Lett.* **2002**, *12*, 1599.

[923] Bigot, A.; Dau, M. E. T. H.; Zhu, J. *J. Org. Chem.* **1999**, *64*, 6283.

[924] Alexander, K.; Cook, S.; Gibson, C. L.; Kennedy, A. R. *J. Chem. Soc., Perkin Trans. 1* **2001**, 1538.

[925] Falck-Pedersen, M. L.; Undheim, K. *Tetrahedron* **1996**, *52*, 7761.

[926] Han, Y.; Liao, S.; Qiu, W.; Cai, C.; Hruby, V. J. *Tetrahedron Lett.* **1997**, *38*, 5135.

[927] Yuan, W.; Hruby, V. J. *Tetrahedron Lett.* **1997**, *38*, 3853.

[928] Moore, S. B.; Grant, M.; Rew, Y.; Bosa, E.; Fabbri, M.; Kumar, U.; Goodman, M. *J. Peptide Res.* **2005**, *66*, 404.

[929] Boteju, L. W.; Wegner, K.; Qian, X.; Hruby, V. J. *Tetrahedron* **1994**, *50*, 2391.

[930] Boteju, L. W.; Wegner, K.; Hruby, V. J. *Tetrahedron Lett.* **1992**, *33*, 7494.

[931] Liu, D.-G.; Gao, Y.; Wang, X.; Kelly, J. A.; Burke, T. R., Jr *J. Org. Chem.* **2002**, *67*, 1448.

[932] Andersen, R. J.; Coleman, J. E.; Piers, E.; Wallace, D. J. *Tetrahedron Lett.* **1997**, *38*, 317.

[933] Chen, H. G.; Beylin, V. G.; Marlatt, M.; Leja, B.; Goel, O. P. *Tetrahedron Lett.* **1992**, *33*, 3293.

[934] McNamara, L. M. A.; Andrews, M. J. I.; Mitzel, F.; Siligardi, G.; Tabor, A. B. *J. Org. Chem.* **2001**, *66*, 4585.
[935] Beylin, V. G.; Chen, H. G.; Dunbar, J.; Goel, O. P.; Harter, W.; Marlatt, M.; Topliss, J. G. *Tetrahedron Lett.* **1993**, *34*, 953.
[936] Lin, J.; Liao, S.; Hruby, V. J. *Tetrahedron Lett.* **1998**, *39*, 3117.
[937] Kaczmarek, K.; Zabrocki, J.; Lachwa, M.; Lipkowski, A. W. In *Peptides 1998*; Bajusz, S., Hudecz, F., Eds.; Akadémiai Kiadó: Budapest, 1999; p 668.
[938] Nadin, A.; Sánchez López, J. M.; Owens, A. P.; Howells, D. M.; Talbot, A. C.; Harrison, T. *J. Org. Chem.* **2003**, *68*, 2844.
[939] Prasad, C. V. C.; Vig, S.; Smith, D. W.; Gao, Q.; Polson, C. T.; Corsa, J. A.; Guss, V. L.; Loo, A.; Barten, D. M.; Zheng, M.; Felsenstein, K. M.; Roberts, S. B. *Bioorg. Med. Chem. Lett.* **2004**, *14*, 3535.
[940] Hanessian, S.; Papeo, G.; Fettis, K.; Therrien, E.; Viet, M. P. T. *J. Org. Chem.* **2004**, *69*, 4891.
[941] Poullenec, K. G.; Kelly, A. T.; Romo, D. *Org. Lett.* **2002**, *4*, 2645.
[942] Butcher, J. W.; Liverton, N. J.; Selnick, H. G.; Elliot, J. M.; Smith, G. R.; Tebben, A. J.; Pribush, D. A.; Wai, J. S.; Claremon, D. A. *Tetrahedron Lett.* **1996**, *37*, 6685.
[943] Chan, P. W. H.; Cottrell, I. F.; Moloney, M. G. *J. Chem. Soc., Perkin Trans. 1* **2001**, 3007.
[944] Chan, P. W. H.; Cottrell, I. F.; Moloney, M. F. *Tetrahedron: Asymmetry* **1999**, *10*, 3887.
[945] Devillers, I.; Pevet, I.; Jacobelli, H.; Durand, C.; Fasquelle, V.; Puaud, J.; Gaudillière, B.; Idrissi, M.; Moreu, F.; Wigglesworth, R. *Bioorg. Med. Chem. Lett.* **2004**, *14*, 3303.
[946] Maeng, J.-H.; Funk, R. L. *Org. Lett.* **2001**, *3*, 1125.
[947] Boche, G.; Bosold, F.; Niessner, M. *Tetrahedron Lett.* **1982**, *23*, 3255.
[948] Pagliarin, R.; Papeo, G.; Sello, G.; Sisti, M.; Paleari, L. *Tetrahedron* **1996**, *52*, 13783.
[949] Jommi, G.; Miglierini, G.; Pagliarin, R.; Sello, G.; Sisti, M. *Tetrahedron* **1992**, *48*, 7275.
[950] Yamamoto, Y.; Hatsuya, S.; Yamada, J.-i. *Tetrahedron Lett.* **1989**, *30*, 3445.
[951] Terada, M.; Nakano, M.; Ube, H. *J. Am. Chem. Soc.* **2006**, *128*, 16044.
[952] Bumbek, S.; Lenarsic, R.; Kocevar, M.; Polanc, S. *Synlett* **2001**, 1237.
[953] Comelles, J.; Moreno-Mañas, M.; Pérez, E.; Roglans, A.; Sebastián, R. M.; Vallribera, A. *J. Org. Chem.* **2004**, *69*, 6834.
[954] Comelles, J.; Pericas, À.; Moreno-Mañas, M.; Vallribera, A.; Drudis-Solé, G.; Lledos, A.; Parella, T.; Roglans, A.; García-Granda, S.; Roces-Fernández, L. *J. Org. Chem.* **2007**, *72*, 2077.
[955] Colonna, F. P.; Pitacco, G.; Valentin, E. *J. Chem. Soc., Chem. Commun.* **1975**, 71.
[956] Marigo, M.; Kumuragurubaran, N.; Jørgensen, K. A. *Synthesis* **2005**, 957.
[957] Winkler, F. J.; Stahl, D. *J. Am. Chem. Soc.* **1978**, *100*, 6780.
[958] Kozikowski, A. P.; Greco, M. N. *J. Org. Chem.* **1984**, *49*, 2310.
[959] Fioravanti, S.; Pellacani, L.; Stabile, S.; Tardella, P. A.; Ballini, R. *Tetrahedron* **1998**, *54*, 6169.
[960] Zeifman, Yu. V.; Koshtoyan, S. O.; Knunyants, I. L. *Dokl. Akad. Nauk SSSR* **1970**, *195*, 93; *Engl. Transl.* p 783.
[961] Zeifman, Yu. V.; Rokhlin, E. M.; Utebaev, U.; Knunyants, I. L. *Dokl. Akad. Nauk SSSR* **1976**, *226*, 1337; *Engl. Transl.* p 149.
[962] Parcell, R. F. *Chem. Ind. (London)* **1963**, 1396.
[963] de Santis, M.; Fioravanti, S.; Pellacani, L.; Tardella, P. A. *Eur. J. Org. Chem.* **1999**, 2709.
[964] Tanaka, K.; Mori, Y.; Narasaka, K. *Chem. Lett.* **2004**, 26.
[965] Drefahl, G.; Ponsold, K.; Schönecker, B. *Chem. Ber.* **1964**, *97*, 2014.

CHAPTER 2

DESULFONYLATION REACTIONS

Diego A. Alonso and Carmen Nájera

Department of Organic Chemistry and Institute of Organic Synthesis (ISO), Faculty of Science, University of Alicante, Apartado 99, E-03080 Alicante, Spain

CONTENTS

diego.alonso@ua.es

Organic Reactions, Vol. 72, Edited by Scott E. Denmark et al.

ACKNOWLEDGMENTS

We gratefully acknowledge the guidance and assistance of the editorial staff of *Organic Reactions*, in particular Professor T. V. RajanBabu, for their kind help during the preparation of this chapter.

INTRODUCTION

Over the last thirty years the use of sulfones in organic chemistry has become a very important synthetic strategy, especially for the formation of carbon-carbon single and double bonds,[1–11] enabling the preparation of a wide variety of functionalized molecules including many natural and biologically active compounds.[12] Sulfones can modify the polarity of a molecule by acting as electron-withdrawing groups. Sulfones can also stabilize α-carbanions and function as good leaving groups in elimination reactions.

The sulfone is a versatile functional group comparable to the carbonyl functionality in its ability to activate molecules for further bond construction, the main difference between these two groups being that the sulfone is usually removed once the synthetic objective is achieved. The removal most commonly involves a reductive desulfonylation process with either replacement of the sulfone by hydrogen (Eq. 1), or a process that results in the formation of a carbon-carbon multiple bond when a β-functionalized sulfone, for example a β-hydroxy or β-alkoxy sulfone, is employed (Eq. 2). These types of reactions are the Julia–Lythgoe or Julia–Paris–Kocienski olefination processes. Alkylative desulfonylation (substitution of the sulfone by an alkyl group, Eq. 3), oxidative desulfonylation (Eq. 4), and substitution of the sulfone by a nucleophile (nucleophilic displacement, Eq. 5) are also known. Finally, β-eliminations (Eq. 6) or sulfur dioxide extrusion processes (Eqs. 7, 8 and 9) have become very popular for the preparation of carbon-carbon single and double bonds. A few reviews[13,14] and book chapters[1,15,16,3] have previously covered the different aspects and synthetic applications of the reductive desulfonylation and reductive elimination processes. This chapter deals exclusively with the replacement of the sulfone group by a hydrogen (reductive desulfonylation reactions, Eq. 1), and reductive elimination reactions of the Julia-type substrates (Julia–Lythgoe olefination process, Eq. 2). In the following sections, the full scope and limitations of these reactions, their synthetic applications, and typical experimental conditions are described. The coverage of the literature through most of 2007 is comprehensive, as is the accompanying Tabular Survey.

$R^1CH(R^2)SO_2R^3 \longrightarrow R^1CH(R^2)H$ (Eq. 1)

$R^1CH(OR^2)CH(R^3)SO_2R^4 \longrightarrow R^1CH{=}CHR^3$ (Eq. 2)

$R^1CH(R^2)SO_2R^3 \xrightarrow{E^+} R^1CH(R^2)E$ (Eq. 3)

$R^1CH(R^2)SO_2R^3 \xrightarrow{[Ox]} R^1C(O)R^2$ (Eq. 4)

$R^1CH(R^2)SO_2R^3 \xrightarrow{Nu^-} R^1CH(R^2)Nu$ (Eq. 5)

$R^1CH(H)CH(R^2)SO_2R^3 \xrightarrow{base} R^1CH{=}CHR^2$ (Eq. 6)

$HC(R^1)(R^2)SO_2C(R^3)(R^4)X \xrightarrow{base}$ [episulfone: $R^1R^2C{-}CR^3R^4$, SO_2 bridged] $\xrightarrow{-SO_2} R^1R^2C{=}CR^3R^4$ (Eq. 7)

$HC(R^1)(R^2)SO_2C(R^3)(R^4)X \xrightarrow{-SO_2} R^1CH(R^2)CH(R^3)R^4$ (Eq. 8)

$R^1CH(O^-)CH(R^2)SO_2Ar \longrightarrow [R^1CH(SO_2^-)CH(R^2)OAr] \xrightarrow[-ArO^-]{-SO_2} R^1CH{=}CHR^2$ (Eq. 9)

MECHANISM AND STEREOCHEMISTRY

The method for removal of the sulfone group depends on the other functionalities present in the molecule, and different desulfonylation conditions have been developed to achieve the synthetic objective.

Reductive Desulfonylations

One of the most widely used transformations in sulfone-mediated synthetic reactions is the substitution of the sulfone group by hydrogen (Eq. 1). The reductive $C{-}SO_2$ cleavage can be performed by chemical, electrochemical and photochemical methods, the most commonly used being the chemical reductive desulfonylation method. For this purpose, a wide range of reducing agents and procedures has been developed, and most methods involve an electron-transfer mechanism. The standard chemical reducing agents and procedures fall into three main categories: reductive desulfonylations mediated by active metals and salts, tin hydrides, and transition-metals.

Reductive Desulfonylations by Active Metals and Salts. Electropositive metals such as alkali metals, Mg, Ca, Al, Zn, and Sm, have been used to cleave $C{-}SO_2$ bonds. The alkali metals and their amalgams are the most widely employed. Most reactions proceed through a single-electron-transfer (SET) mechanism. Therefore, the chemical properties of the electron-transfer reagent and the substrate to be reduced, such as the reduction potentials in solution, are very important. The reduction potential of the sulfone derivatives depends principally on the structure of the reduced substance and its solvation energy. Aromatic sulfones react easily with solvated electrons because energetically accessible LUMO levels are often available given the strong electron-withdrawing nature of the sulfone group. This group decreases the energy level of the LUMO of the aromatic moiety allowing a much faster electron transfer. The reduction of diaryl sulfones is complex and, from a synthetic point of view, not as useful as the selective removal of one arenesulfonyl group connected to an aliphatic chain. Studies on the mechanism of this reaction[17] show that in the first step, an electron is transferred to the substrate to form a radical anion, which rapidly dissociates into an arylsulfinate and a radical. The radical is immediately reduced and, after protonation, affords the observed hydrocarbon (Eq. 10).

$$ArSO_2R \xrightarrow{e^-} [ArSO_2R]^{\bullet -} \rightarrow \begin{cases} R^{\bullet} \xrightarrow{e^-,\ S{-}H} RH \\ ArSO_2^- \end{cases} \qquad \text{(Eq. 10)}$$

Functional groups present in the molecule that are sensitive to SET, usually multiple bonds, aromatic rings, and nitro and carbonyl groups situated in the α-position to the SO_2 group, may play the role of electron acceptor and accelerate the $C{-}SO_2$ cleavage. This activation has been broadly used, for example, to perform selective desulfonylations of α-ketosulfones (Eq. 11), with aluminum amalgam usually being the reagent of choice to achieve efficient desulfonylation.[18]

$$R^1SO_2CH_2C(O)R^2 \xrightarrow{SET} [R^1SO_2\text{–}CH_2\text{–}C^{\bullet}(O^-)R^2] \xrightarrow[H^+]{SET} CH_3C(O)R^2 \quad \text{(Eq. 11)}$$

It is also possible to enhance the electron transfer from a metal by the use of certain solvents such as amines, or through the presence in the reaction medium of external additives such as aromatic compounds. Solutions of alkali and alkaline earth metals (Li, Na, and less frequently, Ca) in anhydrous ammonia or low molecular weight amines such as methyl- or ethylamine at low temperatures represent a powerful reductive desulfonylation system.[19] This method, which has been mainly employed for the reductive desulfonylation of α-keto-, allyl- and remote-functionalized sulfones, proceeds via an SET process to afford carbanionic intermediates and sulfinate species. There is no evidence that reduction of the sulfonyl S–O bond occurs, and it appears that the cleavage process normally takes place only after the second electron has been transferred to the molecule. In the reductive desulfonylation with lithium in methylamine of different alkyl cycloalkyl sulfones, the site selectivity in the $C-SO_2$ bond cleavage is a consequence of a process with appreciable carbanionic character that is governed by both electronic and steric factors.[20] Of synthetic relevance is the reduction of alkyl aryl sulfones (Eq. 12), where the site selectivity seems to be dependent on the substrate and the reaction conditions employed, but selective alkyl–S bond cleavage can be achieved by proper selection of the reaction conditions (Eq. 13).[21] With respect to the desulfonylation of allylic sulfones, rearranged alkene products corresponding to the thermodynamically more stable compounds are usually observed (Eq. 14).[22a]

$$ArS(O_2)R^1 \xrightarrow{R^2NH_2/Li} [R^{1-} + ArSO_2Li] \xrightarrow{H_2O} R^1H + ArSO_2H \quad \text{(Eq. 12)}$$

Ts; EtNH₂/Li, –78°, 30 min; O; O; (82%) (Eq. 13)

OH; H; SO₂Ph; H; H; H; OMe; NH₃/Li, –78°; OH; H; H; H; H; OMe; (65%) (Eq. 14)

Alkali metal arene radical anion complexes are useful sources of solvated electrons for reductive desulfonylation reactions.[14] Aromatic compounds such as naphthalene and its derivatives improve the electron-transfer ability of the alkali metals such as Na and Li in solution. The reduction potential of alkali metal-naphthalene, which is the most employed reagent, is close to the value of the ammonia solutions of alkali metals.[22b,22c] These radical anion solutions constitute a good alternative to dissolving metal reductions. The heterogeneous electron-transfer process between the metal and the aromatic compounds is solvent dependent and is usually carried out in aprotic coordinating solvents such as ethers. These solvents stabilize the radical anion complex by forming solvent-separated ion pairs[23–28] especially at low temperatures (Eq. 15).

$$\mathrm{ArH} + \mathrm{M} \xrightleftharpoons{\mathrm{ROR}} \mathrm{ArH}^{\bullet -} + \mathrm{M}^{+} \quad \mathrm{M} = \mathrm{Li, Na, K} \qquad \text{(Eq. 15)}$$

Samarium(II) iodide is a mild and selective single-electron-transfer reagent that has become very popular for reduction of sulfone derivatives[29] due to its propensity to revert to the more stable Sm(III) oxidation state.[30,31] Additives and cosolvents often have a profound effect on reactions mediated by SmI_2. The additives usually fall into one of two classes: proton sources such as MeOH, or donor ligands such as HMPA (hexamethylphosphoric triamide) or DMPU [1,3-dimethyl-3,4,5,6-tetrahydro-2(1*H*)-pyrimidinone]. The role of these ligands is to increase the reducing power of Sm(II) [from −1.33 V to −2.05 V in the presence of four equivalents of HMPA (vs. $Ag/AgNO_3$ in THF)],[32] and it has been thought that proton sources serve not only to protonate basic organometallic intermediates but also to accelerate the reductive desulfonylation process (Eq. 16). Excess of SmI_2 is usually necessary to bring the reaction to completion.

$$\mathrm{RC(O)CH(SO_2Ar)CH_2R'} \xrightarrow{\mathrm{SmI_2}} \left[\mathrm{R\dot{C}(O^-)CH(SO_2Ar)CH_2R'} \xrightarrow{\mathrm{MeOH}} \mathrm{R\dot{C}(OH)CH(SO_2Ar)CH_2R'} \xrightarrow{\mathrm{SmI_2}} \mathrm{R\bar{C}(OH)CH(SO_2Ar)CH_2R'} \right] \xrightarrow{-\mathrm{ArSO_2^-}} \mathrm{RC(OH){=}CHCH_2R'} \rightleftharpoons \mathrm{RC(O)CH_2CH_2R'} \qquad \text{(Eq. 16)}$$

By far the most employed and general method for the reductive desulfonylation of all types of sulfones is the reduction with metal amalgams, particularly sodium amalgam (5–6%) in a buffered alcohol solution employing four equivalents of disodium hydrogen phosphate.[33] This method is based on an early disclosure[34] where several diaryl and alkyl aryl sulfones are reported to undergo reductive desulfonylation by Na/Hg in boiling ethanol to a sulfinic acid and a hydrocarbon. The selective alkyl–SO_2 bond cleavage that occurs with alkyl aryl sulfones makes this reagent appropriate for removing arenesulfonyl groups in the presence of base-sensitive functional groups since the formation of alkoxides is prevented under the buffered conditions. Dialkyl sulfones are not reactive towards this reagent. The method is highly chemoselective because simple sulfones as well as those having a range of other functional groups (isolated multiple bonds, ethers, acetals, epoxides, ketones, carboxylic acids and their derivatives, and a variety of nitrogen and oxygen protecting groups), are smoothly desulfonylated as depicted in Eqs. 17[35] and 18.[36]

Na/Hg, THF/MeOH (75%) (Eq. 17)

$Na/Hg, Na_2HPO_4$, MeOH, 0°, 2 h (85%) (Eq. 18)

Site and stereocontrol of the reductive desulfonylation reactions of allylic and vinylic sulfones depend on the choice of the reducing agents and reaction conditions. With allylic sulfones migration of the double bond to the most stable position is usually observed with all the reagents investigated (Eq. 19).[37] The migration of the double bond takes place with little or no stereocontrol, and different results can be obtained depending on the method employed as depicted in Eq. 20 for vinylic sulfones.[38]

$Na/Hg, NaH_2PO_4$, MeOH, 0° to rt (19%) + (48%) (Eq. 19)

$EtNH_2/Li$, Et_2O, –30° (60%) Z/E 40:60; K–graphite, THF, rt (61%) Z/E 30:70 (Eq. 20)

The reductive desulfonylation of vinylic sulfones has been mainly carried out with sodium and aluminum amalgams, SmI_2, and Mg in *N*,*N*-dimethylformamide (DMF) in the presence of chlorotrimethylsilane (TMSCl). The configuration of the double bond is not necessarily preserved in the desulfonylation of vinylic sulfones when Na/Hg is employed. This reagent is the most widely used when there is no need for stereochemical control as shown in Eqs. 21[39] and 22.[40]

$Na/Hg, Na_2HPO_4$, MeOH, 0°, 2 h (92%) (Eq. 21)

$Na/Hg, NaH_2PO_4$, THF/MeOH, –30°, <1 h (75%) (Eq. 22)

In contrast, reductions using Al/Hg seem to be stereoselective at high reaction temperatures to afford exclusively the more stable E-alkene product. The equilibration of the anionic intermediates under the reaction conditions is responsible for this isomerization. The temperature is a crucial factor in controlling the stereoselectivity of the process, since the geometry of the vinylic sulfone can be maintained when the reduction is carried out at low temperature. Partial scrambling of configuration is observed upon increasing the reaction temperature (Eq. 23).[41]

H SO$_2$Ph; Al/Hg, THF/D_2O; <0° → H D; >0° → H D (Eq. 23)

The stereochemical outcome of the reduction of vinylic sulfones with SmI_2 seems to be additive-dependent and unpredictable as shown in Eq. 24.[42] The exact role of DMPU or HMPA on the stereoselectivity is still not understood. Deuterium labeling studies employing MeOH-d_1 have revealed an important role of this cosolvent as a proton source, and in the control of the stereochemical course of the reaction. Because hydrogen is not abstracted from the solvent (THF), it is probable that the proton is obtained from methyl alcohol to quench the corresponding vinyl anion (Eq. 24). In general, when the reaction is carried out at low temperatures, with short reaction times, and with the proper solvent-additive combination, the desulfonylation of vinylic sulfones is highly stereoselective (Eq. 25).[43] At ambient or higher temperatures the preference for the thermodynamically more stable alkene geometry is usually observed as a consequence of the equilibration of the alkenyl radical (Eq. 26).[42] Under these conditions, reduction of the double bond is occasionally observed.[44]

SO$_2$Ph, Ph, Ph; SmI_2, THF, additive → [Ph, Ph] "H^+" → Ph, Ph

SmI_2 (eq)	Additive	Time	Yield	Z/E
6	MeOH	5 d	(60%)	86:14
8	MeOH	5 d	(90%)	0:100
8	HMPA	10 min	(40%)	48:52
8	DMPU	35 min	(95%)	10:90
8	DMPU/MeOH (1:1)	35 min	(92%)	0:100

(Eq. 24)

DMPU = *N,N'*-dimethylpropyleneurea

SmI_2, THF, HMPA; −20°, 20 min (62%) (Eq. 25)

SmI_2, THF, MeOH; DMPU, rt, 30 min (95%) (Eq. 26)

Styryl aryl sulfones are efficiently transformed into the corresponding E-β-substituted styrenes in a stereoselective manner through a Mg-promoted reduction.[45] The reaction is carried out in polar solvents such as dimethyl sulfoxide (DMSO) or *N*,*N*-dimethylformamide (DMF) and in the presence of TMSCl, which is believed to activate the metal surface and stabilize the anionic intermediates generated by the electron transfer from Mg metal.[45] Under these reaction conditions, a highly stereoselective reductive desulfonylation takes place to give the corresponding E-styrenes (Eq. 27). The reaction is believed to take place through a stabilized radical anion that undergoes a stereoselective elimination of the arylsulfinyl moiety (Eq. 28).

Mg/TMSCl; DMF, rt, 1 d (Eq. 27)

R	Z/E	Yield	Z/E
Me	45:55	(72%)	<1:99
Et	32:68	(81%)	<1:99
n-Pr	31:69	(75%)	2:98
Bn	10:90	(85%)	1:99

(Eq. 28)

Sodium dithionite ($Na_2S_2O_4$) is a useful reducing agent for reactions of vinyl[46] and β-ketosulfones[47] under weakly basic aqueous conditions. The mechanism of the reaction is believed to follow an addition-elimination process in which a syn-addition of the hydrogen sulfinate ion (HSO_2^-) is followed by an anti-elimination of sulfur dioxide and arylsulfinate ion from the intermediate sulfonylsulfinate (Eq. 29).[48,49] The reaction is stereospecific and retention of the configuration of the original vinylic sulfone is observed. With β-ketosulfones, removal of the sulfone group also takes place via an addition-elimination mechanism.[47]

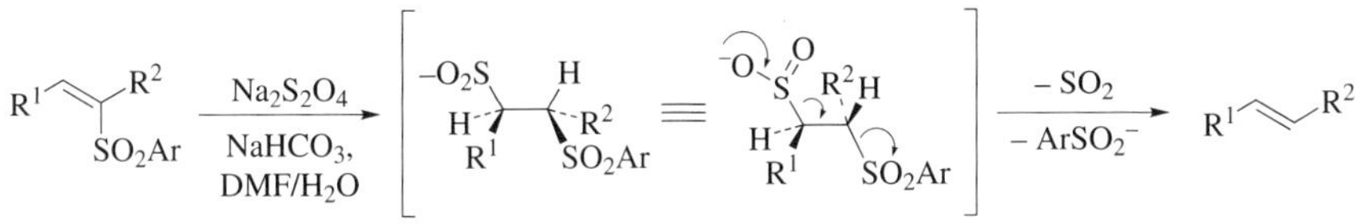

(Eq. 29)

Despite the great amount of interest in reductive desulfonylation reactions, very little research has addressed the stereospecific reductive desulfonylation of chiral α-substituted sulfones. Only limited success has been achieved as shown in Eq. 30.[50] Lithium naphthalenide is used for the stereoselective SET desulfonylation of anomeric sulfones derived from 2-deoxy-D-glucose derivatives.[51–54] The initial homolytic cleavage of the $C-SO_2$ bond generates a σ-radical, which adopts an α-orientation due to stereoelectronic stabilization,[55–57] forcing the anomeric substituent to adopt the β-orientation, an arrangement that is retained through the reduction process (Eq. 31).

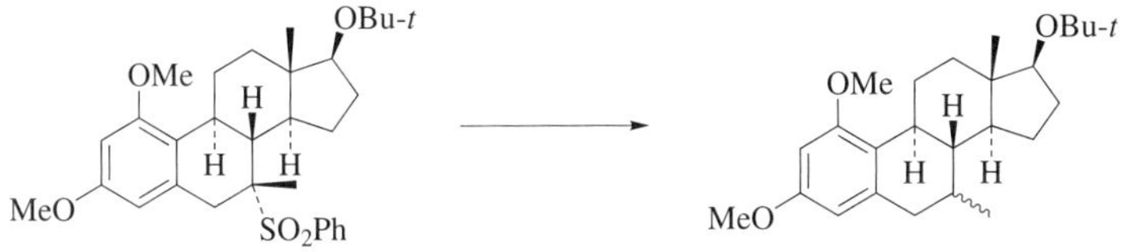

(Eq. 30)

Reducing Agent	Additive	Temp	Time	Solvent	Yield	α/β
Et_2NH/Li	—	–70°	30 min	THF	(—)	72:28
Na/Hg	—	72°	2 h	EtOH	(80%)	63:37
Na/Hg	—	70°	2 h	HMPA	(44%)	45:55
Na/Hg	Na_2HPO_4	50°	1.5 h	$MeOH/C_6H_6$	(78%)	74:26

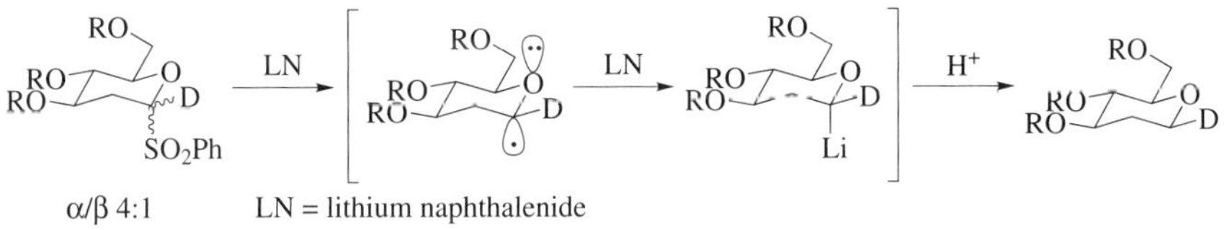

α/β 4:1 LN = lithium naphthalenide

(Eq. 31)

Reductive Desulfonylations by Tin Hydrides. Reductive desulfonylation of allyl, vinyl, and α-functionalized sulfones can be carried out employing tin hydrides. This radical reaction is usually promoted thermally or photochemically and provides organotin derivatives as intermediates which are finally subjected to protonolysis (Eq. 32). Both steps can be carried out in one pot employing catalytic amounts of tin.

$$ArSO_2R^1 \xrightarrow[h\nu \text{ or heat}]{(R^2)_3SnH} R^1Sn(R^2)_3 \xrightarrow{H^+} R^1H \quad \text{(Eq. 32)}$$

α-Keto arylsulfones are easily desulfonylated to the corresponding ketones with tin hydrides in the presence of radical initiators. The reaction mechanism involves the formation of a ketyl-type radical as an intermediate[58,59] that, after β-elimination of a sulfonyl radical, produces the corresponding tin enolate (Eq. 33). In a propagation step, hydrogen transfer from the tin hydride to the sulfonyl radical gives the corresponding sulfinic acid which, in a final step, protonates the tin enolate to furnish the desired desulfonylated product. Tin enolate formation via SET from the tin radical to the keto group has also been proposed.[58,59]

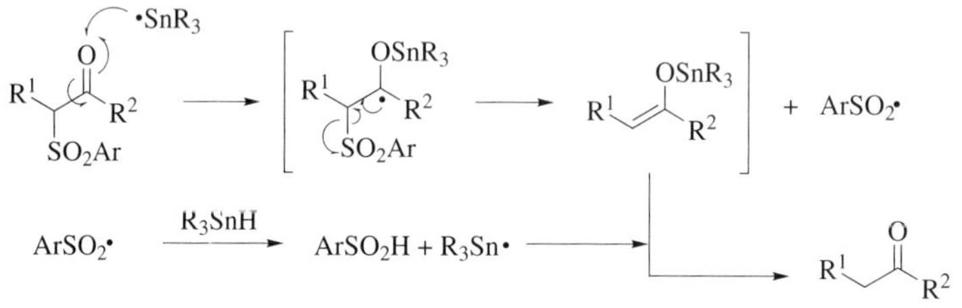

(Eq. 33)

Although stabilized α-keto radical species are also postulated as intermediates in the desulfonylation process,[59] the absence of favored hex-5-enyl radical cyclization processes in the reductive desulfonylation of alkene-containing substrates argues against this possibility.[59,60] Conversely, the tin hydride-mediated reductive desulfonylations of α-sulfonyl phosphonates are suggested to proceed via attack of the tin radical at an oxygen (or sulfur) atom of the sulfonyl group to give a stabilized α-phosphonyl radical intermediate (Eq. 34).[60] This method has been applied to the reductive desulfonylation of π-electron-deficient arylsulfonyl derivatives, such as 2-pyridyl- and 2-pyrimidylsulfonyl compounds; substrates that facilitate the reaction as a result of the strong electron-withdrawing character of the aromatic ring. As a consequence, SET from the tin radical to the electronegative phosphonate system (Eq. 34), followed by sulfinate cleavage, might also lead to the α-phosphonyl radical. Both mechanistic possibilities may be further enhanced by the π-electron-deficient arylsulfonyl moieties.

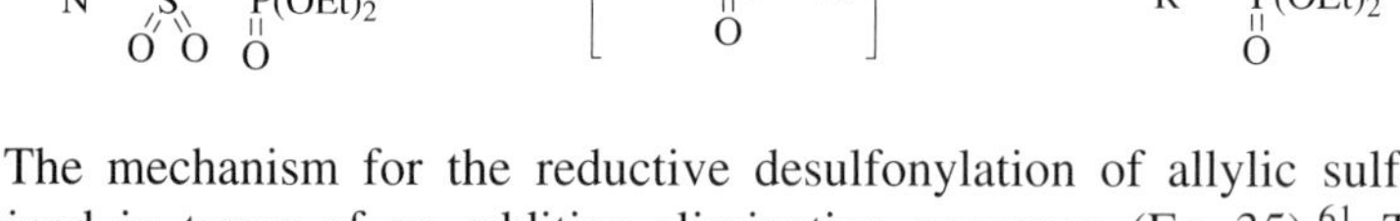

(Eq. 34)

The mechanism for the reductive desulfonylation of allylic sulfones is explained in terms of an addition-elimination sequence (Eq. 35).[61] The process is site selective but not usually stereoselective since allylstannane intermediates are produced as mixtures of stereoisomers. In one exception, α-(hydroxymethyl) allylic sulfones afford, after reaction with n-Bu_3SnH, Z-allyltin intermediates stereoselectively.[62] Coordination between the oxygen and tin atoms in the initial adduct seems to fix the conformation leading to the predominant formation of Z-isomers (Eq. 36).

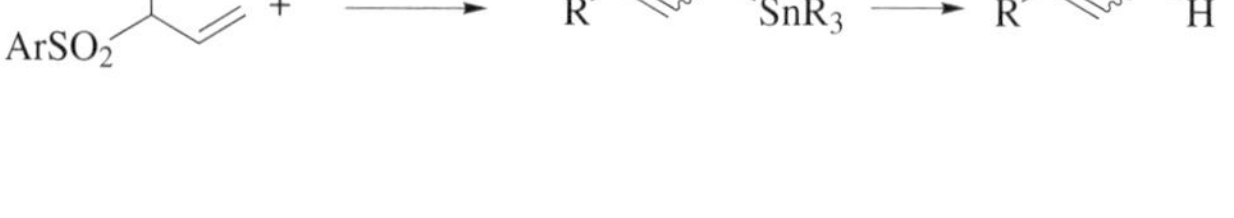

(Eq. 35)

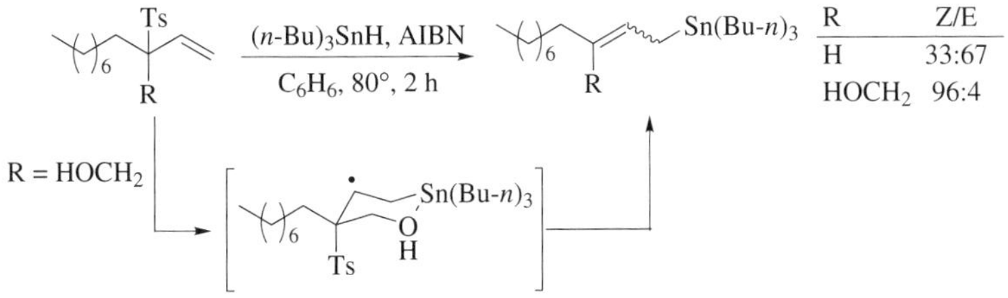

(Eq. 36)

In desulfonylation reactions of vinylic sulfones, the accepted mechanism also involves a free-radical addition-elimination sequence to generate the corresponding vinylstannanes on treatment with tin hydride and a substoichiometric amount of a radical initiator.[63] The stereoselectivity of the reaction has been studied with α-fluorovinylic sulfones,[63,64] showing that the tin-sulfonyl exchange is mostly stereospecific and proceeds with retention of configuration for 2,2-disubstituted derivatives (Eq. 37). Conversely, the 2-monosubstituted analogs equilibrate to E/Z mixtures of (fluorovinyl)stannanes when treated with tributyltin hydride (Eq. 37). A high degree of stereocontrol is observed when the substrate bears a bulky substituent in the β-vinylic position. The radical addition-elimination process is controlled by steric factors preserving the E/Z geometry of the (fluorovinyl)stannanes (Eq. 38).[65]

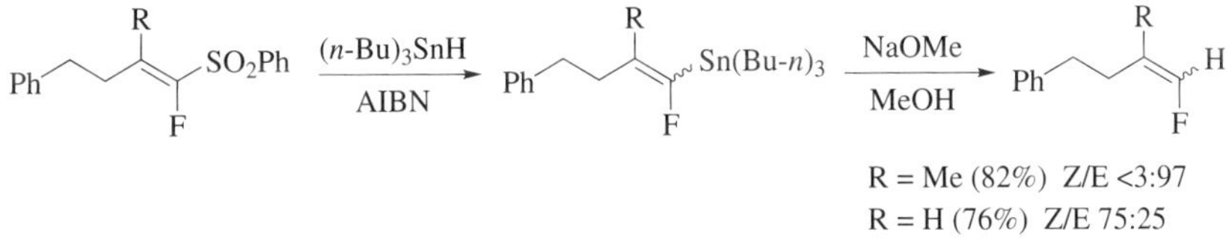

R = Me (82%) Z/E <3:97
R = H (76%) Z/E 75:25

(Eq. 37)

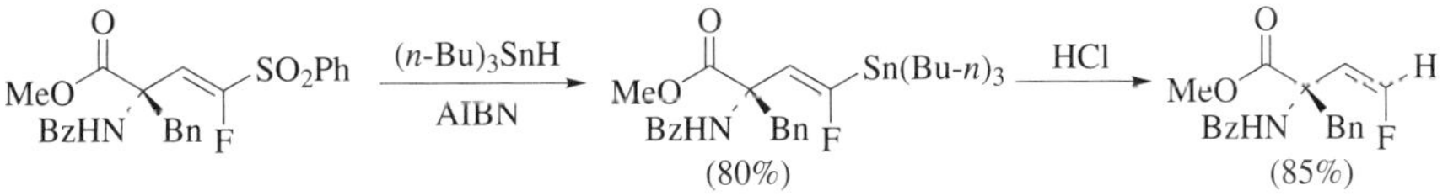

(80%) (85%)

(Eq. 38)

Transition-Metal-Mediated Reductive Desulfonylations. Nickel reagents have been used to remove the sulfonyl group[66,67] under heterogeneous and homogeneous conditions. Although less commonly used, Raney nickel (Ra–Ni) in its different forms can be employed to promote the reductive desulfonylation of organic compounds under heterogeneous conditions. The first step in the process might involve adsorption of the sulfone on the nickel surface through its sulfonyl oxygen atoms. The hydrogen atom that replaces the sulfonyl group generally comes from the large amount of surface-bound hydrogen on the finely divided reagent.[68] The reduction has been postulated to occur through either a radical[69] or an ionic[70] mechanism and experimental results to support both alternatives have been presented. The degree of stereocontrol of the reduction of α-chiral alkyl sulfones is quantified by the resulting diastereomeric ratio of products. The results are sensitive to different factors such as the sulfone structure and the reaction conditions employed.[71] The outcomes range from partial inversion to partial retention so no general conclusions can be drawn about the mechanism or the stereochemical pathway of the reaction.

Homogeneous organonickel reagents prepared by the combination of $LiAlH_4$ with a nickel salt are used in the reductive desulfonylation of alkyl- and α-ketosulfones, and offer an alternative to the Ra–Ni reduction. Nickelocene-lithium aluminum hydride $[(Cp_2NiAlH_2)^- Li^+]_2$,[72] synthesized from nickelocene and $LiAlH_4$ in THF, or a reagent prepared by the combination of $NiBr_2$•DME with two equivalents of PPh_3 and one equivalent of $LiAlH_4$,[73a] are effective reagents for this purpose (Eq. 39). The reaction involves reduction of the S–O bond followed by a desulfurization process. This final C–S cleavage can take place through an oxidative addition reaction of the C–S bond to a Ni(0) complex.[73b,73c] Alternatively, an electron-transfer process can also be envisioned. The intermediates thus obtained might undergo hydrogen abstraction from the nickel hydridic moieties, which are proposed as the active species in the process.

$PhCOCH_2SO_2Ph \xrightarrow[\text{THF, rt, overnight}]{(Cp_2NiAlH_2)^- Li^+} PhCOCH_3$ (63%) (Eq. 39)

Reductive desulfonylations of saturated and unsaturated sulfones can be performed with nickel-containing complex reducing agents (NICRAs) (Eq. 40).[74,75] These complexes, very easily prepared by combining NaH, a sodium alkoxide, and a nickel salt in different ratios, are not very sensitive and are easily handled. During the desulfonylation process, the intermediate formation of the corresponding sulfides is observed.[72] Therefore it has been postulated that the reduction at the sulfur atom might be the first step and that the actual desulfurization takes place on the corresponding thioethers.

$CH_3CH_2(CH_2CH_2)_9SO_2Et \xrightarrow[\text{THF, 63°, 19 h}]{\text{NICRA}} CH_3CH_2(CH_2CH_2)_9H$ (66%) (Eq. 40)

Allylic sulfones can be activated towards nucleophilic attack by conversion into π-allylpalladium complexes (Eq. 41).[76] Although alkyl sulfones can sometimes be reduced in the presence of hydride reagents such as $LiAlH_4$ and diisobutylaluminum hydride (DIBALH),[50,77–81] they are usually resistant to them. On the contrary, allylic sulfones can be desulfonylated with hydride reagents in the presence of metal [generally Pd(0)] complexes as catalysts.[82–85] This protocol is based on the alkylation of allylic sulfones catalyzed by palladium complexes (Eq. 41),[76] which proceeds with overall retention of configuration since both the oxidative addition and the nucleophilic attack involve inversion of configuration (Eq. 42) usually taking place with the nucleophilic attack predominantly at the less hindered position. Nickel and molybdenum have also been identified as efficient catalysts for this transformation.[86,87] The palladium-catalyzed desulfonylation of allylic sulfones is highly site- and stereoselective. Double bond migration and isomerization processes, usually observed when allylic sulfones are subjected to other reductive desulfonylation conditions, are avoided (Eq. 43).[83] The reductive desulfonylation of allylic sulfones employing stoichiometric amounts of $Mo(CO)_6$ in refluxing dioxane has been also reported.[88]

Pd catalyst, MNu, THF

R^1	R^2	R^3	Pd catalyst	MNu	Yield	Yield
H	H	Me	$Pd(PPh_3)_4$	$NaCH(CO_2Me)_2$	(74%)	(0%)
Me	H	H	$Pd(PPh_3)_4$	$NaCH(CO_2Me)_2$	(52%)	(19%)
Me	H	H	$Pd(PPh_3)_4$	$NaCH(SO_2Ph)(CO_2Me)$	(55%)	(18%)
Me	Me	H	$Pd(dppe)_2$	$NaCH(CO_2Me)_2$	(45%)	(35%)
Me	Me	H	$Pd(dppe)_2$	$NaCH(SO_2Ph)(CO_2Me)$	(53%)	(15%)
H	H	Ph	$Pd(PPh_3)_4$	$NaCH(CO_2Me)_2$	(71%)	(0%)

(Eq. 41)

SO_2Ph — $Pd(PPh_3)_4$, inversion 1 → [Nu, Pd^+L_2] — $NaHC(CO_2Me)_2$, inversion 2 → CO_2Me, CO_2Me (64%)

(Eq. 42)

SO_2Ph — $LiBHEt_3$, $Pd(PPh_3)_4$ (5 mol%), THF, 66°, 24 h → Z/E 13:86 + (65%) 99:1

(Eq. 43)

The generally accepted mechanism for Pd-catalyzed allylic desulfonylations is illustrated in Scheme 1. The first step is coordination of the Pd(0) catalyst to the allylic sulfone. Oxidative addition or internal S_N2-type nucleophilic attack of the electron-rich palladium at the allylic position generates a neutral Pd(II) η^3-allyl complex, which leads to a more reactive cationic complex that is finally reduced. The equilibrium between the neutral and the more reactive cationic complexes depends on the nature and concentration of the palladium ligands as well as the counter anions present in solution.

Unsymmetrical π-allyl-Pd complexes usually suffer attack of the hydride nucleophile at the less substituted position in an S_N2-type reaction. However, the site selectivity of the process is controlled by steric and/or electronic effects. The reaction is strongly dependent on the structural features of the substrate and the reaction conditions. Opposite site selectivity is observed when the reduction occurs at the sterically more hindered position via a cationic intermediate (S_N1-type). Very potent nucleophilic hydride sources, such as $LiBHEt_3$ or $LiAlH_4$, may rapidly attack intermediate π-allyl complexes at the less hindered terminal position to give the more substituted alkene, while less effective hydride-transfer reagents ($NaBH_3CN$, $NaBH_4$) attack the π-allyl systems at the site best able to accommodate a positive charge, leading to increasing amounts of the less substituted alkene (Eq. 44).

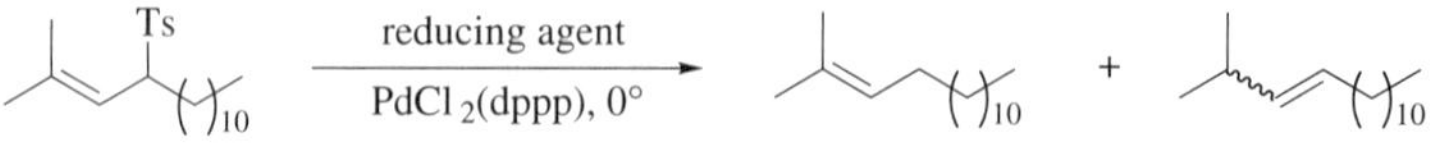

(Eq. 44)

Reducing Agent	Time	Yield	Yield
$NaBH_4$	43 h	(26%)	(7%)
$LiHBEt_3$	1.5 h	(86%)	(0%)

From the constitutional point of view, the employment of formic acid leads to interesting results, since hydride is site selectively transferred to the more hindered position of the allylic sulfone (Eq. 45).[89] Thus, a hydride equivalent generated from formic acid is exceptional in this respect. It is proposed[83] that decarboxylation and hydride transfer is a concerted process in which the hydride site selectively attacks the more substituted (more electropositive) side of the allylic system in a cyclic mechanism (S_Ni transfer of hydride, Scheme 1).

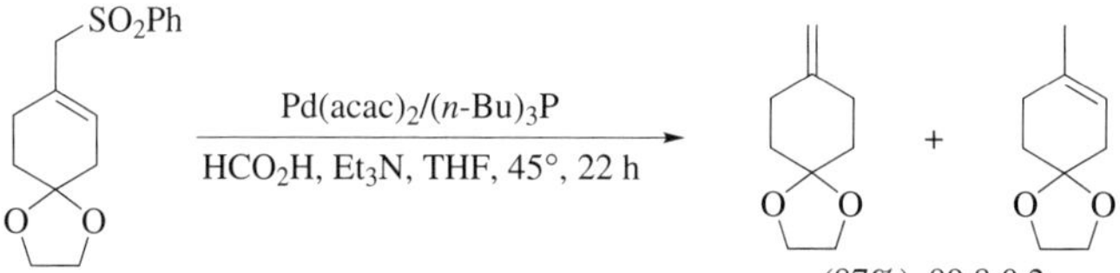

(Eq. 45)

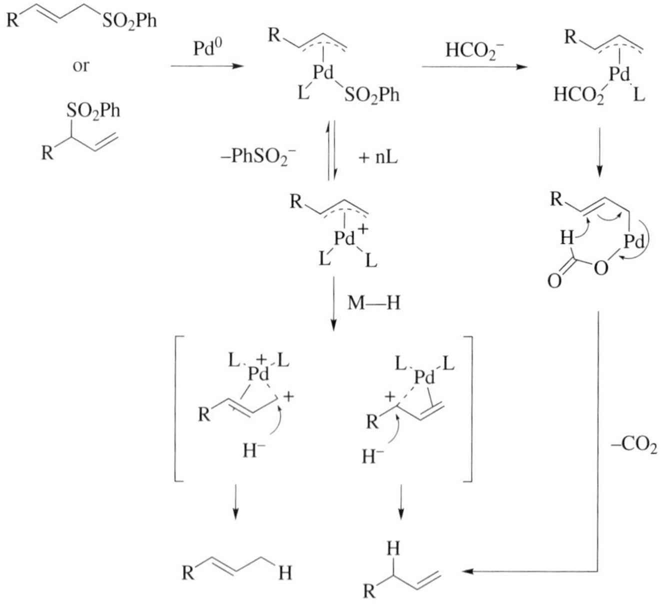

Scheme 1

Other Reducing Agents. 1,4-Dihydropyridines such as 1-benzyl-1,4-dihydronicotinamide (BNAH) are NADH equivalents capable of acting as good electron donors. These reagents have been used in the reductive desulfonylation of α-nitro,[90,91] α-keto,[92] and α-cyanosulfones[92] under sunlight irradiation. Based on the observed experimental results, these reactions seem to proceed via radical-anion species as shown in Scheme 2.[91]

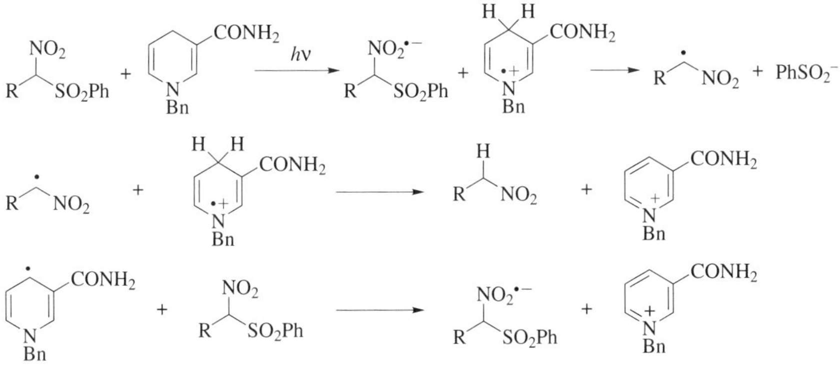

Scheme 2

Reductive desulfonylation of α-nitro sulfones have been carried out employing octylviologen (1,1′-dioctyl-4,4′-bipyridinium dibromide) as an electron-transfer catalyst in a CH_2Cl_2–water two-phase system and in the presence of $Na_2S_2O_4$ (Eq. 46).[93] Octylviologen is reduced by $Na_2S_2O_4$ in the aqueous phase to the cation radical, which after transfer to the organic phase, acts as an SET agent. The method, which is specific for α-nitro sulfones, allows the preparation of nitroalkanes under very mild reaction conditions and is proposed to take place via nitroalkyl radical species.[93]

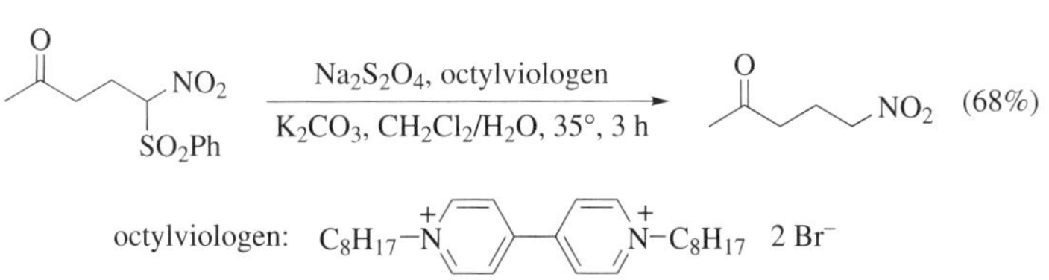

$Na_2S_2O_4$, octylviologen; K_2CO_3, CH_2Cl_2/H_2O, 35°, 3 h (68%) (Eq. 46)

octylviologen: C_8H_{17}–N^+(bipyridinium)N^+–C_8H_{17} 2 Br^-

Reductive Eliminations

Among the different methods for the formation of C–C double bonds, the reductive elimination of β-functionalized (mainly β-hydroxy or β-carboxy) sulfones, is one of the most widely used ones in organic synthesis. The reductive elimination of β-hydroxy sulfones and derivatives is the so-called Julia,[94] or Julia–Lythgoe olefination reaction (Eq. 2). It usually involves a condensation between the anion of an alkyl sulfone and a carbonyl compound to afford a β-hydroxy sulfone (Eq. 47). The metal alkoxide intermediate is typically transformed in situ into a carboxylic or sulfonic ester derivative, which is then reduced with sodium amalgam in methanol. Prior esterification favors the reductive elimination and prevents a possible retro-aldol type process of the alkoxide intermediate.

$$R^1CH(SO_2Ar)R^2 \xrightarrow[\text{2. derivatization}]{\text{1. } R^3COR^4} ArO_2S\text{-}C(R^1)(R^2)\text{-}C(OR)(R^3)R^4 \xrightarrow[\text{MeOH}]{\text{Na/Hg}} R^1R^2C{=}CR^3R^4 \quad \text{(Eq. 47)}$$

$R = H, COR', SO_2R'$

Use of Sodium Amalgam. The reduction classically involves an electron transfer to the sulfone group with loss of the arylsulfinate anion to generate a β-hydroxy or β-carboxy radical, which is further reduced by another equivalent of reducing agent to the alkyl anion. These intermediates are long-lived enough to assume the lowest-energy conformation that undergoes an anti-elimination process to form the E-alkene stereoselectively (Eq. 48). Therefore, the configuration of the alkene product is independent of the intermediate hydroxy sulfone adducts and is strongly influenced by the bulk of the substituents (Eq. 49).[95] This effect is particularly strong when the new double bond is part of a conjugated triene system.

$$n\text{-}C_7H_{15}CH(SO_2Ph)CH(OAc)C_6H_{13}\text{-}n \xrightarrow{\text{Na/Hg}} \left[n\text{-}C_7H_{15}\overset{\ominus}{C}H\text{-}CH(OAc)C_6H_{13}\text{-}n \right] \xrightarrow{-AcO^-} n\text{-}C_7H_{15}CH{=}CHC_6H_{13}\text{-}n \quad \text{(Eq. 48)}$$

dr 60:40 (70%) Z/E 0:100

$$R^1CH(SO_2Ph)CH(OBz)R^2 \xrightarrow[\text{THF/MeOH, }-20^\circ]{\text{Na/Hg}} R^1CH{=}CHR^2 \quad \text{(Eq. 49)}$$

R^1	R^2	Z/E
n-C_7H_{15}	n-C_6H_{13}	20:80
n-C_7H_{15}	i-Pr	10:90
i-Pr	Et_2CH	0:100
1-cyclohexenyl	1-cyclohexenyl	0:100

The Na/Hg reduction of acetoxy sulfones does not follow the originally proposed pathway exclusively. The existence of a different mechanism of reductive elimination was already suggested during a synthesis of vitamin D_4 based on the Julia–Lythgoe olefination.[96] On the basis of deuterium incorporation studies and experimental observations, an alternative mechanism for the reduction of α-acetoxy sulfones with Na/Hg in MeOH has been proposed (Eq. 50).[42] The isolation of the vinylic sulfone intermediate of the reaction as well as the high degree of deuterium incorporation in the final product when the reaction is run in MeOH-d_4 suggest a base-promoted β-elimination of acetate and subsequent reductive desulfonylation. This mechanism allows for the incorporation of deuterium as well as the observed alkene stereoselectivity via equilibration of the vinyl radical. The outcome of a given reduction using Na/Hg may, depending on the structure of the starting β-acetoxy sulfone, follow different reaction mechanisms. It is difficult to determine whether the reaction proceeds exclusively through the vinylic sulfone or through the direct reduction of the acetoxy sulfone. Successful Julia olefinations of substrates with Na/Hg where the formation of the vinylic sulfone intermediate is not possible have been also reported (Eq. 51).[97]

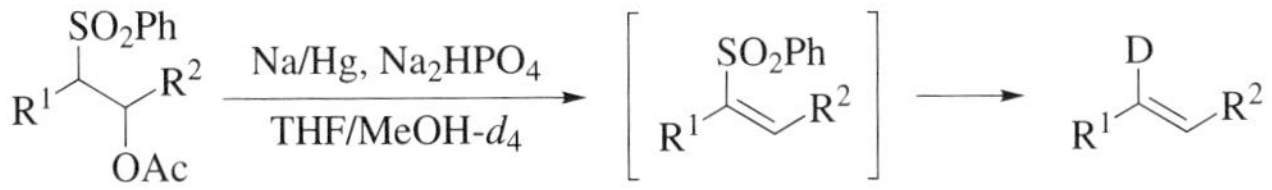

(Eq. 50)

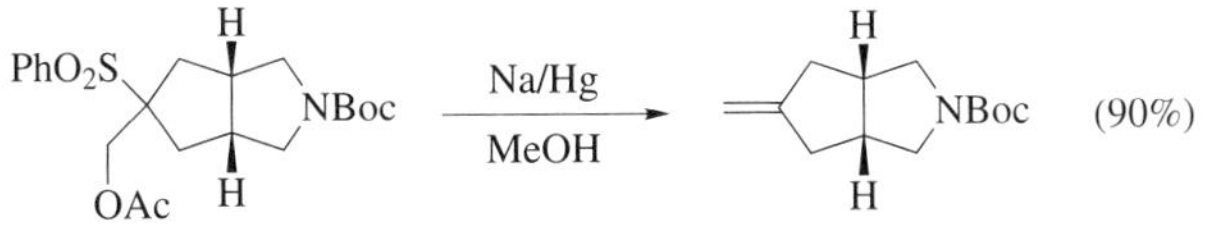

(Eq. 51)

Use of Tin Hydrides. From the mechanistic point of view, the reductive elimination of methyl xanthate derivatives of β-hydroxy sulfones entails an interesting variation of the Julia reaction since an initial fragmentation of the C–O bond through a Barton-McCombie-type radical deoxygenation takes place. Final aryl sulfonyl radical elimination affords the corresponding alkene (Eq. 52).[98,99]

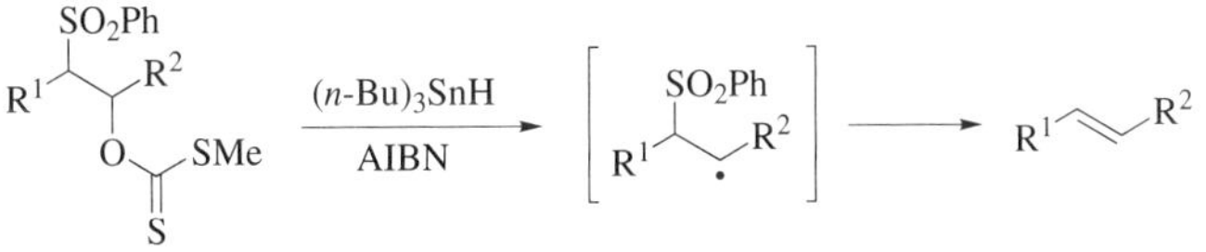

(Eq. 52)

Use of Samarium(II) Iodide. Samarium(II) iodide in the presence of various additives such as HMPA[100] or DMPU[42] is a non-basic alternative to Na/Hg in the classical Julia olefination.[29] The mechanism by which this reaction proceeds depends on the starting sulfone. The enormous rate differences in the formation of the alkene from the β-hydroxy sulfone or its benzoyl derivative strongly suggest different reaction pathways. With β-hydroxy sulfones, a single-electron transfer from SmI_2 to the aromatic sulfone moiety initiates the traditional path via initial C–S bond cleavage (Eq. 53).[101] In contrast, transfer of an electron from SmI_2 to the benzoate function is a much easier process, leading to initial C–O bond fragmentation (Eq. 53).[101]

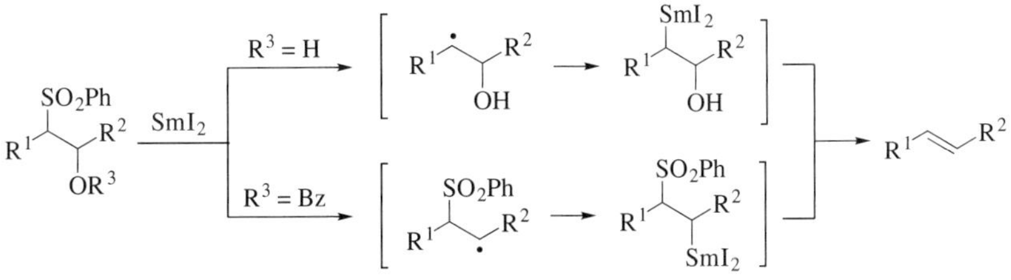

(Eq. 53)

The role of the additive (e.g. HMPA, DMPU) in the process is to increase the reducing power of Sm(II) and is crucial since no elimination reaction is observed in its absence, unless a better single-electron acceptor aromatic sulfone, such as a β-hydroxy imidazolyl sulfone is used (Eq. 54).[102] The employment of imidazolyl sulfones increases the efficiency of the olefination process due to the absence of a retro-aldolization reaction and the change in the nature of the leaving group (probably $HOSmI_2$).

Me—N N SO2 Ph OH; SO2Ph Ph OH; SmI_2 THF, rt; Ph (87%) Z/E 17:83; no reaction (Eq. 54)

In general, better yields are obtained by using SmI_2/HMPA than by the traditional procedure using Na/Hg. Reductions using SmI_2/HMPA afford slightly different E-selectivities, but the diastereoselection is unaffected by the reaction temperature (Eq. 55).[103]

OR OTBDPS SO_2Ph → OTBDPS (Eq. 55)

R	Reducing Agent	Yield	Z/E
H	SmI_2/HMPA	(73%)	25:75
H	Na/Hg	(68%)	32:68
Ac	SmI_2/HMPA	(95%)	24:76
Ac	Na/Hg	(88%)	21:79

SCOPE AND LIMITATIONS

Reductive Desulfonylation

Reductive Desulfonylations by Active Metals and Salts. Relatively large quantities of reducing agent are required because of the stoichiometric nature of the reactions. Further, the separation of products from large amounts of aqueous solutions of metal salts may be laborious and inefficient.

Use of Alkali Metals in Ammonia. The reductive desulfonylation process with solutions of alkali or alkaline earth metals (Li, Na, and less frequently, Ca) in anhydrous ammonia or low molecular weight amines is solvent- and substrate-dependent and the outcome of the desulfonylation may be different depending on the reaction conditions employed. One of the main disadvantages of this method is the strongly basic conditions employed, which are not compatible with base-labile substrates. Incompatibility with diverse protecting groups such as acetamides, as well as various benzyl protecting groups such as benzyloxy (BnO), benzyloxymethoxy (BOMO), and 4-methoxybenzyloxy (PMBO) (Eq. 56) is also a limitation.[104,105] Decreasing the reduction power of the reagent using Ca instead of Li or Na is not sufficient to prevent benzyl deprotections.[106] Other protecting groups for the hydroxy function such as *tert*-butyldimethylsilyl (TBDMS), triisopropylsilyl (TIPS), and methoxymethyl (MOM), as well as acetals, isolated double bonds, carbamates, and epoxides are tolerated.

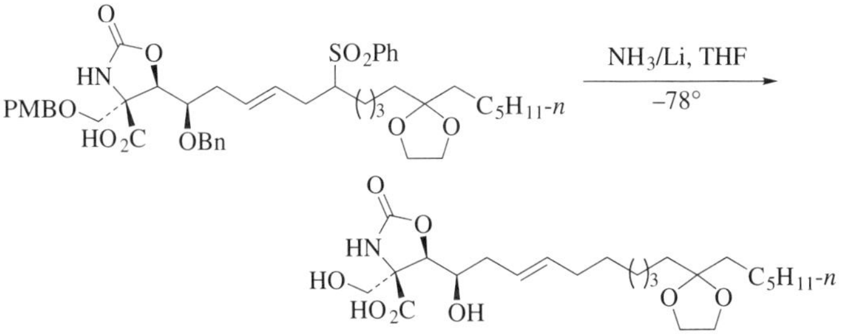

NH_3/Li, THF
−78°

(Eq. 56)

The reduction of allylic sulfones by alkali or alkaline earth metals in ammonia or low molecular weight amines is generally accompanied to varying degrees by regio- and stereochemical problems such as double bond rearrangements and isomerizations (Eq. 57).[107] This drawback is general for the reductive desulfonylation reactions employing activated metals.

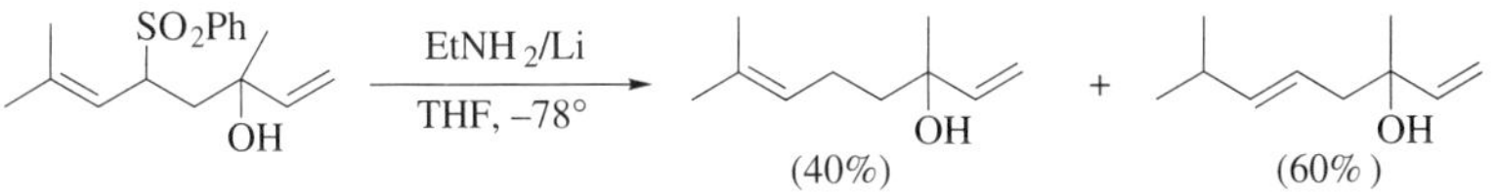

$EtNH_2$/Li
THF, −78°
(40%) (60%)

(Eq. 57)

Furthermore, allylic C−OR bond cleavage of protected allylic alcohols (OR = OMOM, OAc) takes place during reductions with alkaline earth metals in ammonia and forces the selection of different reductive desulfonylation procedures, as depicted in Eq. 58. Reduction with Na/NH_3 practically destroys one of the diastereomeric products by cleavage of the allylic C−O bond. This process is not observed when the reduction is carried out with Mg in MeOH.[108] Alcohol deprotection is a good strategy to avoid C−O cleavage. It has been used, for instance, in the synthesis of all-*trans*-geranylgeraniol from the corresponding tosyl derivative as depicted in Eqs. 59a and 59b.[109] In the case of sulfone derivatives bearing benzyl-protected allylic alcohols, it is possible to avoid the allylic C−O cleavage by proper selection of the solvent. When the reduction is carried out in the presence of MeOH as co-solvent, cleavage of the allylic C−O bond occurs, probably as a consequence of Birch reduction of the benzene ring of the benzyl ether.[110] On the other hand, when the reduction is carried out in the absence of an alcoholic solvent, debenzylation takes place first, which avoids the fragmentation of the allylic C−O bond (Eq. 60).[111]

NH_3/Na	(1%)	(15%)
Mg, MeOH	(46%)	(28%)

(Eq. 58)

$EtNH_2$/Li, Et_2O, –78°, 30 min (70%) (Eq. 59a)

$EtNH_2$/Li, Et_2O, –78°, 75 min (43%) (Eq. 59b)

$EtNH_2$/Li, Et_2O, –78°, 45 min (84%) (Eq. 60)

Use of Metals in Alcoholic Solvents. Magnesium in the presence of mercuric chloride as a catalyst is by far the most employed metal when the reductive desulfonylation is carried out in low molecular weight alcoholic solvents.[112] Only a small number of examples exist where Na or Li in alcoholic solvents is used, mostly for the desulfonylation of alkyl sulfones. The use of Mg in a low molecular weight alcoholic solvent in the presence of mercuric chloride is an extremely convenient desulfonylation method for a wide variety of sulfones (Eq. 61).[113] Although β-ketosulfones are inert towards this reagent,[114] α-sulfonyl esters are efficiently desulfonylated by Mg in MeOH.[115,116]

Mg, $HgCl_2$ (cat), MeOH/THF, rt, 2 h (Eq. 61)

Isomerization of allylic sulfones and reduction of conjugated double bonds (Eq. 62), are possible disadvantages of the use of Mg as an electron-transfer agent.[117] The occurrence of Julia olefination of 1,2-disulfone derivatives is another important drawback when performing reductive desulfonylation on these kinds of substrates (Eq. 63).[118]

Mg, MeOH; rt, 3 h (Eq. 62)

Mg, MeOH; 50°, 4-5 h (21%) + (62%) (Eq. 63)

The functional and protecting group tolerance of the Mg method for reductive desulfonylation is higher compared to those reactions employing alkali metals in ammonia or amine solutions. The cleavage conditions are mild and compatible with most functionalities and protecting groups except for esters and lactones, which may hydrolyze. Hydroxyl protecting groups such as benzyloxy, silyloxy, tetrahydropyranyloxy (THP), and MOM groups as well as a wide variety of functional groups such as carbamates, nitriles, phosphonates, epoxides, acetals, and isolated double bonds, are inert under the reaction conditions (Eq. 64).[119] The reaction is usually carried out in the presence of mercuric chloride as a catalyst. An environmentally benign alternative to the mercury-catalyzed process has been recently reported that employs AcOH as both an activator for the Mg metal and a proton source.[120]

BnO, CN, SO_2Ph — Mg, $HgCl_2$, MeOH/THF; 0° to rt, 2 h → BnO, CN (98%)

(Eq. 64)

Magnesium in MeOH reduces β hydroxy sulfones without producing reductive elimination side products (Eq. 65).[121] The poor leaving group character of the β-hydroxy moiety is key in avoiding elimination since reductive elimination is the main process when attempting the reduction of activated substrates (Eq. 66).[122]

PhO_2S, HO, OTBDPS — Mg, MeOH; 50°, 24 h → OTBDPS, OH (85%) (Eq. 65)

SO_2Ph, Tol, OAc — Mg, $HgCl_2$; EtOH, rt, 1 h → Tol (53%) (Eq. 66)

Use of Lithium Naphthalenide. Lithium arene radical anion complexes are mild and highly effective reagents for the reductive desulfonylation process of functionalized sulfones. These reagents have only rarely been used with vinylic and allylic sulfones. In addition to high yields and their operational simplicity, metal arene radical anion complexes demonstrate high chemoselectivity (Eq. 67).[123]

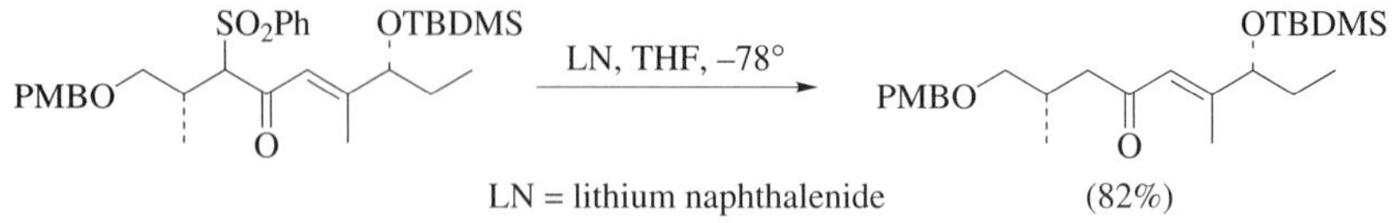

LN, THF, –78°

LN = lithium naphthalenide (82%)

(Eq. 67)

Good tolerance to hydroxyl protecting groups is usually observed at low temperatures (–78°), and it is only when working at higher temperatures that O-debenzylation is detected (Eq. 68).[124]

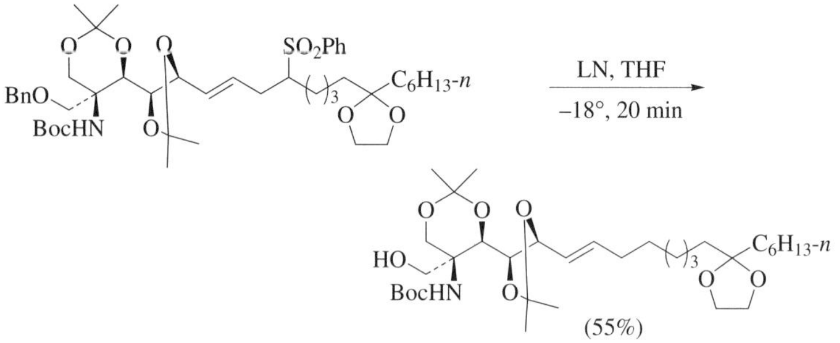

LN, THF, –18°, 20 min (55%)

(Eq. 68)

An interesting feature is that the reductive desulfonylation can be carried out in the presence of thioethers and no desulfurization is observed (Eq. 69).[125] This reagent also tolerates isolated and conjugated double bonds, ketones, acetals, and Boc carbamates. Sulfonamides, however, are not tolerated, and even at low reaction temperatures give the corresponding amines. β-Elimination of arylsulfinates is also observed (Eq. 70).[126]

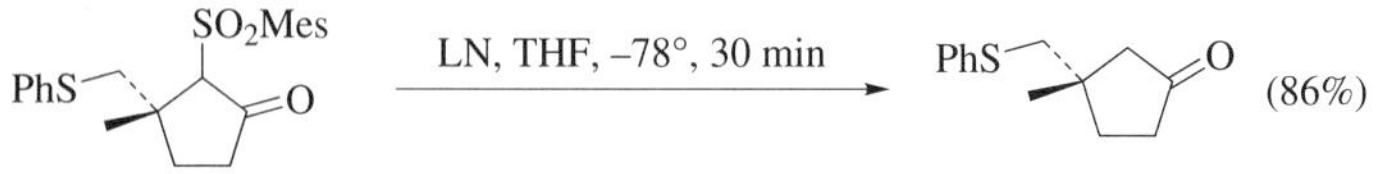

LN, THF, –78°, 30 min (86%) (Eq. 69)

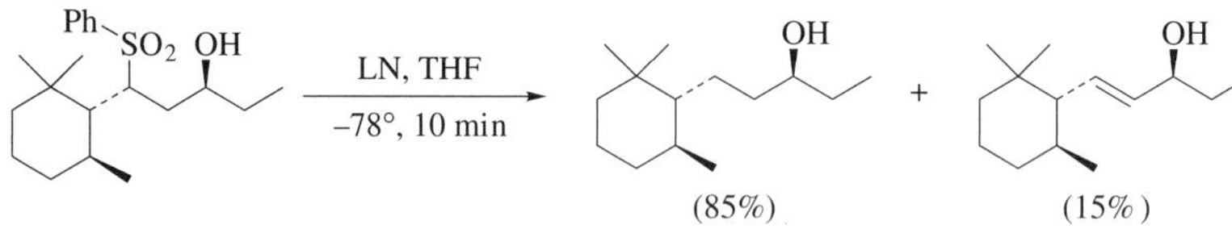

LN, THF, –78°, 10 min (85%) + (15%) (Eq. 70)

The carbanionic reactivity of the anomeric center in carbohydrates has assumed great importance during the last two decades in synthetic carbohydrate and natural product chemistry.[54] A sulfonyl group situated in the anomeric center facilitates removal of the anomeric proton for further functionalization. Additionally, the sulfone group can be replaced by an electrophile through a reductive metalation process. These two features are used for the one-pot stereoselective synthesis of *C*-glycosyl derivatives.[54] For instance, the stereocontrolled desulfonylation with lithium naphthalenide of anomeric glycosyl aryl sulfones derived from 2-deoxy-D-glucose is used in the stereoselective synthesis of *C*-glycosides through the sequence of deprotonation, electrophilic trapping, and in situ reductive desulfonylation followed by proton quenching (Eq. 71).[51–53] Different electrophiles such as alkyl iodides,[52] aldehydes,[52] esters,[52] carbonates,[53] and aziridines[127] have been successfully employed depending on the stereoselectivity of the reductive desulfonylation step on the electrophile (Eq. 71). Experiments using D_2O with alkyl halides and carbonyl compounds show that the anion formed in the desulfonylation step is configurationally stable and the final protonation occurs from the same side as the departing α-*C*-glycosyl sulfone (see Eq. 31). On the contrary, with enolizable carbon substituents, anomeric enolates can form during desulfonylation, which suffer attack by the proton from the sterically less hindered β-face of the pyran ring (Eq. 71).

The reductive cleavage of allylic tosylmethyl ethers with lithium naphthalenide in THF is used to prepare metalated allylic ethers that undergo [2,3] Wittig rearrangements in situ (Eq. 72).[128]

TBDMSO, TBDMSO, TBDMSO, O, SO_2Ph —(1. LDA; 2. E)→ TBDMSO, TBDMSO, TBDMSO, O, E, SO_2Ph —(1. LN, –78° to –100°; 2. H_3O^+)→

TBDMSO, TBDMSO, TBDMSO, O, R + TBDMSO, TBDMSO, TBDMSO, O, R (Eq. 71)

E	R	α/β	Yield
D_2O	D	0:100	(80%)
MeI	Me	0:100	(43%)
PhCHO	PhCHOH	0:100	(74%)
$(MeO)_2CO$	CO_2Me	95:5	(72%)
$PhCO_2Ph$	PhCO	90:11	(72%)

MeO, O, Ts —(LN, THF, TMEDA; –65° to 0°)→ MeO, OH (Eq. 72)

(43%) Z/E 64:36

TMEDA = *N,N,N',N'*-tetramethylethylenediamine

Use of Sodium Amalgam. Reductive desulfonylation with metal amalgams, and particularly Na/Hg (5–6%), is the most widely employed and general method for all types of sulfones even though it requires the handling of substantial quantities of mercury, which is toxic and relatively expensive.

Due to the strongly basic conditions associated with this method, the reduction is usually carried out in buffered methanol or ethanol solutions, which is particularly important for base-sensitive substrates (Eq. 73),[129] and in situations where undesired β-elimination processes should be avoided. The β-elimination of arylsulfinate under Na/Hg reductive desulfonylation conditions was initially avoided by employing mixtures of HMPA/EtOH as the solvent (Eq. 74).[130] The use of disodium hydrogen phosphate[33] is now the most widely used and effective method to control unwanted reactions. However, the β-elimination reaction of the arylsulfinate moiety is sometimes very difficult to avoid completely even under optimal reaction conditions (Eq. 75).[131] This undesirable side-reaction becomes more significant when the newly formed unsaturation is a part of a stabilized conjugated system.

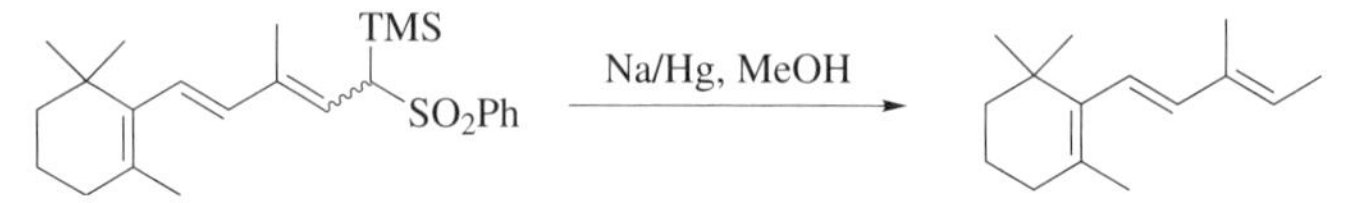

TMS, SO_2Ph — Na/Hg, MeOH → (Eq. 73)

SO_2Ph — Na/Hg, 0°, 1 h → + (Eq. 74)

Solvent		
EtOH	(67%)	(30%)
HMPA/EtOH (9:1)	(85%)	(15%)

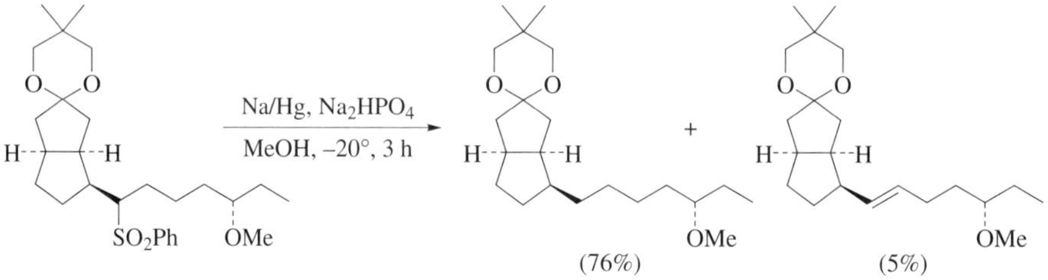

Na/Hg, Na_2HPO_4, MeOH, –20°, 3 h → (76%) + (5%) (Eq. 75)

Another major drawback of this method for reductive desulfonylation of β-alkoxy sulfones is that the reaction is occasionally accompanied by elimination of the β-alkoxy (or hydroxy) group leading to unwanted side products (Eq. 76).[94]

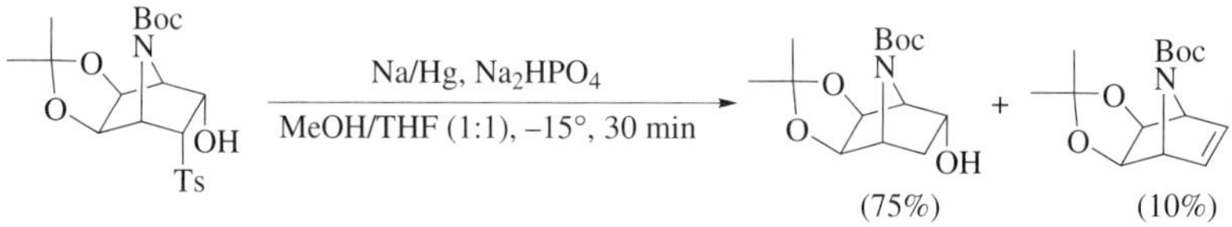

Na/Hg, Na_2HPO_4, MeOH/THF (1:1), –15°, 30 min → (75%) + (10%) (Eq. 76)

The tolerance of hydroxyl protecting groups toward Na/Hg reduction is very high. The latter include all the hydroxyl protecting groups, such as TBDMS, *tert*-butyldiphenylsilyl (TBDPS), TIPS, THP, 2-(trimethylsilyl)ethoxymethyl (SEM), MOM, Bn, and PMB. On the other hand, trimethylsilyl (TMS) protected alcohols and acetates are deprotected as depicted in Eqs. 77[132] and 78[109], respectively. In this latter reaction, the main problem associated with the reductive desulfonylation of the allylic sulfone employing Na/Hg in MeOH is the migration of the allylic double bond to afford mixtures of isomers. Double bond migration is also observed even when the allylic double bond belongs to an α,β-unsaturated system (Eq. 79).[133] The reductive deconjugation in those particular examples is due to the formation, under the basic reaction conditions, of the corresponding dienolate, which undergoes kinetically controlled protonation. Similar results are obtained when using alkali metals in ammonia solutions.[134]

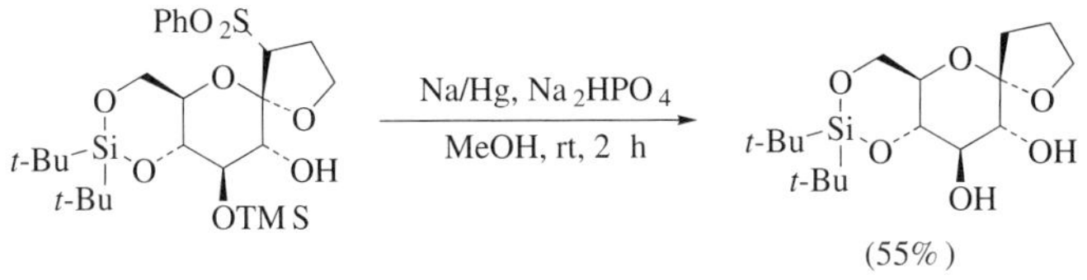

(Eq. 77)

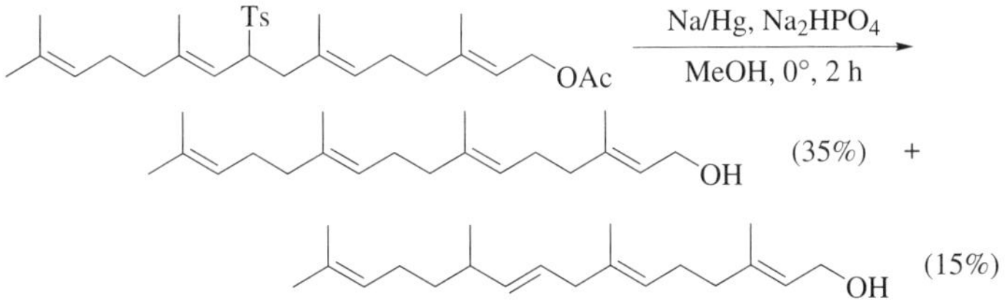

(Eq. 78)

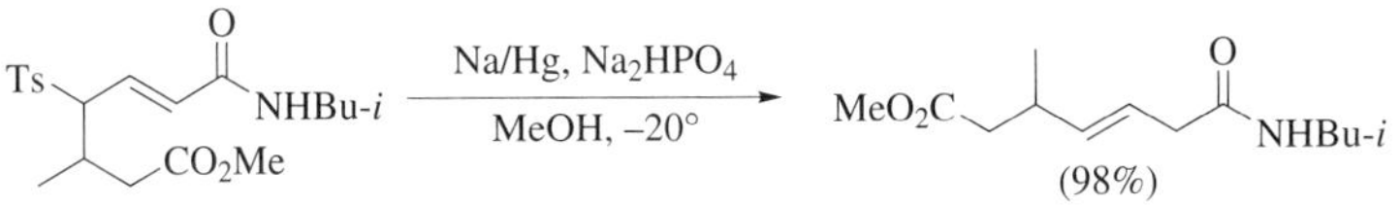

(Eq. 79)

Cleavage of allylic C–O bonds (Eq. 80),[135] reduction of conjugated double bonds,[136,137] and reductive dehalogenations[138,139] occasionally intervene when reducing functionalized sulfones with Na/Hg. These side reactions are dependent on the substrates and reaction conditions, and should not be considered as general limitations.

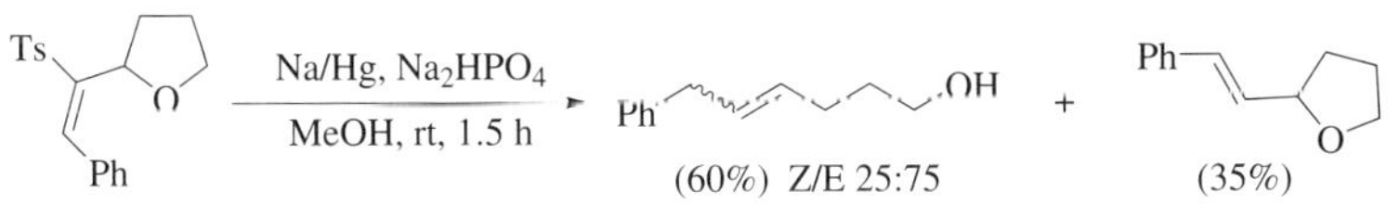

(Eq. 80)

Sodium amalgam buffered with Na_2HPO_4 is also a chemoselective reagent for the desulfonylation of β-oxo sulfones, a frequently used reaction in numerous total syntheses. The example depicted in Eq. 81 comes from the synthesis of an intermediate in the preparation of analogues of migrastatin as anti-metastatic agents.[140] The desulfonylation of this type of sulfone with Na/Hg is a fast, general, and high-yielding process and no problems associated with concomitant reduction of the β-carbonyl group are observed.

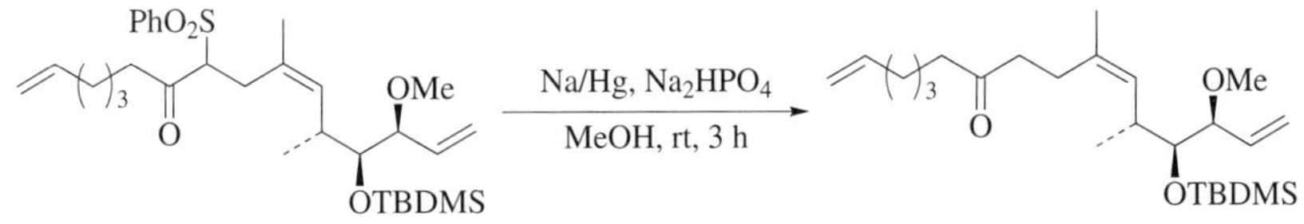

(Eq. 81)

On the other hand, carbonyl reduction is observed in reduction of sulfone derivatives bearing dialkyl ketone moieties with Na/Hg if very long reaction times and room temperature are employed (Eq. 82).[39] This reduction[141] as well as other side-reactions such as pinacol couplings[142] are also observed with substrates bearing aromatic ketones even when low temperature conditions and short reaction times are used (Eq. 83).

COMe, Ts → Na/Hg, Na_2HPO_4 / MeOH, rt, 24 h → OH (90%) (Eq. 82)

SO_2Ph, O, O → Na/Hg, Na_2HPO_4 / MeOH, 0°, 1.5 h → O, OH (59%) (Eq. 83)

A major drawback when reducing alkenyl sulfones with Na/Hg is that the configuration of the double bond is not necessarily preserved. Additionally, reduction of the double bond is occasionally observed as shown in Eq. 84, where a Julia–Lythgoe olefination process is also taking place.[135]

Ts, OMOM, Ph, OH → Na/Hg, Na_2HPO_4 / MeOH, rt, 1.5 h → Ph, OMOM (67%) (Eq. 84)

Use of Aluminum Amalgam. Aluminum amalgam is widely used in the chemoselective reduction of α-sulfonylated carbonyl groups because of the high tolerance shown by this reagent towards other functional groups (Eq. 85).[143] Use of a large excess of toxic mercury is one of the main drawbacks associated with this method.

O, PhO_2S, MeO, O, O, O, H → Al/Hg, THF/H_2O → O, MeO, O, O, O, H (66%) (Eq. 85)

A wide variety of functional groups such as hydroxy and amino groups, esters, amides, carbamates, acetals, thioacetals, and isolated double bonds are tolerated. Sulfonamides, which are labile towards other reductive reagents such as Na/Hg,[144] are unreactive.[145] On the other hand, Al/Hg reduces aromatic nitro compounds to the corresponding anilines, a feature that has been used in the synthesis of the alkaloid dehydroisolongistrobine (Eq. 86).[146] As previously mentioned for Na/Hg desulfonylations, reduction of aromatic ketones to the corresponding benzylic alcohols may also occur. Allylic sulfones usually suffer double bond migration and isomerization. Reduction of the double bond in alkenyl sulfones is also observed when the unsaturation is part of an α,β-unsaturated carbonyl compound (Eq. 87),[147] and with dienyl sulfones.[48]

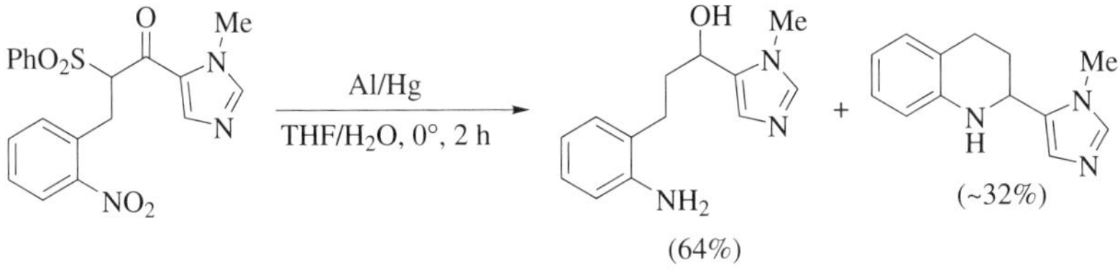

PhO_2S ... Al/Hg, THF/H_2O, 0°, 2 h → (64%) + (~32%) (Eq. 86)

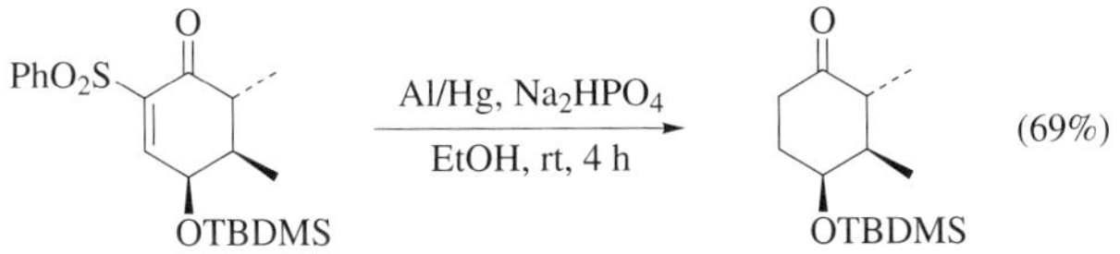

PhO_2S ... OTBDMS — Al/Hg, Na_2HPO_4, EtOH, rt, 4 h → OTBDMS (69%) (Eq. 87)

Under the mild reaction conditions associated with this reducing agent, it is possible to perform reductive desulfonylations of β-hydroxy sulfones without formation of the Julia olefination products (Eq. 88).[148]

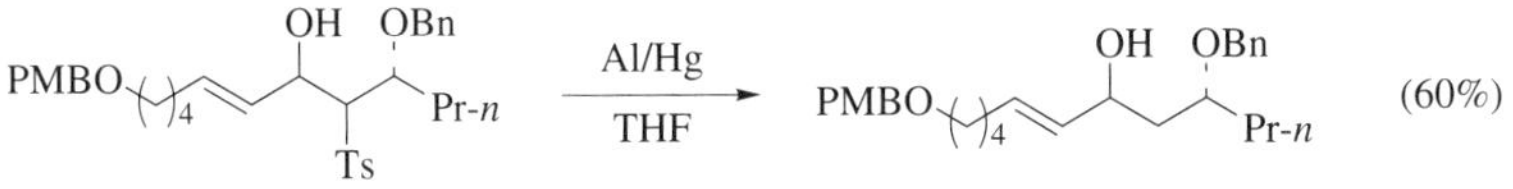

PMBO ... OH OBn, Ts, Pr-*n* — Al/Hg, THF → PMBO ... OH OBn, Pr-*n* (60%) (Eq. 88)

Use of Samarium(II) Iodide. In recent years, samarium(II) iodide has become a popular choice as a single-electron-transfer reagent for mild and selective reductive desulfonylations. This reagent is employed for the reduction of a wide variety of functionalized sulfones, principally β-keto and vinylic sulfones (Eq. 89).[149]

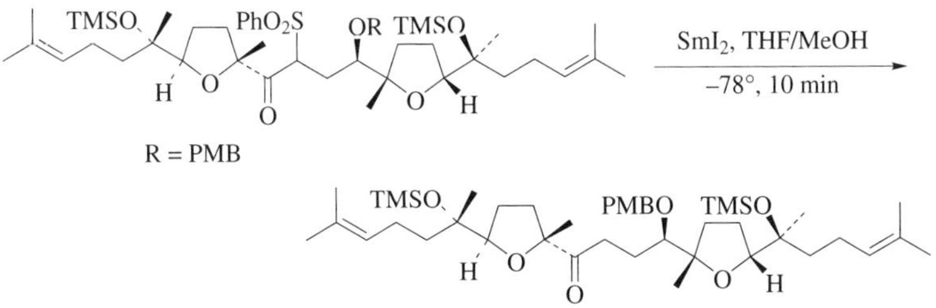

TMSO, PhO_2S, OR TMSO — SmI_2, THF/MeOH, −78°, 10 min →

R = PMB (Eq. 89)

TMSO, PMBO TMSO

The immobilization of sulfones on solid supports has become increasingly popular in organic synthesis. A very interesting solid-phase approach to tetrahydroquinolones using a Merrifield resin-supported sulfone linker that can be cleaved by SmI_2 has been presented (Eq. 90).[150] The same reagent is efficiently employed in a high-throughput fluorous-phase synthesis of nitrogen heterocycle libraries.[151]

C-Glycosides have been stereoselectively synthesized via SmI_2-promoted Barbier reactions between glycosyl pyridylsulfones and carbonyl compounds.[152–159,54] A SmI_2 reduction of glycosyl pyridyl sulfones bearing a silicon-tethered unsaturated group at the C2–OH position is used for the stereospecific synthesis of 1,2-*cis*-*C*-glycosides and *C*-disaccharides such as methyl-α-*C*-isomaltoside in good yield (Eq. 91).[160]

SmI_2, THF, LiCl; rt (Eq. 90)

1. SmI_2, THF
2. TBAB, THF
3. Pd/C, H_2
4. Ac_2O, pyr

(48%) (Eq. 91)

TBAB = tetra(*n*-butyl)ammonium bromide

Samarium(II) iodide is also a good reducing agent for β-hydroxy-functionalized sulfones.[100] Several examples show that Julia olefination can be avoided, at least partially, with these sulfones if SmI_2 is employed (Eq. 92).[100,161] This circumstance, however, is not general and seems to be substrate-dependent (Eq. 93).[103] β-Hydroxy sulfones may be prepared by SmI_2-mediated reductive addition of geminal disulfones to ketones without the concomitant olefination process (Eq. 94).[162] The Julia olefination is the predominant reaction in the reduction of 1,2-disulfonylated compounds (Eq. 95).[100] Another problem arises upon reduction of bromine-containing sulfone derivatives, where reduction of the halide is the principal process (Eq. 96).[163] Such halide reduction is not observed in the reductive desulfonylation of fluorinated compounds with SmI_2.[161]

SmI_2, THF; HMPA, rt, 1 h (53%) + (20%)

(Eq. 92)

OH, SO_2Ph, OTBDMS — SmI_2, THF, HMPA, rt, 1 h → OTBDMS (75%) (Eq. 93)

SO_2Ph, O, PhO_2S — SmI_2, THF, rt → OH, PhO_2S (71%) (Eq. 94)

PhO_2S, SO_2Ph — SmI_2, THF, HMPA, –20°, 30 min → (91%) (Eq. 95)

Br, SO_2Ph — SmI_2, THF, rt → SO_2Ph (65%) (Eq. 96)

Use of Sodium Hydrogen Telluride. Certain α-functionalized-α,β-unsaturated sulfones are desulfonylated using sodium hydrogen telluride in ethanol at room temperature.[164–166a] α-Methylthio-α,β-unsaturated sulfones are reduced to give vinylic thioethers in good yields albeit with moderate selectivities (Eq. 97).[164] On the other hand, α-methylsulfonyl chalcones suffer tandem reduction-desulfonylation in the presence of DMF as cosolvent.[165,166a]

Ar, SO_2Ph, SMe — NaTeH, EtOH, rt, 2-3 h → Ar, SMe (Eq. 97)

Ar	Yield	Z/E
Ph	(82%)	74:26
4-MeC_6H_4	(73%)	74:26
4-$MeOC_6H_4$	(75%)	68:32
3-ClC_6H_4	(78%)	76:24
4-ClC_6H_4	(80%)	68:32
2-furyl	(67%)	72:28

Use of Sodium Dithionite. Sodium dithionite is a mild and inexpensive reducing agent that has numerous applications in organic synthesis.[166b–166e] With respect to the reductive desulfonylation reaction, alkenyl sulfones are readily reduced to alkenes by reaction with $Na_2S_2O_4$ under weakly basic conditions in aqueous DMF at high temperatures. The process gives good yields of alkenes and is stereospecific with retention of the configuration of the original alkenyl sulfone (Eq. 98).[46] This method allows the preparation of monodeuterated alkenes by replacing the water with D_2O.[167] Sodium dithionite affords the corresponding allylic sulfones when reducing conjugated sulfonyl dienes as a consequence of the reduction of the alkenyl sulfone double bond (Eq. 99).[48] In contrast, sulfonyl 1,4-dienes are stereospecifically reduced to the corresponding dienes in good yields (Eq. 100).[48]

$Na_2S_2O_4$, $NaHCO_3$; DMF/H_2O, 120°, 2 h (80%) (Eq. 98)

$Na_2S_2O_4$, $NaHCO_3$; adogen, cyclohexane/H_2O, reflux, 18 h (50%) (Eq. 99)

adogen = methyltrialkyl(C_8–C_{10})ammonium chloride

$Na_2S_2O_4$, $NaHCO_3$; H_2O, reflux, 18 h (65%) (Eq. 100)

Although less commonly employed, $Na_2S_2O_4$ also reduces β-ketosulfones in moderate yields (Eq. 101).[47] In spite of the limited use of $Na_2S_2O_4$ as a reductive desulfonylating agent and the need for high temperatures for a successful reaction, the absence of toxic by-products makes this an attractive "green" reagent and should be considered as a viable alternative to the well-established methods.

$Na_2S_2O_4$, $NaHCO_3$; DMF, 100°, 24 h (50%) (Eq. 101)

Reductive Desulfonylations by Tin Hydrides. Tin hydrides such as n-Bu_3SnH and Ph_3SnH are used to perform small- to large-scale reductive desulfonylations of allyl, alkenyl, and α-functionalized sulfones. The desulfonylation of allylic sulfones is a site selective process but not usually stereoselective since the allylstannane intermediates are produced as mixtures of stereoisomers[61] (Eq. 35). Alkenyl sulfones generate alkenylstannanes that have been used as intermediates in different reactions, such as the palladium-catalyzed cross-coupling with aryl and alkenyl halides, and the tin–lithium exchange and subsequent reaction with electrophiles. In this manner fluoroalkenes[168] and different natural product derivatives such as functionalized glycals[169–171] and nucleic acid analogues[172,173] have been prepared. The reduction of β-oxo sulfones is slow and does not proceed to completion when small amounts of the radical initiator 2,2′-azobis(2-methylpropionitrile) (AIBN) are used. With larger amounts of AIBN, the reduction is complete in minutes (Eq. 102) indicating a short radical chain length.[59] The reduction does not work well with substrates possessing a phenyl ring attached to the carbonyl of the β-keto phenyl sulfone derivative. For these substrates, the alternative use of triphenylstannane renders the process more effective (Eq. 103).[59]

(n-Bu)$_3$SnH, AIBN; toluene, reflux, 40 min (89%) (Eq. 102)

R_3SnH, AIBN; solvent, reflux (Eq. 103)

R_3SnH	Solvent	Time	Yield
$(n\text{-Bu})_3SnH$	toluene	40 min	(16%)
Ph_3SnH	toluene	5 min	(43%)
$(n\text{-Bu})_3SnCl + NaBH_3CN$	*t*-BuOH	1 h	(90%)

The tin hydride species can be genereated in solution using n-Bu_3SnCl in the presence of $NaBH_3CN$, a method that allows the production of tin hydride in low concentration, which is particularly effective for the desulfonylation of β-ketosulfones bearing a phenyl group directly attached to the carbonyl moiety (Eq. 103).[174] However, under these conditions, sterically crowded substrates are not desulfonylated even after prolonged reaction periods and/or using a large excess of the reducing agent.[174]

The reductive desulfonylation of α-sulfonyl esters with tin hydrides only succeeds with π-deficient heterocyclic sulfones (Eq. 104).[175] These sulfones are inert under standard procedures using Al/Hg or Na/Hg, but undergo facile C–S cleavage with tin hydrides. Substitution of n-Bu_3SnD for n-Bu_3SnH gives access to α-deuterated esters.[58] A catalytic version of the reaction is carried out with substoichiometric amounts of tributyltin chloride and an excess of poly(methylhydrosiloxane) (PMHS) in the presence of potassium fluoride. This method has been employed for the synthesis of 2-fluoroalkanoates (Eq. 105).[58]

$(n\text{-Bu})_3SnH$, AIBN; C_6H_6, reflux (Eq. 104)

Z	Y	Time	Yield
CH	CH	48 h	(0%)
CH	N	36 h	(60%)
N	N	1 h	(>99%)

$(n\text{-Bu})_3SnCl$, AIBN; PMHS, KF, toluene/H_2O, reflux (84%) (Eq. 105)

Tri-n-butyltin hydride is unproductive in the reduction of non-activated alkyl phenyl sulfones. Thus, the selective desulfonylation of a β-keto phenyl sulfone in the presence of an alkyl sulfone is possible. Another interesting example of chemoselectivity is seen in the reduction of α-arylsulfonyl phosphonates.[176] Tin hydride reduction of this kind of non-activated sulfone produces C–P fragmentation to afford the corresponding sulfone derivatives (Eq. 106a). In contrast, it is possible to carry out the C–S cleavage (Eq. 106b) employing Na/Hg under standard conditions.[176] Tin hydride reductions also complement the SET reductions of α-nitro sulfones with 1,4-dihydropyridines (Scheme 2), because reduction of the nitro group is the only process observed.

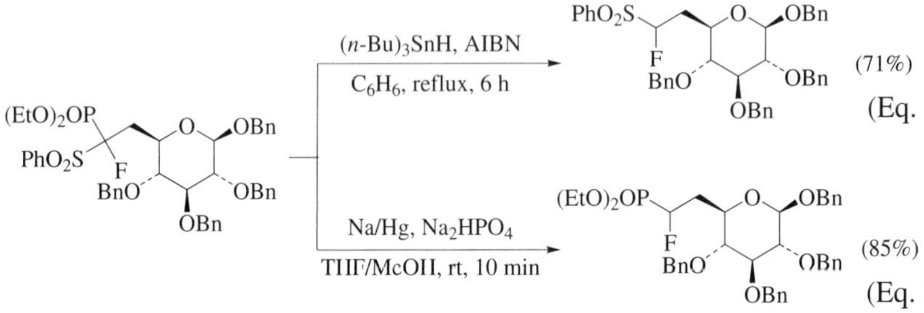

$(n\text{-Bu})_3SnH$, AIBN; C_6H_6, reflux, 6 h (71%) (Eq. 106a)

Na/Hg, Na_2HPO_4; THF/MeOH, rt, 10 min (85%) (Eq. 106b)

Reductive desulfonylation by tin hydride is carried out under neutral conditions, which is particularly applicable for substrates that are labile toward acid or base. Thus, functional groups such as isolated and conjugated double bonds, esters, acetals, nitriles, and epoxides are tolerated in addition to a wide variety of hydroxyl (silyl and benzyl) and amine (carbamates, benzyl) protecting groups (Eq. 107).[177]

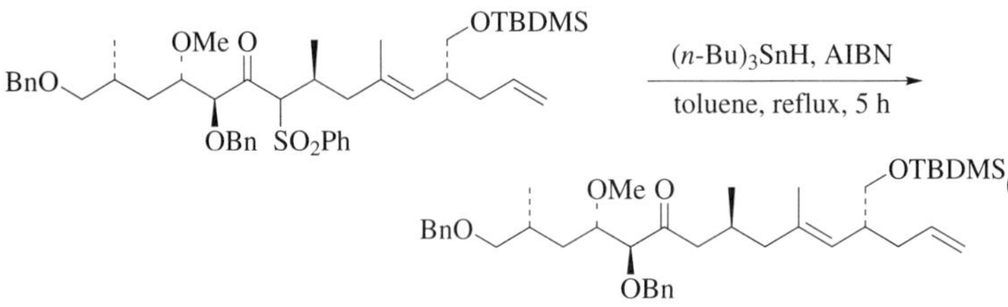

$(n\text{-Bu})_3SnH$, AIBN; toluene, reflux, 5 h (Eq. 107) (77%)

Transition-Metal-Mediated Reductive Desulfonylations. Few conclusions can be drawn with respect to functional group tolerance of Ra–Ni since the use of this reagent is not very common, given the many operational drawbacks that Ra–Ni presents. Among them are the tediousness of the preparation, its pyrophoric nature, the loss of activity on storage, and the difficulty of quantifying the nickel reagent. Thus, the main problem in the use of this reagent is the reproducibility of reactions, especially since detailed experimental conditions and information about the Ra–Ni used are often not reported. In spite of these disadvantages Ra–Ni has been successfully used in the reductive desulfonylation of β-hydroxy sulfones (Eq. 108)[178] and has shown tolerance towards carbonyl compounds and their derivatives such as ketones,[179] acetals,[180] amides,[181] esters,[182] and carbamates,[183] as well as towards nitriles[184].

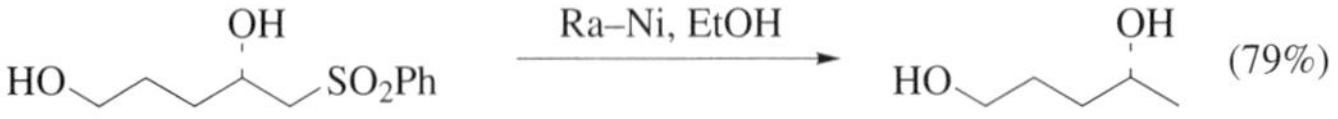

Ra–Ni, EtOH (79%) (Eq. 108)

Homogeneous organonickel reagents are good alternatives to Ra–Ni since they are easily handled and not as sensitive. These reagents, however, have scarcely been used for reductive desulfonylations processes. Not much information is available with respect to functional group tolerance. Some generalizations can be found in the literature, mostly based on the chemoselective desulfurization of thioethers. However, given that sulfones are tolerated in the desulfurization of thioethers employing Ni reagents,[185] these generalizations cannot be extended to the sulfonyl group. β-Ketosulfones seem to be reduced in good yields with nickelocene–lithium aluminum hydride (Eq. 39).[72]

Nickel-containing complex reducing agents prepared in the presence of 2,2′-bipyridine (NICRA-bpy) are employed for the selective desulfonylation of alkenyl sulfones.[75] Results are not very satisfactory because considerable starting material is recovered, reduction of the double bond is observed, and the reaction is not stereospecific (Eq. 109). Replacement of the nitrogen ligand 2,2′-bipyridine with quinoline seems to overcome these problems, although the generality of the method still has to be established (Eq. 110).[75]

SO_2Ph — NICRA–Bpy, DME, 65° → (65%) Z/E 33:67 + (15%) (Eq. 109)

Bpy = 2,2'-bipyridyl

Ph, SO_2Ph — NICRA–quinoline → Ph, H (81%) Z/E 95:5 (Eq. 110)

A novel $Mg/MeOH/NiBr_2$ desulfonylating system has been very recently presented for the reductive desulfonylation of β-sulfonylated aminosugars.[186] Reduction of the nickel halide with a low oxidation potential metal such as magnesium is supposed to produce finely divided Ni(0), which exhibits better catalytic activity than Ra–Ni or Mg/MeOH in the reductive desulfonylation of these aminosugar substrates (Eq. 111).

PhO_2S, NHBn, BnO, O, OMe — Mg, $NiBr_2$ (10-20 mol%), MeOH, rt, 4-6 h → NHBn, BnO, O, OMe (75%) (Eq. 111)

Copper dichloride[187] and titanium tetrachloride[80] have also been used in combination with lithium aluminum hydride for the reduction of alkenyl and aryl sulfones, respectively. The presence of a transition metal such as Ni, Cu, and Ti, in combination with $LiAlH_4$ permits reductive desulfonylations with these reagents under relatively mild conditions. Sulfones are generally resistant to reductions with hydride reagents alone, and very few examples of this type of desulfonylation are found in the literature, mostly reporting the use of DIBALH (Eq. 112),[78,79] or $LiAlH_4$.[50,77]

S, Ts, H — DIBALH, toluene, 50-80°, 5 min → S, H (81%) (Eq. 112)

Since the first report[82,83] of the reductive desulfonylation reaction of allylic sulfones employing a hydride source and a palladium catalyst, a large number of applications in organic synthesis have appeared in the literature. The method is operationally simple and the complications associated with other reductive desulfonylation methods of allylic sulfones, such as overreduction, migration of the double bond, and stereochemical problems are seldom seen under the reaction conditions. Although reductive desulfonylation processes employing a rhodium catalyst in combination with a dihydronicotinamide derivative,[188] and those using stoichiometric amounts of $Mo(CO)_6$,[88] are known, none of the systems are as efficient and general as those employing palladium catalysts. Allylic sulfones can be easily desulfonylated to the corresponding alkenes by $LiHBEt_3$ in the presence of a catalytic amount of $[PdCl_2(dppp)]$ under mild conditions with the preservation of the alkene position and configuration, as depicted in Eq. 113 for the synthesis of squalene.[84] The desulfonylation proceeds through attack at the less substituted terminus of the allyl moiety to give the more substituted alkene (see mechanism in Scheme 1). This method is also used for the synthesis of α,β-unsaturated ketones,[189] allylic and homoallylic alcohols,[82,190] and different natural products such as lavandulol and isolavandulol,[190] and the human redox carrier coenzyme Q_{10}.[85] Of special note is the selective preparation of either allylic or homoallylic alcohols from the same 2-tosyl homoallylic alcohol by an appropriate selection of the reaction conditions (Eq. 114).[190]

Ts

$PdCl_2(dppp)$ (5 mol%)
$LiHBEt_3$, THF, 0°

(94%) Z/E 3:97

(Eq. 113)

Ts, OH, Ph

Pd catalyst, reducing agent
THF, additive

Ph, OH + Ph, OH

$PdCl_2(dppp)$, $LiHBEt_3$, Ph_3SiH, 20°, 3 min — (98%) Z/E 2:98 — (2%)
$PdCl_2(PPh_3)_2$, $LiBH_4$, –15°, 4.5 h — (3%) — (97%) Z/E 9:91

(Eq. 114)

The Pd-catalyzed, $LiHBEt_3$-mediated reductive desulfonylation of allylic sulfones is also used in the ligand-controlled stereoselective synthesis of dienes[191] where it is possible to control the geometry of the diene by a proper selection of the palladium ligand as shown by the distribution of products **1** and **2** (Eq. 115). The method described herein is also applicable to the so-called "integrated chemical processes", which allow the preparation of a wide variety of alkenes by combining alkylation of allylic sulfones and reductive desulfonylation in one pot.[192,193]

The palladium-catalyzed deprotection of allyl-based protecting groups has emerged as an important tool in organic synthesis.[194] Consequently, special care must be taken when an allyl-protected functional group that is susceptible to deprotection under the Pd-catalyzed conditions also exists in the target molecule. During the synthesis of the kinesin motor protein inhibitor adociasulfate 1 (Eq. 116),[195] the allyloxy group is removed during the Pd-catalyzed reductive desulfonylation step. In sharp contrast, competing reduction of the allylic silyl ether functionality also present in the molecule is not observed when the reduction is conducted at low temperatures.

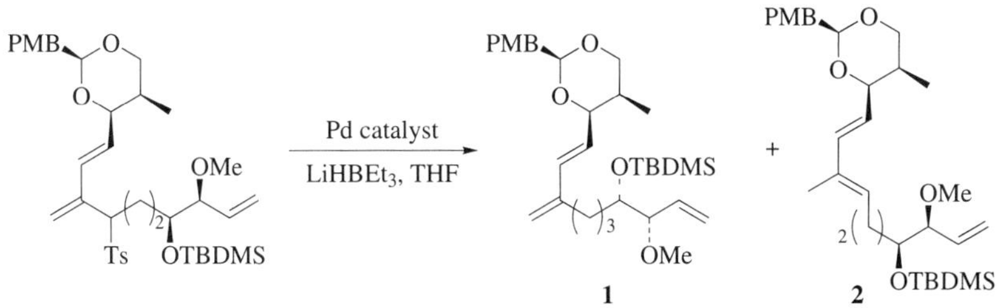

Pd catalyst	Additive	Temp	Time	**1**	**2**
$Pd[P(Bu\text{-}t)_3]_2$	–	45°	4 h	(90%)	(0%)
$(^3\eta\text{-}C_3H_5PdCl)_2$/(1-pyrrolidinyl)$_3$P	–	0°	20 min	(6%)	(84%)
$PdCl_2[(S,S)\text{-}\mathbf{3}]$	LiCl	0°	10 h	(74%)	(10%)

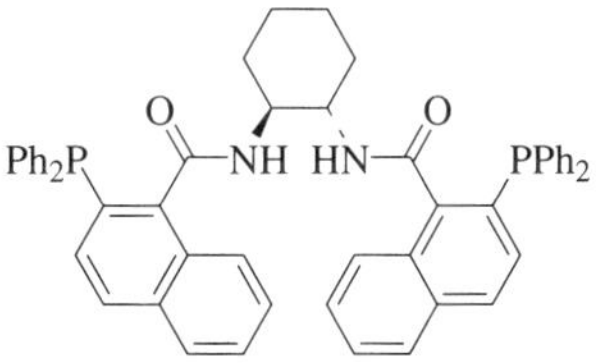

(Eq. 115)

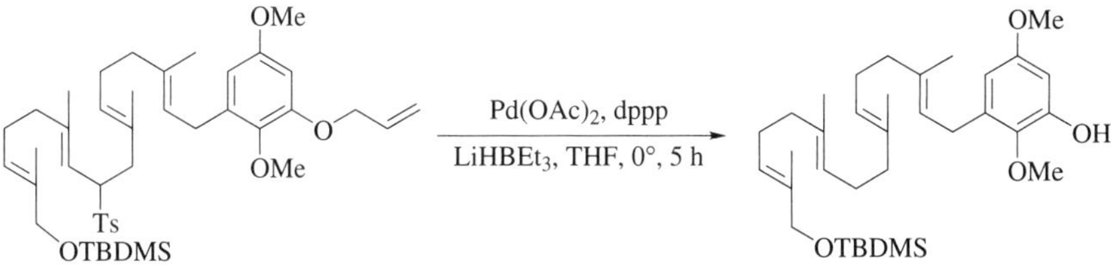

(Eq. 116)

A simple and stereospecific way to carry out the reductive desulfonylation of alkenyl sulfones consists of the use of an excess of a Grignard reagent such as *n*-BuMgCl in the presence of Ni(II) or Pd(II) complexes as catalyst and nitrogen or phosphorus ligands such as 1,4-diazabicyclo[2.2.2]octane (DABCO), $(n\text{-Bu})_3P$, or Ph_3P (Eq. 117).[196,48] Palladium catalysts appear to be superior to nickel catalysts, giving higher yields and stereoselectivities. This is a very important method in sulfone chemistry considering that the reductive desulfonylation of alkenyl sulfones using dissolving metals or metal amalgams is generally not stereospecific.

The process requires very mild reaction conditions when compared to $Na_2S_2O_4$-mediated reductive desulfonylations, although small amounts of products derived from the sulfone substitution by the Grignard reagent are also obtained. With 2,2-disubstituted alkenyl sulfones, major amounts of a dimer are formed in addition to the expected hydrogenolysis product (Eq. 118).[196]

t-BuO_2S — n-BuMgCl, $Pd(acac)_2$ / DABCO, THF, rt → (70%) Z/E 99.5:0.5 (Eq. 117)

Ts — n-BuMgCl / $Pd(acac)_2$, THF, rt → (20%) + (30%) (Eq. 118)

The reductive desulfonylation of sulfonyl 1,3-dienes with Grignard reagents is particularly important since many other reagents such as $Na_2S_2O_4$, Na/Hg, and Al/Hg do not work with these substrates. The yields are moderate, but the reaction is stereospecific, with the addition of Ni catalysts giving slightly better results (Eq. 119).[48]

THPO, SO_2Ph — n-BuMgCl, $Ni(acac)_2$ / THF, 1 h, rt → THPO (49%) 95% E,Z (Eq. 119)

Other Reducing Agents. 1,4-Dihydropyridines have been successfully employed for the reductive desulfonylation of functionalized sulfones. Of special interest is the desulfonylation of α-nitro sulfones with BNAH under sunlight irradiation to give the corresponding nitro compounds in good yields.[90,91] The reaction takes place under mild conditions and tolerates ketones, nitriles, and isolated double bonds (Eq. 120).[91] A photo-induced electron-transfer employing ascorbic acid as electron donor is also an efficient approach for the reductive desulfonylation of β-ketosulfones.[197]

Ph, NO_2, SO_2Ph — BNAH, $h\nu$ / C_6H_6, 42 h → Ph, NO_2 (69%) (Eq. 120)

Reductive Eliminations

Since its original publication,[94] the Julia olefination has become a very important tool in organic synthesis for the site- and stereoselective synthesis of alkenes. The synthetic importance of the process is reflected by its numerous applications in the synthesis of a diverse range of functionalized alkenes such as allylic alcohols,[198] allylic amines,[199–201] homoallylic alcohols,[202] homoallylic amines,[203] and allylsilanes.[204,129] The reaction has also been used as a key step in many natural product syntheses,[12,54] and a solid-phase version of the process has been developed.[205]

In its original form,[94] the Julia reaction consisted of the formation of a carbon–carbon double bond through the coupling of a sulfonyl-stabilized anion and a carbonyl compound to generate a β-hydroxy sulfone, followed by a reductive elimination to afford the alkene (Eq. 47). A subsequent study of its scope and stereochemistry led to improved reaction conditions, which are now widely used.[206] Alternative methods to synthesize the β-hydroxy sulfone intermediates, such as the addition of sulfonyl carbanions to esters with subsequent reduction of the ketone to the β-hydroxy sulfone, are also known (Eq. 121).[207]

OTBDMS; SO_2Ph; N; CO_2Me —LHMDS, THF, 0°→ TBDMSO, H, SO_2Ph, O, N —1. $LiEt_3BH$, THF, –78°; 2. Na/Hg, MeOH, rt→ TBDMSO, H, N (70%) (Eq. 121)

LHMDS = lithium hexamethyldisilazane

One of the main disadvantages of the Julia olefination is that two steps are needed. Many different factors related to the addition and reductive elimination steps have to be considered when attempting a successful olefination. The nature of the coupling partners as well as of the counter ion in the metalated sulfone are both important with respect to the addition step. The addition of metalated sulfones to aldehydes and ketones is reversible and the failure of the process often results from an unfavorable equilibrium at this stage. With easily enolizable carbonyl compounds, use of metalated lithio sulfones can lead to poor yields of the desired product due to competitive deprotonation of the carbonyl compound by the sulfonyl carbanion. The less basic magnesium derivative of the sulfone should therefore be used to overcome this problem.[208–212] Magnesium derivatives can be prepared by warming the sulfone with EtMgBr or, more conveniently, by transmetalation of the lithio derivative with $MgBr_2$. When the lithio or the magnesium derivatives of the sulfone fail, the use of the metalated sulfone in the presence of boron trifluoride may be used successfully (Eq. 122).[213] Occasionally, an appropriate selection of the reaction solvent may help to suppress the enolization.[214] Care must be taken as well with very stable α-sulfonyl carbanions where the reverse reaction can be favored.[215] Stabilization of the sulfone anion by conjugation or chelation with a proximal heteroatom often favors the reverse reaction. Trapping the in situ generated alkoxide, typically with Ac_2O, BzCl, MsCl, or TMSCl, usually shifts the equilibrium to the addition product. Increasing the leaving group character of the hydroxy functionality has a positive effect in the reductive elimination process (Eq. 123).[216]

OMe, O, SO_2Ph, OTBDMS

1. n-BuLi, THF, –78°
2. n-C_5H_{11} CH(OTBDMS) CHO

OMe, O, SO_2Ph, C_5H_{11}-n, TBDMSO, HO, OTBDMS

without BF_3•OEt_2 (<10%)
with BF_3•OEt_2 (90%)

Na/Hg, Na_2HPO_4
THF/MeOH, rt

OMe, O, C_5H_{11}-n, TBDMSO, OTBDMS (76%)

(Eq. 122)

TIPSO, MeO, OMe OMe, S, SO_2Ph, OTBDMS, S, O, O, OH, OTBDMS, OMe

1. $(CF_3CO)_2O$, Py, DMAP, CH_2Cl_2
2. LN, THF, –20°
DMAP = 4-dimethylaminopyridine

OMe OMe, S, OTBDMS, S, OTBDMS (68%) Z/E 29:71

(Eq. 123)

Problems also occur in the metalation of phenyl[15] and imidazolyl sulfones[217] with n-BuLi, especially when the α-protons are sterically hindered. Competitive metalation on the aromatic ring of the sulfone is the only observed process as depicted in Eq. 124.

Me–N, N, SO_2, OMe H, O, OPMB, TMSO, OTBDMS, CO_2H

1. LDA, THF, –78°
2. i-PrCHO

HO, Pr-i, Me–N, N, SO_2, OMe H, O, OPMB, TMSO, OTBDMS, CO_2H

(Eq. 124)

Even with all the potential problems associated with the condensation step, the flexibility of the Julia olefination often offers viable alternatives to reach the synthetic goal. For example, as depicted in Eqs. 125[206] and 126[218,219], a proper selection of the coupling partners circumvents the problems associated with the stability of the lithium derivative (it forms a stable chelate with the proximate oxygen thereby preventing addition to the aldehyde) or the reactivity of the magnesium sulfone, which acts as a reducing agent toward the aldehyde and affords the corresponding vinylic sulfone after a β-hydride elimination process.

TBDMSO SO$_2$Ph M + OBn O O O OHC; M = Li → TBDMSO----Li SO$_2$Ph; M = MgBr → OBn O O O HO + TBDMSO SO$_2$Ph

(Eq. 125)

OBn O O O PhO$_2$S Li; 1. TBDMSO CHO, THF, 0°; 2. MsCl, Py, rt; 3. NH$_3$/Li, –78° → OH O O O TBDMSO (37%)

(Eq. 126)

β-Hydroxy sulfones can be easily transformed into derivatives for radical-mediated reactions such as thiobenzoates,[220] xanthates,[221,222] selenobenzoates,[221,222] and thionocarbonates.[221,222] These substrates may be used in subsequent stereoselective syntheses of alkenes by free radical methods. Methyl xanthates usually give the best results in the olefination reaction, typically using *O*-acyl *N*-hydroxy-2-thiopyridone under visible light irradiation. Alternatively, diphenylsilane in combination with radical initiators such as Et_3B/O_2, benzoyl peroxide, or AIBN may also be used (Eq. 127).[221,222]

S MeS O *n*-C_6H_{13} C_6H_{13}-*n* SO$_2$Ph → *n*-C_6H_{13} C_6H_{13}-*n*

Conditions	Yield
Ph_2SiH_2, AIBN, toluene, 110°	(75%)
Ph_2SiH_2, Et_3B/O_2, C_6H_6, 80°	(55%)
O-acyl *N*-hydroxy-2-thiopyridone, *h*ν, C_6H_6, rt	(85%)

(Eq. 127)

Trisubstituted alkenes are prepared by reductive elimination of β-hydroxy sulfones but, in general, the reverse reaction competes.[214] The reverse reaction is favored when the β-alkoxy sulfone adduct is sterically encumbered. The olefination of ketones to prepare trisubstituted alkenes employing Na/Hg affords moderate yields, unpredictable stereoselectivities, and large amounts of retroaldol products from the intermediate β-alkoxy sulfones. High yields and moderate stereoselectivities of trisubstituted alkenes are obtained by a modification of the Julia–Lythgoe olefination reaction involving the in situ capture of the intermediate β-alkoxy sulfones with a suitable oxophilic electrophile and the employment of SmI_2/HMPA to promote, under neutral conditions, the reductive elimination at low temperatures (Eq. 128).[223] A recent modification of this protocol, using sulfoxides instead of sulfones, is very efficient in the stereoselective preparation of di-, tri-, and tetrasubstituted alkenes.[224,225]

OBz; Bu-*n*; Ph; SO_2Ph → SmI_2, HMPA; THF, –78°, 1 h → Ph; Bu-*n*; (72%) Z/E 29:71 (Eq. 128)

A further complication of the Julia olefination is reductive desulfonylation. This process can intervene, especially in substrates where the anti elimination process is less favored, leading to significant amounts of the corresponding desulfonylated alcohols via carbanion protonation (Eq. 129).[226]

PhO_2S; OH; OH; OTBDMS → Na/Hg, Na_2HPO_4; THF/MeOH, –20° to rt → OH; OTBDMS (65%) + OH; OH; OTBDMS (24%) (Eq. 129)

An improved modification of the reaction employs β-hydroxy imidazolyl sulfones and SmI_2 as the reducing agent in the absence of additives.[102] The reaction is E-selective and no hydroxy group derivatization is needed. No reaction is observed when β-hydroxy phenyl sulfones are reduced under these conditions unless HMPA is employed to improve the reducing ability of SmI_2.[103] Similar results have been observed in the SmI_2-promoted reductive elimination of glycosyl aryl- and heteroaryl sulfones (Eq. 130).[160,227–229] Lithium naphthalenide, however, has been used in the reductive elimination of phenyl glycosyl sulfones in the absence of additives.[230,231] With poor leaving groups in the 2-position of the glycoside, reductive desulfonylation is the major process even with activated heteroarylic sulfonyl glycosides (Eq. 132).[152] D-manno 2-Pyridylsulfonyl derivatives give the corresponding glucal in good yield with in situ generated Cp_2TiCl from Cp_2TiCl_2 and Mn.[232]

R^1	R^2	Reducing Agent	Additive		
β-SO_2Ph	Ac	SmI_2	—	(0%)	(0%)
β-SO_2Ph	Ac	SmI_2	HMPA	(96%)	(0%)
α-SO_2Ph	Ac	SmI_2	—	(<5%)	(0%)
α-SO_2-2-naphthyl	Ac	SmI_2	—	(22%)	(0%)
α-SO_2-2-pyrimidyl	Ac	SmI_2	—	(72%)	(0%)
α-SO_2-2-pyridyl	Ac	SmI_2	—	(94%)	(0%)
α-SO_2-2-pyridyl	TMS	SmI_2	—	(9%)	(91%)
α-SO_2-2-pyridyl	AcO	Cp_2TiCl_2/Mn	—	(70%)	(0%)

(Eq. 130)

Anomeric C–S bonds in glycosyl sulfones can be cleaved by chromium(II) complexes in water/DMF leading to the corresponding glycals.[233] Phenyl sulfones are unreactive under the tested reaction conditions while 2-pyridyl and 2-benzothiazolyl compounds exhibit high reactivities (Eq. 131).

(Eq. 131)

R	Time	
Ph	48 h	(0%)
2-pyridyl	5 h	(>95%)
2-benzothiazolyl	5 h	(>95%)

The mixture SmI_2/HMPA has been applied to the conversion of vicinal bis (sulfonyl) derivatives into the corresponding alkenes.[100] An application of this reaction to the synthesis of a difluoromethylene nucleoside, where conventional difluoromethylation strategies (e.g., Wadsworth–Emmons and Wittig reactions) failed, is shown in Eq. 132.[234]

(46%)

(Eq. 132)

An interesting variation of the Julia olefination is the reductive elimination of 2,3-epoxy sulfones. This reaction, which leads to allylic alcohols,[198] consists of alkylation of a sulfone-stabilized allylic carbanion followed by epoxidation of the double bond and reductive elimination (Eq. 133). The synthesis of disubstituted alkenes is trans-selective and proximate branching increases the stereoselectivity. However, this procedure is not selective for the preparation of trisubstituted alkenes.

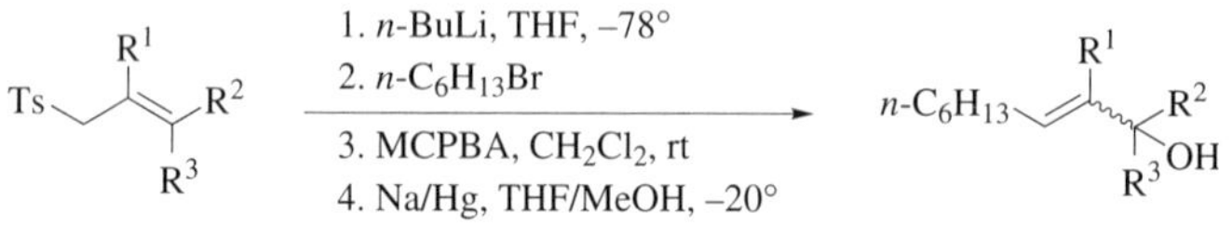
R1
Ts
R2
R3
1. n-BuLi, THF, –78°
2. n-C6H13Br
3. MCPBA, CH2Cl2, rt
4. Na/Hg, THF/MeOH, –20°
R1
n-C6H13
R2
R3
OH

(Eq. 133)

R^1	R^2	R^3	Yield	Z/E
Me	H	H	(70%)	30:70
H	Me	H	(62%)	20:80
H	Me	Me	(89%)	0:100

High yields of functionalized acetylenes and enynes are obtained from unsaturated β-arylsulfonyl enol phosphates,[235–237] enol acetates,[237] and enol carbonates.[235] These compounds, which are obtained from the corresponding β-ketosulfones, are subjected to reductive elimination using Na/Hg, NH_3/Na, or SmI_2 to afford the corresponding alkynes (Eq. 134). Considerable formation of β-ketosulfones by reduction of the starting phenylsulfonyl enol acetates and enol phosphates employing NH_3/Na or Na/Hg is also observed. Careful control of the reaction conditions is necessary to avoid over-reduction of the alkyne to the trans-alkene when using alkali metals in ammonia.[235] The latter process, however, has been elegantly used in a key step of the synthesis of brefeldin A (Eq. 135).[238]

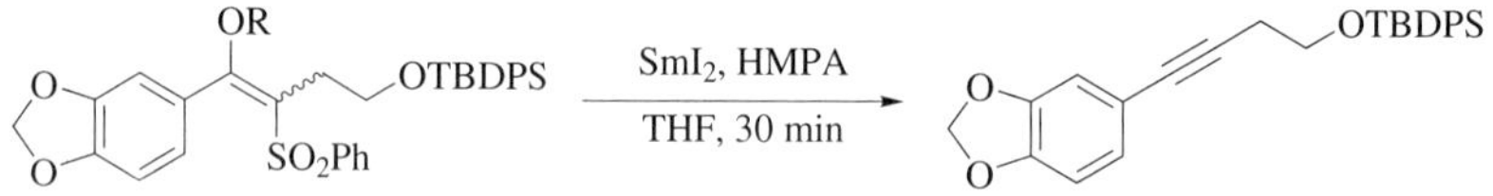
OR
O
O
OTBDPS
SO2Ph
SmI2, HMPA
THF, 30 min
O
O
OTBDPS

R	Yield
Ac	(68%)
$(PhO)_2OP$	(94%)

(Eq. 134)

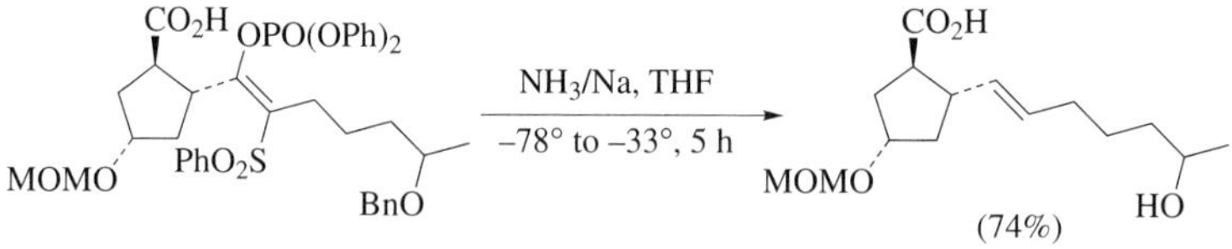
CO2H
OPO(OPh)2
MOMO
PhO2S
BnO
NH3/Na, THF
–78° to –33°, 5 h
CO2H
MOMO
(74%)
HO

(Eq. 135)

1,2-Di(phenylsulfonyl)ethylene is employed as a synthetic equivalent of acetylene in cycloaddition reactions to prepare polycyclic dienes.[239] The high activation due to the presence of two sulfonyl groups promotes the cycloaddition to very unreactive systems. The reductive elimination of the resulting 1,2-disulfones, which is usually carried out with Na/Hg, affords the corresponding alkenes (Eq. 136).[240] A similar method is employed in the synthesis of tetrasubstituted polycyclic alkenes.[241] Alternatively, SmI_2[100] and Mg in MeOH[118] are also successful reagents for this transformation.

1. heat
2. WCl_6, n-BuLi
3. Na/Hg
(78%)
(Eq. 136)

A novel method for the synthesis of alkenes is based on the coupling of aldehydes with dithioacetals to give the corresponding hydroxy thioacetals, which afford vicinal disulfides via reductive phenylthio migration.[242] The syn-diastereomers are the major products from symmetrical compounds while the anti-isomers are obtained with high selectivity with unsymmetrical compounds. Separation of the diastereomers, oxidation to the 1,2-disulfones, and reductive elimination give the corresponding alkenes with moderate stereoselectivities (Eq. 137).[242]

1. n-BuLi
2. RCHO
3. AcCl
4. Et_3SiH, $TiCl_4$

R = n-C_7H_{15}: (67%) syn/anti 88:12 → 1. MCPBA 2. Na/Hg → (98%) Z/E 85:15

R = c-C_6H_{11}: (80%) syn/anti 5:95 → 1. MCPBA 2. Na/Hg → (64%) Z/E 19:81

(Eq. 137)

Other uncommon reductive eliminations of β-halogeno sulfones are carried out using Na/Hg,[243] and especially Mg in MeOH.[244,122] Yields are usually very low with Na/Hg but very high with Mg as depicted in Eq 138.[122]

Na/Hg, THF/MeOH
−20°, 30 min
(30%)
Mg, $HgCl_2$ (cat)
EtOH, rt, 1 h
(95%)
(Eq. 138)

β-Nitro sulfones react with tin radicals to afford alkenes in good yields.[245] The reaction is stereospecific, especially for the formation of α,β-unsaturated nitriles (Eq. 139), since the elimination from the radical intermediate is faster than bond rotation.

$(n\text{-}Bu)_3SnH$, AIBN
C_6H_6, 80°, 2 h
(86%)
(Eq. 139)

A very interesting synthesis of medium-sized cyclic amines has been performed by selective ring cleavage of sulfonylated bicyclic amines.[246] A Julia-type desulfonylation of an activated β-amino sulfone is the key step in this method, which takes place even in the presence of a hydroxyl leaving group in the β position (Eq. 140).

TIPSO H OH SO_2Ph — 1. MeI, CH_2Cl_2, rt; 2. Na/Hg, Na_2HPO_4, MeOH, rt, 6 h → TIPSO OH N Me (67%) (Eq. 140)

APPLICATIONS TO SYNTHESIS OF NATURAL PRODUCTS

The sulfone group has been used in synthesis as an activating group for carbon–carbon single and double bond formation involving reductive desulfonylation or reductive elimination processes. The facile and regioselective generation of carbanions α to the sulfone group enables efficient carbon–carbon single bond construction via alkylation, acylation, and aldol-like reactions. Since the sulfonyl group is also easily removed from the synthetic intermediate, many sulfonyl-containing derivatives have been employed in the preparation of intermediates for the synthesis of a wide variety of functionalized molecules and many natural and biologically active compounds.[3,14,12] Among reactions of sulfones playing an essential role in the synthesis of natural products, alkylation of carbanions and the Julia olefination have become conventional processes. Both reactions usually complement each other in the syntheses of many natural products, as found, for instance, in the synthesis of the secosesquiterpene (−)-anthoplalone.[247] The reductive desulfonylation approach to this compound presents a double-bond site selectivity problem (Eq. 141). This difficulty is solved by using a Julia-type olefination protocol (Eq. 142).

OTBDPS SO_2Ph —
1. *t*-BuOK, TBAI, THF, –78°
2. Br–(cyclopropyl)–CO_2Bu-*t*
3. TBAF, THF, 0° to rt
4. $PdCl_2$(dppp), $LiBHEt_3$, THF, 0°

TBAI = tetra(*n*-butyl)ammonium iodide
TBAF = tetra(*n*-butyl)ammonium fluoride

→ OH CO_2Bu-*t* (35%) + OH CO_2Bu-*t* (11%)

(Eq. 141)

1. n-BuLi, THF, −78°
2. [ketone ester] CO_2Bu-t
3. SmI_2, THF, rt

(82%)

(−)-Anthoplalone

(Eq. 142)

Different routes to the total synthesis of natural products that are based on these two reactions (reductive desulfonylation or reductive elimination) as the key connection steps have been reported. Two different total syntheses[248,249] of the 24-membered macrolide isolated from a deep-sea bacterium, (−)-macrolactin A, involve as the key step a sulfone-mediated C−C coupling, with subsequent reductive desulfonylation[249] or reductive elimination[248].

Reductive Desulfonylation
Reductive Elimination

(−)-Macrolactin A

It is also common to find synthetic routes where both methods are employed in sequence as in the construction of the C18−C34 fragment of the macrolide antibiotic antascomicin A (Eq. 143).[250]

PhO_2S … OPMB

1. n-BuLi, THF, −78°
2. TBDMSO … CHO
3. Ac_2O, TEA, DMAP, CH_2Cl_2, 0° to rt
4. Na/Hg, Na_2HPO_4, MeOH, 0°

TBDMSO … OPMB → PhO_2S … OPMB

(Eq. 143)

1. LHMDS, THF, −78° to 0°
2. [BnO, BnO cyclohexyl iodide, OTBDMS]
3. Na/Hg, Na_2HPO_4, THF/MeOH, 0° to rt

[product: BnO, BnO, OTBDMS, OPMB]

Reductive Desulfonylations in the Synthesis of Natural Products

The first step when using sulfones in the synthesis of natural products consists of the formation of the new C–C bond. This process is normally performed using the sulfone as a nucleophile via the corresponding α-sulfonyl carbanion. Three different strategies are normally employed: alkylation of α-sulfonyl carbanions followed by reductive removal of the sulfonyl group, acylation of α-sulfonyl carbanions followed by reductive removal of the sulfonyl group, and finally, reaction of α-sulfonyl carbanions with activated multiple bonds followed by reductive desulfonylation.

By far, the most widely used method is the alkylation of an α-sulfonyl carbanion followed by reductive removal of the sulfonyl group. Different electrophiles such as alkyl halides, sulfonates, sulfinates, acetates, oxiranes, and electron-deficient multiple bonds are employed for the formation of the new C–C bond. Palladium-catalyzed π-allylic alkylation with α-sulfonyl carbanions is also a commonly used method. After the C–C bond formation, the conditions for the final desulfonylation reaction with the appropriate reagent will depend on the structure of the sulfone intermediate.

Synthesis of (+)-Chatancin. The alkylation of an α-sulfonyl carbanion derived from a γ-alkoxy functionalized sulfone with an allylic bromide and subsequent reductive desulfonylation with Na/Hg constitutes a key step in the synthesis of the marine diterpene (+)-chatancin (Eq. 144).[251]

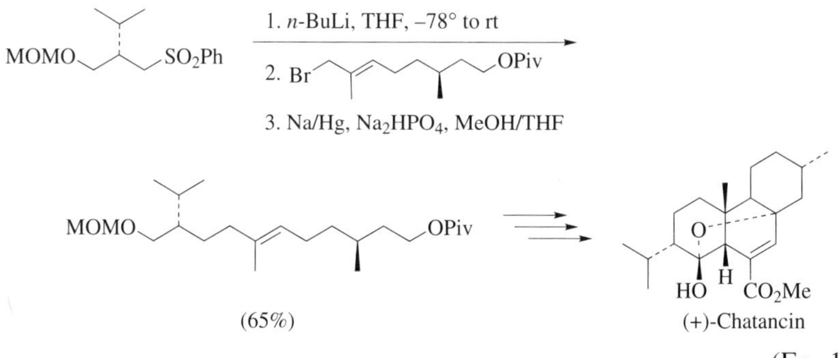

(Eq. 144)

Synthesis of Bacillariolides I-III. Marine oxylipin bacillariolides I-III are synthesized from (*R*)-malic acid, using a common chiral cyclopentane derivative prepared as depicted in Eq. 145.[252] Two consecutive alkylation reactions of lithioallyl sulfone are responsible for the generation of the cyclopentane intermediate. The synthetic route also includes a reductive desulfonylation with Na/Hg in MeOH/THF (Eq. 145).

PhO_2S 1. n-BuLi, THF, –78° 2. MsO … OPMB, OMOM → SO_2Ph, MOMO, H, OPMB, OH (99%)

1. OsO_4, $NaIO_4$, 1,4-dioxane/H_2O, rt
2. CrO_3/H_2SO_4, acetone, 0°
3. Na/Hg, Na_2HPO_4, MeOH/THF

→ MOMO, H, O, OPMB (75%) (Eq. 145)

HO, H, R^2, R^1, H, O

R^1 = (1Z,4Z,7Z,10Z)-trideca-1,4,7,10-tetraenyl, R^2 = H, Bacillariolide I
R^1 = H, R^2 = (1Z,4Z,7Z,10Z)-trideca-1,4,7,10-tetraenyl, Bacillariolide II
R^1 = (1Z)-4-carboxybut-1-enyl, R^2 = H, Bacillariolide III

An asymmetric total synthesis of bacillariolide III is achieved in fifteen linear steps with a good overall yield.[253] The key feature of this synthetic route involves a highly stereoselective construction of a vinyl-substituted bicyclic lactone by an intramolecular Pd(0)-catalyzed π-allylic alkylation with an α-sulfonyl carbanion (Eq. 146).

PhO_2S, O, OCO_2Et → $Pd(PPh_3)_4$, CH_2Cl_2, reflux, 1 h → O, SO_2Ph, H (88%) endo/exo 30:1

Na/Hg, $B(OH)_3$, MeOH, rt, 3 h → O, H → HO, H, O, CO_2H (Eq. 146)

Bacillariolide III

Synthesis of All-*trans*-Geranylgeraniol. The type of alkylation described above for the synthesis of bacillariolide III is widely used in the synthesis of natural products due to the mild reaction conditions and high stereospecificity. The formation of the C–C bond takes place when activated α-sulfonyl carbanions derived from β-ketosulfones, α-sulfonyl sulfones or, less often, allylic sulfones react with the π-allyl palladium complex. In the synthesis of all-*trans*-geranylgeraniol, the α-sulfonyl carbanion adds to the π-allylpalladium complex of 2-(prop-1-en-2-yl)oxirane. Final reductive desulfonylation affords the desired compound, as depicted in Eq. 147.[254]

SO$_2$Ph CO$_2$Me

Pd(PPh$_3$)$_4$, THF, rt, 4 h

SO$_2$Ph MeO$_2$C OH (90%)

1. PhSH, Cs$_2$CO$_3$, DMF, 85°, 3 h
2. Pd(dppe)Cl$_2$, LiEt$_3$BH, THF, 0°, 5 h

OH

all-*trans*-Geranylgeraniol (84%)

(Eq. 147)

Synthesis of (±)-Tacamonine. α-Sulfonyl acetamides are very effective reagents for the synthesis of glutarimides and pyroglutamates, intermediates that are efficiently transformed into a wide variety of alkaloids such as (±)-tacamonine,[255] (±)-pseudoheliotridane,[256] (±)-homopumiliotoxin 223G,[257] (±)-deplancheine,[258] and (±)-yohimbane.[258] As depicted in Eq. 148 for the synthesis of (±)-tacamonine, the alkaloid precursor is obtained through a stepwise [3+3] annulation reaction, that starts with a Michael addition of the α-sulfonyl carbanion to the appropriate activated alkenes followed by ring closure. The precursor thus obtained is then transformed into the target alkaloid in a sequence that involves reductive desulfonylation with Na/Hg.[259]

1. NaH, THF, rt
2. OEt

(82%)

1. LiAlH$_4$, TEA, THF, reflux
2. Na/Hg, MeOH

(75%)

(±)-Tacamonine

(Eq. 148)

Synthesis of (+)-Eurylene. The reaction of α-sulfonyl anions with carboxylic acid derivatives is used as the key step in the construction of various natural products. The resulting β-oxo sulfone intermediate is then further elaborated and/or desulfonylated to afford the desired product. A variety of carboxylic acid derivatives has been used, esters being most often employed, as depicted in Eq. 149 for the synthesis of the triterpene polyether (+)-eurylene.[149]

1. LHMDS, DMPU, THF, –78° to –30°
2. [structure]
3. SmI_2, THF/MeOH, –78°
4. HCl, MeOH

(85%)

(Eq. 149)

(+)-Eurylene

Synthesis of (−)-Azaspiracid-1. A different approach to C–C coupling through β-oxo sulfones consists of the addition of an α-sulfonyl carbanion to an aldehyde followed by oxidation. This reaction sequence has been widely used in the preparation of various natural products such as the marine toxin (−)-azaspiracid-1 (Eq. 150).[260]

1. LDA, THF, –78°
2. [structure]
3. TPAP, NMO, CH_2Cl_2
4. Na/Hg, Na_2HPO_4, THF/H_2O

TPAP = tetrapropylammonium perruthenate
NMO = *N*-methylmorpholine *N*-oxide

(54%)

(Eq. 150)

(–)-Azaspiracid-1

Synthesis of (+)-Rhizoxin. A different strategy is employed in the asymmetric total synthesis of rhizoxin D, where two alkene linkages are established by a modified Julia protocol.[261] The initial β-hydroxy sulfone, obtained after addition of the sulfonyl carbanion to the aldehyde partner, is transformed into the corresponding vinylic sulfone through a sequence of acetylation and elimination. Final reductive desulfonylation of the vinylic sulfone with SmI_2 gives the desired E-alkene as a single geometric isomer. The construction of one of these linkages is depicted in Eq. 151.

1. LHMDS, THF, –78°
2. (aldehyde: MEMO, OTBDMS, OTBDMS, OMe)
3. Ac_2O, DMAP, Py, rt
4. DBU, THF, rt
5. SmI_2, DMPU, MeOH/THF, rt

DBU = 1,8-diazabicyclo[5.4.0]undec-7-ene

(73%)

(+)-Rhizoxin

(Eq. 151)

Synthesis of (+)-Tricycloclavulone. Many syntheses of natural products have been performed employing unsaturated sulfones. These activated substrates are versatile synthetic reagents due to the activating effect of the sulfone moiety that enables them to undergo conjugate additions, cycloadditions, and deprotonation-alkylation sequences by way of the corresponding α-anions. With regard to the 1,4-addition reactions, vinylic and acetylenic sulfones have been used as Michael acceptors in the preparation of different natural products. New C–C and C–heteroatom bonds are formed upon addition of carbon (mainly metal alkyls) and nitrogen nucleophiles to the activated unsaturation followed by intra- or intermolecular reaction of the resulting highly nucleophilic α-sulfonyl carbanion with electrophiles. The synthesis of (+)-tricycloclavulone, an abnormal marine prostanoid isolated from *Clavularia viridis*, provides an example (Eq. 152).[262] The process features a highly efficient SmI_2-promoted reductive desulfonylation of a β-ketosulfone intermediate.

1. Bu_2CuLi, THF
2. TBAF, THF
3. SmI_2, THF

(70%)

(+)-Tricycloclavulone

(Eq. 152)

Synthesis of (−)-Sibirine. Various types of alkaloids have been prepared by conjugate addition of carbon-centered radicals to unsaturated sulfones. This approach is used in the stereoselective synthesis of the *Nitraria* spirocyclic alkaloid (−)-sibirine, where a 6-*exo-trig* radical cyclization to an α,β-unsaturated sulfone leads to the spirocyclic skeleton of the natural product (Eq. 153).[263] The γ-nitrogen-functionalized sulfone so obtained is then desulfonylated under dissolving-metal conditions.

1. $(n\text{-Bu})_3SnH$, AIBN, toluene, reflux, 7 h
2. Na/EtOH, −20° to rt, 2 h
3. $LiAlH_4$, THF, rt, 30 min
4. HF, MeCN, rt, 1.5 h

(−)-Sibirine (47%) (Eq. 153)

Synthesis of (−)-Lasubine II. A reductive desulfonylation with lithium in ammonia is employed in the total synthesis of quinolizidine alkaloid (−)-lasubine II.[264] A conjugate addition of methyl (*S*)-(2-piperidyl)acetate to an acetylenic sulfone, followed by lithium diisopropylamide (LDA)-promoted intramolecular acylation is the key step in the preparation of the quinolizine structure of (−)-lasubine II (Eq. 154).

1. MeOH, heat
2. LDA, THF, −78°

(53%)

1. $NaBH_4$, MeOH
2. Swern oxidation
3. NH_3/Li
4. L-Selectride

(−)-Lasubine II (45%)

(Eq. 154)

Synthesis of (+)-7-Deoxypancratistatin. Different approaches to the total synthesis of (+)-7-deoxypancratistatin have been reported recently because of the promising biological properties shown by this alkaloid. An elegant synthesis of (+)-7-deoxypancratistatin has been achieved from furan and *trans*-1,2-bis(phenylsulfonyl)ethylene (Eq. 155).[265] This synthesis clearly illustrates the utility of alkenyl sulfones as Michael acceptors and dienophiles for cycloaddition reactions.

PhO_2S SO_2Ph CH_2Cl_2, rt, 7 h SO_2Ph SO_2Ph (94%)

1. Li THF/toluene, –78°
2. t-BuO_2H, n-BuLi, THF, –78°
3. Na/Hg, MeOH/THF, –23°

PhO_2S OH (65%)

OH HO OH OH NH O

(+)-7-Deoxypancratistatin

(Eq. 155)

Synthesis of Hesitine Diterpenoid Alkaloids. An efficient enantioselective approach to the hesitine class of the C_{20}-diterpenoid alkaloids involves an intramolecular oxidopyridinium dipolar cycloaddition with a vinylic sulfone as the key transformation as depicted in Eq. 156.[266] Once the sulfonyl group has played its role in the C–C bond formation, it is removed by a Na/Hg-promoted reductive desulfonylation.

N^+ O^- SO_2 Ph toluene, reflux (70%) SO_2Ph H H N H O

1. L-Selectride, –78°, $PhNTf_2$
2. HCO_2H, TEA, $PdCl_2(PPh_3)_2$
3. Na/Hg, Na_2HPO_4, THF/t-BuOH, rt, 30 h

H H N H (48%)

(Eq. 156)

Reductive Eliminations in the Synthesis of Natural Products

Synthesis of (−)-Siccanin. The Julia–Lythgoe olefination provides an important tool in the total synthesis of a number of natural products.[12] The reductive elimination of β-hydroxy sulfones and their derivatives usually involves a condensation between an anion α to an alkyl sulfone and a carbonyl compound to afford a β-hydroxy sulfone. Acylation of the alcohol, followed by reductive elimination affords the corresponding alkene. This reaction has been used in the synthesis of (−)-siccanin, a mold metabolite isolated from the culture of *Helminthosporium siccans*, which exhibits potent antifungal activity (Eq. 157).[267] This example clearly demonstrates the utility of the Julia reductive elimination since other different C–C couplings based on a reductive desulfonylation process or a modified Julia olefination (Eq. 9) failed.

1. n-BuLi, THF, –78°
2. (OHC-chroman, OMe)
3. Ac_2O
4. Na/Hg, Na_2HPO_4

(93%)

(–)-Siccanin

(Eq. 157)

Synthesis of (−)-Laulimalide. Different approaches to the β-hydroxy sulfone moiety needed for the olefination reaction are frequently used in the synthesis of natural products. For instance, a very common strategy consists of carbonyl reduction of the corresponding α-ketosulfone followed by reductive elimination. This sequence is employed in the synthesis of polyhydroxylated indolizidine alkaloids (Eq. 121),[207] (+)-dihydromevinolin,[268] pleraplysillin-1,[269] amphidinolide B,[270] and the novel antitumor agent (−)-laulimalide (Eq. 158).[271]

1. $LiBH_4$, THF, 0° to rt
2. PhCOCl, TEA, DMAP, CH_2Cl_2
3. Mg, $HgCl_2$, EtOH

(44%)

(–)-Laulimalide

(Eq. 158)

Synthesis of (−)-Tricycloillicinone. An elegant synthetic route to the neurotrophic (−)-tricycloillicinone employs a sodium amalgam mediated reductive elimination of a β-alkoxy sulfone obtained by a thermal Claisen rearrangement (Eq. 159).[272]

MeO$_2$C … PhO$_2$S … 1. neat, 165°, 12 h; 2. K$_2$CO$_3$, acetone, heat → (90%)

Na/Hg, MeOH/EtOAc, −20°, 12 h → (87%) → (−)-Tricycloillicinone

(Eq. 159)

Synthesis of (+)-Pseudomonic Acid C. A total synthesis of (+)-pseudomonic acid C employes an n-Bu$_3$SnH-mediated reductive elimination of the methyl xanthate derivative of a β-hydroxy sulfone as one of the key steps (Eq. 160).[99]

SO$_2$Ph, OTBDMS; 1. LDA, THF, −60°; 2. OHC, OTBDPS; 3. CS$_2$, 0°, 10 min; 4. MeI, 0°, 20 min; 5. (n-Bu)$_3$SnH, toluene, reflux, 3.5 h → OTBDMS, OTBDPS (67%)

→ OH, HO, OH, CO$_2$H, 8 — (+)-Pseudomonic Acid C

(Eq. 160)

Synthesis of L-Amiclenomycin. Different β-functionalized sulfones have been employed in natural product synthesis. For example, L-amiclenomycin, an antibiotic isolated from cultures of different *Streptomyces* strains, has been prepared employing in the final steps of the sequence a reductive elimination involving a 1,2-di(phenylsulfonyl) derivative (Eq. 161).[273]

(Eq. 161)

COMPARISON WITH OTHER METHODS

Reductive Decyanations

The use of the nitrile function for C–C bond-forming reactions has increased recently.[274–276] Alkylation of nitriles[277] followed by reductive decyanation[278] is a good alternative to the sulfone alkylation and reductive cleavage. A number of methods perform the reductive decyanation in good yields and stereoselectivities, with the use of dissolving metals being the most popular strategy. The Li/NH_3 or Na/NH_3 system has been principally used for the reduction of tertiary nitriles since primary and secondary nitriles give not only the expected decyanated products, but also the corresponding amines (Eq. 162).[279]

$$RCN \xrightarrow[-33°,\ 10\text{-}15\ \text{min}]{NH_3/Na} RH + RCH_2NH_2$$

R	RH	RCH_2NH_2
Ph_3C	(90-96%)	(0%)
Ph_2CH	(76%)	(11%)
n-$C_{13}H_{27}$	(35%)	(65%)

(Eq. 162)

Potassium is a more general reducing agent and allows decyanation of primary, secondary, and tertiary cyanides in good yields when employed in a mixture of HMPA and *t*-BuOH as solvent,[280] on neutral alumina,[281] or in the presence of dicyclohexano-18-crown-6 in toluene[282,283] as depicted in Eq. 163 for the synthesis of *ent*-cholesterol.[284]

(Eq. 163)

Reductive decyanation of α-aminonitriles with aluminum- or borohydrides such as $LiAlH_4$, $NaBH_4$, BH_3, and $NaBH_3CN$ provides access to amines through an S_N1 mechanism with formation of an iminium ion followed by reduction by the hydride reagent. The process is stereoelectronically controlled and proceeds with high stereoselectivity (Eq. 164).[285]

$NaBH_4$, MeOH; 65°, 1.5 h (85%) (Eq. 164)

Reduction of cyano groups under radical conditions is carried out employing tin hydrides such as n-Bu_3SnH.[286] Unfortunately, the reaction is so far restricted to malonitriles. Samarium iodide is a valid alternative to tin hydrides since in the presence of HMPA it promotes the reductive decyanation of malononitriles and α-cyanoesters in high yields (Eq. 165).[287–289]

SmI_2, HMPA; THF (Eq. 165)

R^1	R^2	R^3	Temp	Yield
Bn	H	NC	0°	(85%)
Bn	Bn	NC	0°	(97%)
$H_2C{=}CH(CH_2)_3$	H	NC	0°	(53%)
Bn	H	EtO_2C	rt	(54%)
Bn	Bn	EtO_2C	rt	(87%)
$H_2C{=}CH(CH_2)_3$	$H_2C{=}CH(CH_2)_3$	EtO_2C	rt	(87%)
$HO(CH_2)_6$	H	EtO_2C	rt	(49%)
n-C_7H_{15}	$EtCO_2(CH_2)_2$	EtO_2C	rt	(75%)
n-C_7H_{15}	$CN(CH_2)_3$	EtO_2C	rt	(85%)
n-C_7H_{15}	$Et_2NCO(CH_2)_2$	EtO_2C	rt	(88%)
n-C_7H_{15}	$Cl(CH_2)_4$	EtO_2C	rt	(61%)

The decyanation reaction is also observed employing Brønsted acids and bases, although the harsh reaction conditions required (very high temperatures and long reaction times) limit the applicability of this method.

Reductive Eliminations

Among the different methods for the synthesis of alkenes, none matches the versatility of carbonyl olefination.[290] It has been extensively studied since the discovery of the Wittig reaction,[291,292] and a wide variety of approaches to transform carbonyl compounds into alkenes have been developed. Besides the classical Julia reaction,[94] the most generally applicable methods for direct olefination of carbonyl compounds include the Wittig,[291,293,294] Horner–Wittig,[295,293,296] Horner–Wadsworth–Emmons,[297,298,293] Peterson,[299–301] Johnson,[302] and Julia–Kocienski[303,304] reactions. All of these methods involve the addition of a metalated substrate to a carbonyl compound followed by elimination, rearrangement, or cleavage to the alkene (Eq. 166).

Reaction	Y
Julia (1973)	$ArSO_2$
Wittig (1953)	R_3P^+
Horner–Wittig (1958)	$R_2P{=}O$
Horner–Wadsworth–Emmons (1961)	$(RO)_2P{=}O$
Peterson (1968)	R_3Si
Johnson (1973)	ArSONMe
Julia–Kocienski (1991)	$HeteroarylSO_2/ArylSO_2$

(Eq. 166)

These olefination reactions can be applied with confidence to the stereoselective synthesis of alkenes. Both isomers of a wide variety of alkenes can be obtained with very high stereoselectivities when suitable reaction conditions are selected. Compared with other methods, the Julia reductive elimination has some advantages. First, sulfones are more readily available and easily purified than the corresponding phosphorus and silicon derivatives. There is a wide range of mild and high-yielding routes to synthesize sulfones.[3] Furthermore, the sulfone group also confers stability and frequently crystalline properties to the substrate.

The Julia reductive elimination is a good choice when trying to prepare mono-, 1,1-di- and E-1,2-disubstituted alkenes. This application has been demonstrated, for example, in the preparation of a key intermediate in the synthesis of calciferol (Eq. 167).[305] The alternative Wittig olefination approach failed due to the difficulties encountered in the preparation of the corresponding sterically hindered phosphonium halide.[305]

1. *n*-BuLi, THF, –78°
2. (aldehyde)
3. Ac_2O, THF, rt
4. Na/Hg, MeOH/EtOAc, –20°
5. KOH, EtOH

(Eq. 167)

A highly stereoselective synthesis of 1,2-disubstituted alkenes can be also accomplished via reductive desulfonylation of 1,2-disubstituted vinylic sulfones since both E- and Z-vinylic sulfones can be stereoselectively prepared by several methods.[8] Although different reducing agents have been used for this purpose (Eqs. 24–28), the reduction of 1,2-disubstituted alkenyl sulfones by $Na_2S_2O_4$ is particularly efficient and highly stereoselective (Eq. 168).

n-Bu / PhO_2S (vinyl sulfone) $\xrightarrow[\text{DMF/H}_2\text{O, 120°, 2 h}]{Na_2S_2O_4,\ NaHCO_3}$ n-Bu (alkene) (Eq. 168)

Z/E
0:100
93:7

Yield	Z/E
(80%)	100:0
(60%)	10:90

With respect to the synthesis of trisubstituted alkenes, high yields but modest selectivities are obtained through the coupling between ketones and primary sulfones and subsequent reductive elimination employing SmI_2 (Eq. 128).[223] On the other hand, and considering that formally it is not a reductive desulfonylation process in the sense of a substitution of the sulfonyl group by hydrogen, the coupling reaction of vinylic sulfones with Grignard reagents catalyzed by nickel or iron complexes is a good alternative for this purpose (Eq. 169).[306]

t-BuO_2S (trisubstituted alkene) $\xrightarrow[\text{THF, rt, 20 h}]{\text{PhMgBr, Fe(acac)}_3}$ Ph (trisubstituted alkene) (60%) Z/E 0:100 (Eq. 169)

No stereocontrolled syntheses of tetrasubstituted alkenes have been reported via reductive eliminations, including the synthesis of tetrasubstituted alkenes using the Horner–Wadsworth–Emmons reaction since it also proceeds with moderate selectivity.[307,308] The stereochemical course of the HWE reaction usually depends on the nature of the phosphonate employed. Bulky substituents at the phosphorus atom and the carbon close to the carbanion favor the formation of the E-alkene. Z-selectivity can be achieved using the Still–Gennari modification[309] and the Ando method.[310] With respect to trisubstituted alkenes, the HWE reaction occurs with moderate to good E-selectivity either using the addition of a phosphonate to a ketone[311] or the reaction of an α-substituted phosphonate with an aldehyde. Electronic and steric effects can modulate these tendencies.[312,313]

The Horner–Wittig reaction[295,293,296] consists of the preparation of alkenes by treatment of a phosphine oxide with base followed by the addition of the carbonyl compound. If a lithium base is used, the intermediate β-hydroxy phosphine oxide diastereomers can be isolated and separated. They can then be treated separately with base to give the corresponding Z- or E-alkenes with high stereoselectivity. The Peterson olefination[299–301] is also a good alternative, allowing for the synthesis of pure syn or anti β-hydroxyalkylsilane intermediates, from which the stereocontrolled preparation of alkenes proceeds.

A new variant of the classical Julia olefination, the Julia–Kocienski olefination, also called modified or one-pot Julia olefination, has recently emerged as a powerful tool for alkene synthesis via the condensation of certain heteroaryl or aryl sulfonyl anions with carbonyl compounds.[303,304] A reaction pathway has been proposed for this reaction involving, after the initial coupling between the metalated sulfone and the carbonyl compound, a Smiles rearrangement,[314] and spontaneous sulfur dioxide elimination (Eq. 170a).[304] This process was originally described with benzothiazol-2-yl (BT) sulfones[303,315] (Eq. 170a) and has been extended to include other types of heteroaryl sulfones such as pyrid-2-yl

(PYR),[315,316] 1-phenyl−1H-tetrazol-5-yl (PT),[317] and 1-*tert*-butyl−1H-tetrazol-5-yl (TBT)[318] sulfones and non-heteroaryl 3,5-bis(trifluoromethyl)phenyl sulfones (BTFP)[319−322] (Eq. 170a). The Julia–Kocienski olefination shares with the Julia reductive elimination the critical dependence that reaction conditions such as counterion, solvent, and temperature have on the yield and the stereochemical outcome of the elimination. Some recent examples demonstrate the utility of this reaction as an alternative to the Julia–Lythgoe olefination as shown for the total synthesis of the alkaloid (−) -spirotryprostatin B[323] where the Julia–Lythgoe reaction affords the alkene in a very poor yield and shows epimerization problems in the final product (Eq. 170b). The Julia–Kocienski olefination using the corresponding 1-phenyl−1H-tetrazol-5-yl (PT) sulfone solves this problem affording the alkene in high yield and without epimerization (Eq. 170b).

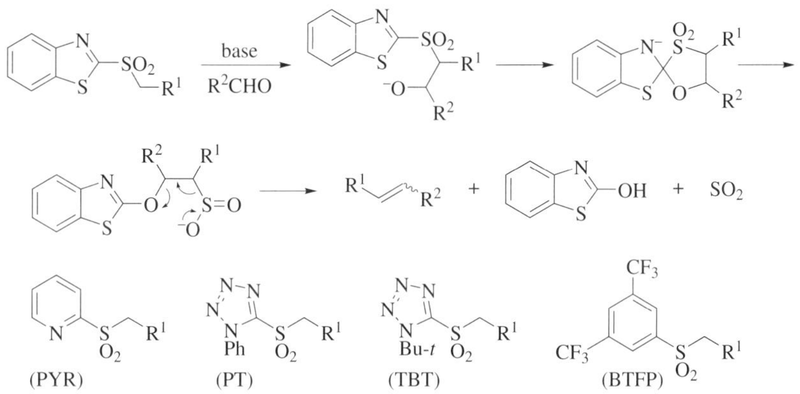

base, R^2CHO

(PYR) (PT) (TBT) (BTFP)

(Eq. 170a)

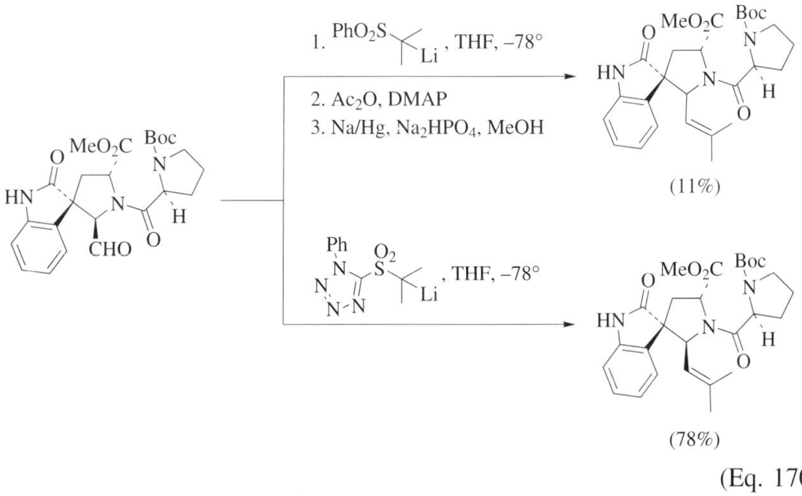

1. PhO_2S–C(Me)$_2$Li, THF, −78°
2. Ac_2O, DMAP
3. Na/Hg, Na_2HPO_4, MeOH

(11%)

THF, −78°

(78%)

(Eq. 170b)

On some occasions, both methods have been used complementarily, as in the synthesis of the C(43)–C(67) subunit of polyketide metabolite amphidinol 3.[324]

Julia-Kocienski
Julia-Lythgoe

Alkene cross-metathesis[325–327] represents an attractive alternative to the carbonyl olefination methods described above. The applicability of the Julia olefination is sometimes limited if highly functionalized substrates are involved. Moreover, it employs functional groups such as aldehydes and ketones that often require protecting-group strategies prior to the olefination process. Cross-metathesis tolerates a wide variety of functional groups, as illustrated for reactions employing the ruthenium-complex catalyst **4**, giving access to functionalized alkenes under very mild reaction conditions that can be used in subsequent synthetic manipulations (Eq. 171).[328]

4 (5 mol%), CH_2Cl_2, 40°

R = Ph, X = BH_3: (63%) Z/E <5:95

R = Ph, X = O=: (90%) Z/E <5:95

R = EtO, X = O=: (87%) Z/E <5:95

(Eq. 171)

4

The major drawback of alkene cross-metathesis is the limited ability to control the chemo- and stereoselectivity of the reaction. High yields of the cross-product are obtained by either stoichiometric control or by the use of functionalized alkenes. When unfunctionalized alkenes are used in the reaction, an excess of one of the alkenes must be used in order to get a synthetically useful yield of the cross-product. Good yields of the cross-product are also obtained by combining a sterically hindered alkene with a readily available one (Eq. 172).[329]

4 (5 mol%)
CH_2Cl_2, 40°, 4 h

(87%) Z/E <5:95

(Eq. 172)

Cross-metathesis of conjugated electron-deficient alkenes such as α,β-unsaturated esters, ketones, aldehydes, and amides often give high cross-product/dimer ratios due to the slow rate of dimerization of these substrates (Eq. 171). When this occurs, the cross-product is dominant even when the reactions are performed with a 1 : 1 stoichiometry of the reactants.[330] When one of the alkene partners homodimerizes slowly, such as happens with electron-deficient and sterically hindered alkenes, the reaction is driven to the cross-product. With respect to the stereochemistry of the reaction, the E-isomer is obtained with electron-deficient alkenes (Eq. 171), and the E/Z ratio may also vary depending on the types of substituents present on the reactants.

Very recently, cross-metathesis has also been employed for the synthesis of functionalized trisubstituted alkenes.[331–335] This method is, however, rather limited and, in general, poor Z/E selectivities are observed (Eq. 173).

BzO + OAc
4 (5 mol%)
CH_2Cl_2, 40°, 12 h

(Eq. 173)

BzO OAc (80%) Z/E 26:74

EXPERIMENTAL CONDITIONS

General

With the exception of the reductive desulfonylations employing sodium dithionite and Zn, these reactions are carried out under an inert atmosphere employing anhydrous solvents due to the high reactivity of the reducing agents with water and moist air. All the reagents are commercially available but they are usually freshly prepared prior to their use. In general, the reactions are carried out employing an excess of the reducing agent.

Reductions with Active Metals and Salts

Reductive desulfonylations and reductive eliminations can be performed employing alkali or alkaline earth metals in ammonia or low molecular weight amines. Lithium or sodium in ammonia or ethylamine are very effective systems at 0°, typically for reductions with ethylamine, or lower temperatures such as −33° or −78°. Very short reaction times must be used to minimize side reactions. The reaction is occasionally carried out in the presence of low molecular weight alcohols or ethers as cosolvents (conditions of the Birch reduction) such as ethanol, *tert*-butanol, tetrahydrofuran, and diethyl ether. Lithium–amine solutions are more sensitive to catalytic decomposition than lithium in ammonia,[336] so purified solvents are mandatory. The persistence of the deep blue color produced by lithium or sodium metals in ammonia (solvated electrons) is used to judge when the reaction is completed. The mixture is then quenched, typically with ammonium chloride, sodium benzoate, or dienes such as isoprene or 1,3-butadiene, and then the reaction is warmed to room temperature in order to remove the ammonia. *Reductive desulfonylations and reductive eliminations employing alkali or alkaline earth metals in ammonia present a high fire hazard and should be conducted in a properly functioning chemical fume hood away from flammable solvents. Ammonia is a corrosive gas with a pungent odor. Alkali metals react violently with water or even moist air to generate hydrogen, which can then be ignited by the heat of the reaction.*

The majority of the reductions with metals in low molecular weight alcoholic solvents are carried out with magnesium in methanol, ethanol, or mixtures of these solvents with tetrahydrofuran or ethyl acetate to improve substrate solubility. The temperature of the reaction depends on the sulfone derivative and ranges from low temperature to reflux conditions, with room temperature usually preferred. Addition of catalytic amounts of mercury dichloride makes the reaction more efficient, which avoids using a large excess of activated magnesium and high temperatures. Upon completion, excess magnesium is typically destroyed with dilute acid.

From the experimental point of view, reductive desulfonylations with alkali metal arene radical anion complexes require a large excess of the radical anion, very short reaction times at low temperatures, and must be run under an inert atmosphere. Sodium or lithium naphthalenides in tetrahydrofuran at −78° or lower temperatures are typical reaction conditions. Tetrahydrofuran solutions of lithium naphthalenide are dark green. This color is lost when the substrate is added and restored once the reaction is finished. Upon completion, the excess reagent is quenched with a saturated aqueous solution of ammonium chloride or low molecular alcohols such as methanol or ethanol.

Sodium amalgam containing 2–6% sodium is the most commonly employed reagent for the reductive desulfonylation and reductive elimination processes. Sodium amalgam can be prepared and freshly used by the addition of Hg(0) to ribbons of sodium metal.[337,338a] Normally, Na/Hg is used in large excess and a solution of the sulfone to be reduced in methanol or tetrahydrofuran–methanol mixtures is added at low temperature (−40° to 0°) to a suspension of Na/Hg and Na_2HPO_4 in the same solvent. Buffering the reaction with NaH_2PO_4 avoids side reactions with base-labile compounds. Sodium amalgam is a commercially available (3%, 4%, 5%, 10%, and 20% in Na) air and moisture sensitive compound. *Mercury and mercury compounds are poisonous and teratogenic. They should be used in a properly functioning chemical fume hood. Mercury cannot be destroyed but it can be removed from aqueous solutions by using ion-exchange resins or by amalgamation with iron.*

Aluminum amalgam reductions are generally carried out in aqueous THF solutions (typically 10% aqueous THF) at higher temperatures (from 0° to reflux conditions) than those generally employed for Na/Hg desulfonylations. Aluminum amalgam is not commercially available but is easily prepared by immersion of aluminum foil into a 2% aqueous solution of $HgCl_2$.[18]

Samarium(II) iodide is a deep blue air-sensitive compound. Therefore, all manipulations involving this reagent must be carried out under an inert atmosphere. It does not react significantly with water over several hours and is less reactive towards other protic solvents such as alcohols. Therefore, reductions with SmI_2 are usually carried out in tetrahydrofuran or tetrahydrofuran–methanol mixtures under low or room temperature conditions. Additives such as HMPA or DMPU often have a profound effect on reactions mediated by SmI_2 since they improve the reducing power of Sm(II). Excess SmI_2 is usually necessary for the completion of the reactions. The vast majority of reactions employing SmI_2 are carried out in tetrahydrofuran, and since SmI_2 is conveniently generated in this solvent, the in situ preparation of the reducing reagent is particularly useful and highly recommended. Oxidation of samarium metal with organic dihalides (usually diiodomethane) is typically the method of choice.[31] The solution thus obtained can be stored for long periods of time without a decrease in Sm(II) concentration if kept under an inert atmosphere and in the presence of a small amount of Sm metal. Due to the intense interest in this reagent, SmI_2 is now commercially available as a 0.1 M solution in tetrahydrofuran. *Hexamethylphosphoramide is a potent carcinogen, and thus must be handled with extreme care. It can be hydrolyzed by refluxing in concentrated HCl to dimethylamine and phosphoric acid.*

Aqueous conditions are employed for the reductive desulfonylations with $Na_2S_2O_4$. Mixtures of dimethylformamide–water, and less often tetrahydrofuran–water or cyclohexane–water, temperatures in the range of 80–120°, and excess base ($NaHCO_3$) are commonly used.

Reductions with Tin Hydrides

Reductive desulfonylations stoichiometric in tin are performed employing *n*-Bu_3SnH in toluene at reflux and in the presence of catalytic amounts of AIBN as radical initiator. Ph_3SnH has been occasionally used as a hydride source when *n*-Bu_3SnH is not effective. In the catalytic version of the reaction, *n*-Bu_3SnCl is used as a tin hydride precursor employing either poly(methylhydrosiloxane) (PMHS) or $NaBH_3CN$ as hydride reagents and AIBN as the radical initiator. When using PMHS the reaction is performed in mixtures of toluene and water at reflux and in the presence of potassium fluoride. Desulfonylations using the system *n*-$Bu_3SnCl/NaBH_3CN$ are performed in *tert*-butanol as solvent under reflux conditions. *Tin hydrides are irritants and toxic and should be handled with care in a fume hood.*

Transition-Metal-Mediated Reductive Desulfonylations

Raney nickel catalysts are designated as W-1, W-2, W-3, W-4, W-5, W-6 and W-7 according to their hydrogen content, which depends on the method of preparation. The most employed catalyst for the reductive desulfonylation process is W-2, although in many experimental procedures the type of catalyst is not specified, making any generalization difficult. Under typical reaction conditions, the Ra–Ni reagent is used in large excess and is added as an alcoholic or aqueous suspension to a solution of the derivative to be reduced. It is usually stored as an alcoholic suspension, or occasionally in water, ether, methylcyclohexane, or dioxane. In order to obtain reproducible results it is desirable to use freshly prepared or recently purchased reagent since with aging the catalyst suffers deactivation due to hydrogen loss. The reduction is usually carried out in EtOH under reflux conditions for long periods of time (10–24 h), unless ultrasound is used. Other solvents such as methanol at room temperature, 1,4-dioxane at reflux, tetrahydrofuran, and ethyl acetate are used as well. *Raney nickel ignites on contact with air and should never be allowed to dry.*

The reductive desulfonylation reactions employing homogeneous organonickel reagents such as nickelocene-lithium aluminum hydride are carried out in THF at room temperature employing an excess of the reagent. The reducing agent is prepared in situ by mixing nickelocene with $LiAlH_4$ in tetrahydrofuran at room temperature. Reductions with nickel-containing complex reducing agents (NICRAs) are performed in tetrahydrofuran or 1,2-dimethoxyethane at 65°. The reagents are also freshly prepared before use, by mixing $Ni(OAc)_2$, degreased NaH, and *t*-AmOH in tetrahydrofuran or 1,2-dimethoxyethane.[74] External ligands such as 2,2′-bipyridyl and triphenylphosphine are added in some cases. Nickel-containing complex reducing agents have been designated according to the stoichiometry employed for their preparation. Thus, a NICRA prepared from NaH, *t*-AmOH, $Ni(OAc)_2$ and the external ligand is abbreviated NICRAL (x/y/z/t) where x/y/z/t is the molar ratio NaH/*t*-AmOH/$Ni(OAc)_2$/L.[338b]

With respect to the palladium-catalyzed reductive desulfonylations of allylic sulfones, [$PdCl_2$(dppp)] is the preferred catalyst in combination with superhydride ($LiHBEt_3$) in THF solutions, usually working under low temperature (0–4°) to room temperature conditions. *Lithium triethylborohydride is supplied as a 1 M solution in tetrahydrofuran and is corrosive and flammable. Handle and store under an inert atmosphere in a cool dry place. Use the solution in a fume hood and avoid contact with skin.*

Transition-metal-catalyzed stereoselective reductions of vinylic sulfones with Grignard reagents are achieved with excess *n*-BuMgCl in tetrahydrofuran at room temperature.[196,48] Better yields and selectivities are obtained with palladium catalysts [$Pd(acac)_2$] than with nickel complexes such as [$Ni(acac)_2$], especially if external ligands such as DABCO, triethylamine (TEA), or $(n\text{-Bu})_3P$ are used.[196,48] Nickel catalysts are used more often for reducing sulfonyl-1,3-dienes than palladium catalysts. When using this method, it is very important to remove the catalyst before isolation of the products in order to avoid isomerization of the alkene moiety during solvent evaporation. *Tetrahydrofuran solutions of n-BuMgCl are highly flammable, sensitive to moisture, and cause burns.*

EXPERIMENTAL PROCEDURES

Reductive Desulfonylations

PhO$_2$S — NH_3/Li, THF, –78°, 30 min — (97%)

(1*R*,4*S*,5*S*)-4,6,6-Trimethyl-4-vinylbicyclo[3.1.1]heptan-2-one (Desulfonylation of a β-Ketosulfone).[339] Anhydrous liquid NH_3 (80 mL) distilled from Li wire was stirred and cooled at −78° as a solution of (1*R*, 4*S*, 5*S*)-4,6,6-trimethyl-3-(phenylsulfonyl)-4-vinylbicyclo[3.1.1]-heptan-2-one (1.42 g, 4.5 mmol) in THF (7 mL) was added. After brief stirring, Li wire (103 mg, 0.02 g-atom), cut into small pieces, was added, and stirring was continued for an additional 30 minutes. Excess solid NH_4Cl was added cautiously, and most of the NH_3 was allowed to evaporate at room temperature. Water was added, and the product was extracted with Et_2O. Removal of the solvent gave an oily residue which was chromatographed on silica gel (4 : 1 hexane/Et_2O) to give the title product as an oil (792 mg, 97%): $[\alpha]_D^{21}$ + 87.9° (*c* 2.15, $CHCl_3$); IR (film) 3070, 1710, 1640, 915 cm^{-1}; ^{1}H NMR (90 MHz, $CDCl_3$) δ 1.08 (s, 3H), 1.23 (s, 3H), 1.38 (s, 3H), 1.4–1.7 (s, 1H), 1.9–2.1 (m, 1H), 4.92 (d, *J* = 17.1 Hz, 1H), 4.99 (d, *J* = 9.9 Hz, 1H), 5.78 (dd, *J* = 17.1, 9.9 Hz, 1H). Anal. Calcd. for $C_{12}H_{18}O$: C, 80.85; H, 10.18. Found C, 80.64; H, 9.92.

PhO$_2$S SO$_2$Ph, BnO OBn — Mg, MeOH, rt, 6 h — BnO OBn (77%)

(1*S*,2*S*)-1,2-Bis(benzyloxy)cyclopentane (Desulfonylation of an α-Functionalized Sulfone).[340] To a solution of (1*S*, 2*S*)-1,2-bis(benzyloxy)-4,4-bis(phenylsulfonyl)cyclopentane (13.92 g, 24.7 mmol) in MeOH (625 mL) at 50° under a nitrogen atmosphere was added activated Mg (4.33 g, 178 mmol). Once evolution of hydrogen began, the heating source was removed and the reaction was maintained over a period of 6 hours by the addition of two supplementary portions of Mg (2 x 4.33 g). It was occasionally necessary to cool the reaction mixture in a 15° water bath during this time. After all the Mg had reacted, the cloudy gray solution was concentrated, diluted with H_2O (300 mL), and then acidified with concentrated HCl at 0° until all the Mg salts were dissolved. The resulting clear solution was extracted with Et_2O (3 × 200 mL). The combined ethereal extracts were washed with 1 M KOH (3 × 200 mL) and saturated NaCl (200 mL), dried ($MgSO_4$), and concentrated to give a pungent oil. Column chromatography (20 : 1 pentane/Et_2O) afforded the title product as a clear fragrant oil (5.37 g, 19.0 mmol, 77%): $[\alpha]_D^{20}$ + 32.18° (*c* 5.6, $CHCl_3$); IR (CH_2Cl_2) 3030, 2950, 1500, 1450, 1360, 1340, 1220, 1100 cm^{-1}; ^1H NMR (200 MHz, $CDCl_3$) δ 1.60–1.82 (m, 4H), 1.87–2.10 (m, 2H), 3.92–4.03 (m, 2H), 4.52 (d, *J* = 12.5 Hz, 4H), 7.34 (s, 10H). Anal. Calcd. for $C_{19}H_{22}O_2$: C, 80.82; H, 7.85. Found: C, 80.42; H, 8.03.

Ts OTBDPS n-$C_{12}H_{23}$ OTIPS O NHBoc —LN, THF / –78°, 20 min→ n-$C_{12}H_{23}$ OTBDPS OTIPS O NHBoc (93%)

(2*S*, 3*S*)-2-*tert*-Butoxycarbonylamino-3-*tert*-butyldiphenylsilyloxy-1-triisopropylsilyloxyoctadecan-4-one (Desulfonylation of a β-Ketosulfone).[341] To a solution of lithium naphthalenide, prepared from naphthalene (64 mg, 0.50 mmol) in THF (1.0 mL) and Li wire (4.3 mg, 0.63 mmol), was added a solution of (2*S*, 3*S*)-2-*tert*-butoxycarbonylamino-3-*tert*-butyldiphenylsilyloxy-5-(*p*-toluenesulfonyl)-1-triisopropylsilyloxyoctadecan-4-one (120 mg, 0.12 mmol) in THF (0.50 mL) via cannula. The mixture was stirred for 20 minutes at −78°, treated with saturated NH_4Cl (0.50 mL), and poured into H_2O (10 mL). After extraction with Et_2O, the organic layer was washed with brine, dried over anhydrous $MgSO_4$, and concentrated to give a residue that was purified by silica gel column chromatography. Elution with EtOAc/hexane (0 : 100, then 5 : 95) afforded the title product as a colorless oil (93 mg, 93%): $[\alpha]_D^{24}$ + 4.9° (*c* 1, $CHCl_3$); IR ($CHCl_3$) 3445, 1709, 1501 cm^{-1}; ^1H NMR (270 MHz, $CDCl_3$) δ 0.88 (t, *J* = 6.5 Hz, 3H), 0.94–1.32 (m, 63H), 2.08 (dt, *J* = 17.6, 5.8 Hz, 1H), 2.30 (dt, *J* = 17.6, 6.4 Hz, 1H), 3.69 (dd, *J* = 9.9, 8.0 Hz, 1H), 3.77 (dd, *J* = 9.9, 5.8 Hz, 1H), 4.08 (m, 1H), 4.39 (d, *J* = 4.8 Hz, 1H), 4.82 (d, *J* = 9.0 Hz, 1H), 7.30–7.66 (m, 10H); ^{13}C NMR (100 MHz, $CDCl_3$) δ 11.92 (3C), 14.1, 18.0, 19.6, 22.7, 27.1, 28.3, 29.0, 29.4, 29.7, 31.6, 32.0, 39.3, 55.0, 62.0, 77.9, 79.3, 127.6, 127.7, 129.86, 129.90, 133.0, 133.1, 135.9, 136.0, 155.4, 209.4; HRMS–FAB (*m*/*z*): $[M + Na]^+$ calcd for $C_{48}H_{83}NNaO_5Si_2$, 832.5708; found, 832.5697. Anal. Calcd for $C_{48}H_{85}NO_5Si_2$: C, 70.97; H, 10.55; N, 1.72. Found: C, 71.37; H, 10.15; N, 1.58.

TBDMSO PhO$_2$S TBDMSO O OTBDMS O Si O t-Bu Bu-t —SmI_2, THF/MeOH / –78° to rt, 1.5 h→ TBDMSO TBDMSO O OTBDMS O Si O t-Bu Bu-t (89%)

(2S,7S,8R,9S,12R)-7,9-Di[(*tert*-butyldimethylsilyl)oxy]-12-[(*tert*-butyldimethylsilyl)oxymethyl]-2-[(4S,6R)-2,2-di-*tert*-butylsilylene-6-methyl-1,3-dioxan-4-yl]-8-methyltetradecan-5-one (Desulfonylation of a β-Ketosulfone).[342] A flame-dried flask under argon was charged with Sm (900 mg, 6.00 mmol). The flask was evacuated to high vacuum for 15 minutes and was then refilled with argon. This process was repeated three times. Freshly distilled THF (30 mL) and diiodomethane (0.244 mL, 3.00 mmol) were added with vigorous stirring at room temperature, and the dark blue solution was stirred for 1 hour. This stock solution of samarium diiodide could be stored for 3 months under argon.

To a solution of (2S, 7S, 8R, 9S, 12R)-7,9-di[(*tert*-butyldimethylsilyl)oxy]-12-[(*tert*-butyldimethylsilyl)oxymethyl]-2-[(4S, 6R)-2,2-di-*tert*-butylsilylene-6-methyl-1,3-dioxan-4-yl]-8-methyl-4-phenylsulfonyltetradecan-5-one (14.5 mg, 0.0145 mmol) in THF (1.6 mL) and MeOH (0.8 mL) under argon at −78° was added a freshly prepared 1 M solution of samarium diiodide in THF (0.580 mL, 0.0580 mmol). The reaction flask was covered with foil, and the dark blue solution was stirred for 30 minutes at −78°. The solution was left to warm to room temperature during 1 hour and then diluted with Et_2O (20 mL). The ethereal solution was washed with saturated K_2CO_3 solution (20 mL), and the aqueous wash was extracted three times with Et_2O (20 mL). The combined ethereal extracts were dried ($MgSO_4$), and the solvent was removed under reduced pressure. Chromatography of the residue on silica gel, with gradient elution from 2–5% EtOAc in hexane, gave the title compound as a colorless oil (11.0 mg, 89%): $[\alpha]_D^{22}$ + 19.8° (c 0.85, $CHCl_3$); IR (neat) 2963, 2932, 2896, 2860, 1715, 1476, 1386, 1257, 1103, 840 cm^{-1}; ^1H NMR (400 MHz, $CDCl_3$) δ −0.01 (s, 3H), 0.03 (s, 6H), 0.06 (s, 6H), 0.07 (s, 3H), 0.83 (d, J = 7.0 Hz, 3H), 0.84–0.91 (m, 6H), 0.86 (s, 9H), 0.89 (s, 9H), 0.90 (s, 9H), 1.00 (s, 18H), 1.15–1.36 (m, 5H), 1.29 (d, J = 7.0 Hz, 3H), 1.37–1.53 (m, 5H), 1.60–1.67 (m, 1H), 1.70–1.81 (m, 1H), 2.00–2.08 (ddd, J = 16.0, 10.0, 6.0 Hz, 1H), 2.34–2.54 (m, 2H), 2.56–2.68 (ddd, J = 20.0, 16.0, 4.0 Hz, 1H), 3.48 (ddd, J = 15.0, 10.0, 6.0 Hz, 2H), 3.83 (q, J = 6.0 Hz, 1H), 4.00 (m, 1H), 4.21 (q, J = 6.0 Hz, 1H), 4.39 (ddd, J = 12.0, 6.0, 2.0 Hz, 1H); ^{13}C NMR (100 MHz, $CDCl_3$) δ − 5.5, −5.4, −4.5, −4.4, −4.2, −3.6, 9.9, 11.1, 13.9, 18.0, 18.1, 18.3, 20.8, 21.3, 23.3, 23.5, 25.8, 25.9, 26.0, 26.8, 27.3, 32.2, 32.7, 38.9, 41.9, 42.3, 42.4, 47.9, 65.1, 67.7, 70.1, 71.4, 72.1, 209.9; MS–CI m/z: M^+ 858, 844, 802, 728, 670, 630, 596, 538, 498, 471, 359, 269, 227, 199, 147, 115. HRMS–CI (m/z): calcd for $C_{46}H_{98}O_6Si_4$–C_4H_9, 801.5739; found, 801.5738.

HO–(chain)–SO_2Ph —[6% Na/Hg, Na_2HPO_4; MeOH, −10° to rt, 14 h]→ HO–(chain) (90%)

(S)-4-Methylnon-8-en-1-ol (Desulfonylation of a Non-Functionalized Sulfone).[343] A solution of (R)-4-methyl-6-(phenylsulfonyl)non-8-en-1-ol (0.24 g, 0.8 mmol) in dry MeOH (3 mL) was added to a stirred suspension of Na/Hg [freshly prepared from Na (0.37 g, 16.1 mmol) and Hg (6.2 g, 30.9 mmol)] and

Na_2HPO_4 (2.28 g, 16.1 mmol) in MeOH (10 mL) under argon. The reaction progress was monitored by TLC (ca. 14 h). The mixture was then filtered and the filter cake was washed with Et_2O. The combined filtrate and washings were evaporated at room temperature under vacuum. The residue was treated with H_2O (40 mL) and extracted with Et_2O (3 × 15 mL). The ethereal phase was washed with H_2O (15 mL) and brine (15 mL), dried ($MgSO_4$), and concentrated under vacuum at room temperature. The residue was purified by flash chromatography (silica gel, 2 : 1 light petroleum ether/Et_2O) to give the title compound as a colorless oil (0.11 g, 90%): R_f 0.19 (4 : 1 light petroleum/Et_2O); $[\alpha]_D^{23}$ −1.8° (*c* 1.23, $CHCl_3$); IR (film) 3400–3200, 3078, 2920, 2860, 1640, 1405, 1373, 1055, 990, 905 cm^{-1}; ^1H NMR (300 MHz, $CDCl_3$) δ 0.88 (d, J = 6.3 Hz, 3H), 1.07–1.68 (m, 7H), 1.96–2.07 (m, 3H), 3.60 (t, J = 6.3 Hz, 2H), 4.90–5.04 (m, 2H), 5.81 (ddt, J = 17.0, 10.5, 6.6 Hz, 1H); ^{13}C NMR (75 MHz, $CDCl_3$) δ19.6, 26.4, 30.3, 32.6, 32.9, 34.2, 36.5, 63.3, 114.2, 139.1; MS–EI m/z: M^+156, 123, 112, 97, 95, 82, 81, 70, 69, 55, 41. Anal. Calcd for $C_{10}H_{20}O$: C, 76.85; H, 12.90. Found: C, 76.71; H, 13.03.

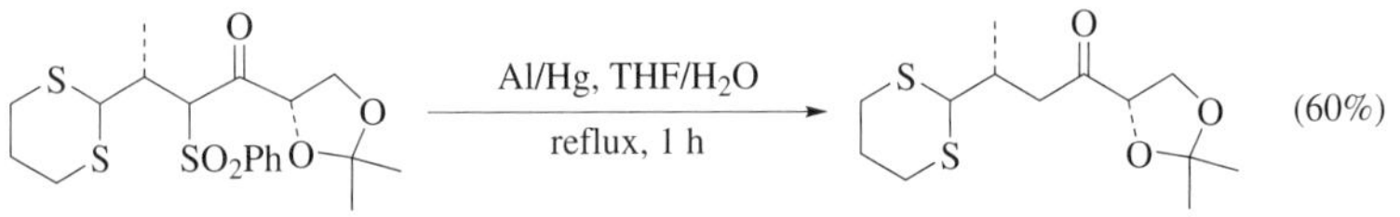

Al/Hg, THF/H_2O, reflux, 1 h (60%)

(3*R*)-1-[(4*S*)-2,2-Dimethyl-1,3-dioxolan-4-yl]-3-(1,3-dithian-2-yl)butan-1-one (Desulfonylation of a β-Ketosulfone).[344] A solution of mercury (II) chloride (60.3 g, 222 mmol) in water (1.2 L) was added to a vigorously stirred suspension of aluminum powder (11.9 g, 449 mmol) in water (50 mL). The supernatant was decanted and the amalgam washed with methanol (3 × 50 mL) followed by THF (3 × 50 mL). A suspension of the amalgam in THF (50 mL) was poured through a funnel into a solution of (3*R*)-1-[(4*S*)-2,2-dimethyl-1,3-dioxolan-4-yl)-3-(1,3-dithian-2-yl)-2-(phenylsulfonyl)butan-1-one (4.8 g, 11.1 mmol) in THF (70 mL). The reaction vessel was fitted with a reflux condenser, and water (5 mL) was added. After approximately 5 minutes, the reaction mixture began to reflux. Stirring was continued for 1 hour, and the mixture was then filtered through a pad of Celite and sand on a sintered-glass funnel. The solids were rinsed with EtOAc (300 mL), and the filtrate was washed with water (200 mL) and brine (200 mL), dried over $MgSO_4$, filtered, and concentrated. Flash chromatography (3 : 1 hexanes/EtOAc) provided the title product (1.9 g, 60%) as a colorless oil: $[\alpha]_D^{23}$ −15° (*c* 1.7, $CHCl_3$); IR ($CHCl_3$) 3345, 1670, 1590, 1110 cm^{-1}; ^1H NMR (500 MHz, DMSO–d_6) δ 0.90 (d, J = 7.0 Hz, 3H), 0.97 (s, 9H), 1.40 (s, 3H), 1.48 (dq, J = 14.0, 7.0 Hz, 1H), 1.51–1.58 (m, 1H), 1.52 (s, 3H), 1.84 (t, J = 7.0 Hz, 2H), 1.98 (q, J = 7.0 Hz, 2H), 2.25 (sextet, J = 7.0 Hz, 1H), 3.52 (dt, J = 10.1, 7.0 Hz, 1H), 3.56 (dt, J = 10.1, 7.0 Hz, 1H), 3.89 (t, J = 5.2 Hz, 2H), 4.39 (t, J = 5.2 Hz, 1H), 5.06 (t, J = 7.0, 1H), 5.21 (dt, J = 5.2, 1.2 Hz, 1H), 7.44 (m, 6H), 7.59 (m, 4H); HRMS–CI (NH_3) (m/z): $[M + H]^+$ calcd for $C_{13}H_{23}O_3S_2$, 291.1088; found, 291.1063.

Ph Ts / O — Zn, NH_4Cl / THF, rt, 2 h → Ph / O (92%)

1-Cyclohexylidenyl-3-phenylpropan-2-one (Desulfonylation of a β-Ketosulfone).[345] To a solution of 1-cyclohexylidenyl-3-phenyl-3-(*p*-toluenesulfonyl)propan-2-one (1.02 g, 2.77 mmol) in THF (15 mL) was added activated Zn (400 mg) and saturated aqueous NH_4Cl (15 mL). The mixture was stirred vigorously at room temperature for 2 hours and then diluted with EtOAc and filtered. The filtrate was washed with $NaHCO_3$ and brine, dried, and evaporated. Purification by flash chromatography (9 : 1 hexane/Et_2O) of the residue afforded the title product as a viscous colorless liquid (546 mg, 92%): IR (neat) 2932, 1688, 1613 cm^{-1}; ^1H NMR (200 MHz, $CDCl_3$) δ 1.57 (m, 6H), 2.13 (m, 2H), 2.80 (m, 2H), 3.69 (s, 2H), 6.00 (s, 1H), 7.26 (m, 5H); ^{13}C NMR (50 MHz, $CDCl_3$) δ 16.1, 27.7, 28.6, 29.7, 37.9, 51.3, 120.2, 126.5, 128.4, 129.2, 135.0, 162.7, 198.1. Anal. Calcd for $C_{15}H_{18}O$: C, 84.07; H, 8.47. Found: C, 84.10; H, 8.42.

Ph O Ts N Pr-*i* O — $Na_2S_2O_4$, $NaHCO_3$ / DMF/H_2O, 100°, 1 d → Ph O N Pr-*i* O (40%)

***N*-Isopropyl-γ-(2-phenyl-2-oxoethyl)-γ-butyrolactam (Desulfonylation of a β-Ketosulfone).**[346] To a solution of *N*-isopropyl-γ-[2-phenyl-2-oxo-1-(*p*-toluenesulfonyl)ethyl]-γ-butyrolactam (80 mg, 0.2 mmol) in DMF (4 mL) and water (2 mL) was added $Na_2S_2O_4$ (102 mg, 0.5 mmol) and $NaHCO_3$ (42 mg, 0.5 mmol). The mixture was stirred for 1 day at 100°, was cooled to rt, H_2O was added and the mixture was extracted with EtOAc (3 × 20 mL). The organic layer was dried (Na_2SO_4) and concentrated under vacuum (15 Torr) to give a residue that was chromatographed (silica gel, hexane/EtOAc) to afford the pure title product (19 mg, 40%): R_f 0.40 (EtOAc); IR (neat) 1670 cm^{-1}; ^1H NMR (300 MHz, $CDCl_3$) δ 1.29, 1.30 (2d, $J = 6.9$ Hz, 6H), 1.71 (m, 1H), 2.30 (m, 1H), 2.49 (m, 2H), 3.20 (dd, $J = 17.2, 9.5$ Hz, 1H), 3.32 (dd, $J = 17.2, 3.5$ Hz, 1H), 4.16 (m, 1H), 4.35 (m, 1H), 7.50, 7.61, 7.94 (3m, 5H); ^{13}C NMR (75 MHz, $CDCl_3$) δ 19.9, 21.6, 25.9, 30.2, 43.6, 44.5, 53.4, 128.0, 128.8, 133.6, 136.7, 174.9, 197.6; MS–EI m/z: M^+ 245, 217, 202, 126, 125, 110, 105, 84, 77, 55, 51, 43, 42, 41; HRMS–EI (m/z): calcd for $C_{15}H_{19}NO_2$, 245.1416; found, 245.1413.

MeO O BzHN SO_2Ph F — $(n\text{-Bu})_3SnH$, AIBN / C_6H_6, reflux, 24 h (80%) → MeO O BzHN $Sn(Bu\text{-}n)_3$ F

— HCl / reflux, 17 h (89%) → HO O ClH_3N H F

(Z)-α-(2′-Fluoro)vinylalanine Hydrochloride (Desulfonylation of a Vinylic Sulfone).[65] Argon was bubbled into a solution of (*S*, *E*)-methyl 2-benzamido-4-fluoro-2-methyl-4-(phenylsulfonyl)but-3-enoate (41 mg, 0.10 mmol) in benzene (1 mL) for 2 minutes. Tributyltin hydride (64 mg, 020 mmol) and AIBN (2.0 mg, 0.01 mmol) were then added under an argon atmosphere. The reaction mixture was heated under reflux for 24 hours, concentrated, and the residue was chromatographed (hexane to 90 : 10 hexane/EtOAc) to give the tri-*n*-butylstannyl vinylalaninate derivative (45 mg, 80%): IR (film) 1741, 1525 cm^{-1}; ^1H NMR (500 MHz, $CDCl_3$) δ 0.85 (m, 9H), 0.99 (m, 6H), 1.30 (m, 6H), 1.50 (m, 6H), 1.81 (s, 3H), 3.78 (s, 3H), 5.38 (d, $J = 56.0$ Hz, 1H), 7.40 (m, 2H), 7.46 (d, $J = 7.0$ Hz, 1H), 7.49 (bs, 1H), 7.77 (m, 2H); ^{13}C NMR 125 MHz, ($CDCl_3$) δ 8.7, 10.1, 13.5, 13.6, 24.3, 26.8, 27.0, 27.2, 28.6, 28.7, 28.8, 52.9, 57.9, 58.0, 124.2, 127.0, 127.1, 128.4, 131.4, 134.6, 165.9, 172.2, 173.9, 174.7; ^{19}F NMR (470 MHz, $CDCl_3$) δ −94.99 (d, $J = 55.0$ Hz). Anal. Calcd for $C_{25}H_{40}NO_3FSn$: C, 55.58; H, 7.46; N, 2.59. Found: C, 54.85; H, 7.34; N, 2.54.

A suspension of the tri-*n*-butylstannyl vinylalaninate (46.4 mg, 0.1 mmol) in 6 N HCl (2 mL) was refluxed for 17 hours. Following sequential extraction with CH_2Cl_2 and EtOAc, the aqueous layer was evaporated under vacuum and mild heating (40°) to give the title product salt (13.4 mg, 89%): ^1H NMR (500 MHz, D_2O) δ 1.73 (s, 3H), 5.21 (dd, $J = 44.0$, 5.0 Hz, 1H), 6.77 (dd, $J = 82.0$, 5.0 Hz, 1H); ^{19}F NMR (470 MHz, $CDCl_3$) δ −117.85 (dd, $J = 82.0$, 43.0 Hz); HRMS–FAB (m/z): $[M + H]^+$ calcd for $C_5H_9FNO_2$,134.0617; found, 134.0616.

$$\text{2-pyrimidinyl-}SO_2\text{-}CF(CH_3)\text{-}P(O)(OEt)_2 \xrightarrow[\text{toluene, } H_2O\text{, reflux, 3 h}]{(n\text{-Bu})_3SnCl\text{, AIBN, PMHS, KF}} CH_3CHF\text{-}P(O)(OEt)_2 \quad (82\%)$$

Diethyl 1-Fluoroethylphosphonate (Desulfonylation of an α-Functionalized Sulfone).[60] Nitrogen was bubbled through a solution of diethyl 1-fluoro-1-(pyrimidin-2-ylsulfonyl)ethylphosphonate (117 mg, 0.36 mmol), *n*-Bu_3SnCl (18 mg, 0.015 mL, 0.054 mmol), and AIBN (14 mg, 0.09 mmol) in toluene (3 mL) for 15 minutes. The solution was heated at reflux for 3 hours and PMHS (0.15 mL) and KF [(42 mg, 0.72 mmol) in H_2O (0.3 mL)] were added in three equal portions, immediately after the boiling point was reached, after 1 hour, and after 2 hours. Three extra portions of AIBN (14 mg, 0.09 mmol) in toluene (0.2 mL) were added via syringe after 45 minutes, 1.5 hours, and 2 hours. The volatiles were evaporated, and the residue was partitioned ($EtOAc/NaHCO_3/H_2O$). The organic layer was washed with brine, dried ($MgSO_4$), evaporated, and chromatographed (70–20% hexane/EtOAc) to give diethyl 1-fluoroethylphosphonate (54 mg, 82%): ^{19}F NMR (376.4 MHz, CCl_3F) δ −202.38 (ddq, $J = 76.0$, 46.8, 24.4 Hz); ^{31}P NMR (161.9 MHz, H_3PO_4) δ 19.87 (dm, $J = 75.2$, 7.2 Hz); MS–APCI m/z: $[M + H]^+$185.

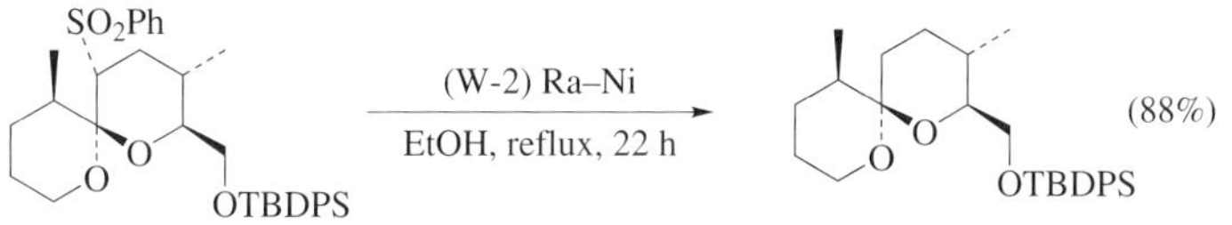

(W-2) Ra–Ni, EtOH, reflux, 22 h (88%)

(2*S*,3*S*,6*R*,11*R*)-3,11-Dimethyl-2-[[[(1,1-dimethylethyl)diphenylsilyl]oxy]methyl]-1,7-dioxaspiro[5.5]undecane (Desulfonylation of a β-Functionalized Sulfone).[180] To a solution of (2*S*,3*S*,5*R*,6*R*,11*R*)-3,11-dimethyl-2-[[[(1,1-dimethylethyl)diphenylsilyl]oxy]methyl]-5-(phenylsulfonyl)-1,7-dioxaspiro[5.5]undecane (2.83 g, 4.77 mmol) in EtOH (100 mL) was added a suspension of (W-2) Ra–Ni (42.0 g) in EtOH (100 mL). After the mixture was heated to reflux with vigorous stirring for 22 hours, Et_2O (100 mL) was added and the mixture refluxed again for 30 minutes. Insoluble material was removed by filtration through a pad of Celite, and the bed was washed with Et_2O (200 mL). The combined filtrate was concentrated under vacuum, and the residue obtained was purified by silica gel flash chromatography (benzene) and subsequent crystallization from MeCN to give the title compound as colorless needles (1.90 g, 88%): mp 82–84°; $[\alpha]_D^{22}$ + 40.9° (*c* 1.80, $CHCl_3$); IR (KBr) 2925, 2870, 1105 cm^{-1}; ^1H NMR (270 MHz, $CDCl_3$) δ 0.85 (d, $J = 5.8$ Hz, 3H), 0.99 (d, $J = 6.5$ Hz, 3H), 1.00–1.89 (m, 10H), 1.04 (s, 9H), 3.37 (ddd, $J = 9.0$, 5.3, 2.5 Hz, 1H), 3.53 (ddd, $J = 11.3$, 3.1, 1.6 Hz, 1H), 3.68 (dt, $J = 11.3$, 2.3 Hz, 1H), 3.73 (dd, $J = 10.6$, 5.3 Hz, 1H), 3.81 (dd, $J = 10.6$, 2.5 Hz, 1H), 7.35–7.80 (m, 10H); ^{13}C NMR (67.8 MHz, $CDCl_3$) δ 16.8, 17.6, 19.2, 26.4, 26.6, 27.5, 27.8, 30.6, 31.8, 38.8, 59.8, 64.9, 75.9, 97.9, 127.7, 129.5, 133.8, 134.0, 135.7; HRMS–EI (*m/z*): calcd for $C_{28}H_{40}O_3Si$, 452.2747; found, 452.2729.

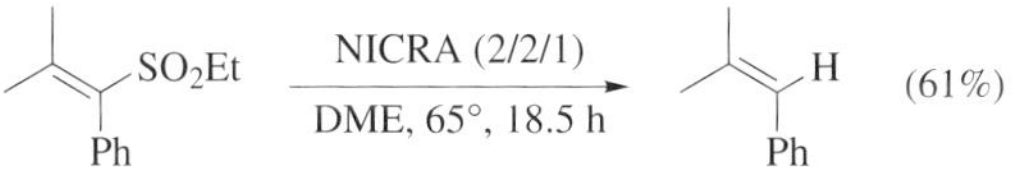

NICRA (2/2/1), DME, 65°, 18.5 h (61%)

1-Phenyl-2-methyl-1-propene (Desulfonylation of a Vinylic Sulfone).[75] *tert*-Amyl alcohol (20 mmol) in anhydrous DME (10 mL) was added dropwise to a suspension of NaH (40 mmol) and $Ni(OAc)_2$ (10 mmol) in refluxing anhydrous DME (30 mL). After 2 hours stirring at the same temperature the NICRA (2/2/1) was formed and ready for use. A solution of 1-(ethylsulfonyl)-1-phenyl-2-methyl-1-propene (224 mg, 1 mmol) in DME (10 mL) was then added dropwise and the reaction mixture was stirred for 18.5 hours at reflux. The excess NaH was carefully destroyed by dropwise addition of EtOH at room temperature. Analysis by GC of the crude reaction mixture showed 1-phenyl-2-methyl-1-propene as the major reaction product (95%) together with small amounts of 1-phenyl-2-methylpropane (5%). After classical work-up, the residue was purified by flash chromatography on silica gel (EtOAc/hexane) to give 1-phenyl-2-methyl-1-propene as a colorless oil (80.5 mg, 61%).

SO_2Me $\xrightarrow[\text{THF, rt, overnight}]{[Cp_2NiAlH_2]^- Li^+}$ (54%)

2-Methylnaphthalene (Desulfonylation of a Non-Functionalized Sulfone).[72] A solution of nickelocene (0.23 g, 1.2 mmol) in THF (20 mL) was added under argon to $LiAlH_4$ (0.046 g, 1.2 mmol) and the resulting solution was stirred at room temperature for 15 minutes. A solution of 2-(methylsulfonylmethyl)naphthalene (0.14 g, 0.61 mmol) in THF (10 mL) under argon was added and the mixture was stirred overnight at room temperature. Water was added and after stirring for 20 minutes, the mixture was filtered, and the filtrate was extracted with Et_2O. The combined organic extracts were dried ($MgSO_4$) and filtered, and the filtrate was evaporated in vacuo to give the title product (0.065 g, 54%).

MeO, OMe, SO_2Ph $\xrightarrow[\text{THF, }-78^\circ\text{ to }20^\circ\text{, 8 h}]{LiAlH_4, TiCl_4}$ MeO, OMe (87%)

9-Isopropyl-1,3-dimethoxy-4,7,12-trimethylbenzo[*a*]heptalene (Desulfonylation of an Aryl Sulfone).[80] Titanium tetrachloride (0.26 mL, 2.4 mmol) was added dropwise at -78° to anhydrous THF (8 mL) under an argon atmosphere. A 1 M solution of $LiAlH_4$ (7.1 mL, 7.1 mmol) in THF was then added slowly, upon which a dark gray suspension formed, which was left to warm to -10° within 3 hours. The mixture was cooled again to -78° and a solution of 9-isopropyl-1,3-dimethoxy-4,7,12-trimethyl-2-(phenylsulfonyl)benzo[*a*]heptalene (0.090 g, 0.184 mmol) in THF (4 mL) was added slowly under argon. After 0.5 hours at -78°, the temperature was raised within 2 hours to room temperature and stirring was continued for an additional 2 hours. The still dark gray mixture was added slowly to a saturated solution of NH_4Cl (150 mL), and the mixture was stirred for about 1.5 hours. After extraction with EtOAc (3 × 50 mL), the organic layer was washed with H_2O (50 mL), brine (50 mL), and dried (Na_2SO_4). Evaporation of the solvent under vacuum left a solid, which was purified by flash chromatography (SiO_2, 70 g, 4 : 1 hexane/EtOAc) to give the pure title product as a yellow crystalline powder (0.056 g, 87%): mp 132.5–132.9° (Et_2O/hexane); R_f (3 : 1 hexane/EtOAc) 0.75; ^1H NMR (300 MHz, $CDCl_3$) δ 1.15, 1.16 (2d, $J = 6.9$ Hz, 6H), 1.56 (s, 3H), 1.72 (s, 3H), 2.22 (s, 3H), 2.58 (septet, $J = 6.9$ Hz, 1H), 3.68 (s, 3H), 3.84 (s, 3H), 5.74 (s, 1H), 6.26 (d, $J = 12.0$ Hz, 1H), 6.34 (dd, $J = 11.8$, 1.2 Hz, 1H), 6.44 (d, $J = 11.8$ Hz, 1H), 6.62 (s, 1H), 6.99 (d, $J = 12.0$ Hz, 1H); ^{13}C NMR (75.5 MHz, $CDCl_3$) δ 11.2, 16.7, 19.0, 22.8, 23.1, 34.6, 56.0, 57.3, 98.6, 116.6, 121.3, 122.1, 127.3, 128.3, 129.7, 130.2, 132.2, 133.3, 135.6, 136.3, 138.0, 146.3, 154.2, 156.8.

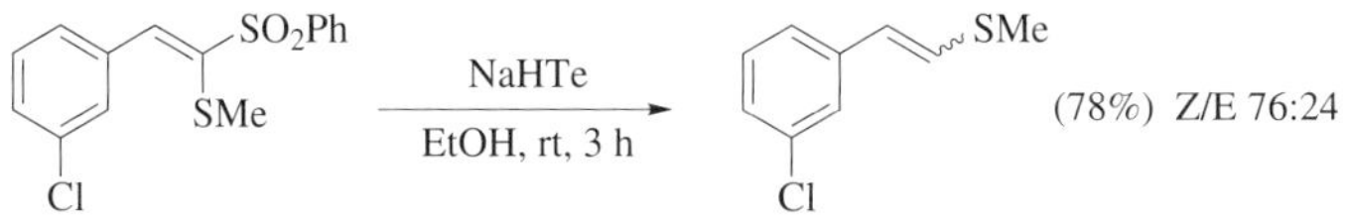

Methyl (β-3-Chlorophenylethenyl) Sulfide (Desulfonylation of a Vinylic Sulfone).[164] To a solution of NaHTe, prepared from Te (1.3 g, 10 mmol), and $NaBH_4$ (0.9 g, 24 mmol) in EtOH (20 mL) under a nitrogen atmosphere, was added a solution of (*E*)-α-methylthio-β-(3-chlorophenyl)ethenyl phenyl sulfone (1.3 g, 4 mmol) in EtOH (30 mL). The mixture was stirred at room temperature for 3 hours, quenched with water (30 mL), and kept open to air to precipitate the Te powder. After 1 hour, the mixture was filtered and the filtrate was extracted with Et_2O (3 × 30 mL). The combined ethereal solution was dried ($MgSO_4$) and concentrated to give the crude product, which was purified by column chromatography on silica gel using benzene as eluent to afford the pure title product as a colorless oil (0.57 g, 78%, Z/E = 76 : 24): IR (neat) 1600, 1592, 1482, 830, 788, 770, 672, 560 cm^{-1}; ^1H NMR (90 MHz, $CDCl_3$) Z-isomer: δ 2.36 (s, 3H), 6.09, 6.33 (2d, J = 11.0 Hz, 2H), 7.16–7.52 (m, 4H); E-isomer: δ 6.21, 6.83 (2d, J = 15.4 Hz, 2H).

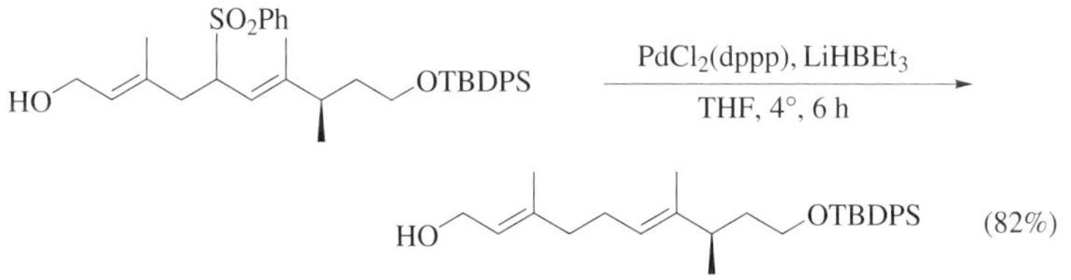

(*R*, 2*E*, 6*E*)-10-(*tert*-Butyldiphenylsilyloxy)-3,7,8-trimethyldeca-2,6-dien-1-ol (Desulfonylation of an Allylic Sulfone).[347] To a solution of (*R*, 2*E*, 6*E*)-10-(*tert*-butyldiphenylsilyloxy)-3,7,8-trimethyl-5-(phenylsulfonyl)deca-2,6-dien-1-ol (5.74 g, 9.71 mmol) and palladium chloride/1,3-bis(diphenylphosphano)propane complex (767 mg, 1.30 mmol) in dry THF (100 mL) was added a solution of lithium triethylhydroborate (1.0 M in THF, 29.0 mL, 29.0 mmol) at 0° under argon. The mixture was stirred at 4° for 6 hours, then was diluted with 10% aqueous NaCN solution and extracted with diethyl ether. The extracts and the organic layer were combined, washed with water and brine, dried with $MgSO_4$, and concentrated under reduced pressure. The residue was chromatographed on silica gel (80 g, 40 : 1 hexane/EtOAc) to give the title product (3.57 g, 82%) as a colorless oil: $[\alpha]_D^{22}$ + 0.628° (*c* 1.0, $CHCl_3$); IR (film) 3345 cm^{-1}; ^1H NMR (500 MHz, $CDCl_3$) δ 1.07 (d, J = 6.7 Hz, 3H), 1.36 (s, 3H), 1.46 (s, 3H), 1.82 (m, 1H), 2.06 (m, 1H), 2.55 (m, 1H), 2.58 (dd, J = 17.7, 7.9 Hz, 1H), 2.81 (m, 4H), 2.95 (dd, J = 17.7, 4.6 Hz, 1H), 3.79 (dd, J = 8.6, 5.5 Hz, 1H), 4.07 (d, J = 4.8 Hz, 1H), 4.16 (dd, J = 8.6, 7.7 Hz, 1H), 4.41 (dd, J = 7.7, 5.5 Hz, 1H); ^{13}C NMR (62.8 MHz, $CDCl_3$) δ 17.6, 24.9, 26.0, 30.2, 30.4, 33.1, 42.9, 53.9, 66.3, 80.2, 110.9, 209.2. Anal. Calcd for $C_{29}H_{42}O_2Si$: C, 77.28; H, 9.39. Found: C, 77.10; H, 9.43.

SO$_2$Ph

n-BuMgCl, Ni(acac)$_2$

THF, 1 h, rt

(51%)

98.5% E,E

(51%) 96% E,Z

(3E, 5Z)-Dodecadiene (Desulfonylation of a Vinylic Sulfone).[48] A mixture of (3E, 5Z)-5-(phenylsulfonyl)dodeca-3,5-diene (153 mg, 0.5 mmol) and Ni(acac)$_2$ (2.6 mg, 0.01 mmol) was purged three times with nitrogen before adding anhydrous THF (2.5 mL). The mixture was stirred at room temperature for 0.25 hour and a 1 M solution of n-BuMgCl in THF (1 mL, 1 mmol) was added dropwise. The resulting pale blue solution was poured over a mixture of saturated aqueous ammonium chloride and ice. The mixture was extracted five times with pentane and the combined organic layers were washed 5 times with H_2O. After elution over a column of silica gel, the solvent was distilled through a glass-bead column to yield (3E, 5Z)-dodecadiene (44 mg, 51%) contaminated with small amounts (4%) of the EE–isomer: ^{1}H NMR (250 MHz, $CDCl_3$) δ 0.91 (m, 3H), 1.04 (t, J = 7.5 Hz, 3H), 1.22–1.45 (m, 8H), 2.08–2.24 (m, 4H), 5.35 (m, 1H), 5.64 (br dt, J = 15.0, 6.7 Hz, 1H), 6.01 (br t, J = 11.0, 1H), 6.37 (br ddd, J = 15.0, 11.0, 1.5 Hz, 1H); MS–EI m/z: M^+166, 137, 123, 109, 95, 82, 81, 67; HRMS–EI (m/z): calcd for $C_{12}H_{22}$, 166.1721; found, 166.1721.

Reductive Eliminations

OH NHBoc OBn O O SO$_2$Ph

6% Na/Hg, Na$_2$HPO$_4$

MeOH, 0°, 3 h

NHBoc OBn O O

(80%) Z/E 23:77

(2R, 5R)-1-Benzyloxy-2-[(*tert*-butoxycarbonyl)amino]-5,6-isopropylidene-dioxyhex-3-ene (Reductive Elimination of a β-Hydroxysulfone).[201] To a solution of (2S, 5S)-6-benzyloxy-5-[(*tert*-butoxycarbonyl)amino]-1,2-isopropylidenedioxy-4-(phenylsulfonyl)hexan-3-ol (4.55 g, 8.5 mmol) in HPLC grade MeOH (70 mL) containing Na_2HPO_4 (12.1 g, 85 mmol) was added 6% Na/Hg (25 g, 65 mmol) at 0°. The mixture was stirred at this temperature for 3 hours. Mercury was removed by decanting the reaction mixture and the MeOH was evaporated. The residue was diluted in H_2O (200 mL) and extracted with EtOAc (3x100 mL). The organic extracts were washed successively with H_2O (2 × 100 mL) and brine (100 mL), dried over Na_2SO_4, and evaporated. Flash chromatography of the residue (3 : 1 heptane/EtOAc) provided two alkenes (2.54 g, 80%): 1.95 g (77%) of the E-isomer and 0.59 g (23%) of the Z-isomer.

E-isomer: $[\alpha]_D^{20}$ −7.3° (c 2.0, $CHCl_3$); IR (neat) 3348, 3030, 2982, 2934, 2869, 1715, 1511, 1498, 1455, 1391, 1368, 1247 cm^{-1}; ^{1}H NMR (300 MHz, $CDCl_3$) δ 1.38, 1.41 (2s, 6H), 1.44 (s, 9H), 3.45–3.54 (m, 2H), 3.56 (t, J = 8.0 Hz, 1H), 4.07 (dd, J = 8.0, 6.1 Hz, 1H), 4.35 (br s, 1H), 4.44–4.58 (m, 3H), 4.92

(br s, 1H), 5.64 (ddd, $J = 15.6, 7.1, 1.2$ Hz, 1H), 5.83 (dd, $J = 15.6, 5.1$ Hz, 1H), 7.30–7.33 (m, 5H); ^{13}C NMR (75.5 MHz, $CDCl_3$) δ 26.0, 26.7, 28.4, 51.4, 69.5, 72.0, 73.2, 76.6, 79.6, 109.4, 127.7, 127.8, 128.5, 128.9, 132.4, 137.9, 155.4; MS–CI m/z: $[M+H]^+$ 378, 278, 264. Anal. Calcd for $C_{21}H_{31}NO_5$: C, 66.8; H, 8.3; N, 3.7. Found: C, 66.5; H, 8.7; N, 3.6.

Z-isomer: $[\alpha]_D^{20} + 3.7°$ (c 2.0, $CHCl_3$); IR (neat) 3347, 3030, 2982, 2933, 2969, 1715, 1511, 1498, 1455, 1391, 1368, 1247 cm^{-1}; ^{1}H NMR (300 MHz, $CDCl_3$) δ 1.36, 1.41 (2s, 6H), 1.44 (s, 9H), 3.45–3.61 (m, 2H), 4.18 (dd, $J = 8.1$, 6.2 Hz, 1H), 4.49, 4.57 (2d, $J = 11.9$ Hz, 2H), 4.55–4.65 (m, 1H), 4.90–5.10 (m, 2H), 5.33 (dd, $J = 11.0, 8.7$ Hz, 1H), 5.68 (dd, $J = 11.0, 10.4$ Hz, 1H), 7.30–7.35 (m, 5H); ^{13}C NMR (75.5 MHz, $CDCl_3$) δ 25.9, 26.9, 28.5, 48.0, 69.8, 72.2, 72.3, 73.4, 79.6, 109.4, 127.8, 127.9, 128.5, 129.8, 130.3, 131.9, 137.9, 155.1. MS–CI m/z: $[M+H]^+$ 378, 278, 264; HRMS–CI (m/z): calcd for $C_{21}H_{32}NO_5$, 378.2280; found, 378.2282.

TABULAR SURVEY

Tables 1–8 are organized by substrate and cover the reductive desulfonylation reactions of non-functionalized sulfones, α-functionalized sulfones, β-functionalized sulfones, remote functionalized sulfones, β-oxo sulfones (and β-oxo equivalents), allylic sulfones, and vinylic sulfones, respectively. Table 8 covers the reductive elimination (Julia–Lythgoe olefination) of β-functionalized sulfones. In general, a polyfunctionalized sulfone substrate will be ordered according to the following substitution classification: β-oxo, allyl, vinyl > α > β > remote. For example, an α-substituted vinylic sulfone will be found in Table 7. Entries in Tables 1–8 are ordered by increasing carbon count of the compound. Protecting groups are included in the carbon count. For a particular carbon count, entries are ordered according to increasing hydrogen count. The tables contain all examples that could be found in the literature through September 2007.

Abbreviations used in the tables are as follows:

Ac	acetyl
AIBN	2,2′-azobis(2-methylpropionitrile)
All	allyl
Alloc	allyloxycarbonyl
Bn	benzyl
BNAH	1-benzyl-1,4-dihydronicotinamide
Boc	*tert*-butoxycarbonyl
BOM	benzyloxymethyl
bpy	2,2′-bypiridyl
Bz	benzoyl
CAN	ceric ammonium nitrate
Cbz	benzyloxycarbonyl
$C_{10}H_7$	naphthyl
$C_{10}H_8$	naphthalene

Cp	cyclopentadienyl
DABCO	1,4-diazabicyclo[2.2.2]octane
DEIPS	diethylisopropylsilyl
DIBALH	diisobutylaluminum hydride
DMAN	1-(dimethylamino)naphthalene
DMAP	4-(dimethylamino)pyridine
DME	1,2-dimethoxyethane
DMF	*N*,*N*-dimethylformamide
DMPM	3,4-dimethoxyphenylmethyl
DMPU	*N*,*N*′-dimethylpropyleneurea
DMSO	dimethyl sulfoxide
dppe	1,2-bis(diphenylphosphino)ethane
dppp	1,3-bis(diphenylphosphino)propane
dr	diastereomeric ratio
EDTA	ethylenediaminetetraacetic acid
HMPA	hexamethylphosphoric triamide
LDA	lithium diisopropylamide
LDTBB	lithium 4,4′-di-*tert*-butylbiphenylide
LHMDS	lithium hexamethyldisilazane
MCPBA	3-chloroperbenzoic acid
MEM	2-methoxyethoxymethyl
Mes	mesityl
MOM	methoxymethyl
MP	4-methoxyphenyl
MR	Merrifield resin
Ms	methanesulfonyl or mesyl
MTM	methylthiomethyl
NADH	reduced nicotinamide adenine dinucleotide
NICRA	nickel-containing complex reducing agent
NICRA (x/y/z/t)	nickel-containing complex reducing agents (NaH/*t*-AmONa/Ni(OAc)$_2$/external ligand)
NMO	*N*-methylmorpholine *N*-oxide
OcV^{2+}	octylviologen (1,1′-dioctyl-4,4′-bipyridinium)
Piv	pivaloyl
PMB	4-methoxybenzyl
PMHS	poly(methylhydrosiloxane)
PNAH	1-propyl-1,4-dihydronicotinamide
PNB	4-nitrobenzyl
PNBz	4-nitrobenzoyl
PPTS	pyridinium 4-toluenesulfonate
Py	pyridine
Ra–Ni	Raney nickel
rt	room temperature
SEM	2-(trimethylsilyl)ethoxymethyl
TBAF	tetra(*n*-butyl)ammonium fluoride

TBAI	tetra(*n*-butyl)ammonium iodide
TBDMS	*tert*-butyldimethylsilyl
TBDPS	*tert*-butyldiphenylsilyl
TEA	triethylamine
Teoc	2-(trimethylsilyl)ethoxycarbonyl
TES	triethylsilyl
THF	tetrahydrofuranyl
THP	2-tetrahydropyranyl
TIPS	triisopropylsilyl
TMEDA	N,N,N',N'-tetramethylethylenediamine
TMS	trimethylsilyl
TMSCl	trimethylsilyl chloride
Tol	tolyl
TPAP	tetrapropylammonium perruthenate
Tr	trityl
Ts	4-toluenesulfonyl or tosyl

TABLE 1. REDUCTIVE DESULFONYLATION OF NON-FUNCTIONALIZED SULFONES

Substrate	Conditions	Product(s) and Yield(s) (%)	Refs.
C_{10} (structure, SO_2Me)	$EtNH_2$/Li, 0°, 1.5 h	(structure) (31)	348
C_{10-25} $R^1CH(R^2)SO_2R^3$	NICRA (x/y/z/t), THF, 65°	$R^1CH(R^2)H$	75

R^1	R^2	R^3	x/y/z/t	Time	Yield (%)
n-C_6H_{13}	Me	Et	7/2/1/0	21 h	(80)
Ph	Ph	Et	7/2/1/0	21 h	(80)
n-$C_{12}H_{25}$	H	Et	5/2/1/0	19 h	(60)
n-$C_{12}H_{25}$	H	n-$C_{12}H_{25}$	7/2/1/0	16.5 h	(55)

Substrate	Conditions	Product(s) and Yield(s) (%)	Refs.
C_{12} (2-naphthyl)CH_2SO_2Me	$NiBr_2$•DME, Ph_3P, $LiAlH_4$, rt, 1 d	2-methylnaphthalene (92)	72
C_{13-25} (bicyclic, n, R^1, R^2, SO_2Ph)	Na/Hg, Na_2HPO_4, THF/MeOH, –20°, 11 h	(bicyclic, n, R^1, R^2)	

n	R^1	R^2	Time	Yield (%)	Refs.
1	H	H	11 h	(80)	349
2	H	H	11 h	(78.5)	349
2	MeO	H	11 h	(91)	349
2	MeO	$H_2C{=}CH_2CH_2$	(—)	(95)	350
2	MeO	$(EtO)_2CHCH_2$	(—)	(80)	350
2	MeO	(E)-$Me_2C{=}CH(CH_2)_2C(Me){=}CHCH_2$	(—)	(93)	350

C_{14}

SO$_2$Ph | Mg, MeOH, 50° | (68) | 118

C_{14-17}

R, SO$_2$C$_6$H$_4$Cl-4 | Na/Hg, EtOH, reflux, 15 h | | 351

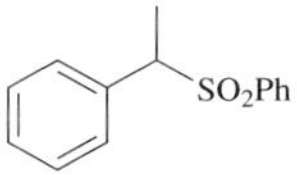

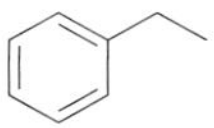

R	
Me	(70)
n-Bu	(55)

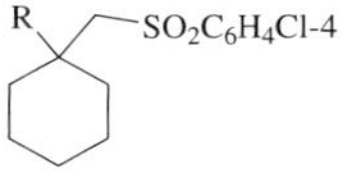

C_{14-20}

R[1], R[2], R[3], R[4], SO$_2$C$_6$H$_4$Cl-4 | Na/Hg, EtOH, reflux | R[1], R[2], R[3], R[4] | 351

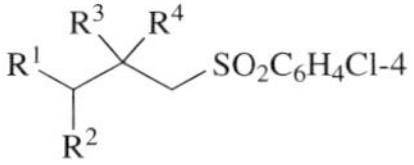

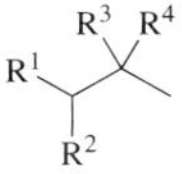

R^1	R^2	R^3	R^4	Time	
n-Pr	H	Me	Me	15 h	(70)
H	H	Ph	H	5 h	(99)
n-C_5H_{11}	H	Me	H	4 h	(85)
Ph	Me	Me	H	18 h	(100)
n-Pr	H	Ph	H	15 h	(92)
n-C_5H_{11}	H	n-Bu	H	12 h	(72)
Ph	Me	n-Bu	H	15 h	(80)

TABLE 1. REDUCTIVE DESULFONYLATION OF NON-FUNCTIONALIZED SULFONES (Continued)

Substrate	Conditions	Product(s) and Yield(s) (%)	Refs.
C_{14-20}			
PhO_2S–CH_2–$C(R^2)(R^3)$–CH=CH–R^1 (cis)	Na/Hg, Na_2HPO_4, MeOH, rt	CH_3–$C(R^2)(R^3)$–CH=CH–R^1 (cis)	352

R^1	R^2	R^3	Yield (%)
t-Bu	H	H	(82)
n-Bu	Me	H	(98)
s-Bu	Me	H	(95)
n-Bu	Me	Me	(83)
n-Bu	Ph	H	(83)
s-Bu	Ph	H	(87)

Substrate	Conditions	Product(s) and Yield(s) (%)	Refs.
C_{14-26}			
1-R^1-2-R^2-cyclohexene bearing R^3 and SO_2Ph at C-4	Na/Hg, Na_2HPO_4, THF/MeOH, –20°, 5.5 h	1-R^1-2-R^2-4-R^3-cyclohexene	349, 350

R^1	R^2	R^3	Yield (%)
Me	Me	H	(76)
H	$Me_2C{=}CH(CH_2)_2$	H	(92)
H	Me	$EtO_2C(CH_2)_4$	(81)
Me	Me	Bn	(96)
Me	Me	$PhS(CH_2)_2$	(83)
Me	Me	$4\text{-TolCH(Me)}(CH_2)_3$	(93)

Substrate	Conditions	Product(s) and Yield(s) (%)	Refs.
C_{15}			
3,5-(MeO)$_2C_6H_3CH_2SO_2Ph$	Mg, $HgCl_2$, EtOH, rt, 2 h	3,5-(MeO)$_2C_6H_3CH_3$ (98)	114
C_{16}			
3-($PhSO_2CH_2$)-6-methyl-4,5,6,7-tetrahydrobenzofuran	Na/Hg	3,6-dimethyl-4,5,6,7-tetrahydrobenzofuran (—)	353

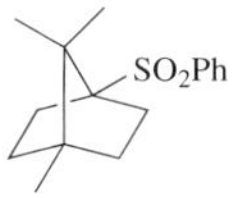	Ra-Ni, EtOH, reflux, 17 h	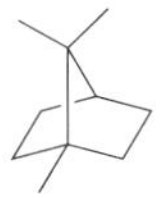 (61)	69
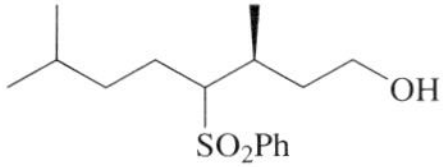	$H_2N(CH_2)_2NH_2$/Li, pentane, rt	 (62)	354
C_{17}			
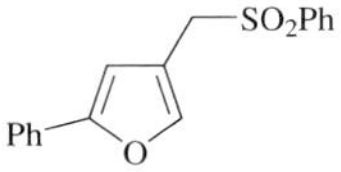	Na/Hg	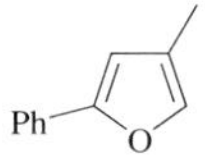 (—)	353
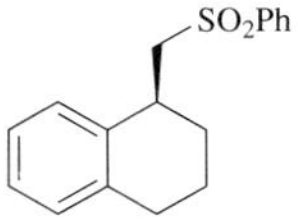	Na/Hg, Na_2HPO_4, THF/MeOH, rt, 40 h	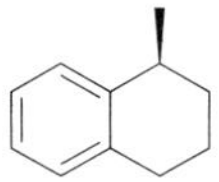 (95)	355
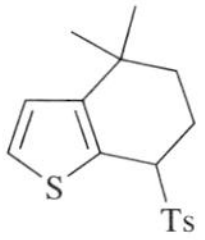	DIBALH, toluene, 50-80°, 5 min	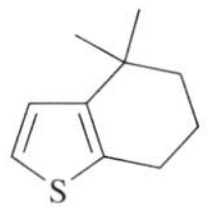 (72)	78
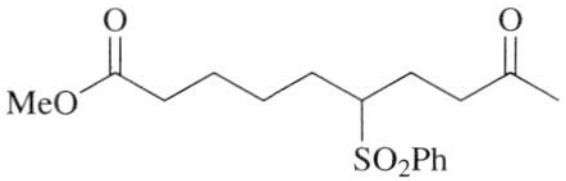	Na/Hg, Na_2HPO_4, MeOH, 0°	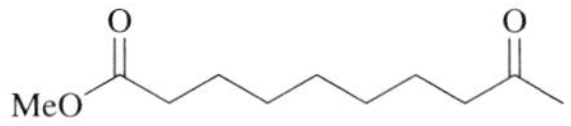 (60)	356
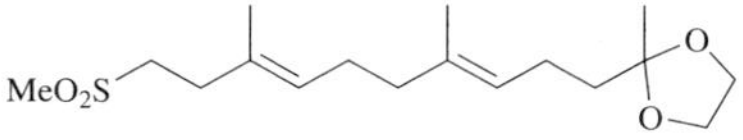	1. $EtNH_2$/Li, 0°, 1.5 h 2. H_3O^+	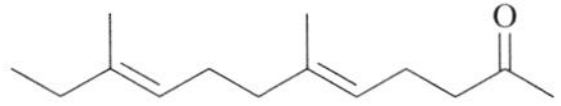 (86)	348

TABLE 1. REDUCTIVE DESULFONYLATION OF NON-FUNCTIONALIZED SULFONES (Continued)

Substrate	Conditions	Product(s) and Yield(s) (%)	Refs.
C_{17-19} R^1 R^2 SO_2Ph	Mg, $HgCl_2$, EtOH, rt, 2 h	R^1 R^2 H; R^1 = H, R^2 = H (99); R^1 = Me, R^2 = H (100); R^1 = Me, R^2 = Me (100)	114
C_{17-24} OMe SO_2Ph N R	Na/Hg, Na_2HPO_4, MeOH, rt	OMe N R; R = H, Time 2.5 h (94); R = (*E*)-$TBDMSOCH_2$, Time 6 h (93); R = (*Z*)-$TBDMSOCH_2$, Time 1 h (99)	357
C_{18} H H SO_2Ph	Na/Hg, Na_2HPO_4, MeOH, rt, 40 h	H H (92)	358
O O PhO_2S	Na/Hg, Na_2HPO_4, THF/MeOH, –25°	O O (94)	359
C_{19} SO_2Ph	Na/Hg, Na_2HPO_4, THF/MeOH, –20°, 5 h	(97)	349
O O SO_2Ph	Na/Hg, Na_2HPO_4, MeOH/DME, –25°, 3 h	O O (90)	360

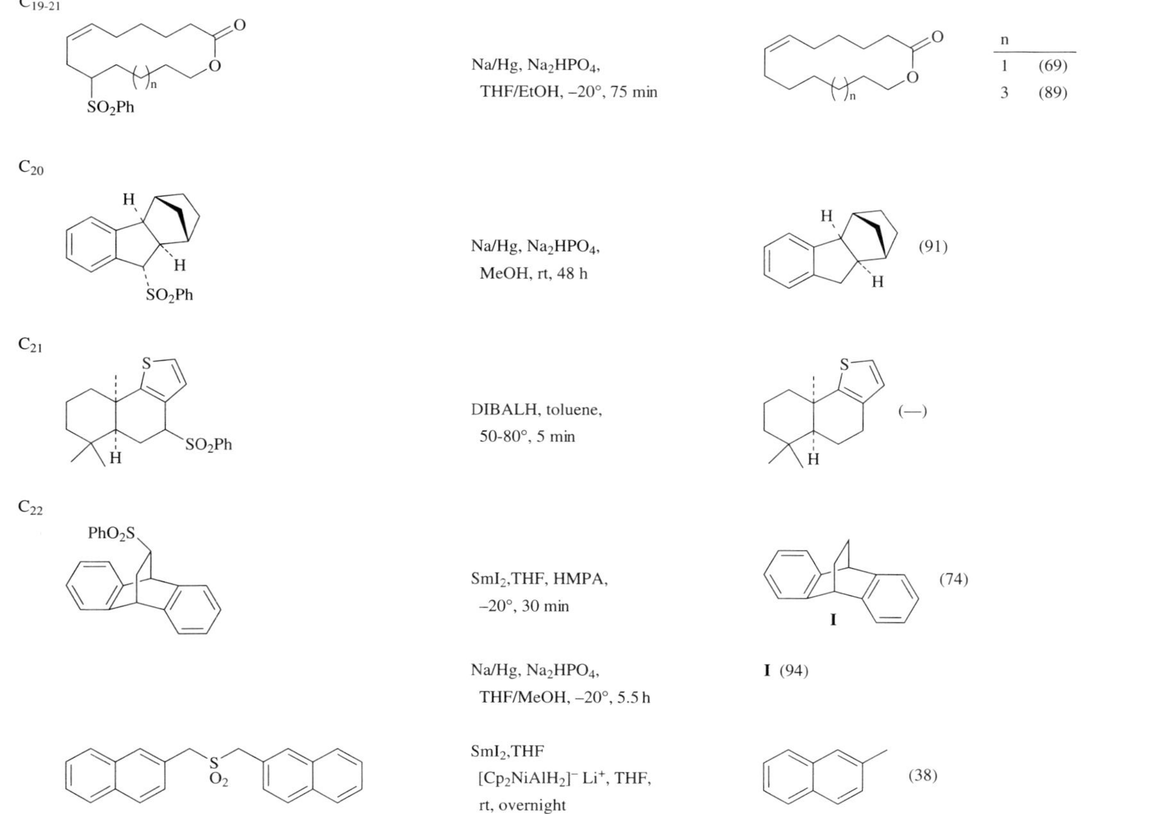

Substrate	Conditions	Product(s) and Yield(s) (%)	Refs.
C_{19-21}	Na/Hg, Na_2HPO_4, THF/EtOH, –20°, 75 min	n = 1 (69); n = 3 (89)	359
C_{20}	Na/Hg, Na_2HPO_4, MeOH, rt, 48 h	(91)	361
C_{21}	DIBALH, toluene, 50-80°, 5 min	(—)	79
C_{22}	SmI_2,THF, HMPA, –20°, 30 min	**I** (74)	100
	Na/Hg, Na_2HPO_4, THF/MeOH, –20°, 5.5 h	**I** (94)	349
	SmI_2,THF $[Cp_2NiAlH_2]^-$ Li^+, THF, rt, overnight	(38)	72

TABLE 1. REDUCTIVE DESULFONYLATION OF NON-FUNCTIONALIZED SULFONES (Continued)

	Substrate	Conditions	Product(s) and Yield(s) (%)	Refs.
C_{22}	SO$_2$Ph	Na/Hg, Na_2HPO_4, MeOH	(95)	136
C_{22-25}	Ts, R^1, R^2, N H	See table.	R^1, R^2, N H	362
C_{23}	SO$_2$Ph, OAc	Na/Hg, HMPA/EtOH, 0°, 1 h	OH (60)	130
	SO$_2$Ph	Na/Hg, EtOH, 0°, 4 h	(89)	343

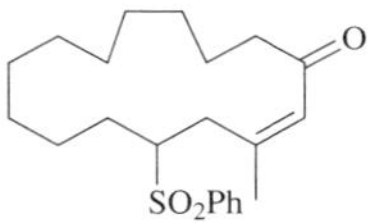

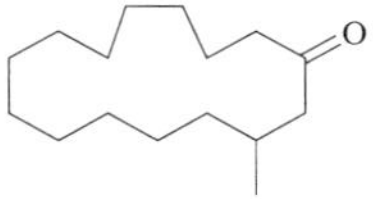

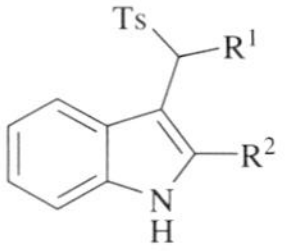

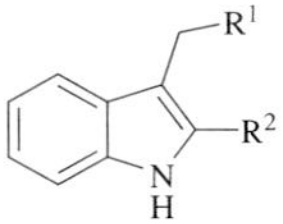

R^1	R^2	Reagents	Solvent	Temp	Time	
n-C_5H_{11}	Me	(n-Bu)$_3$SnH, AIBN	toluene	reflux	3 h	(85)
Ph	H	(n-Bu)$_3$SnH, AIBN	toluene	reflux	4 h	(86)
Ph	H	Na/Hg, Na_2HPO_4	EtOH	rt	2 h	(61)
Ph	Me	Na/Hg, Na_2HPO_4	EtOH	rt	4 h	(62)
4-$NO_2C_6H_4$	Me	(n-Bu)$_3$SnH, AIBN	toluene	reflux	8 h	(56)
$PhCH_2CH_2$	Me	(n-Bu)$_3$SnH, AIBN	toluene	reflux	8 h	(77)
$PhCH_2CH_2$	Me	$LiAlH_4$	THF	rt	2 h	(85)

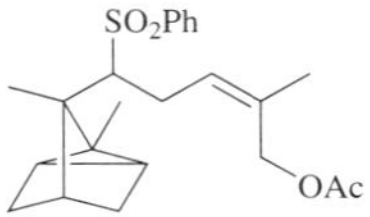

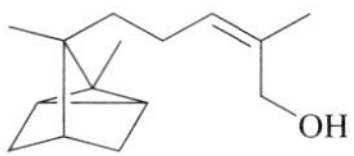

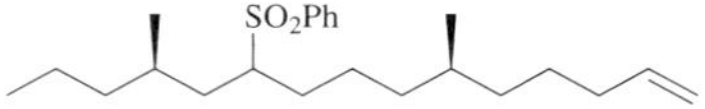

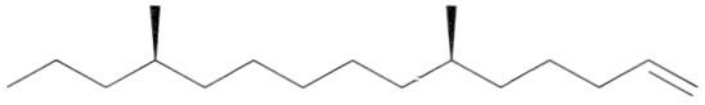

Substrate	Conditions	Product(s) and Yield(s) (%)	Refs.
C_{24}			
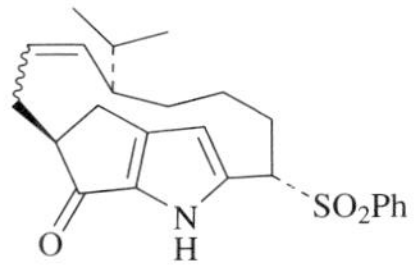	Na/Hg, Na_2HPO_4, THF/MeOH, 0°, 3 h	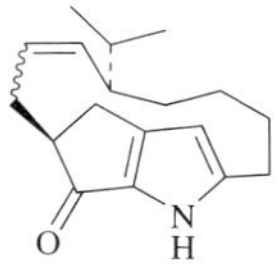 (90)	363
C_{25}			
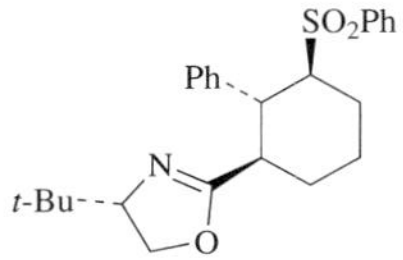	Na/Hg, Na_2HPO_4, EtOH, 0° to rt, 12 h	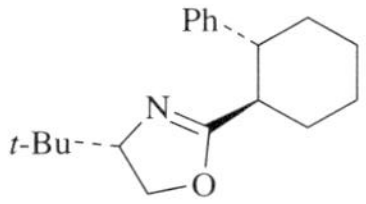 (85)	364
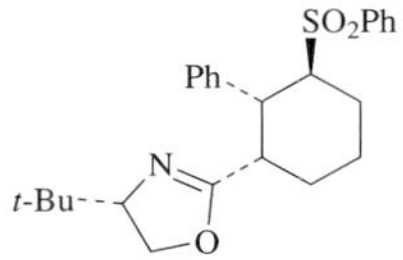	Na/Hg, Na_2HPO_4, EtOH, 0° to rt, 12 h	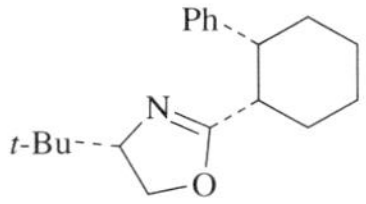 (62)	364
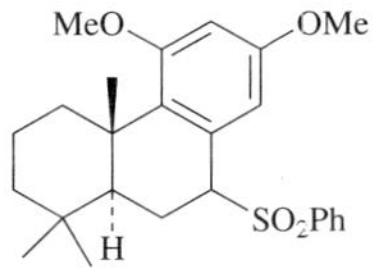	DIBALH, toluene, 50-80°, 5 min	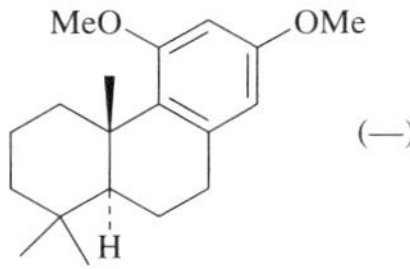 (—)	79
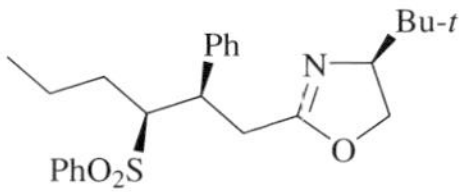	Na/Hg, Na_2HPO_4, EtOH, 0° to rt, 12 h	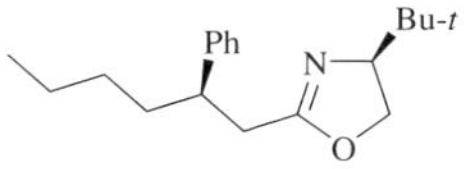 (86)	364
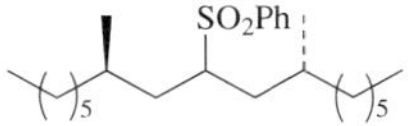	Na/Hg, Na_2HPO_4, EtOH, rt, 6 h	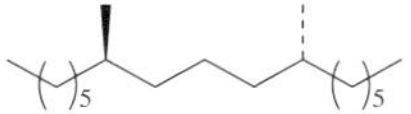 (33)	365

TABLE 1. REDUCTIVE DESULFONYLATION OF NON-FUNCTIONALIZED SULFONES (Continued)

	Substrate	Conditions	Product(s) and Yield(s) (%)	Refs.
C_{28}	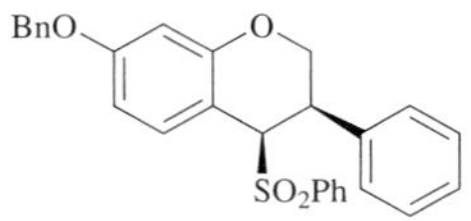 BnO, O, SO_2Ph	Na/Hg, Na_2HPO_4, MeOH/THF, rt	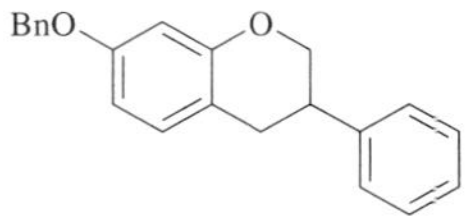 BnO, O (80)	366
	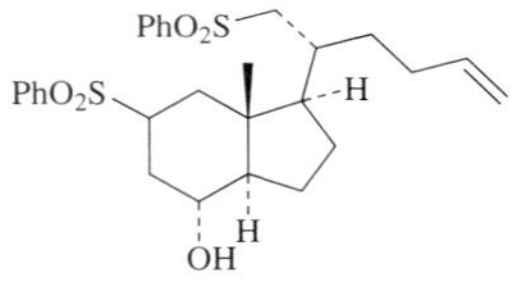 PhO_2S, PhO_2S, H, OH, H	Na/Hg, Na_2HPO_4, MeOH, rt, 1 h	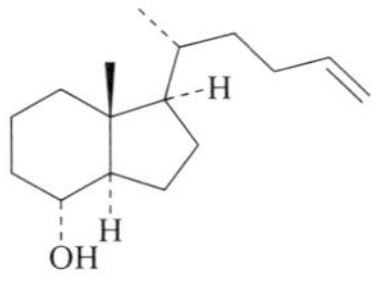 H, OH, H (49)	367
$C_{28\text{-}31}$	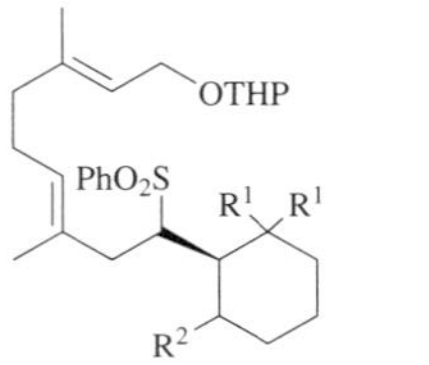 OTHP, PhO_2S, R^1, R^1, R^2	Reagents, Na_2HPO_4, rt	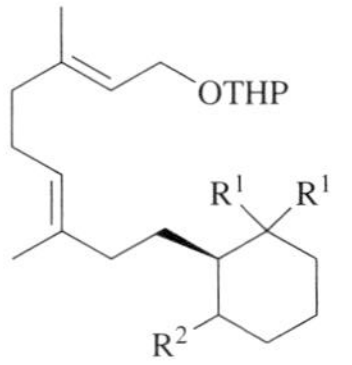 OTHP, R^1, R^1, R^2 (—)	368
C_{29}	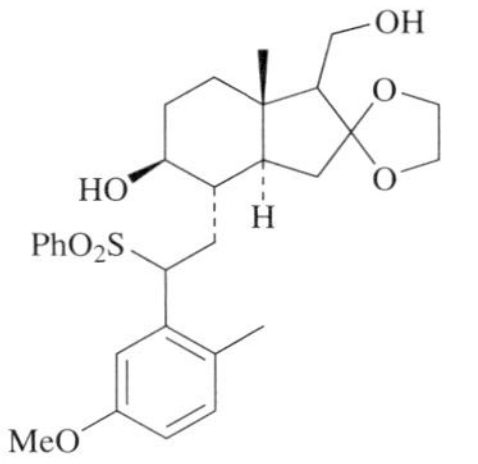 OH, O, O, HO, H, PhO_2S, MeO	Na, EtOH/THF, –20°, 2 h	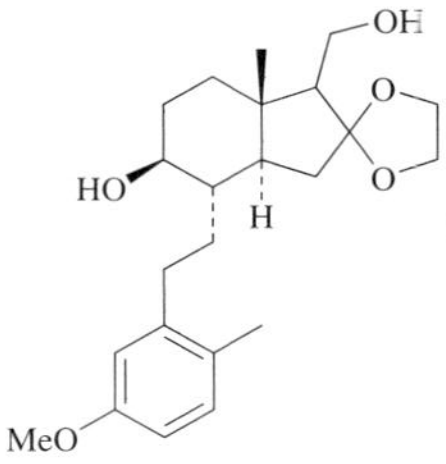 OH, O, O, HO, H, MeO (89)	369

R^1	R^2
H	H
Me	$H_2C=$

Reagents	Solvent	Time
Na/Hg	MeOH	2 h
Na	THF/EtOH	16 h

Substrate	Conditions	Product(s) and Yield(s) (%)	Refs.
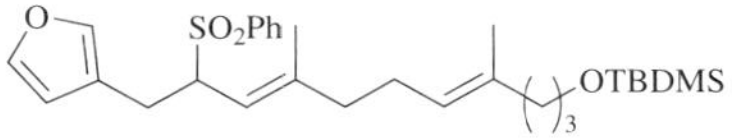	Na, THF/*i*-PrOH, 0°	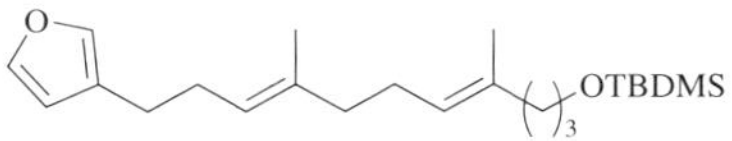 (76)	370
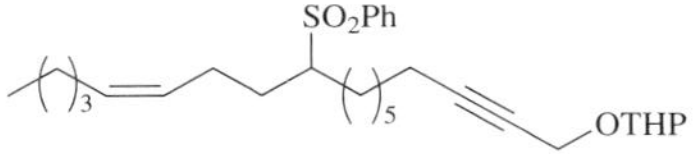	Na/Hg, Na_2HPO_4, MeOH, rt, 21 h	 (88)	371
$C_{29\text{-}33}$			
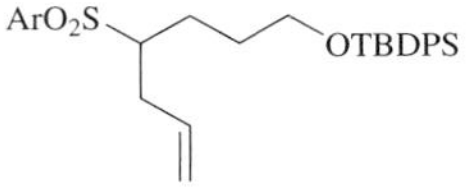	Na/Hg, Na_2HPO_4, MeOH, 10°, 0.5 h	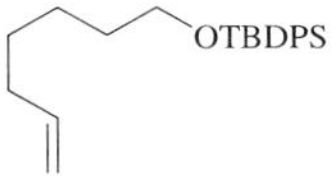	372
C_{30}			
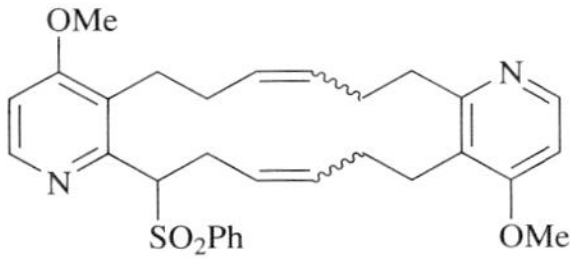	Al/Hg, THF/H_2O, 0°, 1.5 h	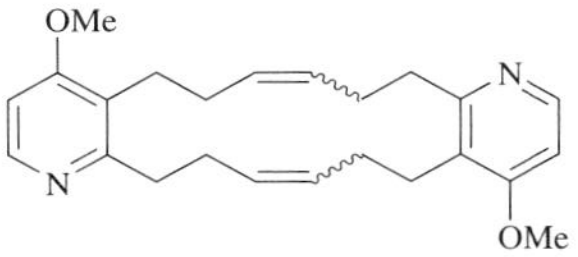 (81)	357
	Ra-Ni, MeOH, rt, 20 h	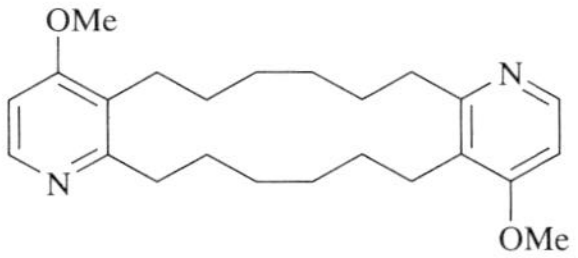 (65)	357
C_{32}			
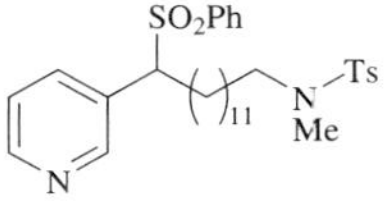	Na/$C_{10}H_8$, DME, –40°	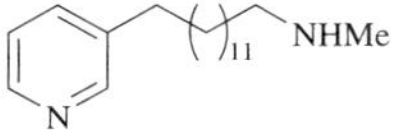 (100)	373

Ar	Time	
4-FC_6H_4	0.5 h	(97)
Ph	1 h	(88)
4-MeC_6H_4	1 h	(75)
2-naphthyl	0.5 h	(93)

TABLE 1. REDUCTIVE DESULFONYLATION OF NON-FUNCTIONALIZED SULFONES (*Continued*)

Substrate	Conditions	Product(s) and Yield(s) (%)	Refs.
C_{32}			
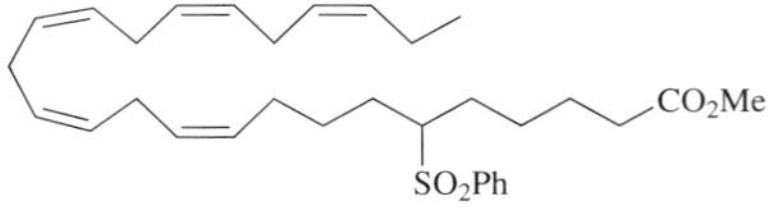 CO_2Me, SO_2Ph	Na/Hg, Na_2HPO_4, MeOH, –20°	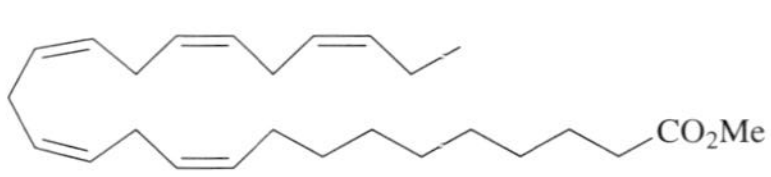 CO_2Me (—)	374
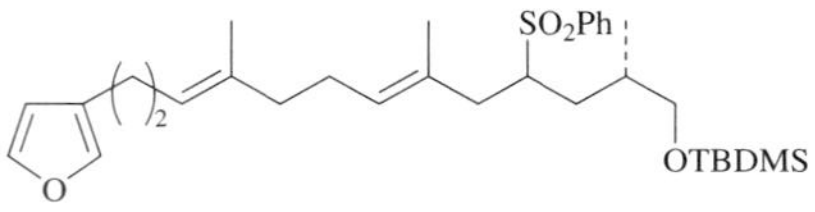 SO_2Ph, OTBDMS	Na, THF/*i*-PrOH	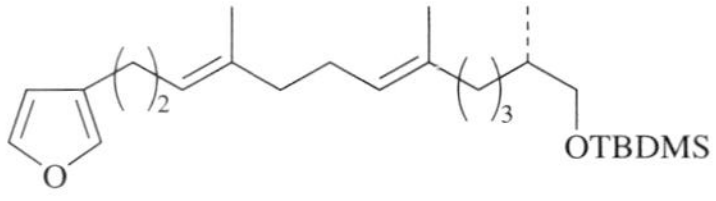 OTBDMS (76)	370
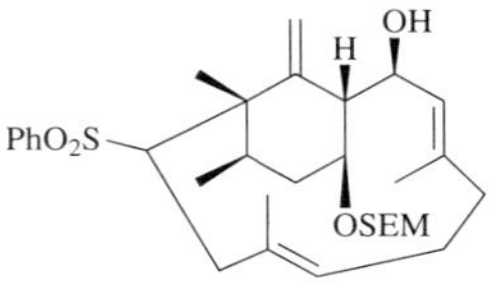 PhO_2S, H, OH, OSEM	NH_3/Na, THF/EtOH, –60°	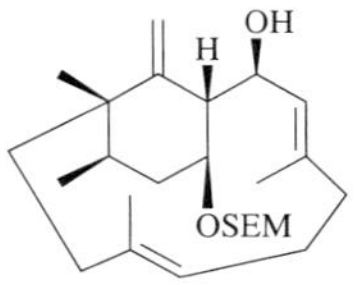 H, OH, OSEM (—)	375
C_{33}			
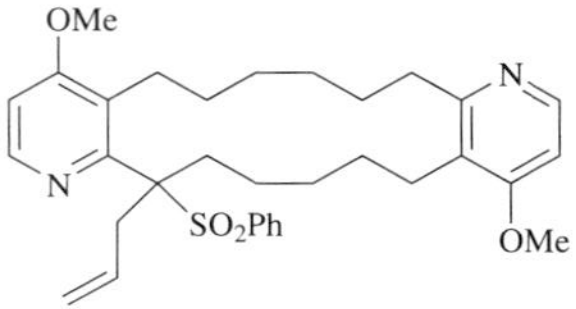 OMe, N, SO_2Ph, OMe	Na/Hg, Na_2HPO_4, THF/MeOH, rt, 12 h	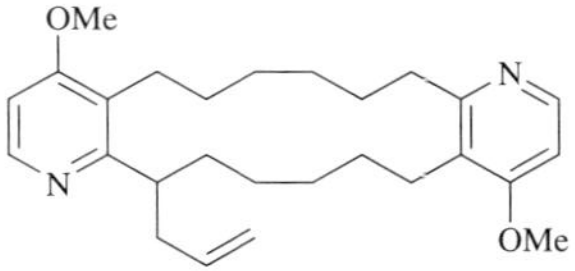 OMe, N, OMe (92)	357
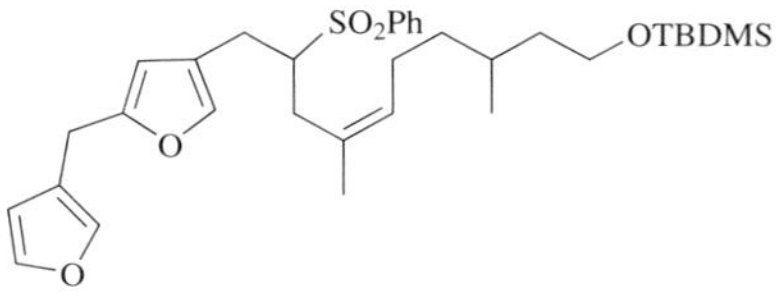 SO_2Ph, OTBDMS	Na/Hg, Na_2HPO_4, MeOH	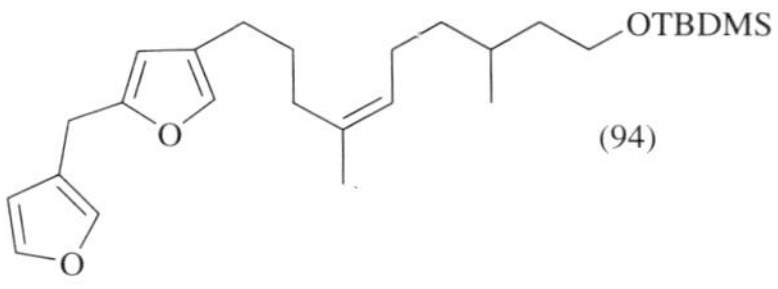 OTBDMS (94)	376

Substrate	Conditions	Product(s) and Yield(s) (%)	Refs.
C_{35}			
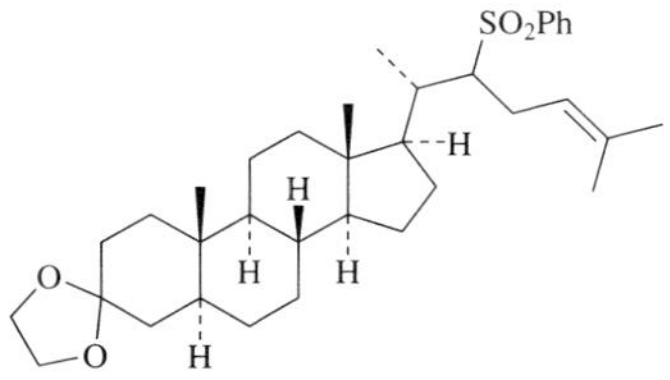 SO$_2$Ph; H	Na/Hg, Na_2HPO_4, MeOH/DME, rt, 30 min	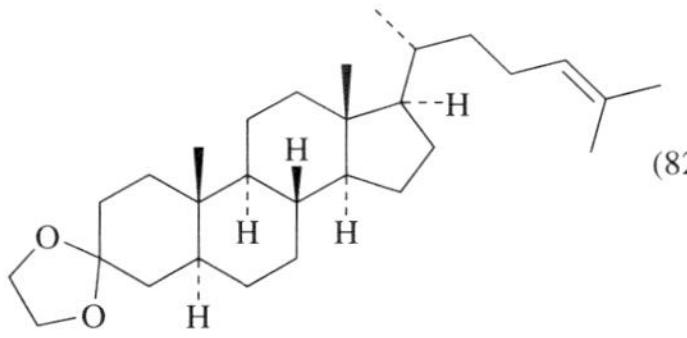 (82)	377
C_{36}			
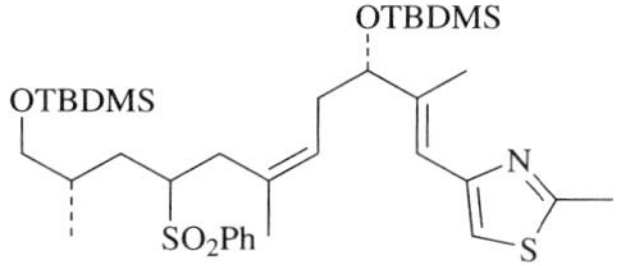 OTBDMS; OTBDMS; SO$_2$Ph; N; S	Na/Hg, Na_2HPO_4, MeOH/THF, –15° to rt, 2 h 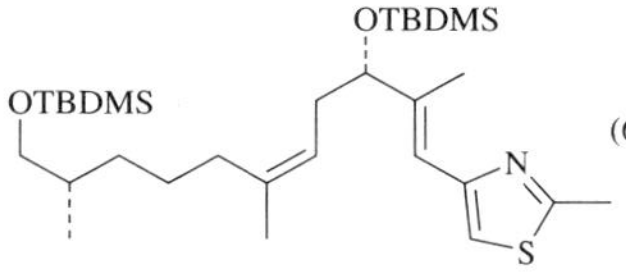	OTBDMS; OTBDMS; N; S (65)	378
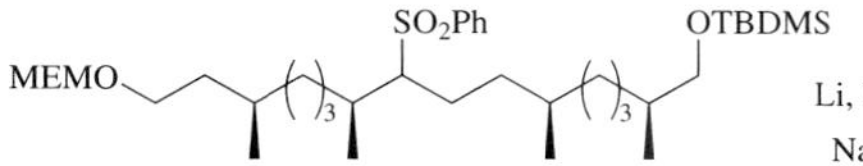 MEMO; SO$_2$Ph; OTBDMS; ( )$_3$; ( )$_3$	Li, HMPA, THF/*t*-BuOH, Na_2HPO_4, ultrasound, 0°, 2 h	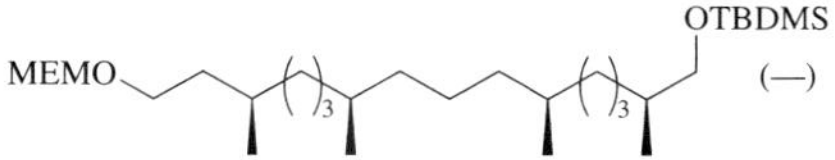 MEMO; OTBDMS; ( )$_3$; ( )$_3$ (—)	379
$C_{37\text{-}39}$			
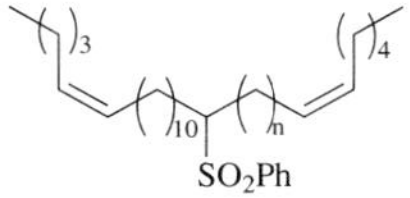 ( )$_3$; ( )$_{10}$; ( )$_n$; ( )$_4$; SO$_2$Ph	SmI_2, THF/HMPA, 30°, 1 h	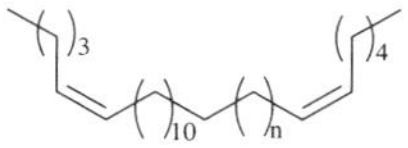 ( )$_3$; ( )$_{10}$; ( )$_n$; ( )$_4$ n = 7 (74) n = 9 (76)	380

TABLE 1. REDUCTIVE DESULFONYLATION OF NON-FUNCTIONALIZED SULFONES (*Continued*)

Substrate	Conditions	Product(s) and Yield(s) (%)	Refs.
C_{38}			
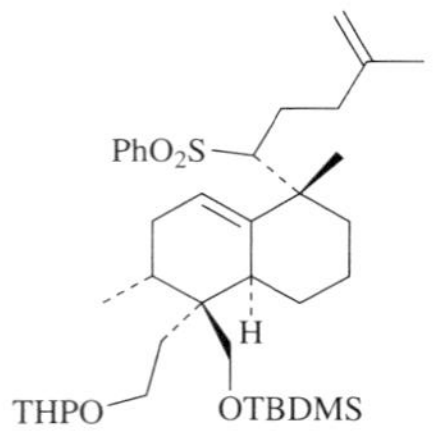 PhO$_2$S; THPO; OTBDMS; H	Na/Hg, Na_2HPO_4, MeOH, rt, 2 h	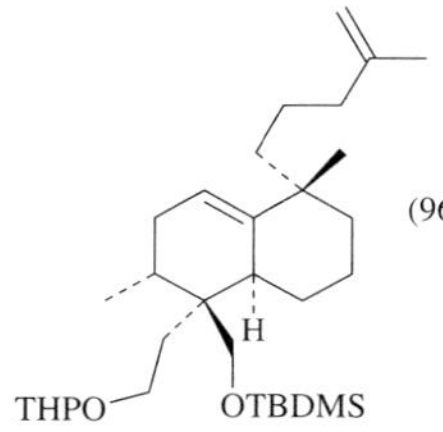 THPO; OTBDMS; H (96)	381
C_{39}			
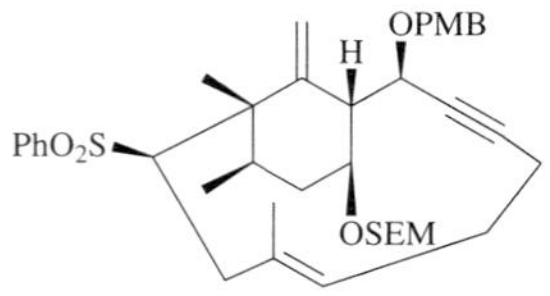 PhO$_2$S; OPMB; H; OSEM	Na/Hg, Na_2HPO_4, MeOH/THF, 0°	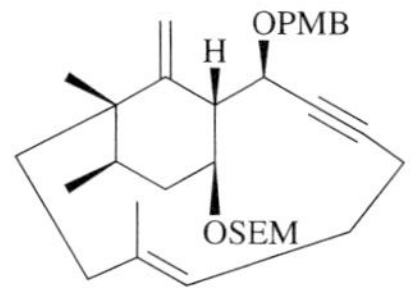 OPMB; H; OSEM (93)	375
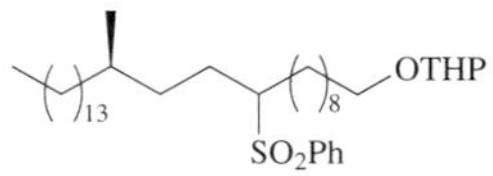 ( )$_{13}$; ( )$_8$; OTHP; SO$_2$Ph	Na/Hg, Na_2HPO_4, EtOH, rt, 36 h	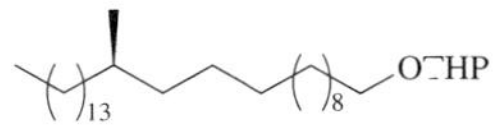 ( )$_{13}$; ( )$_8$; OTHP (78)	382
$C_{39\text{-}40}$			
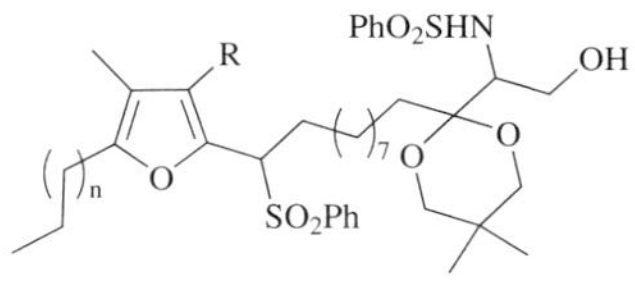 R; ( )$_n$; O; SO$_2$Ph; ( )$_7$; PhO$_2$SHN; OH; O; O	Na/Hg, Na_2HPO_4, MeOH, 65°	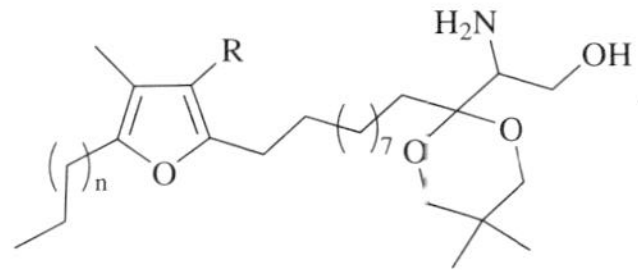 R; ( )$_n$; O; ( )$_7$; H$_2$N; OH; O; O R = Me, n = 1 (56) R = H, n = 3 (60)	383

Carbons	Reactant	Conditions	Product(s) and Yield(s) (%)	Refs.
C_{42}	OMe, OTBDMS, SPh, N, SO_2Ph, MeO, N	Na/Hg, Na_2HPO_4, MeOH, rt, 2 h	OMe, OTBDMS, SPh, N, N, MeO (66)	357
	CO_2Me, SEMO, OSEM, SO_2Ph, O	Na/Hg, Na_2HPO_4, MeOH, rt, 5 h	CO_2Me, SEMO, OSEM, O (55)	384
C_{47}	O, O, O, SO_2Ph, BocHN, BnO, O, $( )_3$, O, O, $( )_5$	Li, $C_{10}H_8$, THF, –18°, 10 min	O, O, O, BocHN, HO, O, $( )_3$, O, O, $( )_5$ (52)	105
C_{48}	PhO_2S, TBDPSO, TBDMSO, OMOM	Na/Hg, Na_2HPO_4, MeOH, rt	TBDPSO, TBDMSO, OMOM (82)	385

TABLE 1. REDUCTIVE DESULFONYLATION OF NON-FUNCTIONALIZED SULFONES (*Continued*)

Substrate	Conditions	Product(s) and Yield(s) (%)	Refs.
C_{48}			
TBDPSO ... OTBDMS, SO_2Ph	Na/Hg, Na_2HPO_4, MeOH, rt, 2 h	TBDPSO ... OTBDMS (86)	379
C_{53}			
TBDMSO, OTBDMS, H, SO_2Ph, OTES	Na/Hg, Na_2HPO_4, THF/MeOH, 5°, 3 h	TBDMSO, OTBDMS, H, OTES (78)	386
C_{53-57}			
OMe, PhO_2S, NH, OMe, O, TBDMSO, OMe, OR^1, OR^3, OR^2, R^4	Na/Hg, Na_2HPO_4, MeOH, –20°	OMe, NH, OMe, O, TBDMSO, OMe, OR^1, OR^3, OR^2, R^4	

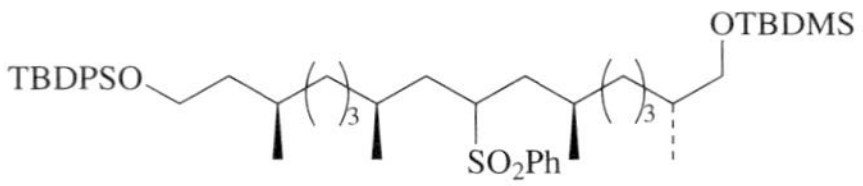

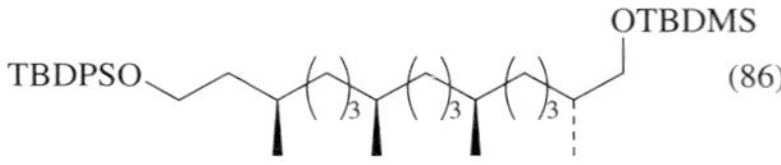

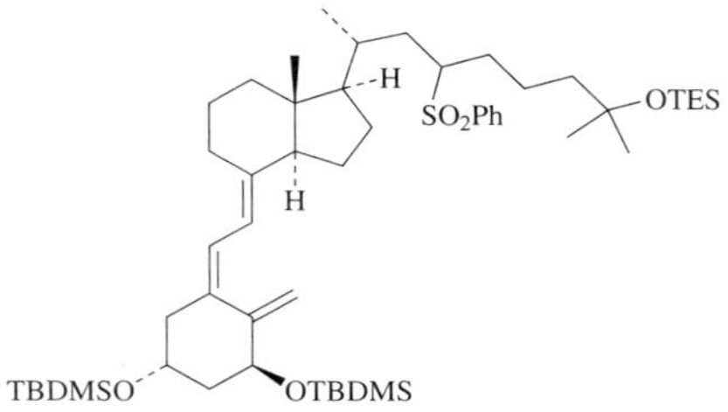

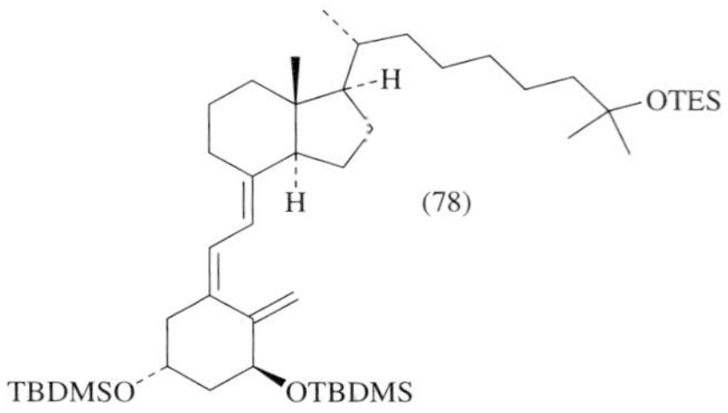

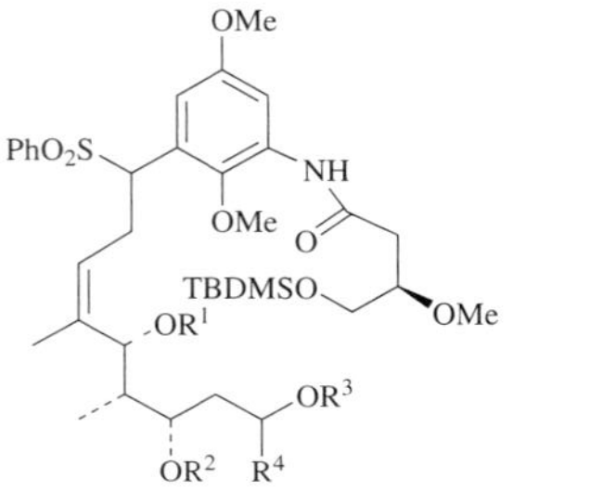

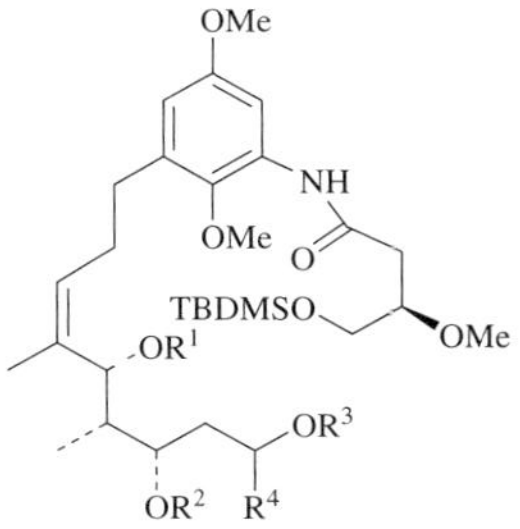

R^1	R^2	R^3	R^4	Time	Product(s) and Yield(s) (%)	Refs.
TBDMS	TIPS	Me	MeO	0.5 h	(98)	387
TIPS	TBDMS	TBDMS	H	1.5 h	(96)	388

TABLE 2. REDUCTIVE DESULFONYLATION OF α-FUNCTIONALIZED SULFONES

Substrate	Conditions	Product(s) and Yield(s) (%)	Refs.
C_{9-15}			
$R^1C(NO_2)(R^2)SO_2Ar$	See table.	$R^1C(NO_2)(R^2)H$	

Ar	R^1	R^2	Reagents	Solvent	Temp	Time	Yield (%)	Refs.
Ph	Me	Me	BNAH	DMF	rt	6 h	(95)	90
4-Tol	Me	Me	$Na_2S_2O_4$, OcV^{2+}, K_2CO_3	CH_2Cl_2/H_2O	35°	3 h	(65)	60
Ph	EtO_2CCH_2	H	$Na_2S_2O_4$, OcV^{2+}, K_2CO_3	CH_2Cl_2/H_2O	35°	3 h	(60)	60
Ph	Et	$NCCH_2CH_2$	BNAH	DMF	rt	6 h	(75)	90
Ph	Et	$MeCOCH_2CH_2$	BNAH	DMF	rt	6 h	(72)	90
4-Tol	—$(CH_2)_5$—		$Na_2S_2O_4$, OcV^{2+}, K_2CO_3	CH_2Cl_2/H_2O	35°	3 h	(55)	60
Ph	H	n-C_6H_{13}	BNAH	DMF	rt	24 h	(55)	90
Ph	Bn	H	$Na_2S_2O_4$, OcV^{2+}, K_2CO_3	CH_2Cl_2/H_2O	35°	3 h	(62)	60
Ph	Bn	H	BNAH, *hν*	DMF	rt	42 h	(62)	91
Ph	2-Tol	H	BNAH, *hν*	DMF	rt	42 h	(61)	91
Ph	Me	n-C_6H_{13}	BNAH	DMF	rt	8 h	(65)	90
4-Tol	Bn	H	$Na_2S_2O_4$, OcV^{2+}, K_2CO_3	CH_2Cl_2/H_2O	35°	3 h	(98)	60
Ph	$MeCOCH_2CH_2$	$MeCOCH_2CH_2$	$Na_2S_2O_4$, OcV^{2+}, K_2CO_3	CH_2Cl_2/H_2O	35°	3 h	(50)	60
Ph	n-C_8H_{17}	H	$Na_2S_2O_4$, OcV^{2+}, K_2CO_3	CH_2Cl_2/H_2O	35°	3 h	(76)	60

TABLE 2. REDUCTIVE DESULFONYLATION OF α-FUNCTIONALIZED SULFONES (*Continued*)

Substrate	Conditions	Product(s) and Yield(s) (%)	Refs.
C_{9-21}			
$R^1R^2C(PO(OR^3)_2)(SO_2Ar)$	$(n\text{-}Bu)_3SnH$, AIBN, toluene, reflux, 4 h	$R^1R^2C(PO(OR^3)_2)(H)$ **I**	60
$(n\text{-}Bu)_3SnCl$ (cat.), AIBN, PMHS, KF, toluene/H_2O, reflux, 7 h		**I**	60

$(n\text{-}Bu)_3SnH$, AIBN, toluene, reflux, 4 h:

Ar	R^1	R^2	R^3	Yield (%) of **I**
2-pyrimidyl	H	F	Et	(45)
2-pyrimidyl	Me	F	Et	(61)
2-pyrimidyl	Me	H	Et	(56)
2-pyridyl	Me	F	Et	(48)
2-pyridyl	Me	H	Et	(32)
2-pyrimidyl	Ph	F	*i*-Pr	(80)
2-pyrimidyl	Ph	H	*i*-Pr	(88)
2-pyridyl	Ph	F	*i*-Pr	(40)
2-pyridyl	Ph	H	*i*-Pr	(45)
2-pyrimidyl	2-naphthyl	F	*i*-Pr	(—)
2-pyrimidyl	2-naphthyl	H	*i*-Pr	(78)

$(n\text{-}Bu)_3SnCl$ (cat.), AIBN, PMHS, KF, toluene/H_2O, reflux, 7 h:

Ar	R^1	R^2	R^3	Yield (%) of **I**
2-pyrimidyl	H	F	Et	(91)
2-pyrimidyl	Me	F	Et	(82)
2-pyrimidyl	Me	H	Et	(60)
2-pyridyl	Me	F	Et	(—)
2-pyrimidyl	Ph	H	Et	(—)
2-pyrimidyl	Ph	F	*i*-Pr	(94)
2-pyridyl	Ph	F	*i*-Pr	(73)
2-pyridyl	Ph	H	*i*-Pr	(55)
2-pyrimidyl	2-naphthyl	F	*i*-Pr	(92)
2-pyrimidyl	2-naphthyl	H	*i*-Pr	(81)

Substrate	Conditions	Product(s) and Yield(s) (%)	Refs.
C_{14-16}			
$RCH(OH)CF_2SO_2Ph$	Na/Hg, Na_2HPO_4, MeOH, –20° to –10°, 1 h	$RCH(OH)CF_2H$	389
		R = Ph (79)	
		R = (*E*)-PhCH=CH (86)	
		R = $PhCH_2CH_2$ (84)	
C_{14-19}			
$R^1R^2C(PO(OEt)_2)(SO_2Ph)$	Mg, $HgCl_2$, EtOH/THF, rt, 12 h	$R^1R^2C(PO(OEt)_2)H$	390
R^1R^2 = —$(CH_2)_3$—		(96)	
R^1R^2 = —(*Z*)-CH_2CH=$CHCH_2$—		(97)	
R^1R^2 = —$(CH_2)_4$—		(95)	
R^1R^2 = —$CHMe(CH_2)_3$—		(95)	
R^1R^2 = —$(CH_2)_5$—		(94)	
R^1R^2 = 1,2-phenylenebis(methylene)		(96)	

TABLE 2. REDUCTIVE DESULFONYLATION OF α-FUNCTIONALIZED SULFONES (*Continued*)

Substrate		Conditions				Product(s) and Yield(s) (%)	Refs.
C_{14-20}							
$R^1R^2C(OH)CF_2SO_2Ph$		See table.				$R^1R^2C(OH)CF_2H$	
R^1	R^2	Reagents	Solvent	Temp	Time		
H	4-ClC_6H_4	Mg, AcOH/NaOAc	DMF/H_2O	rt	3 h	(83)	120
H	Ph	10% Na/Hg, Na_2HPO_4	MeOH	–20° to 0°	1 h	(79)	391
H	*c*-C_6H_{11}	Mg, AcOH/NaOAc	DMF/H_2O	rt	3 h	(84)	120
Me	*n*-C_5H_{11}	Mg, AcOH/NaOAc	DMF/H_2O	rt	3 h	(83)	120
H	*n*-C_6H_{13}	10% Na/Hg, Na_2HPO_4	MeOH	–20° to 0°	2 h	(76)	391
H	4-$MeOC_6H_4$	Mg, AcOH/NaOAc	DMF/H_2O	rt	3 h	(86)	120
Me	Ph	Mg, AcOH/NaOAc	DMF/H_2O	rt	3 h	(80)	120
Me	Ph	10% Na/Hg, Na_2HPO_4	MeOH	–20° to 0°	1.5 h	(79)	391
H	*n*-C_7H_{15}	Mg, AcOH/NaOAc	DMF/H_2O	rt	3 h	(89)	120
H	(*E*)-PhCH=CH	Mg, AcOH/NaOAc	DMF/H_2O	rt	3 h	(88)	120
H	(*E*)-PhCH=CH	10% Na/Hg, Na_2HPO_4	MeOH	–20° to 0°	1.5 h	(84)	391
H	$PhCH_2CH_2$	10% Na/Hg, Na_2HPO_4	MeOH	–20° to 0°	2 h	(86)	391
H	2-naphthyl	Mg, AcOH/NaOAc	DMF/H_2O	rt	3 h	(91)	120
Ph	Ph	10% Na/Hg, Na_2HPO_4	MeOH	–20° to 0°	2 h	(82)	391
n-C_6H_{13}	*n*-C_6H_{13}	10% Na/Hg, Na_2HPO_4	MeOH	–20° to 0°	2 h	(91)	391
C_{14-22}							
t-Bu–S(O)–NH–CH(R)–CF_2SO_2Ph		1. Na/Hg, Na_2HPO_4, MeOH, –15°, 1 h 2. HCl				$Cl^-H_3N^+$–CH(R)–CF_2H	392
						R	
						Et (70)	
						i-Pr (72)	
						2-furyl (88)	
						t-Bu (94)	
						4-ClC_6H_4 (82)	
						Ph (83)	
						4-$MeCC_6H_4$ (96)	
						2-naphthyl (97)	

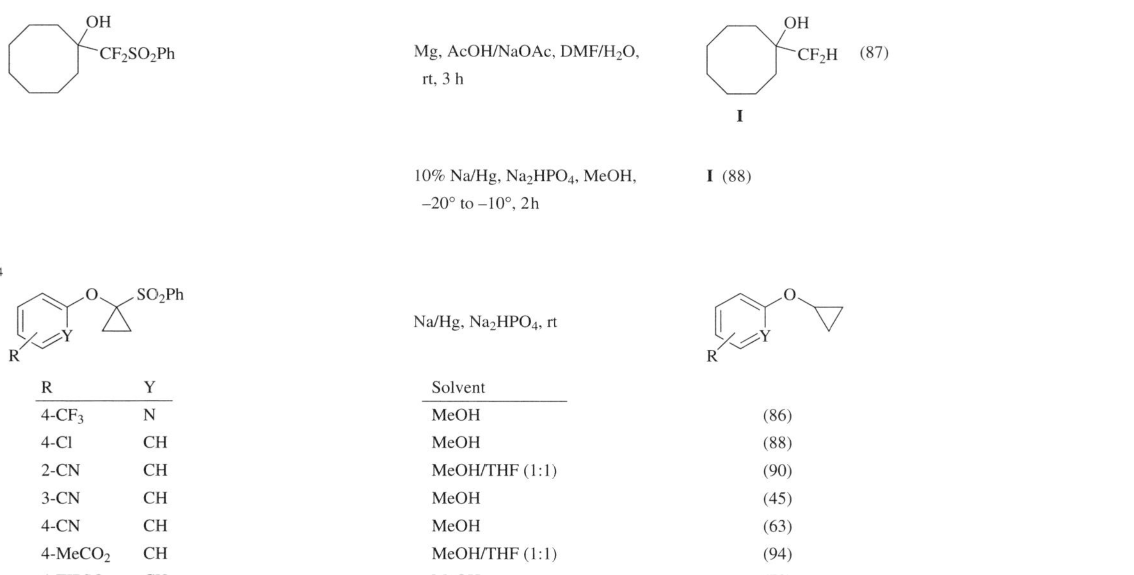

Reactant	Conditions	Product(s) and Yield(s) (%)	Refs.
C_{15} (1-(difluoro(phenylsulfonyl)methyl)cyclooctanol, OH, CF_2SO_2Ph)	Mg, AcOH/NaOAc, DMF/H_2O, rt, 3 h	**I** (OH, CF_2H) (87)	120
	10% Na/Hg, Na_2HPO_4, MeOH, –20° to –10°, 2h	**I** (88)	391
C_{15-24} (aryloxy-cyclopropyl SO_2Ph; R, Y)	Na/Hg, Na_2HPO_4, rt	(aryloxy)cyclopropane (R, Y)	393

R	Y	Solvent	Yield
4-CF_3	N	MeOH	(86)
4-Cl	CH	MeOH	(88)
2-CN	CH	MeOH/THF (1:1)	(90)
3-CN	CH	MeOH	(45)
4-CN	CH	MeOH	(63)
4-$MeCO_2$	CH	MeOH/THF (1:1)	(94)
4-TIPSO	CH	MeOH	(79)

TABLE 2. REDUCTIVE DESULFONYLATION OF α-FUNCTIONALIZED SULFONES (Continued)

Substrate		Conditions				Product(s) and Yield(s) (%)	Refs.
C_{15-32}							
$PhO_2S(R^1)C(R^2)SO_2Ph$		See table.				$R^1CH(SO_2Ph)R^2$	
R^1	R^2	Reagents	Solvent	Temp	Time		
$—(CH_2)_2—$		$Li/C_{10}H_8$	THF	–78°	5 min	(0)	394
$—(CH_2)_2—$		SmI_2	THF	rt	5-15 min	(0)	394
$—(CH_2)_3—$		$Li/C_{10}H_8$	THF	–78°	5 min	(97)	394
$—(CH_2)_3—$		SmI_2	THF	rt	5-15 min	(82)	394
$—CH_2CH=CHCH_2—$		$Li/C_{10}H_8$	THF	–78°	5 min	(68)	394
$—CH_2CH=CHCH_2—$		SmI_2	THF	rt	5-15 min	(87)	394
$—(CH_2)_5—$		$Li/C_{10}H_8$	THF	–78°	5 min	(85)	394
$PhCH_2CH_2$	H	Mg	MeOH	50°	—	(84)	118
$PhCH_2CH_2$	Me	Mg	MeOH	50°	4 h	(81)	118
(*E*)-$PhCH_2CH=CH$	H	$Li/C_{10}H_8$	THF	–78°	5 min	(85)	394
$Ph(CH_2)_2CH_2$	H	$Li/C_{10}H_8$	THF	–78°	5 min	(85)	394
$Ph(CH_2)_2CH_2$	Me	$Li/C_{10}H_8$	THF	–78°	5 min	(94)	394
(*R*)-$TBDPSOCH_2CHMe$	H	$Li/C_{10}H_8$	THF	–78°	5 min	(84)	394
(*R*)-$TBDPSOCH_2CHMe$	H	SmI_2	THF	rt	5-15 min	(82)	394
C_{15-45}							
$RC(F)(PO(OEt)_2)SO_2Ph$		Na/Hg, Na_2HPO_4, MeOH/THF, rt, 10-20 min				$RC(F)(H)PO(OEt)_2$	176
R							
i-Bu						(79)	
2,2-dimethyl-1,3-dioxolan-4-ylmethyl						(89)	
(*S*)-3,7-dimethyloct-6-enyl						(80)	

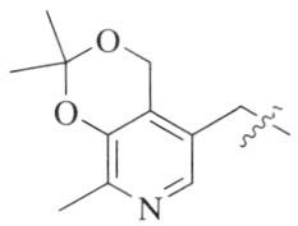

(71)

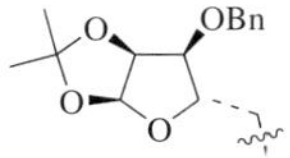

(80)

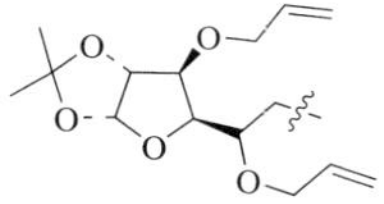

(78)

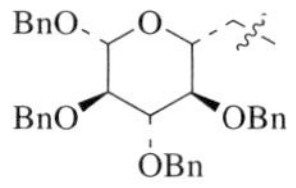

(85)

C_{16}

Substrate	Conditions	Product(s) and Yield(s) (%)	Refs.
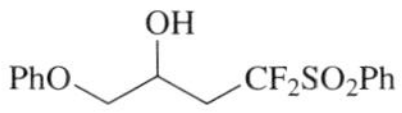 PhO, OH, CF_2SO_2Ph	Mg, HOAc, NaOAc, DMF, rt	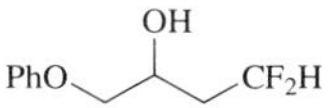 PhO, OH, CF_2H (83)	395
R^1, SO_2Ph, N, R^2	Na/Hg	R^1, N, R^2	138
		R^1 | R^2 |	
		Me | Bn | (~100)	
		Ph | Et | (~100)	
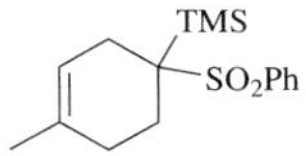 TMS, SO_2Ph	Na/Hg, Na_2HPO_4, MeOH, –20°	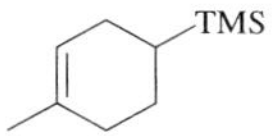 TMS (79)	350

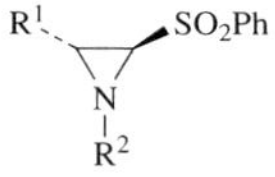

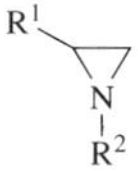

TABLE 2. REDUCTIVE DESULFONYLATION OF α-FUNCTIONALIZED SULFONES (*Continued*)

Substrate	Conditions	Product(s) and Yield(s) (%)	Refs.
C_{16-22}			
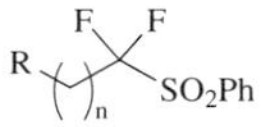 $R(CH_2)_nCF_2SO_2Ph$	Na/Hg, Na_2HPO_4, MeOH, –20° to 0°, 1 h	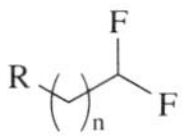 $R(CH_2)_nCHF_2$ R / n PhO / 3 (91) PhO / 4 (88) Ph / 4 (87) 4-$MeOC_6H_4$ / 4 (80) Ph / 5 (90) Ph / 6 (85) Ph_2CH / 2 (89)	396
C_{17}			
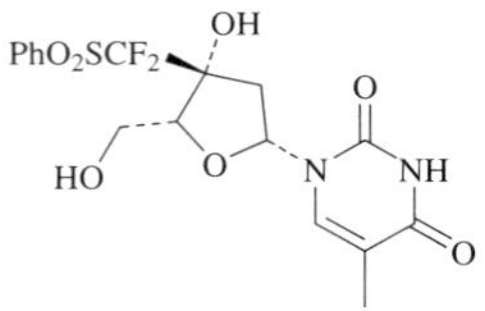 3'-OH, 3'-CF_2SO_2Ph thymidine nucleoside (HO, OH, PhO_2SCF_2, O, N, NH, O)	SmI_2, THF/HMPA, rt, 1 h	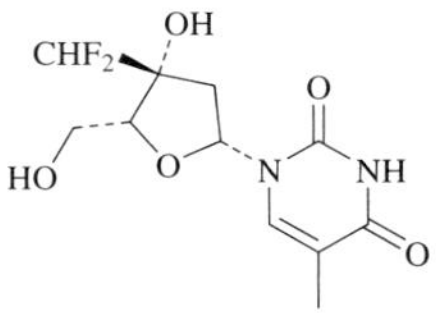 3'-OH, 3'-CHF_2 thymidine nucleoside (HO, OH, CHF_2, O, N, NH, O) (48)	161
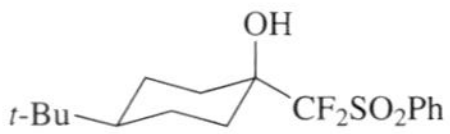 4-*t*-Bu-1-OH-1-CF_2SO_2Ph-cyclohexane (OH axial)	Na/Hg, Na_2HPO_4, MeOH, –20° to –10°, 2 h	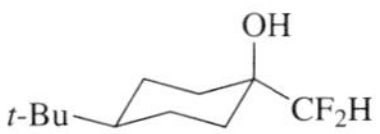 4-*t*-Bu-1-OH-1-CF_2H-cyclohexane (85)	391
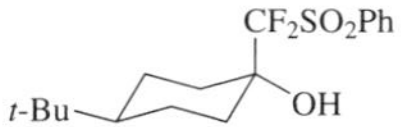 4-*t*-Bu-1-CF_2SO_2Ph-1-OH-cyclohexane (CF_2SO_2Ph axial)	Na/Hg, Na_2HPO_4, MeOH, –20° to –10°, 2 h	 4-*t*-Bu-1-CF_2H-1-OH-cyclohexane (88)	391
C_{17-18}			
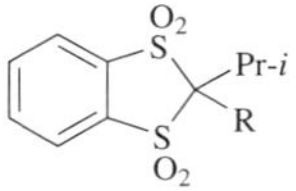 2-R-2-Pr-*i*-1,3-benzodithiole 1,1,3,3-tetraoxide	Mg, MeOH, 50°, 4 h	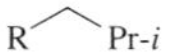 RCH_2Pr-*i* R *c*-$C_6H_{11}CH_2$ (50) $PhCH_2CH_2$ (72)	397

PhO_2S–C(Br)–$(CH_2)_n$–C(Br)–SO_2Ph | SmI_2, THF, 2 h | PhO_2S–bicyclic iodide | 163

n	Temp	
1	rt	(61)
2	reflux	(~100)

C_{18-20}

R–CH(SO_2Ph)–OBu-*t* | See table. | R–CH_2–OBu-*t* | 398

R	Reagents	Solvent	Temp	Time	
$PhCH_2$	DIBALH	toluene	0°	3 h	(100)
$Ph(CH_2)_3$	Na/Hg	EtOH	–25°	overnight	(92)
$Ph(CH_2)_3$	$NaBH_4$	*i*-PrOH/THF	0° to rt	—	(87)
$Ph(CH_2)_3$	DIBALH	toluene	0°	3 h	(100)
$Ph(CH_2)_3$	$LiAlH_4$	Et_2O	0° to rt	3 h	(100)
n-C_9H_{19}	DIBALH	toluene	0°	3 h	(97)

C_{19}

HO, SO_2Ph, SO_2Ph | Mg, MeOH, 0°, 1 h; reflux overnight | HO (45) | 399

C_{21-22}

CO_2Et, PhO_2S, N, O, n | Ra-Ni, EtOH, ultrasound, rt, 5 min | CO_2Et, H, N, O, n | 181

n	dr	
1	80:20	(~100)
2	50:50	(80)

C_{22}

HN, O, N, O, O, O, F, $PO(OEt)_2$, O_2S, N | (*n*-Bu)$_3$SnH, AIBN, C_6H_6, reflux, 48 h | HN, O, N, O, O, O, F, $PO(OEt)_2$ (61) | 175

TABLE 2. REDUCTIVE DESULFONYLATION OF α-FUNCTIONALIZED SULFONES (*Continued*)

Substrate	Conditions	Product(s) and Yield(s) (%)	Refs.
C_{22-28}			
$PhO_2S(SO_2Ph)C(R)F$	Mg, MeOH, 0°	RCH_2F	400
		R: (*E*)-PhCH=CHCH$_2$ (74); PhCH$_2$CHPh (76); Ph$_2$CHCH$_2$CH$_2$ (81)	
C_{22-29}			
RCH(NHBoc)CF(SO_2Ph)$_2$	Reagents, MeOH	RCH(NHBoc)CH$_2$F	401

R	Reagents	Temp	Time	Yield (%)
i-Pr	Mg	0°	2 h	(75)
2-furyl	Mg	0°	2 h	(81)
t-Bu	Mg	0°	2 h	(26)
3-ClC$_6$H$_4$	Mg	0°	2 h	(82)
4-ClC$_6$H$_4$	Mg	0°	2 h	(88)
Ph	Mg	0°	2 h	(84)
Ph	Na/Hg, Na$_2$HPO$_4$	–20° to –10°	1 h	(92)
c-C$_6$H$_{11}$	Mg	0°	2 h	(87)
4-MeOC$_6$H$_4$	Mg	0°	2 h	(80)
Me(CH$_2$)$_6$	Mg	0°	2 h	(83)
PhCH$_2$CH$_2$	Mg	0°	2 h	(87)
2-naphthyl	Mg	0°	2 h	(85)

Substrate	Conditions	Product(s) and Yield(s) (%)	Refs.
C_{23}			
PhCH$_2$CH(NHBn)CH$_2$CF$_2$SO$_2$Ph	Mg, HOAc, NaOAc, DMF, rt	PhCH$_2$CH(NHBn)CH$_2$CF$_2$H (81)	395

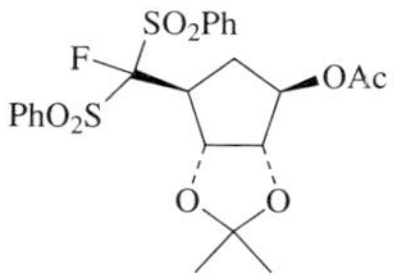	Mg, $NiBr_2$, MeOH, –30°, 8 h	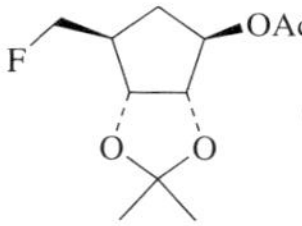 (61)	402
C_{23-30}			
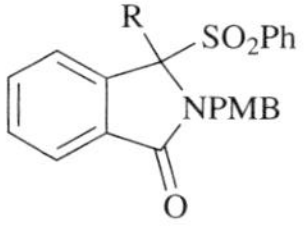	Ra-Ni, EtOH, ultrasound, rt, 5 min	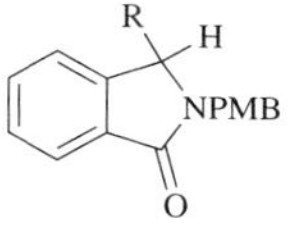 (—) R: Me; $MeC(O)CH_2CH_2$; $MeO_2CCH_2CH_2$; $PhCH_2CH_2$	181
C_{25}			
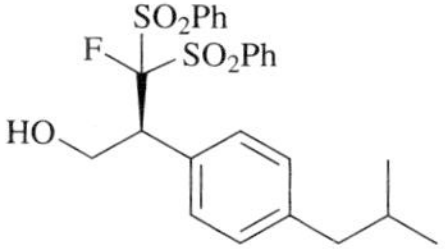	Mg, MeOH, 0°, 2 h	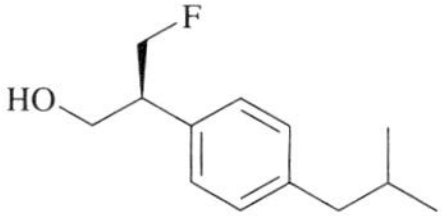 (87)	402
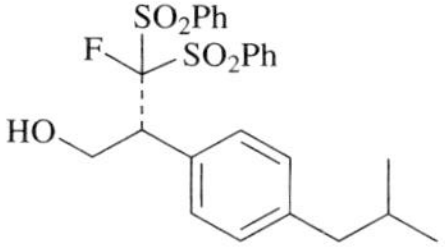	Mg, MeOH, 0°, 2 h	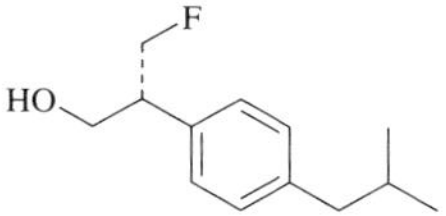 (85)	402
C_{27}			
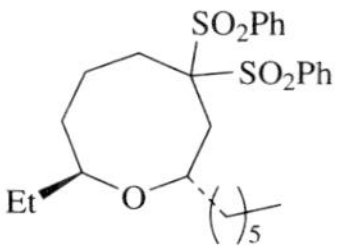	Na/Hg, Na_2HPO_4, MeOH, THF, rt	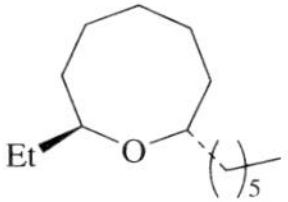 (82)	403

TABLE 2. REDUCTIVE DESULFONYLATION OF α-FUNCTIONALIZED SULFONES (*Continued*)

	Substrate	Conditions	Product(s) and Yield(s) (%)	Refs.
C_{30}	OH, F, F, SO_2Ph, H, H, H, AcO	Na/Hg, Na_2HPO_4, MeOH/THF, −20° to 0°, 2h	OH, CF_2H, H, H, H, HO (93)	391
C_{32}	OPMB, PhO_2S, PhO_2S	Mg, MeOH/THF, 50°, 3 h	OPMB (79)	404
C_{34}	O, $CFRSO_2Ph$, t-Bu, S, N, H, NBn_2, Bn	See table.	O, CHFR, t-Bu, S, N, H, NBn_2, Bn	405
	PhO_2S, F, F, HO, H, H, H, H, $()_3$	Na/Hg, Na_2HPO_4, MeOH, −20° to 0°, 3h	CHF_2, HO, H, H, H, H, $()_3$ (89)	391

R	Reagents	Solvent	Temp	Time	
F	Mg, AcOH, NaOAc	DMF	rt	3 h	(93)
H	Na/Hg, Na_2HPO_4	MeOH	−20° to 0°	4 h	(65)

Substrate	Conditions	Product(s) and Yield(s) (%)	Refs.
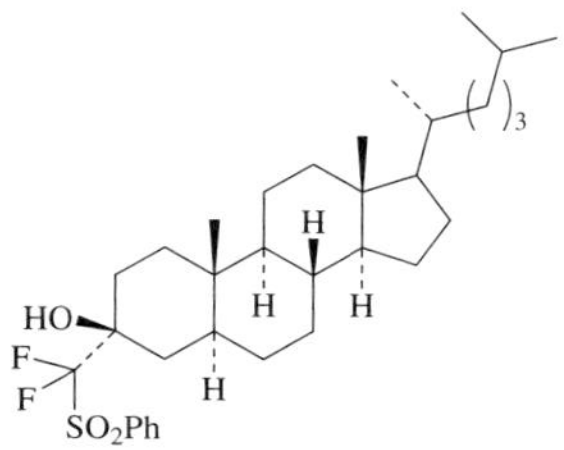 HO, F, F, SO_2Ph, H, H, H, H, ( )$_3$	Na/Hg, Na_2HPO_4, MeOH/THF, −20° to 0°, 3h	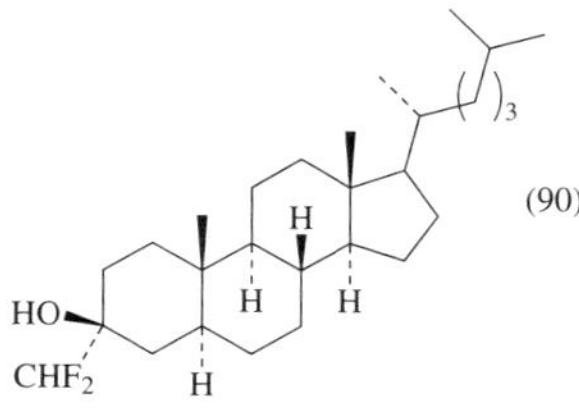 HO, CHF_2, H, H, H, H, ( )$_3$ (90)	391
C_{36}			
TBDPSO, PhO_2S, SO_2Ph, HO	LDTBB, THF, −78°	TBDPSO, HO (75)	406
C_{40}			
F, PhO_2S, SO_2Ph, H, H	Mg, MeOH, 0°	F, H, H (67)	400

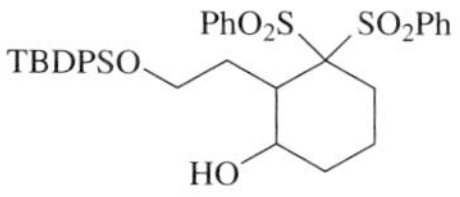

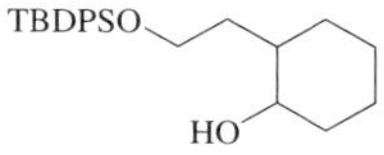

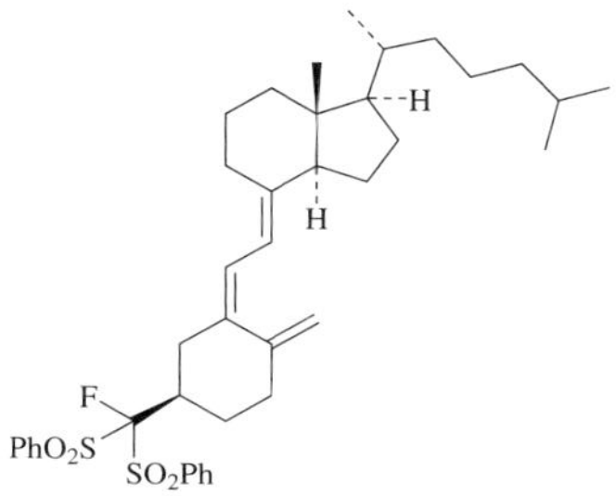

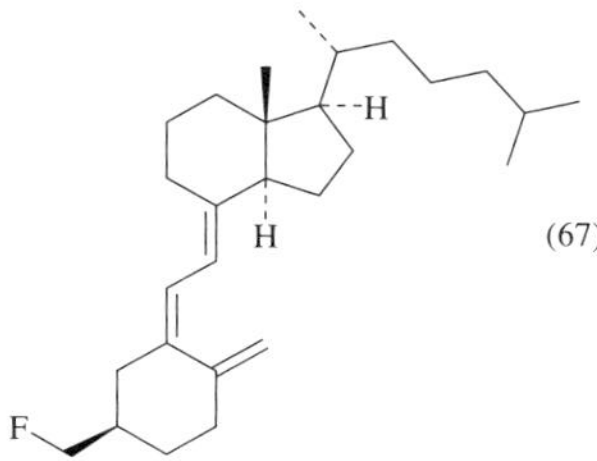

TABLE 2. REDUCTIVE DESULFONYLATION OF α-FUNCTIONALIZED SULFONES (*Continued*)

Substrate	Conditions	Product(s) and Yield(s) (%)	Refs.
C_{53}			
THPO, PhO_2S, PhO_2S, CO_2Me, $Sn(Bu\text{-}n)_3$	Mg, MeOH, rt, 2 h	THPO, CO_2Me, $Sn(Bu\text{-}n)_3$ (—)	407

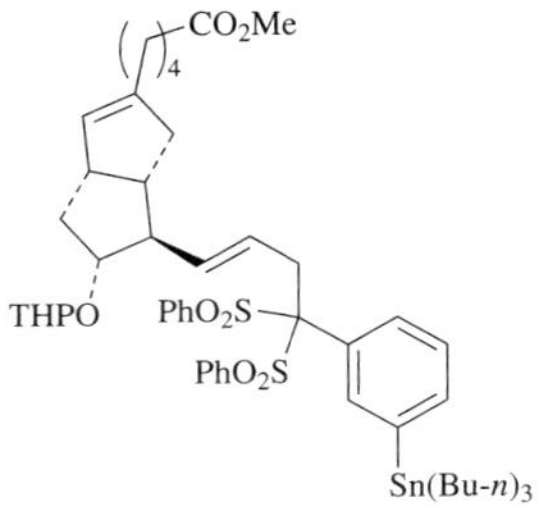

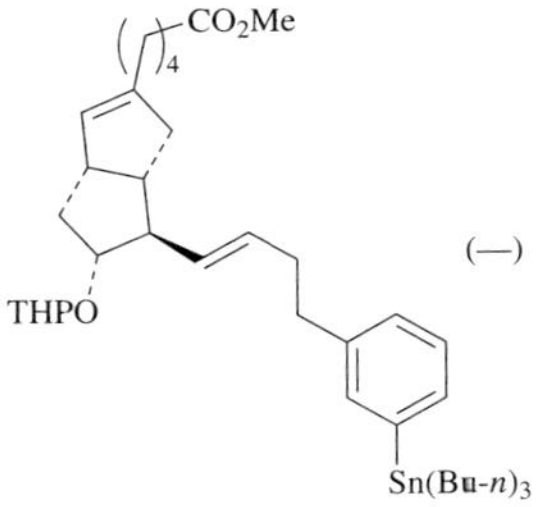

TABLE 3. REDUCTIVE DESULFONYLATION OF β-FUNCTIONALIZED SULFONES

Substrate	Conditions	Product(s) and Yield(s) (%)	Refs.
$C_{10\text{-}13}$			
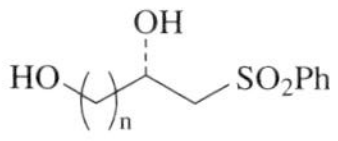 HO–$(CH_2)_n$–CH(OH)–CH_2SO_2Ph	Ra-Ni, MeOH, rt, 24 h	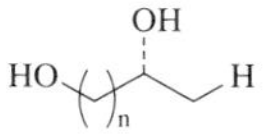 HO–$(CH_2)_n$–CH(OH)–CH_2H; n = 2 (—); 3 (79); 4 (—); 5 (—)	178
$C_{12\text{-}18}$			
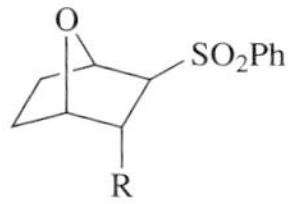 7-oxabicyclo, SO_2Ph, R	Na/Hg	 I + II	408

R	Solvent	Buffer	Temp	Time	**I + II**	**I:II**
H	MeOH	$NaH_2PO_4•H_2O$	–20°	24 h	(48)	100:0
PhO_2S	MeOH	$NaH_2PO_4•H_2O$	rt	24 h	(48)	90:10
PhO_2S	EtOH	$NaH_2PO_4•H_2O$	rt	24 h	(48)	90:10
PhO_2S	DMF	$NaH_2PO_4•H_2O$	rt	5 h	(70)	93:7
PhO_2S	DMF/MeOH (80:20)	—	rt	12 h	(48)	96:4

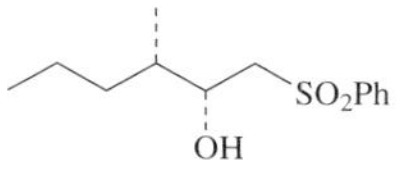

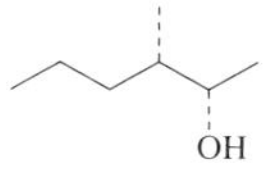

Substrate	Conditions	Product(s) and Yield(s) (%)	Refs.
C_{13}			
7-oxabicyclo, OMe, SO_2Ph	Na/Hg, $NaH_2PO_4•H_2O$, MeOH, rt, 6 h	OMe (48) + (4)	408
OH, SO_2Ph	Na/Hg, Na_2HPO_4, MeOH, rt, 4 h	OH (—)	409

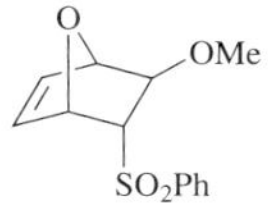

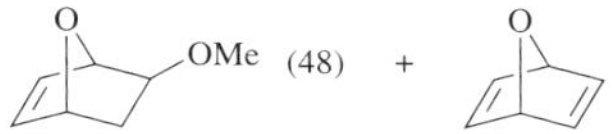

TABLE 3. REDUCTIVE DESULFONYLATION OF β-FUNCTIONALIZED SULFONES (*Continued*)

Substrate	Conditions	Product(s) and Yield(s) (%)	Refs.
$C_{13\text{-}17}$			
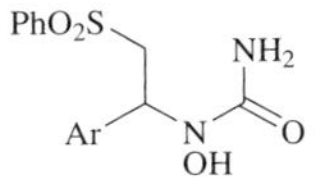 PhO_2S, Ar, N, OH, NH_2, O	Reagents, MeOH	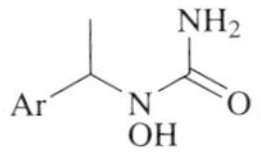 Ar, N, OH, NH_2, O	410
Ar	Reagents / Temp / Time		
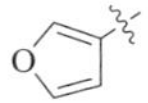 3-furyl	Mg / reflux / 3 h	(87)	
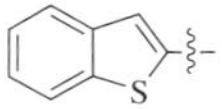 2-benzothienyl	Na/Hg, Na_2HPO_4 / rt / 4 h	(81)	
C_{14}			
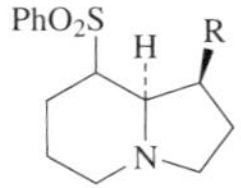 PhO_2S, H, R, N	Na/Hg, MeOH, rt	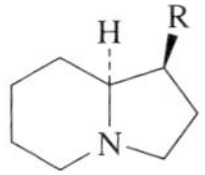 H, R, N	
R	Time		
H	2 h	(71)	411
OH	—	(81)	412
C_{15}			
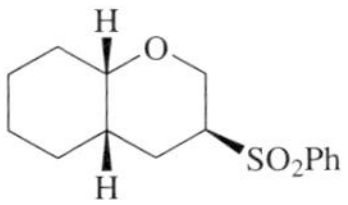 H, O, H, SO_2Ph	Ra-Ni, EtOH, reflux, 9 h	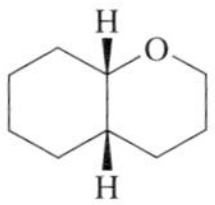 H, O, H (60)	182
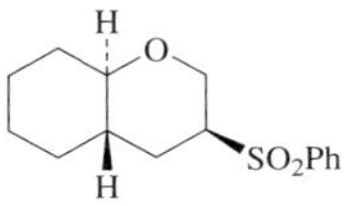 H, O, H, SO_2Ph	Ra-Ni, EtOH, reflux, 10 h	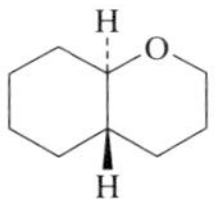 H, O, H (65)	182

C15-16

R SO2Ph (cyclopropane)

Na/Hg, EtOH, reflux, 12 h

R	
HO (1-hydroxyhexyl)	(90)
Bn	(75)

413

Ts, H (indolizidinone)

Na/Hg, Na_2HPO_4, MeOH, 0°, 1.5 h

R	
H	(80)
$H_2C=$	(94)

414

C16

H, O, CO_2Me, SO_2Ph

Ra-Ni, EtOH, reflux, overnight

H, O, CO_2Me (73)

182

BocHN, OH, SO_2Ph

Ra-Ni, EtOH, reflux

BocHN, OH (73)

183

C16-24

R^1, R^2, R^3, R^4, R^5, SO_2MR^a

SmI_2, LiCl, THF, rt

R^1, R^2, R^3, R^4, R^5

150

R^1	R^2	R^3	R^4	R^5	Time	
H	H	morpholine-4-carbonyl	H	Me	—	(—)
F	F	t-$BuCO_2$	H	Me	23 h	(27)
i-Pr	H	t-$BuCO_2$	H	Me	—	(31)
i-Pr	H	t-$BuCO_2$	$H_2C=CHCH_2$	Me	—	(16)
H	H	t-$BuCO_2$	H	Bn	—	(—)
i-Pr	H	t-$BuCO_2$	$Me_2C=CHCH_2$	Me	—	(12)

TABLE 3. REDUCTIVE DESULFONYLATION OF β-FUNCTIONALIZED SULFONES (*Continued*)

Substrate	Conditions	Product(s) and Yield(s) (%)	Refs.
C_{16-27} SO$_2$Ph, R^2, N, R^1	Na/Hg, Na_2HPO_4, MeOH, −10° to 0°	R^2, N, R^1 R^1 / R^2 / Time: C(O)Me / *n*-Pr / 2 h (62); Boc / *n*-$C_{11}H_{23}$ / 5.5 h (56)	43
C_{17} SO$_2$Ph, OH, N, Boc	Na/Hg, Na_2HPO_4, MeOH, 0°	OH, N, Boc (—)	415
C_{17} HO, Ts	Na/Hg, MeOH, 0°, 2.5 h	HO (72) cis:trans = 2:1	416
SO$_2$Ph, TMS	Na/Hg, Na_2HPO_4, THF/MeOH, −20°	TMS (92)	350
C_{17-24} HO, SO$_2$Ph, OR, N, Boc	Na/Hg, Na_2HPO_4, MeOH, 0°, 2 h	HO, OR, N, Boc R: H (96); Bn (96)	417, 418

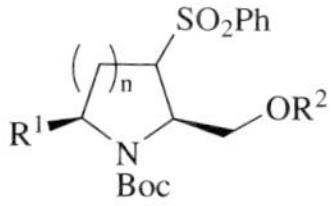

R^1, SO_2Ph, OR^2, N, Boc, n

Na/Hg, Na_2HPO_4, MeOH

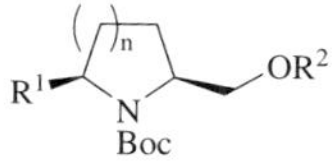

R^1, OR^2, N, Boc, n

n	R^1	R^2	Temp	Time	Yield	Refs.
1	$HOCH_2$	H	0°	1 h	(93)	419
1	H	THP	0° to rt	3 h	(80-85)	420
2	H	THP	0° to rt	3 h	(80-85)	420
3	H	THP	0° to rt	3 h	(80-85)	420
4	H	THP	0° to rt	3 h	(80-85)	420

C_{18}

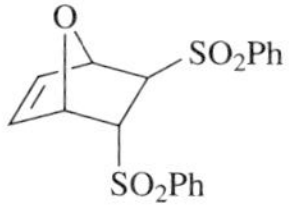

O, SO_2Ph, SO_2Ph

Na/Hg, $NaH_2PO_4•H_2O$

O, **I** + O, **II**

408

Solvent	Temp	Time	**I + II**	**I:II**
MeOH	rt	24 h	(40-50)	50:50
MeOH	reflux	1 h	(46)	65:35
EtOH	rt	24 h	(30-40)	40:60
DMF	rt	24 h	(25)	>95:5

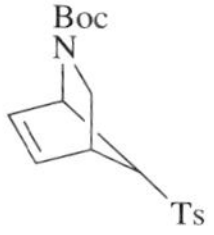

Boc, N, Ts

Na/Hg, NaH_2PO_4, Na_2HPO_4, MeOH, rt, 12 h

Boc, N (33)

421

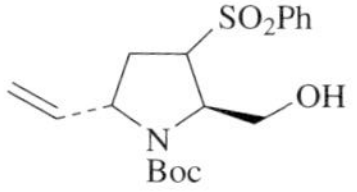

SO_2Ph, OH, N, Boc

Na/Hg, Na_2HPO_4, MeOH, 0°, 1 h

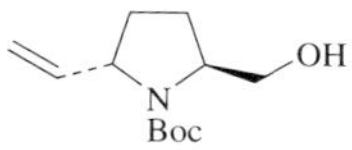

OH, N, Boc (90)

419

TABLE 3. REDUCTIVE DESULFONYLATION OF β-FUNCTIONALIZED SULFONES (Continued)

Substrate	Conditions	Product(s) and Yield(s) (%)	Refs.
C_{18} 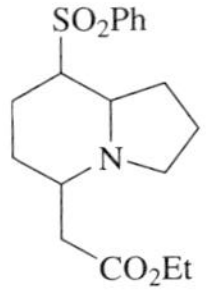 SO_2Ph … N … CO_2Et	Na/Hg, MeOH, rt, 25 h	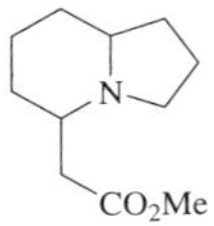 N … CO_2Me (38)	422
C_{19} 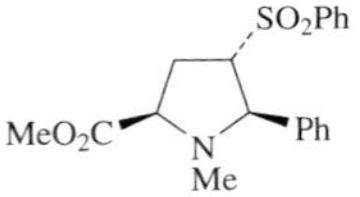 SO_2Ph, MeO_2C, N Me, Ph	Na/Hg, Na_2HPO_4, MeOH/THF, rt, 5 h	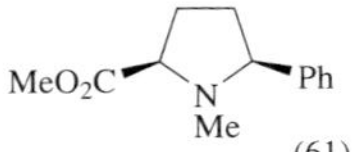 MeO_2C, N Me, Ph (61) 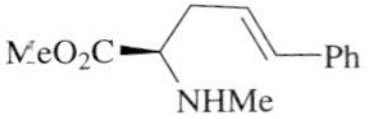 + MeO_2C, NHMe, Ph (29)	423
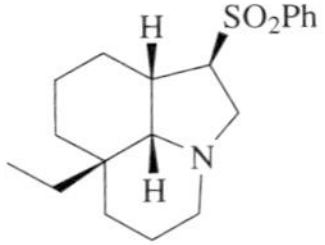 H, SO_2Ph, N, H	Na/Hg, Na_2HPO_4, MeOH, 0°	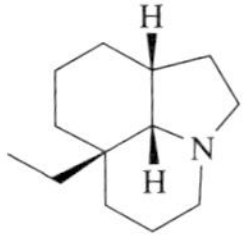 H, N, H (57)	424
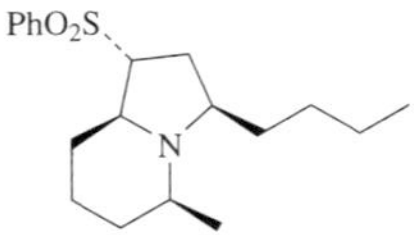 PhO_2S, N	$Na/C_{10}H_8$, THF, rt, 5 min	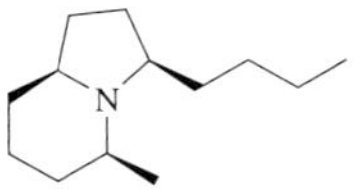 N (55)	425
C_{19-24} 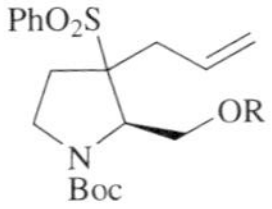 PhO_2S, OR, N Boc R / cis:trans H / 6:94 H / 89:11 THP / 6:94	Na/Hg, Na_2HPO_4, MeOH, rt, 1 h	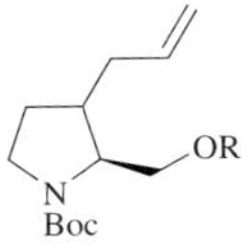 OR, N Boc cis:trans 16:84 (—) 0:100 (80) 63:37 (—)	420

C_{20}

TMSO, MeO_2C, Cl, SO_2Ph

Na/Hg, Na_2HPO_4, THF/MeOH, –40°, 3 h

TMSO, MeO_2C (—)

426

C_{20-27}

Ph, O, SO_2Ph, R, O, O, *, OMe

Mg, MeOH, reflux

Ph, O, R, O, O, *, OMe

427

*	R	Time	
S	H_2N	3.5 h	(—)
S	N	4 h	(91)
S	O N	7 h	(85)
R	N	5 h	(42)
R	O N	1 h	(47)
S	BnHN	11 h	(90)
R	BnHN	2 h	(76)

TABLE 3. REDUCTIVE DESULFONYLATION OF β-FUNCTIONALIZED SULFONES (*Continued*)

Substrate	Conditions	Product(s) and Yield(s) (%)	Refs.
C_{20-29} 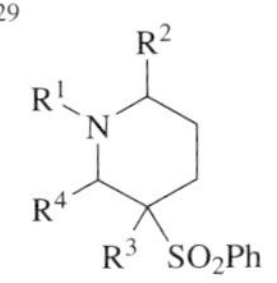 (piperidine: R^1 on N, R^2, R^4, R^3, SO_2Ph)	Na/Hg, MeOH	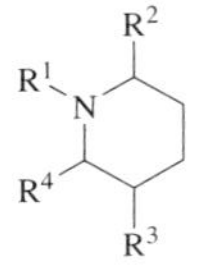 (piperidine: R^1, R^2, R^3, R^4)	

R^1	R^2	R^3	R^4	Buffer	Temp	Time	Yield (%)	Refs.
Me	=O	H	(*E*)-PhCH=CH	Na_2HPO_4	—	—	(93)	428
Me	H	H	(*E*)-PhCH=CH	—	—	—	(—)	428
Bn	=O	H	$BnO(CH_2)_3$	—	rt	3 h	(88)	422
Bn	=O	Me	$BnO(CH_2)_3$	—	rt	3 h	(51)	422

Substrate	Conditions	Product(s) and Yield(s) (%)	Refs.
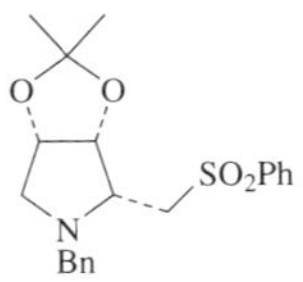 C_{21} (N-Bn pyrrolidine acetonide, SO_2Ph)	Na/Hg, MeOH	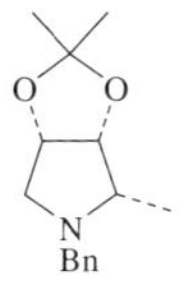 (77)	429
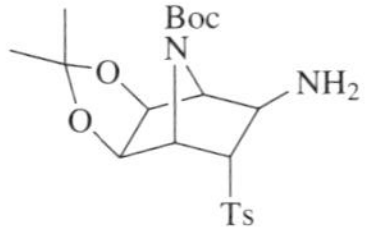 (Boc-N bicyclic acetonide, NH_2, Ts)	Na/Hg, Na_2HPO_4, MeOH/THF, −15°, 1.5 h	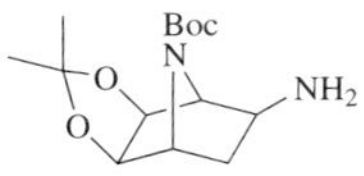 (Boc-N bicyclic, NH_2) (60)	430
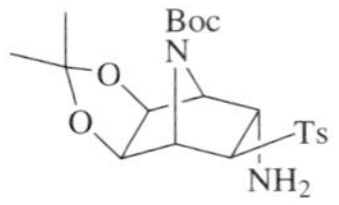 (Boc-N bicyclic acetonide, Ts, NH_2)	Na/Hg, Na_2HPO_4, MeOH/THF, −15°, 1.5 h	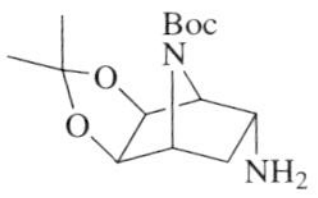 (Boc-N bicyclic, NH_2) (72)	430

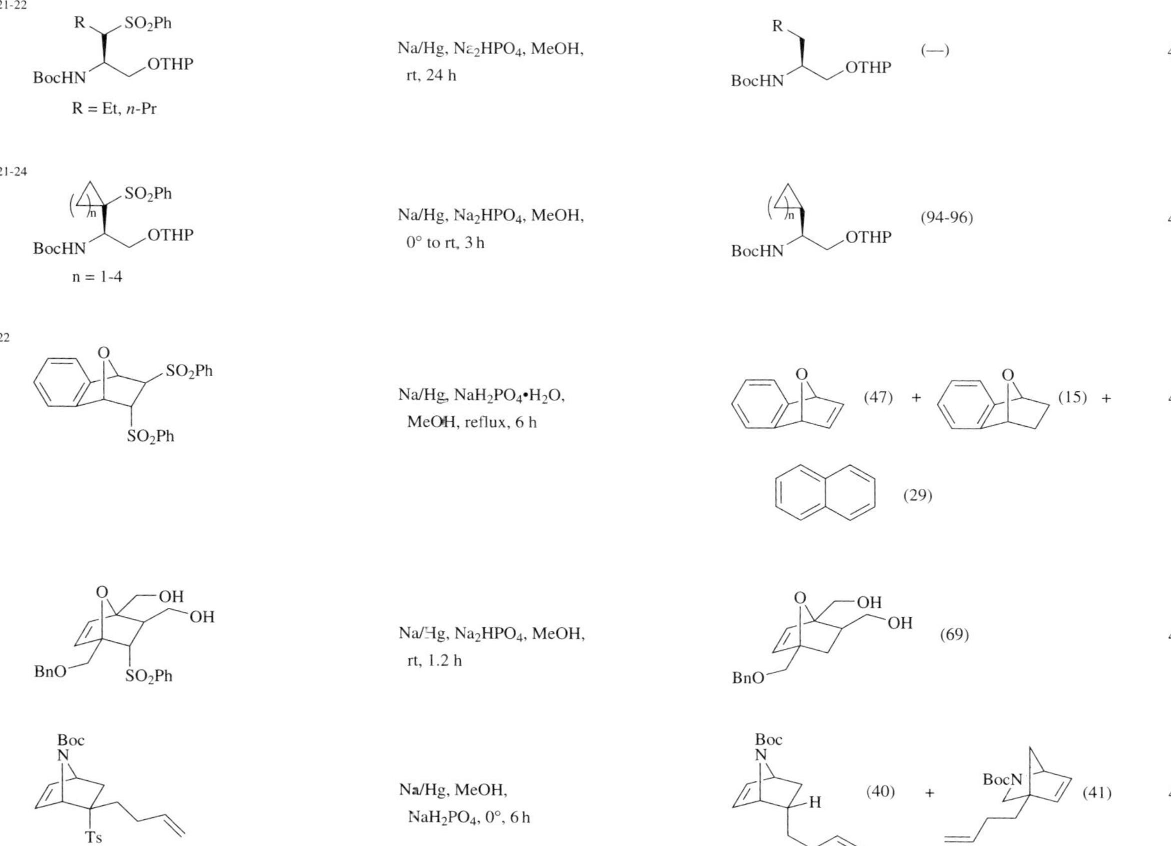

Substrate	Conditions	Product(s) and Yield(s) (%)	Refs.
C_{21-22} R = Et, *n*-Pr	Na/Hg, Na_2HPO_4, MeOH, rt, 24 h	(—)	431
C_{21-24} n = 1-4	Na/Hg, Na_2HPO_4, MeOH, 0° to rt, 3 h	(94-96)	420
C_{22}	Na/Hg, NaH_2PO_4•H_2O, MeOH, reflux, 6 h	(47) + (15) + (29)	408
	Na/Hg, Na_2HPO_4, MeOH, rt, 1.2 h	(69)	432
	Na/Hg, MeOH, NaH_2PO_4, 0°, 6 h	(40) + (41)	433

TABLE 3. REDUCTIVE DESULFONYLATION OF β-FUNCTIONALIZED SULFONES (*Continued*)

Substrate	Conditions	Product(s) and Yield(s) (%)	Refs.
C_{22}			
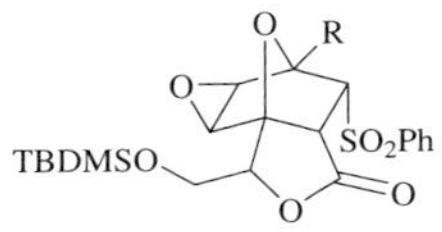 TBDMSO, O, R, SO_2Ph, O, O	Na/Hg, MeOH/THF, rt, 1.5 h	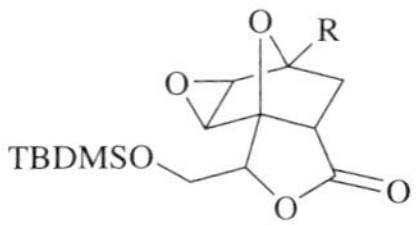 TBDMSO, O, R, O, O R: Me (70); MeO (69)	434
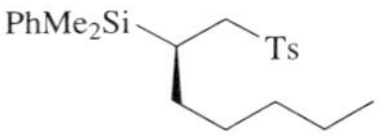 $PhMe_2Si$, Ts	Na/Hg, EtOH, reflux, 12 h	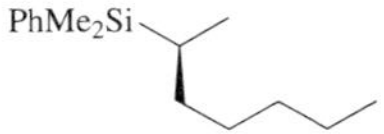 $PhMe_2Si$ (91)	435
C_{24}			
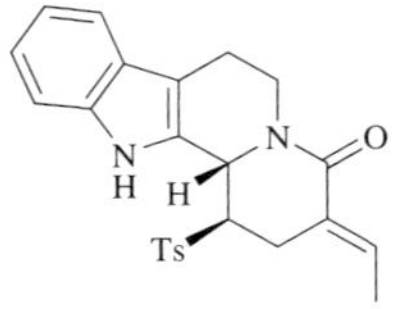 N H, H, N, O, Ts	Na/Hg, MeOH, rt, 2 h	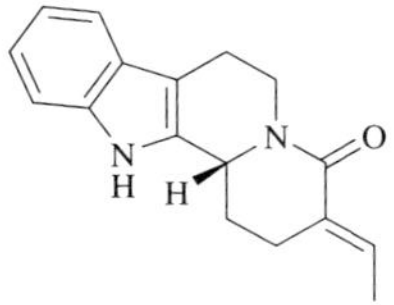 N H, H, N, O (—)	383
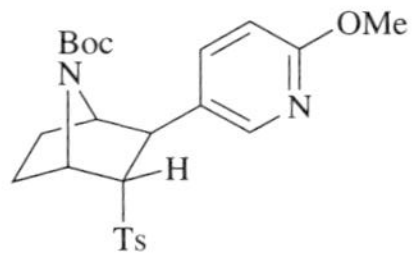 Boc N, OMe, N, H, Ts	Na/Hg, THF/MeOH	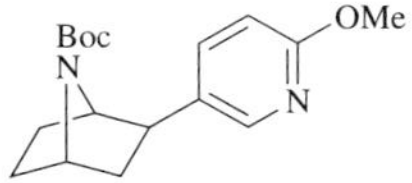 Boc N, OMe, N (58)	139

C_{24-27}

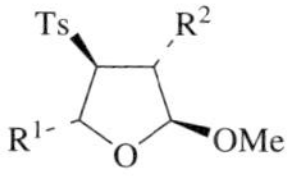

Substrate: 2-methoxy-3-R^2-4-Ts-5-R^1-tetrahydrofuran → Mg, NiX_2, MeOH, 0° to rt, 4-5 h → 2-methoxy-3-R^2-5-R^1-tetrahydrofuran (186)

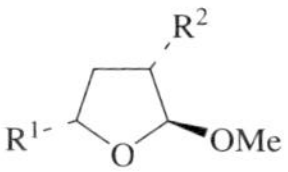

R^1	R^2	NiX_2 (mol%)	Yield (%)
$BnOCH_2$	 morpholino	$NiCl_2$ (20)	(31)
$BnOCH_2$	"	$NiBr_2$ (20)	(50)
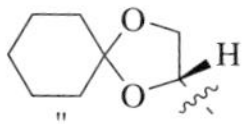 1,4-dioxaspiro[4.5]decan-2-yl	"	$NiCl_2$ (20)	(36)
"	"	$NiBr_2$ (20)	(50)
"	"	NiI_2 (20)	(22)
$BnOCH_2$	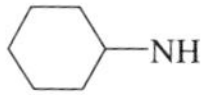 cyclohexyl-NH	$NiCl_2$ (10)	(52)
"	"	$NiBr_2$ (10)	(69)
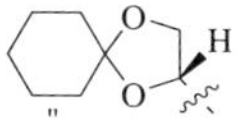 1,4-dioxaspiro[4.5]decan-2-yl	"	$NiCl_2$ (10)	(60)
"	"	$NiBr_2$ (10)	(70)
$BnOCH_2$	BnNH	$NiCl_2$ (10)	(61)
$BnOCH_2$	BnNH	$NiBr_2$ (10)	(75)
$BnOCH_2$	BnNH	NiI_2 (10)	(53)
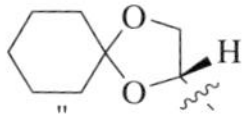 1,4-dioxaspiro[4.5]decan-2-yl	BnNH	$NiCl_2$ (10)	(65)
"	BnNH	$NiBr_2$ (10)	(73)
"	BnNH	NiI_2 (10)	(53)

TABLE 3. REDUCTIVE DESULFONYLATION OF β-FUNCTIONALIZED SULFONES (*Continued*)

Substrate	Conditions	Product(s) and Yield(s) (%)	Refs.
C_{24-28} PhO$_2$S, R, NHBoc (acetonide)	Na/Hg, MeOH, rt, 2 h	I + II	

R	PhO$_2$S*	NHBoc*	Time	I[b]	II	Refs.
MOMO-cyclopropyl	±	S	2 h	(75)	(24)	436
PhCH$_2$	R	S	1 h	(72)	(24)	437
PhCH$_2$	R	R	1 h	(35)	(31)	437
PhCH$_2$	S	R	1 h	(56)	(22)	437
PhCH$_2$	S	S	0.5 h	(65)	(21)	437
TBDMSO-cyclopropyl	R	S	1 h	(75)	(21)	436
TBDMSO-cyclopropyl	S	S	1 h	(75)	(21)	436

Substrate	Conditions	Product(s) and Yield(s) (%)	Refs.
C_{25-29} Ph, OR, SO$_2$Ph	See table.	I + II + III	438

R
Ac
Ac
TBDMS

Reagents	Solvent	Temp
Na/Hg, Na$_2$HPO$_4$	MeOH	rt
Ra-Ni	EtOH	reflux
Ra-Ni	EtOH	reflux

R	I	II	III
H	(36)	(61)	(0)
Ac	(0)	(0)	(74)
TBDMS	(0)	(0)	(84)

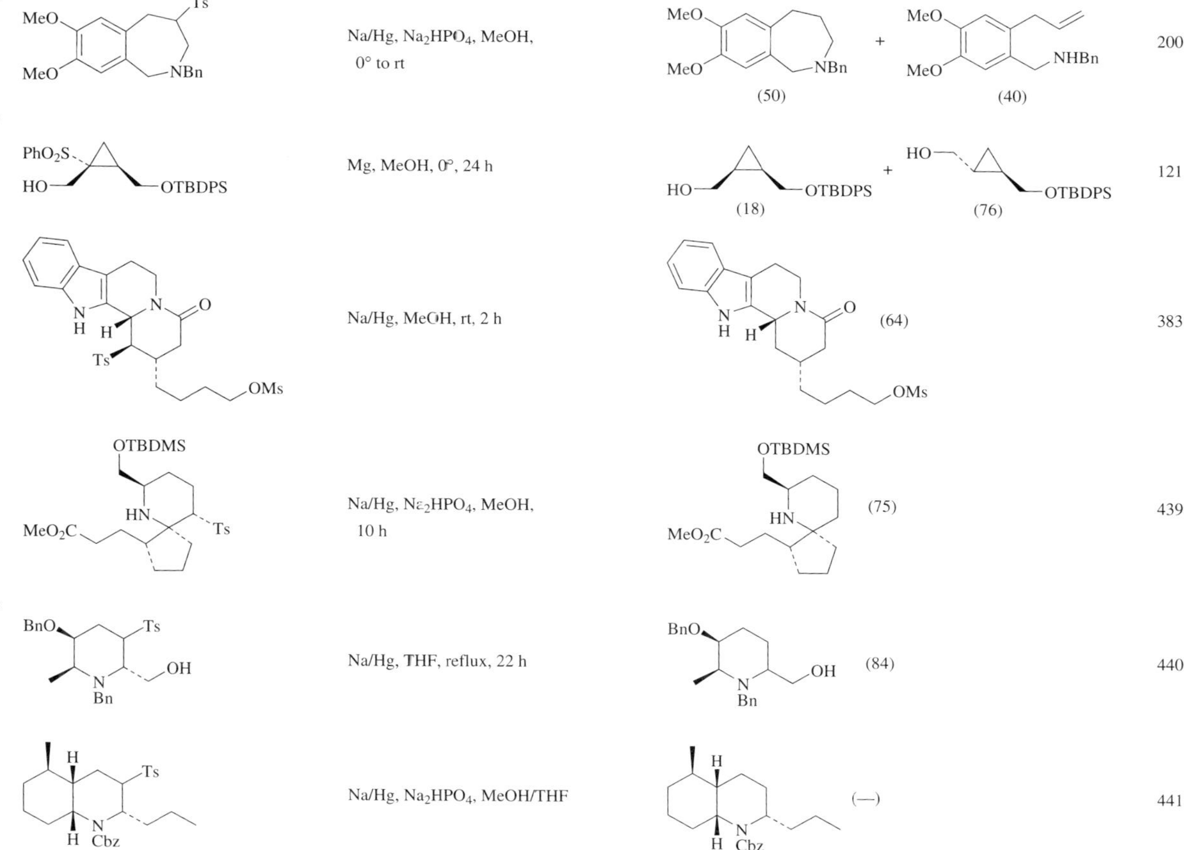

Carbons	Conditions	Product(s) and Yield(s) (%)	Refs.
C_{26}	Na/Hg, Na_2HPO_4, MeOH, 0° to rt	(50) + (40)	200
C_{27}	Mg, MeOH, 0°, 24 h	(18) + (76)	121
	Na/Hg, MeOH, rt, 2 h	(64)	383
	Na/Hg, Na_2HPO_4, MeOH, 10 h	(75)	439
C_{28}	Na/Hg, THF, reflux, 22 h	(84)	440
	Na/Hg, Na_2HPO_4, MeOH/THF	(—)	441

TABLE 3. REDUCTIVE DESULFONYLATION OF β-FUNCTIONALIZED SULFONES (*Contiaued*)

Substrate	Conditions	Product(s) and Yield(s) (%)	Refs.
C_{28} 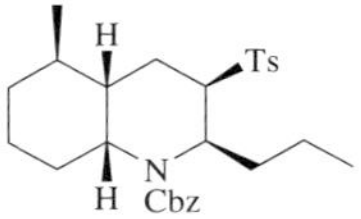 H, Ts, N, H, Cbz	Na/Hg, Na_2HPO_4, MeOH/THF	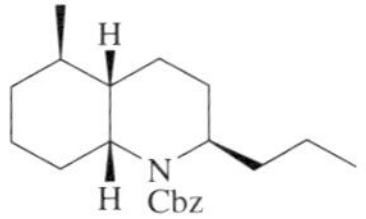 H, N, H, Cbz (—)	441
C_{29} 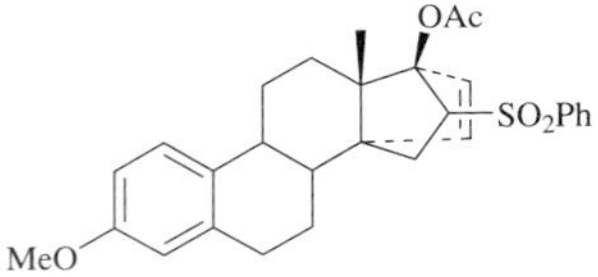 OAc, SO_2Ph, MeO	SmI_2, THF/HMPA, –20°, 70 min	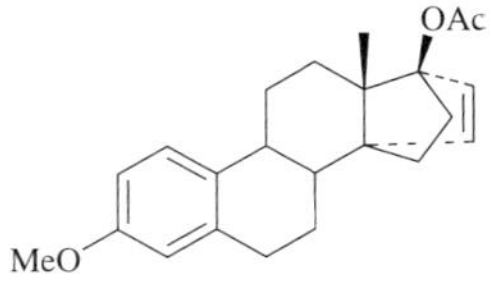 OAc, MeO (70)	100
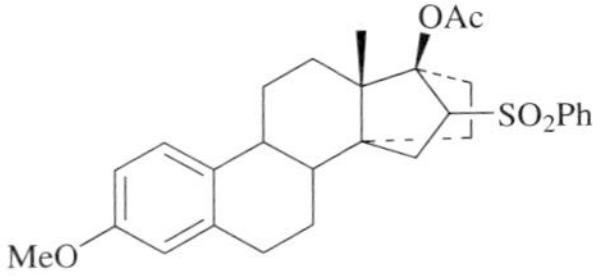 OAc, SO_2Ph, MeO	SmI_2, THF/HMPA, –20°, 90 min	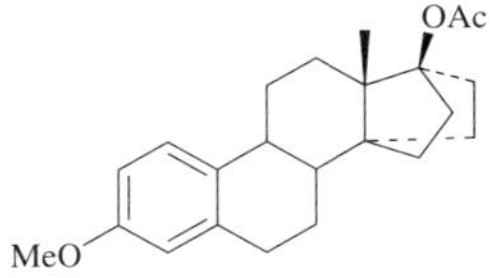 OAc, MeO (87)	100
C_{29-31} 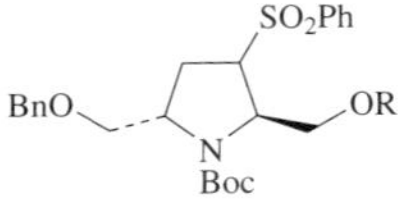 SO_2Ph, BnO, OR, N, Boc	Na/Hg, Na_2HPO_4, MeOH, 0°, 2 h	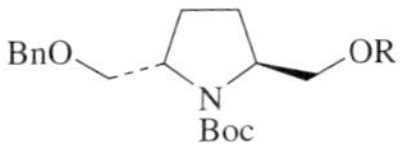 BnO, OR, N, Boc; R: THP (87), Bn (78)	36
C_{30} 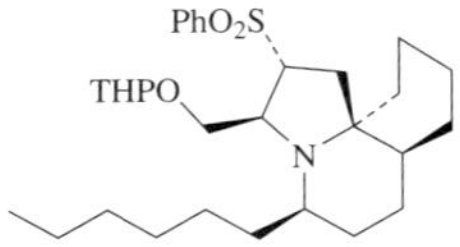 PhO_2S, THPO, N	NH_3/Na, Et_2O, –40°, 3 h	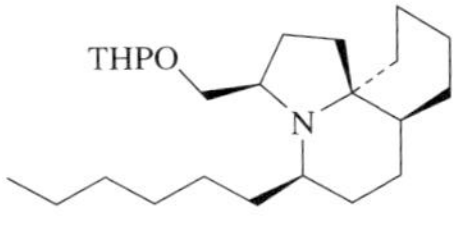 THPO, N (97)	442

Substrate	Conditions	Product(s) and Yield(s) (%)	Refs.
C_{31}			
Ph_3P, O, Ts, OH	Na/Hg, MeOH, 2 h	Ph_3P, O, OH (70)	443
PhO_2S, Ph, N, Pr-*i*, PPh_2	SmI_2, THF/MeOH, DMPU, –20°, 1.5 h	Ph, N, Pr-*i*, PPh_2 (91)	444
MeO, OMe, O, TBDMSO, N, O, Ph, SO_2Ph	Na/Hg, EtOH, reflux	MeO, OMe, O, TBDMSO, N, O, Ph (66)	445

[a] MR stands for Merrifield resin.

[b] The absolute stereochemical designation of the product is opposite that of the starting material due to substituent priority changes.

TABLE 4. REDUCTIVE DESULFONYLATION OF REMOTE-FUNCTIONALIZED SULFONES

Substrate	Conditions	Product(s) and Yield(s) (%)	Refs.
C_{12} 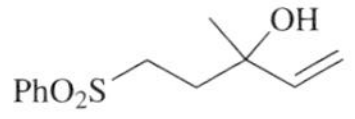 OH, PhO_2S	Na/Hg, MeOH, rt, overnight	 OH (58)	107
C_{13} 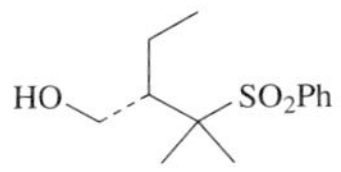 HO, SO_2Ph	Mg, $HgCl_2$, EtOH, rt, overnight	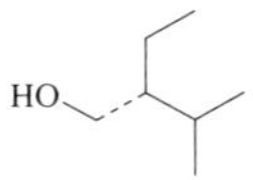 HO (78)	446
C_{14} 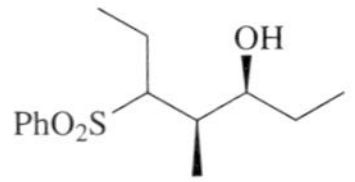 OH, PhO_2S	Na/Hg, Na_2HPO_4, MeOH, rt, 6 h	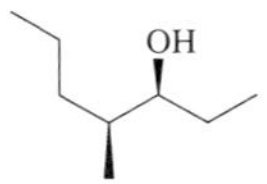 OH (80)	447
C_{15} 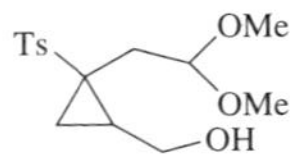 Ts, OMe, OMe, OH	Na/Hg, Na_2HPO_4, MeOH, 0° to rt	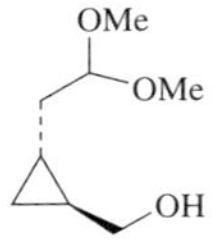 OMe, OMe, OH (46)	448
C_{15-16} 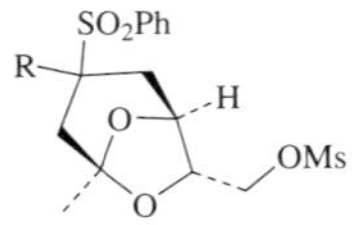 SO_2Ph, R, O, O, H, OMs; R = H, Me	Na/EtOH, THF, –20°, 2 h	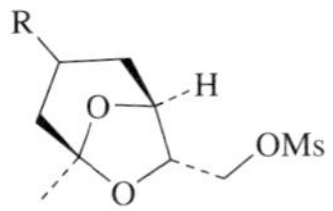 R, O, O, H, OMs (—)	449
C_{15-17} 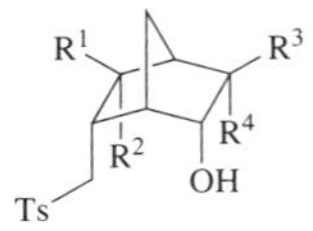 R^1, R^2, R^3, R^4, OH, Ts	Na/Hg, Na_2HPO_4, MeOH, rt	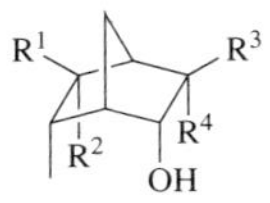 R^1, R^2, R^3, R^4, OH	450

R^1	R^2	R^3	R^4	Time	
H	H	H	H	16 h	(77)
H	H	Me	H	16 h	(80)
H	H	Me	Me	2 h	(41)
Me	Me	H	H	2 h	(71)

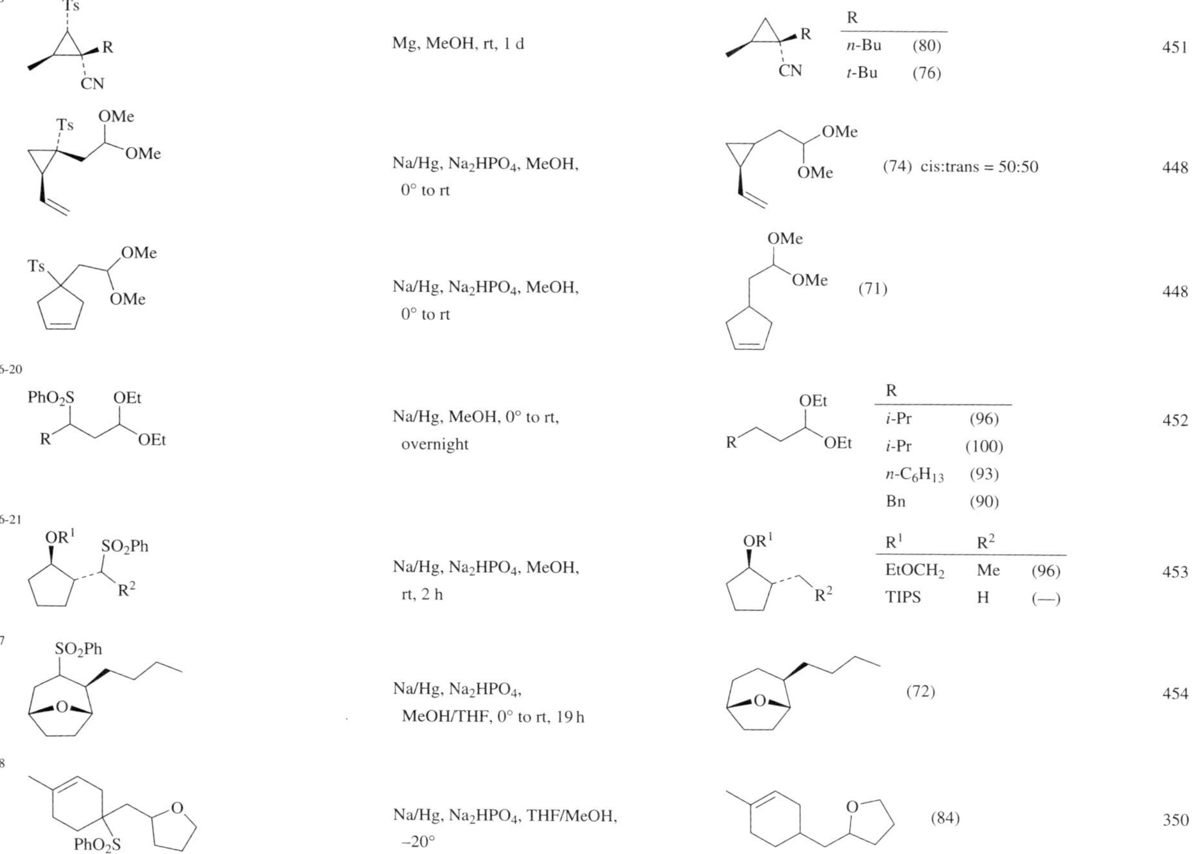

Carbon No.	Conditions	Product(s) and Yield(s) (%)	Refs.
C_{16}	Mg, MeOH, rt, 1 d	R: *n*-Bu (80); *t*-Bu (76)	451
	Na/Hg, Na_2HPO_4, MeOH, 0° to rt	(74) cis:trans = 50:50	448
	Na/Hg, Na_2HPO_4, MeOH, 0° to rt	(71)	448
C_{16-20}	Na/Hg, MeOH, 0° to rt, overnight	R: *i*-Pr (96); *i*-Pr (100); *n*-C_6H_{13} (93); Bn (90)	452
C_{16-21}	Na/Hg, Na_2HPO_4, MeOH, rt, 2 h	R^1 / R^2: $EtOCH_2$ / Me (96); TIPS / H (—)	453
C_{17}	Na/Hg, Na_2HPO_4, MeOH/THF, 0° to rt, 19 h	(72)	454
C_{18}	Na/Hg, Na_2HPO_4, THF/MeOH, –20°	(84)	350

TABLE 4. REDUCTIVE DESULFONYLATION OF REMOTE-FUNCTIONALIZED SULFONES (*Continued*)

	Substrate	Conditions	Product(s) and Yield(s) (%)	Refs.
C_{18}	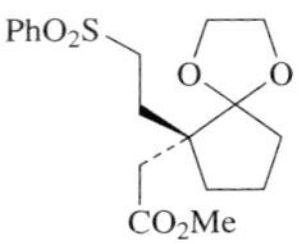 PhO_2S … CO_2Me (dioxolane spiro-cyclopentane)	Na/Hg, MeOH	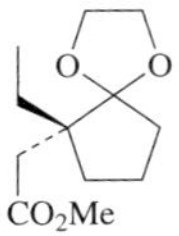 CO_2Me … (—)	455
	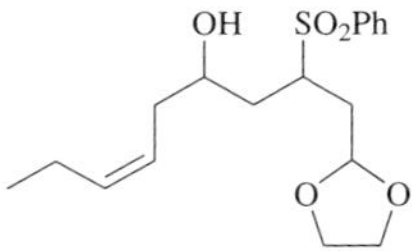 OH, SO_2Ph (dioxolane)	Na/Hg, EtOH, 3 h, rt	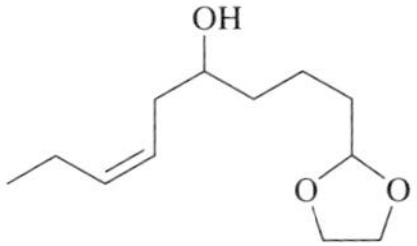 OH … (—)	456
	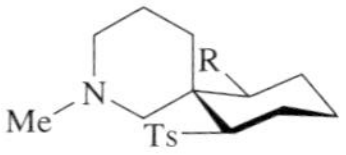 Me–N, R, Ts; R = OH	Na/Hg, Na_2HPO_4, MeOH, –20°, 24 h	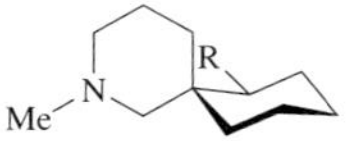 Me–N, R (95)	457
C_{18-22}	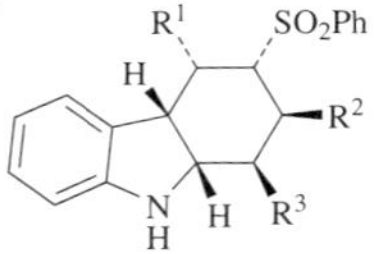 R^1, SO_2Ph, R^2, R^3, H, N–H	Na/Hg, Na_2HPO_4, 0° to rt	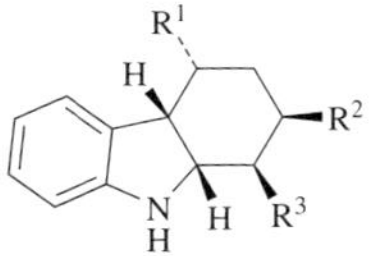 R^1, R^2, R^3, H, N–H	

R^1	R^2	R^3	Solvent	Time	Yield (%)	Refs.
H	H	H	—	—	(—)	458
H	$NCCH_2$	Me	MeOH/THF (1:1)	4 h	(90)	459
Me	$NCCH_2$	Me	MeOH	1 h	(94)	459

	Substrate	Conditions	Product(s) and Yield(s) (%)	Refs.
C_{19}	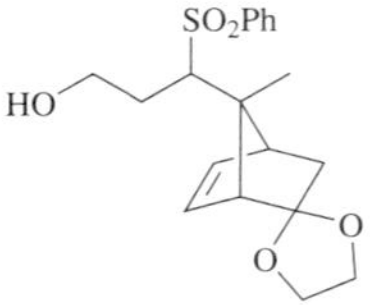 HO, SO_2Ph (dioxolane)	Na/Hg, Na_2HPO_4, MeOH, rt	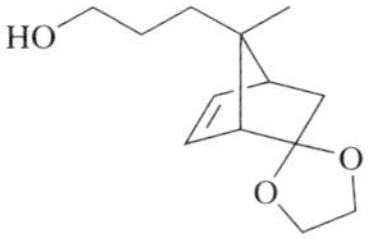 HO … (—)	460

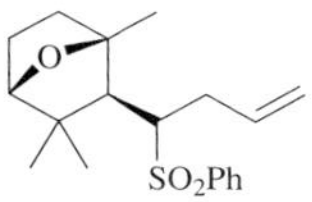

Substrate	Conditions	Product(s) and Yield(s) (%)	Refs.
SO_2Ph	Na/Hg, Na_2HPO_4, EtOH, rt, 6 h	(37) + (—)	461
OH SO_2Ph OTBDMS	Na/Hg, EtOH	OH OTBDMS (92)	342
C_{19-26} R^1 R^2 SO_2Ph O	Na/Hg, Na_2HPO_4, THF/MeOH, rt, 2 h	MeO_2C R^2 O **I** + OH R^2 O **II** + R^2 HO OH R^2 O O **III**	142
C_{20} Ts OMe OMe	Na/Hg, Na_2HPO_4, MeOH, 0° to rt	OMe MeO (83)	448

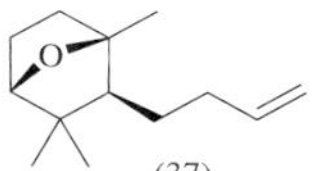

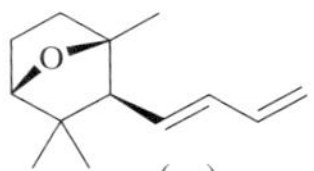

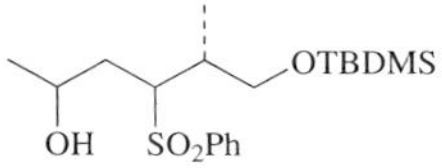

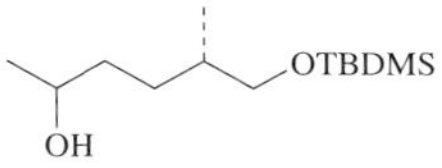

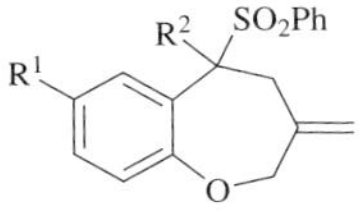

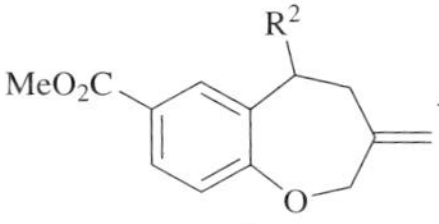

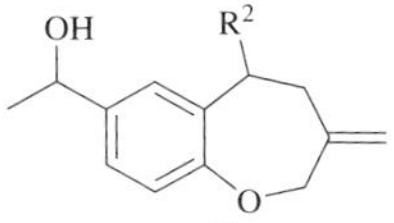

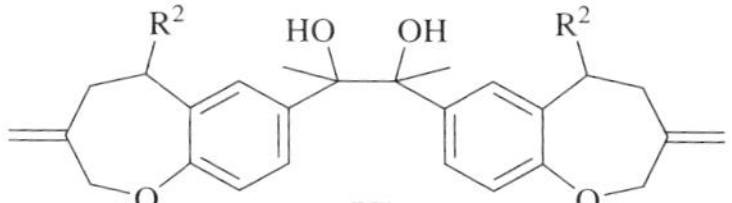

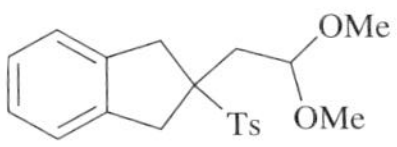

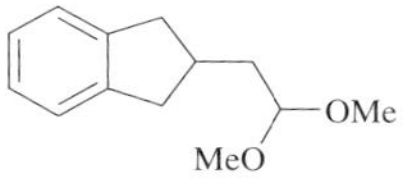

R^1	R^2	**I**	**II + III**
CO_2Me	H	(100)	(0)
C(O)Me	H	(0)	(83)
CO_2Me	Bn	(100)	(0)
C(O)Me	Bn	(0)	(81)

TABLE 4. REDUCTIVE DESULFONYLATION OF REMOTE-FUNCTIONALIZED SULFONES (*Continued*)

Substrate	Conditions	Product(s) and Yield(s) (%)	Refs.
C_{20}			
SO_2Ph; N, Bn	Na/Hg, EtOH, rt	N, Bn (93)	462
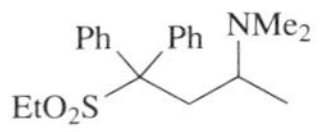 Ph, Ph, NMe_2, EtO_2S	Ra-Ni, EtOH, reflux, 6 h	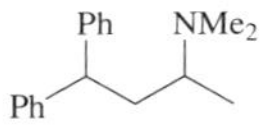 Ph, NMe_2, Ph (—)	463
C_{21}			
PhO_2S, H, O, H, 5	Na/Hg, Na_2HPO_4, MeOH, rt, 12 h	H, O, H, 5 (96)	464
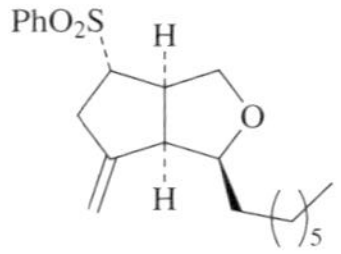 PhO_2S, H, O, H, 5	Na/Hg, Na_2HPO_4, MeOH, rt, 12 h	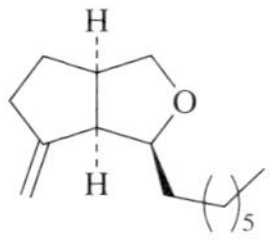 H, O, H, 5 (60)	464
C_{22}			
MOMO, MeO, H, N, N, OMe, SO_2Ph	Na/Hg, MeOH, rt, 3 h	MOMO, MeO, H, N, N, OMe (72)	465

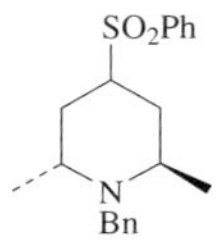

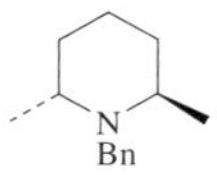

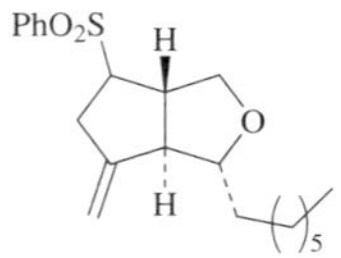

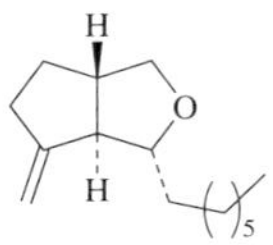

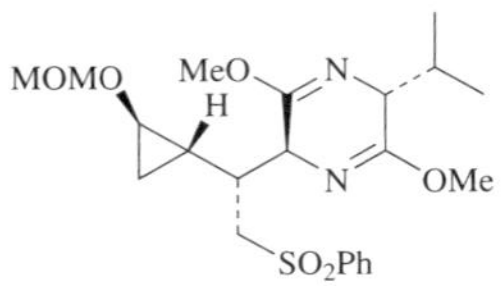

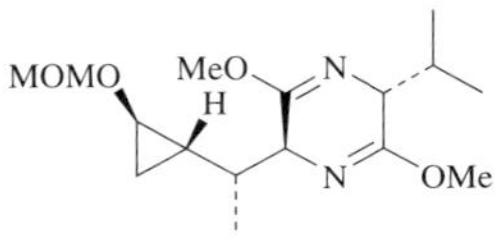

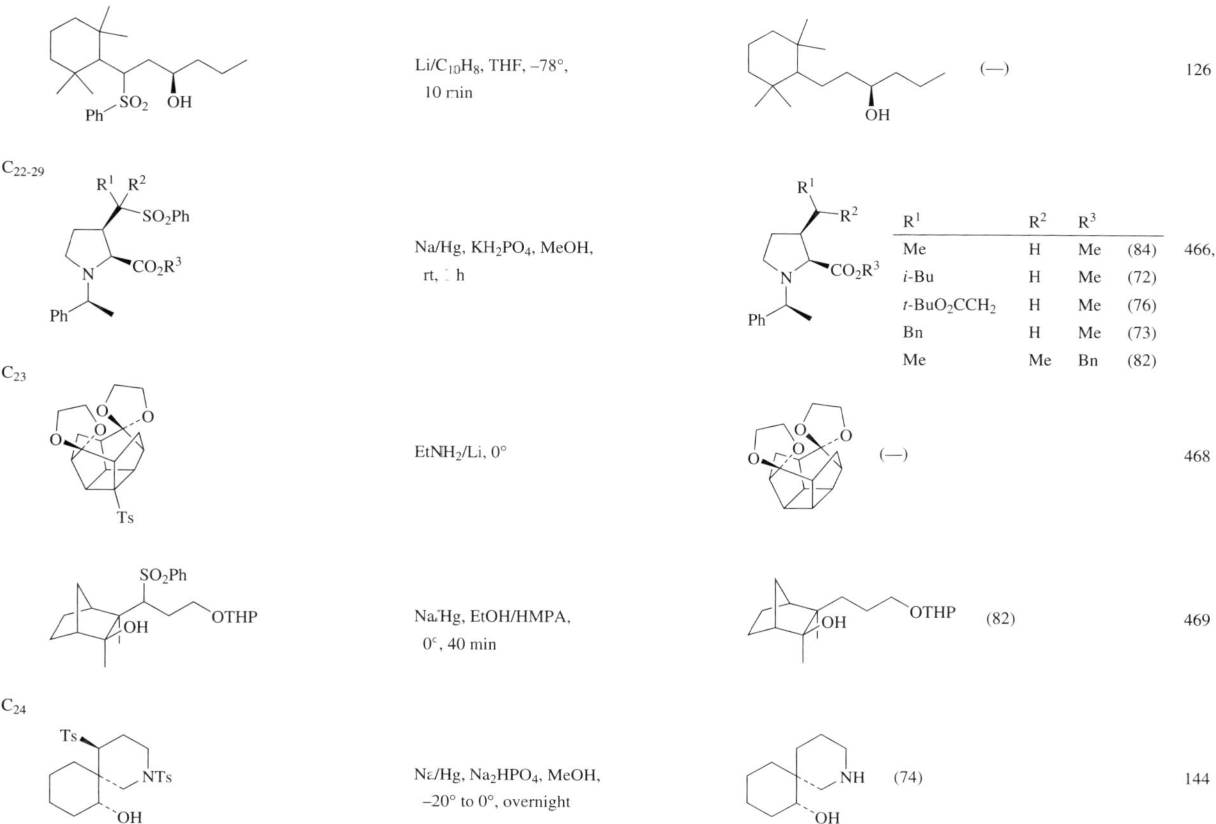

Substrate	Conditions	Product(s) and Yield(s) (%)	Refs.
	$Li/C_{10}H_8$, THF, –78°, 10 min	(—)	126
C_{22-29}	Na/Hg, KH_2PO_4, MeOH, rt, [illegible] h		466, 467
C_{23}	$EtNH_2$/Li, 0°	(—)	468
	Na/Hg, EtOH/HMPA, 0°, 40 min	(82)	469
C_{24}	Na/Hg, Na_2HPO_4, MeOH, –20° to 0°, overnight	(74)	144

R^1	R^2	R^3	
Me	H	Me	(84)
i-Bu	H	Me	(72)
t-BuO_2CCH_2	H	Me	(76)
Bn	H	Me	(73)
Me	Me	Bn	(82)

TABLE 4. REDUCTIVE DESULFONYLATION OF REMOTE-FUNCTIONALIZED SULFONES (*Continued*)

Substrate	Conditions	Product(s) and Yield(s) (%)	Refs.
C_{25}			
Bn, SO_2Ph, OH	Na/Hg, Na_2HPO_4, MeOH, 0°, 3 h	Bn, OH (87)	141
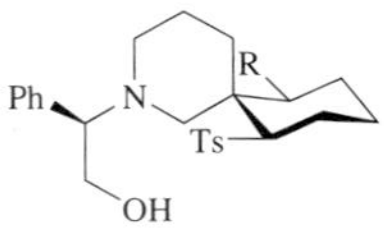 Ph, N, R, Ts, OH R = OH	Na/Hg, Na_2HPO_4, MeOH, –20°, 24 h	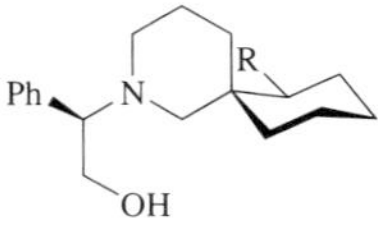 Ph, N, R, OH (95)	457
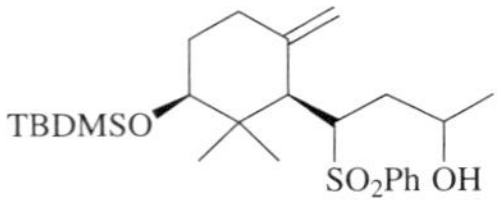 TBDMSO, SO_2Ph OH	Na/Hg, Na_2HPO_4, EtOH, reflux, 4 h	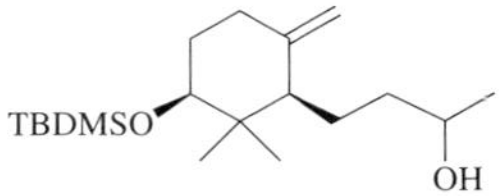 TBDMSO, OH (85)	470, 461
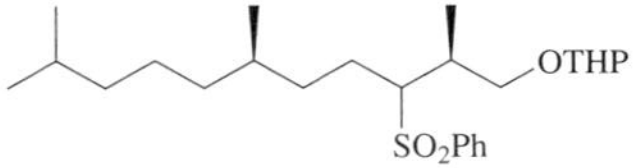 OTHP, SO_2Ph	Na/Hg, EtOH, rt, 48 h	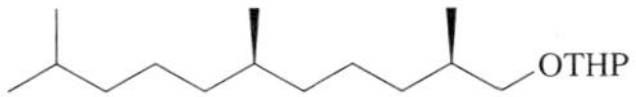 OTHP (99)	471
$C_{25\text{-}27}$			
OR, H, H, H, SO_2Ph, MeO	SmI_2, THF/HMPA, 1.5 h	OR, H, H, H, MeO R | Temp | H | 22° | (50) Ac | –20° | (52)	100

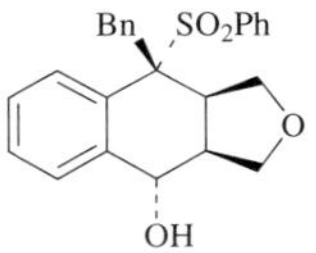

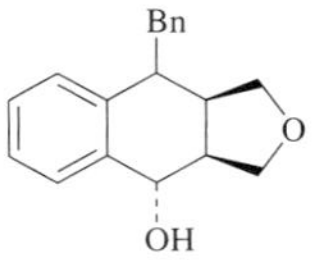

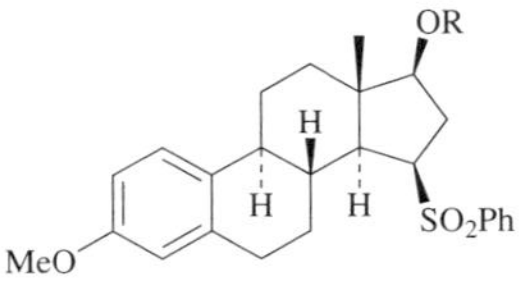

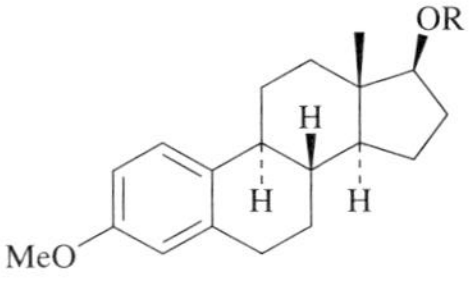

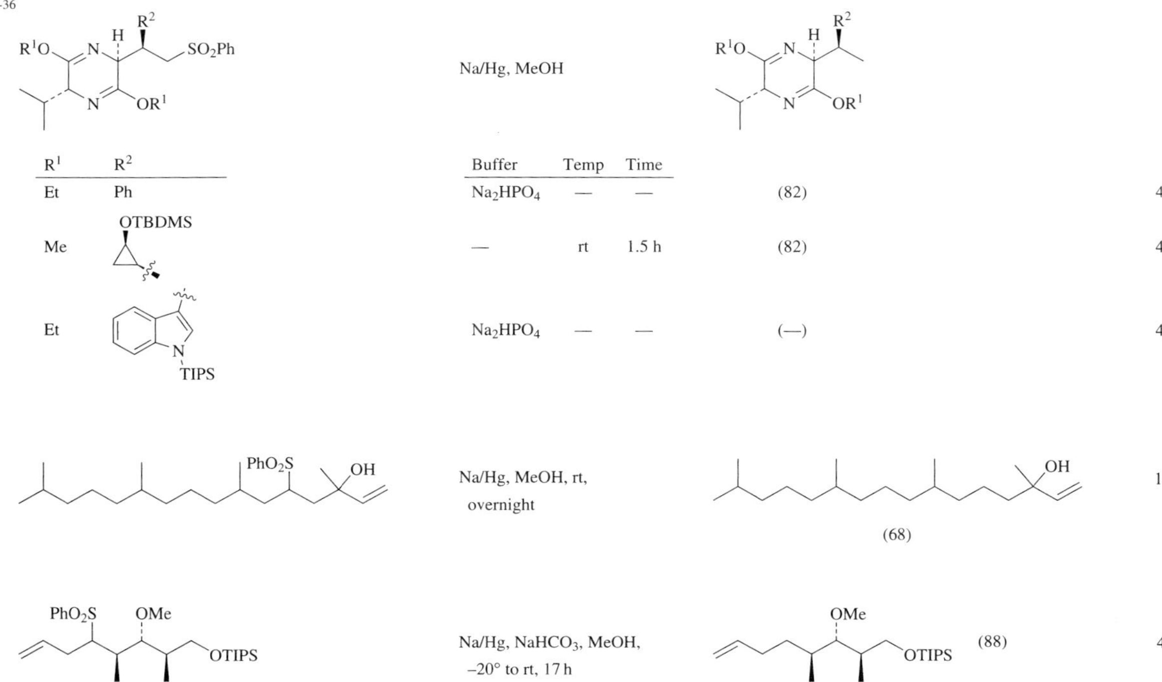

R^1	R^2	Buffer	Temp	Time	Yield	Ref.
Et	Ph	Na_2HPO_4	—	—	(82)	472
Me	OTBDMS-cyclopropyl	—	rt	1.5 h	(82)	473
Et	N-TIPS-indol-3-yl	Na_2HPO_4	—	—	(—)	472

TABLE 4. REDUCTIVE DESULFONYLATION OF REMOTE-FUNCTIONALIZED SULFONES (*Continued*)

Substrate	Conditions	Product(s) and Yield(s) (%)	Refs.
C_{26}			
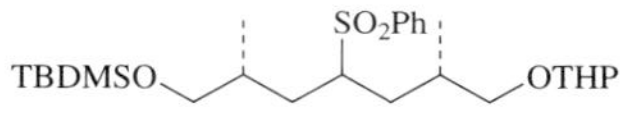 TBDMSO, SO_2Ph, OTHP	Na/Hg, EtOH, 0° to rt, 25 h	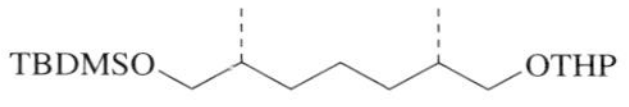 TBDMSO, OTHP (73)  + TBDMSO, OTHP (5)	475
C_{26-27}			
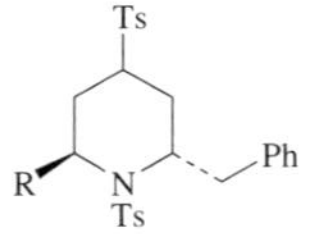 Ts, R, N, Ts, Ph	Na/Hg, THF/MeOH, reflux, 12 h	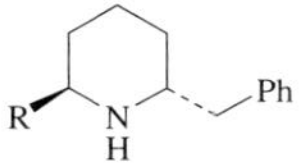 R, N, H, Ph; R: H (97), Me (67)	476
C_{27}			
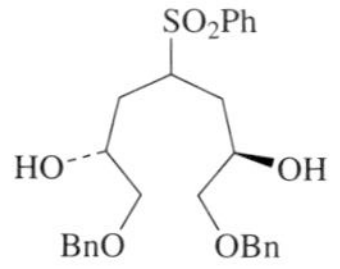 SO_2Ph, HO, OH, BnO, OBn	Na/Hg, EtOH, rt	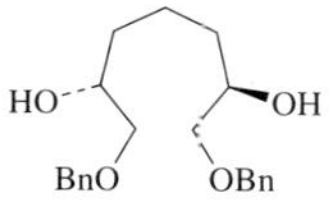 HO, OH, BnO, OBn (91)	462
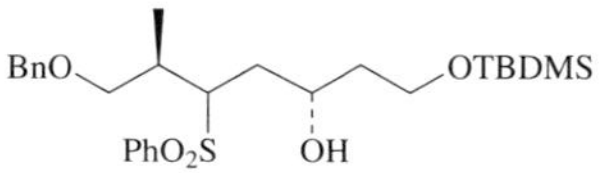 BnO, PhO_2S, OH, OTBDMS	Na/Hg, EtOH	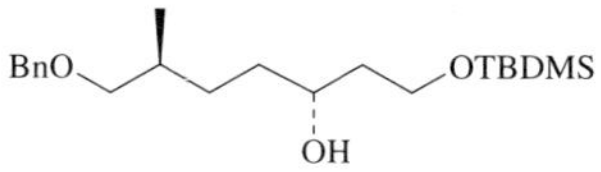 BnO, OH, OTBDMS (71)	477

Substrate	Conditions	Product(s) and Yield(s) (%)	Refs.
OH, SO_2Ph	Na/Hg, MeOH, reflux, 1.5 h	OH (99)	478
TIPSO, MeO, OH, SO_2Ph	Mg, MeOH, 50°, 45 min	TIPSO, MeO, OH (88)	479
OMOM, OTBDMS, OH, SO_2Ph	Na/Hg, Na_2HPO_4, MeOH, 0°, 2 h; rt, overnight	OMOM, OTBDMS, OH (75)	480
C_{28}			
SO_2Ph, PMBO	Na/Hg, Na_2HPO_4, MeOH, 0°, 30 min	PMBO (68)	481

TABLE 4. REDUCTIVE DESULFONYLATION OF REMOTE-FUNCTIONALIZED SULFONES (*Continued*)

Substrate	Conditions	Product(s) and Yield(s) (%)	Refs.
C_{28}			
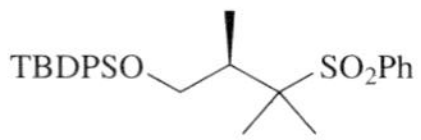 TBDPSO, SO_2Ph	Mg, $HgCl_2$, EtOH, rt, overnight	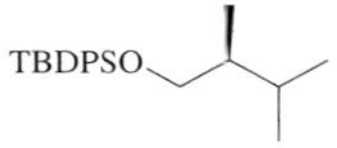 TBDPSO (—)	446
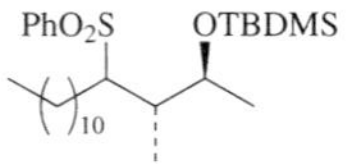 PhO_2S, OTBDMS, 10	Ra-Ni (W-2), 1,4-dioxane, reflux, 2 h	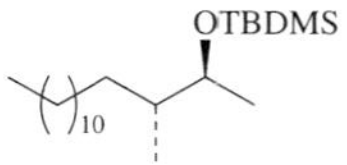 OTBDMS, 10 (73)	482
C_{29}			
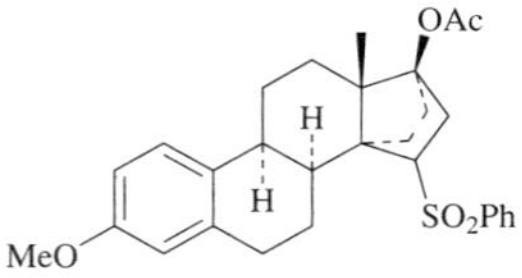 OAc, H, H, MeO, SO_2Ph	Reagents, THF	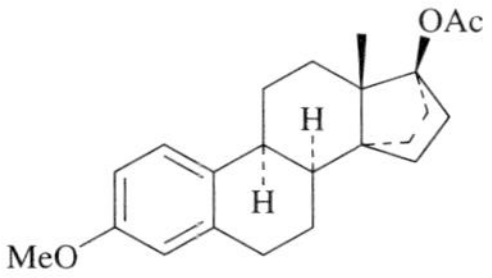 OAc, H, H, MeO	483
	Reagents / Temp / Time		
	SmI_2, HMPA / –20° / 4 h	(89)	
	NH_3/Na / –33° / 2 h	(53)	
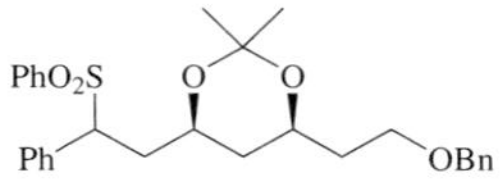 PhO_2S, Ph, O, O, OBn	Na/Hg, MeOH, 0° to rt, 14 h	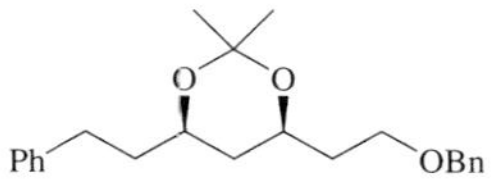 Ph, O, O, OBn (—)	484

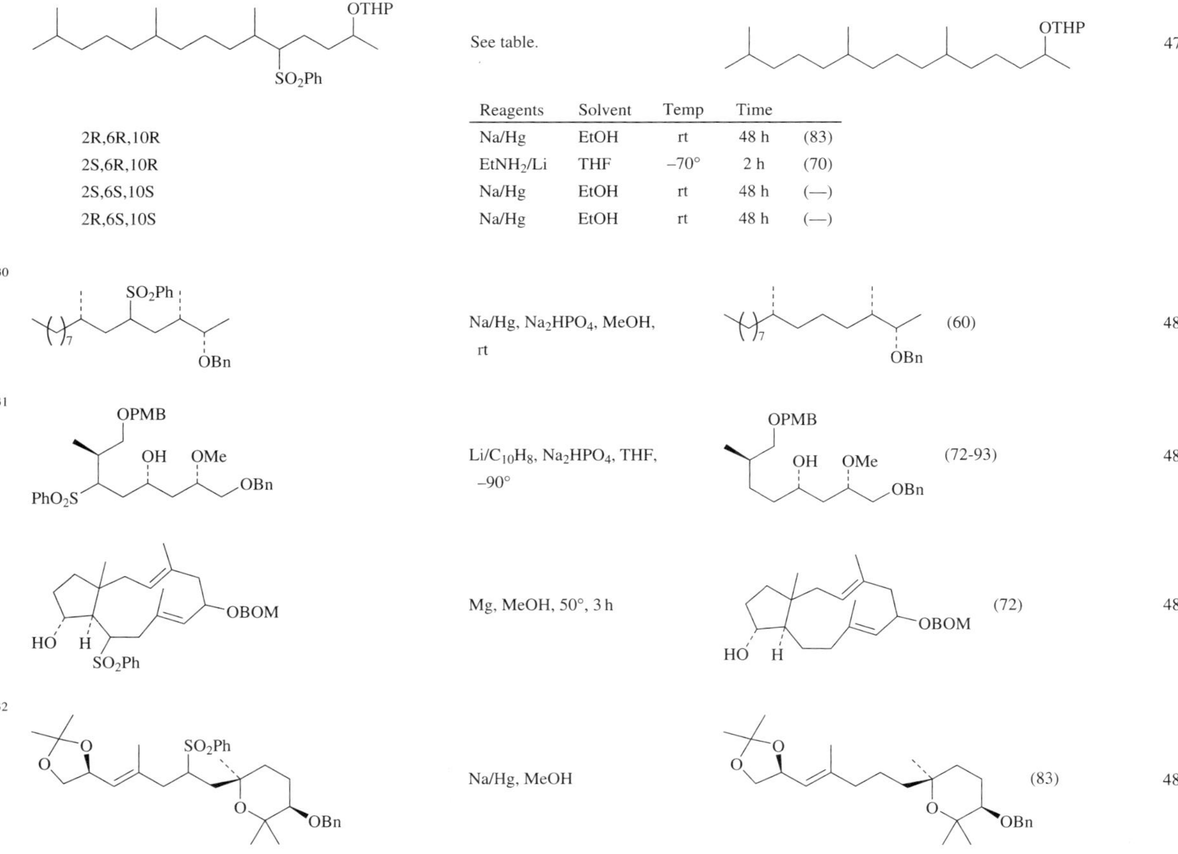

Reagents	Conditions	Product(s) and Yield(s) (%)	Refs.
2R,6R,10R; 2S,6R,10R; 2S,6S,10S; 2R,6S,10S	See table.		471
C30	Na/Hg, Na_2HPO_4, MeOH, rt	(60)	485
C31	Li/$C_{10}H_8$, Na_2HPO_4, THF, –90°	(72-93)	486
	Mg, MeOH, 50°, 3 h	(72)	487
C32	Na/Hg, MeOH	(83)	488

Reagents	Solvent	Temp	Time	
Na/Hg	EtOH	rt	48 h	(83)
$EtNH_2$/Li	THF	–70°	2 h	(70)
Na/Hg	EtOH	rt	48 h	(—)
Na/Hg	EtOH	rt	48 h	(—)

TABLE 4. REDUCTIVE DESULFONYLATION OF REMOTE-FUNCTIONALIZED SULFONES (*Continued*)

Substrate	Conditions	Product(s) and Yield(s) (%)	Refs.
C_{33}			
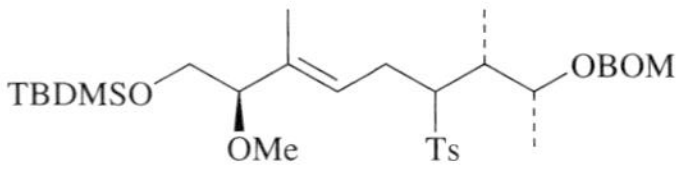 TBDMSO, OMe, Ts, OBOM	Na/Hg, Na_2HPO_4, MeOH, rt, 2 h	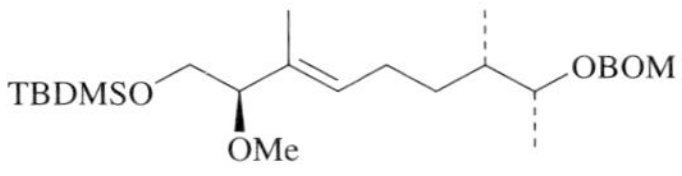 TBDMSO, OMe, OBOM (87)	489
C_{34}			
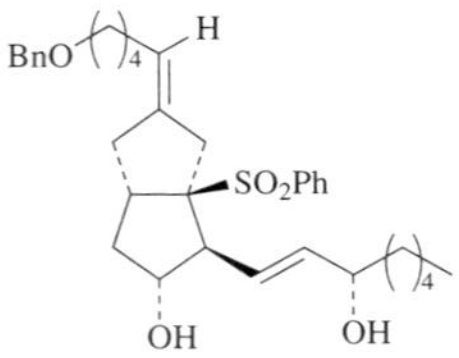 BnO, ( )4, H, SO_2Ph, OH, OH, ( )4	NH_3/Li, *t*-BuOH, THF, –78°, 15 min	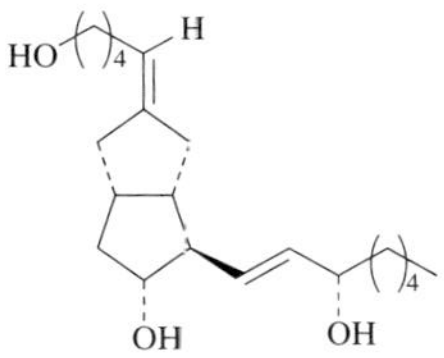 HO, ( )4, H, OH, OH, ( )4 (88)	490
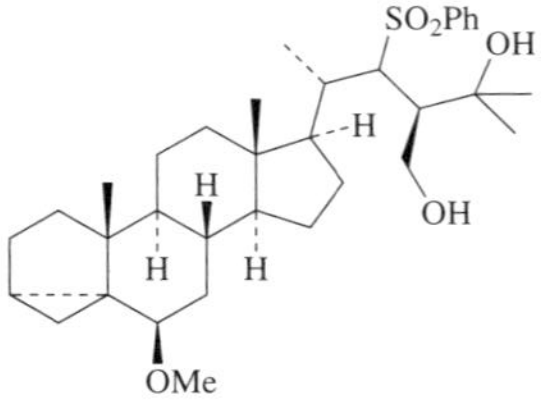 SO_2Ph, OH, H, OH, H, H, H, OMe	NH_3/Li, THF/EtOH, –78°, 25 min	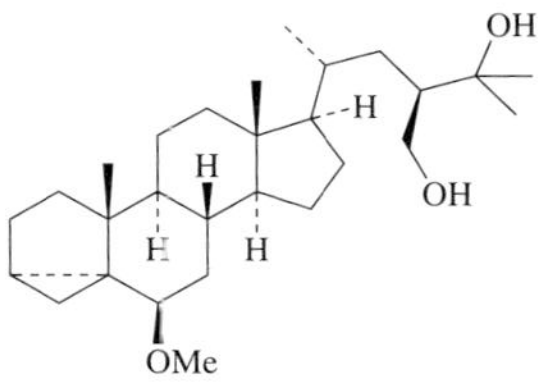 OH, H, OH, H, H, H, OMe (77)	491
C_{35}			
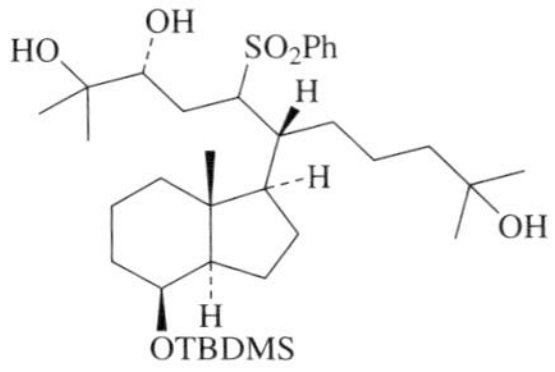 HO, OH, SO_2Ph, H, H, OH, H, OTBDMS	Na/Hg, THF/MeOH, rt, 24 h	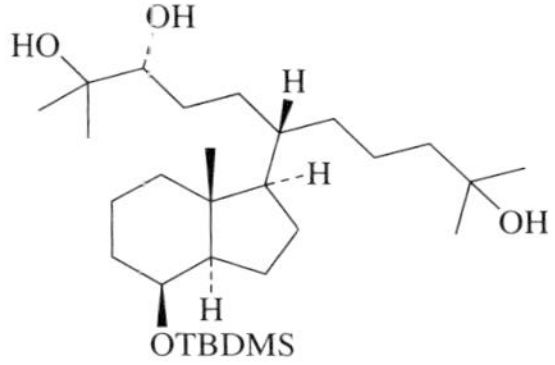 HO, OH, H, H, OH, H, OTBDMS (73)	492

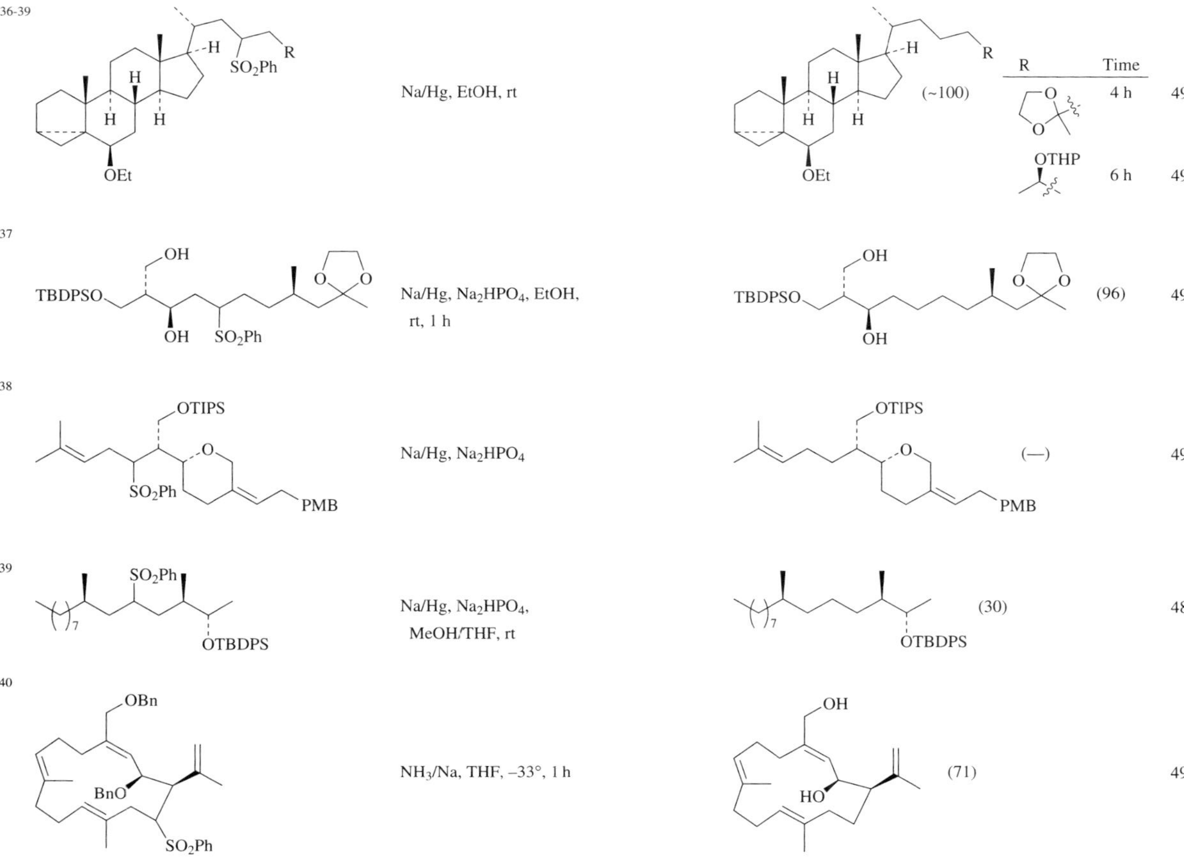

Carbon	Conditions	Product(s) and Yield(s) (%)	Refs.
C_{36-39}	Na/Hg, EtOH, rt	(~100); R = 2-methyl-1,3-dioxolan-2-yl, Time 4 h	493
		R = CH(OTHP)CH$_3$, Time 6 h	494
C_{37}	Na/Hg, Na_2HPO_4, EtOH, rt, 1 h	(96)	495
C_{38}	Na/Hg, Na_2HPO_4	(—)	496
C_{39}	Na/Hg, Na_2HPO_4, MeOH/THF, rt	(30)	485
C_{40}	NH_3/Na, THF, –33°, 1 h	(71)	497

TABLE 4. REDUCTIVE DESULFONYLATION OF REMOTE-FUNCTIONALIZED SULFONES (*Continued*)

Substrate	Conditions	Product(s) and Yield(s) (%)	Refs.
C_{41} 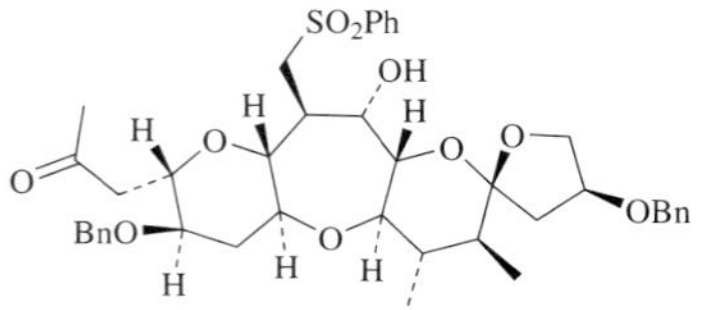	Na/Hg, Na_2HPO_4, MeOH, rt, 4.5 h	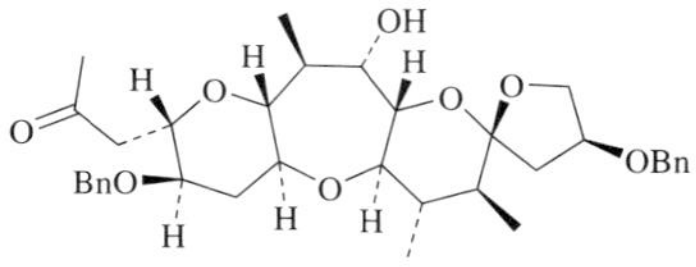 (82)	498, 499
C_{42} 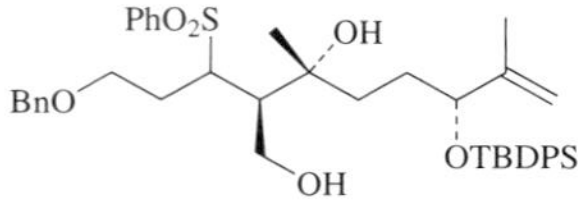	Na/Hg, Na_2HPO_4, EtOH, rt, 1 h	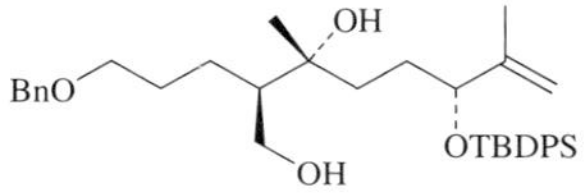 (—)	500
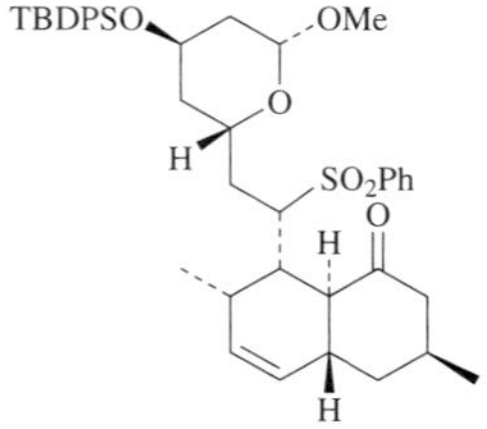	Na/Hg, MeOH, 2 h	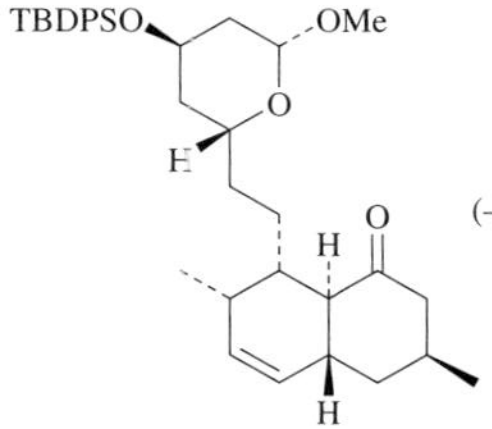 (—)	501
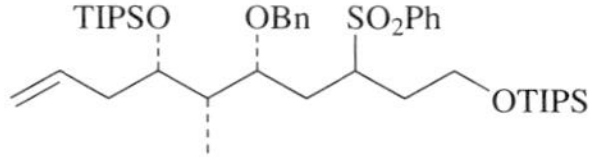	NH_3/Li, THF, –78°	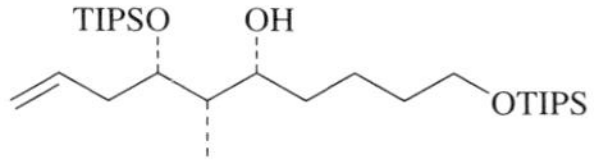 (85)	502
C_{43} 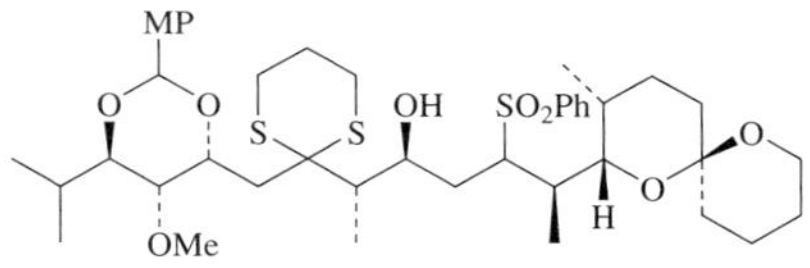	Na/Hg	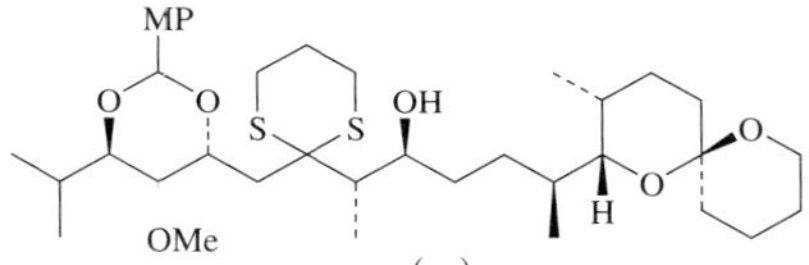 (—)	503

Reactant	Conditions	Product(s) and Yield(s) (%)		Refs.
C_{43}				
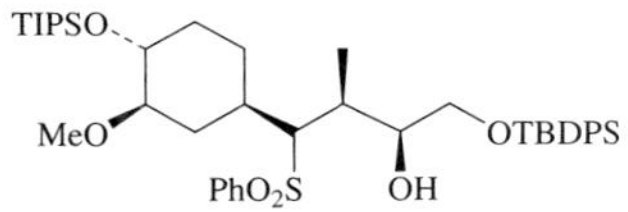 TIPSO, MeO, PhO_2S, OH, OTBDPS	Na/Hg, Na_2HPO_4, MeOH	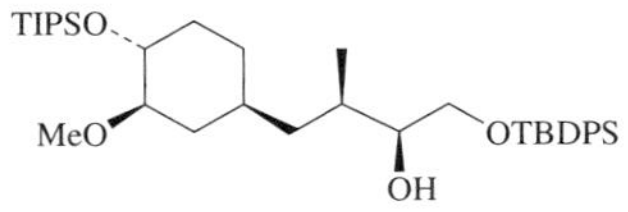 TIPSO, MeO, OH, OTBDPS	(—)	504
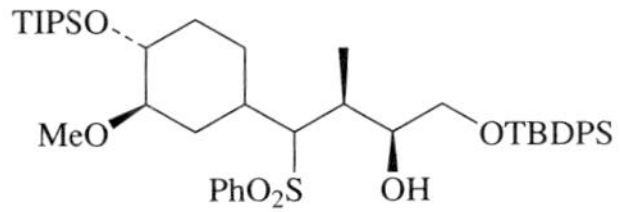 TIPSO, MeO, PhO_2S, OH, OTBDPS	Na/Hg, Na_2HPO_4, MeOH/THF, rt, 45 min	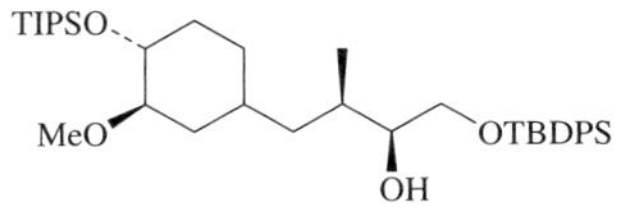 TIPSO, MeO, OH, OTBDPS	(—)	344
C_{45}				
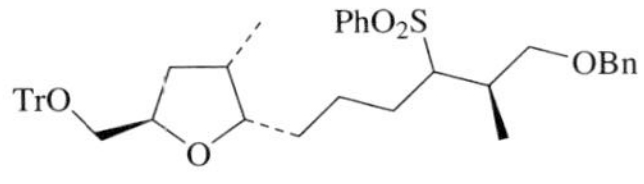 TrO, PhO_2S, OBn	Na/Hg, EtOH, 0° to rt, 2-3 h	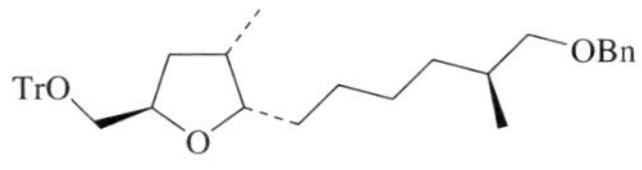 TrO, OBn	(83)	505
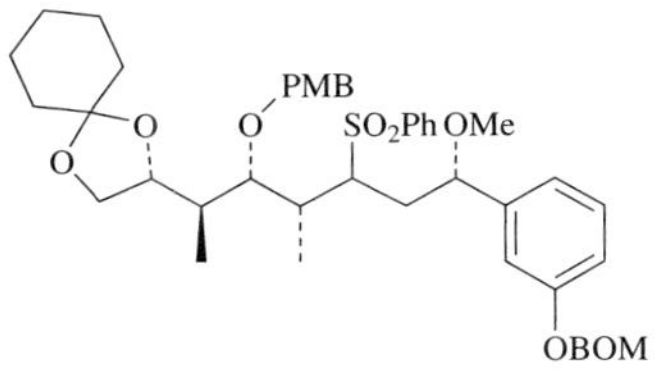 PMB, SO_2Ph, OMe, OBOM	Na/Hg, Na_2HPO_4, MeOH, rt	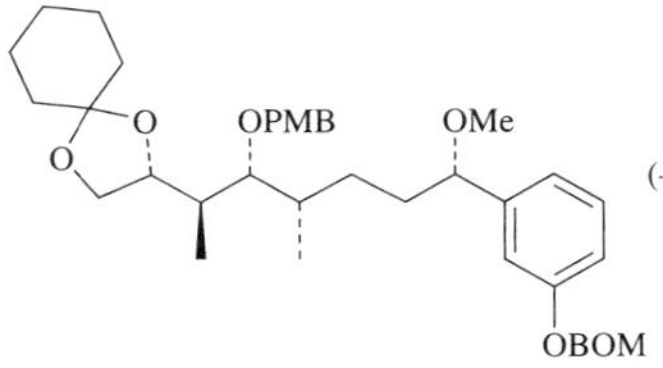 OPMB, OMe, OBOM	(—)	506
C_{51}				
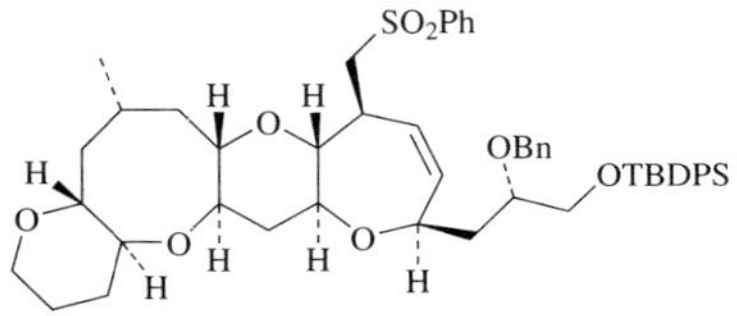 SO_2Ph, OBn, OTBDPS	Na/Hg, Na_2HPO_4, MeOH, rt	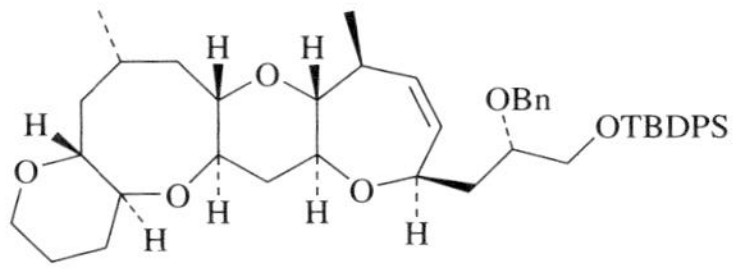 OBn, OTBDPS	(98)	507

TABLE 4. REDUCTIVE DESULFONYLATION OF REMOTE-FUNCTIONALIZED SULFONES (*Continued*)

	Substrate	Conditions	Product(s) and Yield(s) (%)	Refs.
C_{52} 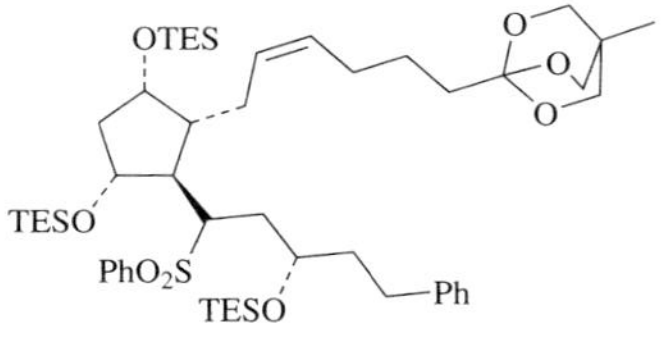	OTES, TESO, PhO_2S, TESO, Ph	Na/Hg, Na_2HPO_4, MeOH, 0° to rt, 100 min	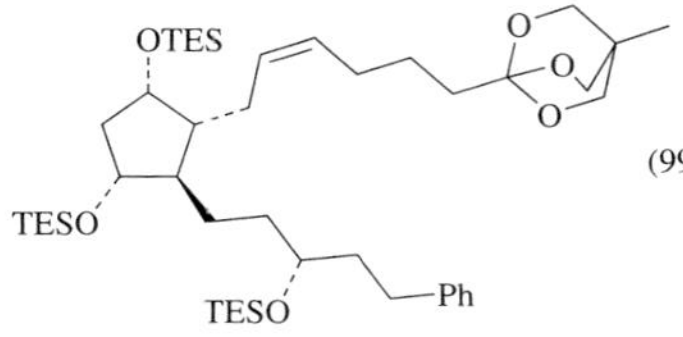 OTES, TESO, TESO, Ph (99)	508
C_{53} 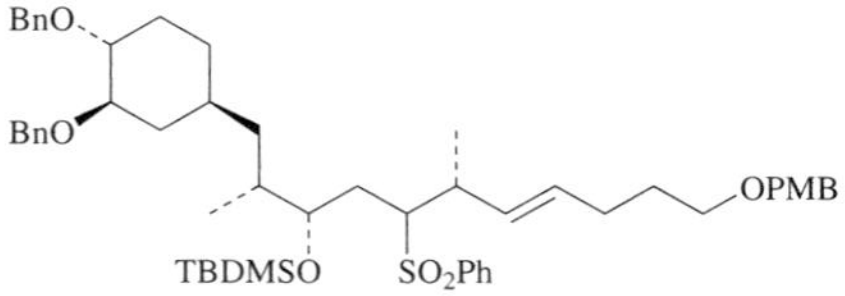	BnO, BnO, TBDMSO, SO_2Ph, OPMB	Na/Hg, Na_2HPO_4, THF/MeOH, 0° to rt, overnight	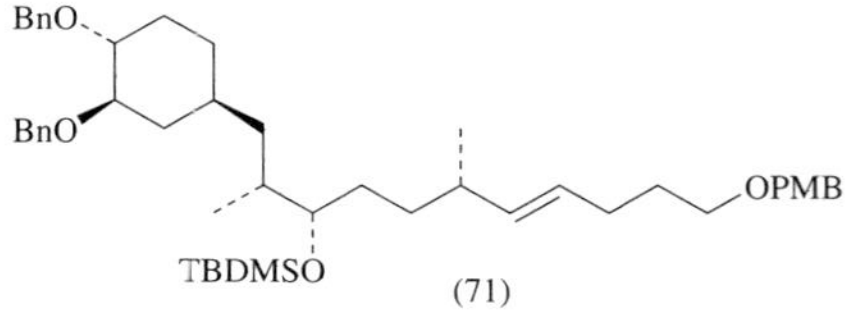 BnO, BnO, TBDMSO, OPMB (71)	250
	OTES, SO_2Ph, H, TBDMSO, OTBDMS	Na/Hg, Na_2HPO_4, THF/MeOH, 5°, 3 h	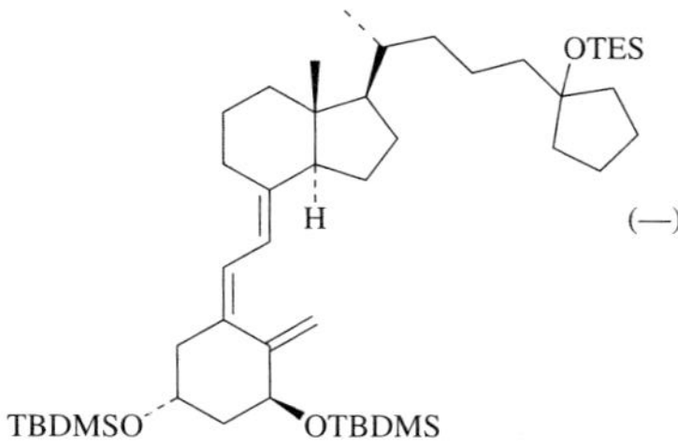 OTES, H, TBDMSO, OTBDMS (—)	386

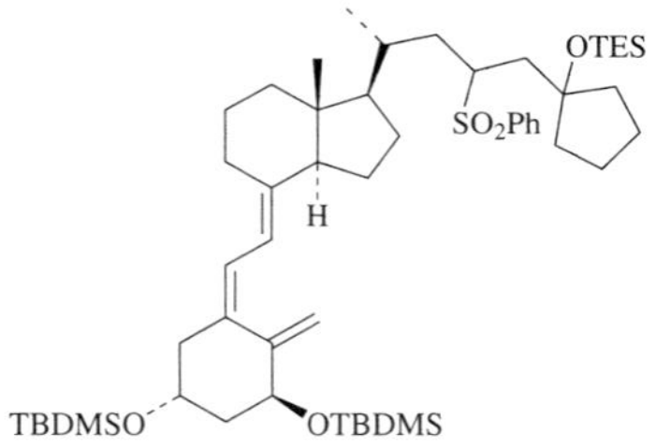

	Reactant	Conditions	Product(s)	Refs.
C_{59}	(structure: TBDPSO, SO_2Ph, OBn, OTBDPS)	Na/Hg, MeOH, rt	(structure: TBDPSO, OBn, OTBDPS) (91)	509
C_{61}	(structure: TBDMSO, TBDMSO, TBDMSO, PhO_2S, HO, OMOM, MeO, OMe)	Li/DTBB, THF, –78°, 2 h	(structure: TBDMSO, TBDMSO, TBDMSO, HO, OMOM, MeO, OMe) (78)	510

TABLE 5. REDUCTIVE DESULFONYLATION OF β-OXO-FUNCTIONALIZED SULFONES

Substrate	Conditions	Product(s) and Yield(s) (%)	Refs.
C_{7-16} 2-(SO_2R^2)-4-R^1-cyclohexanone	Reagents, THF/H_2O, rt	4-R^1-cyclohexanone	
		R^1 = H, R^2 = Me, Reagents = Al/Hg (89)	18
		R^1 = t-Bu, R^2 = Ph, Reagents = Zn, NH_4Cl (97)	511
C_8 1,2-bis(SO_2Me)-1,2-bis($CONH_2$)cyclobutane	Ra-Ni, H_2O	1,2-bis($CONH_2$)cyclobutane (60-70)	512
C_{9-21} $ArCOCH(R^1)SO_2R^2$	See table.	$ArCOCH_2R^1$	

Ar	R^1	R^2	Reagents	Solvent	Temp	Time	Yield (%)	Refs.
Ph	H	Me	$TiCl_4/Zn$	THF	rt	2 h	(75)	513
Ph	H	Me	Sm, $HgCl_2$	THF/H_2O	rt	5-6 h	(48)	44
Ph	Me	Me	Al/Hg	THF/H_2O	65°	2.5 h	(57)	514
Ph	Et	Me	$TiCl_4/Zn$	THF	rt	2 h	(76)	513
4-BrC_6H_4	H	Ph	Sm, $HgCl_2$	THF/H_2O	rt	5-6 h	(65)	44
Ph	H	Ph	$TiCl_4/Zn$	THF	rt	2 h	(83)	513
Ph	H	Ph	Sm, $HgCl_2$	THF/H_2O	rt	5-6 h	(65)	44
4-BrC_6H_4	H	4-Tol	Sm, $HgCl_2$	THF/H_2O	rt	5-6 h	(70)	44
4-BrC_6H_4	H	4-Tol	$TiCl_4/Zn$	THF	rt	2 h	(75)	513
Ph	Me	Ph	Sm, $HgCl_2$	THF/H_2O	rt	5-6 h	(66)	44
Ph	4-$ClC_6H_4CH_2$	Me	Sm, $HgCl_2$	THF/H_2O	rt	5-6 h	(52)	44
Ph	Bn	Me	Sm, $HgCl_2$	THF/H_2O	rt	5-6 h	(60)	44

Ph	Bn	Me	$TiCl_4$/Zn	THF	rt	2 h	(82)	513
Ph	Et	Ph	Sm, $HgCl_2$	THF/H_2O	rt	5-6 h	(66)	44
Ph	Et	Ph	$TiCl_4$/Zn	THF	rt	2 h	(74)	513
Ph	*c*-Pr	Ph	Sm, $HgCl_2$	THF/H_2O	rt	5-6 h	(67)	44
Ph	*n*-Pr	Ph	$TiCl_4$/Zn	THF	rt	2 h	(87)	513
Ph	4-$ClC_6H_4CH_2$	Ph	$TiCl_4$/Zn	THF	rt	2 h	(82)	513
Ph	4-$ClC_6H_4CH_2$	Ph	Sm, $HgCl_2$	THF/H_2O	rt	5-6 h	(62)	44
Ph	Bn	Ph	Sm, $HgCl_2$	THF/H_2O	rt	5-6 h	(66)	44
Ph	Bn	Ph	$TiCl_4$/Zn	THF	rt	2 h	(92)	513

C_{11}

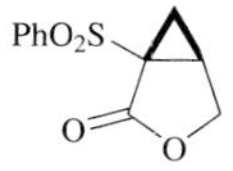

PhO_2S (bicyclic lactone) — Mg, MeOH, 50° — (bicyclic lactone) (75) — 115

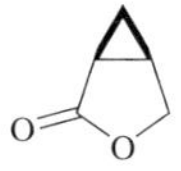

C_{11-19}

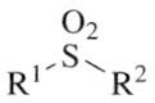

$R^1S(O_2)R^2$ — See table. — R^1H

R^1	R^2	Reagents	Solvent	Temp	Time		
MeC(O)CHMe	4-Tol	$Na_2S_2O_4$, $NaHCO_3$	DMF/H_2O	100°	24 h	(34)	47
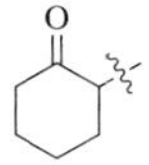 2-oxocyclohexyl	Ph	$Na_2S_2O_4$, $NaHCO_3$	DMF/H_2O	100°	24 h	(51)	47
"	Ph	SmI_2	THF/MeOH	–78° to rt	—	(88)	515
4-$BrC_6H_4C(O)CH_2$	Ph	$Na_2S_2O_4$, $NaHCO_3$	DMF/H_2O	100°	24 h	(44)	47
$PhC(O)CH_2$	4-Tol	$Na_2S_2O_4$, $NaHCO_3$	DMF/H_2O	100°	24 h	(65)	47
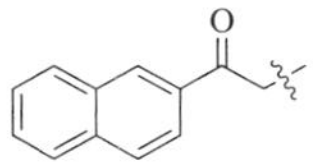 2-naphthyl-C(O)CH_2	4-Tol	$Na_2S_2O_4$, $NaHCO_3$	DMF/H_2O	100°	24 h	(44)	47

TABLE 5. REDUCTIVE DESULFONYLATION OF β-OXO-FUNCTIONALIZED SULFONES (*Continued*)

Substrate	Conditions	Product(s) and Yield(s) (%)	Refs.
C_{12}			
H, OH, O, Ms	Al/Hg, THF/H_2O, 65°, 2.5 h	H, OH, O (89)	514
$C_{12\text{-}17}$			
PhO_2S, R^1, R^2, R^3, O, O, *, *	See table.	R^1, R^2, R^3, O, O	

R^1	*	R^2	R^3	*	Reagents	Solvent	Temp	Time	Yield (%)	Refs.
H	—	H	Et	S	Al/Hg	THF/H_2O	reflux	1 h	(75)	516
H	—	H	Et	R	Na/Hg, Na_2HPO_4	MeOH	0°	3 h	(65)	517
H	—	Me	Me	R	Na/Hg, Na_2HPO_4	MeOH	rt	1.5 h	(90)	518, 519
H	S	Me	Me	R	Na/Hg, Na_2HPO_4	MeOH	rt	2 h	(87)	518, 519
H	—	Me	*i*-Pr	R	Na/Hg, Na_2HPO_4	MeOH	rt	1.5 h	(72)	518, 519
H	S	Me	*i*-Pr	R	Na/Hg, Na_2HPO_4	MeOH	rt	2 h	(72)	518, 519
Me	—	Me	*i*-Pr	R	Na/Hg, Na_2HPO_4	MeOH	rt	3 h	(68)	518, 519
Me	S	Me	*i*-Pr	R	Na/Hg, Na_2HPO_4	MeOH	rt	3 h	(92)	518, 519
H	S	Me	*n*-C_6H_{13}	R	Na/Hg, Na_2HPO_4	MeOH	rt	1 h	(93)	519

C_{12-22}

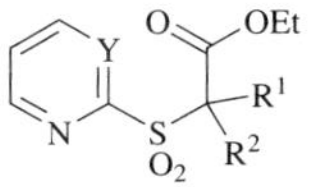

Reagents, reflux 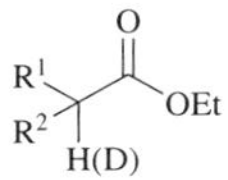 — 58, 175

R^1	R^2	Y	Reagents	Solvent	Time	
CH_2=CH(CH_2)$_2$	F	N	(*n*-Bu)$_3$SnH, AIBN	toluene	30 min	(81)
CH_2=CH(CH_2)$_2$	H	N	(*n*-Bu)$_3$SnH, AIBN	toluene	30 min	(81-91)
n-Bu	F	N	(*n*-Bu)$_3$SnD, AIBN	toluene	30 min	(91)[a]
n-Bu	F	N	(*n*-Bu)$_3$SnH, AIBN	toluene	1 h	(95)
n-Bu	H	N	(*n*-Bu)$_3$SnD, AIBN	toluene	2 h	(90)[a]
n-Bu	H	N	(*n*-Bu)$_3$SnH, AIBN	toluene	30 min	(95)
CH_2=CH(CH_2)$_3$	F	N	(*n*-Bu)$_3$SnH, AIBN	toluene	30 min	(91)
CH_2=CH(CH_2)$_3$	F	N	(*n*-Bu)$_3$SnCl (cat.), AIBN, PHMS, KF	toluene/H_2O	3 h	(84)
CH_2=CH(CH_2)$_3$	H	N	(*n*-Bu)$_3$SnCl (cat.), AIBN, PHMS, KF	toluene/H_2O	3 h	(89)
CH_2=CH(CH_2)$_3$	H	N	(*n*-Bu)$_3$SnH, AIBN	toluene	1 h	(84)
n-Bu	F	CH	(*n*-Bu)$_3$SnH, AIBN	toluene	28 h	(60)
n-Bu	H	CH	(*n*-Bu)$_3$SnH, AIBN	toluene	30 min	(91)
CH_2=CH(CH_2)$_4$	F	N	(*n*-Bu)$_3$SnH, AIBN	toluene	30 min	(85)
CH_2=CH(CH_2)$_4$	H	N	(*n*-Bu)$_3$SnH, AIBN	toluene	30 min	(88)
n-C_6H_{13}	F	N	(*n*-Bu)$_3$SnH, AIBN	toluene	30 min	(88)
n-C_6H_{13}	F	N	(*n*-Bu)$_3$SnCl (cat.), AIBN, PHMS, KF	toluene/H_2O	3 h	(86)
n-C_6H_{13}	H	N	(*n*-Bu)$_3$SnCl (cat.), AIBN, PHMS, KF	toluene/H_2O	3 h	(85)
$EtO_2C(CH_2)_5$	F	N	(*n*-Bu)$_3$SnH, AIBN	toluene	30 min	(88)
$EtO_2C(CH_2)_5$	H	N	(*n*-Bu)$_3$SnH, AIBN	toluene	30 min	(81)
$EtO_2C(CH_2)_5$	H	N	(*n*-Bu)$_3$SnCl (cat.), AIBN, PHMS, KF	toluene/H_2O	3 h	(88)
TBDMSO(CH_2)$_8$	F	N	(*n*-Bu)$_3$SnH, AIBN	toluene	30 min	(77)

TABLE 5. REDUCTIVE DESULFONYLATION OF β-OXO-FUNCTIONALIZED SULFONES (*Continued*)

Substrate	Conditions	Product(s) and Yield(s) (%)	Refs.
C_{12-28}			
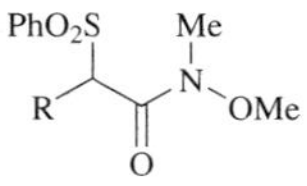 PhO_2S, Me, R, N, OMe, O	Na/Hg, Na_2HPO_4, MeOH, 0°, 1.5 h	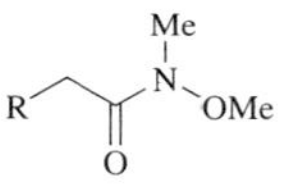 Me, R, N, OMe, O	520
R			
—CH_2CH_2—		(73)	
HC≡CCH_2		(71)	
H_2C=$CHCH_2$		(70)	
n-Bu		(69)	
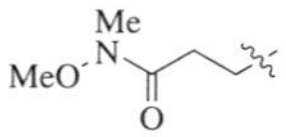 Me, MeO, N, O		(73)	
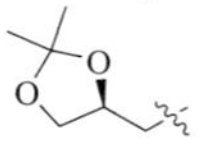 O, O		(74)	
Bn		(70)	
$PhOCH_2CH_2$		(82)	
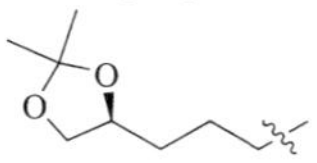 O, O		(72)	
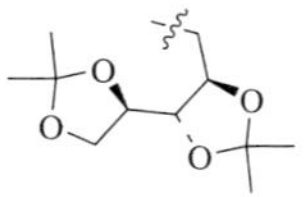 O, O, O, O		(73)	
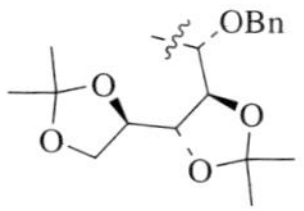 OBn, O, O, O, O		(70)	

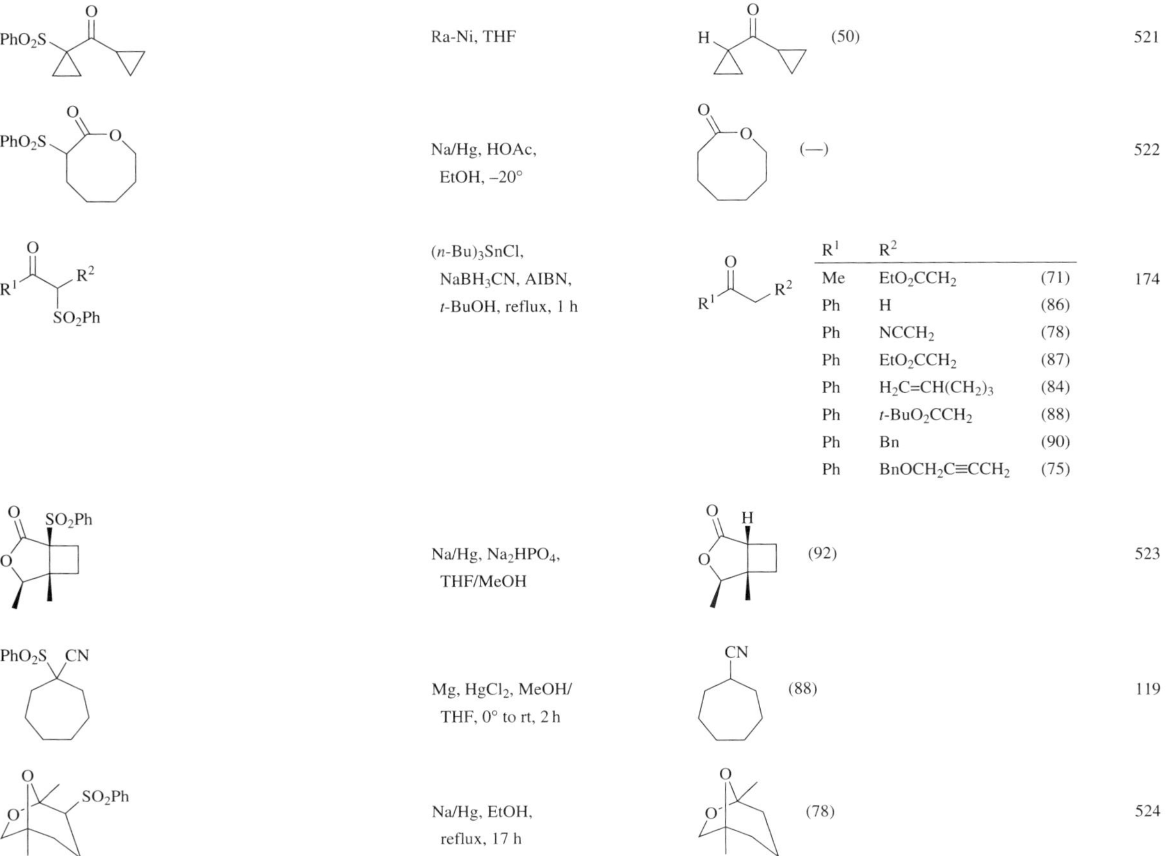

Substrate	Conditions	Product(s) and Yield(s) (%)	Refs.
C_{13}	Ra-Ni, THF	(50)	521
	Na/Hg, HOAc, EtOH, –20°	(—)	522
$C_{13\text{-}25}$	$(n\text{-Bu})_3SnCl$, $NaBH_3CN$, AIBN, *t*-BuOH, reflux, 1 h		174
C_{14}	Na/Hg, Na_2HPO_4, THF/MeOH	(92)	523
	Mg, $HgCl_2$, MeOH/THF, 0° to rt, 2 h	(88)	119
	Na/Hg, EtOH, reflux, 17 h	(78)	524

R^1	R^2	
Me	EtO_2CCH_2	(71)
Ph	H	(86)
Ph	$NCCH_2$	(78)
Ph	EtO_2CCH_2	(87)
Ph	$H_2C{=}CH(CH_2)_3$	(84)
Ph	$t\text{-}BuO_2CCH_2$	(88)
Ph	Bn	(90)
Ph	$BnOCH_2C{\equiv}CCH_2$	(75)

TABLE 5. REDUCTIVE DESULFONYLATION OF β-OXO-FUNCTIONALIZED SULFONES (*Continued*)

	Substrate	Conditions	Product(s) and Yield(s) (%)	Refs.
C_{14} 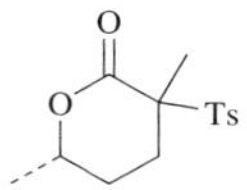	[structure: lactone with Ts]	1. Na/Hg, Na_2HPO_4, MeOH, 0°, 3 h 2. TsOH, C_6H_6, reflux	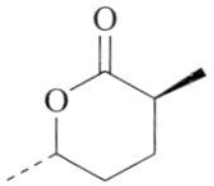 [structure] (33) 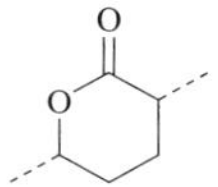 + [structure] (24)	517
C_{14-28} 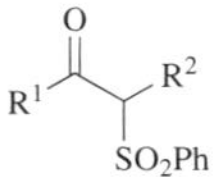	$R^1C(O)CH(R^2)SO_2Ph$	See table.	$R^1C(O)CH_2R^2$	525

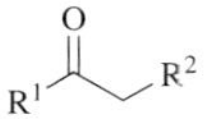

R^1	R^2	Reagents	Solvent	Temp	Time	
EtNH	*n*-Bu	Na/Hg, Na_2HPO_4	MeOH	rt	2 h	(93)
1-piperidinyl	Et	Na/Hg, Na_2HPO_4	MeOH	rt	2 h	(86)
1-azetidinyl	$H_2C{=}CH(CH_2)_3$	SmI_2	THF/MeOH	–78°	—	(92)
1-azetidinyl	*t*-BuO_2CCH_2	SmI_2	THF/MeOH	–78°	—	(91)
1-piperidinyl	*t*-BuO_2CCH_2	SmI_2	THF/MeOH	–78°	—	(96)
1-indolinyl	$H_2C{=}CH(CH_2)_2$	Na/Hg, Na_2HPO_4	MeOH	rt	2 h	(89)
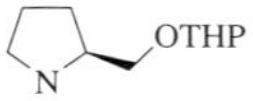 [structure: 2-(OTHP-methyl)pyrrolidin-1-yl]	Et	SmI_2	THF/MeOH	–78°	—	(97)
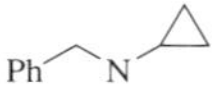 [structure: PhCH₂N(cyclopropyl)]	*n*-Bu	Na/Hg, Na_2HPO_4	MeOH	rt	2 h	(97)
1-azetidinyl	$MeO_2C(CH_2)_{10}$	SmI_2	THF/MeOH	–78°	—	(97)
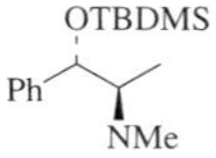 [structure: Ph, OTBDMS, NMe]	H	Na/Hg, Na_2HPO_4	MeOH	rt	2 h	(76)
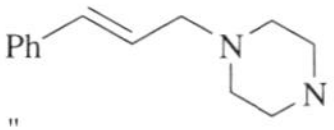 [structure: cinnamyl piperazinyl]	*n*-Bu	SmI_2	THF/MeOH	–78°	—	(90)
"	*i*-Bu	SmI_2	THF/MeOH	–78°	—	(85)

OTBDMS (pyrrolidine)	Bn	Na/Hg, Na_2HPO_4	MeOH	rt	2 h	(85)		
Ph, N, N (piperazine)	Bn	SmI_2	THF/MeOH	–78°	—	(73)		
C_{15} Ts, H, O, N		Na/Hg, Na_2HPO_4, MeOH, rt, 2 h				H, O, N (86)		256
H, OH, Ts, N, O		Na/Hg, Na_2HPO_4, MeOH, –15°, 1 h				H, OH, N, O (75)		526
C_{15-17} CN, SO_2Ar, O		SmI_2, THF/MeOH, –78°, 5 min				CN, O	Ar: 2,4-$Me_2C_6H_3$ (51); 2,4,6-$Me_3C_6H_2$ (59); 1-naphthyl (48)	527
C_{15-18} O, O, R, SO_2Ph		Na/Hg, MeOH, –50° to –20°, 2h				O, O, R		137

R	
(*Z*)-EtCH=CHCH$_2$	(90)
Ph	(77)
c-C_6H_{11}	(77)
n-C_6H_{13}	(90)
(*E*)-PhCH=CH	(63)[b]
$PhCH_2CH_2$	(71)

TABLE 5. REDUCTIVE DESULFONYLATION OF β-OXO-FUNCTIONALIZED SULFONES (*Continued*)

Substrate		Conditions			Product(s) and Yield(s) (%)	Refs.
$C_{15\text{-}24}$ $R^1CH(SO_2Ph)C(O)R^2$		Al/Hg			$R^1CH_2C(O)R^2$	
R^1	R^2	Solvent	Temp	Time		
n-Pr	$(OCH_2CH_2O)CHCH_2$	*n*-PrOH/H_2O	rt	3 h	(89)	528
n-C_5H_{11}	$HOCH_2CH_2$	THF/H_2O	reflux	2-6 h	(92)	529
$H_2C{=}CHCH_2CH_2$	$(OCH_2CH_2O)CHCH_2$	*n*-PrOH/H_2O	rt	3 h	(85)	528
(*Z*)-$MeCH_2CH{=}CHCH_2$	$HO(CH_2)_3$	THF/H_2O	reflux	2-6 h	(—)	529
n-C_5H_{11}	$HO(CH_2)_3$	THF/H_2O	reflux	2-6 h	(85)	529
n-C_5H_{11}	$MeCH(OH)CH_2$	THF/H_2O	reflux	2-6 h	(72)	529
n-C_7H_{15}	$HOCH_2CH_2$	THF/H_2O	reflux	2-6 h	(99)	529
(*Z*)-$MeCH_2CH{=}CHCH_2$	$MeC(O)(CH_2)_2C(O)$	THF/H_2O	reflux	2-6 h	(90)	530
n-C_5H_{11}	$MeC(O)(CH_2)_2C(O)$	THF/H_2O	reflux	2-6 h	(92)	530
n-C_6H_{13}	$(OCH_2CH_2O)CHCH_2$	*n*-PrOH/H_2O	rt	3 h	(92)	528
n-C_7H_{15}	$HOCH_2CHMe$	THF/H_2O	reflux	2-6 h	(70)	529
n-C_6H_{13}	$(OCH_2CH_2O)C(Me)CH_2$	*n*-PrOH/H_2O	rt	3 h	(—)	528
n-C_7H_{15}	$HO(CH_2)_4$	THF/H_2O	reflux	2-6 h	(85)	529
n-C_8H_{17}	$(OCH_2CH_2O)CHCH_2$	*n*-PrOH/H_2O	rt	3 h	(98)	528
n-C_7H_{15}	$HO(CH_2)_5$	THF/H_2O	reflux	2-6 h	(73)	529
t-$BuO_2C(CH_2)_7$	$(OCH_2CH_2O)CHCH_2$	*n*-PrOH/H_2O	rt	3 h	(—)	528
$C_{15\text{-}25}$ $CF_3C(O)CH(R)SO_2Ph$		See table.			$CF_3C(O)CH_2R$	531

R	Z:E	Reagents	Solvent	Temp	
$MeCH_2CH{=}CH(CH_2)_2$	70:30	Al/Hg	THF/H_2O	reflux	(54)
$MeCH_2CH{=}CH(CH_2)_2$	70:30	SmI_2, HMPA	THF	–20°	(58)
n-C_8H_{17}	—	Al/Hg	THF/H_2O	reflux	(62)
$MeCH_2CH{=}CH(CH_2)_4$	20:80	Al/Hg	THF/H_2O	reflux	(60)
$MeCH_2CH{=}CH(CH_2)_4$	90:10	Al/Hg	THF/H_2O	reflux	(68)
$MeCH_2CH{=}CH(CH_2)_4$	90:10	SmI_2, HMPA	THF	–20°	(73)
$Me(CH_2)_3CH{=}CH(CH_2)_3$	90:10	Al/Hg	THF/H_2O	reflux	(70)
$Ph(CH_2)_3$	—	Al/Hg	THF/H_2O	reflux	(68)
$H_2C{=}CH(CH_2)_9$	—	Al/Hg	THF/H_2O	reflux	(75)
$Me(CH_2)_{11}$	—	Al/Hg	THF/H_2O	reflux	(60)
$Me(CH_2)_3CH{=}CH(CH_2)_{10}$	95:5	Al/Hg	THF/H_2O	reflux	(68)

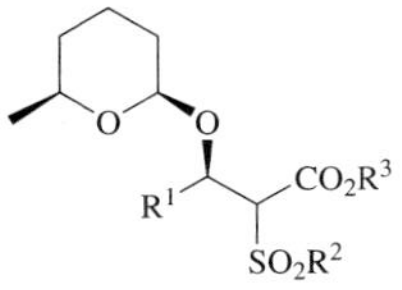

O, R^1, CO_2R^3, SO_2R^2

SmI_2, THF, –78°, 12 h

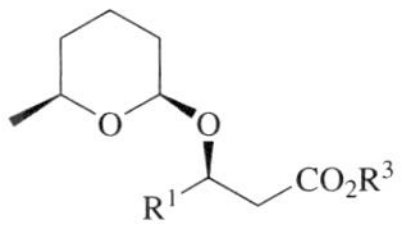

O, O, R^1, CO_2R^3

(—)

R^1	R^2	R^3	
i-Pr	Me	Et	532
Ph	Me	Me	
4-BrC_6H_4	Me	Et	
Ph	Me	Et	
4-$MeOC_6H_4$	Me	Me	
i-Pr	Ph	Et	
Ph	Me	t-Bu	
Ph	Ph	Me	
c-C_6H_{11}	Ph	Me	
i-Pr	Ph	t-Bu	
Ph	Ph	Et	
c-C_6H_{11}	Ph	Et	
Ph	Ph	t-Bu	
c-C_6H_{11}	Ph	t-Bu	

TABLE 5. REDUCTIVE DESULFONYLATION OF β-OXO-FUNCTIONALIZED SULFONES (*Continued*)

Substrate	Conditions	Product(s) and Yield(s) (%)	Refs.
C_{16} PhO_2S, CO_2Me	Na/Hg, Na_2HPO_4, MeOH, 0°, 1 h	CO_2Me (88)	33, 348
Ts, OH, O, N	Na/Hg, Na_2HPO_4, MeOH, rt, 2 h	OH, O, N (88)	257
O, O, SO_2Ph	Na/Hg, Na_2HPO_4, EtOH, –20°	O, O (—)	522
$C_{16\text{-}17}$ PhO_2S, O, O, n	Na/Hg	O, O, n n: 1 (60-70); 2 (60-70)	533
$C_{16\text{-}21}$ R, SO_2Ph, N, O, OMOM	Na/Hg, Na_2HPO_4, MeOH, rt, 2 h	R, N, O, OMOM R: Me (98); Ph (77)	534

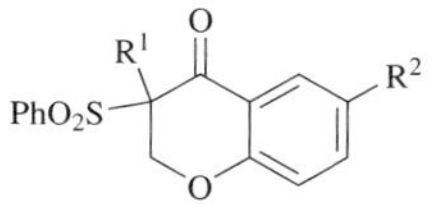

Substrate	Conditions	Product(s) and Yield(s) (%)	Refs.
PhO_2S, R^1 chromanone, R^2	Zn, AcOH, reflux, 1 h	R^1 chromanone, R^2	

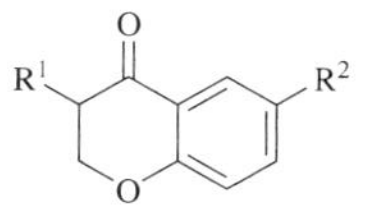

R^1	R^2	Yield (%)	Refs.
Me	H	(75)	535
Me	Me	(76)	535
H_2C=$CHCH_2$	H	(74)	535
Et	Me	(72)	535
HC≡CCH_2	Me	(74)	536
H_2C=$CHCH_2$	Me	(78)	535
Cl	Ph	(80)	536
H	Ph	(80)	536
Me	Ph	(82)	536
MeO	Ph	(84)	536
Me	Bn	(75)	535
H	(*E*)-PhCH=$CHCH_2$	(72)	535

C_{17}

Substrate	Conditions	Product(s) and Yield(s) (%)	Refs.
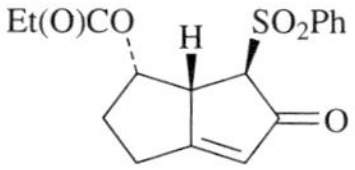 Et(O)CO, H, SO_2Ph, =O	Zn, NH_4Cl, THF/H_2O, rt	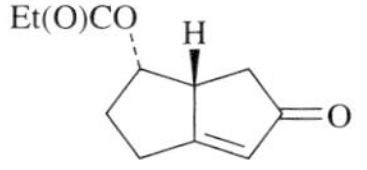 Et(O)CO, H, =O (94)	537
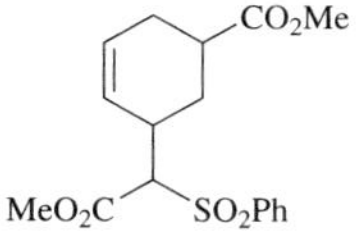 CO_2Me, MeO_2C, SO_2Ph	Na/Hg, Na_2HPO_4, MeOH, 0°, 1 h	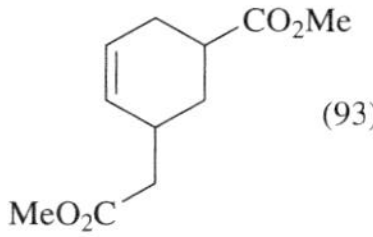 CO_2Me, MeO_2C (93)	33
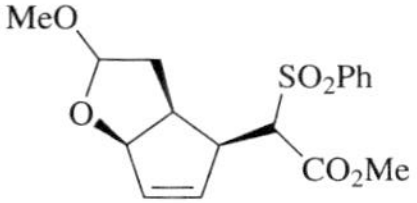 MeO, O, SO_2Ph, CO_2Me	Mg, KH_2PO_4, MeOH, rt, 3 h	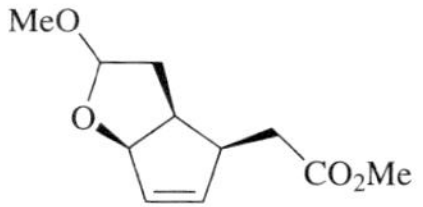 MeO, O, CO_2Me (—)	538

TABLE 5. REDUCTIVE DESULFONYLATION OF β-OXO-FUNCTIONALIZED SULFONES (*Continued*)

Substrate	Conditions	Product(s) and Yield(s) (%)	Refs.
C_{17} 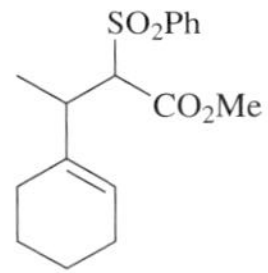 SO_2Ph CO_2Me	NH_3/Ca, Et_2O, –78° to –33°, 15 min	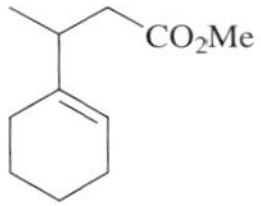 CO_2Me (100)	348
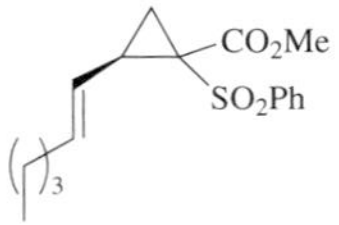 CO_2Me SO_2Ph $(\,)_3$	Na/Hg, Na_2HPO_4, rt	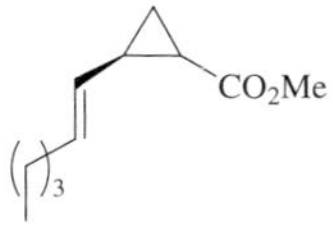 CO_2Me $(\,)_3$ (—)	539
$C_{17\text{-}18}$ 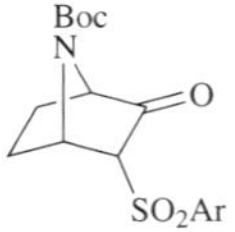 Boc N O SO_2Ar	Al/Hg, THF/H_2O	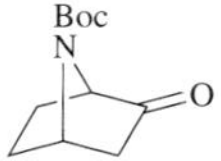 Boc N O; Ar: Ph (60); 4-Tol (60)	540; 541
$C_{17\text{-}21}$ 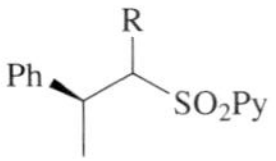 R Ph SO_2Py	Zn, NH_4Cl	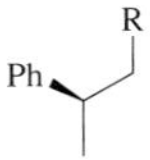 R Ph; R: CO_2Et (—); C(O)Ph (—)	542
$C_{17\text{-}29}$ 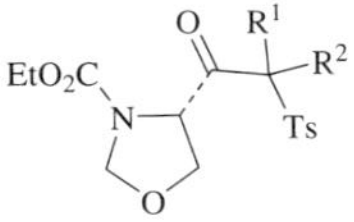 EtO_2C N O O R^1 R^2 Ts	Al/Hg, THF/H_2O, reflux	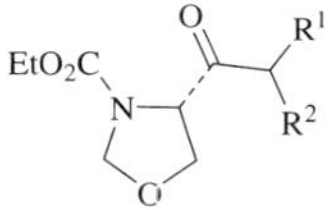 EtO_2C N O O R^1 R^2; R^1 / R^2: Me / Me (78); H / n-$C_{14}H_{29}$ (68)	543

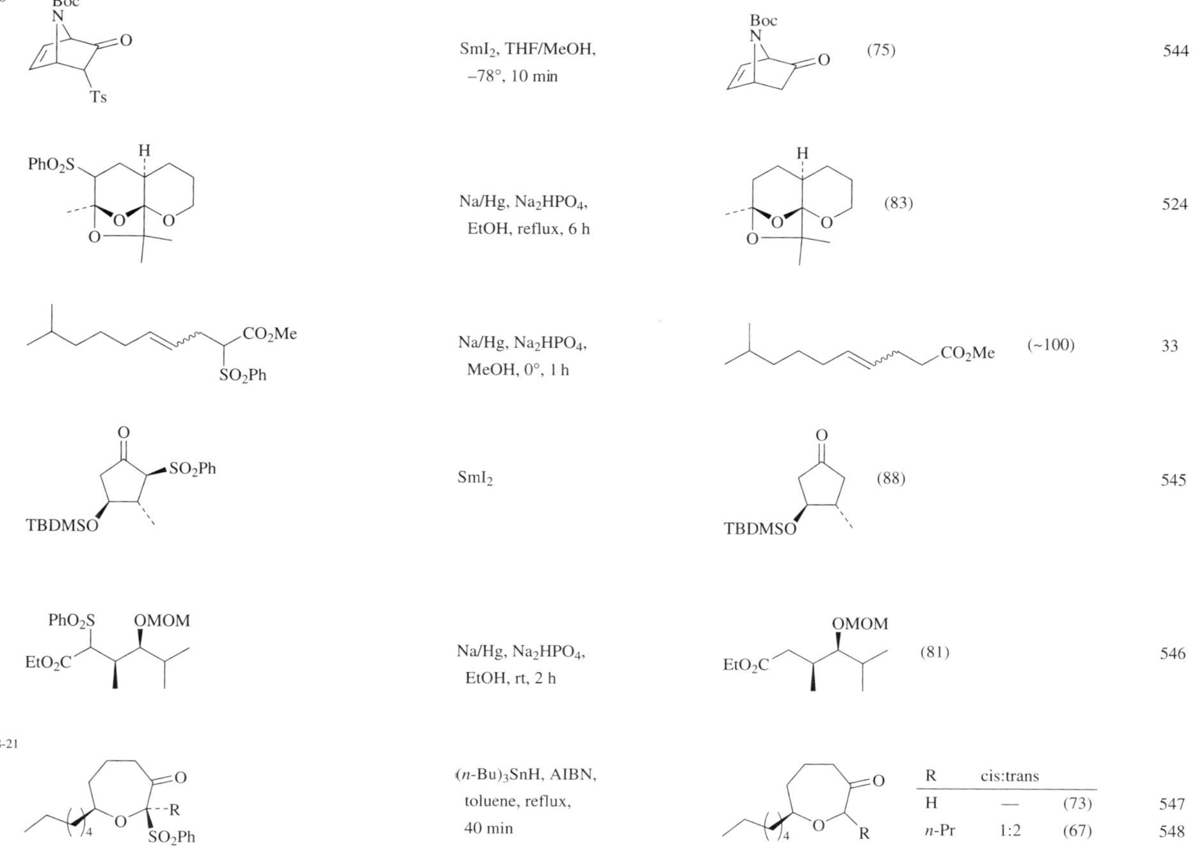

Substrate	Conditions	Product(s) and Yield(s) (%)	Refs.
C_{18}	SmI_2, THF/MeOH, –78°, 10 min	(75)	544
	Na/Hg, Na_2HPO_4, EtOH, reflux, 6 h	(83)	524
	Na/Hg, Na_2HPO_4, MeOH, 0°, 1 h	(~100)	33
	SmI_2	(88)	545
	Na/Hg, Na_2HPO_4, EtOH, rt, 2 h	(81)	546
C_{18-21}	$(n\text{-}Bu)_3SnH$, AIBN, toluene, reflux, 40 min	R = H, cis:trans —, (73)	547
		R = n-Pr, cis:trans 1:2, (67)	548

TABLE 5. REDUCTIVE DESULFONYLATION OF β-OXO-FUNCTIONALIZED SULFONES (*Continued*)

	Substrate	Conditions	Product(s) and Yield(s) (%)	Refs.
C_{19} 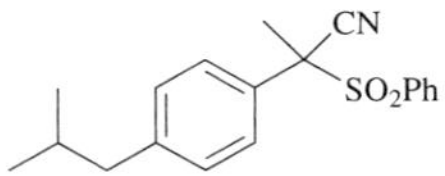	CN, SO_2Ph	Ra-Ni	CN (—)	184
	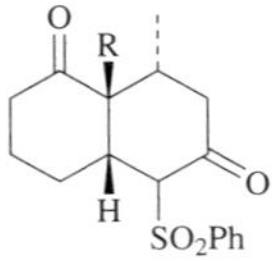 O, R, O, H, SO_2Ph; R = CO_2Me	Al/Hg, THF/H_2O, 65°, 3 h	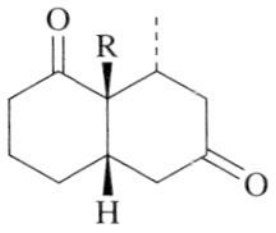 O, R, O, H (48)	549
	MeO_2C, H, O, SO_2Ph, H	Al/Hg, THF/H_2O, 65°, 3 h	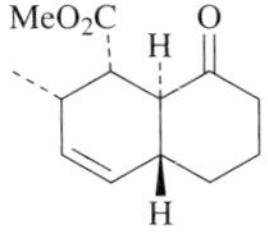 MeO_2C, H, O, H (—)	501
	O, H, O, H, O, O, H, SO_2Ph	Al/Hg, THF/H_2O, 65°, 3 h	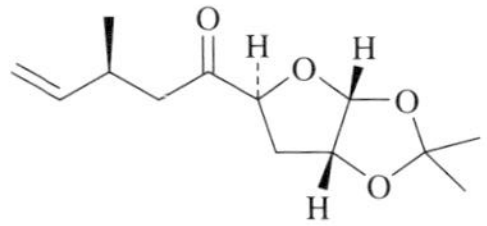 O, H, O, H, O, O, H (86)	550
	CO_2Et, SO_2Ph, OH, N, Boc	Na/Hg, EtOH, AcOH, –10°, 5 h	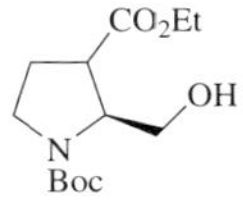 CO_2Et, OH, N, Boc (55)	551
	SO_2Ph, TBDMSO, CO_2Et, O	Al/Hg, THF/H_2O, rt	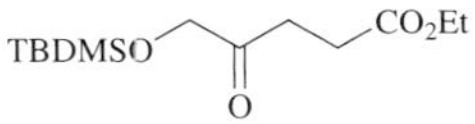 TBDMSO, CO_2Et, O (97)	552

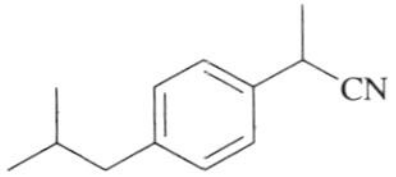

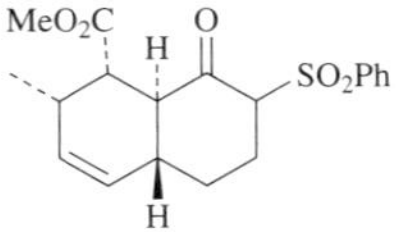

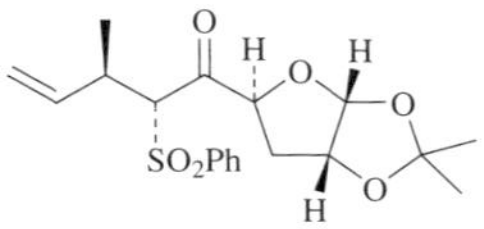

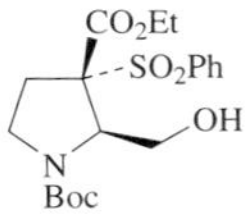

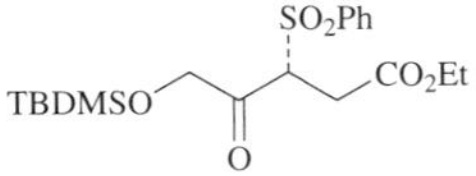

C_{19-21}

Ph$_3$SnH, AIBN, toluene, reflux

R^1	R^2	Time		
H	H	10 min	(79)	59
H	Me	5 min	(91)	
Me	Me	5 min	(91)	

Al/Hg, Na_2HPO_4, MeOH

R		
EtO_2CCH_2	(87)	553
MeO_2C–CH(Me)–NH–C(O)–CH_2–	(72)	

C_{19-33}

SmI_2, THF, rt

R^1	R^2	R^3	R^4	R^5	Additive		
H	MeO	H_2C=CHCH$_2$	MR	Bn	DMPU	(—)	554
(TMS)C≡C	H	Me	n-$C_8F_{17}(CH_2)_2$	n-Pr	—	(88)	151
Br	H	Bn	n-$C_8F_{17}(CH_2)_2$	n-Pr	—	(98)	151
4-Tol	H	Me	n-$C_8F_{17}(CH_2)_2$	n-Pr	—	(94)	151
thiophen-2-yl	H	Bn	n-$C_8F_{17}(CH_2)_2$	n-Pr	—	(82)	151
pyridin-3-yl	H	Bn	n-$C_8F_{17}(CH_2)_2$	n-Pr	—	(72)	151

C_{20}

Al/Hg, THF/H_2O, rt, overnight

(—) 555

TABLE 5. REDUCTIVE DESULFONYLATION OF β-OXO-FUNCTIONALIZED SULFONES (*Continued*)

	Substrate	Conditions	Product(s) and Yield(s) (%)	Refs.
C_{20}	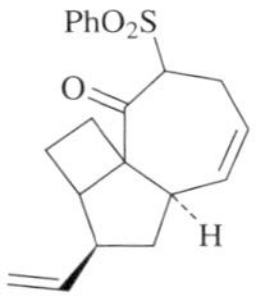 PhO_2S, O, H	Na/Hg, MeOH, –50°, 5 h	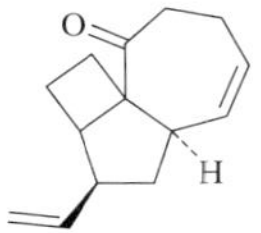 O, H (80)	556
	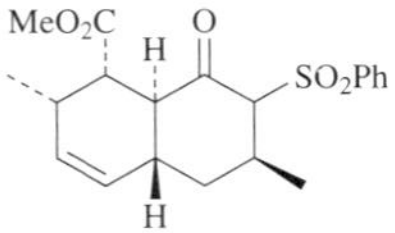 MeO_2C, O, H, SO_2Ph, H	Al/Hg, THF/H_2O, 3 h	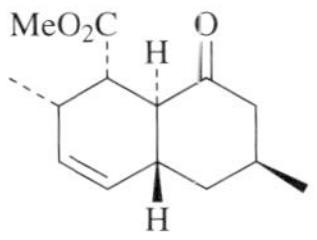 MeO_2C, H, O, H (—)	501
	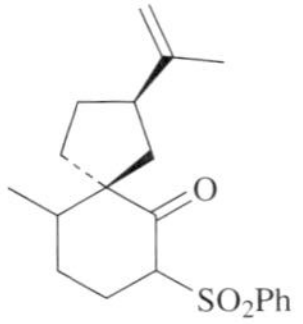 O, SO_2Ph	Al/Hg, THF/H_2O	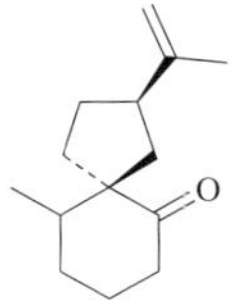 O (—)	557
	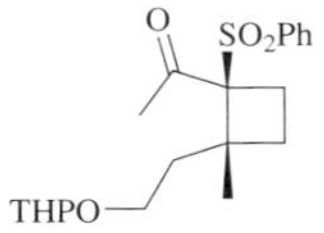 O, SO_2Ph, THPO	Na/Hg, Na_2HPO_4, MeOH	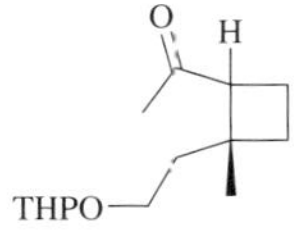 O, H, THPO Temp / Time / cis:trans –40° / 3 h / 50:50 / (92) rt / 0.5 h / 9:91 / (—)	523
	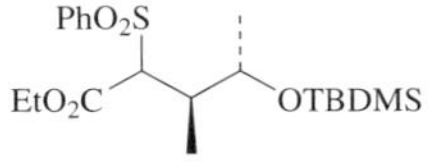 PhO_2S, EtO_2C, OTBDMS	NH_3/Li, Et_2O, –78° to –30°, 20 min	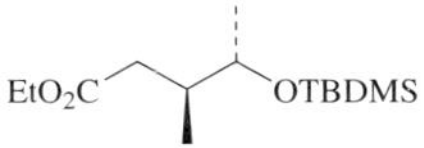 EtO_2C, OTBDMS (—)	558

C_{20-21}

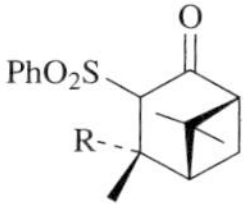

PhO_2S, R (structure) — NH_3/Li, THF, –78°, 30 min

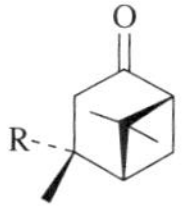

R	
$H_2C=CHCH_2CH_2$	(84)
$H_2C=C(Me)CH_2CH_2$	(88)
$H_2C=CH(CH_2)_2CH_2$	(79)

339

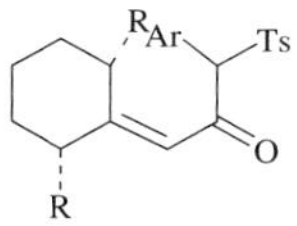

R, Ar, Ts (structure) — Zn, THF, NH_4Cl, rt, 2 h

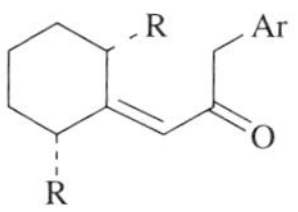

R	Ar	
H	3-furyl	(80)
H	3-thienyl	(80)
Me	3-furyl	(42)

345

C_{20-27}

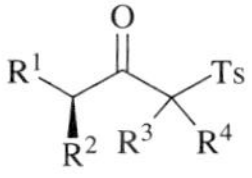

R^1, R^2, R^3, R^4, Ts (structure) — Al/Hg, THF/H_2O, reflux, 2 h — R^1, R^2, R^3, R^4 (structure)

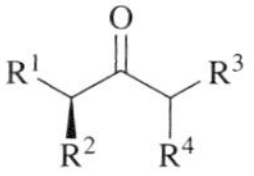

559

R^1	R^2	R^3	R^4	
i-Bu	EtO_2CNH	H	$H_2C=CHCH_2$	(65)
Bn	AcO	H	Et	(56)
i-Bu	EtO_2CNH	Et	Et	(72)
Bn	AcO	H	$H_2C=CHCH_2$	(76)
Bn	AcO	H	EtO_2CCH_2	(75)
Bn	EtO_2CNH	H	$H_2C=CHCH_2$	(60)
i-Bu	EtO_2CNH	$H_2C=CHCH_2$	$H_2C=CHCH_2$	(50)
Me	EtO_2CNH	H	*t*-$BuO_2C(CH_2)_4$	(68)
Bn	EtO_2CNH	H	EtO_2CCH_2	(74)
Bn	AcO	$H_2C=CHCH_2$	$H_2C=CHCH_2$	(86)
Bn	EtO_2CNH	H	$EtO_2C(CH_2)_2$	(70)
Bn	AcO	H	Bn	(80)
Bn	EtO_2CNH	H	Bn	(50)

TABLE 5. REDUCTIVE DESULFONYLATION OF β-OXO-FUNCTIONALIZED SULFONES (*Continued*)

	Substrate	Conditions	Product(s) and Yield(s) (%)	Refs.
C_{21}	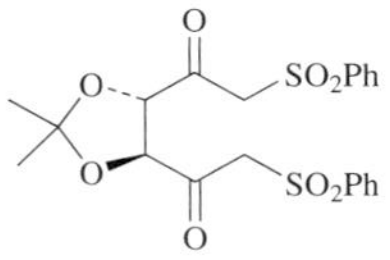 SO_2Ph, SO_2Ph, O, O	Al/Hg, THF/H_2O, 0°, 4 h	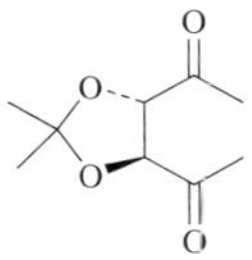 O, O (73)	560
	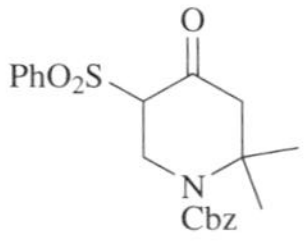 PhO_2S, O, N, Cbz	(*n*-Bu)$_3$SnH, AIBN, toluene, reflux, 10 min	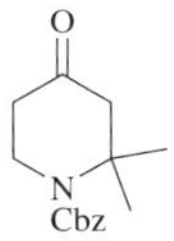 O, N, Cbz (—)	561
	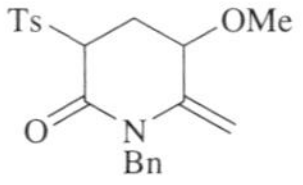 Ts, OMe, O, N, Bn	Na/Hg, Na_2HPO_4, MeOH, rt, 2 h	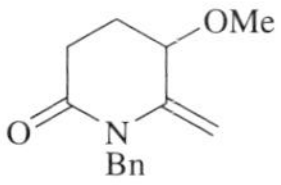 OMe, O, N, Bn (90)	562
	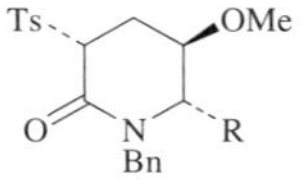 Ts, OMe, O, N, Bn, R	Na/Hg, Na_2HPO_4, MeOH, rt, 2 h	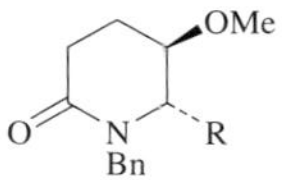 OMe, O, N, Bn, R R: MeO (87); $HOCH_2$ (—)	257
	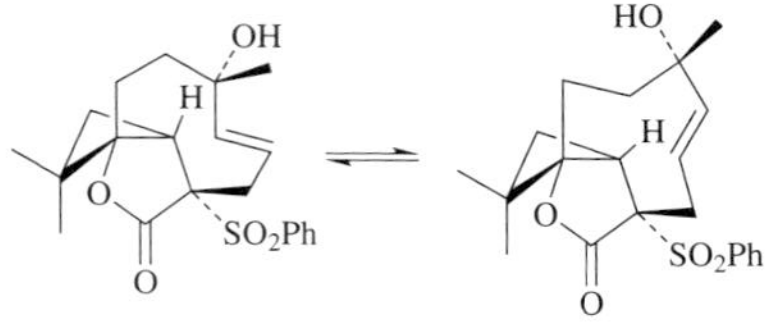 OH, H, O, O, SO_2Ph ⇌ HO, H, O, O, SO_2Ph	Na/Hg, KH_2PO_4	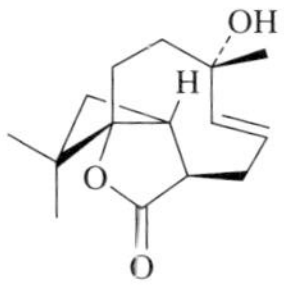 OH, H, O, O (74)	563
	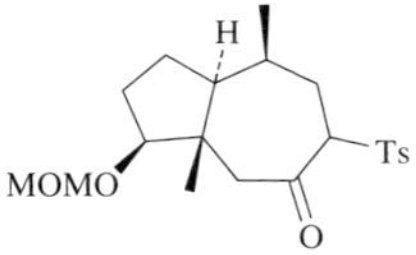 H, MOMO, O, Ts	Na/Hg, NaH_2PO_4, MeOH	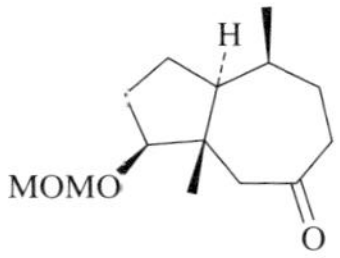 H, MOMO, O (39)	564

Substrate	Conditions	Product(s) and Yield(s) (%)	Refs.
C_{21-22} TBDMSO, Ts, Pr-*n*, O, n	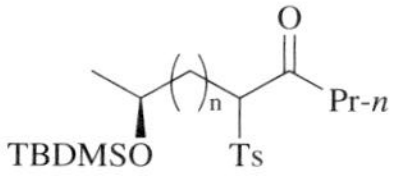 Na/Hg, Na_2HPO_4, MeOH, 0°, 3.5 h	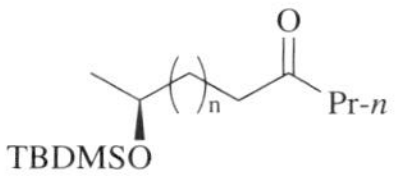 TBDMSO, Pr-*n*, O, n n: 1 (85); 2 (84)	565
C_{22} O, SO_2(naphthyl-1), CN	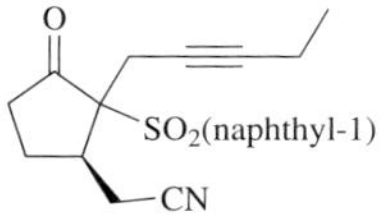 SmI_2, THF/MeOH, –78°, 5 min	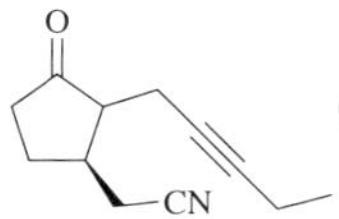 O, CN (100)	527
O, O, OMe, O, O, Ph, SO_2Ph	$(n\text{-Bu})_3SnH$, AIBN, toluene, reflux, 15 min	O, O, OMe, O, O, Ph (83)	59
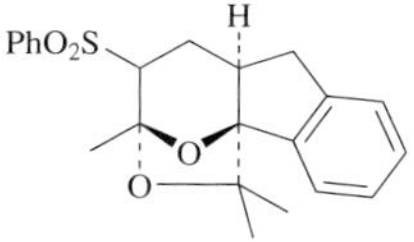 PhO_2S, H, O, O	Na/Hg, EtOH, reflux, 12 h	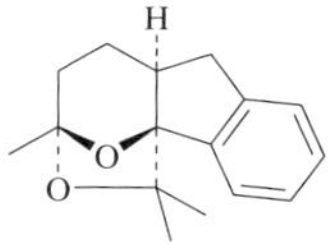 H, O, O (68)	435
C_{22-24} PhO_2S, O, R^1, R^2, O	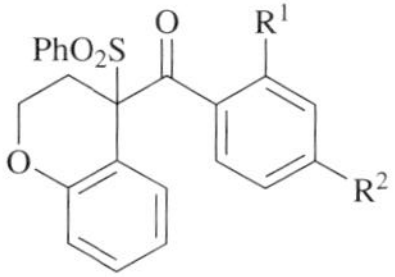 Ra-Ni (W-3), EtOAc, rt, 4 h	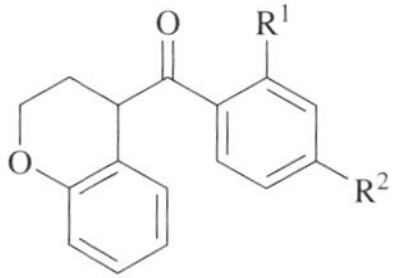 O, R^1, R^2, O R^1 / R^2: HO / H (98); MeO / H (99); MeO / MeO (99)	566
C_{22-30} O, Ts, R, N, CHO	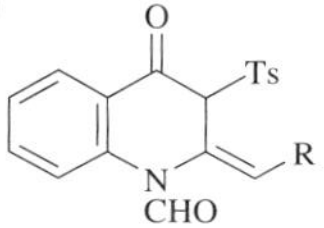 Al/Hg, MeOH/THF, rt, 1.5 h	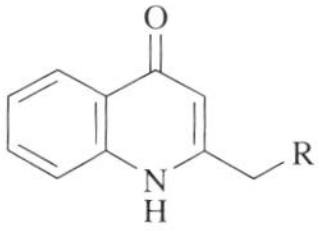 O, R, N, H R: *n*-Bu (68); 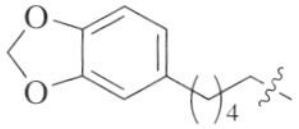 O, O, 4 (68)	567, 568

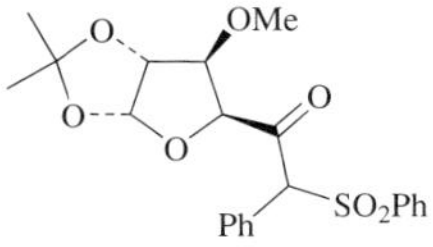

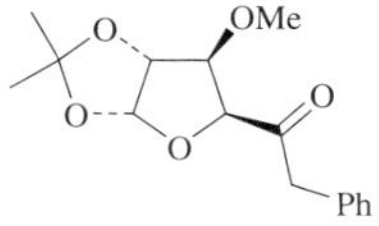

TABLE 5. REDUCTIVE DESULFONYLATION OF β-OXO-FUNCTIONALIZED SULFONES (*Continued*)

Substrate	Conditions	Product(s) and Yield(s) (%)	Refs.
C_{23}			
PhO_2S, t-Bu, t-Bu, Si, O, O, O, O, OH, OH	Na/Hg, Na_2HPO_4, MeOH, rt, 2 h	t-Bu, t-Bu, Si, O, O, O, O, OH, OH (41)	132
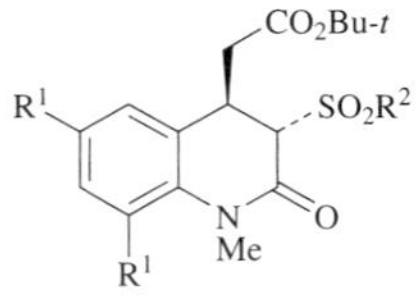 R^1, R^1, CO_2Bu-t, SO_2R^2, N, Me, O	SmI_2, LiCl, THF, rt	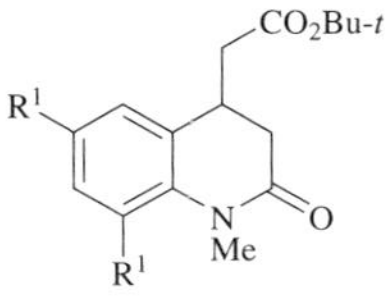 R^1, R^1, CO_2Bu-t, N, Me, O R^1 / R^2 / Time: F / Bn / 18 h (45) H / Bn / 18 h (57) H / MR / 20 h (35)	150
C_{23-27}			
Ph, EtO_2C, Ts, R	Mg, $HgCl_2$, MeOH, rt	Ph, EtO_2C, R (—) R: H Me Et (*E*)-MeCH=$CHCH_2$	116
C_{24}			
CO_2Et, SO_2Ph, OTHP, N, Boc	Na/Hg, AcOH, EtOH, –10°, 5 h	CO_2Et, OTHP, N, Boc (55)	551
C_{24-35}			
PhO_2S, R^1, O, Y, OR^2, NHCbz	Na/Hg, Na_2HPO_4, MeOH, rt	R^1, O, Y, OR^2, NHCbz	569

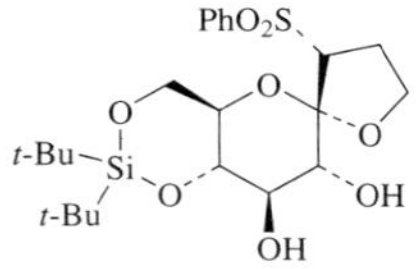

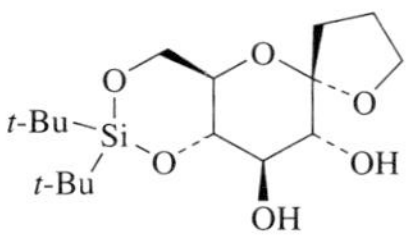

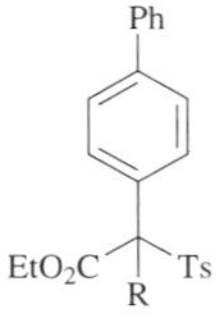

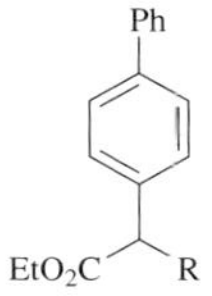

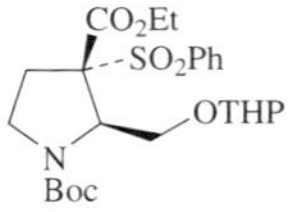

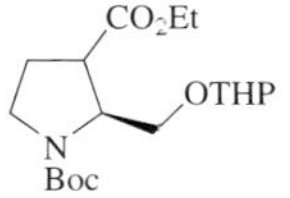

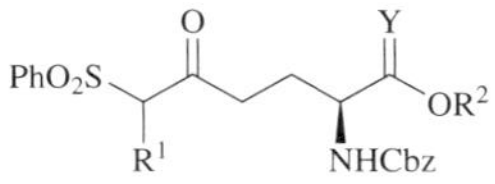

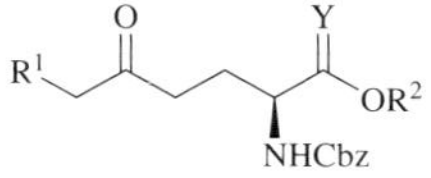

R^1	R^2	Y	Time	
H	*t*-Bu	O	75 min	(58)
H	EtO(Me)CH	H_2	45 min	(63)
H	TBDMS	H_2	50 min	(65)
n-C_9H_{19}	*t*-Bu	O	75 min	(85)
n-C_9H_{19}	EtO(Me)CH	H_2	75 min	(92)
n-C_9H_{19}	TBDMS	H_2	75 min	(92)

C_{25}

Substrate	Conditions	Product(s) and Yield(s) (%)	Refs.
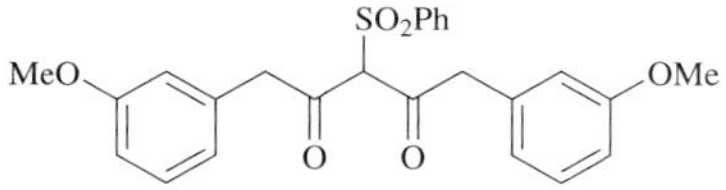 MeO, SO_2Ph, O, O, OMe	Zn	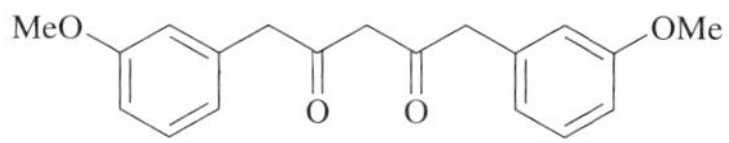 MeO, O, O, OMe (—)	570
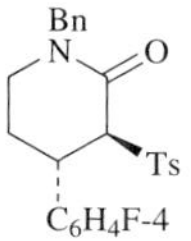 Bn, N, O, Ts, C_6H_4F-4	Na/Hg, Na_2HPO_4, MeOH, rt, 2 h	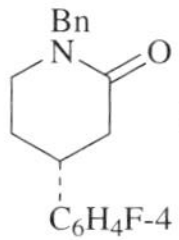 Bn, N, O, C_6H_4F-4 (90)	255
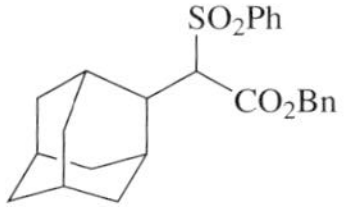 SO_2Ph, CO_2Bn	Mg, $HgCl_2$, EtOH, rt, 2 h	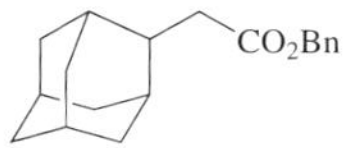 CO_2Bn (98)	114
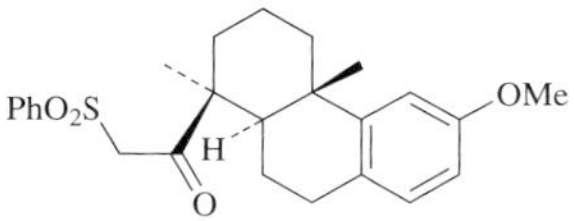 PhO_2S, O, H, OMe	(*n*-Bu)$_3$SnH, AIBN, toluene, reflux, 30 min	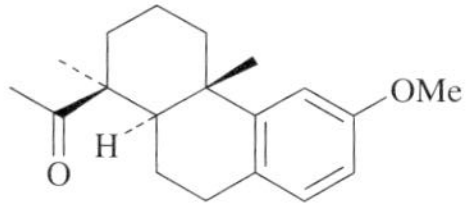 O, H, OMe (96)	59
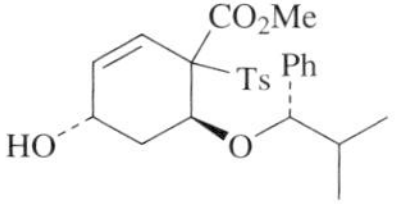 CO_2Me, Ts, Ph, HO, O	Al/Hg, THF/H_2O, 110°, 6 h	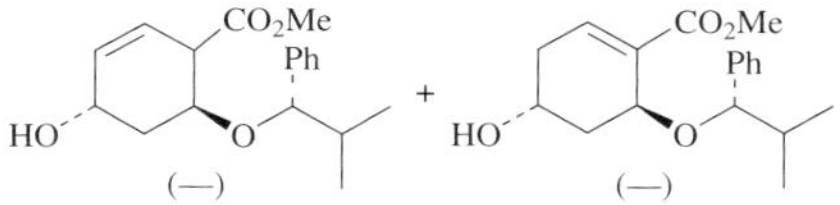 CO_2Me, Ph, HO, O (—) + CO_2Me, Ph, HO, O (—)	571

TABLE 5. REDUCTIVE DESULFONYLATION OF β-OXO-FUNCTIONALIZED SULFONES (*Continued*)

Substrate	Conditions	Product(s) and Yield(s) (%)	Refs.
C_{25} 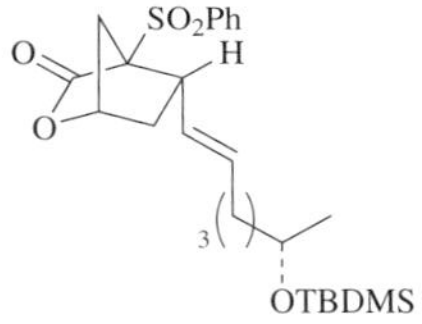 (structure: SO_2Ph, H, OTBDMS)	Na/Hg, $B(OH)_3$, MeOH, rt, 1 h	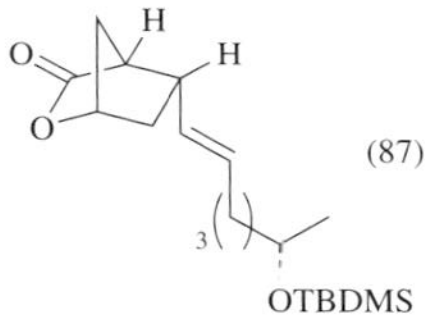 (structure: H, H, OTBDMS) (87)	572
C_{25-29} 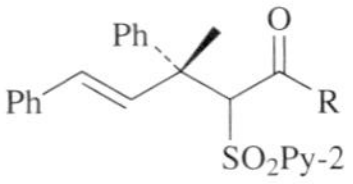 (structure: Ph, Ph, O, R, SO_2Py-2)	Zn, NH_4Cl, THF/H_2O, rt	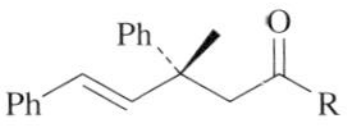 (structure: Ph, Ph, O, R) R: EtO (92); Ph (99)	323
C_{25-32} 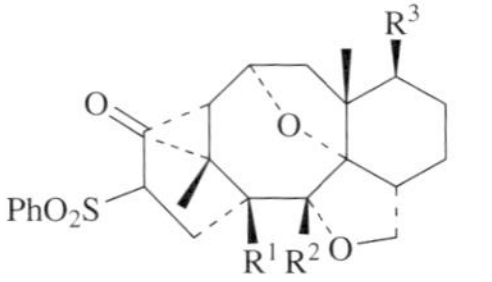 (structure: O, PhO_2S, O, R^1, R^2, R^3, O)	NH_3/Na, THF, –78° to –33°, 1 h	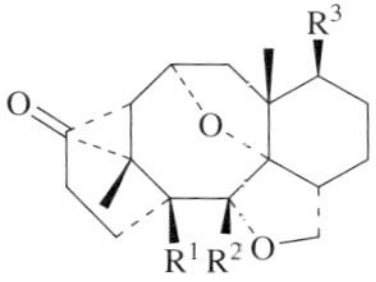 (structure: O, O, R^1, R^2, R^3, O)	573, 574
R^1 = H, R^2 = H, R^3 = H		(95)	
R^1 = HO, R^2 = H, R^3 = H		(100)	
R^1 = H, R^2 = MeO, R^3 = TBDMSO		(—)	
C_{25-41} 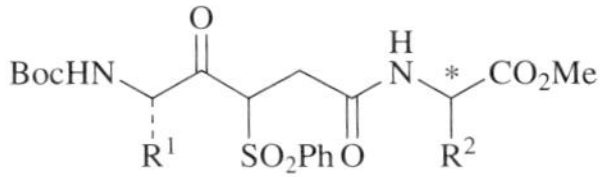 (structure: BocHN, O, R^1, SO_2Ph, O, H, N, *, CO_2Me, R^2)	SmI_2, THF/MeOH, –78° to rt, 20 min	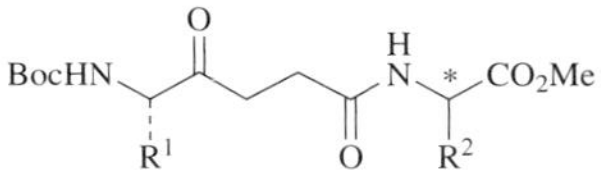 (structure: BocHN, O, R^1, O, H, N, *, CO_2Me, R^2)	575

R^1	R^2	*	
i-Pr	*i*-Pr	S	(66)
i-Pr	*i*-Pr	R	(70)
Bn	*i*-Pr	S	(43)
Bn	*i*-Pr	R	(47)
i-Pr	Bn	S	(61)
i-Pr	Bn	R	(73)
i-Pr	4-$HOC_6H_4CH_2$	S	(55)
Bn	Bn	S	(44)
Bn	Bn	R	(33)
Bn	4-$HOC_6H_4CH_2$	S	(36)
4-$PMBOC_6H_4CH_2$	*i*-Pr	S	(41)
4-$PMBOC_6H_4CH_2$	*i*-Pr	R	(34)
4-$PMBOC_6H_4CH_2$	Bn	S	(54)
4-$PMBOC_6H_4CH_2$	Bn	R	(27)
4-$PMBOC_6H_4CH_2$	4-$HOC_6H_4CH_2$	S	(47)

C_{26}

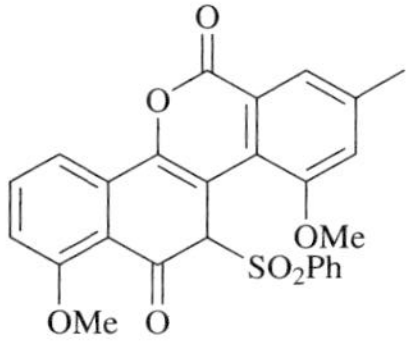

Substrate	Conditions	Product (yield)	Ref.
[structure: OMe, O, SO_2Ph, OMe, O]	$(n\text{-}Bu)_3SnCl$, AIBN, $NaBH_3CN$, *t*-BuOH, reflux, 1 h	[structure: OMe, OH, OMe, O] (94)	576
[structure: Ts, OMe, O, N, Bn, Ph]	Na/Hg, Na_2HPO_4, MeOH, rt, 2 h	[structure: OMe, O, N, Bn, Ph] (90)	562
[structure: N, H, O, H, Ts]	Na/Hg, Na_2HPO_4, MeOH, rt, 12 h	[structure: N, H, O, H] (95)	577

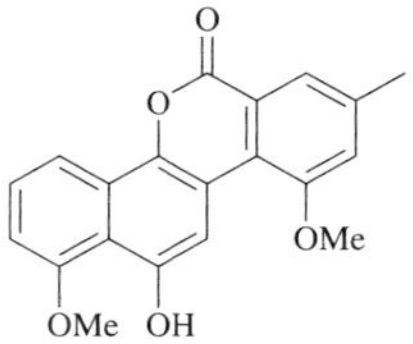

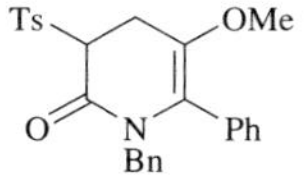

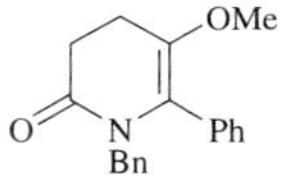

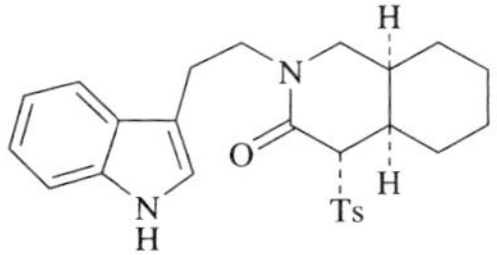

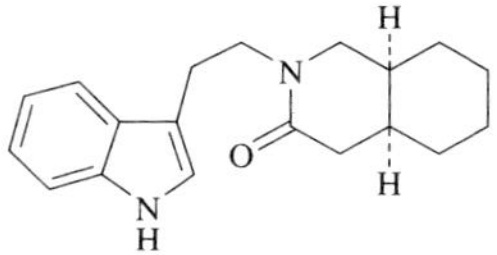

TABLE 5. REDUCTIVE DESULFONYLATION OF β-OXO-FUNCTIONALIZED SULFONES (*Continued*)

Substrate	Conditions	Product(s) and Yield(s) (%)	Refs.
C_{26} t-BuO_2C, SO_2Bn, N Me, O	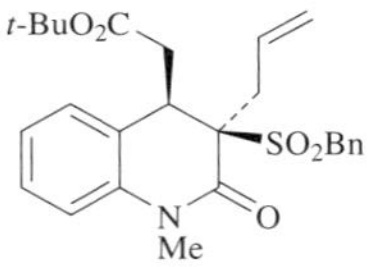 SmI_2, LiCl, THF, rt, 16 h	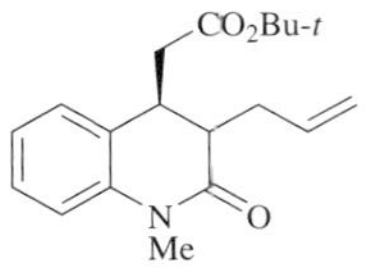 CO_2Bu-t, N Me, O; Additive / cis:trans: — 50:50 (—); t-BuOH 83:17 (79)	150
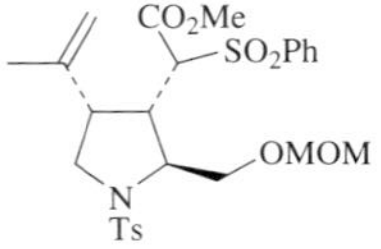 CO_2Me, SO_2Ph, OMOM, N Ts	Na/Hg, Na_2HPO_4, MeOH, 0°	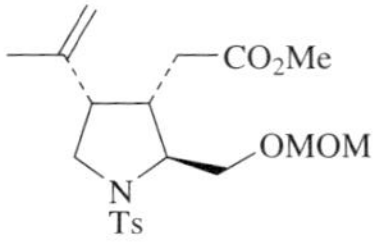 CO_2Me, OMOM, N Ts (60)	578
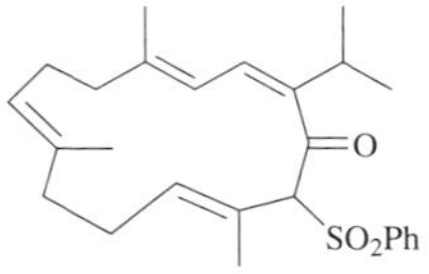 O, SO_2Ph	SmI_2, THF	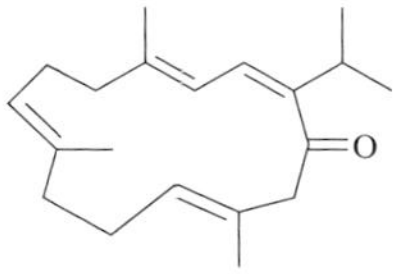 O (90)	579
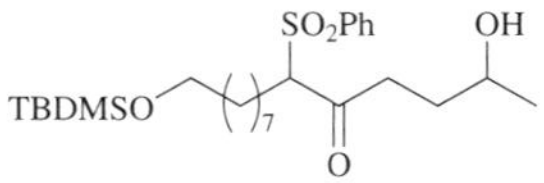 TBDMSO, ( )$_7$, SO_2Ph, O, OH	Na/Hg, MeOH, rt, 24 h	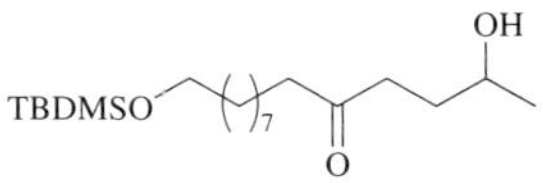 TBDMSO, ( )$_7$, O, OH (66)	580
C_{27} TBDPSO, SO_2Ph, CN	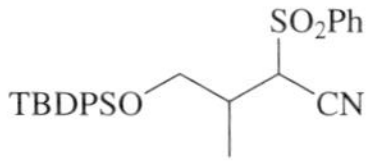 Mg, $HgCl_2$, THF/MeOH, 0° to rt, 2 h	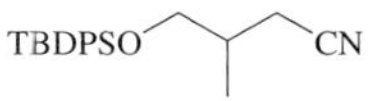 TBDPSO, CN (96)	119
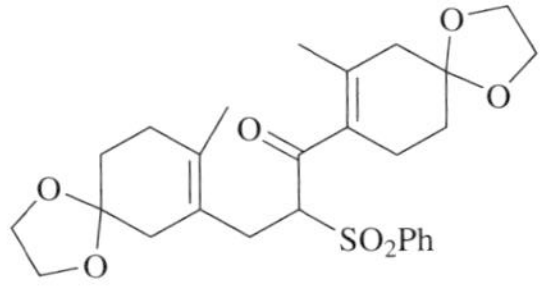 O, O, O, O, O, SO_2Ph	Al/Hg, THF/H_2O, 0°, 30 min	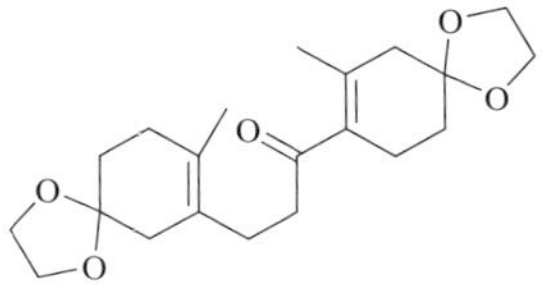 O, O, O, O, O (68)	581

Reactant	Conditions	Product(s) and Yield(s) (%)	Refs.
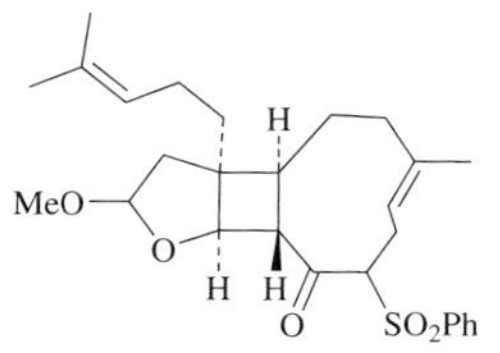 MeO, O, H, H, O, SO_2Ph	Al/Hg, THF/H_2O, rt, 16 h	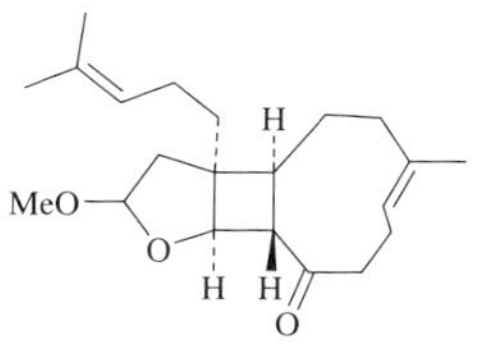 MeO, O, H, H, O (88)	582
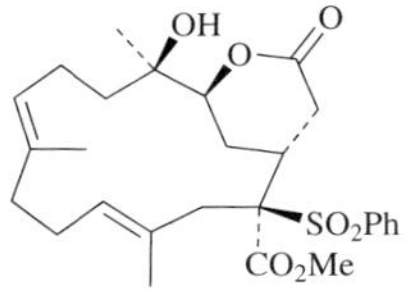 OH, O, O, SO_2Ph, CO_2Me	Na/Hg, Na_2HPO_4, THF/MeOH, –78°, 30 min	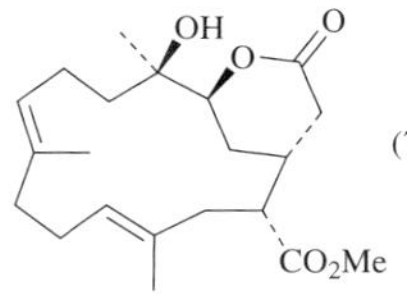 OH, O, O, CO_2Me (76)	583
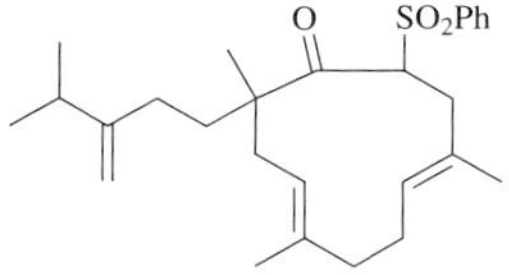 O, SO_2Ph	Al/Hg, THF/H_2O, 70°, 5 h	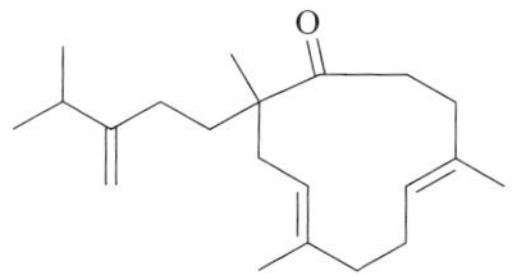 O (96)	584
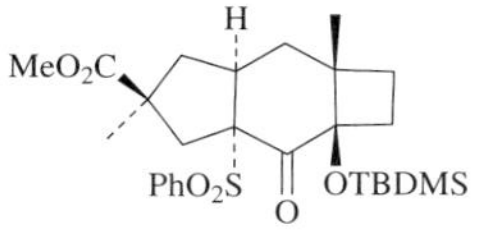 MeO_2C, H, PhO_2S, O, OTBDMS	Al/Hg, Na_2HPO_4, MeOH	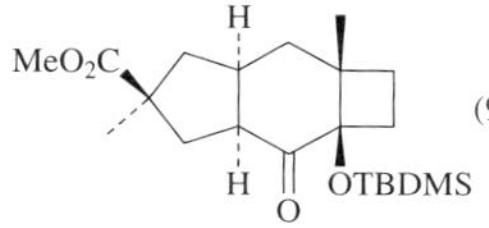 MeO_2C, H, H, O, OTBDMS (93)	585
C_{27-33}			
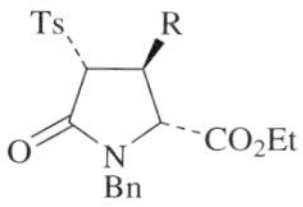 Ts, R, O, N, Bn, CO_2Et R = 4-FC_6H_4, 4-ClC_6H_4, Ph, 3-(c-C_5H_9O)C_6H_3OMe-4	Na/Hg, Na_2HPO_4	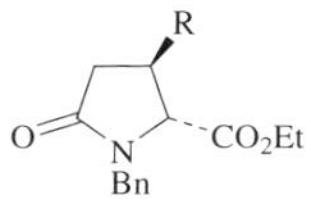 R, O, N, Bn, CO_2Et (—)	586

TABLE 5. REDUCTIVE DESULFONYLATION OF β-OXO-FUNCTIONALIZED SULFONES (*Continued*)

Substrate	Conditions	Product(s) and Yield(s) (%)	Refs.
C_{28} 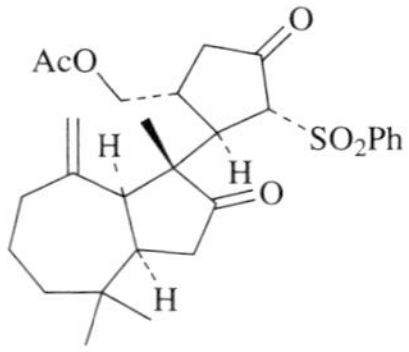 AcO, O, SO_2Ph, H, H, O, H	SmI_2, THF/MeOH, –78°, 20 min	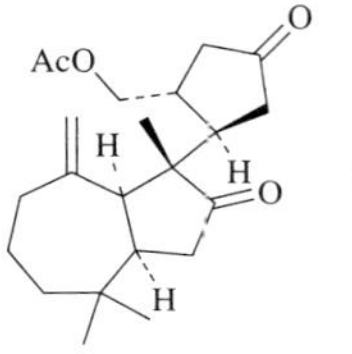 AcO, O, H, H, O, H (79)	587
	1. SmI_2, THF, –78°, 15 min 2. Ac_2O, DMAP, THF, –78°, 30 min	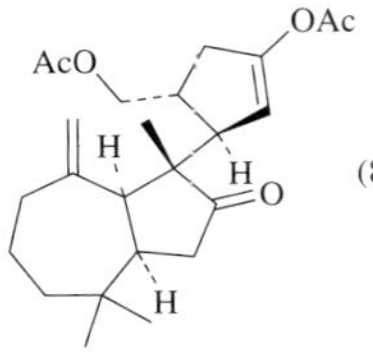 AcO, OAc, H, H, O, H (88)	587
C_{29} 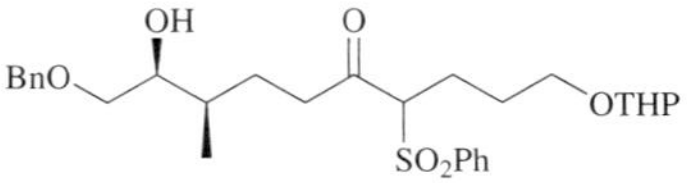 BnO, OH, O, SO_2Ph, OTHP	Al/Hg, THF/H_2O, rt, 1 d	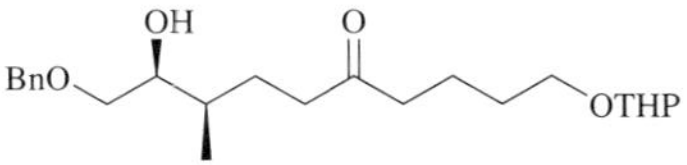 BnO, OH, O, OTHP (—)	588
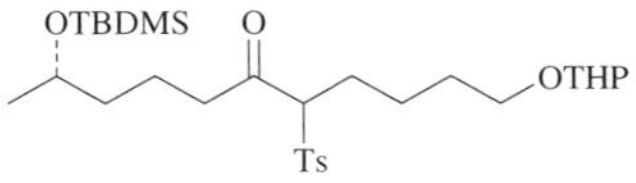 OTBDMS, O, Ts, OTHP	Na/Hg, Na_2HPO_4, MeOH	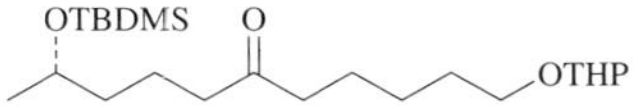 OTBDMS, O, OTHP (87)	589
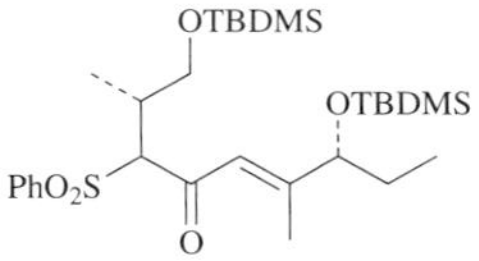 OTBDMS, OTBDMS, PhO_2S, O	Li, $C_{10}H_8$, THF, –78°	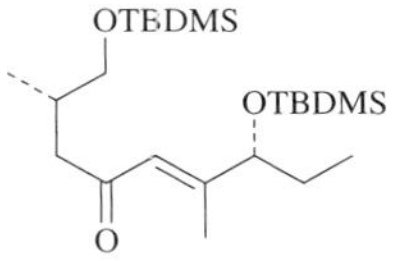 OTBDMS, OTBDMS, O (70)	123

	Substrate	Conditions	Product(s) and Yield(s) (%)	Refs.
C_{30}	SO_2Ph, CO_2Me, MeO, H, H, H	NH_3/Ca, reflux, 10 min	CO_2Me, MeO, H, H, H (61)	590
	PhO_2S, O, H, O, O	Al/Hg	O, H, O, O (38)	579
C_{31}	O, O, O, H, O, SO_2Ph, O, O	Al/Hg, THF/H_2O, 75°, 1 h	O, O, O, H, O, O, O (75)	591
	O, O, O, H, O, SO_2Ph, O, O	Al/Hg, THF/H_2O, 75°, 1 h	O, O, O, H, O, O, O (—)	592

TABLE 5. REDUCTIVE DESULFONYLATION OF β-OXO-FUNCTIONALIZED SULFONES (*Continued*)

	Substrate	Conditions	Product(s) and Yield(s) (%)	Refs.
C_{32} 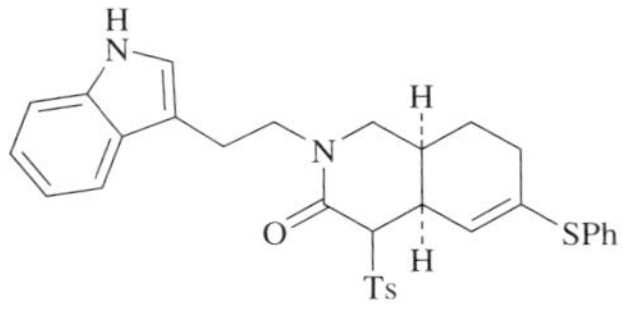	H N; N; H; H; O; Ts; SPh	Na/Hg, Na_2HPO_4, MeOH, rt, 1 h	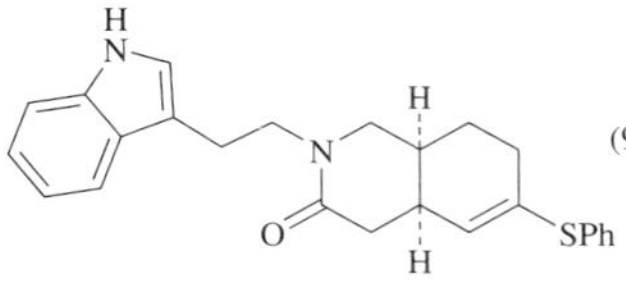 H N; N; H; H; O; SPh (91)	593
	OHC; O; SO_2Ph; OTBDPS	SmI_2, THF, rt, 1 h	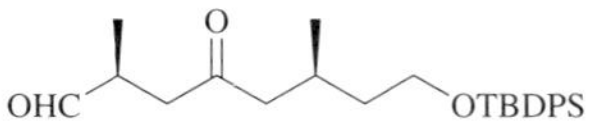 OHC; O; OTBDPS (—)	594
	NHCbz; O; OH; SO_2Ph; $(\ )_9$	Na/Hg, Na_2HPO_4, MeOH, –10°, 4 h	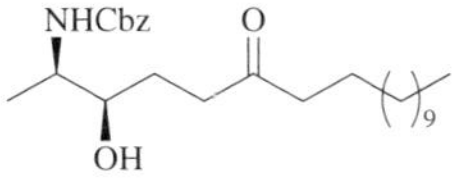 NHCbz; O; OH; $(\ )_9$ (68)	595
	O; R; BocN; SO_2Ph; O	Al/Hg, THF/H_2O, rt, overnight	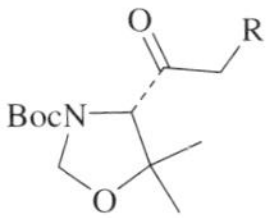 O; R; BocN; O — R: (*E*)-$Me(CH_2)_{10}CH{=}CHCH_2$ (86)	596
			R: *n*-$C_{14}H_{29}$ (85)	597
C_{33} 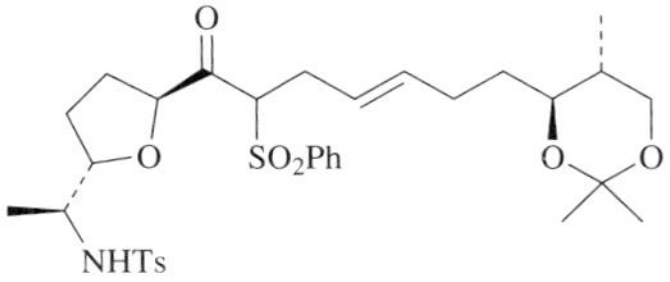	O; O; SO_2Ph; O; O; NHTs	Al/Hg, THF/*n*-PrOH/ H_2O, 40°, 3 h	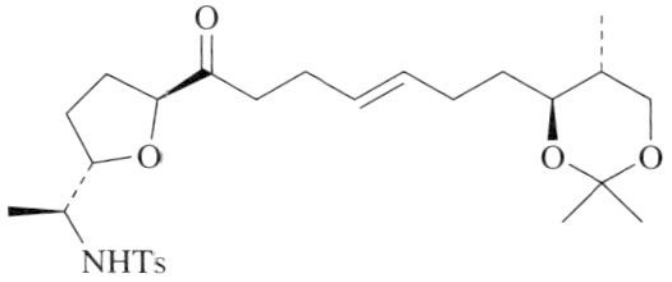 O; O; O; O; NHTs (80)	145

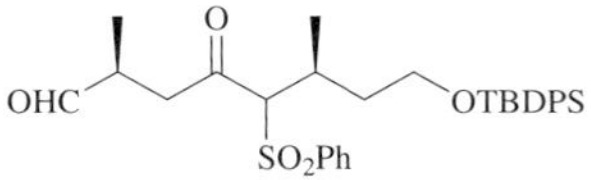

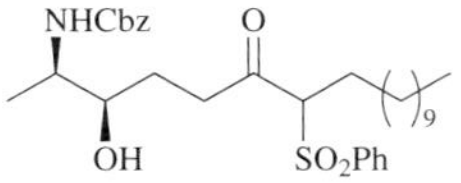

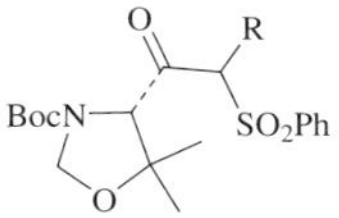

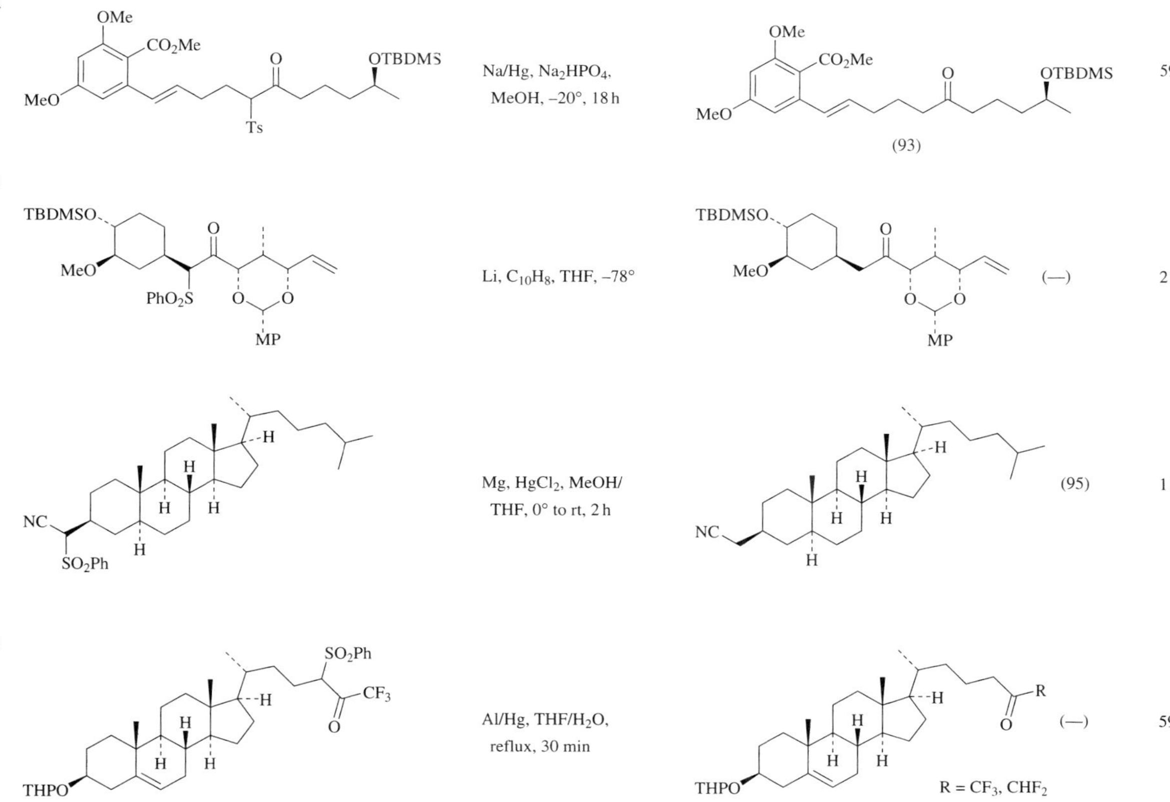

Carbons	Conditions	Yield (%)	Refs.
C_{34}	Na/Hg, Na_2HPO_4, MeOH, –20°, 18 h	(93)	598
C_{35}	Li, $C_{10}H_8$, THF, –78°	(—)	216
	Mg, $HgCl_2$, MeOH/THF, 0° to rt, 2 h	(95)	119
C_{37}	Al/Hg, THF/H_2O, reflux, 30 min	(—) $R = CF_3, CHF_2$	599

TABLE 5. REDUCTIVE DESULFONYLATION OF β-OXO-FUNCTIONALIZED SULFONES (*Continued*)

	Substrate	Conditions	Product(s) and Yield(s) (%)	Refs.
C_{37}	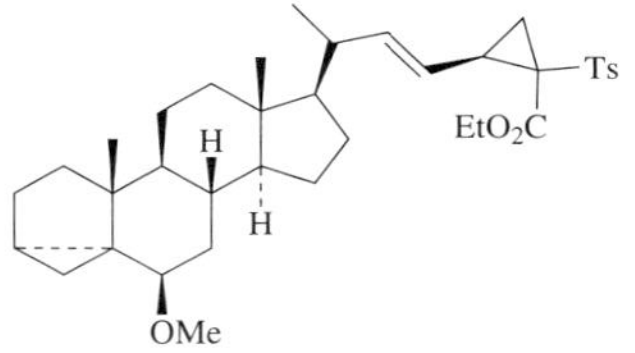 Ts, EtO_2C, H, H, OMe	Na/Hg, Na_2HPO_4, MeOH, rt, 2 h	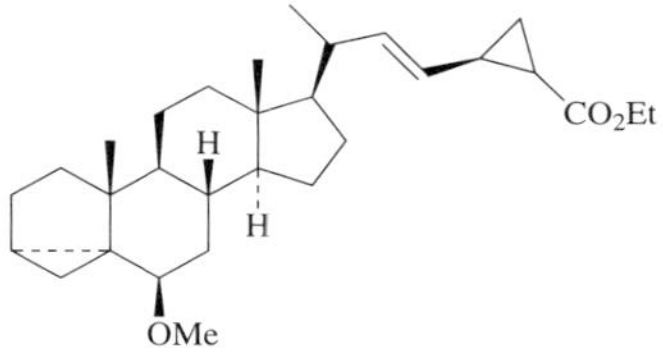 CO_2Et, H, H, OMe (—)	600
C_{38}	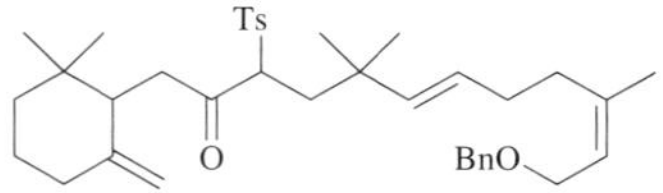 Ts, O, BnO	Na/Hg, EtOH, rt, 1 h	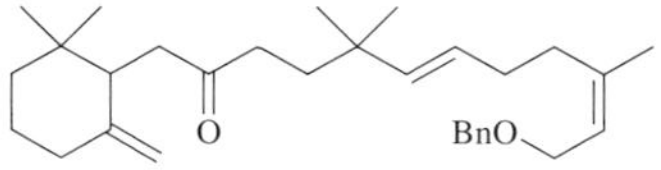 O, BnO (77)	601
C_{39}	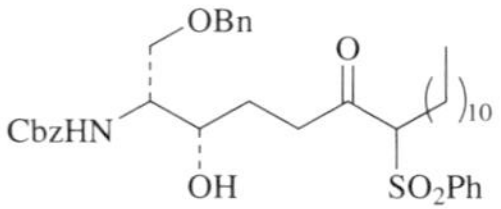 OBn, O, CbzHN, $()_{10}$, OH, SO_2Ph	Na/Hg, Na_2HPO_4, MeOH, –10°	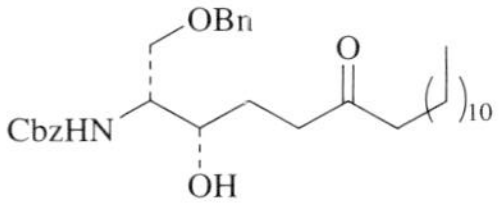 OBn, O, CbzHN, $()_{10}$, OH (95)	602
C_{40}	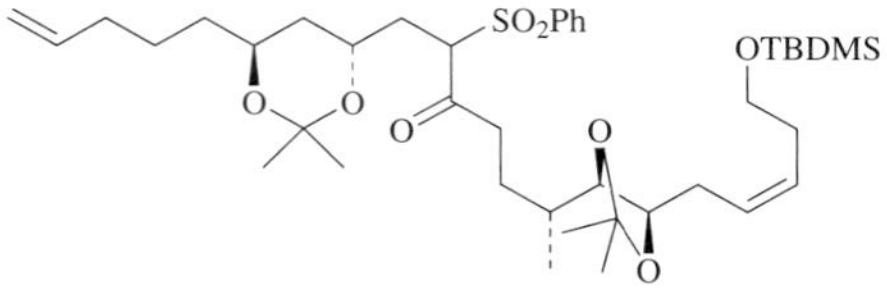 SO_2Ph, OTBDMS, O, O, O, O, O	Na/Hg, Na_2HPO_4, MeOH, 0°	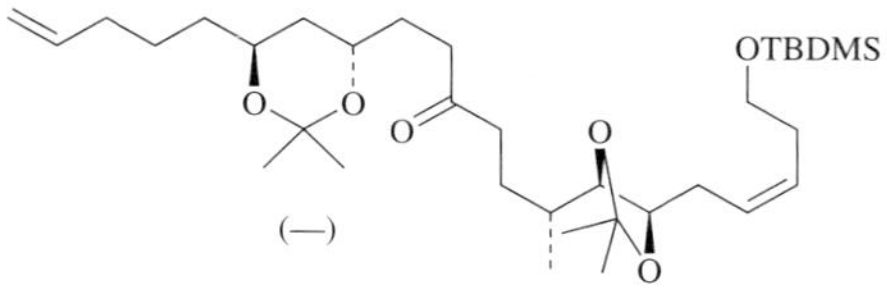 OTBDMS, O, O, O, O, O (—)	603
	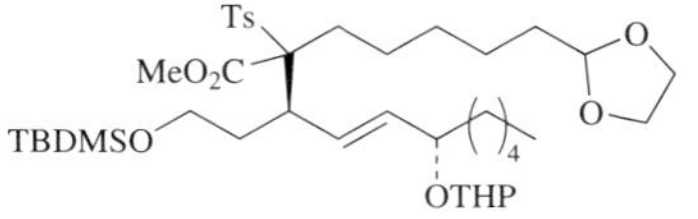 Ts, MeO_2C, O, O, TBDMSO, $()_4$, OTHP	Na/Hg, Na_2HPO_4, EtOH, rt, 12 h	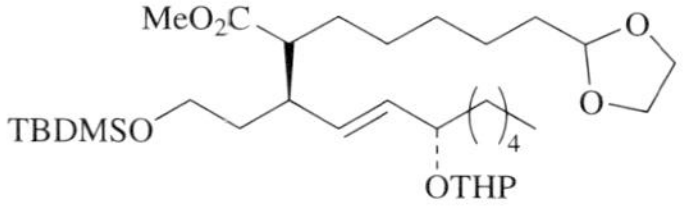 MeO_2C, O, O, TBDMSO, $()_4$, OTHP (93)	604

Carbons	Substrate	Conditions	Product(s) and Yield(s) (%)	Refs.
		Na/Hg, Na_2HPO_4, MeOH, rt, 1 h	(66)	605
C_{40-57}	R = H, $TBDPSOCH_2$	Al/Hg, THF/H_2O, rt, overnight	(—)	555
C_{41}		SmI_2, THF/MeOH, –78°	(93)	606
C_{42}		Li/$C_{10}H_8$, THF, –78°, 1 h	(60)	607
C_{44}		SmI_2, THF/MeOH, –78° to rt, 15 min	(95)	608

TABLE 5. REDUCTIVE DESULFONYLATION OF β-OXO-FUNCTIONALIZED SULFONES (*Continued*)

Substrate	Conditions	Product(s) and Yield(s) (%)	Refs.
C_{44} 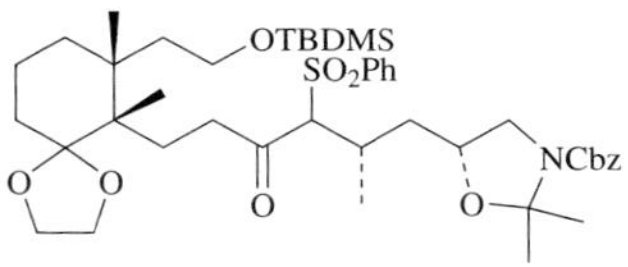 OTBDMS, SO_2Ph, NCbz	Na/Hg, Na_2HPO_4, MeOH, rt, 2 h	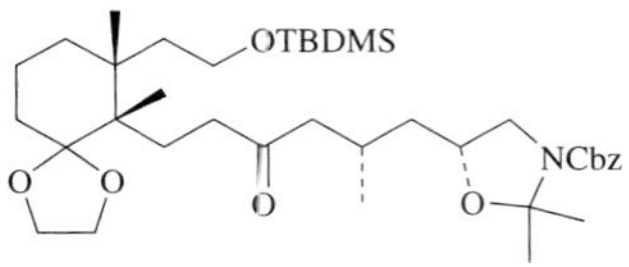 OTBDMS, NCbz (87)	609
C_{45} 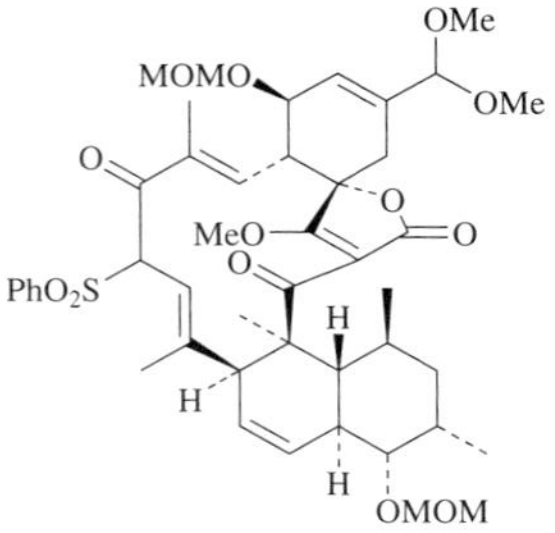 OMe, OMe, MOMO, PhO_2S, MeO, H, H, H, OMOM	Al/Hg, THF/H_2O, 3.5 h	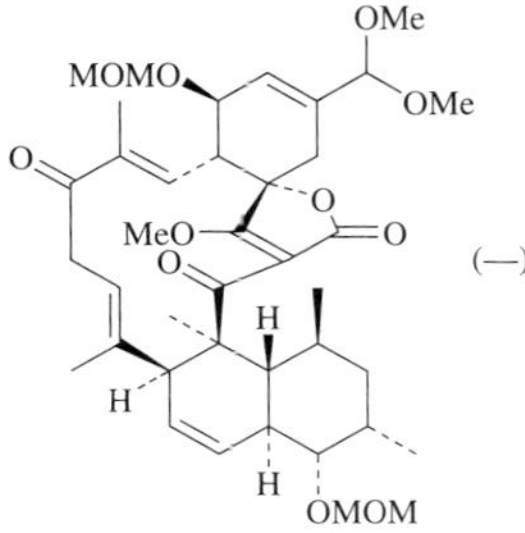 OMe, OMe, MOMO, MeO, H, H, H, OMOM (—)	610
C_{45-46} 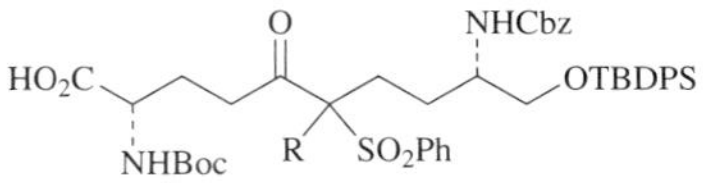 HO_2C, NHBoc, NHCbz, OTBDPS, R, SO_2Ph; R = H, Me	Ra-Ni, EtOH, reflux	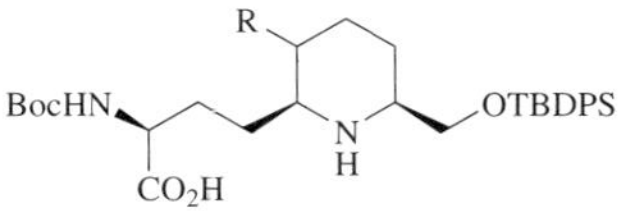 R, BocHN, CO_2H, N H, OTBDPS (—)	179
C_{47-63} 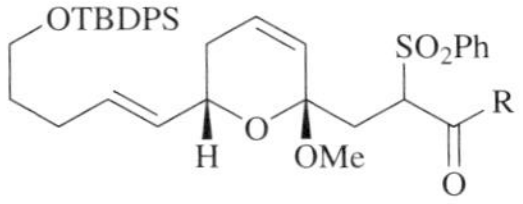 OTBDPS, H, OMe, SO_2Ph, R	Na/Hg, Na_2HPO_4, THF/MeOH, −10°	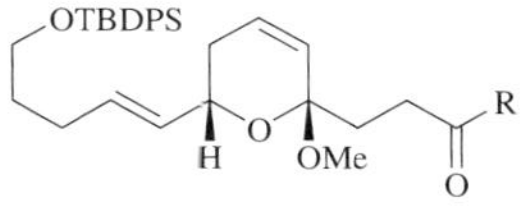 OTBDPS, H, OMe, R	

	R	Time		
	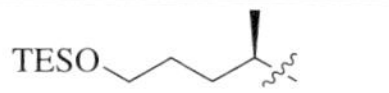 TESO (R group structure)	75 min	(—)	611
	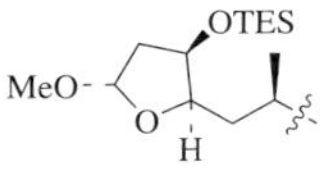 MeO-, OTES, O, H (R group structure)	45 min	(88)	611
	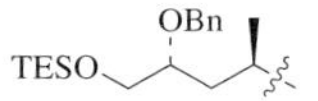 TESO, OBn (R group structure)	2 h	(—)	611
	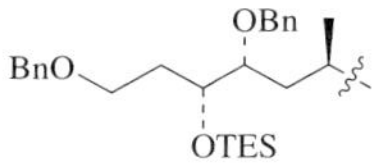 BnO, OBn, OTES (R group structure)	—	(82)	260
C_{48}	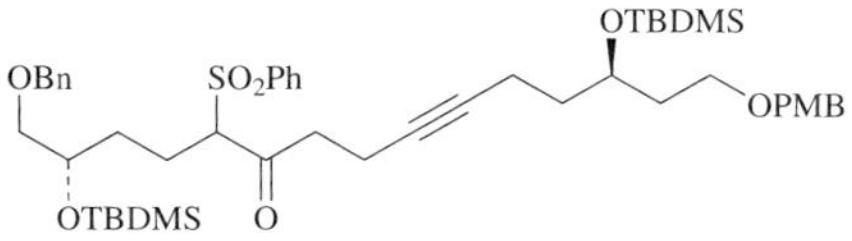 OBn, OTBDMS, SO$_2$Ph, O, OTBDMS, OPMB	Al/Hg, THF/H_2O, rt, 23 h	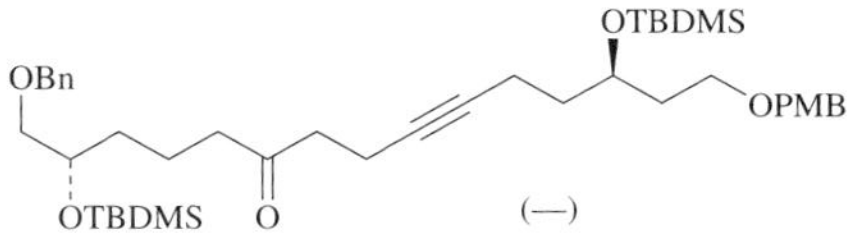 OBn, OTBDMS, O, OTBDMS, OPMB (—)	612
C_{49}	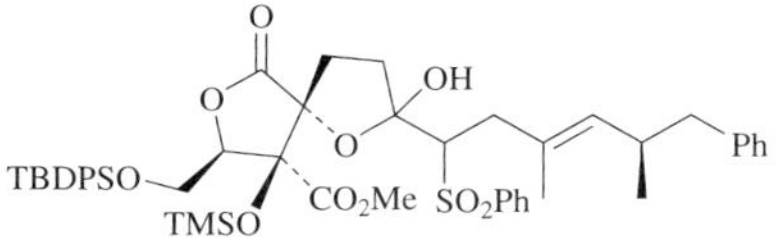 O, O, OH, O, TBDPSO, TMSO, CO_2Me, SO_2Ph, Ph	Al/Hg, THF/HMPA/H_2O, rt, 3 h	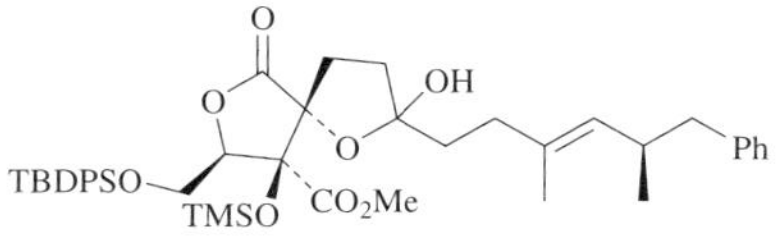 O, O, OH, O, TBDPSO, TMSO, CO_2Me, Ph (—)	613
C_{50}	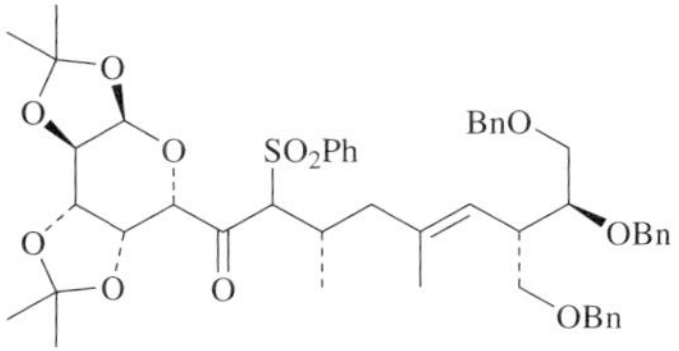 O, O, O, O, O, SO_2Ph, O, BnO, OBn, OBn	(n-Bu)$_3$SnH, AIBN, toluene, reflux, 40 min	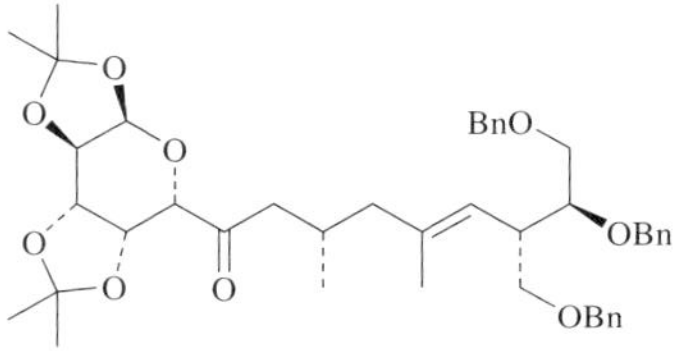 O, O, O, O, O, O, BnO, OBn, OBn (68)	59

TABLE 5. REDUCTIVE DESULFONYLATION OF β-OXO-FUNCTIONALIZED SULFONES (*Continued*)

	Substrate	Conditions	Product(s) and Yield(s) (%)	Refs.
C_{52}	EtO, OEt, OR, O, OR, OR, SO_2Ph, OR R = TBDMS	Na/Hg, Na_2HPO_4, MeOH, rt, 1 h	EtO, OEt, OR, O, OR, OR, OR (—)	614
	TBDMSO, SO_2Ph, O, TBDMSO, OTBDMS, O, Si, O, t-Bu, Bu-t	SmI_2, THF/MeOH, –78°	TBDMSO, O, TBDMSO, OTBDMS, O, Si, O, t-Bu, Bu-t (89)	615
C_{53}	TBDPSO, ( )$_3$, OMOM, O, OPNBz, SO_2Ph	SmI_2, THF/MeOH, –78°, 20 min	TBDPSO, ( )$_3$, OMOM, O, OPNBz (51)	180
	TBDPSO, PhO_2S, OTBDMS, OBn, OTBDMS, O	Na/Hg, Na_2HPO_4, MeOH, rt, 1 h	TBDPSO, OTBDMS, OBn, OTBDMS, O (68)	616
C_{55}	BOMO, O, SO_2Ph, H, H, H, H, H, BOMO, OBOM	$(n\text{-Bu})_3SnH$, AIBN, toluene, reflux, 30 min	BOMO, O, H, H, H, H, H, BOMO, OBOM (83)	59

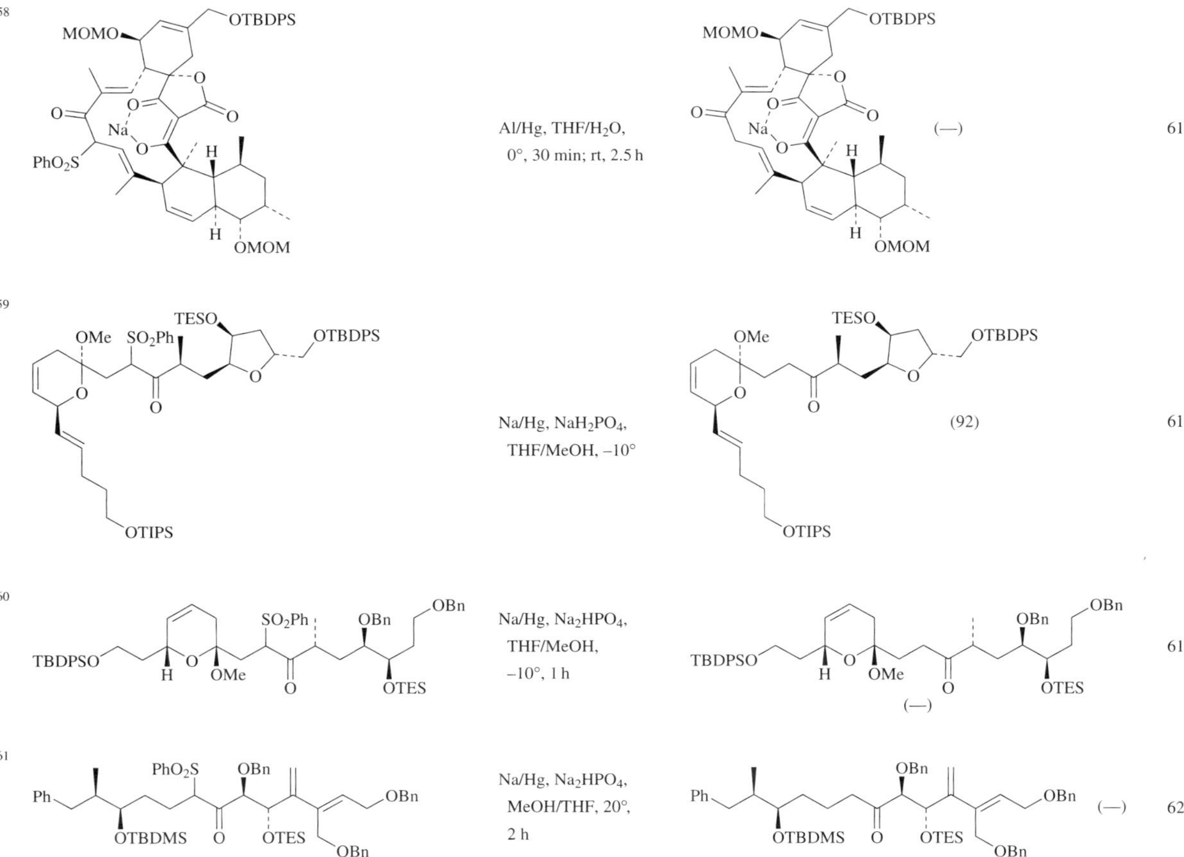

	Conditions	Yield (%)	Ref.
C_{58}	Al/Hg, THF/H_2O, 0°, 30 min; rt, 2.5 h	(—)	617
C_{59}	Na/Hg, NaH_2PO_4, THF/MeOH, –10°	(92)	618
C_{60}	Na/Hg, Na_2HPO_4, THF/MeOH, –10°, 1 h	(—)	619
C_{61}	Na/Hg, Na_2HPO_4, MeOH/THF, 20°, 2 h	(—)	620

TABLE 5. REDUCTIVE DESULFONYLATION OF β-OXO-FUNCTIONALIZED SULFONES (*Continued*)

Substrate	Conditions	Product(s) and Yield(s) (%)	Refs.
C_{64}			
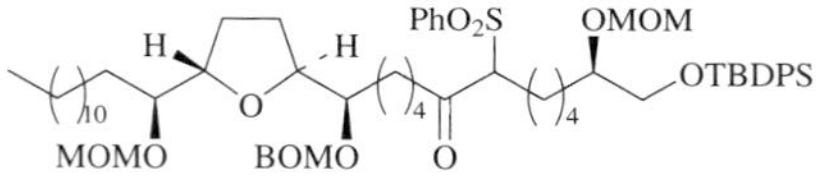 H, H, PhO_2S, OMOM, OTBDPS, 10, 4, 4, MOMO, O, BOMO, O	Na/Hg, Na_2HPO_4, MeOH, 0°, 3 h	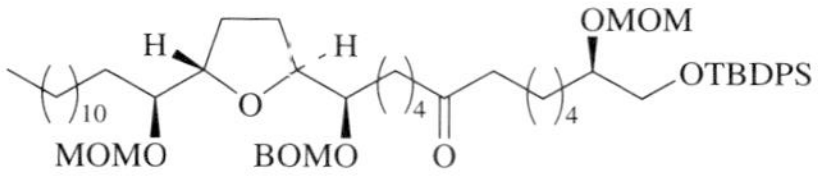 H, H, OMOM, OTBDPS, 10, 4, 4, MOMO, O, BOMO, O (94)	621
C_{71}			
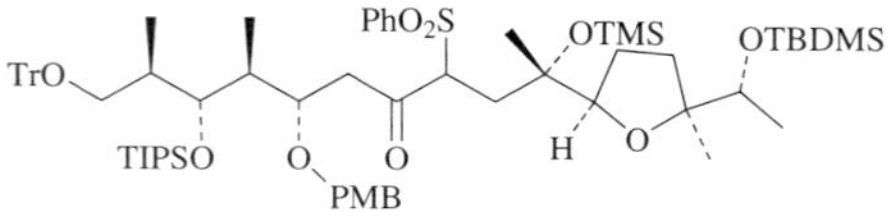 TrO, PhO_2S, OTMS, OTBDMS, TIPSO, O, PMB, O, H, O	SmI_2, THF/MeOH, −78°, 30 min	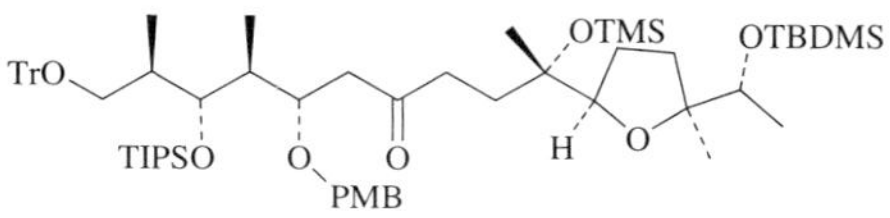 TrO, OTMS, OTBDMS, TIPSO, O, PMB, O, H, O (—)	622

[a] The product has 95% deuterium incorporation.

[b] The alkene is reduced under these conditions.

TABLE 6. REDUCTIVE DESULFONYLATION OF ALLYL SULFONES

Substrate	Conditions	Product(s) and Yield(s) (%)	Refs.
C_{9-12}			
$CH_2{=}C(R)CH_2SO_2Ph$	$(n\text{-}Bu)_3SnH$, AIBN, C_6H_6	$CH_2{=}C(R)CH_2Sn(Bu\text{-}n)_3$	
		R: Cl, Temp 80° (—)	623
		R: CN, Temp 80° (—)	623
		R: TMS, Temp 65° (90)	624
C_{10}			
$CH_2{=}CHCH(CH_3)SO_2Ph$	$(n\text{-}Bu)_3SnH$, AIBN	$CH_2{=}CHCH(CH_3)Sn(Bu\text{-}n)_3$ (—)	625
C_{13}			
4-methylcyclohex-3-enyl–$C({=}CH_2)CH_2SO_2CH_2Ac$	Al/Hg, THF/H_2O, 20°, 4 h	limonene (58)	626
C_{13-17}			
2-R^3-3-[$C(R^1)(R^2)SO_2Ph$]cyclopent-2-enone	1. $(n\text{-}Bu)_3SnH$, AIBN, toluene, 110°, 5 h 2. KF	2-R^3-3-[$CH(R^1)R^2$]cyclopent-2-enone	
R^1 = H, R^2 = H, R^3 = Me		(85-95)	627
R^1 = H, R^2 = Me, R^3 = Me		(85-95)	627
R^1 = Me, R^2 = Me, R^3 = Me		(85-95)	627
R^1 = H, R^2 = H, R^3 = (Z)-$MeCH_2CH{=}CHCH_2$		(—)	628
C_{15}			
$PhCH{=}CHCH_2SO_2Ph$	$RhCl(PPh_3)_3$, PNAH, $LiClO_4$, CH_3CN, 70°, 17 h	$PhCH{=}CHCH_3$ (79) + $PhCH_2CH{=}CH_2$ (5.5)	188

TABLE 6. REDUCTIVE DESULFONYLATION OF ALLYL SULFONES (*Continued*)

Substrate	Conditions	Product(s) and Yield(s) (%)	Refs.
$C_{15\text{-}18}$ 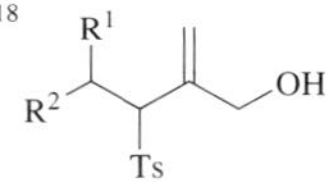 R[1] R[2] Ts OH	Na/Hg, Na_2HPO_4, MeOH, 0° to rt	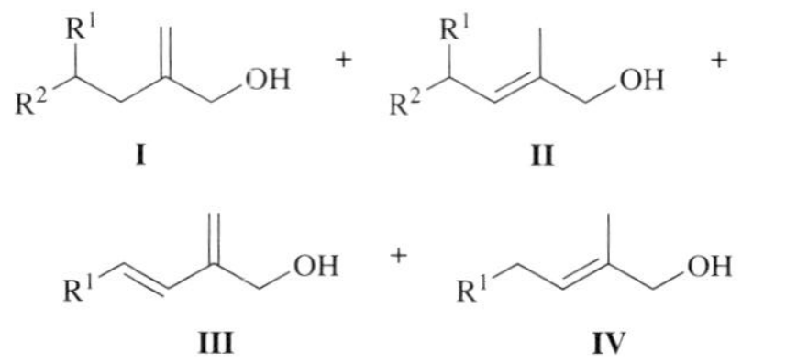 R[2] R[1] OH **I** + R[2] R[1] OH **II** + R[1] OH **III** + R[1] OH **IV**	629
$C_{15\text{-}21}$ R PhO$_2$S O O	$(n\text{-}Bu)_3SnH$, AIBN, toluene, reflux, 2 h	R O O	630

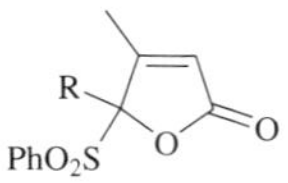

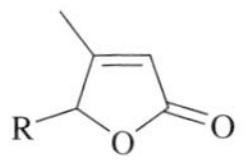

R^1	R^2	I	II	III	IV
i-Pr	HO	(19)	(21)	(8)	(0)
Ph	HO	(28)	(18)	(0)	(12)
H	*t*-BuO_2CCH_2	(33)	(32)	(0)	(0)

R	Yield (%)
(*E*)-MeCH=CHCH$_2$	(69)
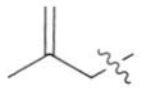	(88)
n-Bu	(75)
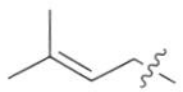	(72)
Bn	(80)
n-C_8H_{17}	(80)
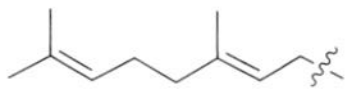	(77)

TMS OR[1] R[2] SO$_2$Ph

$(n\text{-Bu})_3SnH$, AIBN, C_6H_6, 65°, 2 h

$(n\text{-Bu})_3Sn$ TMS OR[1] R[2]

Z:E = 60:40

R^1	R^2		
H	Et	(70)	624
H	*i*-Pr	(80)	
Bn	Me	(70)	

$C_{15\text{-}24}$

SO$_2$Ar, R[1], R[2], R[3]

See table.

R^1	R^2	R^3	Ar	Reagents	Solvent	Temp	Time		
H	H	$H_2C{=}CHCH_2$	Ph	Na/Hg, Na_2HPO_4	MeOH	0°	1.5 h	(92)	631
H	H	*i*-Pr	4-Tol	Na	EtOH/THF	rt	18 h	(44)	632
H	H	*n*-Bu	4-Tol	Na	EtOH/THF	rt	18 h	(10)	632
H	H	Bn	4-Tol	Na	EtOH/THF	rt	18 h	(73)	632
Ph	EtO	H	Ph	Na/$C_{10}H_8$	THF, *n*-$PrNH_2$	–78°	—	(66)	633
Me	EtO	Bn	Ph	Na/$C_{10}H_8$	THF, *n*-$PrNH_2$	–78°	—	(55)	633
i-Pr	EtO	Bn	Ph	Na/$C_{10}H_8$	THF, *n*-$PrNH_2$	–78°	—	(52)	633

C_{16}

Ts

Na/Hg, Na_2HPO_4, MeOH, 0°, 1 h

I + **II**

I + II (94), **I:II** = 42:58 — 39

Ts

Na/Hg, Na_2HPO_4, MeOH

(92) — 631

TABLE 6. REDUCTIVE DESULFONYLATION OF ALLYL SULFONES (*Continued*)

Substrate	Conditions	Product(s) and Yield(s) (%)	Refs.
C_{16}	Li/NH_3, THF, rt, 5 min	(—)	134
C_{16-17}	Na/Hg, Na_2HPO_4, MeOH, –20° to –5°	R: 3-pyridyl (60); Ph (45)	634, 635
C_{16-22} (SO_2R)	Pd catalyst, $LiHBEt_3$, THF, 0°	**I** + **II** + **III**	85
C_{17} (SO_2Ph, Bu-*t*)	$Pd(acac)_2$, *n*-Bu_3P, HCO_2H, TEA, C_6H_6, 45°, 22 h	Bu-*t* (98.7) + Bu-*t* (1.3)	89

R	Pd catalyst	Time	**I + II + III**	**I:II:III**
Me	$PdCl_2(dppp)$	1 h	(84)	98:2:0
Me	$PdCl_2(dppb)$	7 h	(92)	96:3:1
4-Tol	$PdCl_2(dppp)$	0.5 h	(87)	97:2:1
4-Tol	$PdCl_2(dppb)$	0.5 h	(86)	>99:0:0

C_{18}

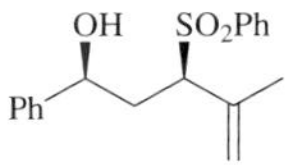

Substrate	Conditions	Product(s) and Yield(s) (%)	Refs.
OH, SO_2Ph, Ph	Mg, $HgCl_2$, EtOH, rt, 2 h	OH, Ph (98)	114
PhO_2S, F, F, O	Mg, $HgCl_2$, EtOH/THF, 0°	F, F, O (75) + F, F, O (15)	636
H, OH, SO_2Ph	Na/Hg, Na_2HPO_4, MeOH, –10° to rt, 1 h	H, OH (71)	33
SO_2CH_2Ac	Al/Hg, THF/H_2O, 20°, 4 h	(71) Z:E = 1:3	626

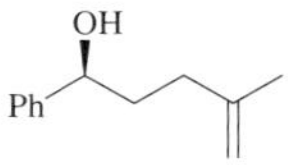

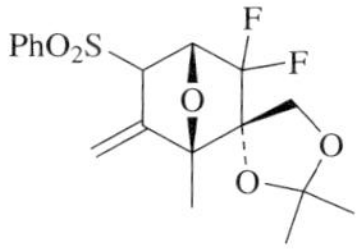

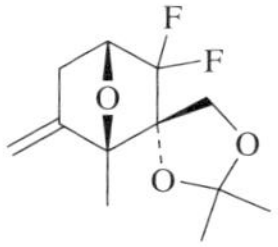

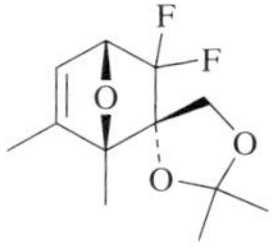

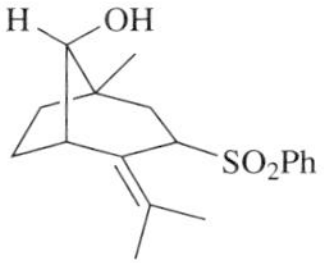

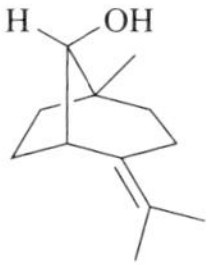

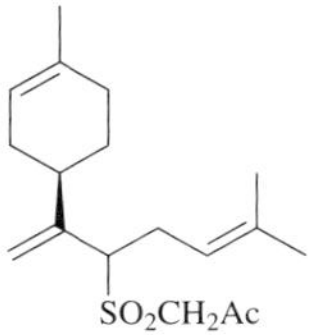

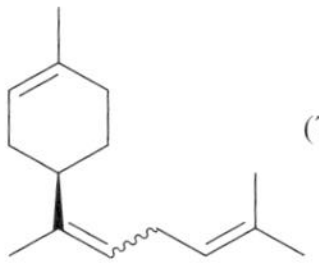

$C_{18\text{-}25}$

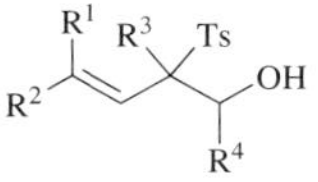

R^1, R^2, R^3, Ts, OH, R^4 — $Pd(PPh_3)_4$, $NaBH_4$ — R^1, R^2, R^3, OH, R^4 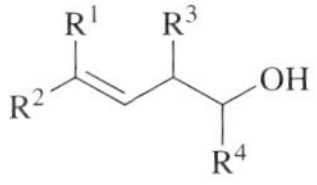 — 82

R^1	R^2	R^3	R^4	Solvent	Temp	Time	
H	H	Bn	H	THF/*i*-PrOH (2:1)	0°	1 h	(91)
Me	Me	Bn	H	THF/*i*-PrOH/EtOH (5:2:2)	20°	5 h	(72)
H	Ph	H	Ph	THF/*i*-PrOH (2:1)	–35° to 0°	6 h	(75)
H	Ph	Bn	H	THF/*i*-PrOH (2:1)	0°	2 h	(91)
H	Ph	H	*n*-C_8H_{17}	THF/*i*-PrOH (2:1)	rt	0.5 h	(81)

TABLE 6. REDUCTIVE DESULFONYLATION OF ALLYL SULFONES (*Continued*)

Substrate	Conditions	Product(s) and Yield(s) (%)	Refs.
C_{19} 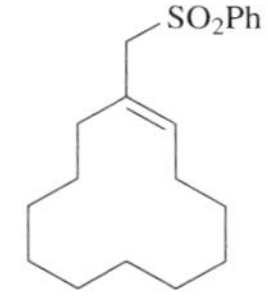 SO_2Ph	$Pd(acac)_2$, n-Bu_3P, HCO_2H, TEA, THF, 45°, 22 h	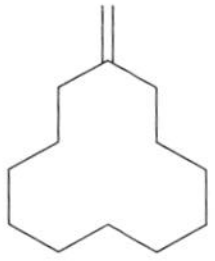 (82) 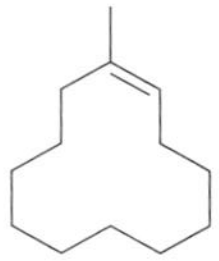 + (18)	89
C_{19-26} 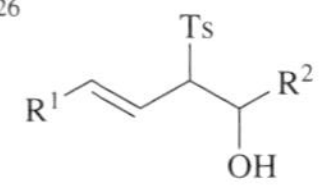 Ts, R^1, R^2, OH	Pd catalyst, hydride, THF	 R^1 … R^2 OH **I** + R^1 … R^2 OH **II**	190

R^1	R^2	Pd Catalyst/Additive	Hydride	Temp	Time	**I + II**	**I:II**	Z:E[a]
Bn	Me	$PdCl_2(dppp)/Ph_3SiH$	$LiHBEt_3$	20°	3 min	(100)	93:7	4:96
Bn	Me	$PdCl_2(PPh_3)_2$	$LiBH_4$	–4°	2.5 h	(100)	7:93	11:89
Me	n-C_8H_{17}	$PdCl_2(dppp)/Ph_3SiH$	$LiHBEt_3$	20°	20 min	(98)	92:8	2:98
Me	n-C_8H_{17}	$PdCl_2(PPh_3)_2$	$LiBH_4$	–18°	4 h	(96)	3:97	10:90
Bn	i-Pr	$PdCl_2(dppp)/Ph_3SiH$	$LiHBEt_3$	20°	15 min	(78)	98:2	4:96
Bn	i-Pr	$PdCl_2(PPh_3)_2$	$LiBH_4$	0°	4 h	(98)	16:84	24:76
i-Pr	$PhCH_2CH_2$	$PdCl_2(dppp)/Ph_3SiH$	$LiHBEt_3$	20°	45 min	(99)	93:7	2:98
i-Pr	$PhCH_2CH_2$	$PdCl_2(PPh_3)_2$	$LiBH_4$	–20°	7 h	(87)	3:97	<1:99
Bn	$PhCH_2CH_2$	$PdCl_2(dppp)/Ph_3SiH$	$LiHBEt_3$	20°	3 min	(94)	96:4	4:96
Bn	$PhCH_2CH_2$	$PdCl_2(PPh_3)_2$	$LiBH_4$	–45°	18 h	(~100)	1:99	11:89
Bn	n-C_8H_{17}	$PdCl_2(dppp)/Ph_3SiH$	$LiHBEt_3$	20°	3 min	(92)	99:1	2:98
Bn	n-C_8H_{17}	$PdCl_2(PPh_3)_2$	$LiBH_4$	–5°	3 h	(83)	1:99	14:86

Substrate	Conditions	Product(s) and Yield(s) (%)	Refs.
C_{20} SO_2Et, N, O, O	$LiHBEt_3$, $PdCl_2(dppp)$, THF, 0°, 0.5 h	N, O, O (84)	637

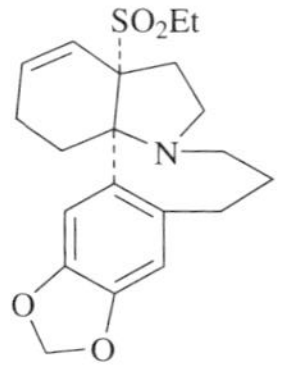

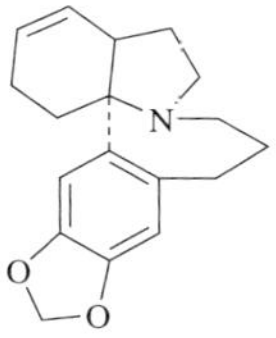

HO, BocN, Ts — Na/Hg, B(OH)$_3$, MeOH, reflux, 12 h — HO, BocN (74) — 638

C_{20-24}

R^2, Ts, R^1 — PdCl$_2$(dppp), LiHBEt$_3$, THF, 0° — R^2, R^1 **I** + R^2, R^1 **II** — 85

R^1	R^2	Time	I + II	I:II
Bn	H	0.5 h	(89)	0:>99
H	Bn	0.5 h	(84)	>99:0
PhCH$_2$CH$_2$	H	0.5 h	(99)	0:>99
H	PhCH$_2$CH$_2$	0.5 h	(89)	>99:0
n-C$_{11}$H$_{23}$	H	1 h	(91)	31:69
H	n-C$_{11}$H$_{23}$	40 min	(97)	>99:0

C_{21-22}

Ts, R, NHBu-*i*, O — Na/Hg, Na$_2$HPO$_4$, MeOH, –20° to rt — R, NHBu-*i*, O — 133

R	
n-C$_5$H$_{11}$CHOH	(75)
Bn	(—)

C_{21-24}

SO$_2$Ph, R — See table — R — 129

R	Reagents	Solvent	Temp		Z:E
H	Na/Hg	MeOH	—	(—)	—
TMS	Na, DMAN	Et$_2$NH/THF	–85°	(84)	13:87

TABLE 6. REDUCTIVE DESULFONYLATION OF ALLYL SULFONES (*Continued*)

Substrate	Conditions	Product(s) and Yield(s) (%)	Refs.
C_{21-27} [structure: PhO$_2$S, R^1, R^2, n]	Li/EtNH$_2$	[structure: R^1, R^2, n]	
n / R^1 / R^2	Temp / Time		
1 / CD_3 / $HOCH_2$	— / —	(—)	639
1 / Me / $HOCH_2$	–78° / 1 h	(92)	348
2 / CD_3 / $HOCH_2$	— / —	(—)	639
2 / Me / $HOCH_2$	–78° / 1 h	(98)	348
2 / Me / MeO_2C	–78° / 1 h	(98)	640
C_{22} [structure: Ph, Ph, SO_2Ph]	$Mo(CO)_6$, reflux, 21 h	[structures] I + II + III	88
	Solvent	I / II / III	
	dioxane	(11) / (4) / (15)	
	dioxane/H_2O	(32) / (12) / (0)	
[structure: Ts, R, CN]	Na/Hg, Na_2HPO_4, MeOH	[structure: R, CN]	33
		R / Temp / Time	
		n-Bu / –20° / 4 min (49)	
		s-Bu / –10° / 20 min (58)	
C_{23} [structure: Ts, EtO_2C, N, Bn]	Na/Hg, NaH_2PO_4, MeOH, 0° to rt	[structure: EtO_2C, N, Bn] (9) + [structure: EtO_2C, N, Bn] (46)	37

Na/Hg, Na_2HPO_4, MeOH, –10° to rt, 7 min (66) 117

C_{23-28}

$Pd(PPh_3)_4$, $LiHBEt_3$, THF, 0°

I + II + III

84

R^1	R^2	R^3	Time	I + II + III	I:II:III
H	Ph	Bn	0.5 h	(94)	94:0:6
Me	Me	n-$C_{11}H_{23}$	1.5 h	(86)	>99:0:0
H	Ph	n-$C_{11}H_{23}$	40 min	(~100)	84:0:16
$Me_2C{=}CHCH_2CH_2$	Me	n-$C_{11}H_{23}$	5 h	(83)	91:9:0
Me	$Me_2C{=}CHCH_2CH_2$	n-$C_{11}H_{23}$	1.5 h	(98)	94:6:0

C_{23-48}

$PdCl_2(dppp)$, $LiHBEt_3$, THF

641

n	R	Temp	Time	
1	OH	0°	1.5 h	(90)
1		0° to rt	16 h	(97)
2	"	0° to rt	16 h	(95)
2		0° to rt	16 h	(88)

TABLE 6. REDUCTIVE DESULFONYLATION OF ALLYL SULFONES (*Continued*)

Substrate	Conditions	Product(s) and Yield(s) (%)	Refs.
C_{24}			
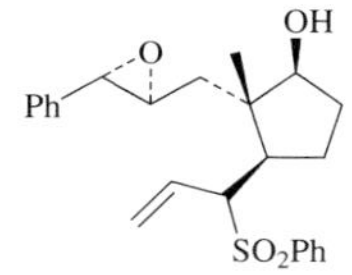 Ph, O, OH, SO_2Ph	$(n\text{-Bu})_3SnH$, AIBN, C_6H_6, 80°, 2 h	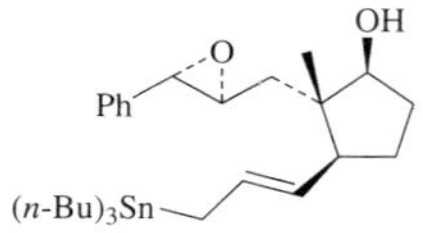 Ph, O, OH, $(n\text{-Bu})_3Sn$ (67)	642
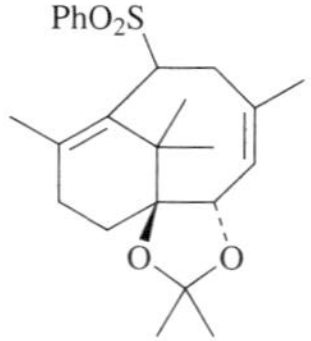 PhO_2S, O, O	Na/NH_3, THF, −10°, 5 min	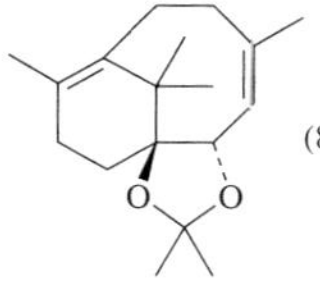 O, O (87)	643
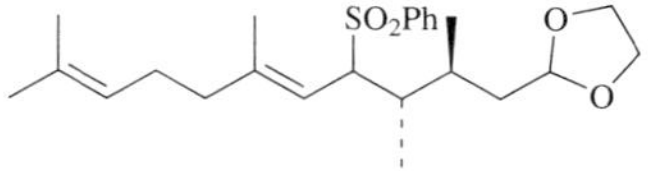 SO_2Ph, O, O	$PdCl_2(dppp)$, $LiHBEt_3$, THF, 3 h	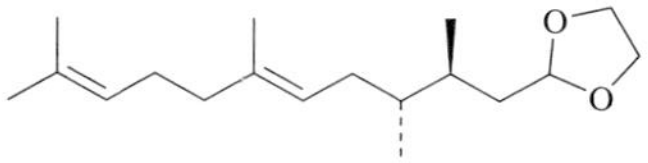 O, O (74)	644
$C_{25\text{-}37}$			
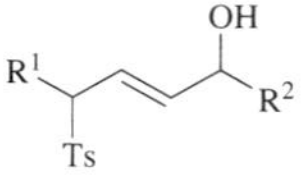 R^1, Ts, OH, R^2 R^1 / R^2: n-C_8H_{17} / Ph; n-$C_{18}H_{37}$ / i-Pr; n-$C_{18}H_{37}$ / n-C_8H_{17}	1. $Pd(PPh_3)_4$, TEA, $ClCH_2CH_2Cl/MeOH$, reflux, 7.5 h 2. TsOH, MeOH/ $CHCl_3$, rt, overnight	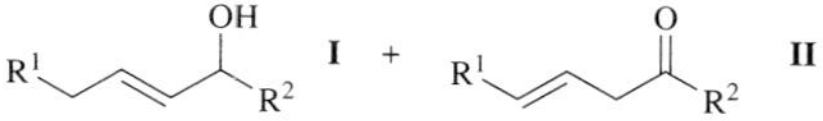 R^1, OH, R^2 **I** + R^1, O, R^2 **II** **I** / **II**: (59) / (0); (71) / (11); (65) / (7)	189
C_{26}			
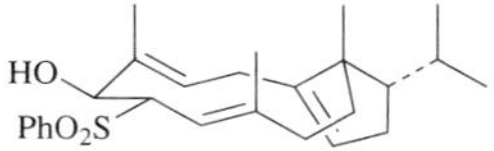 HO, PhO_2S	Na/Hg, Na_2HPO_4, MeOH, 22°, 1 h	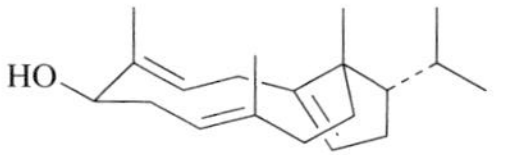 HO (65)	645

Reactant	Conditions	Product(s) and Yield(s) (%)	Refs.
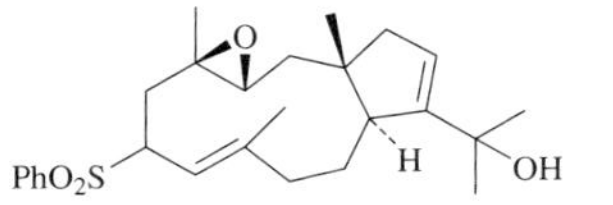 PhO$_2$S	Na/Hg, Na_2HPO_4, MeOH/THF, 0°, 1 h	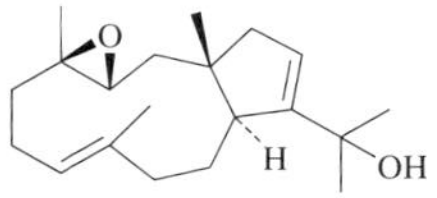 H, OH (83)	646
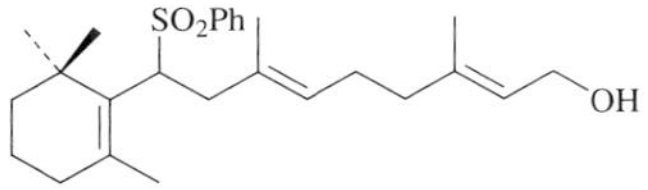 SO_2Ph, OH	Li/$EtNH_2$, Et_2O, –78°	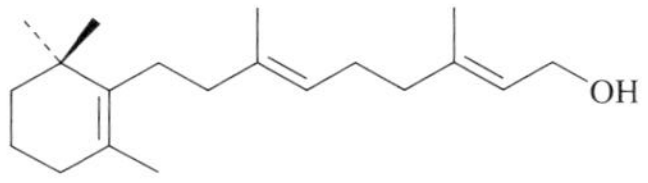 OH (76)	647
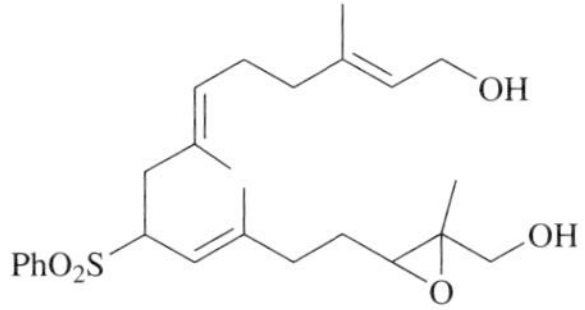 PhO$_2$S, OH, OH	Li/$EtNH_2$, THF, –78°, 3.5 h	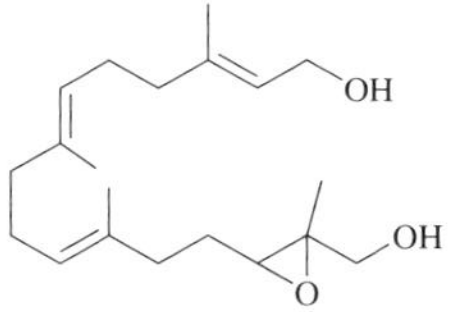 OH, OH (78)	648
C_{27} Ph, Ph, H, SO_2Ph	Na/Hg, Na_2HPO_4, MeOH, rt, 4 h	Ph, Ph (52) + Ph, Ph (35)	649
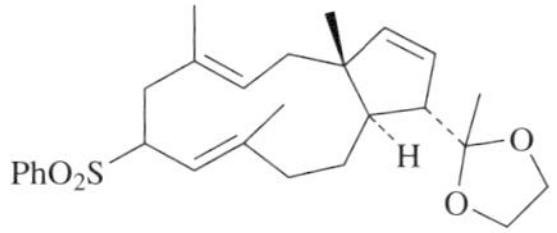 PhO$_2$S, H, O, O	Na/Hg, Na_2HPO_4, MeOH/THF, 0°, 1 h	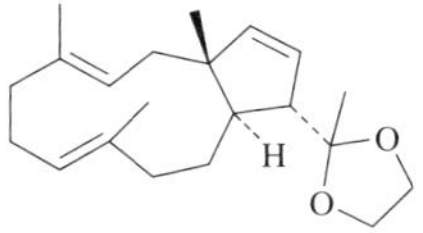 H, O, O (86)	646
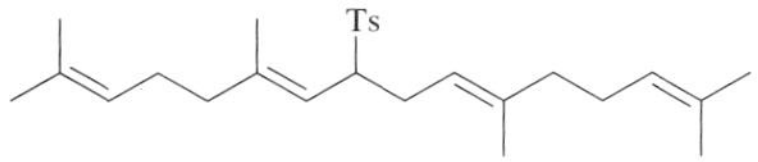 Ts	Li/$EtNH_2$, 0°, 30 min	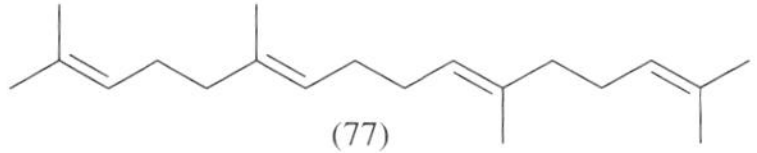 (77)	650

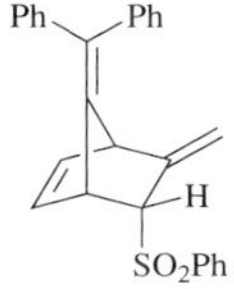

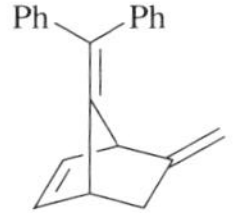

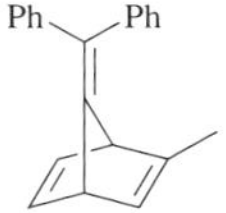

TABLE 6. REDUCTIVE DESULFONYLATION OF ALLYL SULFONES (*Continued*)

Substrate	Conditions	Product(s) and Yield(s) (%)	Refs.
C_{27} Ts	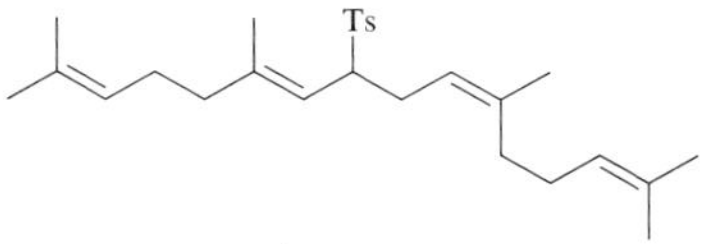 Li/$EtNH_2$, 0°, 30 min	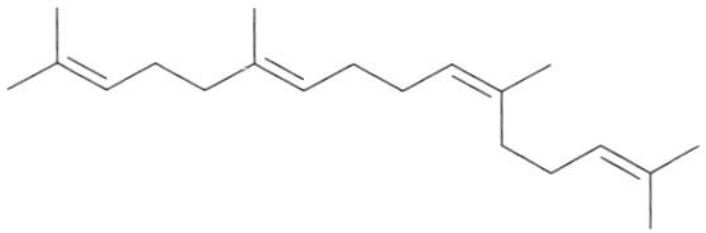 (82)	650
C_{29} O, O, O, HO, SO_2Ph	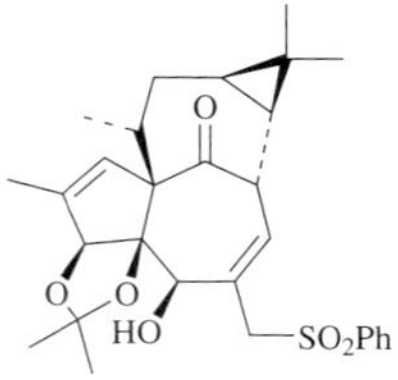 Na/Hg, Na_2HPO_4, MeOH, –20° to –10°, 20 min	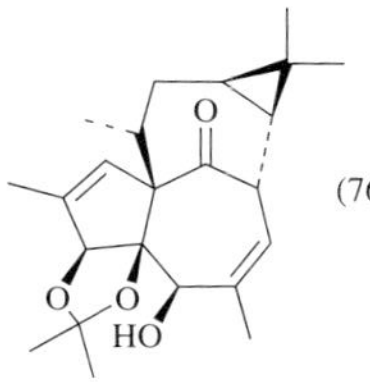 O, O, O, HO (76)	651
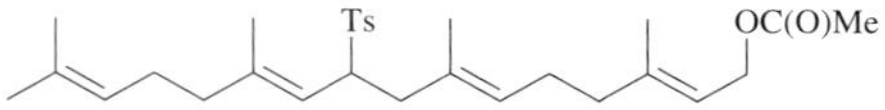 Ts, OC(O)Me	Na/Hg, Na_2HPO_4, MeOH, 0°, 2 h	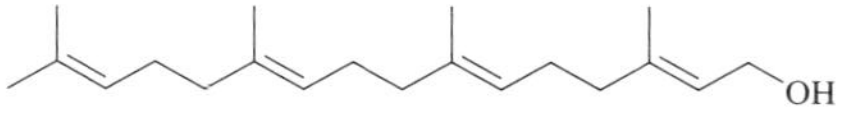 OH (35) + OH (15)	109
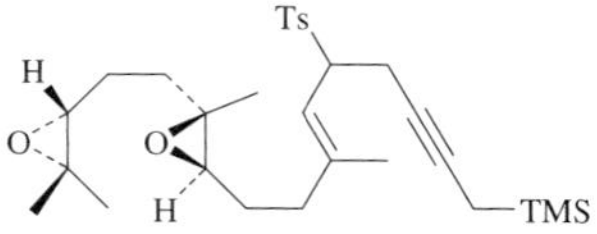 H, O, O, H, Ts, TMS	$PdCl_2$(dppp), $LiHBEt_3$, THF, 0°, 60 min	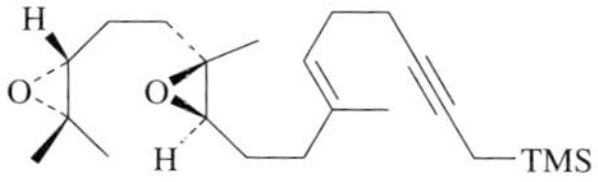 H, O, O, H, TMS (63)	652
C_{30} SO_2Ph, O, O	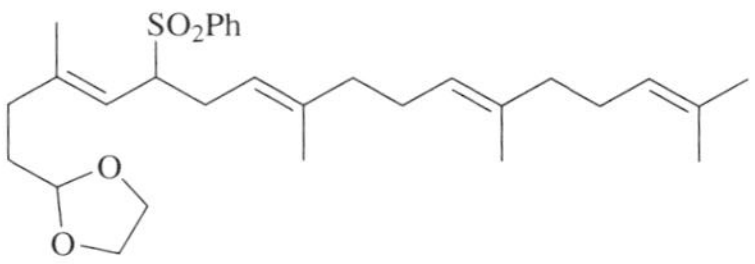 Li/$EtNH_2$, –78°	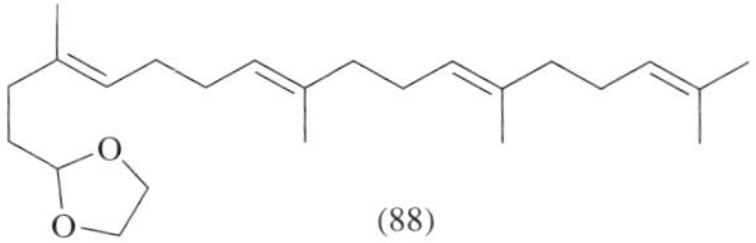 O, O (88)	653

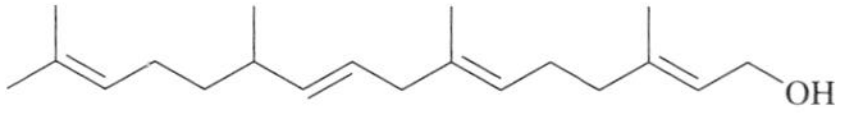

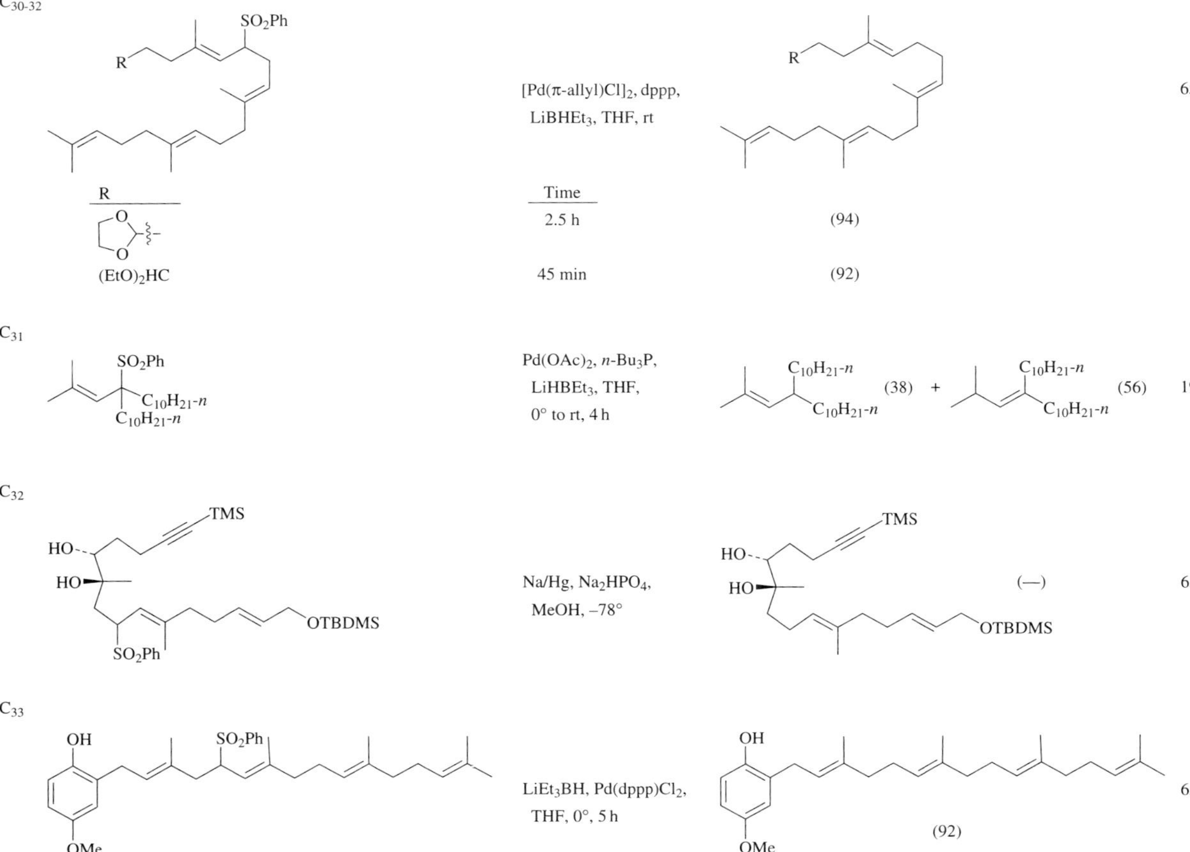
C_{30-32}
SO_2Ph
R
$[Pd(\pi\text{-allyl})Cl]_2$, dppp,
$LiBHEt_3$, THF, rt
654a
R
Time
2.5 h
(94)
$(EtO)_2HC$
45 min
(92)
C_{31}
SO_2Ph
$C_{10}H_{21}$-*n*
$C_{10}H_{21}$-*n*
$Pd(OAc)_2$, *n*-Bu_3P,
$LiHBEt_3$, THF,
0° to rt, 4 h
$C_{10}H_{21}$-*n*
$C_{10}H_{21}$-*n*
(38)
+
$C_{10}H_{21}$-*n*
$C_{10}H_{21}$-*n*
(56)
193
C_{32}
TMS
HO
HO
OTBDMS
SO_2Ph
Na/Hg, Na_2HPO_4,
MeOH, –78°
TMS
HO
HO
OTBDMS
(—)
654b
C_{33}
OH
SO_2Ph
OMe
$LiEt_3BH$, $Pd(dppp)Cl_2$,
THF, 0°, 5 h
OH
OMe
(92)
655

TABLE 6. REDUCTIVE DESULFONYLATION OF ALLYL SULFONES (*Continued*)

Substrate	Conditions	Product(s) and Yield(s) (%)	Refs.
C_{35} OH, SO_2Ph, Pr-*i*, H, H, H, H, OMe	Li/NH_3, –78°	OH, Pr-*i*, H, H, H, H, OMe (65)	22a
C_{36} ( )$_2$, SO_2Ph, O, O	Na/Hg, MeOH, –10°	( )$_2$, O, O (80-82)	656
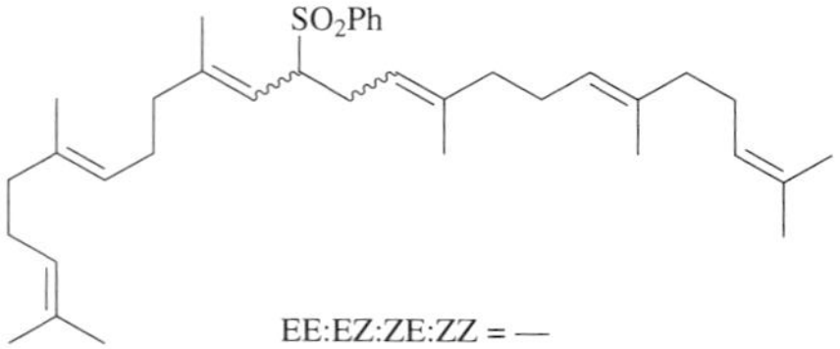 SO_2Ph EE:EZ:ZE:ZZ = —	$Mo(CO)_6$, dioxane, reflux, 21 h	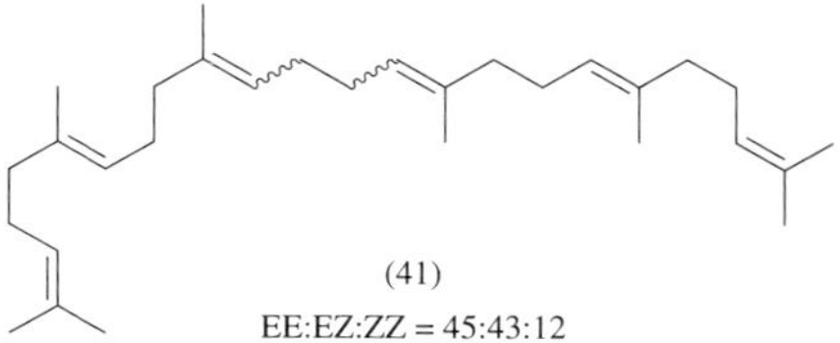 (41) EE:EZ:ZZ = 45:43:12	88
C_{38} PMBO, O, SO_2Ph, HO	$Pd(OAc)_2$, dppp, $NaBH_4$, DMSO, overnight	PMBO, O, HO (77)	657

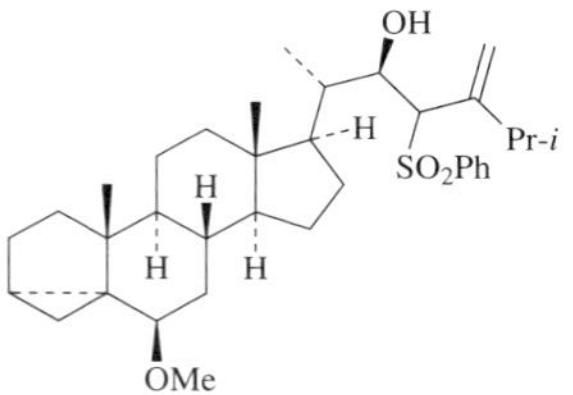

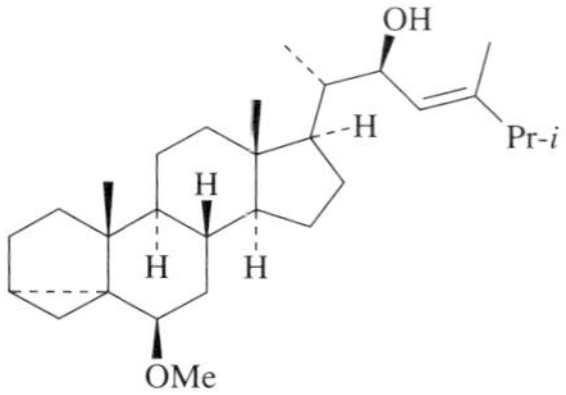

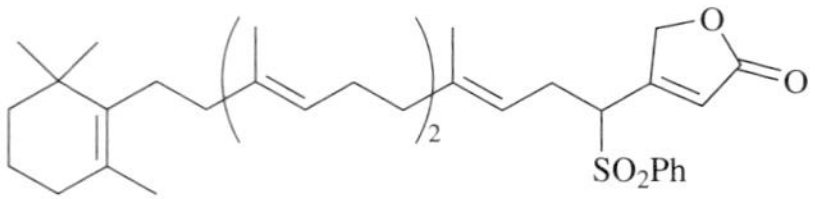

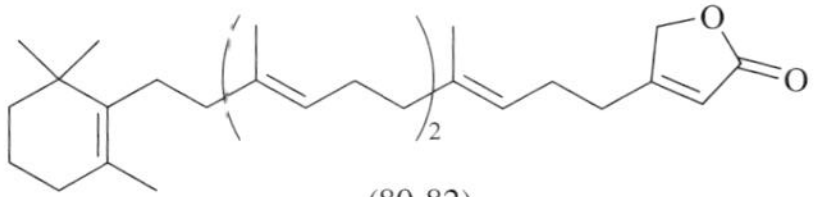

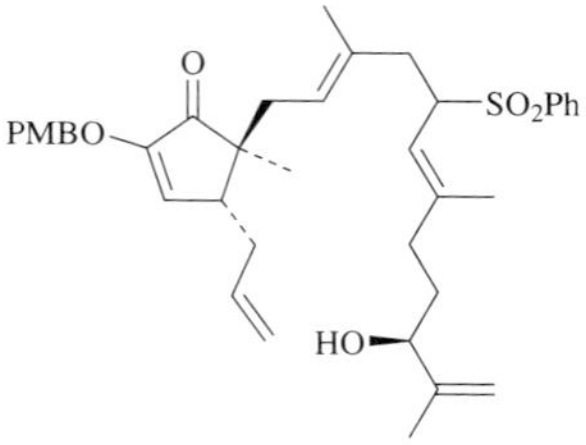

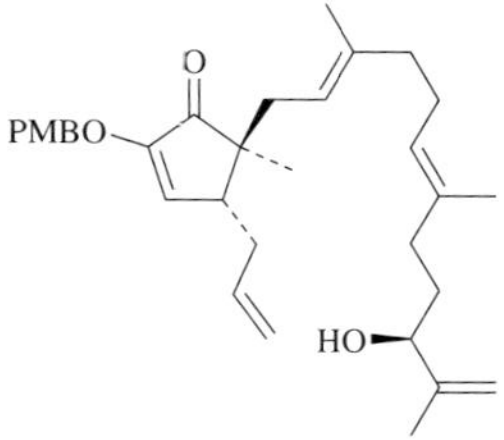

Reactant	Conditions	Product(s) and Yield(s) (%)	Refs.
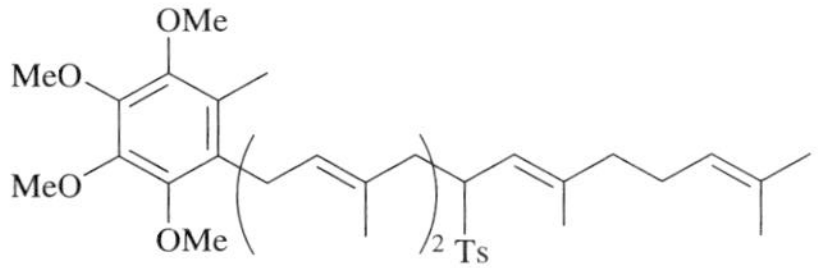	$PdCl_2$(dppp), $LiHBEt_3$, THF, 0°, 5 h	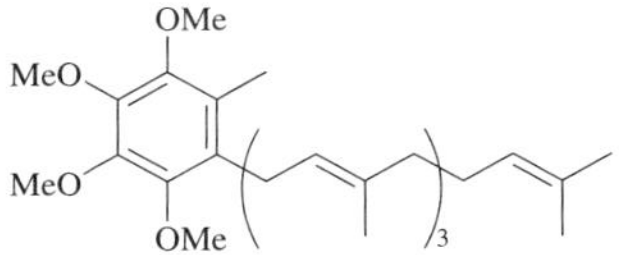 (92)	658
C_{39}			
	$PdCl_2$(dppe), $LiHBEt_3$, THF, 0°, 1 h	(91)	659
C_{40}			
	Na/Hg, Na_2HPO_4, THF/MeOH, –78° to rt	(80)	660
C_{42}			
	$PdCl_2$(dppp), $LiHBEt_3$, THF, 0°, 8 h	(94) Z:E = 9:91	661

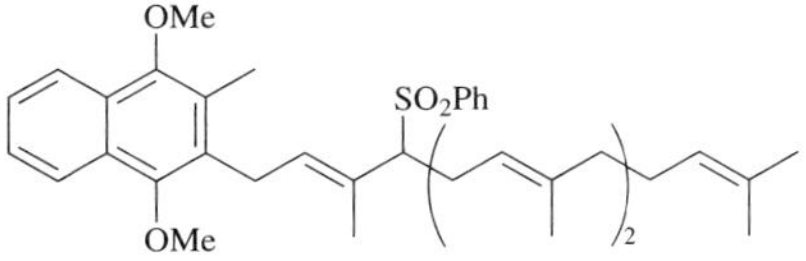

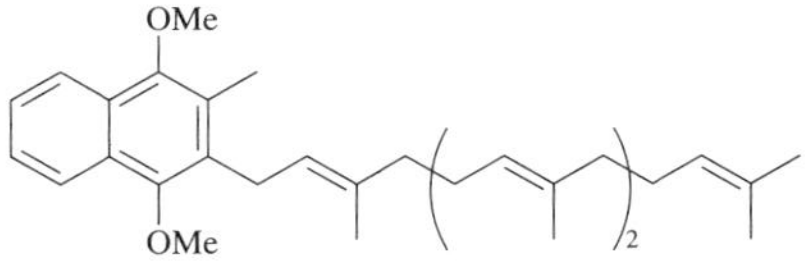

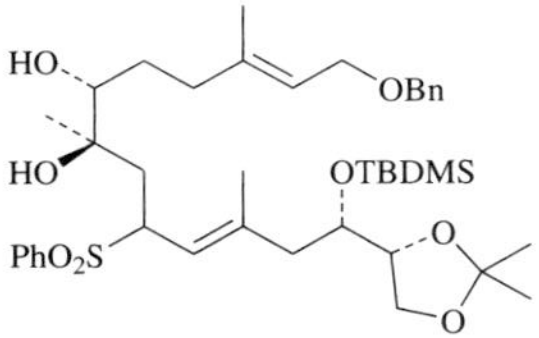

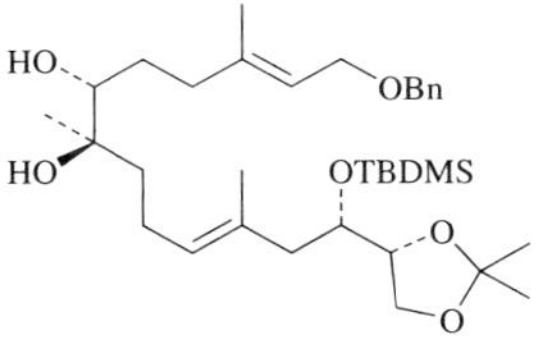

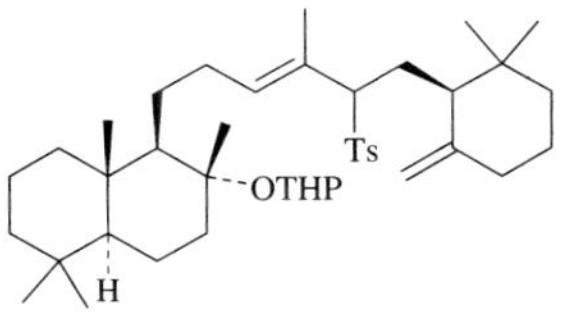

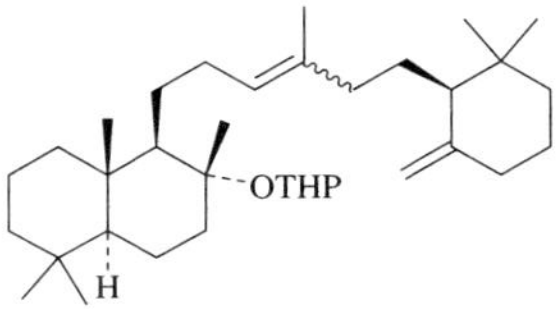

TABLE 6. REDUCTIVE DESULFONYLATION OF ALLYL SULFONES (*Continued*)

Substrate	Conditions	Product(s) and Yield(s) (%)	Refs.
C_{42-45} THPO, SO_2Ph, OSEM, R	$PdCl_2(dppp)$, $LiHBEt_3$, THF, 4°	THPO, OSEM, R	662
R	Time		
$HO(CH_2)_3$	7 h	(93)	
$Me_2C{=}CH(CH_2)_2$	5 h	(83)	
C_{43} BnO, SO_2Ph, H, 2	$Li/EtNH_2$, Et_2O, –78°	HO, H, 2 (83)	663
OTBDMS, SO_2Ph, OTBDMS, OMe, OH	Na/Hg, Na_2HPO_4, MeOH	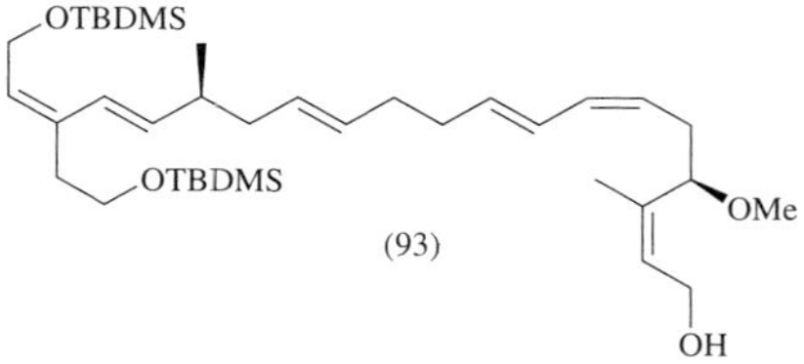 OTBDMS, OTBDMS, OMe, OH (93)	251
C_{45} TBDPSO, O, SO_2Ph	$LiHBEt_3$, $PdCl_2(dppf)$, THF, rt, 6 h	TBDPSO, O (—)	664

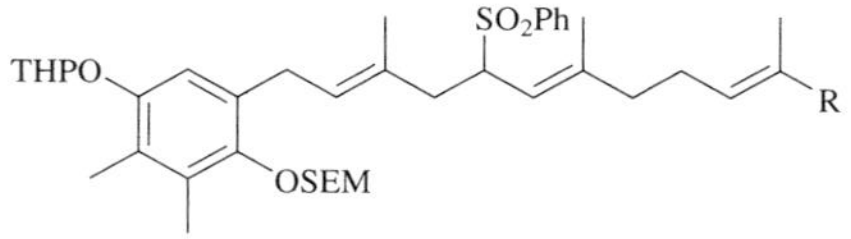

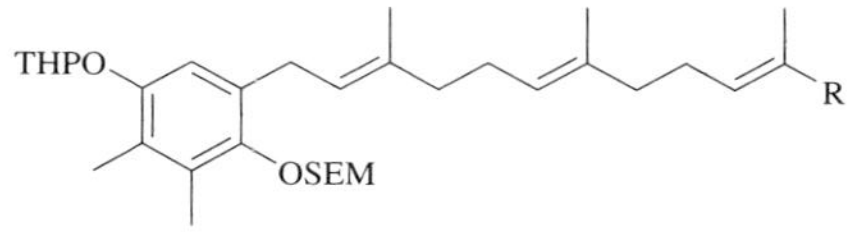

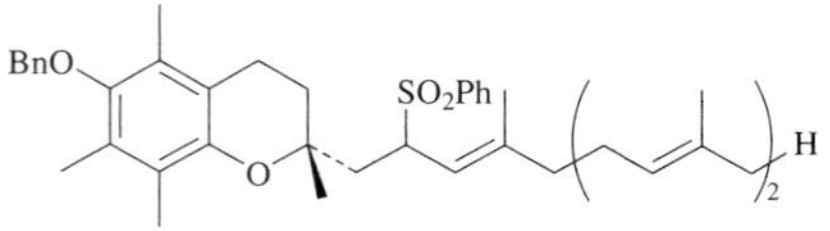

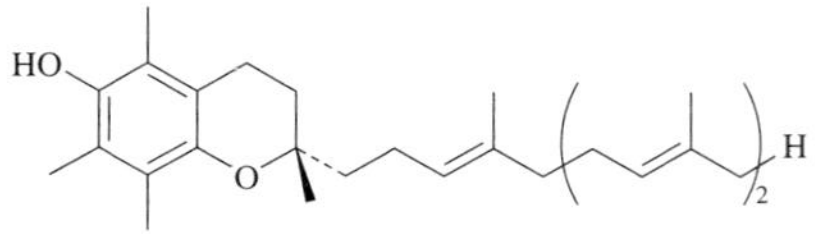

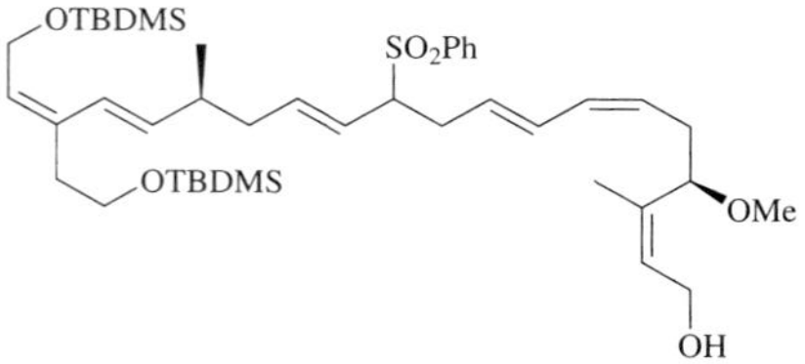

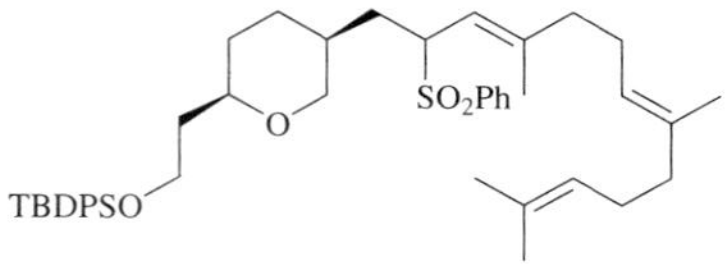

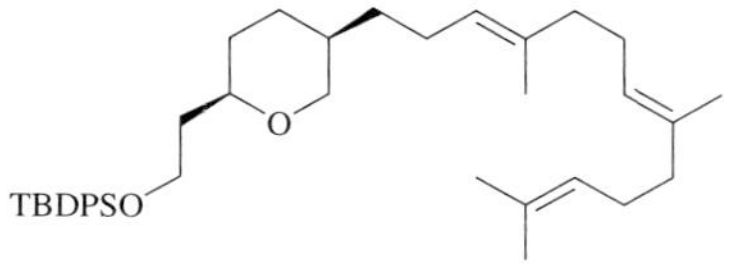

Carbons	Substrate	Conditions	Product(s) and Yield(s) (%)	Refs.
C_{46}	MeO, OMe, OTBDPS, SO_2Ph, O, O	Na/Hg, Na_2HPO_4, MeOH, 26°, 6 h	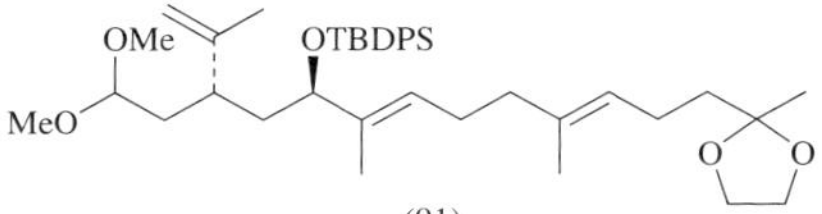 MeO, OMe, OTBDPS, O, O (91)	665
C_{47}	PhO_2S, CO_2Bn, OBn, H	Na/Hg, Na_2HPO_4, MeOH/THF, 0°, 0.5 h; rt, 16 h	CO_2Bn, OBn, H (—)	666
C_{67}	OMe, MeO, MeO, OMe, SO_2Ph, 8	$PdCl_2(dppe)$, $LiHBEt_3$, THF, 0°, 1 h	OMe, MeO, MeO, OMe, 8 (77)	659

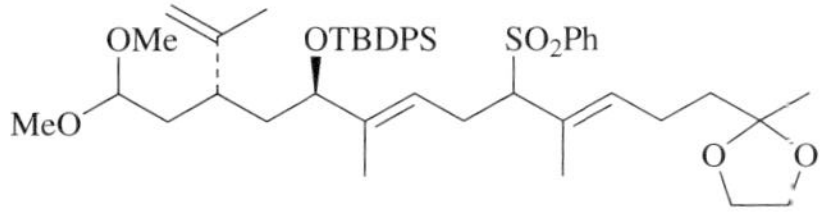

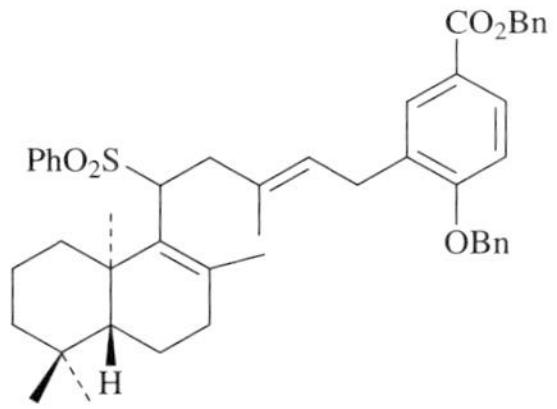

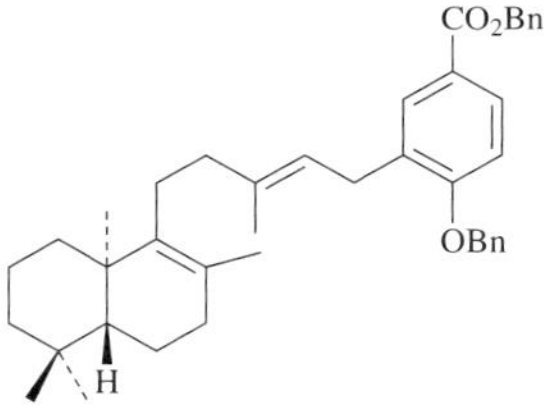

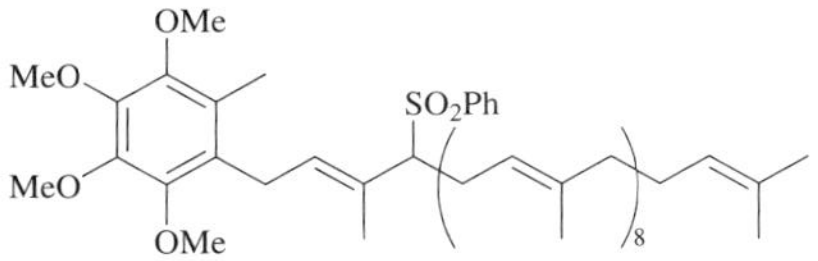

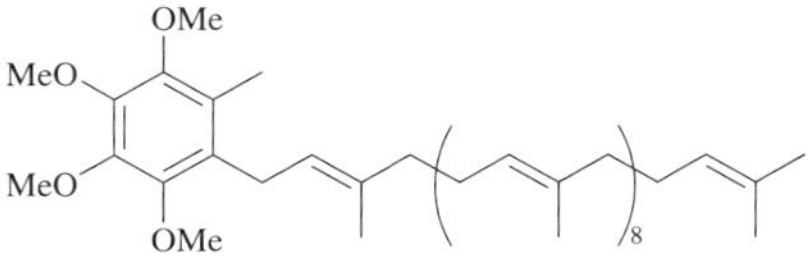

TABLE 6. REDUCTIVE DESULFONYLATION OF ALLYL SULFONES (*Continued*)

Substrate	Conditions	Product(s) and Yield(s) (%)	Refs.
C_{70-80} 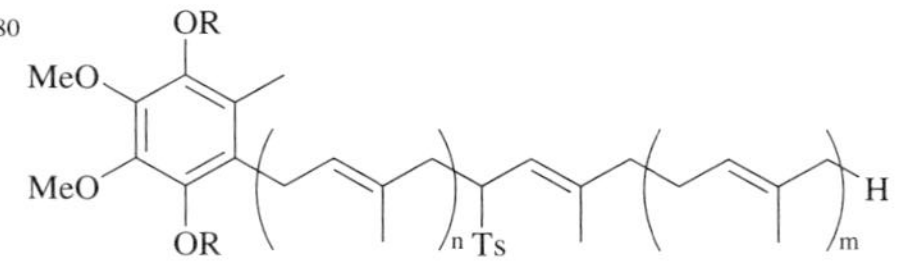 MeO, MeO, OR, OR, Ts, n, m, H	See table.	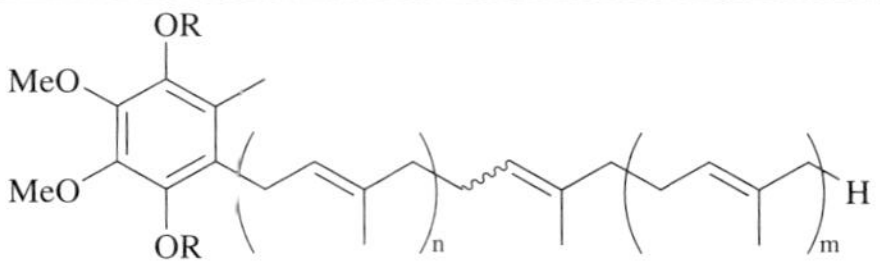 MeO, MeO, OR, OR, n, m, H	
C_{81} OTES, 4, SO_2Ph, 3, OTBDMS	Na/Hg, Na_2HPO_4, MeOH/THF, rt, 2.5 h	OTES, 4, 3, OTBDMS (92)	662
C_{88} H, 2, PhO_2S, PhO_2S, 3, OH, TIPSO	$PdCl_2$(dppp), $LiHBEt_3$, THF, 0°, 6 h	H, 2, 3, OH, TIPSO (51)	670

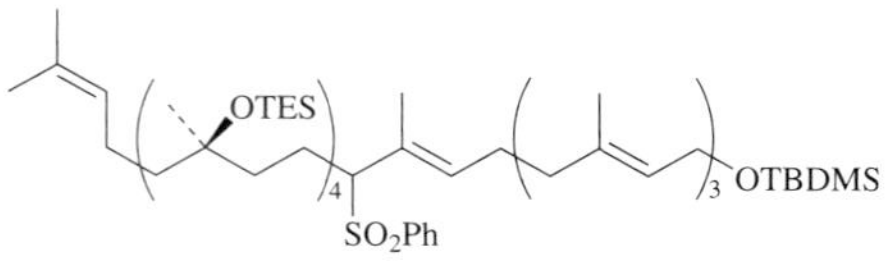

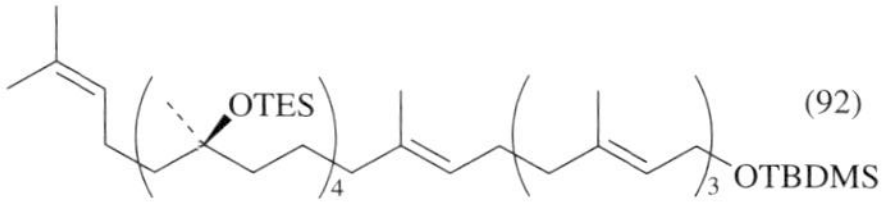

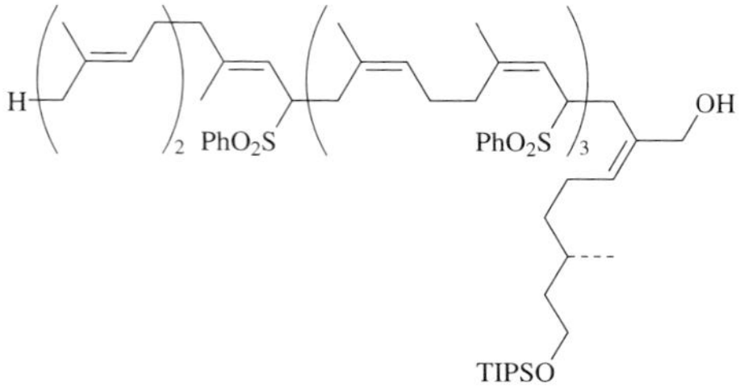

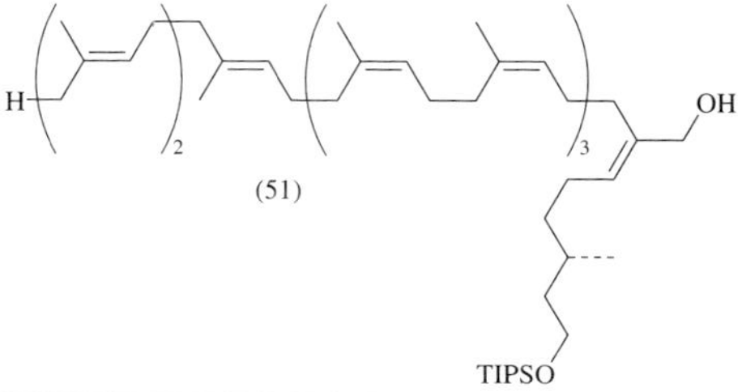

Table for the C_{70-80} entry:

R	n	m	Reagents	Solvent	Temp	Time	Yield (%)	Z:E	Refs.
Bn[b]	7	0	Li/$EtNH_2$	—	–30° to –20°	—	(—)	0:100	667
MOM	1	8	$PdCl_2$(dppp), $LiBHEt_3$	THF	0°	2 h	(94)	5:95	85
MOM	1	8	Na/$C_{10}H_8$	THF	–78°	0.5 h	(99)	0:100	668
Bn[b]	7	1	Li/$EtNH_2$	—	–30° to –20°	—	(—)	0:100	667
Bn[b]	1	8	Li/$C_{10}H_8$	THF	–78°	2 h	(77)	0:100	669
Bn[b]	7	2	Li/$EtNH_2$	—	–30° to –20°	—	(—)	0:100	667

[a] The Z:E ratio is of the major product.

[b] The benzyl group is removed under the reaction conditions.

TABLE 7. REDUCTIVE DESULFONYLATION OF VINYL SULFONES

Substrate			Conditions	Product(s) and Yield(s) (%)	Refs.
$C_{7\text{-}19}$					
2-(SO_2R^3)-5-R^1-1-R^2-pyrrole			$(n\text{-Bu})_3SnH$, AIBN, C_6H_6, reflux	2-R^1-1-R^2-pyrrole	671
R^1	R^2	R^3	Time		
MeC(O)	H	Me	17 h	(58)	
MeC(O)	Me	Me	5 h	(64)	
H	Bn	Me	22 h	(73)	
CHO	Bn	Me	5 h	(40)	
PhC(O)	Me	Me	9 h	(52)	
MeC(O)	Bn	Me	6 h	(62)	
H	Bn	Ph	7 h	(84)	
PhC(O)	Bn	Me	14 h	(36)	
$C_{10\text{-}15}$					
$H(SO_2Ph)C{=}CR^1R^2$			$(n\text{-Bu})_3SnH$, AIBN, xylene, 140°	$H_2C{=}CR^1R^2$	672
R^1	R^2				
Me	Me			(58)	
H	2-furyl			(72)	
H	4-ClC_6H_4			(78)	
H	Ph			(75)	
H	4-MeOC_6H_4			(65)	

TABLE 7. REDUCTIVE DESULFONYLATION OF VINYL SULFONES (*Continued*)

Substrate	Conditions	Product(s) and Yield(s) (%)	Refs.
C_{11-34}			
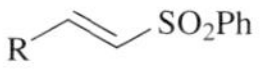 R–CH=CH–SO_2Ph	Na/Hg, NaH_2PO_4, MeOH/THF	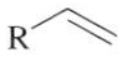 R–CH=CH_2	673
R	Temp		
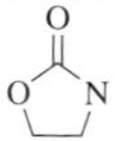 [oxazolidin-2-one, N-linked]	–30°	(86)	
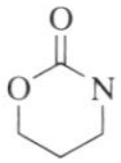 [1,3-oxazinan-2-one, N-linked]	–30°	(13)	
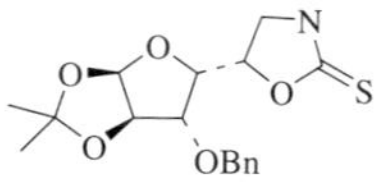 [1,2-O-isopropylidene-3-O-Bn furanose bearing oxazolidine-2-thione]	rt	(64)	
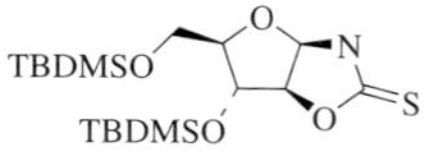 [TBDMSO, TBDMSO furano-fused oxazolidine-2-thione]	rt	(76)	
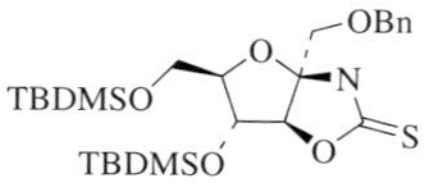 [TBDMSO, TBDMSO, CH_2OBn furano-fused oxazolidine-2-thione]	rt	(98)	
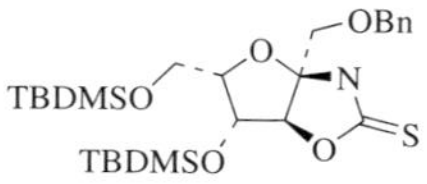 [TBDMSO, TBDMSO, CH_2OBn furano-fused oxazolidine-2-thione, epimer]	rt	(96)	

C_{12}

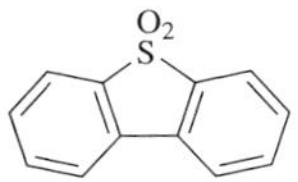

Substrate	Conditions	Product(s) and Yield(s) (%)	Refs.
(dibenzothiophene S,S-dioxide)	NICRA-bpy (4/2/1/2), DME, 63°, 18 h	(biphenyl) (77)	74
(diphenyl sulfone, $PhSO_2Ph$)	Ra-Ni, EtOH, reflux, 6 h	(benzene) (65)	674

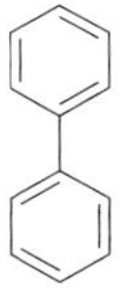

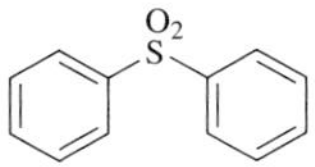

C_{13}

Substrate	Conditions	Product(s) and Yield(s) (%)	Refs.
(4-OMe-3-SO_2Ph-1-Me-pyridin-2-one)	$Na_2S_2O_4$, $NaHCO_3$, TBAI, toluene/H_2O, 90°	(4-OMe-1-Me-pyridin-2-one) (93)	675

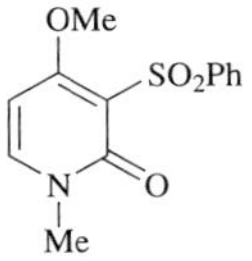

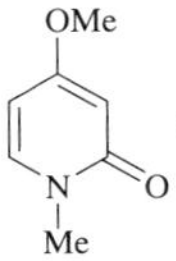

C_{13-18}

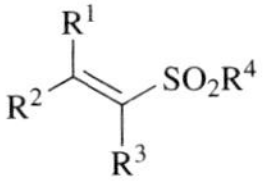

$R^1R^2C{=}C(R^3)SO_2R^4$ → NICRA (2/2/1), DME, 65° → $R^1R^2C{=}C(R^3)H$ **I** + $R^1R^2CH{-}CH(R^3)H$ **II** (75)

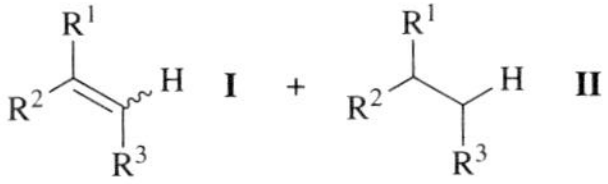

R^1	R^2	R^3	R^4	Time	**I**	Z:E	**II**
H	*n*-Bu	Me	Ph	20 h	(62)	33:67	(14)
H	—$(CH_2)_{10}$—		Et	18 h	(50)	30:70	(0)
H	Ph	Me	Ph	2 h	(77)	73:27	(3)
Me	Me	Ph	Ph	16 h	(84)	—	(6)
H	—$(CH_2)_{10}$—		Ph	3.25 h	(58)	42:58	(0)

TABLE 7. REDUCTIVE DESULFONYLATION OF VINYL SULFONES (*Continued*)

Substrate				Conditions			Product(s) and Yield(s) (%)		Refs.
$C_{13\text{-}22}$ (R¹C(SO₂Ar)=CH~R²)				Reagents, rt			R¹CH=CH~R²		
Ar	R¹	R²	Z:E	Reagents	Solvent	Time		Z:E	
Ph	MeS	2-furyl	<1:99	NaTeH	EtOH	2-3 h	(67)	72:28	164
Ph	MeS	3-ClC_6H_4	<1:99	NaTeH	EtOH	2-3 h	(78)	76:24	164
Ph	MeS	4-ClC_6H_4	<1:99	NaTeH	EtOH	2-3 h	(80)	68:32	164
Ph	MeS	Ph	<1:99	NaTeH	EtOH	2-3 h	(82)	74:26	164
4-Tol	MeS	4-FC_6H_4	<1:99	Mg, TMSCl	DMSO	1 d	(72)	<1:99	45
4-Tol	MeS	4-ClC_6H_4	<1:99	Mg, TMSCl	DMSO	1 d	(71)	<1:99	45
4-Tol	MeS	3-ClC_6H_4	<1:99	Mg, TMSCl	DMSO	1 d	(73)	<1:99	45
4-Tol	MeS	2-ClC_6H_4	<1:99	Mg, TMSCl	DMSO	1 d	(63)	<1:99	45
Ph	MeS	4-MeC_6H_4	<1:99	NaTeH	EtOH	2-3 h	(73)	74:26	164
Ph	MeS	4-$MeOC_6H_4$	<1:99	NaTeH	EtOH	2-3 h	(75)	68:32	164
4-Tol	MeS	Ph	<1:99	Mg, TMSCl	DMSO	1 d	(68)	<1:99	45
4-Tol	Me	Ph	45:55	Mg, TMSCl	DMF	1 d	(72)	<1:99	45
4-Tol	MeS	4-$MeOC_6H_4$	<1:99	Mg, TMSCl	DMSO	1 d	(57)	<1:99	45
4-Tol	MeS	4-MeC_6H_4	<1:99	Mg, TMSCl	DMSO	1 d	(67)	<1:99	45
4-Tol	Et	Ph	32:68	Mg, TMSCl	DMF	1 d	(81)	<1:99	45
4-Tol	*n*-Pr	Ph	31:69	Mg, TMSCl	DMF	1 d	(75)	2:98	45
4-Tol	Bn	Ph	10:90	Mg, TMSCl	DMF	1 d	(85)	1:99	45
C_{14} (Z)-PhSO₂CH=CHSO₂Ph				$(n\text{-}Bu)_3SnH$, CH_2Cl_2, rt, 5 min			(Z)-PhSO₂CH=CH₂ (H)	(~100)	676

C_{14-25}

PhO_2S–C(R^1)=CH–R^2 → $Na_2S_2O_4$ (x eq) → R^1–CH=CH–R^2

R^1	R^2	x	Base (eq)	Solvent	Temp	Time		Z:E	
n-C_5H_{11}	Me	6	Na_2CO_3 (12)	C_6H_{12}/H_2O	80°	3 h	(88)	0.5:99.5	46
Me	*n*-C_5H_{11}	3	$NaHCO_3$ (6)	DMF/H_2O	120°	1.5 h	(52)	0.5:99.5	46
n-C_6H_{13}	Me	3	$NaHCO_3$ (6)	DMF/H_2O	120°	1.5 h	(82)	0:100	46
Bn	MeO	—	—	DMF/H_2O	100°	—	(61)	0:100	677
n-C_6H_{13}	Et	3	$NaHCO_3$ (6)	DMF/H_2O	120°	1.5 h	(65)	0:100	46
n-C_6H_{13}	Et	4	Na_2CO_3 (12)	C_6H_{12}/H_2O	80°	3 h	(55)	0:100	46
n-C_7H_{15}	Me	3	$NaHCO_3$ (6)	DMF/H_2O	120°	1.5 h	(62)	0:100	46
n-C_8H_{17}	Me	3	$NaHCO_3$ (6)	DMF/H_2O	120°	1.5 h	(74)	0:100	46
i-Pr	TBDMSC≡C	4	$NaHCO_3$ (6)	DMF/THF/H_2O	50°	4 h	(50)	<5:95	678
i-Pr	TBDMSC≡C	4	$NaHCO_3$ (6)	DMF/THF/H_2O	50°	18 h	(60)	<5:95	678
n-Bu	THPO$(CH_2)_4$	—	$NaHCO_3$ (—)	DMF/H_2O	80°	7 h	(—)	0:100	679
n-C_6H_{13}	TBDMSC≡C	6	$NaHCO_3$ (12)	DMF/THF/H_2O	60°	2 h	(57)	3:97	678
n-C_6H_{13}	TBDMSC≡C	6	$NaHCO_3$ (12)	THF/H_2O	80°	2 h	(44)	2:98	678
PhCH(OH)	TBDMSC≡C	4	$NaHCO_3$ (6)	DMF/H_2O	120°	1 h	(39)	33:67	678
PhCH(OH)	TBDMSC≡C	6	$NaHCO_3$ (12)	DMF/THF/H_2O	80°	1 h	(55)	6:94	678
n-C_5H_{11}	(*E*)-THPO$(CH_2)_3$CH=CHCH$_2$	—	$NaHCO_3$ (—)	DMF/H_2O	80°	7 h	(—)	0:100	679
n-C_5H_{11}	THPO$(CH_2)_7$	—	$NaHCO_3$ (—)	DMF/H_2O	80°	7 h	(—)	0:100	679

TABLE 7. REDUCTIVE DESULFONYLATION OF VINYL SULFONES (*Continued*)

Substrate		Conditions	Product(s) and Yield(s) (%)	Refs.
C_{14-32}				
R~CH=CH–SO_2Ph		Al/Hg, 0° to rt, 48 h	R–CH=CH_2	680
R	Z:E	Solvent		
2-methyl-1,3-dioxocyclopentan-2-yl	100:0	THF	(47)[a]	
$(EtO_2C)_2C(Et)$–	100:0	CH_3CN	(81)	
(polycyclic N-Ts indole lactam ketone)	50:50	THF	(41)	
C_{15}				
(8-Ts-bicyclic enaminone)		Na/Hg, Na_2HPO_4, THF/MeOH, rt	(bicyclic enaminone) (77)	40
C_{15-16}				
(bicyclic diene, bis-SO_2Bu-*t*)		SmI_2, THF/MeOH, –70°	(bicyclic diene, H, SO_2Bu-*t*) n = 1 (100); n = 2 (85)	681

$C_{15\text{-}17}$

Ar–CH=C(CN)(SO$_2$Ph) — SmI_2, THF/MeOH, 60° — Ar–CH$_2$CH$_2$–CN — 44

Ar	Time	Yield
2,6-$Cl_2C_6H_3$	8 h	(70)
4-ClC_6H_4	4 h	(70)
2-ClC_6H_4	6 h	(60)
4-BrC_6H_4	4 h	(68)
Ph	6 h	(67)
4-MeC_6H_4	4 h	(83)
3-MeC_6H_4	4.5 h	(67)
4-$MeOC_6H_4$	4.5 h	(62)
4-$Me_2NC_6H_4$	5 h	(82)

$C_{15\text{-}18}$

R–N (7-azabicyclo, Ts) — Na/Hg, Na_2HPO_4 — R–N (7-azabicyclo)

R	Solvent	Temp	Time	Yield	Refs.
CO_2Me	MeOH/THF	–78° to rt	—	(—)	682
Boc	EtOAc/*t*-BuOH (1:1)	0° to rt	24 h	(55)	683

PhO$_2$S (N-Ac, R, R tetrahydropyridine) — Na/Hg, Na_2HPO_4, THF/MeOH, rt, 12 h — N-Ac, R, R tetrahydropyridine — 561

R	Yield
Me	(91)
—$(CH_2)_5$—	(93)

TABLE 7. REDUCTIVE DESULFONYLATION OF VINYL SULFONES (*Continued*)

Substrate	Conditions	Product(s) and Yield(s) (%)	Refs.
C_{15-21}			
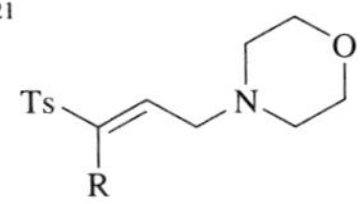 Ts, R, N, O	SmI_2, THF, HMPA, –20°	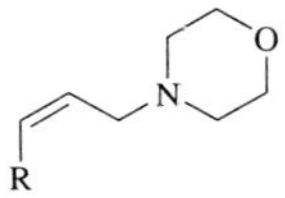 R, N, O	684
R		Z:E	
Me		(42) 88:12	
$H_2C{=}CHCH_2$		(85) 99:1	
Bn		(75) 94:6	
C_{16-20}			
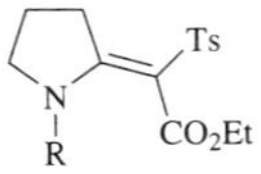 N, R, Ts, CO_2Et	Na/Hg, Na_2HPO_4, MeOH/THF, rt	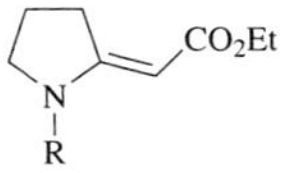 N, R, CO_2Et	40
R		R	
Me		Me (86)	
$AcO(CH_2)_3$		$HO(CH_2)_3$ (87)	
C_{17}			
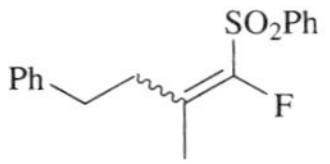 Ph, SO_2Ph, F	1. $(n\text{-Bu})_3SnH$, AIBN, C_6H_6, reflux 2. NaOMe, MeOH, reflux	 Ph, F	63
Z		(76) E	
E		(82) Z	
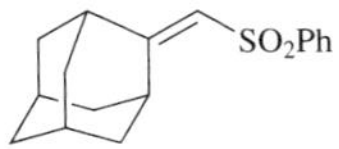 SO_2Ph	Mg, $HgCl_2$, EtOH, rt, 2 h	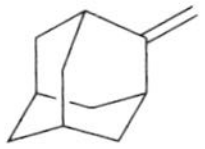 (99)	114

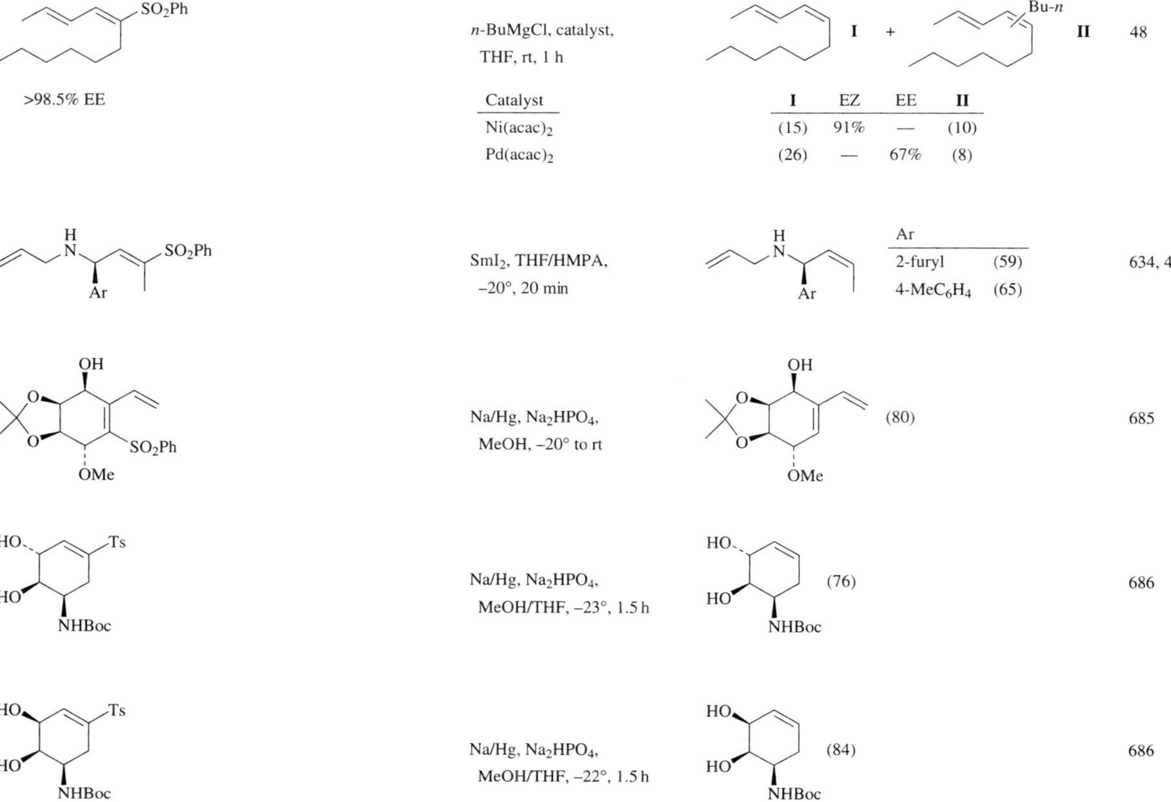

Reactant	Conditions	Product(s) and Yield(s) (%)	Refs.
>98.5% EE	n-BuMgCl, catalyst, THF, rt, 1 h	I + II	48
	Catalyst: $Ni(acac)_2$	I (15), EZ 91%, EE —, II (10)	
	Catalyst: $Pd(acac)_2$	I (26), EZ —, EE 67%, II (8)	
$C_{17\text{-}20}$	SmI_2, THF/HMPA, –20°, 20 min	Ar = 2-furyl (59); Ar = 4-MeC_6H_4 (65)	634, 41
C_{18}	Na/Hg, Na_2HPO_4, MeOH, –20° to rt	(80)	685
	Na/Hg, Na_2HPO_4, MeOH/THF, –23°, 1.5 h	(76)	686
	Na/Hg, Na_2HPO_4, MeOH/THF, –22°, 1.5 h	(84)	686

TABLE 7. REDUCTIVE DESULFONYLATION OF VINYL SULFONES (*Continued*)

	Substrate	Conditions	Product(s) and Yield(s) (%)	Refs.
C_{18}	HO, HO, Ts, NHBoc	Na/Hg, Na_2HPO_4, MeOH/THF, –22°, 1.5 h	HO, HO, NHBoc (70)	686
	SO_2Ph >99% EE	n-BuMgCl, Ni(acac)$_2$, THF, rt, 1 h	(51) >85% EZ + Bu-n (12)	48
	Ts, OMOM, t-Bu, OH	Na/Hg, Na_2HPO_4, MeOH, 0° to rt, 1.5 h	OH, t-Bu, OMOM (73) Z:E = 25:75	135
	SO_2Ph Z:E = 1:99	n-BuMgCl, Ni(acac)$_2$, THF, rt, 1 h	(70) Z:E = 97.5:2.5	48
C_{18-25}	R^1, SO_2Ph, R^2	See table	R^1, R^2	48

R^1	R^2	EE	EZ	ZE
Et	n-C_6H_{13}	98.5%	1%	0.5%
Et	$THPO(CH_2)_8$	98.7%	1%	0.3%
Me	$THPO(CH_2)_8$	99.5%	—	—

Conditions		EZ	EE	ZE	ZZ
n-BuMgCl, Ni(acac)$_2$/(n-Bu)$_3$P, THF, rt, 1 h	(51)	96%	4%	0	0
n-BuMgCl, Ni(acac)$_2$, THF, rt, 1 h	(35)	2%	5%	93%	0
$Na_2S_2O_4$, $NaHCO_3$, H_2O, reflux, 18 h	(65)	0	0	>97%	<3%

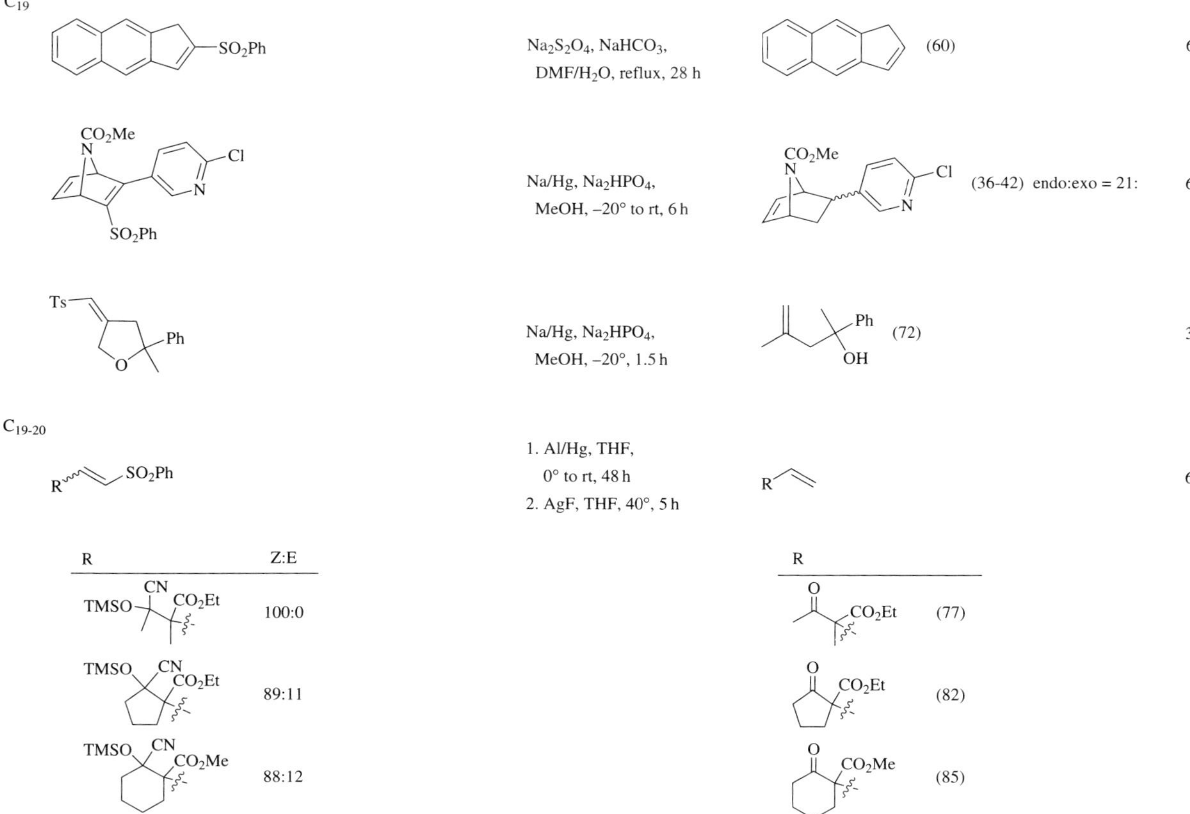

Substrate	Conditions	Product(s) and Yield(s) (%)	Refs.
C_{19}			
SO$_2$Ph	$Na_2S_2O_4$, $NaHCO_3$, DMF/H_2O, reflux, 28 h	(60)	687
CO_2Me, N, Cl, N, SO_2Ph	Na/Hg, Na_2HPO_4, MeOH, –20° to rt, 6 h	CO_2Me, N, Cl, N (36-42) endo:exo = 21:	688
Ts, O, Ph	Na/Hg, Na_2HPO_4, MeOH, –20°, 1.5 h	Ph, OH (72)	39
C_{19-20}			
R, SO_2Ph	1. Al/Hg, THF, 0° to rt, 48 h 2. AgF, THF, 40°, 5 h	R	680

R	Z:E
TMSO, CN, CO_2Et	100:0
TMSO, CN, CO_2Et	89:11
TMSO, CN, CO_2Me	88:12

R	
O, CO_2Et	(77)
O, CO_2Et	(82)
O, CO_2Me	(85)

TABLE 7. REDUCTIVE DESULFONYLATION OF VINYL SULFONES (*Continued*)

Substrate	Conditions	Product(s) and Yield(s) (%)	Refs.
C_{19-29}			
MeO_2C, R, HN, Ph, O, SO_2Ph	Al/Hg, THF/H_2O, 70°, 4 h	MeO_2C, R, HN, Ph, O; R = Me (80)	689
		R = Me (—)	690
		R = $HO_2CCH_2CH_2$ (—)	690
		R = *i*-Bu (—)	690
		R = $PhC(O)NH(CH_2)_4$ (—)	690
C_{20}			
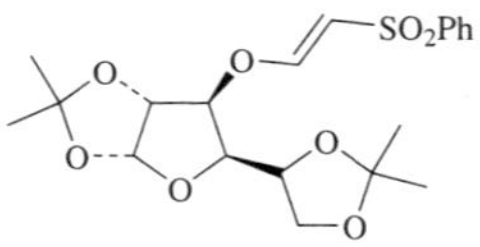 SO_2Ph	Na/Hg, NaH_2PO_4, MeOH/THF, rt, 1-2 h	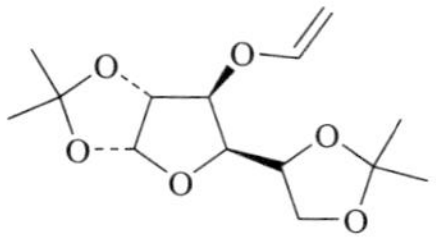 (60)	691
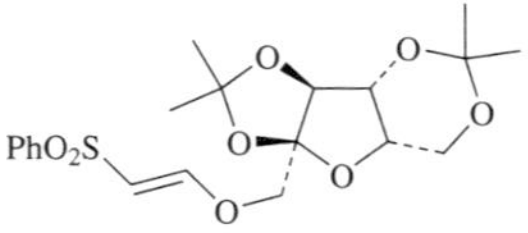 PhO_2S	Na/Hg, NaH_2PO_4, MeOH/THF, rt, 1-2 h	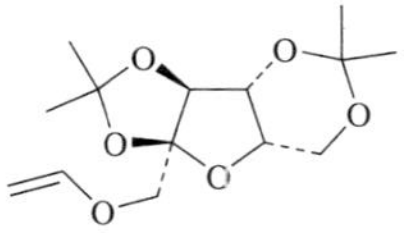 (65)	691
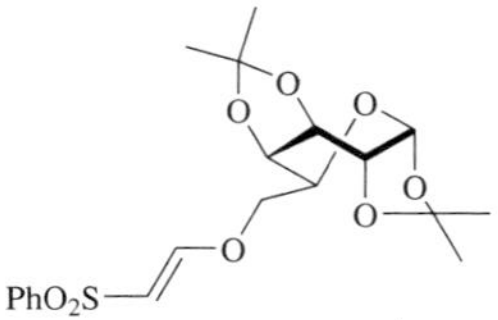 PhO_2S	Na/Hg, NaH_2PO_4, MeOH/THF, rt, 1-2 h	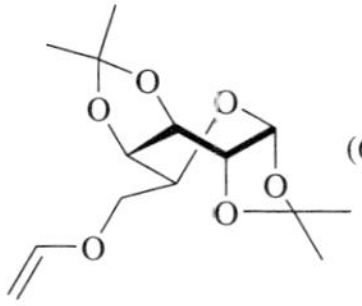 (66)	691
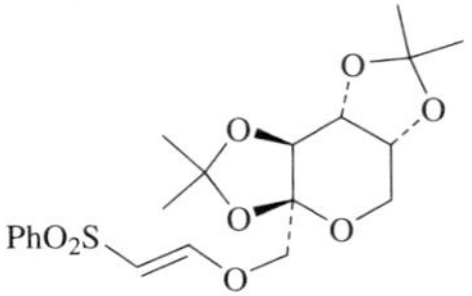 PhO_2S	Na/Hg, NaH_2PO_4, MeOH/THF, rt, 1-2 h	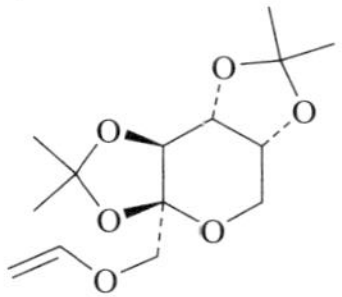 (82)	691

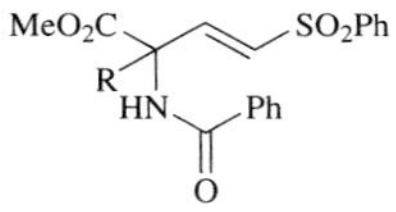

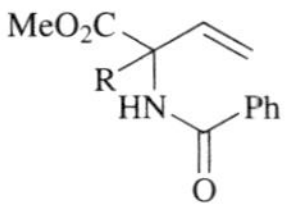

PhO_2S

See table.

691

Reagents	Solvent	Temp	Time	
Na/Hg, NaH_2PO_4	MeOH/THF	rt	1-2 h	(94)
Mg	MeOH	—	—	(64)

PhO_2S

See table.

691

Reagents	Solvent	Temp	Time	
Na/Hg, NaH_2PO_4	MeOH/THF	rt	1-2 h	(55)
Mg	MeOH	—	—	(65)

PhO_2S

Na/Hg, NaH_2PO_4,
MeOH/THF, rt, 1-2 h

(65)

691

TABLE 7. REDUCTIVE DESULFONYLATION OF VINYL SULFONES (*Continued*)

Substrate	Conditions	Product(s) and Yield(s) (%)	Refs.
C_{20-26} R¹C(SO₂Ph)=CHR² Z:E = <1:99	SmI_2, DMPU, THF, MeOH, rt, 30 min	R¹CH=CHR²	42

R^1	R^2	Yield (%)	Z:E
Me_2C=CH	(*E*)-PhCH=CH	(94)	17:83
Me_2C=CH	$PhCH_2CH_2$	(89)	0:100
$PhCH_2CH_2$	Me_2C=CH	(95)	17:83
n-C_7H_{15}	n-C_6H_{13}	(69)	20:80
$PhCH_2CH_2$	Ph	(85)	0:100
$PhCH_2CH_2$	(*E*)-PhCH=CH	(94)	0:100
geranyl-type (Me₂C=CHCH₂CH₂C(Me)=CH–)	(*E*)-PhCH=CH	(70)	33:67
"	geranyl-type (Me₂C=CHCH₂CH₂C(Me)=CH–)	(78)	14:86

Substrate	Conditions	Product(s) and Yield(s) (%)	Refs.
C_{20-40} PhO₂S(R¹)C=CR²R³	THF, reflux, 3 h	R¹CH=CR²R³	187

R^1	R^2	R^3	Reagents	Yield (%)
Ph	4-ClC_6H_4	H	Al/Hg, $HgCl_2$	(90)
H	Ph	Ph	Al/Hg, $HgCl_2$	(90)
H	Ph	Ph	$LiAlH_4$/$CuCl_2$ (1:2)	(60)
Ph	Ph	H	Al/Hg, $HgCl_2$	(90)
Ph	Ph	Ph	Al/Hg, $HgCl_2$	(85)
Ph	Ph	Ph	$LiAlH_4$	(40)
Ph	Ph	Ph	$LiAlH_4$/$CuCl_2$ (1:2)	(65)
Ph	cholest-4-en-3-ylidene[b]		Al/Hg, $HgCl_2$	(80)[b]
Ph	cholest-4-en-3-ylidene[b]		$LiAlH_4$/$CuCl_2$ (1:2)	(60)[b]

Substrate	Conditions	Product(s) and Yield(s) (%)	Refs.
C_{21}	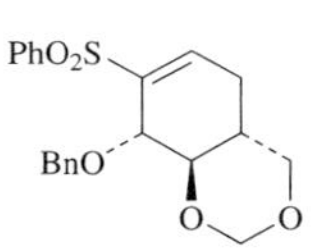 Na/Hg, Na_2HPO_4, MeOH, –20° to rt, 4 h	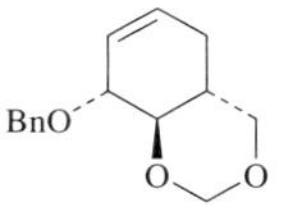 (75)	692
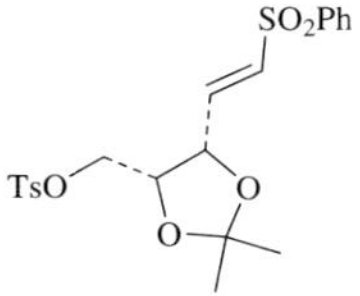	SmI_2, DMPU, THF	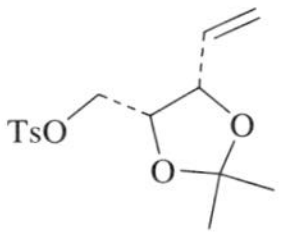 (—)	429
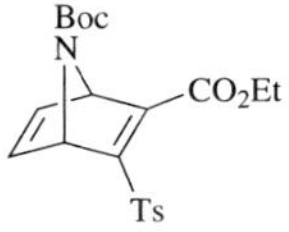	1. $(n\text{-Bu})_3SnH$, THF, rt, 1 h 2. $NaBH_4$, MeOH, 0°, 2 h	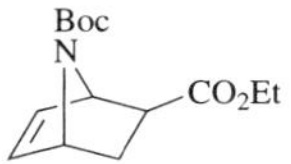 (55)	433
$C_{21\text{-}37}$	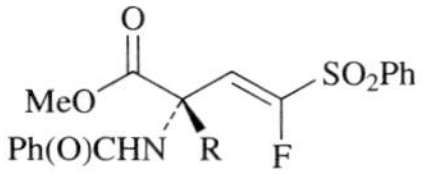 1. $(n\text{-Bu})_3SnH$, AIBN, C_6H_6, reflux, 24 h 2. HCl (6 N), reflux, 17 h	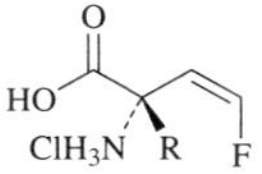	65
R: MeO_2CCH_2		(26)	
R: $PhC(O)NH(CH_2)_4$		(48)	
R: 3-TBDMSO$C_6H_4CH_2$		(85)	
R: 3,4-$(TBDMSO)_2C_6H_3CH_2$		(74)	
C_{22}	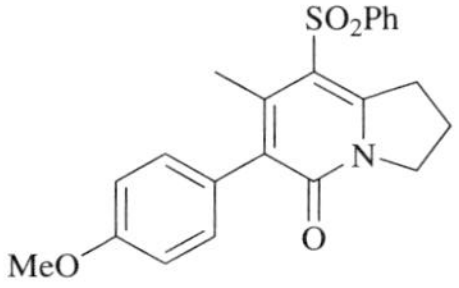 Ra-Ni, EtOH, 65°, 5 h	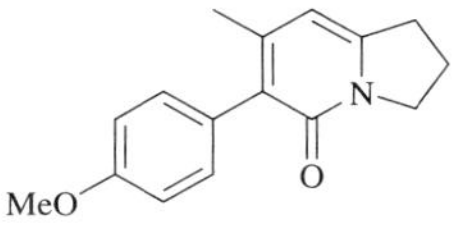 (90)	693

TABLE 7. REDUCTIVE DESULFONYLATION OF VINYL SULFONES (*Continued*)

Substrate	Conditions	Product(s) and Yield(s) (%)	Refs.
C_{22}			
MeO, O, O, Ph, SO_2Ph	Na/Hg, NaH_2PO_4, MeOH/THF, rt, 1-2 h	MeO, O, O, Ph (84)	691
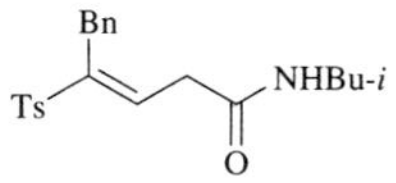 Bn, Ts, NHBu-*i*, O	Na/Hg, Na_2HPO_4, MeOH, –20° to rt	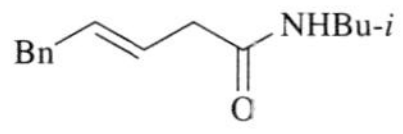 Bn, NHBu-*i*, O (—)	133
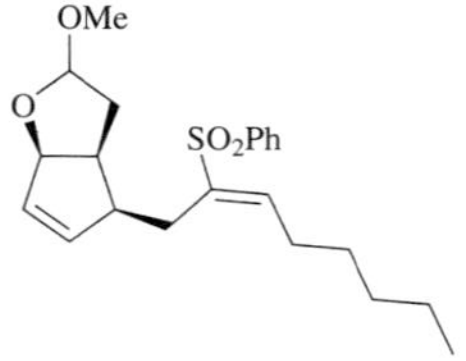 OMe, O, SO_2Ph	$Na_2S_2O_4$, $NaHCO_3$, H_2O/EtOH, reflux, 2 h	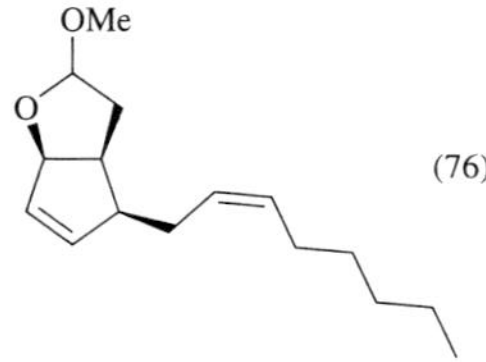 OMe, O (76)	694
C_{23}			
MeO, O, O, Ph, SO_2Ph	Na/Hg, NaH_2PO_4, MeOH/THF, rt, 1-2 h	MeO, O, O, Ph (84)	691
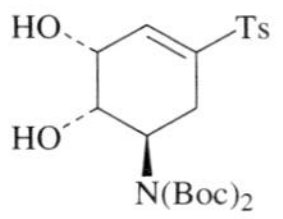 HO, HO, Ts, $N(Boc)_2$	Na/Hg, Na_2HPO_4, MeOH/THF, –12°, 1.5 h	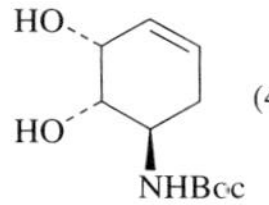 HO, HO, NHBoc (49)	686

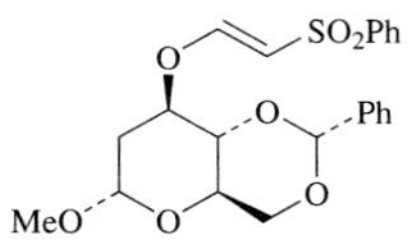

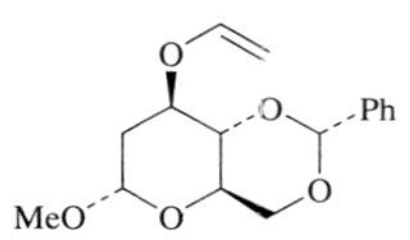

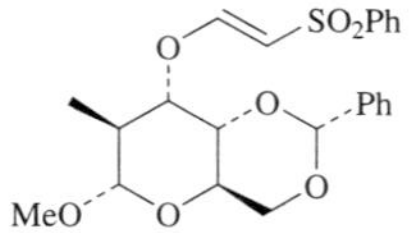

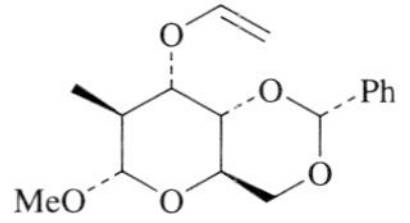

C_{23-25}

PhO$_2$S, Ar, N, S, O, 4-Tol

Na/Hg

Ar, N, S, O, 4-Tol

Ar	
3-pyridyl	(—)
4-MeC$_6$H$_4$	(—)

695

C_{24}

SO$_2$Bu-*t*, *t*-BuO$_2$S

SmI$_2$, THF/MeOH, –70°

H, *t*-BuO$_2$S (96)

681

C_{24-28}

R^{4}O, SO$_2$Ph, OR^4, R^2, R^3, R^2, R^1

R^1	R^2	R^3	R^4
H	Me	H	H
H	Me	H	Me
Me	Me	H	Me
Me	H	*i*-Pr	H
Me	H	*i*-Pr	Me

LiAlH$_4$, additive, THF

R^{4}O, OR^4, R^2, R^3, R^2, R^1

Additive	Temp	Time	
MeLi/LiBr	reflux	50 h	(68)
TiCl$_4$	–78° to rt	2.5 h	(84)
TiCl$_4$	reflux	50 h	(71)
MeLi/LiBr	–78° to rt	6 h	(76)
TiCl$_4$	–78° to rt	1.5 h	(91)

80

C_{25}

MeO$_2$C, EtO$_2$C, NH, O, Ts, OAc

Mg, MeOH, rt, 1 h

MeO$_2$C, EtO$_2$C, NH, O, OH (70)

696

TABLE 7. REDUCTIVE DESULFONYLATION OF VINYL SULFONES (*Continued*)

	Substrate	Conditions	Product(s) and Yield(s) (%)	Refs.
C_{25} 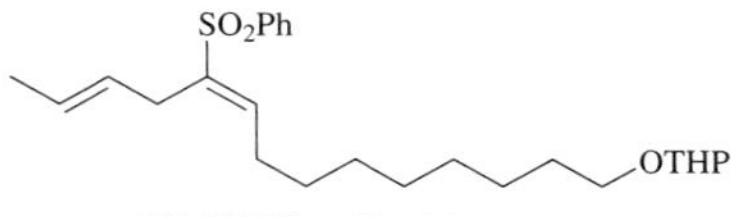	SO_2Ph; OTHP; EE:EZ:ZE = >98:<1:1	1. *n*-BuMgCl, Ni(acac)$_2$, THF, rt, 1 h 2. HCl, MeOH 3. Ac_2O, TEA, DMAP	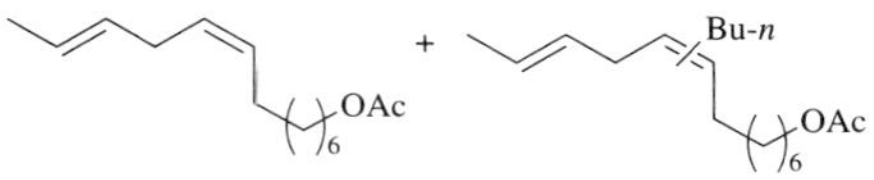 OAc (24) >90% ZE + Bu-*n*, OAc (11)	48
C_{26} 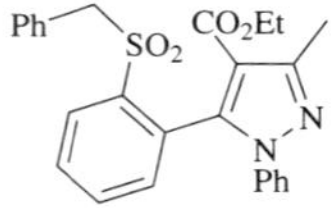	Ph, SO_2, CO_2Et, N, N, Ph	Ra-Ni, EtOH, reflux, 9 h	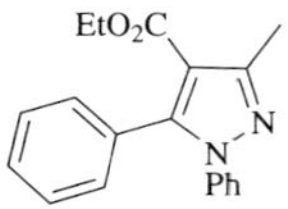 EtO_2C, N, N, Ph (—)	697
	MeO_2C, MeO_2C, NAc, N, Me, Ts, OAc	1. Na/$C_{10}H_8$, THF, –78°, 5 min 2. Ac_2O, Py, –78° to rt, overnight	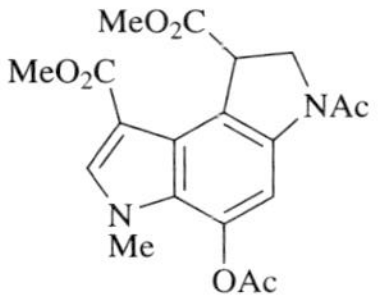 MeO_2C, MeO_2C, NAc, N, Me, OAc (50)	698
C_{27} 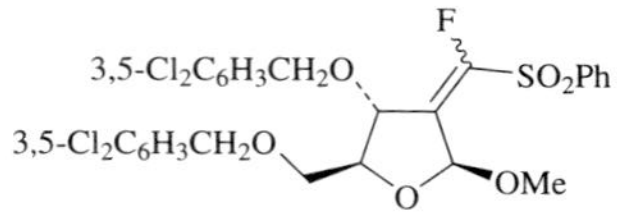	F, SO_2Ph, 3,5-$Cl_2C_6H_3CH_2O$, 3,5-$Cl_2C_6H_3CH_2O$, O, OMe	(*n*-Bu)$_3$SnH, AIBN, C_6H_6	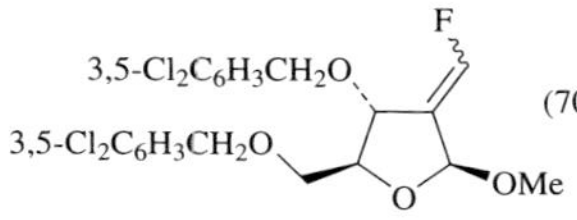 F, 3,5-$Cl_2C_6H_3CH_2O$, 3,5-$Cl_2C_6H_3CH_2O$, O, OMe (70)	699
C_{29} 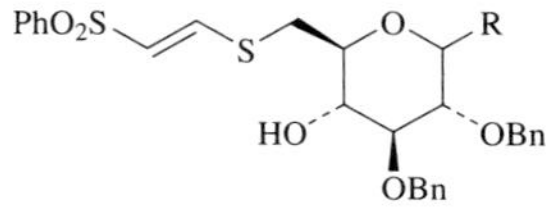	PhO_2S, S, O, R, HO, OBn, OBn	Na/Hg, NaH_2PO_4, MeOH/THF, rt	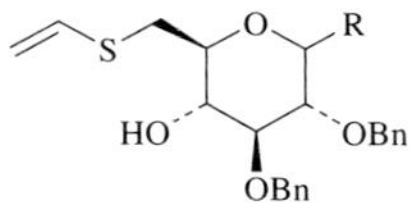 S, O, R, HO, OBn, OBn R α-MeO (59) β-MeO (56)	700

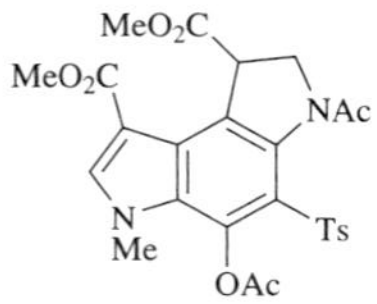

	Substrate	Conditions	Product(s) and Yield(s) (%)	Refs.
C_{30}	(structure) $(i\text{-}Pr)_2Si$, O, Si, O, $(Pr\text{-}i)_2$, O, N, H, OEt, SO_2Ph, F	1. $(n\text{-}Bu)_3SnH$, AIBN, C_6H_6 2. See table	(structure) R^1O, R^2O, O, N, H, NH_2, F	63, 64

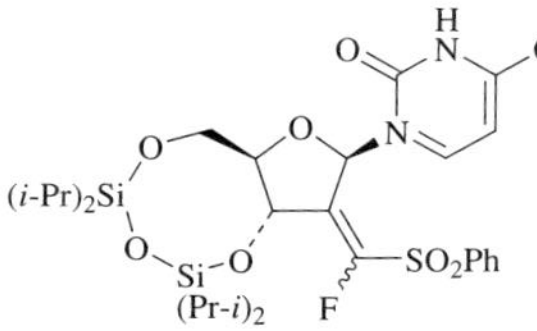

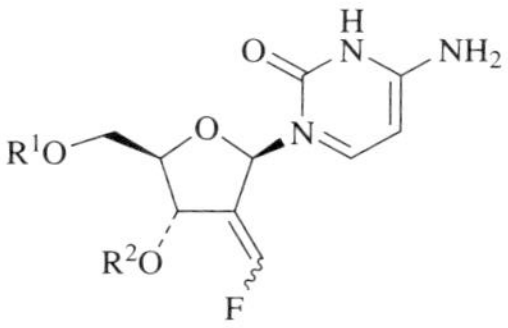

Z:E
100:0
100:0
0:100

Step 2	R^1	R^2		Z:E
NH_3, MeOH	—$Si(i\text{-}Pr)_2OSi(i\text{-}Pr)_2$—		(70)	0:100
CsF, NH_3, MeOH, 50°, 24 h	H	H	(41)	0:100
CsF, NH_3, MeOH, 50°, 24 h	H	H	(46)	100:0

	Substrate	Conditions	Product(s) and Yield(s) (%)	Refs.
$C_{30\text{-}32}$	(structure) TBDMSO, N, PhO_2S, n n = 2, 4	Na/Hg, KH_2PO_4, MeOH/THF, –30°, 1 h	(structure) TBDMSO, N, n (—) EE:EZ = 89:11	701
C_{31}	(structure) BnO, OBn, O, SO_2Ph, O, O, O	Na/Hg, NaH_2PO_4, MeOH/THF, rt, 1-2 h	(structure) BnO, OBn, O, O, O, O (93)	691
	(structure) HO, O, O, H, OBn, Ts	Na/Hg, Na_2HPO_4, MeOH, 0°	(structure) HO, O, O, H, OBn (—)	702

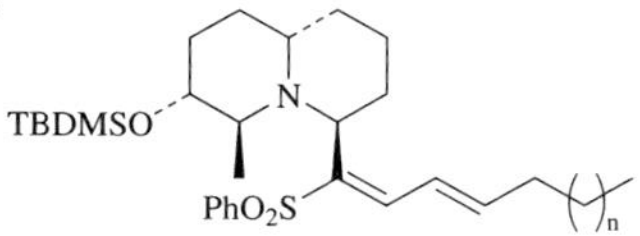

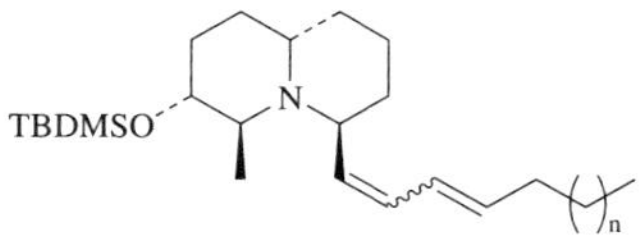

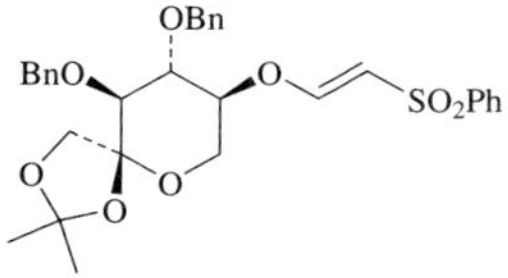

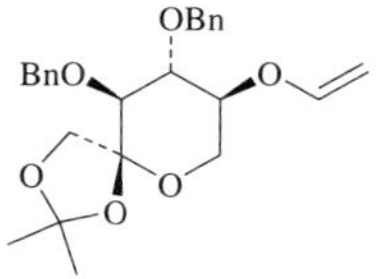

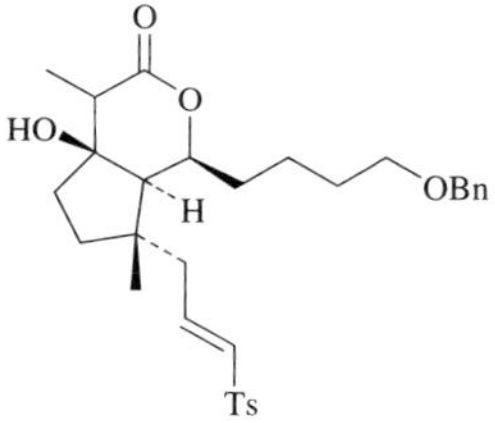

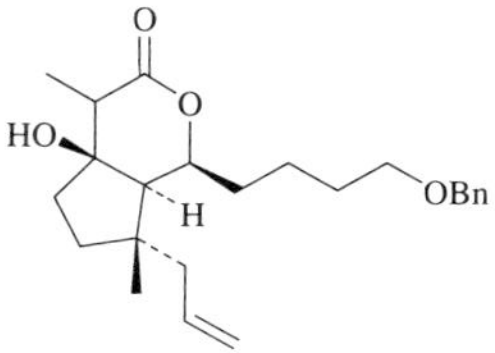

TABLE 7. REDUCTIVE DESULFONYLATION OF VINYL SULFONES (*Continued*)

	Substrate	Conditions	Product(s) and Yield(s) (%)	Refs.
C_{33} 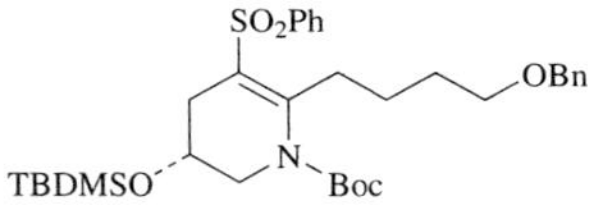	SO$_2$Ph; OBn; TBDMSO; N; Boc	Ra-Ni, EtOH, 70°, 20 h	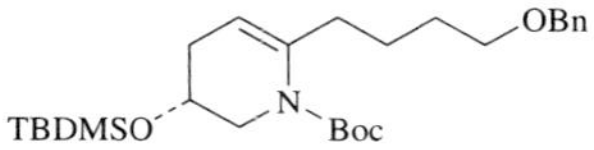 OBn; TBDMSO; N; Boc (100)	701
C_{35-37} 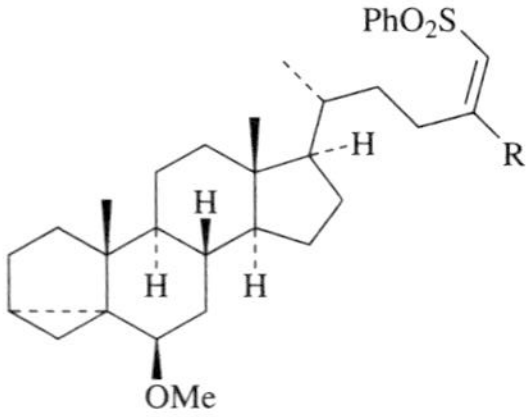	PhO$_2$S; H; R; H; H; H; OMe	Na/Hg, Na_2HPO_4, THF/MeOH, rt, 20 h	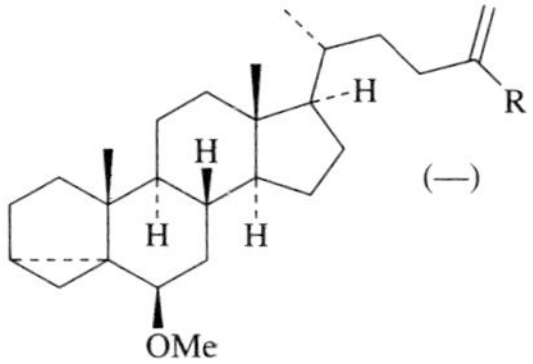 H; R; H; H; H; OMe (—); R: *i*-Pr, (*S*)-MeCH$_2$CHMe, Et$_2$CH	703
C_{40-51} 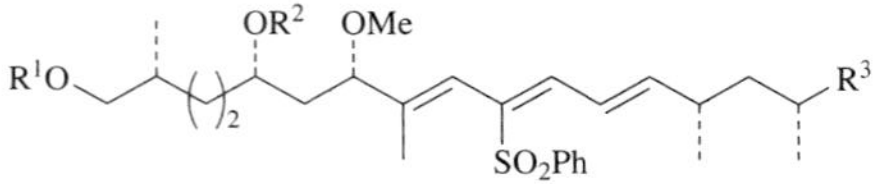	R^{1}O; OR^2; OMe; ( )$_2$; R^3; SO$_2$Ph	Na/Hg, KH_2PO_4, MeOH/THF, –20°, 1 h	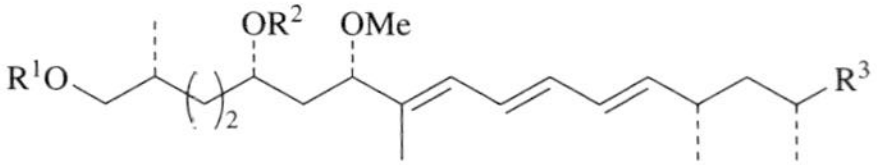 R^{1}O; OR^2; OMe; ( )$_2$; R^3	
	R^1 = TES, R^2 = TBDMS, R^3 = HOCH$_2$		(—)	704
	R^1 = TBDPS, R^2 = Bn, R^3 = CHO		(85)	705
C_{43} 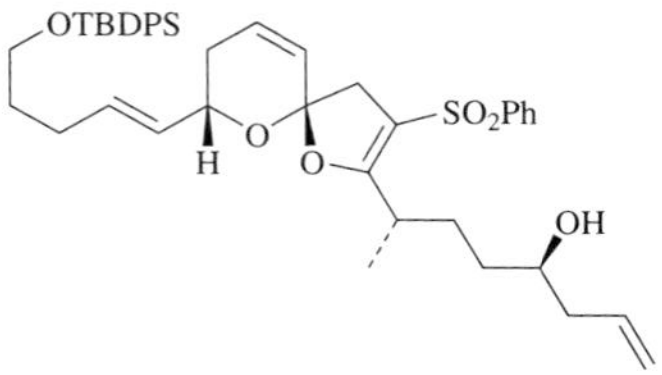	OTBDPS; H; O; O; SO$_2$Ph; OH	Na/Hg, Na_2HPO_4, MeOH/THF, –10°, 20 min; 0°, 1 h	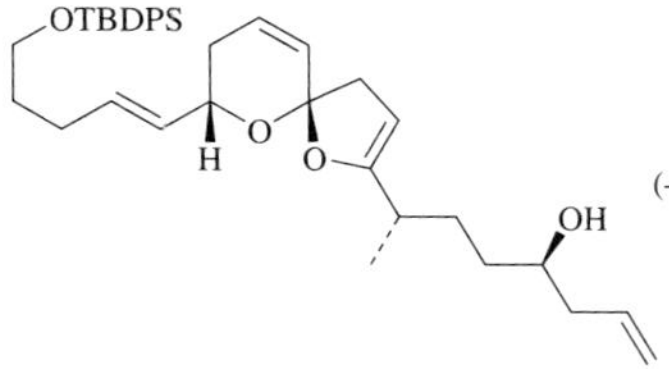 OTBDPS; H; O; O; OH (—)	706, 611

C_{61}	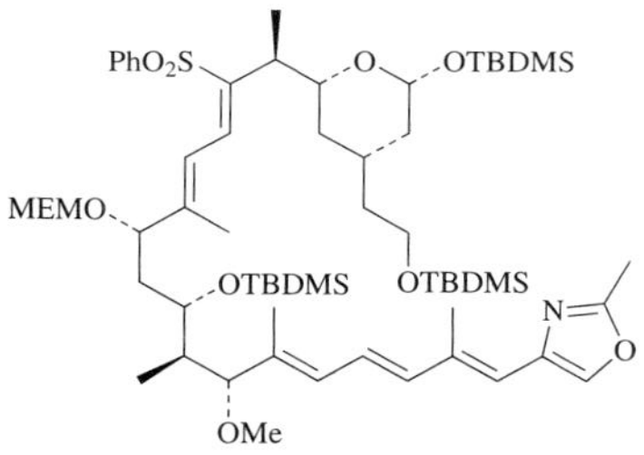 PhO$_2$S … MEMO … OTBDMS … OTBDMS … OTBDMS … OMe	SmI_2, DMPU, THF/MeOH, rt	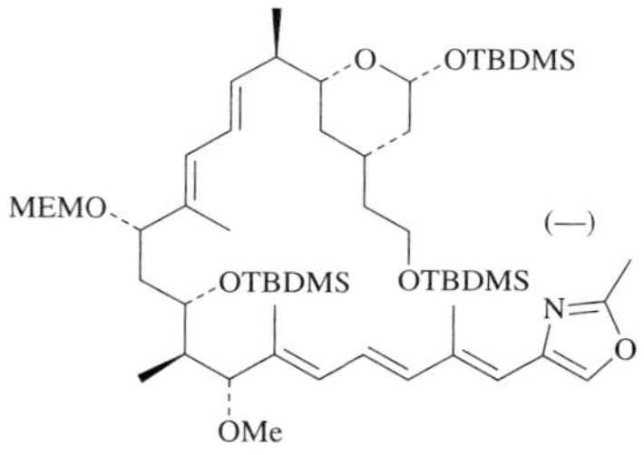 MEMO … OTBDMS … OTBDMS … OTBDMS … OMe (—)	261

[a] One of the carbonyl groups is reduced to the alcohol under the reaction conditions.

[b] A mixture of isomers is produced.

TABLE 8. REDUCTIVE ELIMINATION

Substrate	Conditions	Product(s) and Yield(s) (%)	Refs.
C_{10-20} SO$_2$Ph	Na/Hg, Na_2HPO_4, MeOH, rt	R: H (48); MeO_2C (72); (60) OH	707
C_{11} SO_2Ph	Na/Hg, Na_2HPO_4, MeOH, rt	OH (69)	707
C_{12} SO_2Ph	Na/Hg, Na_2HPO_4, MeOH, rt	OH (63)	707
C_{13} OMe SO_2Ph	Na/Hg, NaH_2PO_4•H_2O	**I** + OMe **II**	408

Solvent	Temp	Time	**I + II**	**I:II**
DMF	rt	24 h	(73)	66:34
MeCN/MeOH (2:1)	0°	6 h	(77)	39:61

C_{13-20}

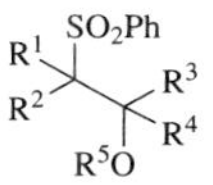

Na/Hg, EtOH, rt — (—) Z:E = — — 94

R^1	R^2	R^3	R^4	R^5
Me	Me	Me	Me	Ms
H	H	—$(CH_2)_5$—		Ms
Me	H	H	*n*-Bu	Ms
H	H	H	Ph	Ms
Me	H	H	*n*-Bu	Ac
Me	H	H	$Me_2C{=}CH(CH_2)_2$	Ms
Me	H	H	*n*-Bu	Ts

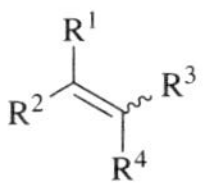

C_{14}

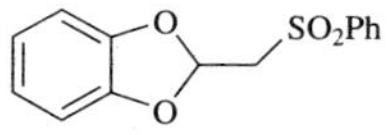

Na/Hg, Na_2HPO_4, MeOH, rt 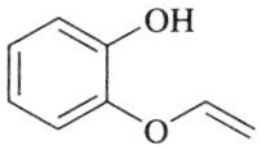 — (90) — 707

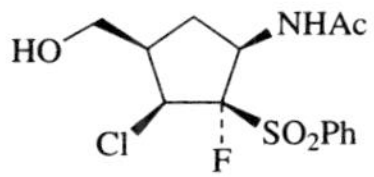

Mg, $HgCl_2$, EtOH/THF 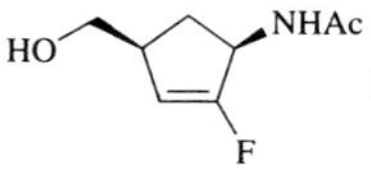 — (65) — 244

C_{14-19}

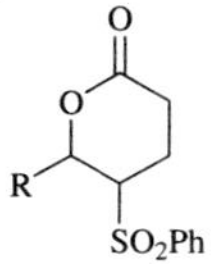

Na/Hg, MeOH, 0° to rt — 708

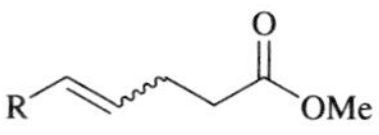

R		Z:E
n-Pr	(85)	20:80
(*E*)-MeCH=CH	(80)	20:80
t-Bu	(81)	3:97
Ph	(70)	20:80
n-C_7H_{15}	(83)	20:80
(*E*)-PhCH=CH	(56)	20:80

TABLE 8. REDUCTIVE ELIMINATION (*Continued*)

Substrate	Conditions	Product(s) and Yield(s) (%)	Refs.
C_{14-21} 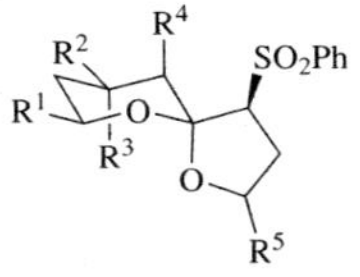	Na/Hg, Na_2HPO_4, MeOH, rt	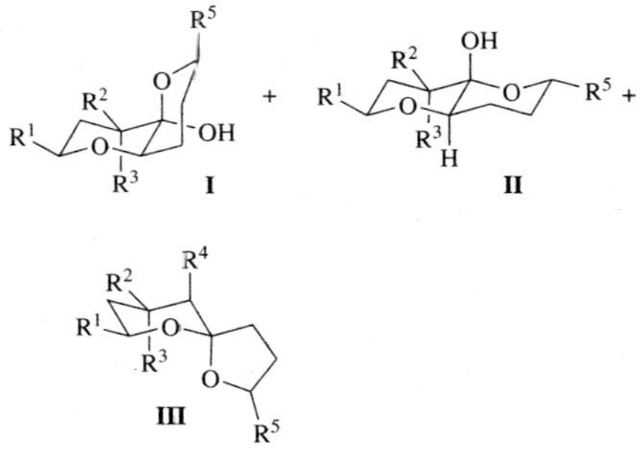 I + II + III	709

R^1	R^2	R^3	R^4	R^5	**I + II + III**	**I:II:III**
H	H	H	O=	H	(64)	66:34:0
H	H	H	HO	H	(73)	0:0:100
H	H	H	O=	Me	(98)	45:55:0
H	H	H	HO	Me	(60)	0:0:100
Me	H	H	O=	Me	(60)	50:50:0
H	Me	Me	O=	H	(85)	95:5:0
Me	H	H	HO	Me	(55)	0:0:100
H	H	H	O=	Ph	(80)	43:57:0
Me	H	H	O=	Ph	(76)	40:60:0

C_{14-23}

See table.

Ar	R^1	R^2	R^3	R^4	Reagents	Solvent	Temp	Time		Z:E	
4-Tol	Me	Et	NC	Me	Na_2S	DMF	rt	3 h	(76)	50:50	245
4-Tol	Me	Et	NC	Me	NaTeH	EtOH	rt	30 min	(83)	37:63	245
4-Tol	Et	Me	NC	Me	$(n\text{-Bu})_3SnH$, AIBN	C_6H_6	80°	2 h	(87)	99:1	245
4-Tol	Et	Me	NC	Me	Na_2S	DMF	rt	3 h	(70)	50:50	245
4-Tol	Et	Me	NC	Me	NaTeH	EtOH	rt	30 min	(83)	64:36	245
4-Tol	Me	Me	(lactone structure)		$(n\text{-Bu})_3SnH$, AIBN	C_6H_6	80°	2 h	(72)	—	710
4-Tol	—$(CH_2)_5$—		NC	Me	$(n\text{-Bu})_3SnH$, AIBN	C_6H_6	80°	2 h	(75)	—	245
4-Tol	Me	Me	EtO_2C	Et	$(n\text{-Bu})_3SnH$, AIBN	C_6H_6	80°	2 h	(81)	—	710
4-Tol	Me	Et	EtO_2C	Et	$(n\text{-Bu})_3SnH$, AIBN	C_6H_6	80°	2 h	(85)	1:99	245
4-Tol	Et	Me	EtO_2C	Et	$(n\text{-Bu})_3SnH$, AIBN	C_6H_6	80°	2 h	(83)	99:1	245
Ph	$BzOCH_2$	Me	H	Et	$(n\text{-Bu})_3SnH$, AIBN	C_6H_6	80°	0.5 h	(80)	11:89	245
Ph	$BzOCH_2$	Me	Et	H	$(n\text{-Bu})_3SnH$, AIBN	C_6H_6	80°	0.5 h	(78)	73:27	245
Ph	$BzOCH_2$	Me	H	*i*-Pr	$(n\text{-Bu})_3SnH$, AIBN	C_6H_6	80°	0.5 h	(88)	4:96	245
Ph	$BzOCH_2$	Me	*i*-Pr	H	$(n\text{-Bu})_3SnH$, AIBN	C_6H_6	80°	0.5 h	(88)	84:16	245
Ph	$BzOCH_2$	Me	H	$n\text{-}C_5H_{11}$	$(n\text{-Bu})_3SnH$, AIBN	C_6H_6	80°	0.5 h	(82)	8:92	245
Ph	$BzOCH_2$	Me	$n\text{-}C_5H_{11}$	H	$(n\text{-Bu})_3SnH$, AIBN	C_6H_6	80°	0.5 h	(84)	84:16	245
Ph	$BzOCH_2$	Me	H	$n\text{-}C_6H_{13}$	$(n\text{-Bu})_3SnH$, AIBN	C_6H_6	80°	0.5 h	(86)	5:95	245
Ph	$BzOCH_2$	Me	$n\text{-}C_6H_{13}$	H	$(n\text{-Bu})_3SnH$, AIBN	C_6H_6	80°	0.5 h	(86)	80:20	245

TABLE 8. REDUCTIVE ELIMINATION (*Continued*)

Substrate	Conditions	Product(s) and Yield(s) (%)	Refs.
C_{14-24} 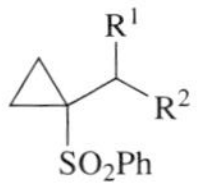	See table.	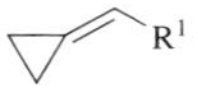	122

R^1	R^2	Reagents	Solvent	Temp	Time	
H_2C=CH	AcO	Na/Hg	THF/MeOH	–20°	30 min	(26)
4-Tol	Cl	Na/Hg	THF/MeOH	–20°	30 min	(30)
4-Tol	Cl	Mg, $HgCl_2$	EtOH	rt	1 h	(95)
4-Tol	AcO	Na/Hg	THF/MeOH	–20°	30 min	(53)
n-$C_{10}H_{21}$	Cl	Mg, $HgCl_2$	EtOH	rt	1 h	(89)
n-$C_{10}H_{21}$	AcO	Na/Hg	THF/MeOH	–20°	30 min	(50)
n-$C_{10}H_{21}$	AcO	Mg, $HgCl_2$	EtOH	rt	1 h	(48)
4-Tol	BzO	Na/Hg	THF/MeOH	–20°	30 min	(35)
4-Tol	TsO	Na/Hg	THF/MeOH	–20°	30 min	(75)

Substrate	Conditions	Product(s) and Yield(s) (%)	Refs.
C_{14-31} 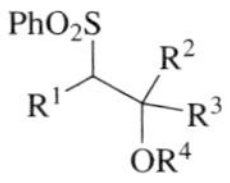	SmI_2, additive, THF	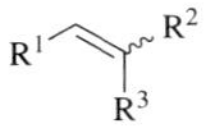	223

R^1	R^2	R^3	R^4	Additive	Temp	Time		Z:E
Me	Me	n-Bu	H	HMPA	0°	1 h	(69)	—
Me	Me	n-Bu	Bz	HMPA	–78°	1 h	(73)	—
n-C_6H_{13}	Me	$PhCH_2CH_2$	H	HMPA	0°	1 h	(66)	45:55
Me	Me	$PhCH_2CH_2$	Bz	HMPA	–84°	1 h	(84)	37:63
n-Bu	—$(CH_2)_2CH(t\text{-Bu})(CH_2)_2$—		Bz	HMPA	–78° to rt	15 h	(85)	—
BnO	H	(E)-PhCH=CH	Bz	DMPU	rt	2 h	(82)	37:63
BnO	H	(E)-PhCH=C(Me)	Bz	DMPU	rt	2 h	(91)	50:50

Substrate	Conditions	Product(s) and Yield(s) (%)	Refs.
C_{15-17} 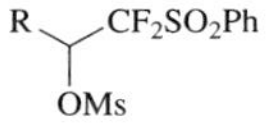 R–CH(OMs)–CF_2SO_2Ph	Na/Hg, Na_2HPO_4, MeOH, –40° to –20°, 1 h	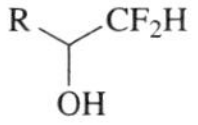 R–CH(OH)–CF_2H; R = 4-BrC_6H_4 (70); Ph (60); $PhCH_2CH_2$ (84)	389
C_{15-19} R^1C(=CH_2)–C(R^2)(Ts)–CH_2OH 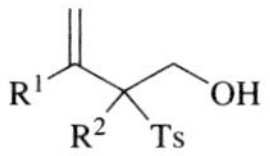	(n-Bu)$_3$SnH, AIBN, C_6H_6, reflux, 2 h	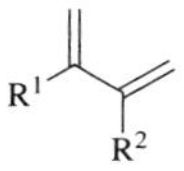 R^1C(=CH_2)–C(=CH_2)R^2; R^1 = H, R^2 = n-Bu (53); Me, Me_2C=$CHCH_2$ (85); H, Me_2C=$CH(CH_2)_2$ (62); H, n-C_6H_{13} (63); H, Bn (100); H, n-C_8H_{17} (92)	62
C_{15-25} TMS–CH_2–CH(SO_2Ph)–C(OMs)R^1R^2 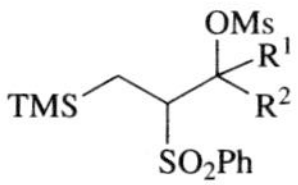	Na/Hg, Na_2HPO_4, MeOH, 0°, 1 h	 TMS–CH_2–CH=CR^1R^2 (—) Z:E = —; R^1, R^2 = Me, Me; —$(CH_2)_3$—; —$(CH_2)_4$—; Me_2C=CH, H; Et, Et; —$(CH_2)_5$—; Ph, H; Ph, Ph	204
C_{16} bis(oxabicyclic) sulfone, H, SO_2Ph 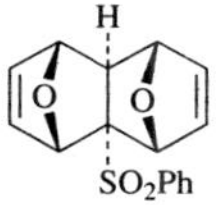	Na/Hg, Na_2HPO_4, MeOH, –20°	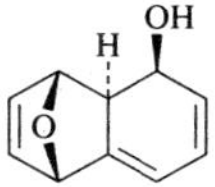 oxabicyclic dienol, H, OH (65)	711
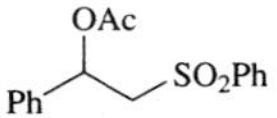 PhCH(OAc)CH_2SO_2Ph	Mg, $HgCl_2$, EtOH, rt, 2 h	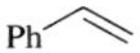 PhCH=CH_2 (98)	712

TABLE 8. REDUCTIVE ELIMINATION (*Continued*)

Substrate	Conditions	Product(s) and Yield(s) (%)	Refs.
C_{16}			
PhO$_2$S	SmI$_2$, HMPA, THF, rt, 45 min	OH (77)	713
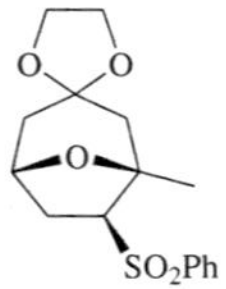 SO$_2$Ph	SmI$_2$, HMPA, THF, rt, 15 min	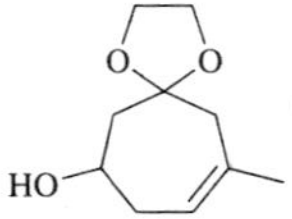 HO (80)	713
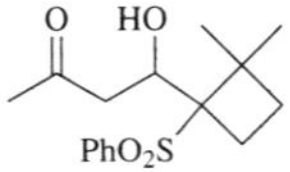 O HO PhO$_2$S	Na/Hg, MeOH, rt, 2 h	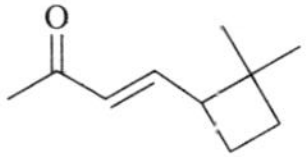 O (25-30)	416
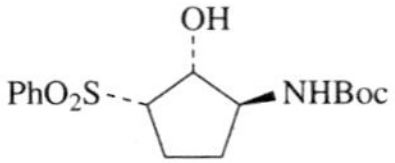 OH PhO$_2$S NHBoc	Na/Hg, Na$_2$HPO$_4$, THF/MeOH	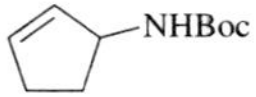 NHBoc (72)	183
C_{16-21}			
R^1 AcO AcO O SO$_2$R^2 OAc	[Cr^{II}(EDTA)$^{2-}$], DMF/H$_2$O, rt, 36 h	R^1 AcO AcO O	233

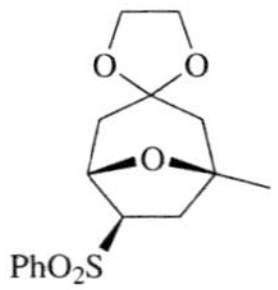

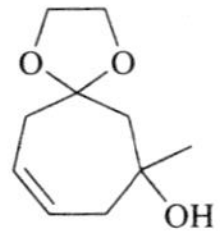

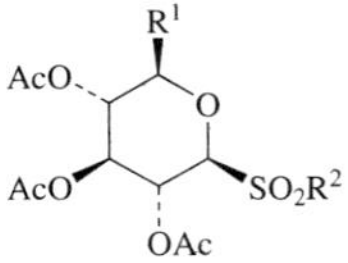

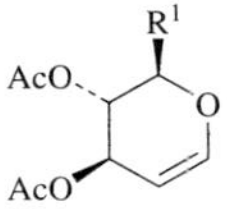

R^1	R^2	
H	2-pyridyl	(80)
AcOCH$_2$	2-pyridyl	(>95)
AcOCH$_2$	2-benzothiazolyl	(>95)

$C_{16\text{-}22}$ 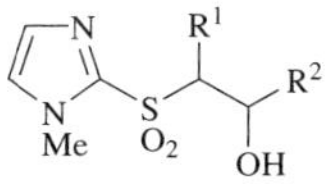	SmI_2, THF, rt	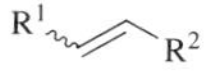		102

R^1	R^2		Z:E
$Me_2C{=}CH$	Ph	(87)	83:17
$Me_2C{=}CH$	(*E*)-PhCH=CH	(78)	100:0
$PhCH_2CH_2$	$Me_2C{=}CH$	(84)	83:17
$PhCH_2CH_2$	$PhCH_2CH_2$	(55)	75:25

C_{17} 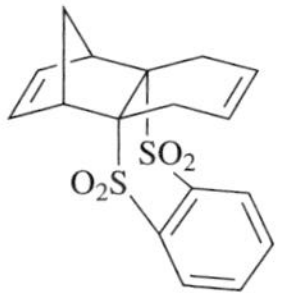	Na/Hg, NaH_2PO_4, MeOH, rt		(—)	241
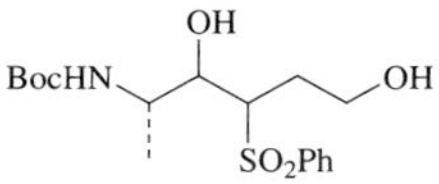	Na/Hg, Na_2HPO_4, MeOH	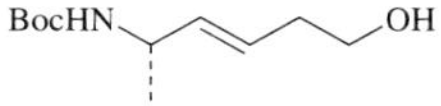	(82)	553
$C_{17\text{-}18}$ 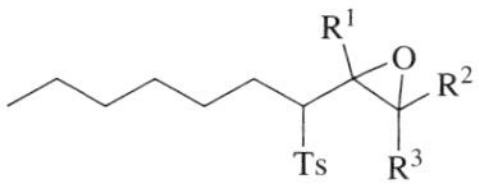	Na/Hg, THF/MeOH, –20°	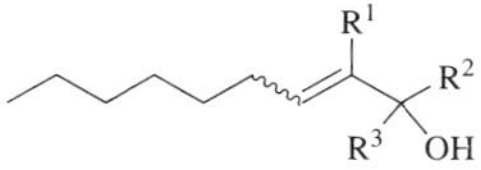		198

R^1	R^2	R^3		Z:E
Me	H	H	(70)	30:70
H	Me	H	(62)	20:80
H	Me	Me	(89)	0:100

TABLE 8. REDUCTIVE ELIMINATION (*Continued*)

Substrate	Conditions	Product(s) and Yield(s) (%)	Refs.
C_{17-19}	SmI_2, HMPA, THF, rt		713

R^1	R^2	R^3	R^4	n	Time	Yield (%)
H	H	H	H	2	10 min	(77)
H	H	H	Me	1	10 min	(84)
H	H	Me	H	1	15 min	(79)
H	MeO_2C	H	H	1	10 min	(86)
MeO_2C	H	H	H	1	10 min	(81)
Me	MeO_2C	H	H	1	10 min	(81)
MeO_2C	Me	H	H	1	20 min	(79)

Substrate	Conditions	Product(s) and Yield(s) (%)	Refs.
C_{17-29}	$Li/C_{10}H_8$, THF, –78°	R: Me (85); Bn (95); Bz (82)	714
C_{18}	Na/Hg, NaH_2PO_4, MeOH, rt	(—)	241
	Na/Hg, Na_2HPO_4, MeOH, 0° to rt, 1.5 h	(45)	135

Reactant	Conditions	Product(s) and Yield(s) (%)	Refs.
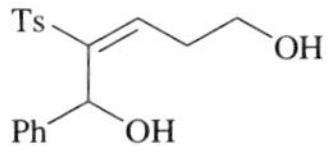 Ts, OH, Ph, OH	Na/Hg, Na_2HPO_4, MeOH, 0° to rt, 1.5 h	 Ph, OH (69)	135
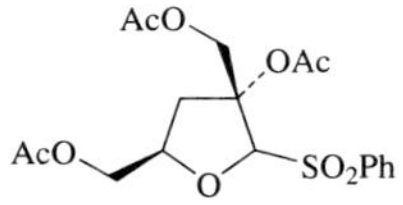 AcO, OAc, AcO, O, SO_2Ph	SmI_2, THF, HMPA, rt, 1 h	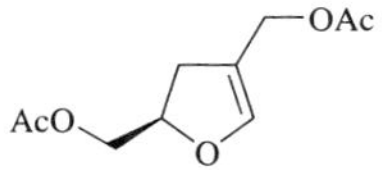 OAc, AcO, O (—)	715
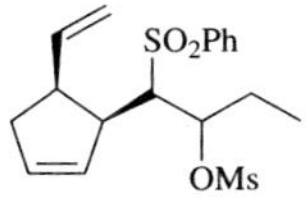 SO_2Ph, OMs	Na/Hg, Na_2HPO_4, MeOH, rt, 2 h	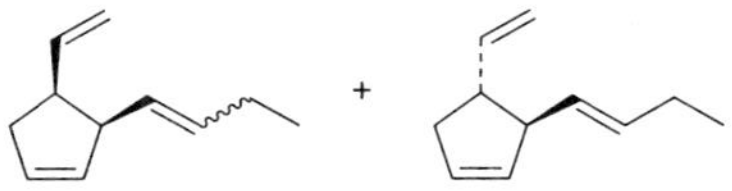 (30) Z:E = 17:83 + (5)	716
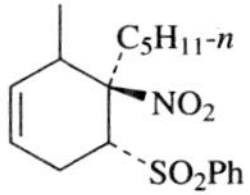 C_5H_{11}-*n*, NO_2, SO_2Ph	(*n*-Bu)$_3$SnH, AIBN, C_6H_6, 80°, 2 h	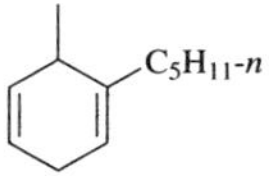 C_5H_{11}-*n* (90)	717
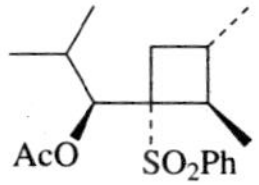 AcO, SO_2Ph	Na, EtOH/THF, 10-15°, 1.5 h	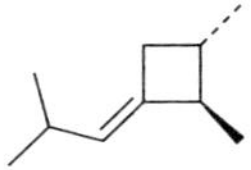 (90)	416
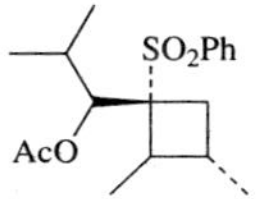 SO_2Ph, AcO	Na, THF/EtOH, 10-15°, 1.5 h	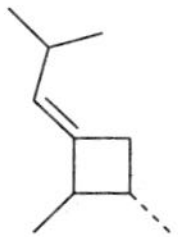 (70)	416

TABLE 8. REDUCTIVE ELIMINATION (*Continued*)

Substrate	Conditions	Product(s) and Yield(s) (%)	Refs.
C_{18-20} 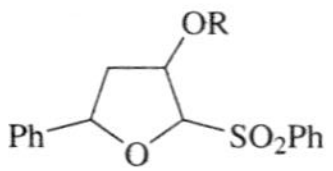 OR; Ph; O; SO_2Ph	See table.	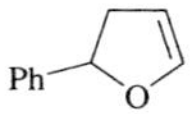 Ph; O	718
R: Ac	Reagents: SmI_2, HMPA; Solvent: THF	(46)	
R: N; N; S (imidazolyl thiocarbonyl)	Reagents: $(n\text{-}Bu)_3SnH$, AIBN; Solvent: toluene, heat	(52)	
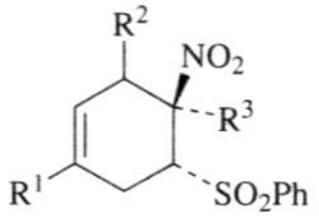 R^2; NO_2; R^3; R^1; SO_2Ph	$(n\text{-}Bu)_3SnH$, AIBN	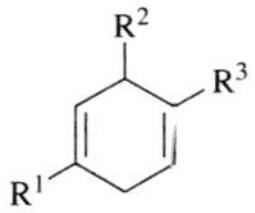 R^2; R^3; R^1	719

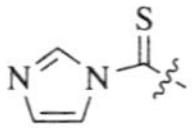

R^1	R^2	R^3	Solvent	Temp	Time	Yield (%)
$Me_2C{=}CHCH_2CH_2$	H	H	toluene	110°	0.5 h	(63)
H	Me	$n\text{-}C_5H_{11}$	C_6H_6	80°	2 h	(79)
H	$n\text{-}C_7H_{15}$	H	toluene	110°	0.5 h	(79)
H	$n\text{-}C_7H_{15}$	Me	C_6H_6	80°	2 h	(76)

Substrate	Conditions	Product(s) and Yield(s) (%)	Refs.
C_{19} 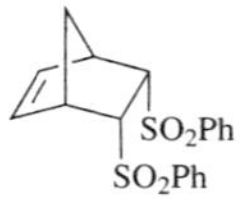 SO_2Ph; SO_2Ph	Na/Hg, NaH_2PO_4, MeOH, rt, overnight	 (65)	239
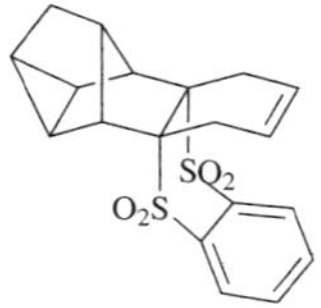 SO_2; O_2S	Na/Hg, NaH_2PO_4, MeOH, rt	 (—)	241

Substrate	Conditions	Product(s) and Yield(s) (%)	Refs.
C_5H_{11}-n, NO_2, SO_2Ph	(n-Bu)$_3$SnH, AIBN, C_6H_6, 80°, 2 h	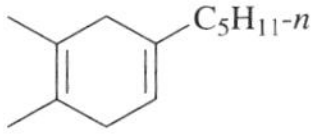 C_5H_{11}-n (81)	717

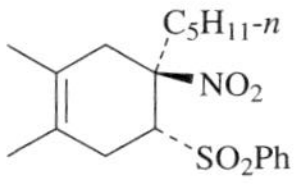

C_{19-35}

Substrate	Conditions	Product(s) and Yield(s) (%)	Refs.
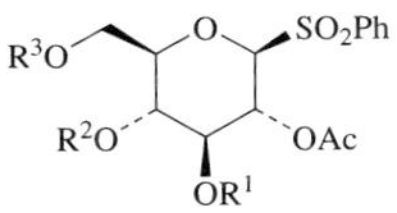 R^3O, O, SO_2Ph, R^2O, OAc, OR^1	SmI_2, HMPA, THF, rt, 15-20 min	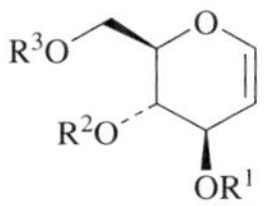 R^3O, O, R^2O, OR^1	229, 228

R^1	R^2	R^3	Yield
—CMe$_2$—		Ac	(98)
Ac	Ac	Ac	(98)
Ac	—CHPh—		(98)
Ac	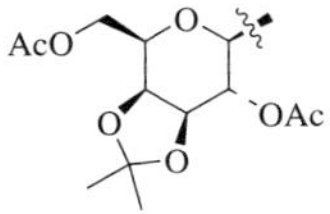 AcO, O, OAc, O, O	Ac	(93)
Ac	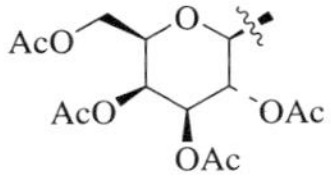 AcO, O, AcO, OAc, OAc	Ac	(95)
Bn	Bn	Bn	(96)

TABLE 8. REDUCTIVE ELIMINATION (*Continued*)

Substrate	Conditions	Product(s) and Yield(s) (%)	Refs.
C_{19-40} R[1]CH(NHTs)CH2CH(SO2Ph)CH(OH)R[2]	Na/Hg, Na_2HPO_4, MeOH, –20° to –15°, 30 min	R[1]CH(NHTs)CH2CH=CHR[2] (—)	203

R^1	R^2	Z:E
Me	Me	20:80
Me	*i*-Pr	10:90
Me	*i*-Bu	20:80
Me	*n*-C_6H_{13}	20:80
Bn	Me	25:75
Bn	*i*-Pr	9:91
Bn	*i*-Bu	20:80
Bn	*n*-C_6H_{13}	25:75
(1-TBDMS-indol-3-yl)methyl	Me	25:75
$TBDPSOCH_2$	Me	25:75
(1-TBDMS-indol-3-yl)methyl	*i*-Pr	12:88
"	*i*-Bu	25:75
$TBDPSOCH_2$	*i*-Pr	10:90
$TBDPSOCH_2$	*i*-Bu	25:75
(1-TBDMS-indol-3-yl)methyl	*n*-C_6H_{13}	25:75
$TBDPSOCH_2$	*n*-C_6H_{13}	25:75

Substrate	Conditions	Product(s) and Yield(s) (%)	Refs.
C_{20}			
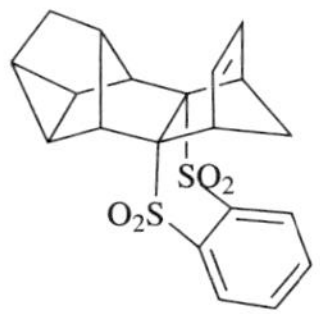 SO_2, O_2S	Na/Hg, NaH_2PO_4, MeOH, rt	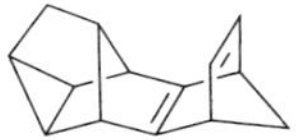 (—)	241
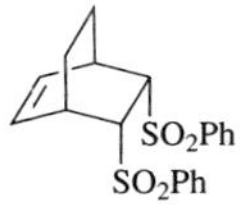 SO_2Ph, SO_2Ph	Na/Hg, NaH_2PO_4, MeOH, rt, overnight	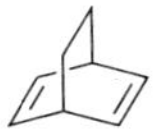 (69)	239
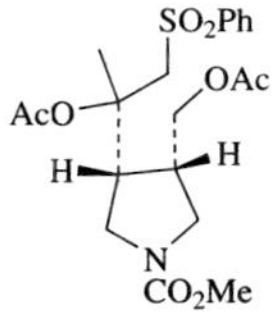 SO_2Ph, OAc, AcO, H, H, N, CO_2Me	Mg, $HgCl_2$, EtOH/THF, rt, 2 h	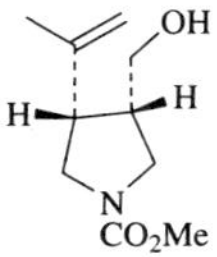 OH, H, H, N, CO_2Me (65)	720
C_{20-24}			
Et, SO_2Ph, SO_2Ph, NHR	Na/Hg, KH_2PO_4, MeOH, rt	Et, NHR	721

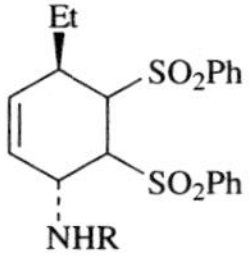

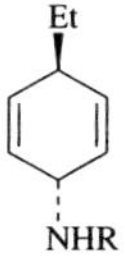

R	Time	
H	2 h	(—)
Alloc	30 min	(80)

TABLE 8. REDUCTIVE ELIMINATION (*Continued*)

Substrate	Conditions	Product(s) and Yield(s) (%)	Refs.
C_{20-28} 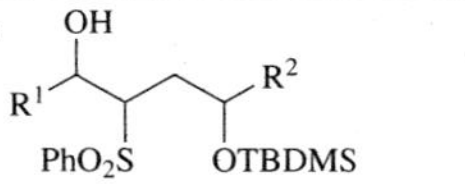 OH, R^1, R^2, PhO_2S, OTBDMS	1. Na/Hg, THF/MeOH, –5° 2. TBAF, THF, 20°	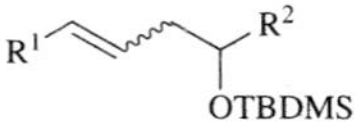 R^1, R^2, OTBDMS	202
R^1 / R^2		Z:E	
i-Pr / Me		(71) 13:87	
Me / Ph		(95) 33:67	
i-Pr / Ph		(87) 12:88	
i-Pr 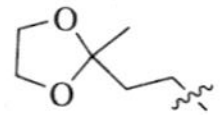 / O, O		(81) 11:89	
i-Pr / $PhCH_2OCH_2$		(75) 13:87	
Ph / Ph		(75) 33:67	
n-C_6H_{13} / Ph		(76) 33:67	
C_{21} 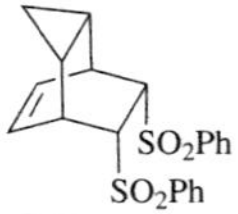 SO_2Ph, SO_2Ph	Na/Hg, NaH_2PO_4, MeOH, rt, overnight	 (61)	239
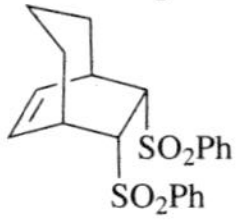 SO_2Ph, SO_2Ph	Na/Hg, NaH_2PO_4, MeOH, rt, overnight	 (84)	239
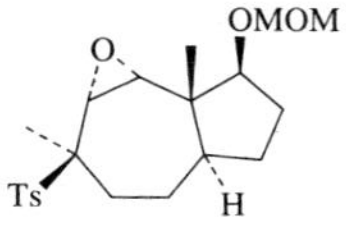 O, OMOM, Ts, H	Na/Hg, THF/MeOH	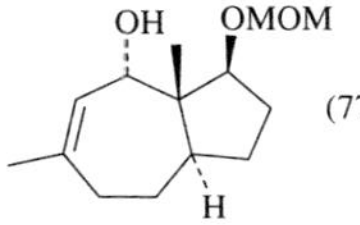 OH, OMOM, H (77)	564
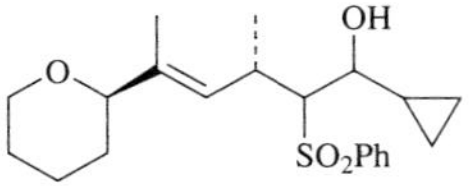 O, OH, SO_2Ph	Na/Hg, Na_2HPO_4, MeOH, 20°	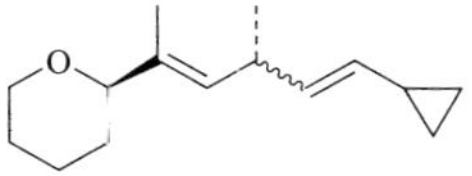 O (70) Z:E = 14:86	722

C_{21-23}

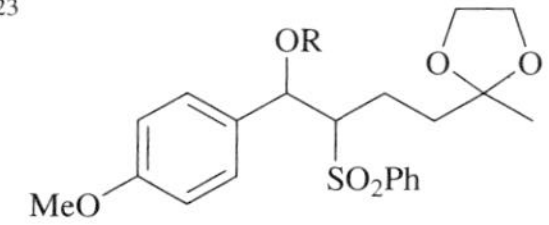

MeO, OR, SO_2Ph (substrate)

R
H
Ac

SmI_2, HMPA, THF, rt

Time
2 h
0.25 h

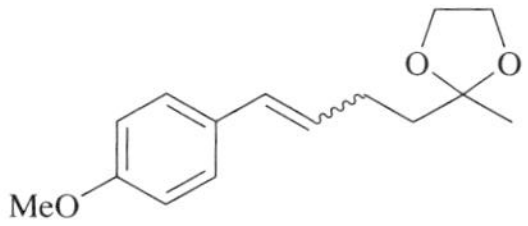

MeO (product)

	Z:E
(59)	12:88
(92)	20:80

103

C_{21-24}

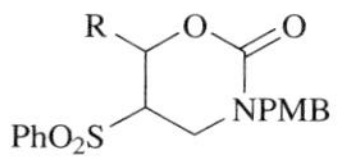

R, O, O, NPMB, PhO_2S (substrate)

R
(*E*)-MeCH=CH
Me_2C=CH
(*E*)-MeCH=C(Me)
Ph

Na/Hg, Na_2HPO_4, MeOH, −15°, 1.5 h

R–NHPMB (product)

	Z:E
(43)	5:95
(54)	4:96
(59)	4:96
(62)	4:96

199

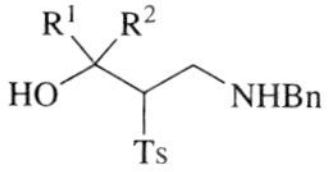

R^1, R^2, HO, NHBn, Ts (substrate)

R^1	R^2
—$(CH_2)_4$—	
t-Bu	H
—$(CH_2)_5$—	
4-$MeOC_6H_4$	H

Na/Hg, Na_2HPO_4, MeOH, 0° to rt

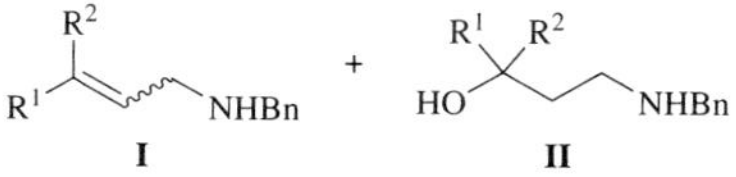

R^1, R^2, NHBn **I** + R^1, R^2, HO, NHBn **II**

I	Z:E	II
(52)	—	(20)
(98)	0:100	(0)
(48)	—	(14)
(61)	33:67	(0)

200

TABLE 8. REDUCTIVE ELIMINATION (*Continued*)

Substrate	Conditions	Product(s) and Yield(s) (%)	Refs.
C_{21-26} HO, OH, OH, R, SO_2Ph	Na/Hg, Na_2HPO_4, MeOH, rt	HO, OH, R **I** + HO, OH, OH, R **II**	723
R: H	Time: 2.5 h	I (—), II (—)	
R: OTHP	Time: 1.5 h	I (64), II (24)	
C_{21-41} $OPO(OR^3)_2$, R^1, R^2, SO_2Ph	Reagents, THF	R^1—≡—R^2 **I** + R^1—CH=CH—R^2 **II**	235

R^1	R^2	R^3	Reagents	Temp	Time	I	II
Ph	*n*-Pr	Et	Na/Hg	rt	20 min	(74)	(0)
c-C_6H_{11}	*n*-Pr	Et	Na/NH_3	–33°	—	(74)	(0)
c-C_6H_{11}	*i*-Pr	Et	Na/NH_3	–33°	—	(78)	(0)
1-adamantyl	H	Ph	Na/NH_3	–33°	—	(51)	(15)
MOMO, CO_2Me (cyclopentyl)	OBz (chain)	Ph	Na/Hg	rt	20 min	(50)	(0)

Substrate	Conditions	Product(s) and Yield(s) (%)	Refs.
C_{22} SO_2Ph, SO_2Ph	Na/Hg, NaH_2PO_4, MeOH, rt, overnight	(61)	239
SO_2Ph, Ph, OAc	Na/Hg, EtOAc/MeOH, –20°, 10 h	Ph (93)	211

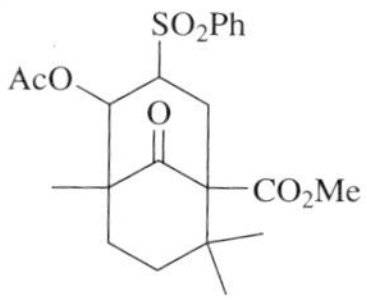

Substrate	Conditions	Product(s) and Yield(s) (%)	Refs.
(structure)	Na/Hg, EtOAc/MeOH, 0°, 3 h	(structure) (57)	724
C_{22-27} (structure)	Na/Hg, EtOAc/MeOH, –20°, 10 h	(structure) (—)	211
C_{22-28} (structure)	See table.	(structure)	

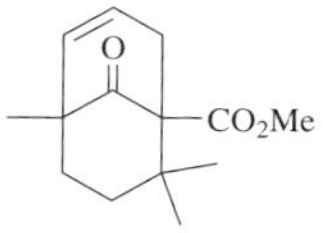

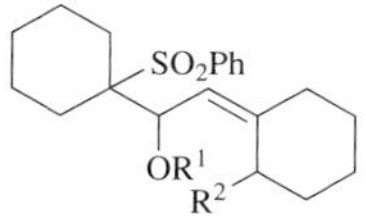

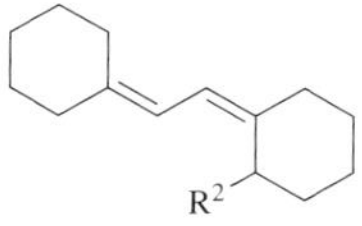

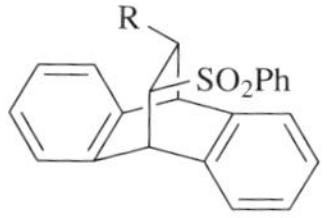

R^1	R^2
Ac	H
Ac	$H_2C=$
Bz	H

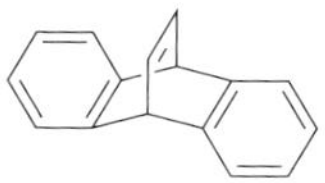

R	Reagents	Solvent	Temp	Time		Refs.
O_2N	$(n\text{-Bu})_3SnH$, AIBN	C_6H_6	80°	2 h	(60)	717
O_2N	$(n\text{-Bu})_3SnH$, AIBN	toluene	110°	0.5 h	(60)	717
PhO_2S	SmI_2	HMPA, THF	–20°	0.5 h	(91)	100
PhO_2S	Na/Hg, NaH_2PO_4	MeOH	–20°	10 h	(91)	239
PhO_2S	Na	toluene	reflux	2 h	(90)	239
PhO_2S	Li/Hg	toluene	rt	14 h	(95)	239

Substrate	Conditions	Product(s) and Yield(s) (%)	Refs.
C_{23} (structure)	Mg, $HgCl_2$, EtOH/THF, ultrasound, 7 d	(structure) (37)	636

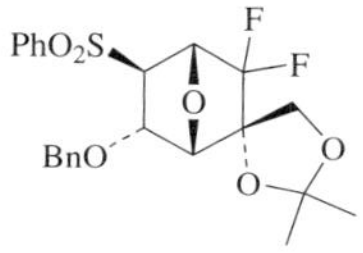

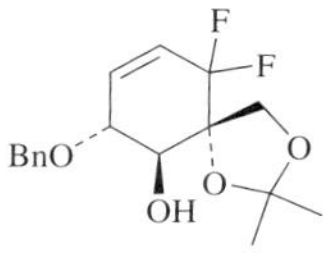

TABLE 8. REDUCTIVE ELIMINATION (*Continued*)

Substrate	Conditions	Product(s) and Yield(s) (%)	Refs.
C_{23}	Na/Hg, EtOAc/MeOH, –20°, 10 h	(—)	211
	Na/Hg, Na_2HPO_4, MeOH, 0°	(88) Z:E = 17:83	725
	Na/Hg, Na_2HPO_4, MeOH, rt, 4 h	(73)	207
	Na/Hg, Na_2HPO_4, MeOH, rt, 6 h	(71)	246
$C_{23\text{-}25}$	Mg, $HgCl_2$, EtOH/THF, 0°, 5 h	R^1 = H, R^2 = H (50); R^1 = Me, R^2 = H (70); R^1 = Me, R^2 = Me (61)	636
$C_{23\text{-}26}$	See table.		726

R
Ac
Ac
Boc

Reagents	Solvent	Temp	Time	
SmI_2, HMPA	THF	—	—	(28)
5% Na/Hg, $B(OH)_3$	THF/MeOH	0° to 20°	1 h	(48)
20% Na/Hg, $B(OH)_3$	THF/MeOH	0° to 20°	1 h	(74)

$C_{23\text{-}34}$

F, SO_2Ph, OBn, OAc, n

Na/Hg, NaH_2PO_4, MeOH/THF, –10° to rt , overnight

F, OBn, n

n		Z:E
3	(80)	75:25
7	(77)	67:33
12	(86)	67:33
14	(46)	29:71

727

C_{24}

Ph, NO_2, SO_2Ph, Ph

(n-Bu)$_3$SnH, AIBN, toluene, 110°, 0.5 h

Ph, Ph (65)

719

OAc, Ts, O

Na, EtOH/THF, –78°, 1 h

O (65) Z:E = 14:86

269

CN, OAc, SO_2Ph

Na/Hg, Na_2HPO_4, THF/EtOAc, –10° to rt, 12 h

CN (55) Z:E = 33:67

728

OAc, CO_2Me, PhO_2S, OTBDMS

Na/Hg, –24°

CO_2Me, OTBDMS (—)

729

TABLE 8. REDUCTIVE ELIMINATION (*Continued*)

Substrate	Conditions	Product(s) and Yield(s) (%)	Refs.
C_{24} TIPSO, H, SO_2Ph, N^+ Me, OH	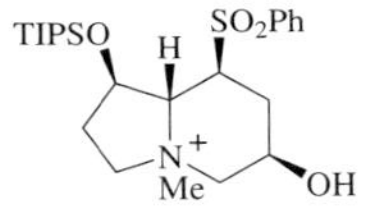 Na/Hg, Na_2HPO_4, MeOH, rt, 5 h	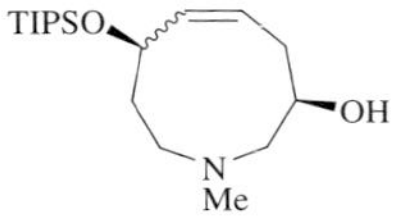 TIPSO, OH, N Me (82) Z:E = 54:46	246
C_{25} Ph, OCS_2Me, PhO_2S, Ph	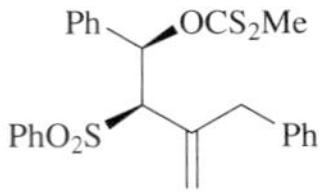 Na/Hg, THF/MeOH, –20°, 3.5 h	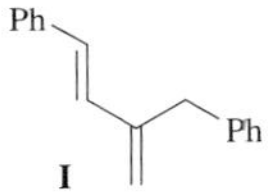 Ph, Ph **I** (>95)	730
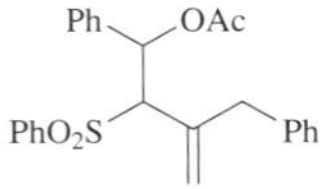 Ph, OAc, PhO_2S, Ph	Na/Hg, THF/MeOH, –20°, 3 h	**I** (50)	730
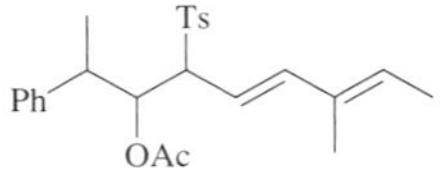 Ts, Ph, OAc	Na/Hg, Na_2HPO_4, MeOH, –20°, 5-7 h	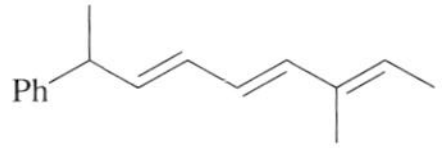 Ph (—)	731
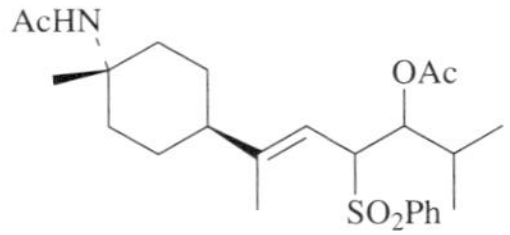 AcHN, OAc, SO_2Ph	Na/Hg, Na_2HPO_4, MeOH, 0°	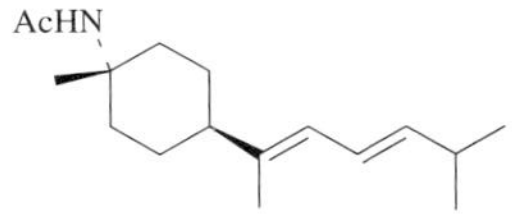 AcHN (—)	725
OMs, TMS, SO_2Ph	Na/Hg	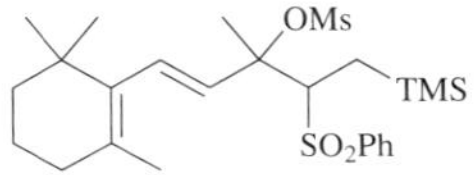 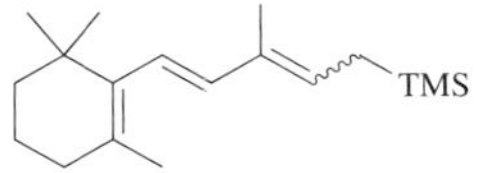 TMS (7) Z:E = 33:67	129
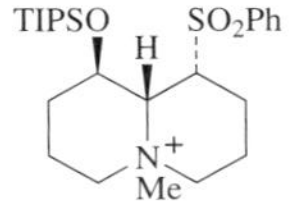 TIPSO, H, SO_2Ph, N^+ Me	Na/Hg, Na_2HPO_4, MeOH, rt, 8 h	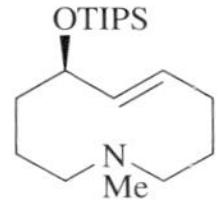 OTIPS, N Me (64)	246

Reactant	Conditions	Product(s) and Yield(s) (%)	Refs.
C_{26} [structure]	Na/Hg, NaH_2PO_4, MeOH, rt	[structure] (—)	241
[structure: OAc, Ph, SO_2Ph]	Na/Hg, THF/MeOH, –20°	[structure: Ph] (56) Z:E = 15:85	355
[structure: PhO_2S, OBz, C_6H_{13}-*n*]	Na/Hg, EtOAc/MeOH, –20°, 10 h	[structure: C_6H_{13}-*n*] (—)	211
[structure: $OPO(NMe_2)_2$, SO_2Ph, Pr-*i*, $()_{10}$]	Na/NH_3, THF, –78°	[structure: Pr-*i*, $()_{10}$] (75)	235
C_{26-33} [structure: MeO, H, H, H, H, PhO_2S, OH, R]	Na/Hg, Na_2HPO_4, MeOH, rt	[structure: MeO, H, H, H, H, R]	732

R	Time		Z:E
[1-methylimidazol-2-yl]	4 h	(62)	32:68
[6-methylpyridin-2-yl]	2 h	(19)	0:100
[N-Boc-6-methylpiperidin-2-yl]	2.5 h	(66)	0:100

TABLE 8. REDUCTIVE ELIMINATION (*Continued*)

Substrate	Conditions	Product(s) and Yield(s) (%)	Refs.
C_{27}			
n-Bu, $OC(O)C_6H_4CF_3$-3, PhO_2S, CO_2Me	Na/Hg, THF/MeOH, –20°, 22 h; 0°, 8 h	n-Bu, CO_2Me (—) Z:E = 38:62	733
OAc, CO_2Et, SO_2Ph, OAc	Na/Hg, MeOH	OAc, CO_2Et (—) Z:E = 14:86	268
OH, NHBoc, OBn, O, O, SO_2Ph	Na/Hg, Na_2HPO_4, MeOH, 0°, 3 h	NHBoc, OBn, O, O (80) Z:E = 20:80	734
C_{27-33}			
NC, SO_2Ph, OBz, R	Na/Hg, THF/MeOH	NC, R	735
R: Me		(—)	
R: 3-furyl-CH_2CH_2–		(—)	
C_{27-45}			
OMe, O, TBDMSO, R^1O, R^2, R^3, PhO_2S	Na/Hg	OMe, O, TBDMSO, R^2, R^3	213

R^1	R^2	R^3	Base	Solvent	Temp	Time		Z:E
H	H	2,2-dimethyl-1,3-dioxolan-4-yl	—	THF/MeOH	–20°	3 h	(—)	—
H	H	n-C_5H_{11}	—	MeOH	rt	3 h	(70)	—
Ms	H	Ph	Na_2HPO_4	THF/MeOH	–20°	3 h	(—)	0:100
H	Me	n-C_6H_{13}	—	MeOH	rt	14 h	(62)	—
PhCO	H	2,2-dimethyl-1,3-dioxolan-4-yl	—	THF/MeOH	–20°	3 h	(—)	—
PhCO	H	Ph	—	THF/MeOH	–20°	3 h	(53)	0:100
Ms	H	$Me(CH_2)_4CH(OTBDMS)$	Na_2HPO_4	MeOH	–40°	—	(67)	—

C_{28}

(OBz, Ts, Ph substrate) — $Na/C_{10}H_8$, THF, –73° — (triene, Ph) (25) — 736

(MeO-pyranone, SO_2Ph, C_8H_{17}-n, OAc substrate) — Na/Hg, Na_2HPO_4, MeOH, –20°, 2 h — (MeO-pyranone, C_8H_{17}-n) (92) Z:E = 36:64 — 737

C_{28-30}

(PhO_2S, R, OBz, OMe OMe, CO_2Me substrate) — See table. — (R, MeO OMe, CO_2Me) — 738

R	Reagents	Solvent	Temp	Time		Z:E
Ph	Na/Hg	MeOH	–20°	~6 h	(82)	18:82
Ph	SmI_2	HMPA, THF	rt	10 min	(75)	64:36
$PhCH_2CH_2$	Na/Hg	MeOH	–20°	~6 h	(76)	22:78
$PhCH_2CH_2$	SmI_2	HMPA, THF	rt	10 min	(78)	50:50

TABLE 8. REDUCTIVE ELIMINATION (*Continued*)

Substrate	Conditions	Product(s) and Yield(s) (%)	Refs.
C_{29-30} 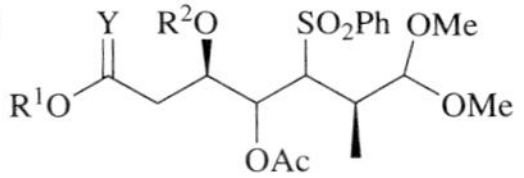	Na/Hg, Na_2HPO_4, EtOAc/MeOH, –50°, 4-8 h	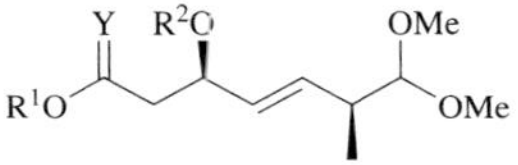	739
R^1 = *t*-BuCO, Y = H_2, R^2 = TBDMS		(—) Z:E 0:100	
R^1 = *t*-Bu, Y = O, R^2 = PMB		(—) Z:E 25:75	
C_{29-31} 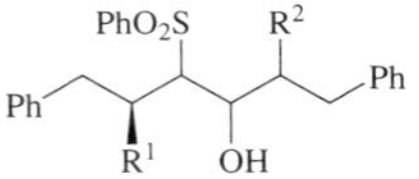	Na/Hg, Na_2HPO_4, MeOH, 0°	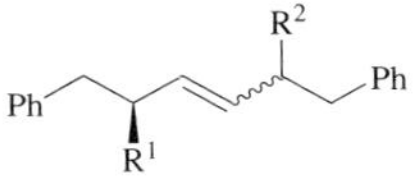	
R^1 = BocNH, R^2 = H	Time: overnight	(—) Z:E 14:86	740
R^1 = H, R^2 = $TBDMSOCH_2$	Time: 4 h	(—) Z:E 12:88	741
C_{29-43} 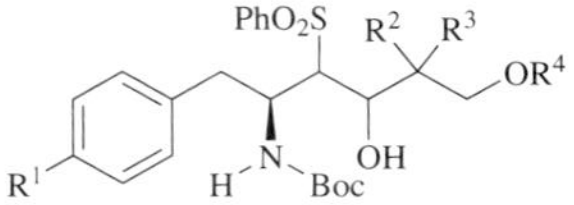	Na/Hg, Na_2HPO_4, MeOH, 0°	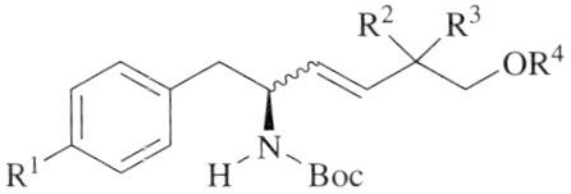 I 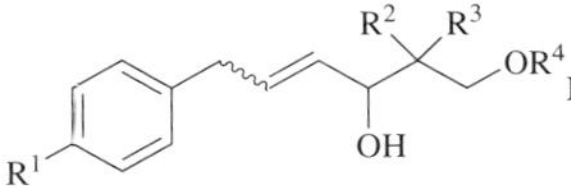 (—) + II (—) Z:E = —	

R^1	R^2	R^3	R^4	Time	I:II	I Z:E	Refs.
H	H	H	TBDMS	overnight	100:0	14:86	742
BnO	(*S*)-Me	H	THP	4 h	100:0	0:100	743
H	Bn	H	TBDMS	overnight	100:0	14:86	742
H	H	Bn	TBDMS	overnight	91:9	14:86	742
H	Bn	Bn	TBDMS	overnight	78:22	0:100	742

Substrate	Conditions	Product(s) and Yield(s) (%)	Refs.
C_{30}			
Ph, Ph, SO_2, O_2S	Na/Hg, NaH_2PO_4, MeOH, rt	Ph, Ph (—)	241
$OPO(OPh)_2$, SO_2Ph	Na/NH_3, THF, –78°	(52) + (6) Z:E = — + (8)	235
SO_2Ph, OH, H, H, H, OMe	Na/Hg, MeOH, rt	H, H, H, H, OMe (77)	744
TBDMSO, OMs, O, O, SO_2Ph, OMOM	Na/Hg, Na_2HPO_4, MeOH, –20°, 1 h	TBDMSO, O, O, OMOM (—)	745

TABLE 8. REDUCTIVE ELIMINATION (*Continued*)

Substrate	Conditions	Product(s) and Yield(s) (%)	Refs.
C_{31}			
OH, SO_2Ph, OTBDPS	SmI_2, HMPA, THF, rt, 0.5 h	OTBDPS (53)	103
TIPSO, H, SO_2Ph, N^+, Bn	Na/Hg, Na_2HPO_4, MeOH, rt, 5 h	TIPSO, N, Bn (75)	246
O, O, H, SO_2Ph, OMTM, MEMO, H, OAc	Na/Hg, Na_2HPO_4, EtOAc/MeOH, –20°, 6.5 h	O, O, H, OMTM, MEMO, H (64)	746
O, OMe, TBDMSO, O, OMe, TMS, SO_2Ph	Na/Hg, $NaHCO_3$, THF/MeOH, –40°, 1.5 h	TBDMSO, CO_2H, TMS, MeO, OMe (—) Z:E = 7:93	747
$C_{31\text{-}33}$			
TBDMSO, OR^1, R^2, SO_2Ph	Reagents, rt	TBDMSO, R^2 Z:E = 0:100	103

R^1	R^2	Reagents	Solvent	Time	
H		SmI_2	HMPA, THF	2 h	(82)
H	"	Na/Hg, Na_2HPO_4	THF/MeOH	2 h	(58)
H		Na/Hg, Na_2HPO_4	THF/MeOH	3 h	(39)
Ac	"	SmI_2	HMPA, THF	1 h	(83)
Ac	"	Na/Hg, Na_2HPO_4	THF/MeOH	2 h	(77)

C_{32}

PhS OMe OMe Ts OAc

Na/Hg, Na_2HPO_4, MeOH, –20°, 5-7 h

PhS OMe OMe (—) 731

PhO$_2$S Bu-*n* OH H OMOM N H Boc

Na/Hg, Na_2HPO_4, MeOH, rt, 2 h

Bu-*n* H OMOM N H Boc (—) 748

$C_{32\text{-}37}$

SO$_2$Ph OR OH BocN O

Na/Hg, Na_2HPO_4, MeOH, –20°, 2 h

OR BocN O 749

R		Z:E
Me	(70)	25:75
TBDMS	(71)	~25:75

TABLE 8. REDUCTIVE ELIMINATION (*Continued*)

	Substrate	Conditions	Product(s) and Yield(s) (%)	Refs.
C_{33} 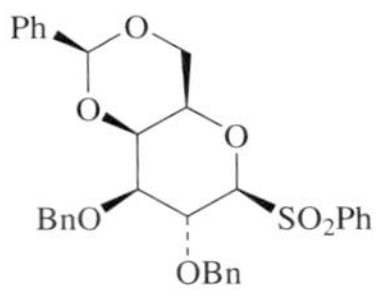	Ph, O, O, O, BnO, OBn, SO_2Ph	$Li/C_{10}H_8$, THF, –78°	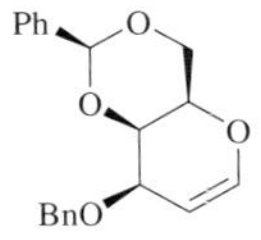 Ph, O, O, O, BnO (30)	714
	Ac, O, OBn, OBn, OBn, N, N, Me, S, O_2, O	SmI_2, THF, 20°	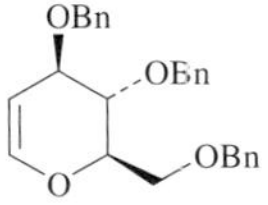 OBn, OBn, OBn, O (76)	160
	Ts, OBz	Na/Hg, THF/MeOH, –20°	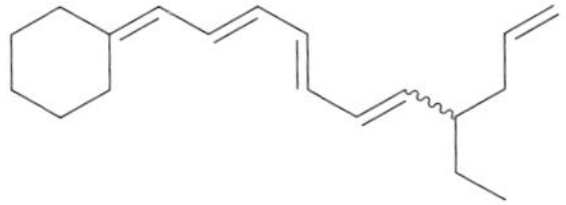 (49) Z:E = 9:91	750
	THPO, SO_2Ph, OTHP, OBz	Na/Hg, Na_2HPO_4, THF/MeOH, –15°	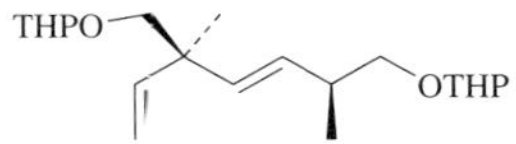 THPO, OTHP (55)	751
	OBz, SO_2Ph, OTBDMS	Na/Hg, THF/MeOH, –30°, 1 h	 OTBDMS (77) Z:E = 10:90	324

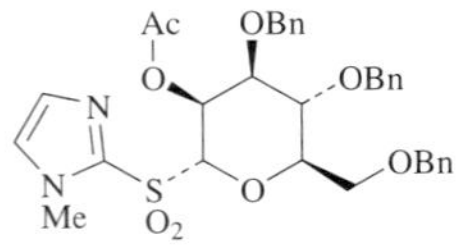

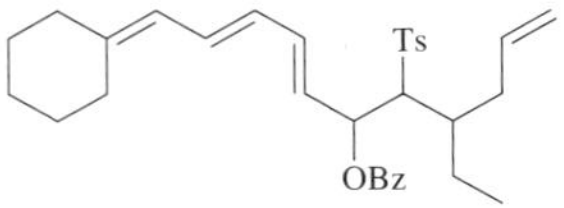

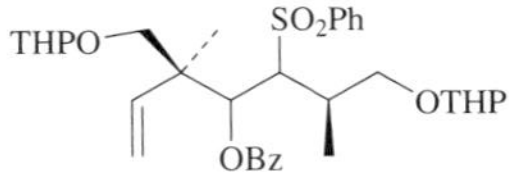

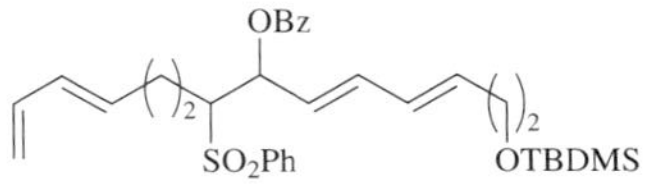

Reactant	Conditions	Product(s) and Yield(s) (%)	Refs.
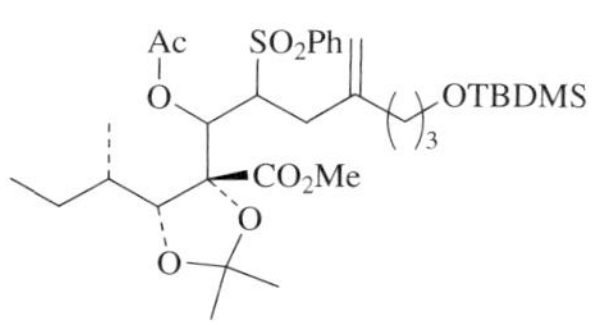 Ac, O, SO_2Ph, CO_2Me, O, O, $(\ )_3$OTBDMS	Na/Hg, EtOAc/MeOH, –40° to –50°, 50 min	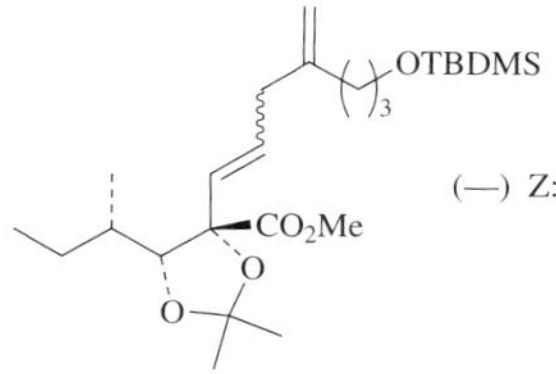 CO_2Me, O, O, $(\ )_3$OTBDMS (—) Z:E = 7:93	752
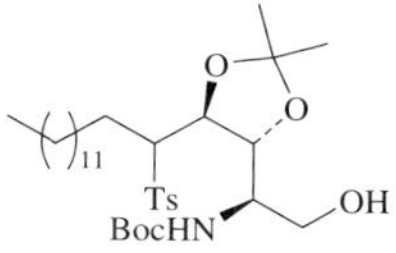 O, O, $(\ )_{11}$, Ts, BocHN, OH	Na/Hg, Na_2HPO_4, MeOH, –20° to rt, 1.5 h	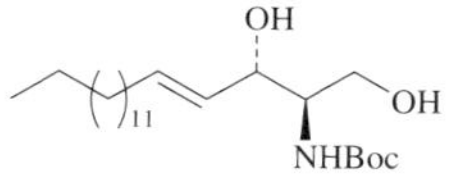 $(\ )_{11}$, OH, OH, NHBoc (60)	753
C_{34-36}			
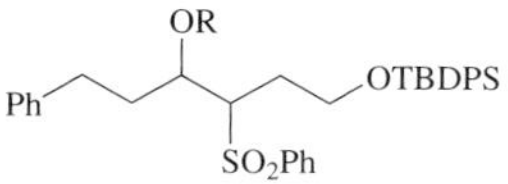 Ph, OR, SO_2Ph, OTBDPS	SmI_2, HMPA, THF	 Ph, OTBDPS	103
R = H	Temp rt, Time 1 h	(30) Z:E 26:74	
R = Ac	Temp rt, Time 0.5 h	(81) Z:E 48:52	
R = Ac	Temp –30° to –10°, Time 3 h	(78) Z:E 22:78	
C_{35}			
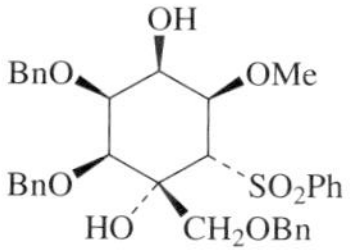 OH, BnO, OMe, BnO, SO_2Ph, HO, CH_2OBn	Na/Hg, Na_2HPO_4, MeOH, –20° to rt, 5 h	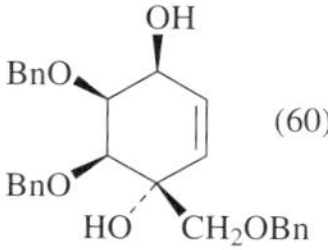 OH, BnO, BnO, HO, CH_2OBn (60)	754

TABLE 8. REDUCTIVE ELIMINATION (*Continued*)

	Substrate	Conditions	Product(s) and Yield(s) (%)	Refs.
C_{35}	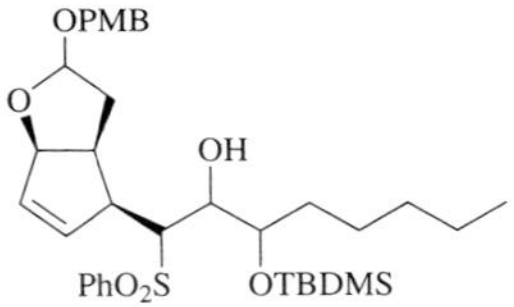	Na/Hg, Na_2HPO_4, MeOH, –40° to –20°, 1.5 h	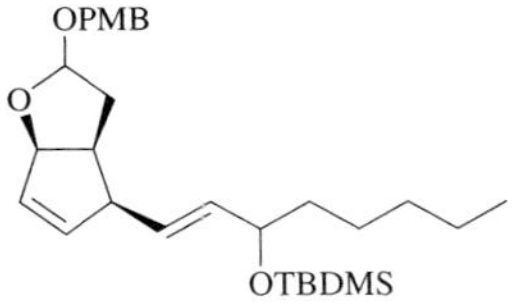 (81)	755
	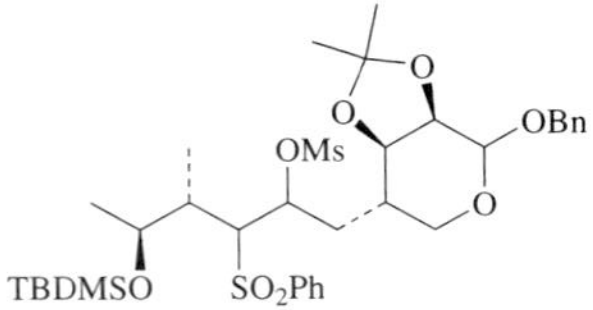	Li/NH_3, THF, –78°	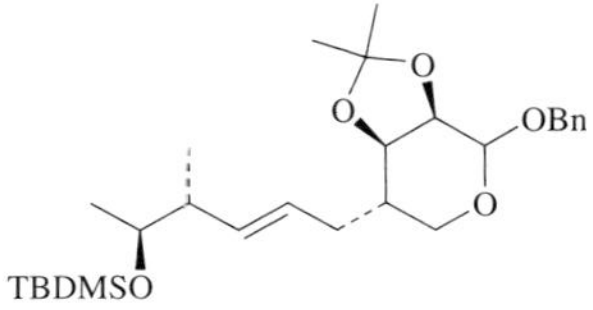 (—)	219
C_{36}	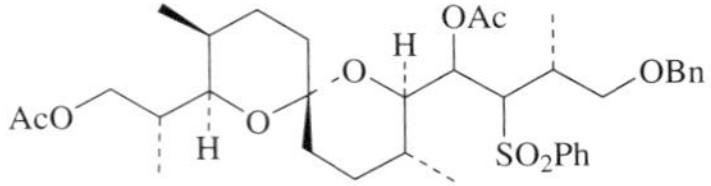	Na/Hg, MeOH, 50°, 1 h	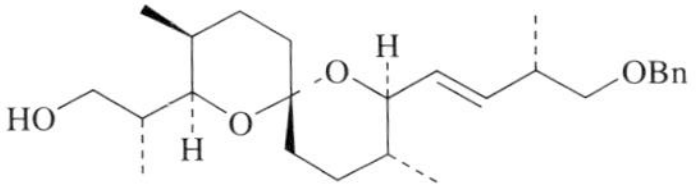 (—)	478
	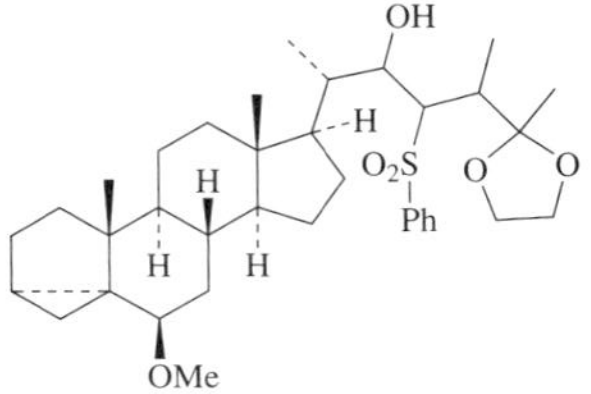	Na/Hg, Na_2HPO_4, MeOH, 4°, 16 h	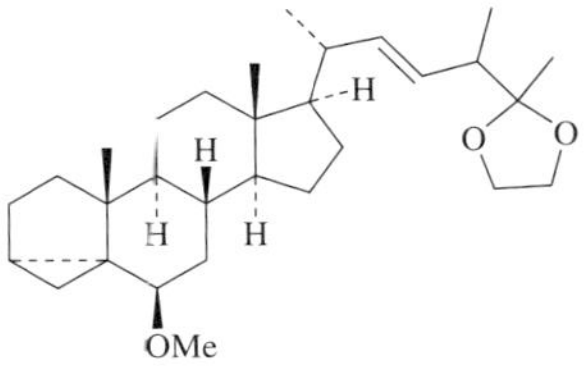 (54)	756
	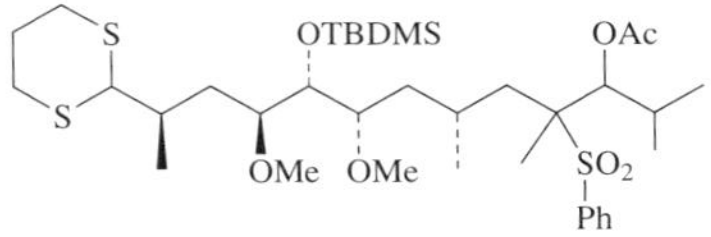	Na/Hg, KH_2PO_4, THF/MeOH, –20°, 12 h	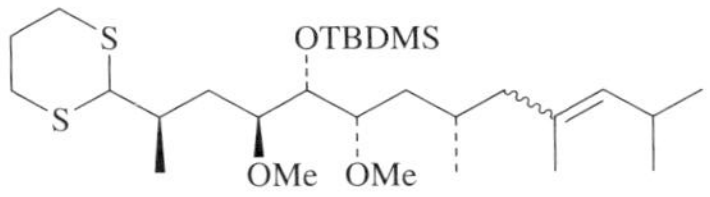 (—) Z:E = 33:67 to 29:71	757

Reactant	Conditions	Product(s) and Yield(s) (%)	Refs.
C_{36-49} R, SO_2Ph, OBz, OMe, H, H	Na/Hg, KH_2PO_4, MeOH, –40°	R, OMe, H, H	758
R (dioxolanyl)		(—)	
R = TBDPSO-acetyl		(—)	
C_{37} TBDMSO, PhO_2S, OAc, MeO, MeO, MeO, OMe, OMe	Mg, Hg_2Cl_2, EtOH, rt, 50 min	TBDMSO, MeO, MeO, MeO, OMe, OMe (—) Z:E = 47:53	759
OMOM, SO_2Ph, N Boc, OAc, NHCbz, OTBDMS	Na/Hg, Na_2HPO_4, MeOH, –20°, 8 h	OMOM, N Boc, NHCbz, OTBDMS (~100) Z:E = 17:83	760
OTBDMS, SO_2Ph, N Teoc, OAc	Na/Hg, Na_2HPO_4, THF/MeOH, –20°, 8 h	OTBDMS, N Teoc (89)	761

TABLE 8. REDUCTIVE ELIMINATION (*Continued*)

Substrate	Conditions	Product(s) and Yield(s) (%)	Refs.
$C_{37\text{-}44}$ 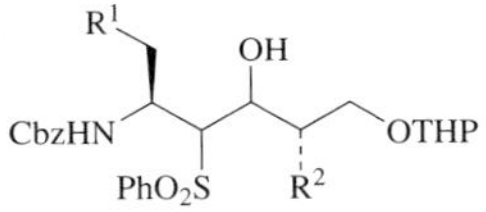 R^1, OH, CbzHN, OTHP, PhO_2S, R^2	Na/Hg, Na_2HPO_4, MeOH, 0°, 2 h	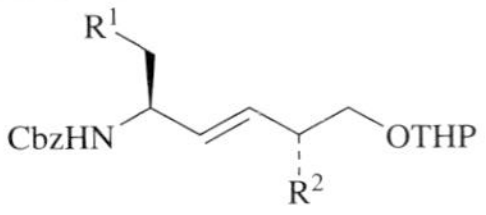 R^1, CbzHN, OTHP, R^2	762
R^1 / R^2: BzO / Boc		(—)	
R^1 / R^2: $(4\text{-}MeOC_6H_4)_2$CHNHCO / *i*-Pr		(95)	
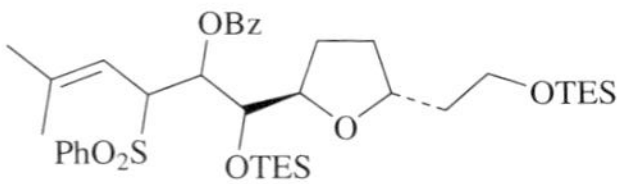 C_{38} OBz, PhO_2S, OTES, O, OTES	Na/Hg, Na_2HPO_4, THF/MeOH, –20°, 1 h	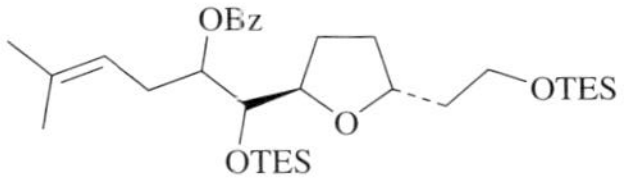 OBz, OTES, O, OTES (67)	763
$C_{38\text{-}41}$ 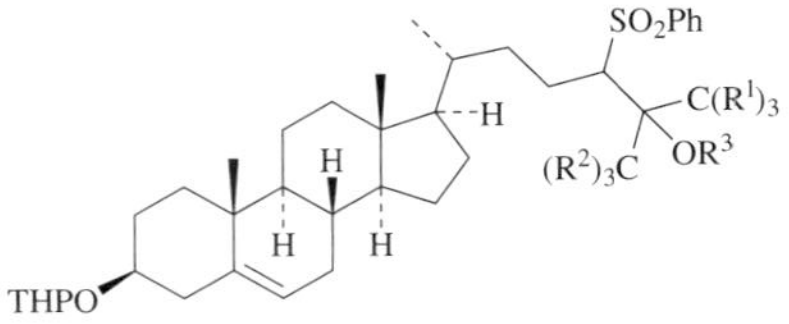 SO_2Ph, $C(R^1)_3$, $(R^2)_3C$, OR^3, H, H, H, H, THPO	Na/Hg, THF/MeOH, rt	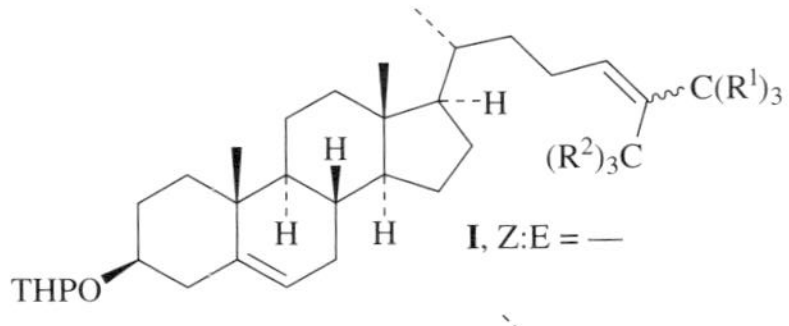 **I** [$C(R^1)_3$, $(R^2)_3C$, H, H, H, H, THPO], **I**, Z:E 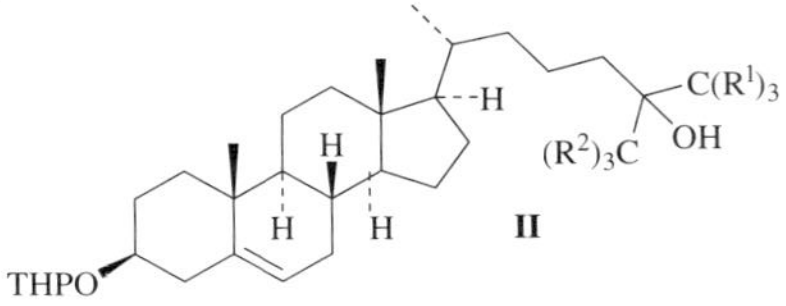 = — + **II** [$C(R^1)_3$, $(R^2)_3C$, OH, H, H, H, H, THPO]	

R^1	R^2	R^3	Additive	Time	I	II	Refs.
D	D	H	—	4 h	(48-52)	(28)	764
F	H	H	Na_2HPO_4	1.5 h	(52)	(14)	599
H	H	TMS	—	4 h	(55)	(31)	764

Reactant	Conditions	Product(s) and Yield(s) (%)	Refs.
C_{39}			
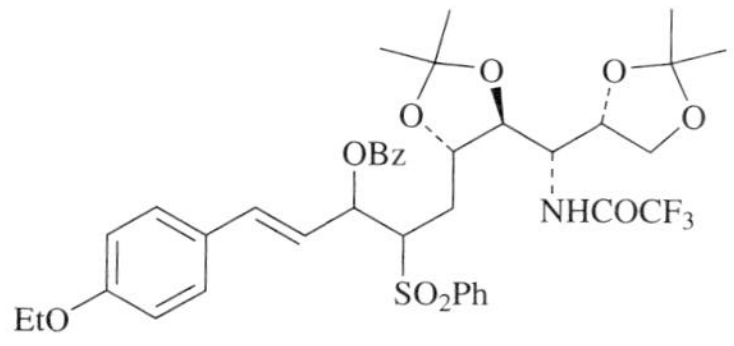 EtO; OBz; SO_2Ph; $NHCOCF_3$	Na/Hg, MeOH	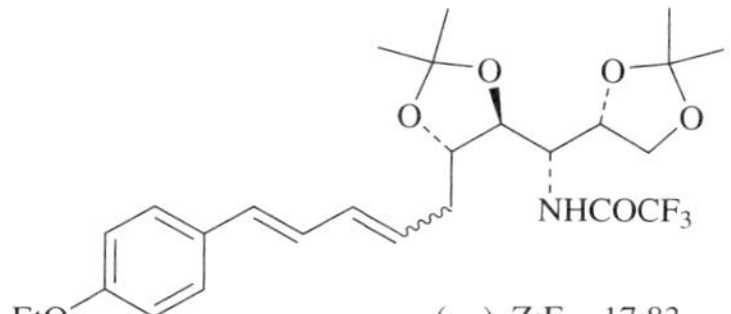 EtO; $NHCOCF_3$ (—) Z:E = 17:83	765
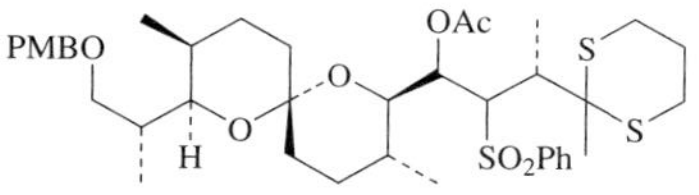 PMBO; H; OAc; SO_2Ph; S; S	Na/Hg, Na_2HPO_4, THF/MeOH, −20°, 1 h	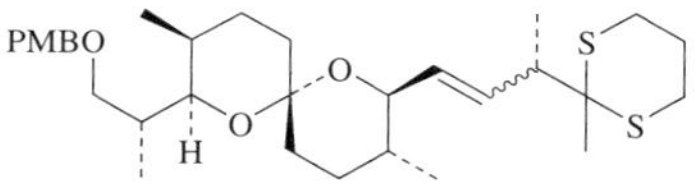 PMBO; H; S; S (—)	766
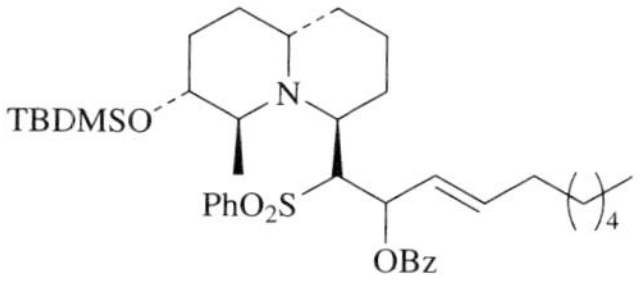 TBDMSO; N; PhO_2S; OBz; $(\ )_4$	Na/Hg, MeOH/EtOAc, −30°	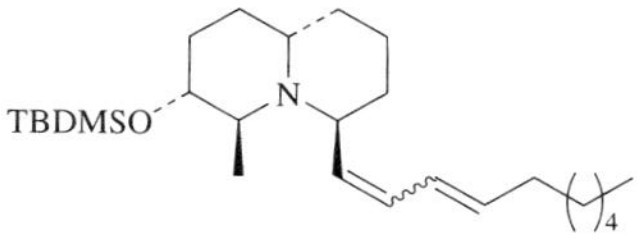 TBDMSO; N; $(\ )_4$ (—) EE:EZ = 80:20	701
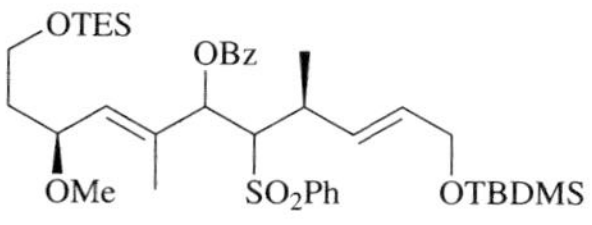 OTES; OMe; OBz; SO_2Ph; OTBDMS	Na/Hg, Na_2HPO_4, MeOH	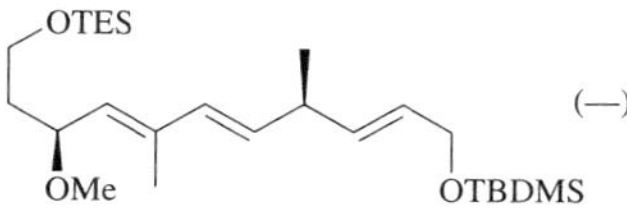 OTES; OMe; OTBDMS (—)	767
C_{40}			
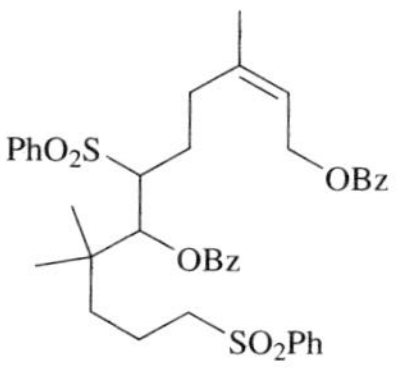 PhO_2S; OBz; OBz; SO_2Ph	Na/Hg, THF/MeOH, −20°, 2 h	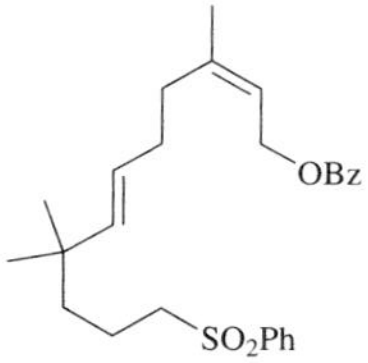 OBz; SO_2Ph (54)	768

TABLE 8. REDUCTIVE ELIMINATION (*Continued*)

Substrate	Conditions	Product(s) and Yield(s) (%)	Refs.
C_{40} 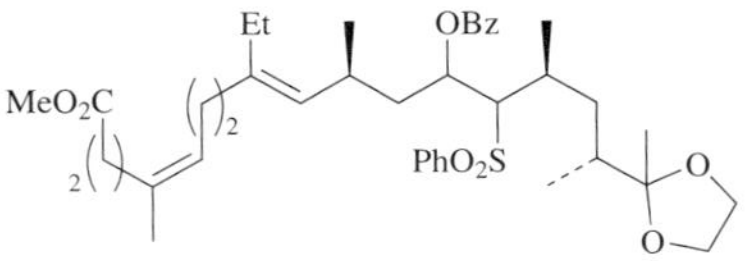 MeO$_2$C, Et, OBz, PhO$_2$S	Na/Hg, EtOAc/MeOH, –30° to –40°, 38h	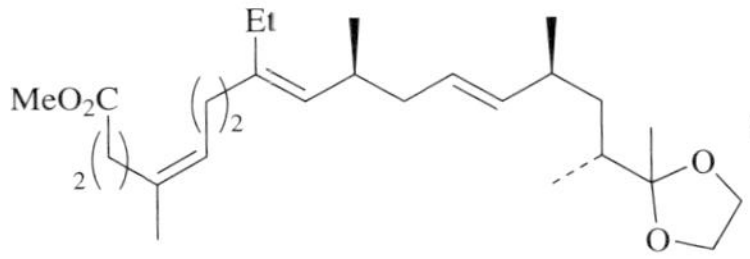 MeO$_2$C, Et (57)	769
C_{41} 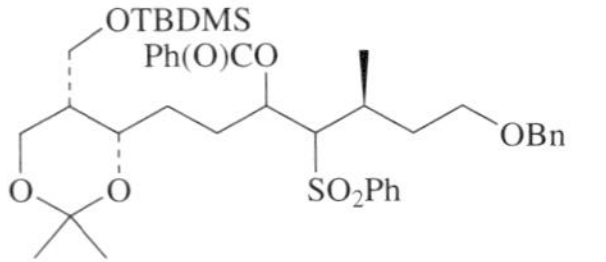 OTBDMS, Ph(O)CO, SO$_2$Ph, OBn	Li/NH$_3$, THF, –78°	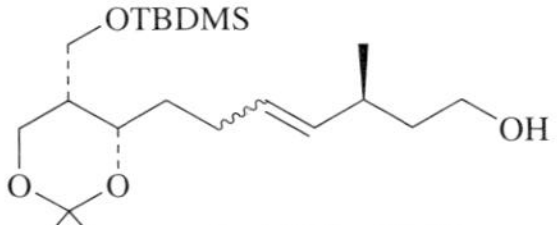 OTBDMS, OH (—) Z:E = 10:90	770
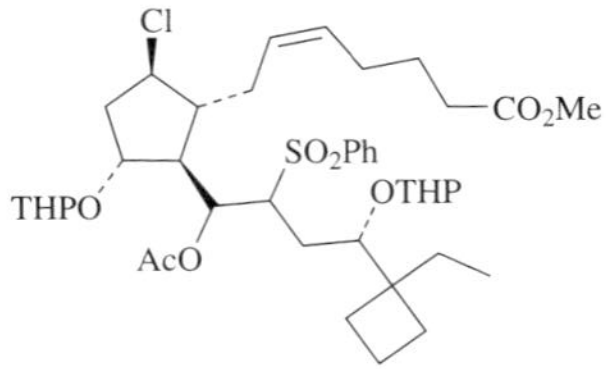 Cl, CO$_2$Me, SO$_2$Ph, OTHP, THPO, AcO	Mg, TMSCl, MeOH, rt, 1 h	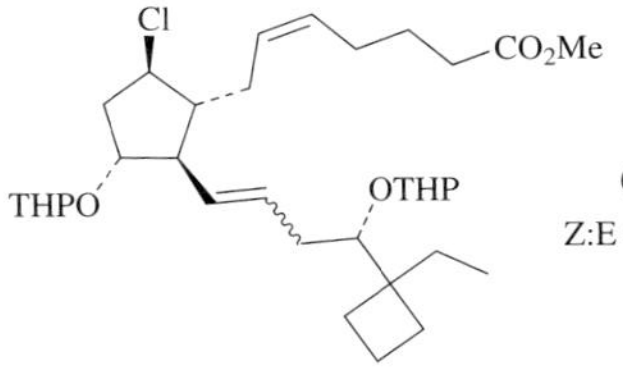 Cl, CO$_2$Me, THPO, OTHP (—) Z:E = 17:83	771
$C_{41\text{-}48}$ 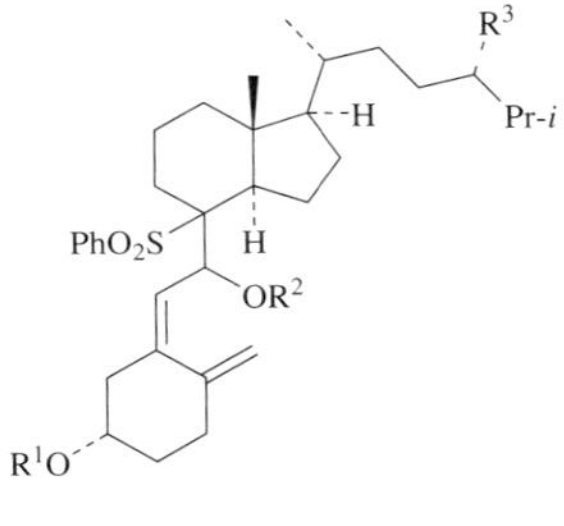 R^3, Pr-*i*, H, PhO$_2$S, H, OR^2, R^{1}O	Reagents, MeOH/THF	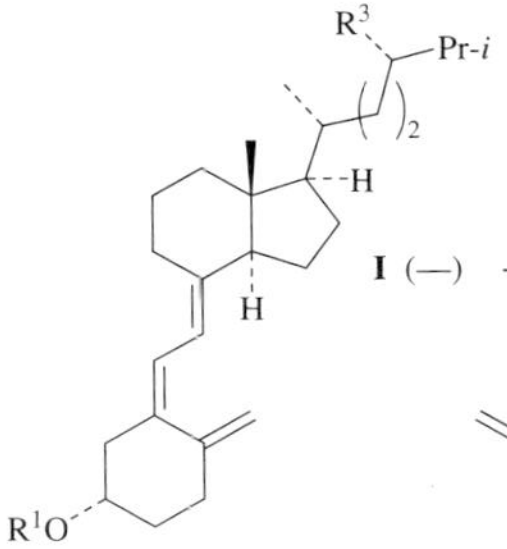 **I** 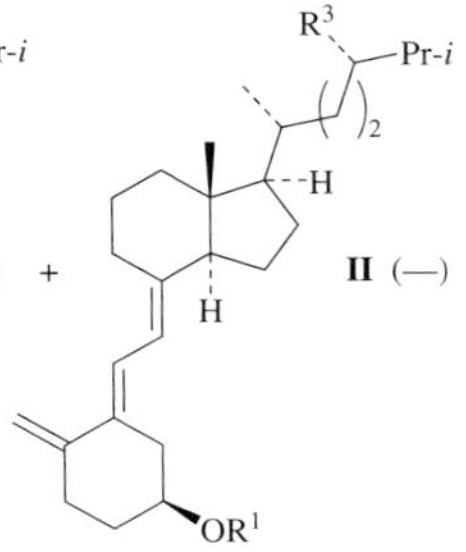 (—) + **II** (—)	

R^1	R^2	R^3	Reagents	Temp	Time	**I:II**	
TBDMS	Ac	H	Na/Hg	–20° to rt	9 h	100:0	772
Bz	Ac	Me	Na/Hg	–20°	—	100:0	96
Bz	TMS	Me	Li/Hg	–20°	3.5 h	100:0	96
Bz	Bz	Me	Na/Hg	–20°	3 h	64:36	96

C_{42}

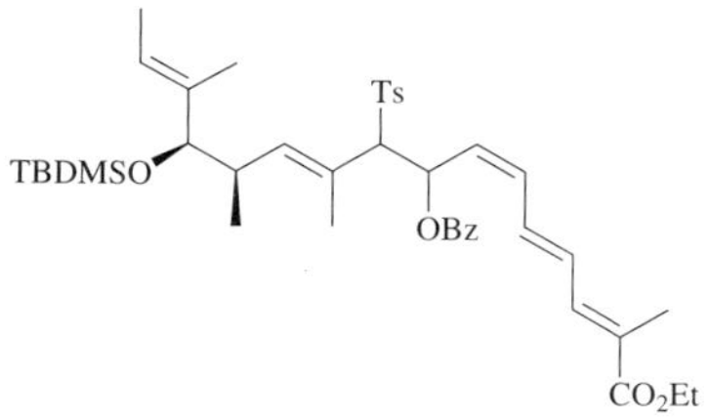

TBDMSO, Ts, OBz, CO_2Et

Na/$C_{10}H_8$, THF, –73°, 20 min

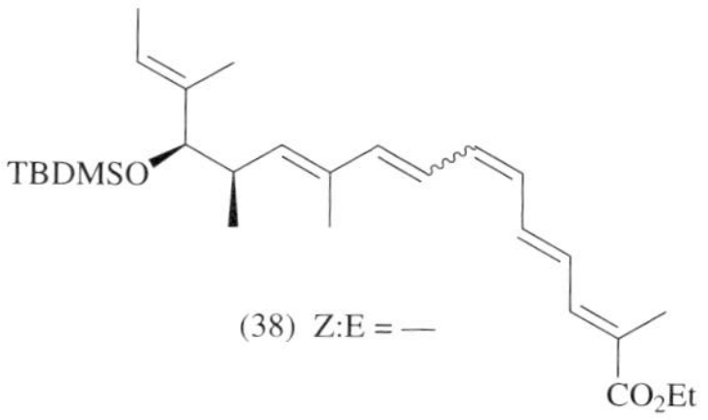

TBDMSO, CO_2Et

(38) Z:E = —

736

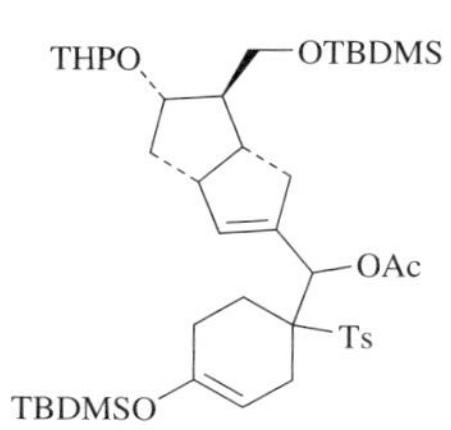

THPO, OTBDMS, OAc, Ts, TBDMSO

Na/Hg, Na_2HPO_4, MeOH, –20°, 2 h

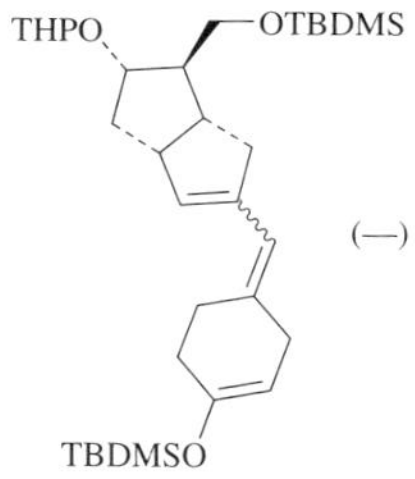

THPO, OTBDMS, TBDMSO

(—) Z:E = 50:50

773

TABLE 8. REDUCTIVE ELIMINATION (*Continued*)

Substrate	Conditions	Product(s) and Yield(s) (%)	Refs.
C_{42-44}			
THPO, PhO_2S, R^2, OR^3, R^4, R^4, OR^1, H, H, H	Na/Hg, Na_2HPO_4, THF/MeOH, 0°	THPO, R^2, OR^3, R^4, R^4, H, H, H	209
R^1 = Ac, R^2 = H, R^3 = MOM, R^4 = CF_3		(73)	
R^1 = H, R^2 = CF_3, R^3 = THP, R^4 = Me		(67)	
C_{43}			
Ph, OAc, NHBoc, Ph, Bn, N, Boc, SO_2Ph	Na/Hg, Na_2HPO_4, MeOH, rt, 12 h	Ph, NHBoc, Ph, Bn, N, Boc (—)	774
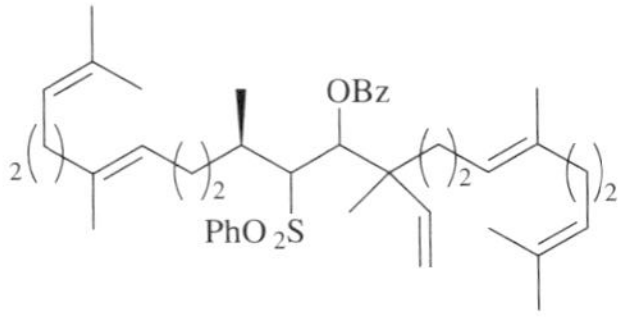 OBz, PhO_2S, ( )$_2$	Na/Hg, Na_2HPO_4, THF/MeOH, –20°	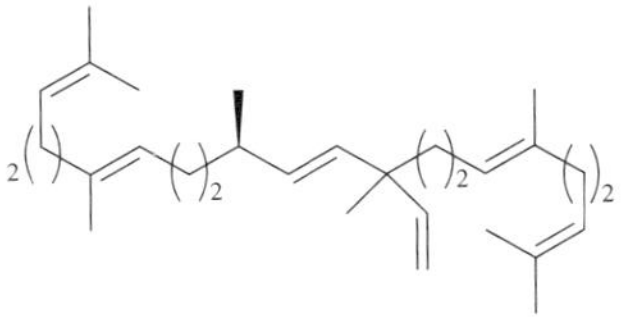 ( )$_2$ (—)	751

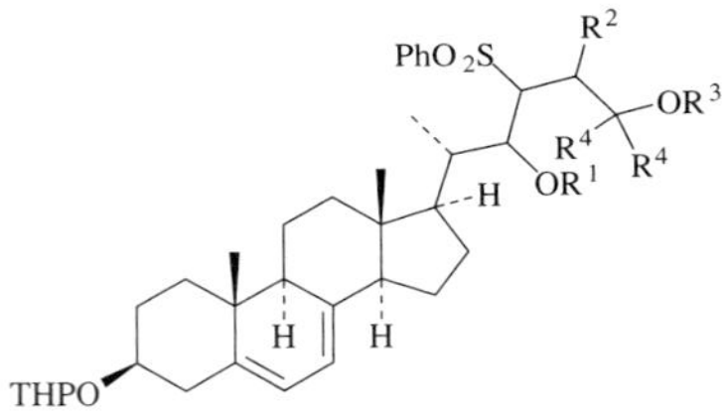

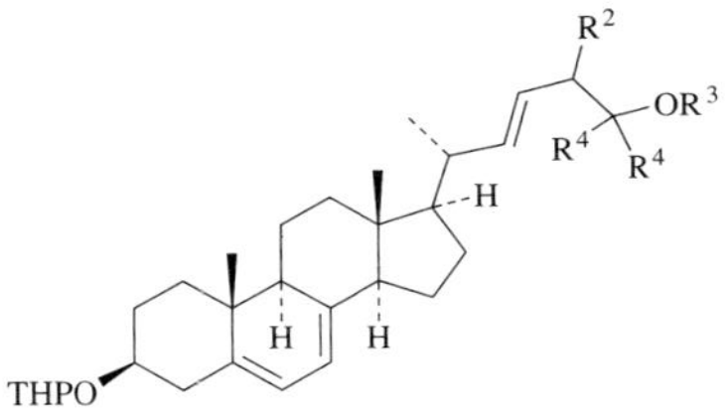

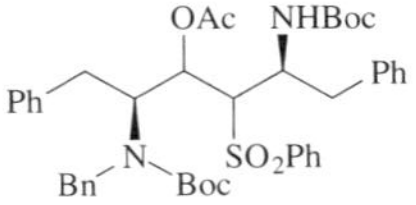

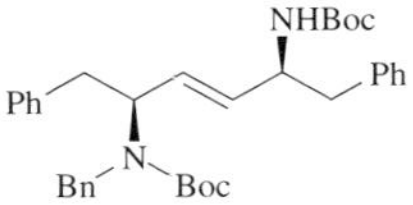

	Reactant	Conditions	Product(s) and Yield(s) (%)	Refs.
C_{44}	TBDPSO, SO_2Ph, OBn, OH, NHBoc	Na/Hg, Na_2HPO_4, MeOH, 0°, 2 h	TBDPSO, OBn, NHBoc (78) Z:E = <1:99 + OR^1, OBn, OR^2, NHBoc	775
			R^1 TBDPS, R^2 H (6); R^1 H, R^2 TBDPS (6)	
	SO_2Ph, OAc, OPMB, TIPSO	Mg, $HgCl_2$, EtOH, rt, 1.5 h	OPMB, TIPSO (—) Z:E = <9:91	270
$C_{44\text{-}49}$	OTBDMS, OTIPS, SO_2Ph, OR	Na/Hg, $NaHCO_3$, THF/MeOH, –35°	OTBDMS, OTIPS	248

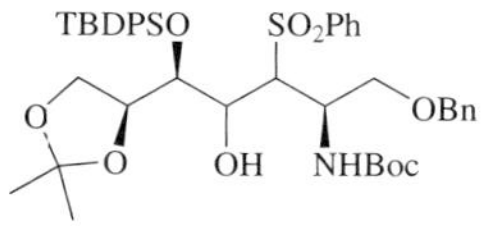

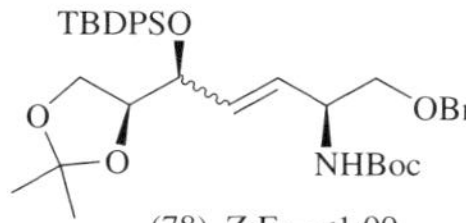

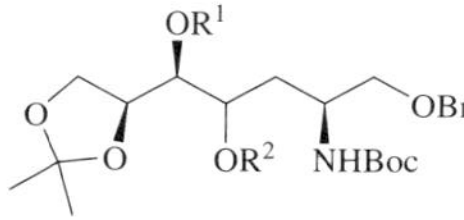

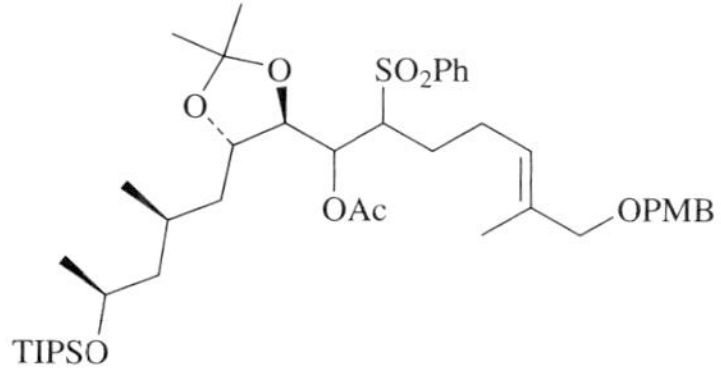

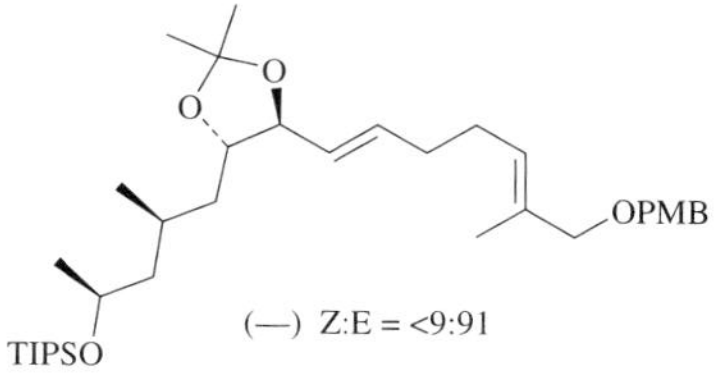

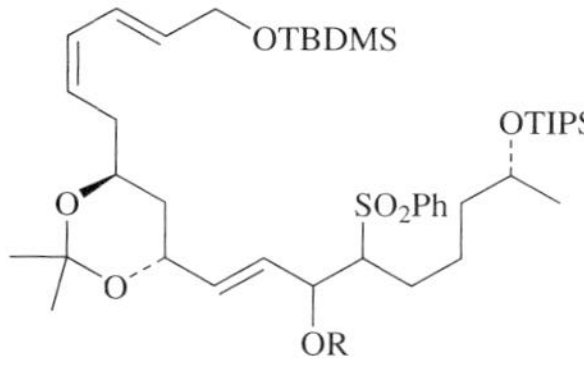

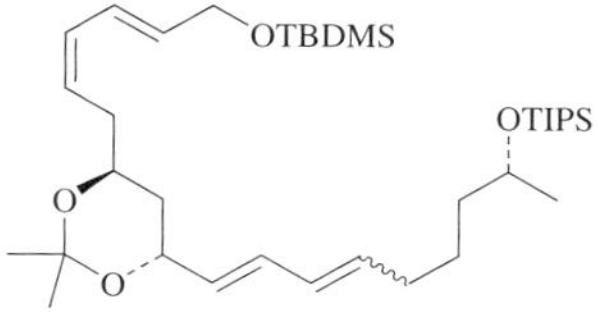

R	Time		Z:E
Ac	—	(—)	36:64
Bz	2 h	(63)	8:92

TABLE 8. REDUCTIVE ELIMINATION (*Continued*)

Substrate	Conditions	Product(s) and Yield(s) (%)	Refs.
C_{45}			
AcO, OH, H, O_2S, Ph, O, O, H, H, N, N, O, NPh, O	Na/Hg, Na_2HPO_4, MeOH, 4°, 15 h	AcO, H, H, H, N, N, O, NPh, O, O, O (44)	756
OBn, OBz, SO_2Ph	NH_3/Na	OBn (93)	776
TBDMSO, OMe, Ph, SO_2, OBn, O, OMe, OH, OTBDMS	Na/Hg, NaH_2PO_4, MeOH, 0°, 20 min	TBDMSO, OMe, OBn, O, OMe, OTBDMS (71)	777, 778
$C_{48\text{-}50}$			
CO_2Me, O, H, OH, Ts, OR, OR, O, Et	Na/Hg, Na_2HPO_4	CO_2Me, O, H, OR, OR, O, Et	

R
TBDMS
Bn

Solvent	Temp	Time		Z:E	Refs.
MeOH	–30°	3 h	(—)	9:91	779
THF/MeOH	–35°	3.5 h	(63)	8:92	780

C_{48-53}

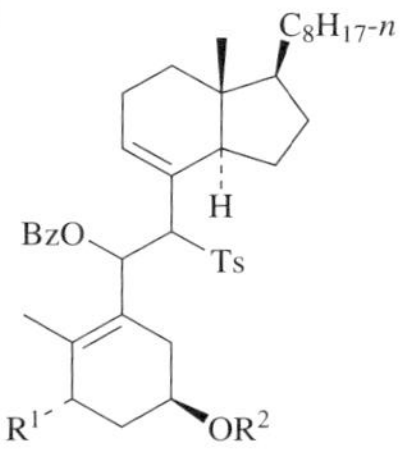

C_8H_{17}-*n*

BzO

Ts

H

R^1

OR^2

Na/Hg, THF/MeOH, −20°

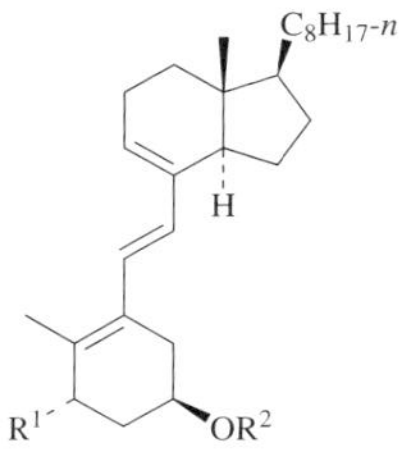

781

R^1	R^2	Time	
H	PhCO	1.5 h	(—)
TBDMSO	TBDMS	2 h	(—)

C_{49}

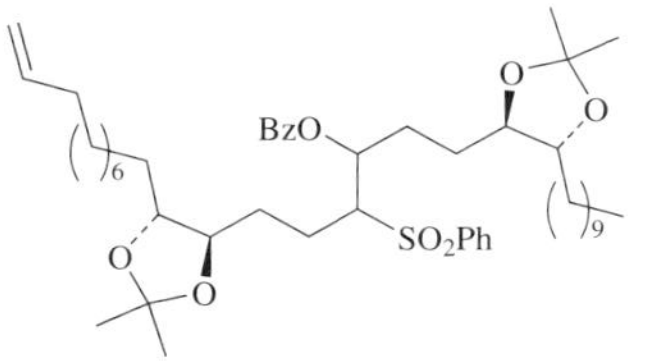

BzO

SO_2Ph

Na/Hg, H_3BO_3, MeOH, rt

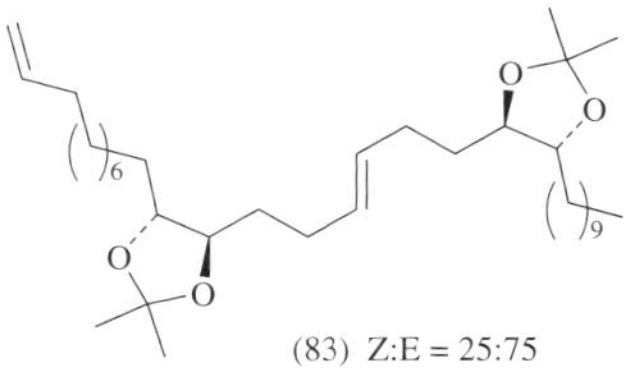

(83) Z:E = 25:75

782

C_{49-53}

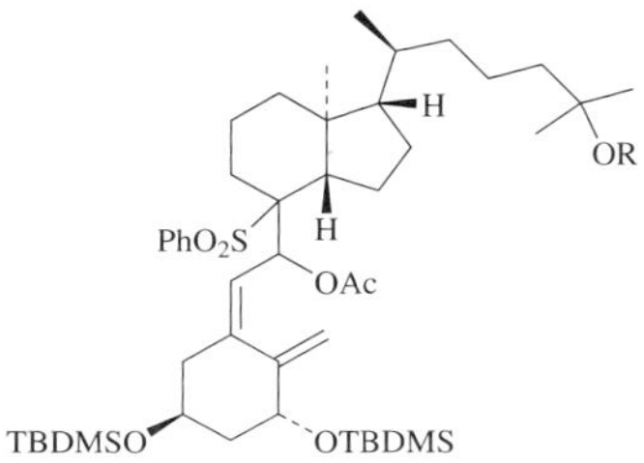

PhO_2S

OAc

H

OR

TBDMSO

OTBDMS

R = mixture of TES and Ac

Na/Hg, Na_2HPO_4, MeOH

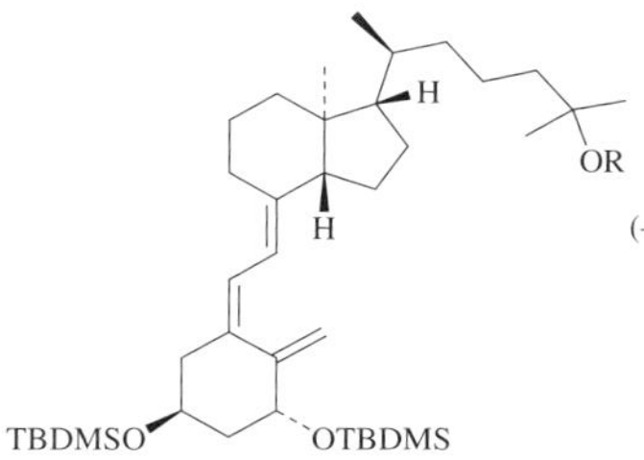

(—)

783

TABLE 8. REDUCTIVE ELIMINATION (*Continued*)

Substrate	Conditions	Product(s) and Yield(s) (%)	Refs.
C_{49-56} (OR[3], R[4], SO[2]Ph, R[1], R[2], TBDMSO, OTBDMS)	Na/Hg, Na_2HPO_4	(R[4], R[1], R[2], TBDMSO, OTBDMS)	

R^1	R^2	R^3	R^4	Solvent	Temp	Time	Yield (%)	Refs.
$H_2C=$	H	H	(dioxane)	MeOH	rt	6 h	(25)	784
H	$H_2C=$	H	OTHF	THF/MeOH	5°	3 h	(55)	386
H	H	Ac	TESO, Et	THF/MeOH	4°	—	(—)	785

Substrate	Conditions	Product(s) and Yield(s) (%)	Refs.
C_{50} (C_8H_{17}-*n*, $(EtO)_2OPO$, Ts, TBDMSO, OTBDMS)	Na/Hg, THF/DMSO, 0°, 6 h	(C_8H_{17}-*n*, TBDMSO, OTBDMS) (56)	236

C_{50-51}

R^1	R^2	R^3
Ac	AcO	Ac
TBDMS	H	H
TBDMS	H	Ms

Na/Hg, Na_2HPO_4, MeOH, 0° to rt, 30 min

R^1	R^2		
H	OH	(98)	786
TBDMS	H	(69)	787
TBDMS	H	(77)	787

C_{51}

Na/Hg, THF/MeOH, –20°, 2 h

(—) 788

Na/Hg, THF/MeOH, –20°, 1 h

(—) Z:E = 63:37 789

TABLE 8. REDUCTIVE ELIMINATION (*Continued*)

Substrate	Conditions	Product(s) and Yield(s) (%)	Refs.
C_{51}			
Ph, OTES, OTES, SO_2, OBn, OMe, OBn, OH	Na/Hg, Na_2HPO_4, MeOH, 0°	OTES, OTES, OBn, OMe, OBn (—)	790
C_{51-91}			
TBDMSO, OMTM, OPiv, OMe, SO_2Ph, R^2, MeO, R^1, OAc	Na/Hg, Na_2HPO_4, MeOH, 0°, 35 h	TBDMSO, OMTM, OPiv, OMe, R^2, MeO, R^1 (—)	791

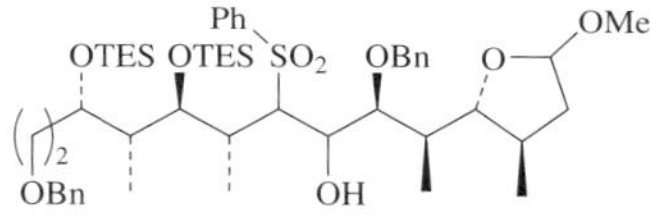

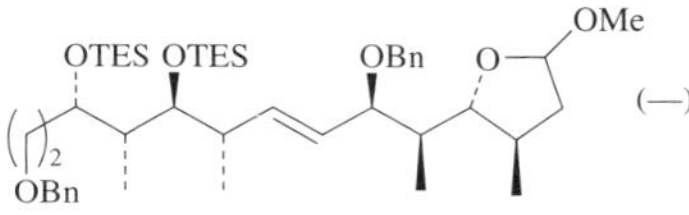

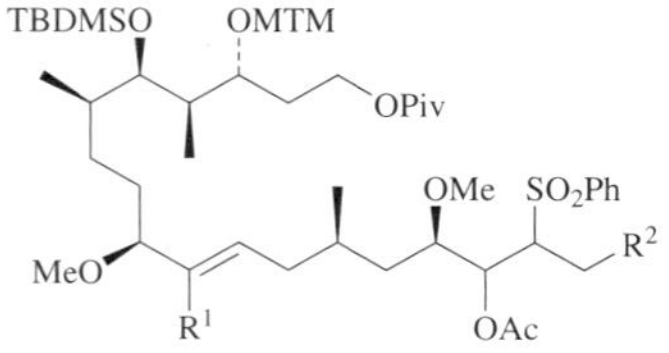

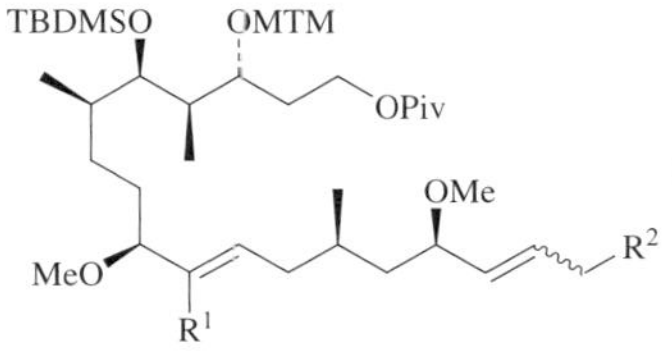

R^1	R^2	Z:E
H	CH_2OTES	10:90
H	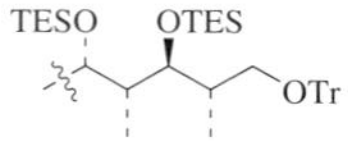 TESO, OTES, OTr	9:91
H	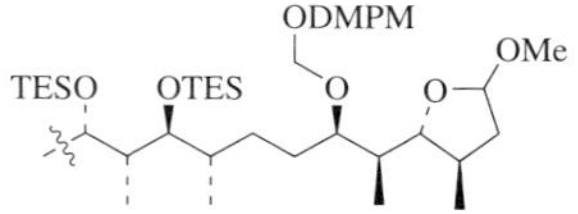 TESO, OTES, ODMPM, O, OMe	9:91
Me	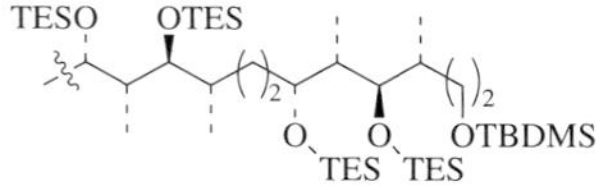 TESO, OTES, O, TES, O, TES, OTBDMS	9:91

Substrate	Conditions	Product	(Yield)	Ref.
C_{52} 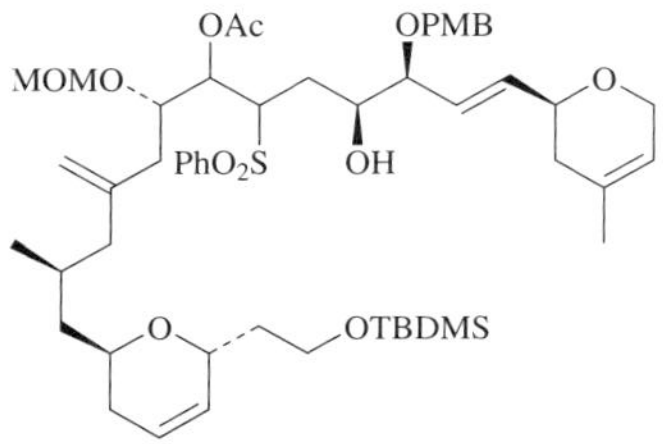	Na/Hg, Na_2HPO_4, MeOH, –20°, 2 h; rt, 0.5 h	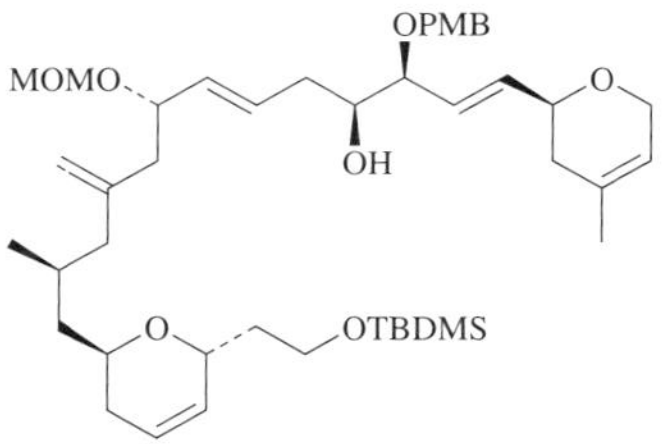	(—) Z:E = 23:77	792
C_{52-55} 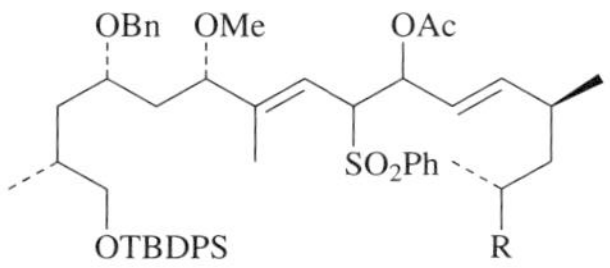	Na/Hg, KH_2PO_4, THF/MeOH, –20°	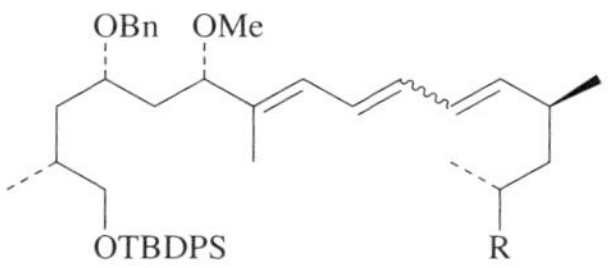		705

R	Time		Z:E
$HOCH_2$	1 h	(65)	0:100
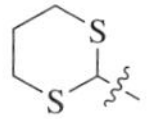 (1,3-dithian-2-yl)	45 min	(85)	40:60

Substrate	Conditions	Product	(Yield)	Ref.
C_{53} 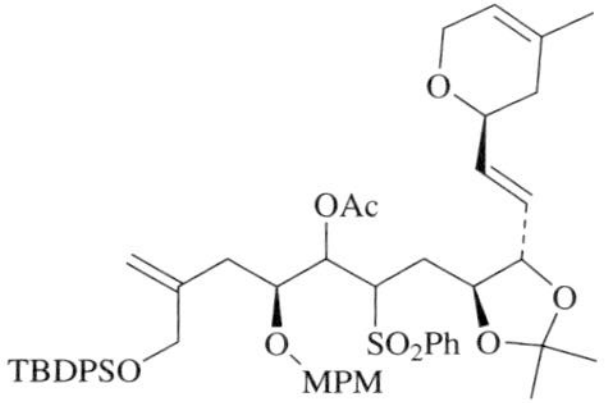	Na/Hg, EtOAc/MeOH, –20°	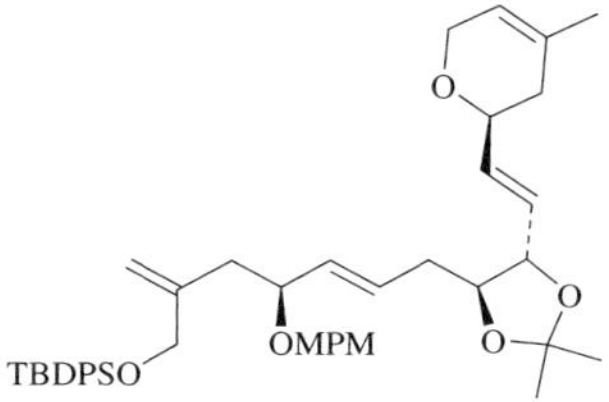	(—)	793

TABLE 8. REDUCTIVE ELIMINATION (*Continued*)

Substrate	Conditions	Product(s) and Yield(s) (%)	Refs.
C_{53}			
OH, O, n-Pr, NH, TBDMSO, OAc, H, O, O, H, SO_2Ph	Na/Hg, KH_2PO_4, THF/MeOH, 0°	OH, O, n-Pr, NH, TBDMSO, H, O, O, H (—) Z:E = 20:80	794
C_{54}			
TBDMSO, OTBDMS, O, AcO, SO_2Ph, H, O, OTBDMS, OPMB	Mg, $HgCl_2$, EtOH/THF, 0° to rt, 5.25 h	TBDMSO, OTBDMS, O, H, O, OTBDMS, OPMB (74) Z:E = <5:95	217
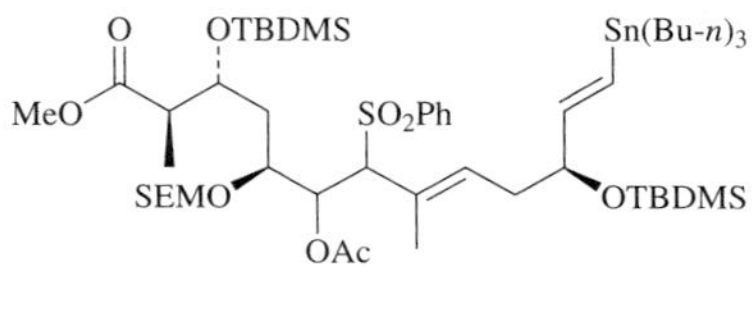 O, OTBDMS, MeO, SEMO, SO_2Ph, OAc, $Sn(Bu\text{-}n)_3$, OTBDMS	Na/Hg, Na_2HPO_4, MeOH/EtOAc	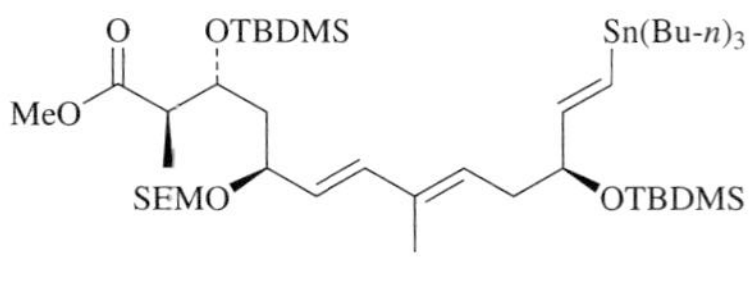 O, OTBDMS, MeO, SEMO, $Sn(Bu\text{-}n)_3$, OTBDMS (—)	795

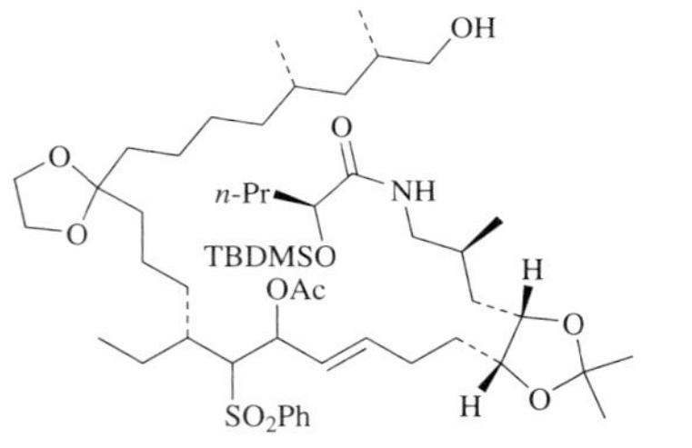

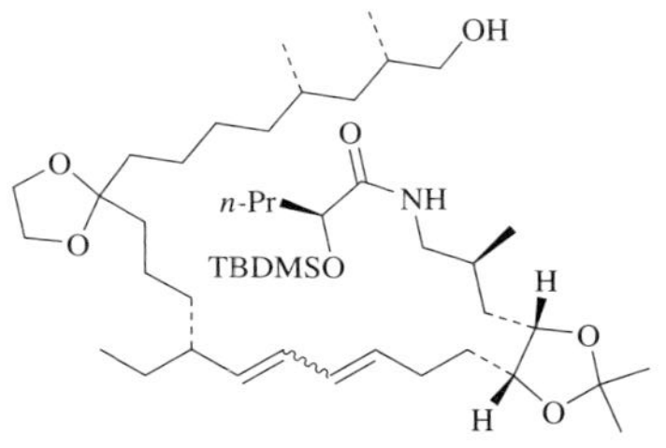

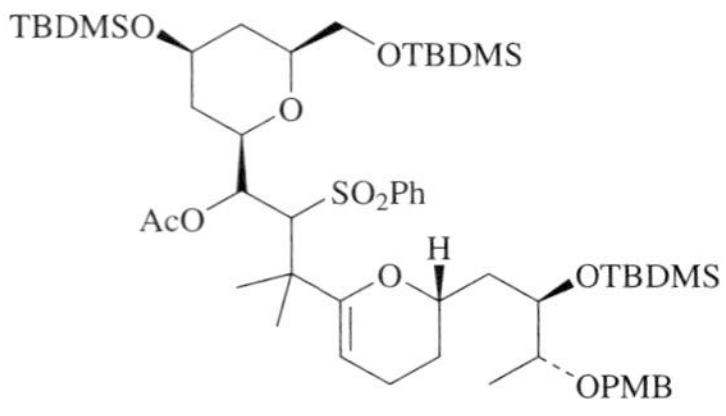

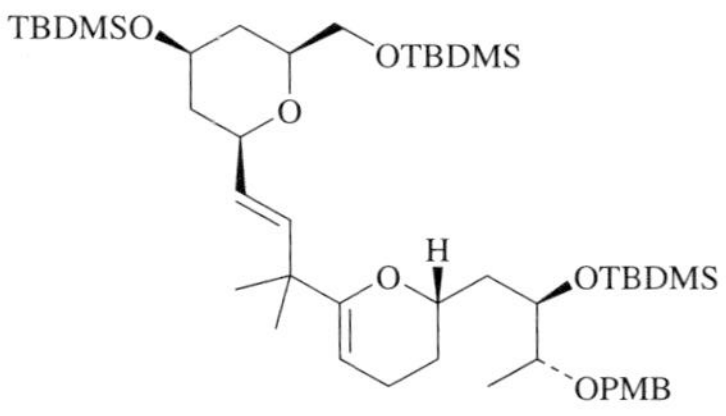

Reactant	Conditions	Product(s) and Yield(s) (%)	Refs.
C_{55}			
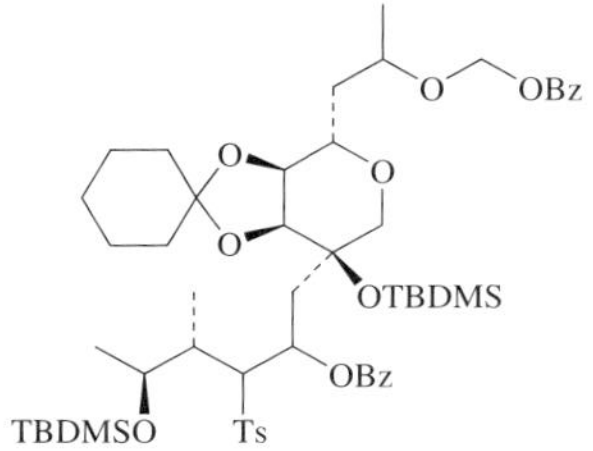	Na/Hg	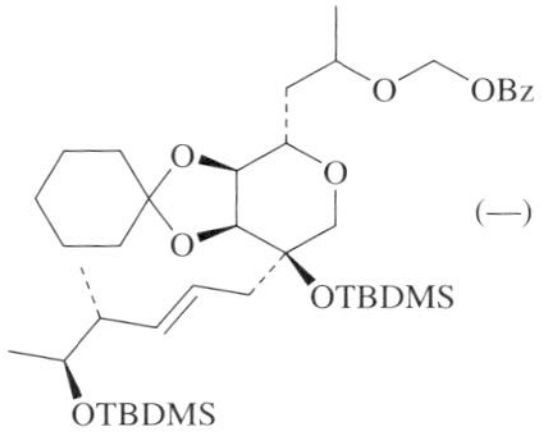 (—)	796
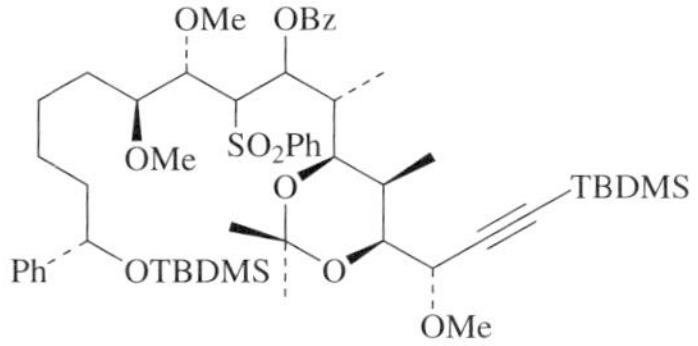	Na/Hg, Na_2HPO_4, MeOH, –30° to –20°, 5h	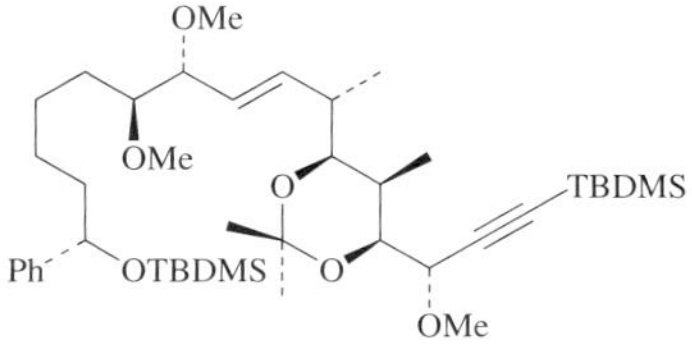 (—)	797
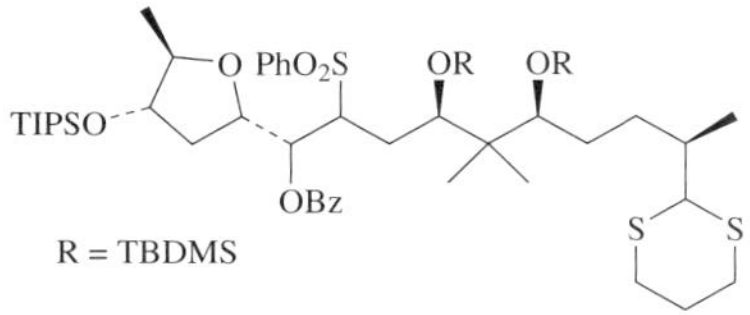 R = TBDMS	Na/Hg, THF/MeOH, –20°, 8 h	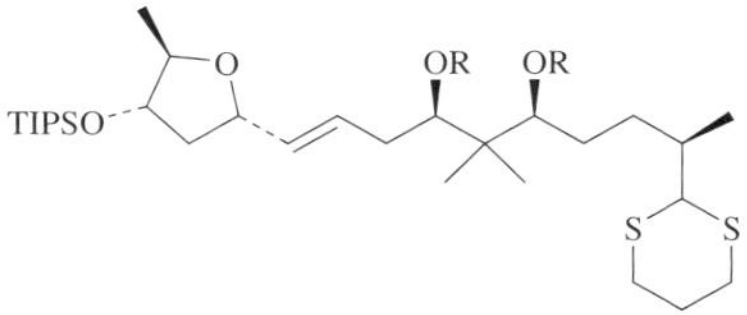 (72)	798
C_{56}			
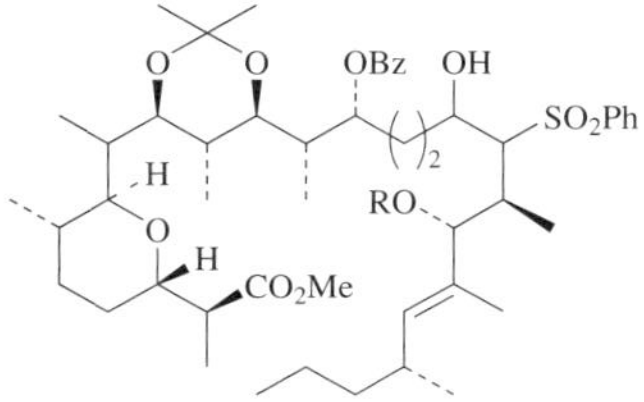 R = TBDMS	Na/Hg, THF/MeOH, –40°, 20 min	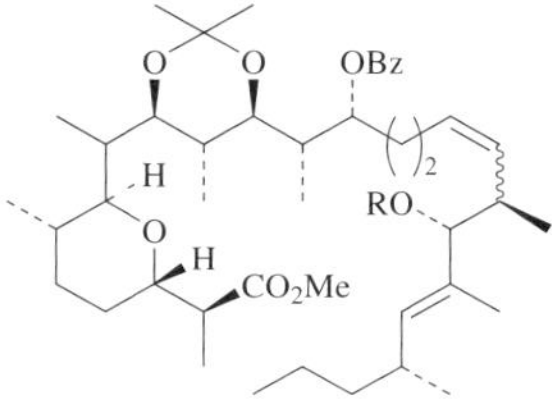 (50) Z:E = 11:89	799

TABLE 8. REDUCTIVE ELIMINATION (*Continued*)

Substrate	Conditions	Product(s) and Yield(s) (%)	Refs.
C_{58}			
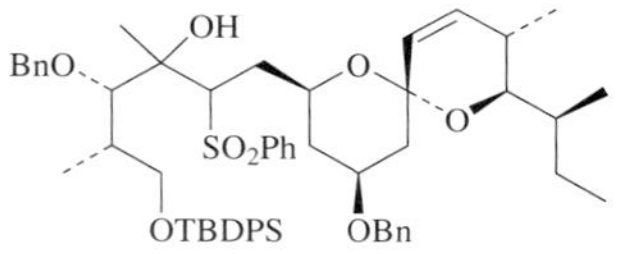 BnO, OH, SO_2Ph, OTBDPS, OBn	Na/Hg, KH_2PO_4, THF/MeOH	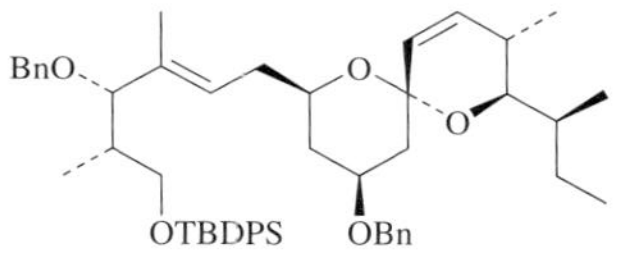 BnO, OTBDPS, OBn (40)	215
C_{59}			
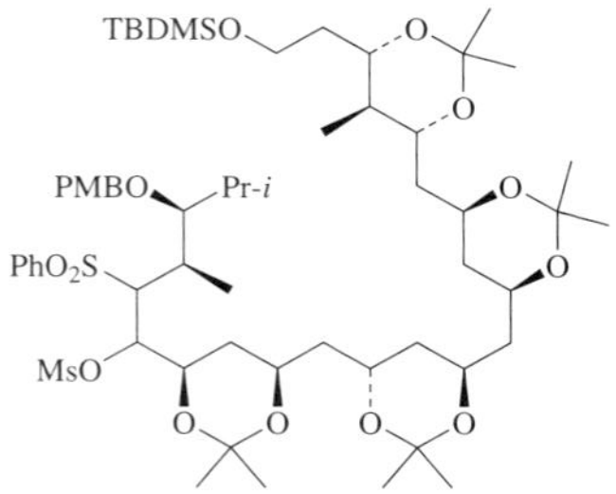 TBDMSO, PMBO, Pr-*i*, PhO_2S, MsO	Na/$C_{10}H_8$	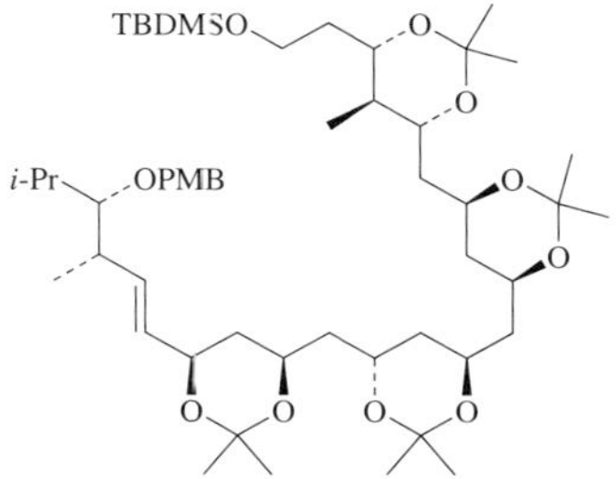 TBDMSO, *i*-Pr, OPMB (—)	570
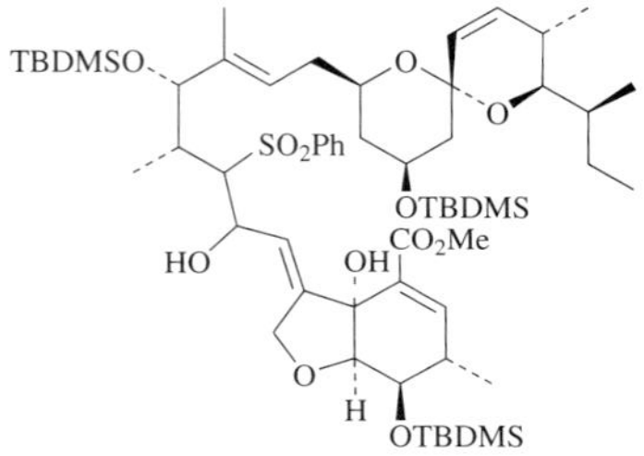 TBDMSO, SO_2Ph, HO, OH, CO_2Me, OTBDMS, H, OTBDMS	Na/Hg, MeOH	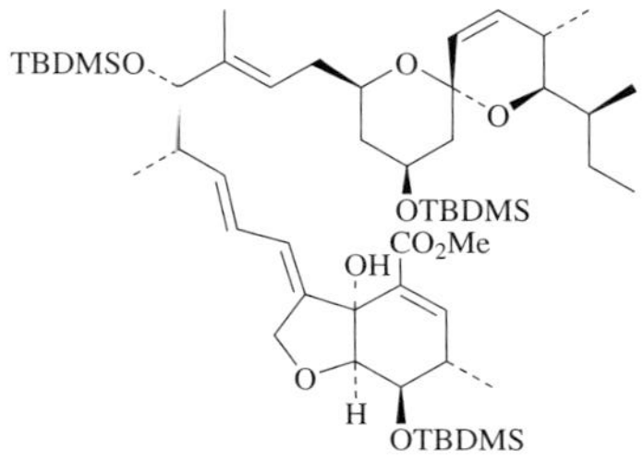 TBDMSO, OH, CO_2Me, OTBDMS, H, OTBDMS (—)	215

Reactant	Conditions	Product(s) and Yield(s) (%)	Refs.
C_{60}			
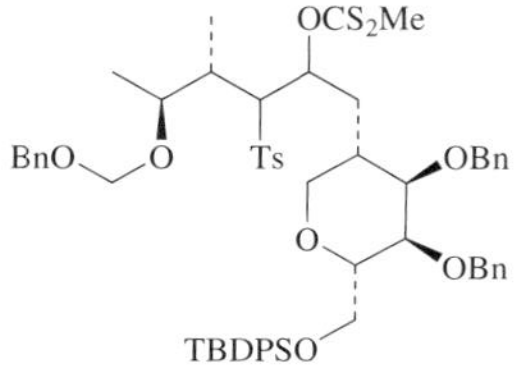	$(n\text{-Bu})_3SnH$, AIBN, toluene, 95°	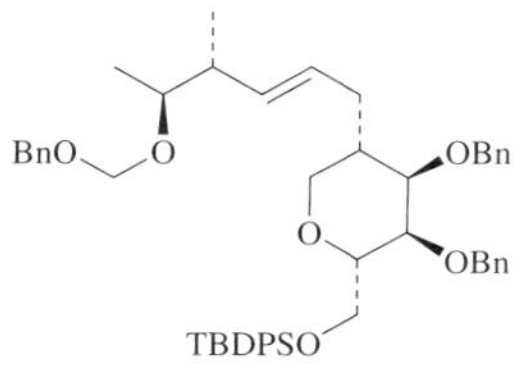 (86)	98
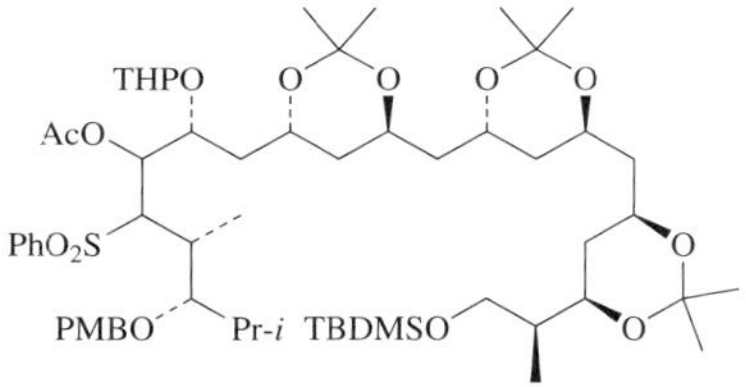	Na/Hg, Na_2HPO_4, THF/MeOH, –20°, 3 h	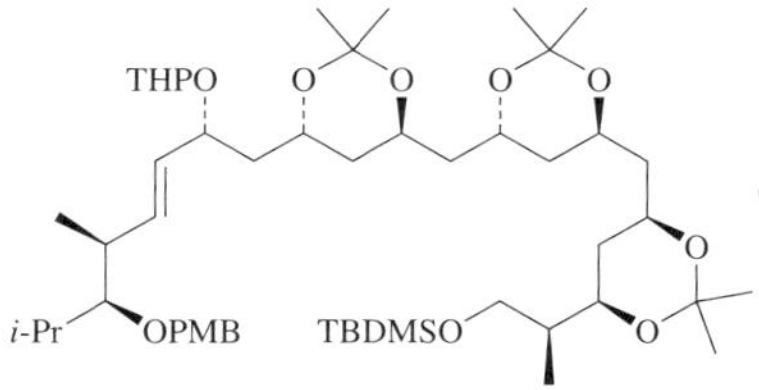 (62)	800
C_{62}			
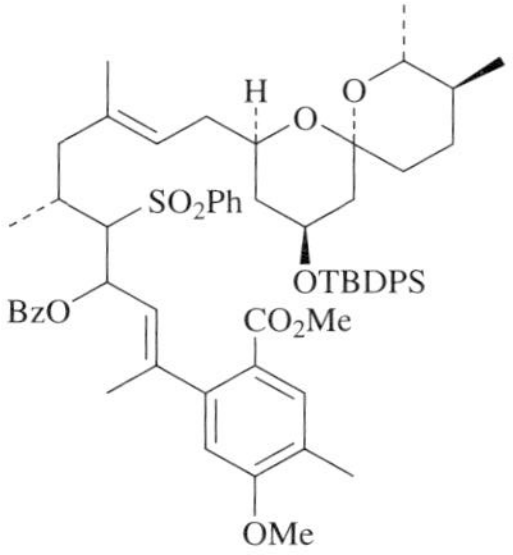	Na/Hg	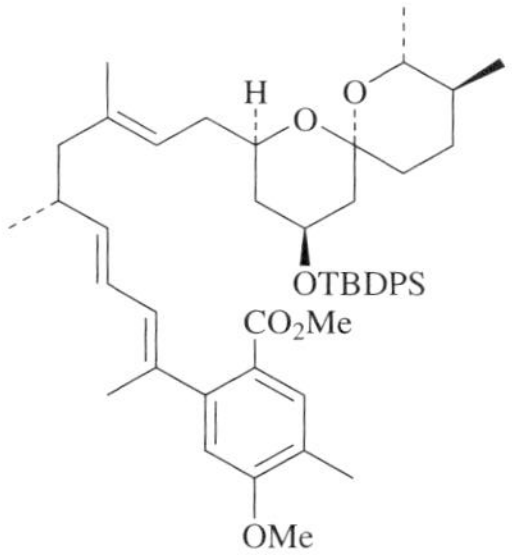 (—)	801
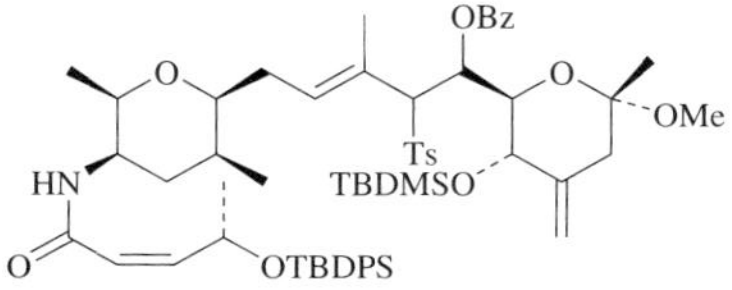	Na/Hg, THF/MeOH, –20°	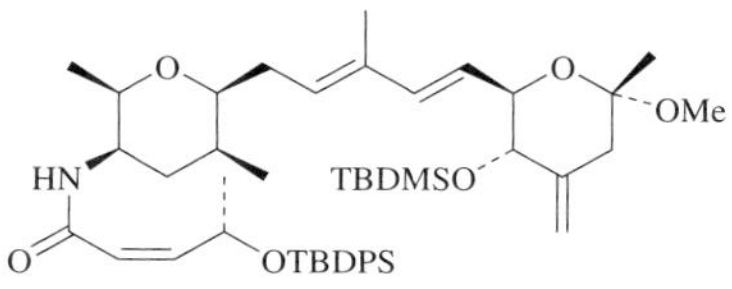 (40)	802

Substrate	Conditions	Product(s) and Yield(s) (%)	Refs.
C_{64} 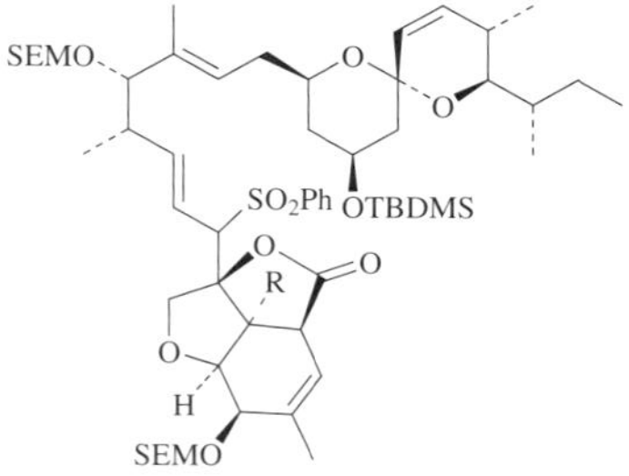 SEMO, SO_2Ph, OTBDMS, R = OTES	Na/Hg, Na_2HPO_4, MeOH, 0°, 8 h	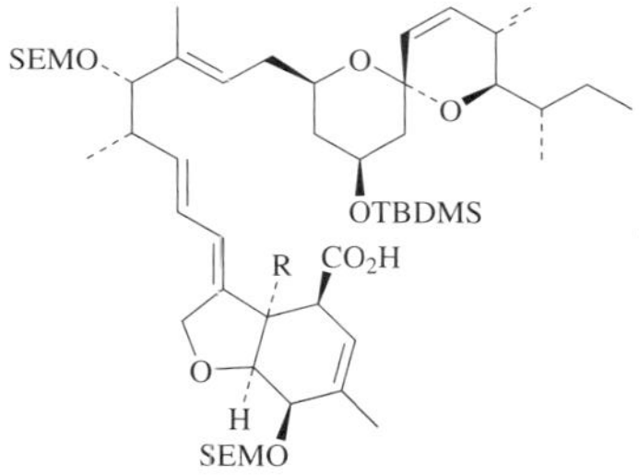 SEMO, OTBDMS, R, CO_2H (58)	803
C_{65} 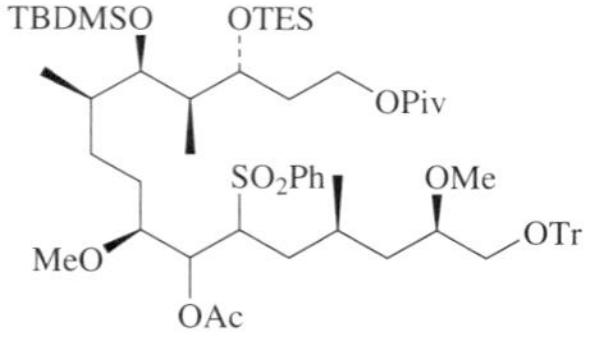 TBDMSO, OTES, OPiv, SO_2Ph, OMe, MeO, OTr, OAc	Na/Hg, Na_2HPO_4, MeOH, 0°, 100 min	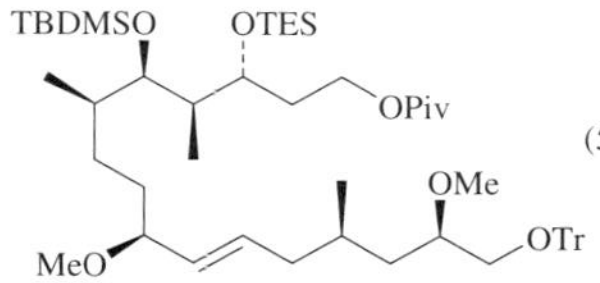 TBDMSO, OTES, OPiv, OMe, MeO, OTr (58) Z:E = 16:84	791
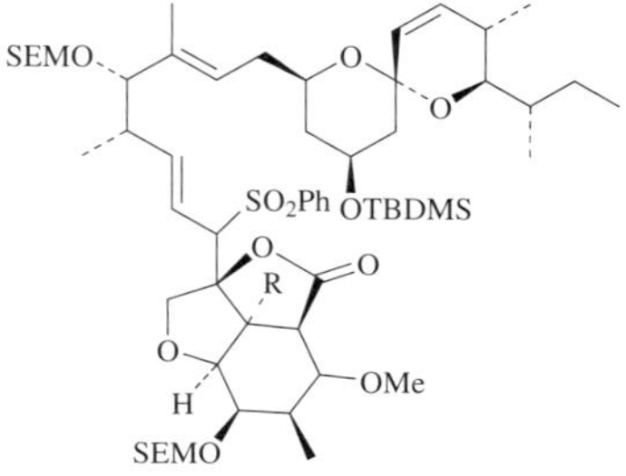 SEMO, SO_2Ph, OTBDMS, OMe, R = OTES	Na/Hg, Na_2HPO_4, MeOH, 0°, 3 h	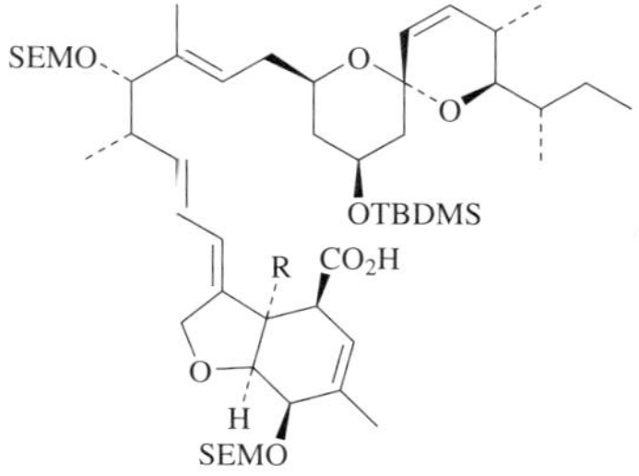 SEMO, OTBDMS, R, CO_2H (71)	803

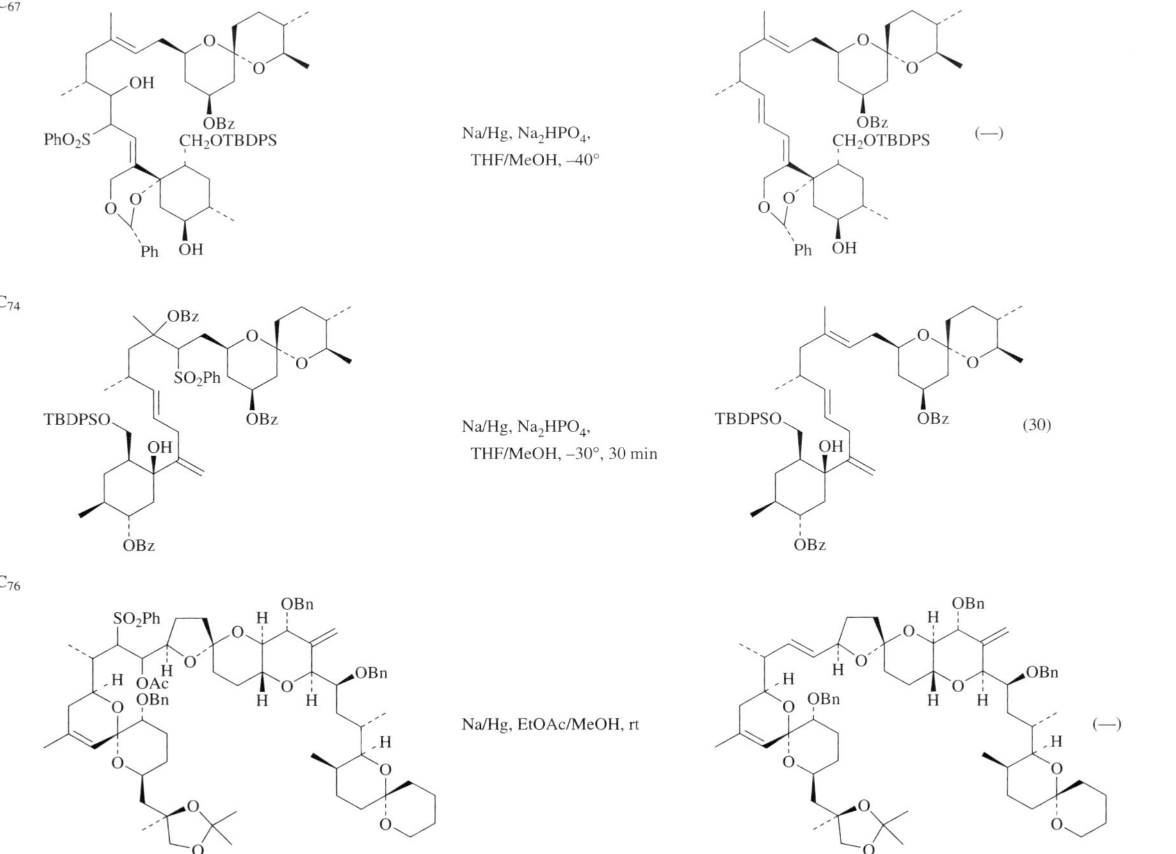

Carbons	Reactant	Conditions	Product(s) and Yield(s) (%)		Refs.
C_{67}		Na/Hg, Na_2HPO_4, THF/MeOH, –40°		(—)	804
C_{74}		Na/Hg, Na_2HPO_4, THF/MeOH, –30°, 30 min		(30)	805
C_{76}		Na/Hg, EtOAc/MeOH, rt		(—)	555

TABLE 8. REDUCTIVE ELIMINATION (*Continued*)

Substrate	Conditions	Product(s) and Yield(s) (%)	Refs.
C_{77}			
TBDMS, Et, PMBO, O, O, AcO, SO$_2$Ph, Mes, TBDPSO, ODMPM	Na/Hg, EtOAc/MeOH, –30°, 5 h	TBDMS, Et, PMBO, O, O, Mes, TBDPSO, ODMPM (82)	806
C_{83}			
NMe$_2$, OTES, H, N, ( )$_2$, O, N, OBz, OMe, OTES, O, PhO$_2$S, (TMSCH$_2$CH$_2$O)$_2$P(O), O, OTBDMS, HO, O, O, OMe	Na/Hg, EtOAc/MeOH, –20°	NMe$_2$, OTES, H, N, ( )$_2$, O, N, OMe, OTES, O, (TMSCH$_2$CH$_2$O)$_2$P(O), O, OTBDMS, HO, O, O, OMe (—)	807

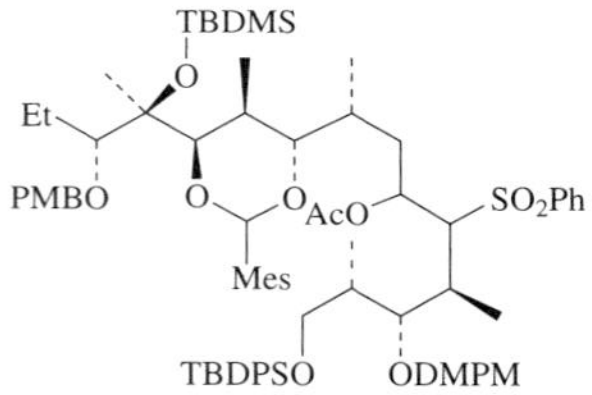

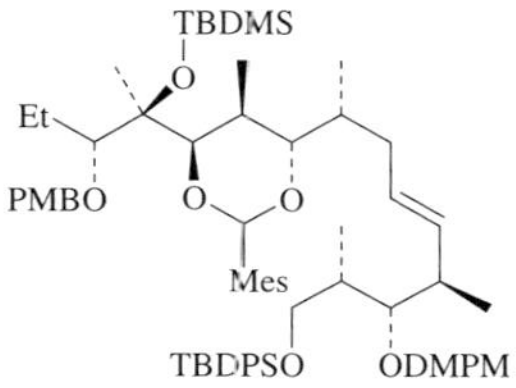

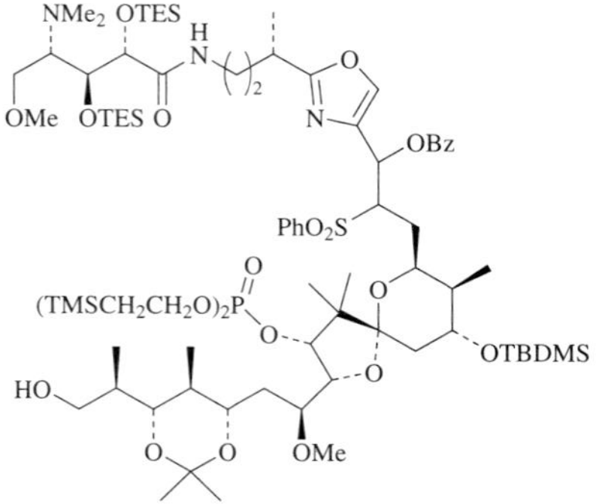

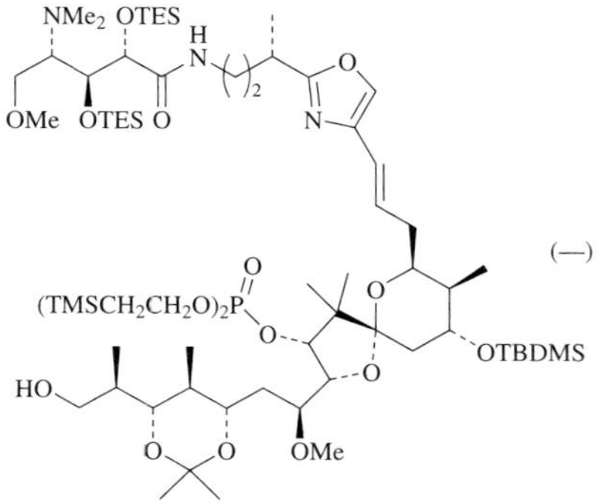

	Reactant	Conditions	Product(s) and Yield(s) (%)	Refs.
C_{86}	MeO, MeO, MeO, OTBDMS, CO_2All, SO_2Ph, BzO, OMe, H, OR, O, TESO, OTES, OTES; R = TBDMS	Na/Hg, Na_2HPO_4, EtOAc/MeOH, –35°	MeO, MeO, MeO, OTBDMS, CO_2All, OMe, H, OR, TESO, OTES, OTES (—)	808
C_{87}	TBDPSO, OTES, SO_2Ph, HO, OMe, H, OTBDMS, TESO, OTBDMS, OTBDPS	Na/Hg, Na_2HPO_4, EtOAc/MeOH, –20°, 4 h; 0°, 2h	TBDPSO, OH, OMe, H, OTBDMS, TESO, OTBDMS, OTBDPS (32)	809

TABLE 8. REDUCTIVE ELIMINATION (*Continued*)

Substrate	Conditions	Product(s) and Yield(s) (%)	Refs.
C_{87}			
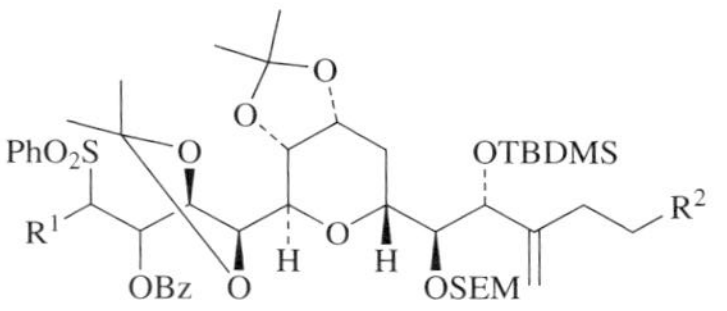 PhO$_2$S, R^1, OBz, O, O, O, H, O, H, OTBDMS, OSEM, R^2; R^1 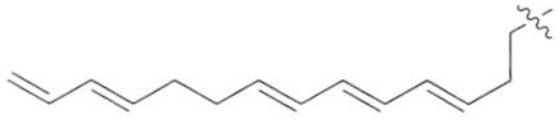 = ; R^2 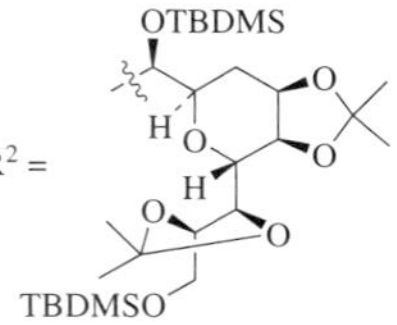 = OTBDMS, H, O, H, O, O, O, O, TBDMSO	Na/Hg, Na_2HPO_4, THF/MeOH, –20°	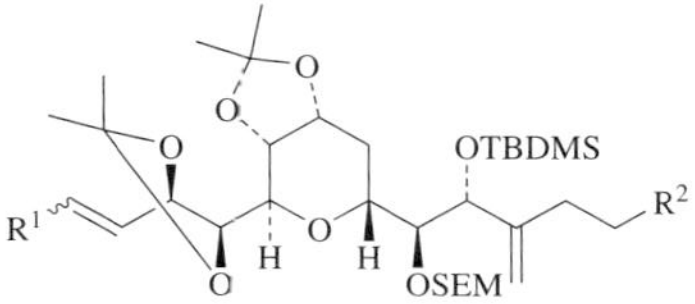 R^1, O, O, O, O, H, O, H, OTBDMS, OSEM, R^2 (72) Z:E = 8:92	810
C_{107}			
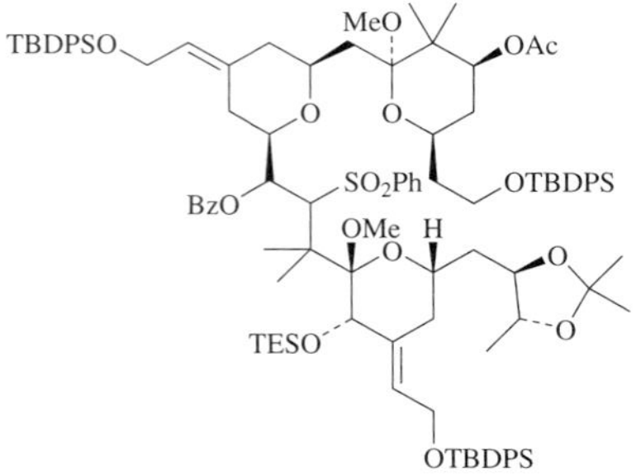 TBDPSO, MeO, OAc, O, O, BzO, SO$_2$Ph, OTBDPS, OMe, O, H, O, O, TESO, OTBDPS	Na/Hg, Na_2HPO_4, EtOAc/MeOH, –20°	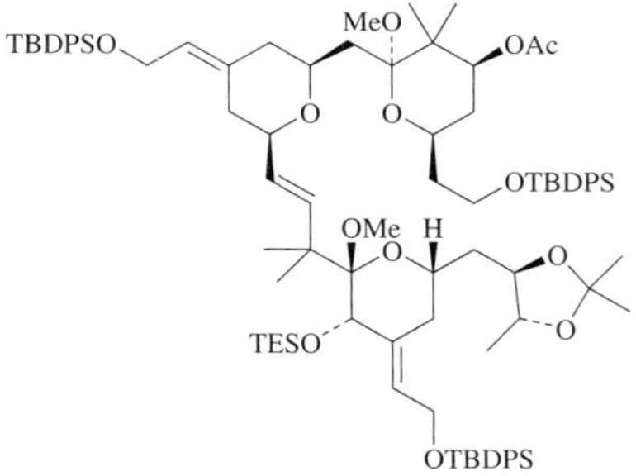 TBDPSO, MeO, OAc, O, O, OTBDPS, OMe, O, H, O, O, TESO, OTBDPS (—)	811

C_{111} 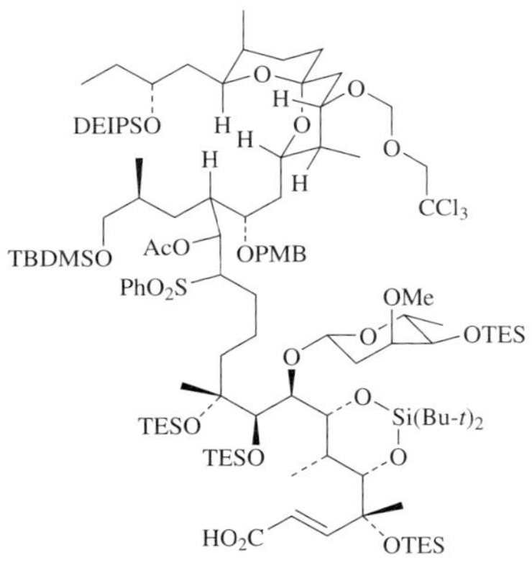	Na/Hg, $NaHCO_3$, THF/MeOH, –40°, 5 min; –30°, 55 min	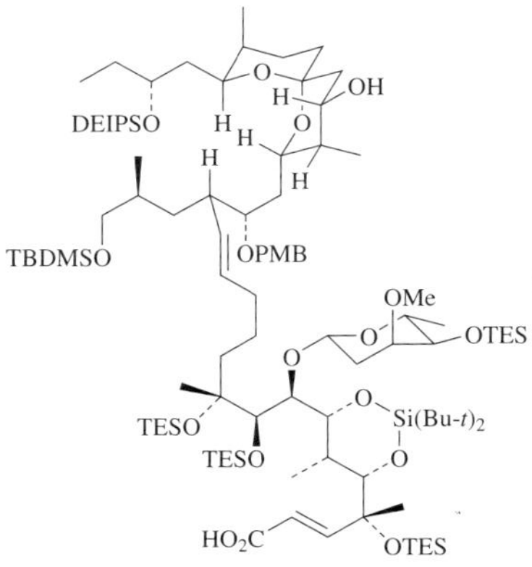	(—)	812

REFERENCES

[1] *The Chemistry of Sulfones and Sulfoxides*; Patai, S.; Rappoport, Z.; Stirling, C., Eds.; Wiley & Sons: Chichester, 1988.
[2] Trost, B. M. *Bull. Chem. Soc. Jpn.* **1988**, *61*, 107.
[3] Simpkins, N. S. *Sulphones in Organic Synthesis*; Pergamon Press: Oxford, 1993.
[4] *The Synthesis of Sulphones, Sulphoxides and Cyclic Sulphides*; Patai, S.; Rappoport, Z., Eds.; Wiley & Sons: Chichester, 1994.
[5] Rayner, C. M. *Contemp. Org. Synth.* **1994**, *1*, 191.
[6] Rayner, C. M. *Contemp. Org. Synth.* **1995**, *2*, 409.
[7] Rayner, C. M. *Contemp. Org. Synth.* **1996**, *3*, 499.
[8] Simpkins, N. S. *Tetrahedron* **1990**, *46*, 6951.
[9] Chinchilla, R.; Nájera, C. *Recent Res. Devel. Org. Chem.* **1997**, *1*, 437.
[10] Nájera, C.; Sansano, J. M. *Recent Res. Devel. Org. Chem.* **1998**, *2*, 637.
[11] Bäckvall, J. E.; Chinchilla, R.; Nájera, C.; Yus, M. *Chem. Rev.* **1998**, *98*, 2291.
[12] Prilezhaeva, E. N. *Russ. Chem. Rev.* **2000**, *69*, 367.
[13] Kice, J. L. *Organic Sulfur Compounds* **1966**, *2*, 115.
[14] Nájera, C.; Yus, C. *Tetrahedron* **1999**, *55*, 10547.
[15] Kocienski, P. J. In *Comprehensive Organic Synthesis*; Trost, B. M.; Fleming, I., Eds.; Pergamon Press: New York, **1991**; Vol. VI, pp 987–1010.
[16] Caubère, P.; Coutrot, P. In *Comprehensive Organic Synthesis*; Trost, B. M.; Fleming, I., Eds.; Pergamon Press: New York, **1991**; Vol. VIII, pp 835–870.
[17] Horner, L.; Neumann, H. *Chem. Ber.* **1965**, *98*, 1715.
[18] Corey, E. J.; Chaykovsky, M. *J. Am. Chem. Soc.* **1964**, *86*, 1639.
[19] Truce, W. E.; Tate, D. P.; Burdge, D. N. *J. Am. Chem. Soc.* **1960**, *82*, 2872.
[20] Truce, W. E.; Frank, F. J. *J. Org. Chem.* **1967**, *32*, 1918.
[21] Grieco, P. A.; Masaki, Y. *J. Org. Chem.* **1975**, *40*, 150.
[22a] Khripach, V. A.; Zhabinskii, V. N.; Zhernosek, E. V. *Tetrahedron Lett.* **1995**, *36*, 607.
[22b] Dewar, M. J. S.; Hashmall, J. A.; Trinajstić, N. *J. Am. Chem. Soc.* **1970**, *92*, 5555.
[22c] Hwu, J. R.; Wein, Y. S.; Leu, Y.-J. *J. Org. Chem.* **1996**, *61*, 1493.
[23] Hogen-Esch, T. E.; Smid, J. *J. Am. Chem. Soc.* **1965**, *87*, 669.
[24] Hogen-Esch, T. E.; Smid, J. *J. Am. Chem. Soc.* **1966**, *88*, 307.
[25] Hogen-Esch, T. E.; Smid, J. *J. Am. Chem. Soc.* **1966**, *88*, 318.
[26] States, R. V.; Szware, M. *J. Am. Chem. Soc.* **1967**, *89*, 6043.
[27] Bank, S.; Bockrath, B. *J. Am. Chem. Soc.* **1971**, *93*, 430.
[28] Bank, S.; Juckett, D. A. *J. Am. Chem. Soc.* **1975**, *97*, 567.
[29] Molander, G. A. *Org. React.* **1994**, *46*, 211.
[30] Namy, J. L.; Girard, P.; Kagan, H. B. *J. Am. Chem. Soc.* **1980**, *102*, 2693.
[31] Girard, P.; Namy, J. L.; Kagan, H. B. *Nouv. J. Chim.* **1977**, *1*, 5.
[32] Shabangi, M.; Flowers, R. A. *Tetrahedron Lett.* **1997**, *38*, 1137.
[33] Trost, B. M.; Arndt, H. C.; Strege, P. E.; Verhoeven, T. R. *Tetrahedron Lett.* **1976**, *17*, 3477.
[34] Dabby, R. E.; Kenyon, J.; Mason, R. F. *J. Chem. Soc.* **1952**, 4881.
[35] Anderson, M. B.; Ranasinghe, M. G.; Palmer, J. T.; Fuchs, P. L. *J. Org. Chem.* **1988**, *53*, 3125.
[36] Wang, Q.; Sasaki, N. A.; Riche, C.; Potier, P. *J. Org. Chem.* **1999**, *64*, 8602.
[37] Alonso, D. A.; Falvello, L. R.; Mancheño, B.; Nájera, C.; Tomás, M. *J. Org. Chem.* **1996**, *61*, 5004.
[38] Savoia, D.; Trombini, C.; Umani-Ronchi, A. *J. Chem. Soc., Perkin Trans. 1* **1977**, 123.
[39] Nájera, C.; Sansano, J. M. *Tetrahedron* **1994**, *50*, 3491.
[40] Michael, J. P.; de Koning, C. B.; Malafetse, T. J.; Yillah, I. *Org. Biomol. Chem.* **2004**, *2*, 3510.
[41] Thyagarajan, B. S.; Majumdar, K. C.; Bates, D. K. *J. Heterocycl. Chem.* **1975**, *12*, 59.
[42] Keck, G. E.; Savin, K. A.; Weglarz, M. A. *J. Org. Chem.* **1995**, *60*, 3194.
[43] Kumareswaran, R.; Hassner, A. *Tetrahedron: Asymmetry* **2001**, *12*, 2001.
[44] Guo, H.; Zhang, Y. *Synth. Commun.* **2000**, *30*, 1879.
[45] Nishiguchi, I.; Matsumoto, T.; Kuwahara, T.; Kyoda, M.; Maekawa, H. *Chem. Lett.* **2002**, 478.
[46] Bremner, J.; Julia, M.; Launay, M.; Stacino, J.-P. *Tetrahedron Lett.* **1982**, *23*, 3265.

[47] Harris, A. R.; Mason, T. J.; Hannah, R. *J. Chem. Res. (S)* **1990**, 218.
[48] Cuvigny, T.; du Penhoat, C. H.; Julia, M. *Tetrahedron* **1987**, *43*, 859.
[49] Julia, M.; Lauron, H.; Stacino, J. P.; Verpeaux, J. N.; Jeannin, Y.; Dromzee, Y. *Tetrahedron* **1986**, *42*, 2475.
[50] Sauer, G.; Junghans, K.; Eder, U.; Haffer, G.; Neef, G.; Wiechert, R.; Cleve, G.; Hoyer, G.-A. *Liebigs Ann. Chem.* **1982**, 431.
[51] Beau, J.-M.; Sinaÿ, P. *Tetrahedron Lett.* **1985**, *26*, 6185.
[52] Beau, J.-M.; Sinaÿ, P. *Tetrahedron Lett.* **1985**, *26*, 6189.
[53] Beau, J.-M.; Sinaÿ, P. *Tetrahedron Lett.* **1985**, *26*, 6193.
[54] Somsák, L. *Chem. Rev.* **2001**, *101*, 81.
[55] Adlington, R M.; Baldwin, J. E.; Basak, A.; Kozyrod, R. P. *J. Chem. Soc., Chem. Commun.* **1983**, 944.
[56] Giese, B.; Dupuis, J. *Angew. Chem., Int. Ed. Engl.* **1983**, *22*, 622.
[57] Baumberger, F.; Vasella, A. *Helv. Chim. Acta* **1983**, *66*, 2210.
[58] Wnuk, S. F.; Rios, J. M.; Khan, J.; Hsu, Y.-L. *J. Org. Chem.* **2000**, *65*, 4169.
[59] Smith, A. B., III; Hale, K. J.; McCauley, J. P., Jr. *Tetrahedron Lett.* **1989**, *30*, 5579.
[60] Wnuk, S. F.; Bergolla, L. A.; Garcia, P. I., Jr. *J. Org. Chem.* **2002**, *67*, 3065.
[61] Ueno, Y.; Aoki, S.; Okawara, M. *J. Am. Chem. Soc.* **1979**, *101*, 5414.
[62] Ueno, Y.; Sano, H.; Auki, S.; Okawara, M. *Tetrahedron Lett.* **1981**, *22*, 2675.
[63] McCarthy, J. R.; Matthews, D. P.; Stemerick, D. M.; Huber, E. W.; Bey, P.; Lippert, B. J.; Snyder, R. D.; Sunkara, P. S. *J. Am. Chem. Soc.* **1991**, *113*, 7439.
[64] McCarthy, J. R.; Huber, E. W.; Le, T.-B.; Laskovics, F. M.; Matthews, D. P. *Tetrahedron* **1996**, *52*, 45.
[65] Berkowitz, D. B.; de la Salud-Bea R.; Jahng, W.-J. *Org. Lett.* **2004**, *6*, 1821.
[66] Hauptmann, H.; Walter, W. F. *Chem. Rev.* **1962**, *62*, 347.
[67] Pettit, G. R.; van Tamelen, E. E. *Org. React.* **1962**, *12*, 356.
[68] Bonner, W. A. *J. Am. Chem. Soc.* **1952**, *74*, 1033.
[69] van Tamelen, E. E.; Grant, E. A. *J. Am. Chem. Soc.* **1959**, *81*, 2160.
[70] Bonner, W. A. *J. Am. Chem. Soc.* **1952**, *74*, 1034.
[71] Grimm, R. A.; Bonner, W. A. *J. Org. Chem.* **1967**, *32*, 3470.
[72] Chan, M.-C.; Cheng, K.-M.; Ho, K. M.; Ng, C. T.; Yam, T. M.; Wang, B. S. L.; Luh, T.-Y. *J. Org. Chem.* **1988**, *53*, 4466.
[73a] Ho, K. M.; Lam, C. H.; Luh, T.-Y. *J. Org. Chem.* **1989**, *54*, 4474.
[73b] Okasada, K.; Macda, M.; Nakamura, Y.; Yamamoto, T.; Yamamoto, A. *J. Chem. Soc., Chem. Commun.* **1986**, 442.
[73c] Wenkert, E.; Shepard, M. E.; McPhail, A. T. *J. Chem. Soc., Chem. Commun.* **1986**, 1390.
[74] Becker, S.; Fort, Y.; Vanderesse, R.; Caubère, P. *Tetrahedron Lett.* **1988**, *24*, 2963.
[75] Becker, S.; Fort, Y.; Caubère, P. *J. Org. Chem.* **1990**, *55*, 6194.
[76] Trost, B. M.; Schmuff, N. R.; Miller, M. J. *J. Am. Chem. Soc.* **1980**, *102*, 5979.
[77] Dufort, N.; Jodoin, B. *Can. J. Chem.* **1978**, *56*, 1779.
[78] Janssen, C. G. M.; van Lier, P. M.; Schipper, P.; Simons, L. H. J. G.; Godefroi, E. F. *J. Org. Chem.* **1982**, *45*, 3159.
[79] Janssen, C. G. M.; Godefroi, E. F. *J. Org. Chem.* **1982**, *47*, 3274.
[80] Abou-Hadeed, K.; Hansen, H.-J. *Helv. Chim. Acta* **2003**, *86*, 4018.
[81] Andel Fattah, M.; El Rayes, S.; Ahmed Soliman, E. S.; Linden, A.; Abou-Hadeed, K.; Hansen, H.-J. *Helv. Chim. Acta* **2005**, *88*, 1085.
[82] Kotake, H.; Yamamoto, T.; Kinoshita, H. *Chem. Lett.* **1982**, 1331.
[83] Hutchins, R. O.; Learn, K. *J. Org. Chem.* **1982**, *47*, 4382.
[84] Mohri, M.; Kinoshita, H.; Inomata, K.; Kotake, H. *Chem. Lett.* **1985**, 451.
[85] Mohri, M.; Kinoshita, H.; Inomata, K.; Kotake, H.; Takagaki, H.; Yamazaki, K. *Chem. Lett.* **1986**, 1177.
[86] Cuvigny, T.; Julia, M. *J. Organomet. Chem.* **1986**, *317*, 383.
[87] Trost, B. M.; Merlic, C. A. *J. Org. Chem.* **1990**, *55*, 1127.
[88] Masuyama, Y.; Yamada, K.; Shimizu, S.; Kurusu, Y. *Bull. Chem. Soc. Jpn.* **1989**, *62*, 2913.

[89] Mandai, T.; Matsumoto, T.; Tsuji, J. *Synlett* **1993**, 113.
[90] Ono, N.; Tamura, R.; Tanikaga, R.; Kaji, A. *J. Chem. Soc., Chem. Commun.* **1981**, 71.
[91] Wade, P. A.; Hinney, H. R.; Amin, N. V.; Vail, P. D.; Morrow, S. D.; Hardinger, S. A.; Saft, M. S. *J. Org. Chem.* **1981**, *46*, 765.
[92] Fujii, M.; Nakamura, K.; Mekata, H.; Oka, S.; Ohno, A. *Bull. Chem. Soc. Jpn.* **1988**, *61*, 495.
[93] Park, K. K.; Lee, C. W.; Choi, S. Y. *J. Chem. Soc., Perkin Trans. 1* **1992**, 601.
[94] Julia, M.; Paris, J.-M. *Tetrahedron Lett.* **1973**, *14*, 4833.
[95] Kocienski, P. J.; Lythgoe, B.; Waterhouse, I. *J. Chem. Soc., Perkin Trans. 1* **1980**, 1045.
[96] Kocienski, P. J.; Lythgoe, B.; Roberts, D. A. *J. Chem. Soc., Perkin Trans. 1* **1979**, 1290.
[97] Flynn, D. L.; Zabrowski, D. L.; Nosal, R. *Tetrahedron Lett.* **1992**, *33*, 7281.
[98] Williams, D. R.; Moore, J. L.; Yamada, M. *J. Org. Chem.* **1986**, *51*, 3916.
[99] Barrish, J. C.; Lee, H. L.; Mitt, T.; Pizzolato, G.; Baggiolini, E. G.; Uskokovic, M. R. *J. Org. Chem.* **1988**, *53*, 4282.
[100] Künzer, H.; Stahnke, M.; Sauer, G.; Wiechert, R. *Tetrahedron Lett.* **1991**, *32*, 1949.
[101] Markó, I. E.; Murphy, F.; Dolan, S. *Tetrahedron Lett.* **1996**, *37*, 2089.
[102] Kende, A. S.; Mendoza, J. S. *Tetrahedron Lett.* **1990**, *31*, 7105.
[103] Ihara, M.; Suzuki, S.; Taniguchi, T.; Tokunaga, Y.; Fukumoto, K. *Synlett* **1994**, 859.
[104] Oishi, T.; Ando, K.; Chida, N. *Chem. Commun.* **2001**, 1932.
[105] Oishi, T.; Ando, K.; Inomiya, K.; Sato, H.; Iida, M.; Chida, N. *Bull. Chem. Soc. Jpn.* **2002**, *75*, 1927.
[106] Hwu, J. R.; Chua, V.; Schroeder, J. E.; Barrans, R. E., Jr.; Khoudary, K. P.; Wang, N.; Wetzel, J. M. *J. Org. Chem.* **1986**, *51*, 4733.
[107] Julia, M.; Uguen, D. *Bull. Soc. Chim. Fr.* **1976**, 513.
[108] Luker, T.; Whitby, R. J. *Tetrahedron Lett.* **1996**, *37*, 7661.
[109] Sato, K.; Inoue, S.; Onishi, A.; Uchida, N.; Minowa, N. *J. Chem. Soc., Perkin Trans. 1* **1981**, 761.
[110] Krapcho, A. P.; Bothner-By, A. A. *J. Am. Chem. Soc.* **1958**, *81*, 3658.
[111] Yee, N. K. N.; Coates, R. M. *J. Org. Chem.* **1992**, *57*, 4598.
[112] Lee, G. H.; Youn, I. K.; Choi, E. B.; Lee, H. K.; Yon, G. H.; Yang, H. C.; Pak, C. S. *Curr. Org. Chem.* **2004**, *8*, 1263.
[113] Sugase, K.; Horikawa, M.; Sugiyama, M.; Ishiguro, M. *J. Med. Chem.* **2004**, *47*, 489.
[114] Lee, G. H.; Choi, E. B.; Lee, E.; Pak, C. S. *Tetrahedron Lett.* **1993**, *34*, 4541.
[115] Benedetti, F.; Berti, F.; Risaliti, A. *Tetrahedron Lett.* **1993**, *34*, 6443.
[116] Costa, A.; Nájera, C.; Sansano, J. M. *J. Org. Chem.* **2002**, *67*, 5216.
[117] Bintz-Giudicelli, C.; Weymann, O.; Uguen, D.; De Cian, A.; Fischer, J. *Tetrahedron Lett.* **1997**, *38*, 2841.
[118] Brown, A. C.; Carpino, L. A. *J. Org. Chem.* **1985**, *50*, 1749.
[119] Lai, J.-Y.; Yu, J.; Hawkins, R. D.; Falck, J. R. *Tetrahedron Lett.* **1995**, *36*, 5691.
[120] Ni, C.; Hu, J. *Tetrahedron Lett.* **2005**, *46*, 8273.
[121] Kazuta, Y.; Matsuda, A.; Shuto, S. *J. Org. Chem.* **2002**, *67*, 1669.
[122] Bernard, A. M.; Frongia, A.; Piras, P. P.; Secci, F. *Synlett* **2004**, 1064.
[123] Smith, P. M.; Thomas, E. J. *J. Chem. Soc., Perkin Trans. 1* **1998**, 3541.
[124] Oishi, T.; Ando, K.; Inomiya, K.; Sato, H.; Lida, M.; Chida, N. *Org. Lett.* **2002**, *4*, 151.
[125] Takano, M.; Umino, A.; Nakada, M. *Org. Lett.* **2004**, *6*, 4897.
[126] Margot, C.; Simmons, D. P.; Reichlin, D.; Skuy, D. *Helv. Chim. Acta* **2004**, *87*, 2662.
[127] Cox, P.; Craig, D.; Ioannidis, S.; Rahn, V. S. *Tetrahedron Lett.* **2005**, *46*, 4687.
[128] Kruse, B.; Brückner, R. *Chem. Ber.* **1989**, 2023.
[129] Chan, T. H.; Labrecque, D. *Tetrahedron Lett.* **1991**, *32*, 1149.
[130] Julia, M.; Ward, P. *Bull. Soc. Chim. Fr.* **1973**, 3065.
[131] Mulzer, J.; Kaselow, U.; Graske, K.-D.; Kühne, H.; Sieg, A.; Martin, H. J. *Tetrahedron* **2004**, *60*, 9599.
[132] Carretero, J. C.; Eugenio de Diego, J.; Hamdouchi, C. *Tetrahedron* **1999**, *55*, 15159.
[133] Caturla, F.; Nájera, C. *Tetrahedron Lett.* **1996**, *37*, 4787.
[134] Lansbury, P. T.; Erwin, R. W.; Jeffery, D. A. *J. Am. Chem. Soc.* **1980**, *102*, 1602.

[135] Caturla, F.; Nájera, C. *Tetrahedron* **1997**, *53*, 11449.
[136] Trost, B. M.; Vincent, J. E. *J. Am. Chem. Soc.* **1980**, *102*, 5680.
[137] Fargeas, V.; Baalouch, M.; Metay, E.; Baffreau, J.; Ménard, D.; Gosselin, P.; Bergé, J.-P.; Barthomeuf, C.; Lebreton, J. *Tetrahedron* **2004**, *60*, 10359.
[138] Gaillot, J.-M.; Gelas-Mialhe, Y.; Vessiere, R. *Chem. Lett.* **1983**, 1137.
[139] Giblin, G. M. P.; Jones, C. D.; Simpkins, N. S. *Synlett* **1997**, 589.
[140] Gaul, C.; Njardarson, J. T.; Shan, D.; Dorn, D. C.; Wu, K.-D.; Tong, W. P.; Huang, X.-Y.; Moore, M. A. S.; Danishefsky, S. J. *J. Am. Chem. Soc.* **2004**, *126*, 11326.
[141] Clayden, J.; Kenworthy, M. N.; Helliwell, M. *Org. Lett.* **2003**, *5*, 831.
[142] Lemaire, P.; Balme, G.; Desbordes, P.; Vors, J.-P. *Org. Biomol. Chem.* **2003**, *1*, 4209.
[143] Takeda, K.; Urahata, M.; Yoshii, E. *J. Org. Chem.* **1986**, *51*, 4735.
[144] Tanner, D.; He, H. M. *Tetrahedron* **1989**, *45*, 4309.
[145] Wagner, H.; Harms, K.; Koert, U.; Meder, S.; Boheim, G. *Angew. Chem., Int. Ed. Engl.* **1996**, *35*, 2643.
[146] Wuonola, M. A.; Woodward, R. B. *Tetrahedron* **1976**, *32*, 1085.
[147] Tatsuta, K.; Yasuda, S.; Kurihara, K.; Tanabe, K.; Shinei, R.; Okonogi, T. *Tetrahedron Lett.* **1997**, *38*, 1439.
[148] Nanda, S. *Tetrahedron Lett.* **2005**, *46*, 3661.
[149] Hioki, H.; Yoshio, S.; Motosue, M.; Oshita, Y.; Nakamura, Y.; Mishima, D.; Fukuyama, Y.; Kodama, M.; Ueda, K.; Katsu, T. *Org. Lett.* **2004**, *6*, 961.
[150] Turner, K. L.; Baker, T. M.; Islam, S.; Procter, D. J.; Stefaniak, M. *Org. Lett.* **2006**, *8*, 329.
[151] McAllister, L. A.; McCormick, R. A.; Brand, S.; Procter, D. J. *Angew. Chem. Int. Ed.* **2005**, *44*, 452.
[152] Mazéas, D.; Skrydstrup, T.; Beau, J.-M. *Angew. Chem., Int. Ed. Engl.* **1995**, *34*, 909.
[153] Jarreton, O.; Skrydstrup, T.; Beau, J.-M. *Chem. Commun.* **1996**, 1661.
[154] Jarreton, O.; Skrydstrup, T.; Beau, J.-M. *Tetrahedron Lett.* **1997**, *38*, 1767.
[155] Andersen, L.; Mikkelsen, L. M.; Beau, J.-M.; Skrydstrup, T. *Synlett* **1998**, 1393.
[156] Urban, D.; Skrydstrup, T.; Beau, J.-M. *J. Org. Chem.* **1998**, *63*, 2507.
[157] Skrydstrup, T.; Jarreton, O.; Mazéas, D.; Urban, D.; Beau, J.-M. *Chem. Eur. J.* **1998**, *4*, 655.
[158] Jarreton, O.; Skrydstrup, T.; Espinosa, J.-F.; Jiménez-Barbero, J.; Beau, J.-M. *Chem. Eur. J.* **1999**, *5*, 430.
[159] Krintel, S. L.; Jiménez-Barbero, J.; Skrydstrup, T. *Tetrahedron Lett.* **1999**, *40*, 7565.
[160] Skrydstrup, T.; Mazéas, D.; Elmouchir, M.; Doisneau, G.; Riche, C.; Chiaroni, A.; Beau, J.-M. *Chem. Eur. J.* **1997**, *3*, 1342.
[161] Serafinowski, P. J.; Barnes, C. L. *Synthesis* **1997**, 225.
[162] Chandrasekhar, S.; Yu, J.; Falck, J. R.; Mioskowski, C. *Tetrahedron Lett.* **1994**, *35*, 5441.
[163] Yoshimatsu, M.; Ohara, M. *Tetrahedron Lett.* **1997**, *38*, 5651.
[164] Huang, X.; Zhang, H.-Z. *Synthesis* **1989**, 42.
[165] Huang, X.; Zhang, H.-Z. *Synth. Commun.* **1989**, *19*, 97.
[166a] Huang, X.; Pi, J.-H. *Synth. Commun.* **1990**, *20*, 2297.
[166b] Louis-André, O.; Gelbard, G. *Bull. Soc. Chim. Fr.* **1986**, 565.
[166c] De Vries, J. G.; Kellog, R. M. *J. Org. Chem.* **1980**, *45*, 4126.
[166d] Adinolfi, M.; Guariniello, L.; Iadonisi, A.; Mangoni, L. *Synlett* **2000**, 1277.
[166e] Park, K. K.; Oh, C. H.; Joung, W. K. *Tetrahedron Lett.* **1993**, *34*, 7445.
[167] Julia, M.; Stacino, J.-P. *Bull. Soc. Chim. Fr.* **1985**, 831.
[168] Chen, C.; Wilcoxen, K.; Zhu, Y.-F.; Kim, K.; McCarthy, J. R. *J. Org. Chem.* **1999**, *64*, 3476.
[169] Lesimple, P.; Beau, J.-M.; Jaurand, G.; Sinaÿ, P. *Tetrahedron Lett.* **1986**, *27*, 6201.
[170] Dubois, E.; Beau, J.-M. *Tetrahedron Lett.* **1990**, *31*, 5165.
[171] Zhang, H.-C.; Brakta, M.; Daves, G. D., Jr. *Tetrahedron Lett.* **1993**, *34*, 1571.
[172] Wnuk, S. F.; Yuan, C.-S.; Borchardt, R. T.; Balzarini, J.; De Clercq, E.; Robins, M. J. *J. Med. Chem.* **1994**, *37*, 3579.
[173] Wnuk, S. F.; Ro, B.-O.; Valdez, C. A.; Lewandowska, E.; Valdez, N. X.; Sacasa, P. R.; Yin, D.; Zhang, J.; Borchardt, R. T.; De Clercq, E. *J. Med. Chem.* **2002**, *45*, 2651.
[174] Giovannini, R.; Petrini, M. *Synlett* **1995**, 973.

[175] Wnuk, S. F.; Robins, M. J. *J. Am. Chem. Soc.* **1996**, *118*, 2519.
[176] Berkowitz, D. B.; Bose, M.; Asher, N. G. *Org. Lett.* **2001**, *3*, 2009.
[177] Maremoto, Y.; Mikami, A.; Kuwabe, S.; Shirahama, H. *Tetrahedron: Asymmetry* **1996**, *7*, 3371.
[178] Sato, T.; Okumura, Y.; Itai, J.; Fujisawa, T. *Chem. Lett.* **1988**, 1537.
[179] Truchot, C.; Wang, Q.; Sasaki, A. *Eur. J. Org. Chem.* **2005**, 1765.
[180] Oikawa, M.; Ueno, T.; Oikawa, H.; Ichihara, A. *J. Org. Chem.* **1995**, *60*, 5048.
[181] Luzzio, F. A.; Zacheri, D. P. *Tetrahedron Lett.* **1998**, *39*, 2285.
[182] Jones, D. N.; Khan, M. A.; Mirza, S. M. *Tetrahedron* **1999**, *55*, 9933.
[183] Gamble, M. P.; Giblin, G. M. P.; Taylor, J. K. *Synlett* **1995**, 779.
[184] Suzuki, H.; Yi, Q.; Inoue, J.; Kusume, K.; Ogawa, T. *Chem. Lett.* **1987**, 887.
[185] Ogura, K.; Yanagisawa, A.; Fujino, T.; Takahashi, K. *Tetrahedron Lett.* **1988**, *29*, 5387.
[186] Das, I.; Pathak, T. *Org. Lett.* **2006**, *8*, 1303.
[187] Pascali, V.; Umani-Ronchi, A. *J. Chem. Soc., Chem. Commun.* **1973**, 351.
[188] Nakamura, K.; Ohno, A.; Oka, S. *Tetrahedron Lett.* **1983**, *24*, 3335.
[189] Inomata, K.; Murata, Y.; Kato, H.; Tsukahara, Y.; Kinoshita, H.; Kotake, H. *Chem. Lett.* **1985**, 931.
[190] Inomata, K.; Igarashi, S.; Mohri, M.; Yamamoto, T.; Kinoshita, H.; Kotake, H. *Chem. Lett.* **1987**, 707.
[191] Li, X.; Lantrip, D.; Fuchs, P. L. *J. Am. Chem. Soc.* **2003**, *125*, 14262.
[192] Orita, A.; Watanabe, A.; Otera, J. *Chem. Lett.* **1997**, 1025.
[193] Orita, A.; Watanabe, A.; Tsuchiya, H.; Otera, J. *Tetrahedron* **1999**, *55*, 2889.
[194] Guibé, F. *Tetrahedron* **1998**, *54*, 2967.
[195] Bogenstätter, M.; Limberg, A.; Overman, L. E.; Tomasi, A. L. *J. Am. Chem. Soc.* **1999**, *121*, 12206.
[196] Fabre, J. L.; Julia, M. *Tetrahedron Lett.* **1983**, *24*, 4311.
[197] Liu, Q.; Han, B.; Liu, Z.; Yang, L.; Liu, Z.-L.; Yu, W. *Tetrahedron Lett.* **2006**, *47*, 1805.
[198] Kocienski, P. J. *Tetrahedron Lett.* **1979**, *20*, 441.
[199] Breuilles, P.; Kaspar, K.; Uguen, D. *Tetrahedron Lett.* **1995**, *36*, 8011.
[200] Alonso, D. A.; Costa, A.; Mancheño, B.; Nájera, C. *Tetrahedron* **1997**, *53*, 4791.
[201] Ermolenko, L.; Sasaki, N. A.; Potier, P. *J. Chem. Soc., Perkin Trans. 1* **2000**, 2465.
[202] Craig, D.; Smith, A. M. *Tetrahedron Lett.* **1990**, *31*, 2631.
[203] Berry, M. B.; Craig, D.; Jones, P. S. *Synlett* **1993**, 513.
[204] Hsiao, C.-N.; Shechter, H. *Tetrahedron Lett.* **1982**, *23*, 1963.
[205] D'herde, J. N. P.; De Clercq, P. J. *Tetrahedron Lett.* **2003**, *44*, 6657.
[206] Kocienski, P. *Phosphorus, Sulfur Silicon Relat. Elem.* **1985**, *24*, 97.
[207] Carretero, J. C.; Arrayás, R. G. *J. Org. Chem.* **1995**, *60*, 6000.
[208] Talke, A.; Kocienski, P. *Tetrahedron* **1990**, *46*, 4503.
[209] Taguchi, T.; Namba, R.; Nakazawa, M.; Nakajima, M.; Nakama, Y.; Kobayashi, Y.; Hara, N.; Ikekawa, N. *Tetrahedron Lett.* **1988**, *29*, 227.
[210] Morzycki, J. W.; Schnoes, H. K.; DeLuca, H. F. *J. Org. Chem.* **1984**, *49*, 2148.
[211] Kocienski, P. J.; Lythgoe, B.; Ruston, S. *J. Chem. Soc., Perkin Trans. 1* **1978**, 829.
[212] Danishefsky, S. J.; Selnick, H. G.; DeNinno, M. P.; Zelle, R. E. *J. Am. Chem. Soc.* **1987**, *109*, 1572.
[213] Achmatowicz, B.; Baranowska, E.; Daniewski, A. R.; Pankowski, J.; Wicha, J. *Tetrahedron* **1988**, *44*, 4989.
[214] Hart, D. J.; Wu, W.-L. *Tetrahedron Lett.* **1996**, *37*, 5283.
[215] Hanessian, S.; Ugolini, A.; Dubé, D.; Hodges, P. J.; André, C. *J. Am. Chem. Soc.* **1986**, *108*, 2776.
[216] Jones, A. B.; Villalobos, A.; Linde, R. G., II; Danishefsky, S. J. *J. Org. Chem.* **1990**, *55*, 2786.
[217] Evans, D. A.; Carter, P. H.; Carreira, E. M.; Charette, A. B.; Prunet, J. A.; Lautens, M. *J. Am. Chem. Soc.* **1999**, *121*, 7540.
[218] Keck, G. E.; Kachensky, D. F.; Enholm, E. J. *J. Org. Chem.* **1984**, *49*, 1462.
[219] Keck, G. E.; Kachensky, D. F.; Enholm, E. J. *J. Org. Chem.* **1985**, *50*, 4317.
[220] Lythgoe, B.; Waterhouse, I. *Tetrahedron Lett.* **1977**, *18*, 4223.

[221] Barton, D. H. R.; Jaszberenyi, J. C.; Tachdjian, C. *Tetrahedron Lett.* **1991**, *32*, 2703.
[222] Barton, D. H. R.; Tachdjian, C. *Tetrahedron* **1992**, *48*, 7109.
[223] Markó, I. E.; Murphy, F.; Kumps, L.; Ates, A.; Touillaux, R.; Craig, D.; Carballares, S.; Dolan, S. *Tetrahedron* **2001**, *57*, 2609.
[224] Satoh, T.; Hanaki, N.; Yamada, N.; Asano, T. *Tetrahedron* **2000**, *56*, 6223.
[225] Pospísil, J.; Pospísil, T.; Markó, I. E. *Org. Lett.* **2005**, *7*, 2373.
[226] Aceña, J. L.; Arjona, O.; León, M.; Plumet, J. *Tetrahedron Lett.* **1996**, *37*, 8957.
[227] De Pouilly, P.; Vauzeilles, B.; Mallet, J.-M.; Sinaÿ, P. *C. R. Hebd. Sceance Acad. Sci., Ser. II* **1991**, *313*, 1391.
[228] De Pouilly, P.; Chénedé, A.; Mallet, J.-M.; Sinaÿ, P. *Bull. Soc. Chim. Fr.* **1993**, *130*, 256.
[229] De Pouilly, P.; Chénedé, A.; Mallet, J.-M.; Sinaÿ, P. *Tetrahedron Lett.* **1992**, *33*, 8065.
[230] Fernández-Mayoralas, A.; Marra, A.; Trumtel, M.; Veyrières, A.; Sinaÿ, P. *Carbohydr. Res.* **1989**, *188*, 81.
[231] Fernández-Mayoralas, A.; Marra, A.; Trumtel, M.; Veyrières, A.; Sinaÿ, P. *Tetrahedron Lett.* **1989**, *30*, 2537.
[232] Hansen, T.; Krintel, S. L.; Daasbjerg, K.; Skrydstrup, T. *Tetrahedron Lett.* **1999**, *40*, 6087.
[233] Micskei, K.; Juhász, Z.; Ratkovic, Z. R.; Somsák, L. *Tetrahedron Lett.* **2006**, *47*, 6117.
[234] Sabol, J. S.; McCarthy, J. R. *Tetrahedron Lett.* **1992**, *33*, 3101.
[235] Bartlett, P. A.; Green, F. R., III; Rose, E. H. *J. Am. Chem. Soc.* **1978**, *100*, 4852.
[236] Lythgoe, B.; Waterhouse, I. *J. Chem. Soc., Perkin Trans. 1* **1980**, 1405.
[237] Ihara, M.; Suzuki, S.; Taniguchi, T.; Tokunaga, Y.; Fukumoto, K. *Tetrahedron* **1995**, *51*, 9873.
[238] Bartlett, P. A.; Green F. R., III *J. Am. Chem. Soc.* **1978**, *100*, 4858.
[239] de Lucci, O.; Lucchini, C.; Pasquato, L.; Modena, G. *J. Org. Chem.* **1984**, *49*, 596.
[240] Cossu, S.; Battaggia, S.; De Lucchi, O. *J. Org. Chem.* **1997**, *62*, 4162.
[241] De Lucchi, O.; Cossu, S. *Eur. J. Org. Chem.* **1998**, 2775.
[242] Sato, T.; Tsuchiya, H.; Otera, J. *Synlett* **1995**, 628.
[243] Fischer, T.; Kunz, U.; Lackie, S. E.; Cohrs, C.; Palmer, D. D.; Christi, M. *Angew. Chem. Int. Ed.* **2002**, *41*, 2969.
[244] Toyota, A.; Nishimura, A.; Kaneko, C. *Tetrahedron Lett.* **1998**, *39*, 4687.
[245] Ono, N.; Kamimura, A.; Kaji, A. *J. Org. Chem.* **1987**, *52*, 5111.
[246] Iradier, F.; Gómez Arrayás, R.; Carretero, J. C. *Org. Lett.* **2001**, *3*, 2957.
[247] Hanessian, S.; Cantin, L.-D.; Andreotti, D. *J. Org. Chem.* **1999**, *64*, 4893.
[248] Marino, J. P.; McClure, M. S.; Holub, D. P.; Comasseto, J. V.; Tucci, F. C. *J. Am. Chem. Soc.* **2002**, *124*, 1664.
[249] Li, S.; Xiao, X.; Yan, X.; Liu, X.; Xu, R.; Bai, D. *Tetrahedron* **2005**, *61*, 11291.
[250] Fuwa, H.; Okamura, Y.; Natsugari, H. *Tetrahedron* **2004**, *60*, 5341.
[251] Shindo, M.; Sugioka, T.; Umaba, Y.; Shishido, K. *Tetrahedron Lett.* **2004**, *45*, 8863.
[252] Miyaoka, H.; Tamura, M.; Yamada, Y. *Tetrahedron Lett.* **1998**, *39*, 621.
[253] Seo, S.-Y.; Jung, J.-K.; Paek, S.-M.; Lee, Y.-S.; Kim, S.-H.; Lee, K.-O.; Suh, Y.-G. *Org. Lett.* **2004**, *6*, 429.
[254] Bouzbouz, S.; Kirschleger, B. *Synlett* **1994**, 763.
[255] Chen, C. Y.; Chang, B. R.; Tsai, M. R.; Chang, M. Y.; Chang, N. C. *Tetrahedron* **2003**, *59*, 9383.
[256] Chang, M.-Y.; Hsu, R.-T.; Tseng, T.-W.; Sun, P. P.; Chang, N. C. *Tetrahedron* **2004**, *60*, 5545.
[257] Chen, B.-F.; Tasi, M.-R.; Yang, C.-Y.; Chang, J.-K.; Chang, N.-C. *Tetrahedron* **2004**, *60*, 10223.
[258] Chang, M.-Y.; Chen, C.-Y.; Chung, W.-S.; Tasi, M.-R.; Chang, N.-C. *Tetrahedron* **2005**, *61*, 585.
[259] Chen, C.-Y.; Chang, M.-Y.; Hsu, R.-T.; Chen, S. T.; Chang, N. C. *Tetrahedron Lett.* **2003**, *44*, 8627.
[260] Zhou, X.-T.; Carter, R. G. *Chem. Commun.* **2004**, 2138.
[261] Keck, G. E.; Wager, C. A.; Wager, T. T.; Savin, K. A.; Covel, J. A.; McLaws, M. D.; Krishnamurthy, D.; Cee, V. J. *Angew. Chem. Int. Ed.* **2001**, *40*, 231.
[262] Ito, H.; Hasegawa, M.; Takenaka, Y.; Kobayashi, T.; Iguchi, K. *J. Am. Chem. Soc.* **2004**, *126*, 4520.
[263] Koreeda, M.; Wang, Y.; Zhang, L. *Org. Lett.* **2002**, *4*, 3329.
[264] Back, T. G.; Hamilton, M. D. *Org. Lett.* **2002**, *4*, 1779.

[265] Aceña, J. L.; Arjona, O.; León, M. L.; Plumet, J. *Org. Lett* **2000**, *2*, 3683.
[266] Peese, K. M.; Gin, D. Y. *Org. Lett.* **2005**, *7*, 3323.
[267] Trost, B. M.; Shen, H. C.; Surivet, J.-P. *J. Am. Chem. Soc.* **2004**, *126*, 12565.
[268] Blackwell, C. M.; Davidson, A. H.; Launchbury, S. B.; Lewis, C. N.; Morrice, E. M.; Reeve, M. M.; Roffey, J. A. R.; Tipping, A. S.; Todd, R. S. *J. Org. Chem.* **1992**, *57*, 5596.
[269] Masaki, Y.; Hashimoto, K.; Serizawa, Y.; Kaji, K. *Chem. Lett.* **1982**, 1879.
[270] Eng, H. M.; Myles, D. C. *Tetrahedron Lett.* **1999**, *40*, 2275.
[271] Ghosh, A. K.; Wang, Y. *Tetrahedron Lett.* **2000**, *41*, 2319.
[272] Pettus, T. R. R.; Chen, X.-T.; Danishefsky, S. J. *J. Am. Chem. Soc.* **1998**, *120*, 12684.
[273] Mann, S.; Carillon, S.; Breyne, O.; Marquet, A. *Chem. Eur. J.* **2002**, *8*, 439.
[274] Schaefer, F. C. In *The Chemistry of the Cyano Group*; Rappoport, Z., Ed.; Interscience: London, **1970**; pp 239–305.
[275] Arseniyadis, S.; Kyler, K. S.; Watt, D. S. *Org. React.* **1984**, *31*, 1.
[276] Collier, S. J.; Langer, P. *Science of Synthesis* **2004**, *19*, 403.
[277] Murahashi, S.-I. *Science of Synthesis* **2004**, *19*, 345.
[278] Mattalia, J.-M.; Marchi-Delapierre, C.; Hazimeh, H.; Chanon, M. *Arkivoc* **2006**, *(iv)*, 90.
[279] Arapakos, P. G.; Scott, M. K.; Huber, F. E., Jr. *J. Am. Chem. Soc.* **1969**, *91*, 2059.
[280] McGrane, P. L.; Livinghouse, T. *J. Am. Chem. Soc.* **1993**, *115*, 11485.
[281] Savoia, D.; Tagliavini, E.; Trombini, C.; Umani-Ronchi, A. *J. Org. Chem.* **1980**, *45*, 3227.
[282] Ohsawa, T.; Kobayashi, T.; Mizuguchi, Y.; Saitoh, T.; Oishi, T. *Tetrahedron Lett.* **1985**, *26*, 6103.
[283] Wender, P. A.; deLong, M. A. *Tetrahedron Lett.* **1990**, *31*, 5429.
[284] Rychnovsky, S. D.; Mickus, D. E. *J. Org. Chem.* **1992**, *57*, 2732.
[285] Posson, H.; Hurvois, J.-P.; Moinet, C. *Synlett* **2000**, 209.
[286] Curran, D. P.; Seong, C. M. *Synlett* **1991**, 107.
[287] Kang, H.-Y.; Hong, W. S.; Cho, Y. S.; Koh, H. Y. *Tetrahedron Lett.* **1995**, *36*, 7661.
[288] Lee, J. C.; Koh, H. Y.; Lee, Y. S.; Kang, H.-Y. *Bull. Korean Chem. Soc.* **1997**, *18*, 783.
[289] Molander, G. A.; Wolfe, J. P. *J. Braz. Chem. Soc.* **1996**, *7*, 335.
[290] Takeda, T. *Modern Carbonyl Olefination: Methods and Applications*; Wiley-VCH: Weinheim, 2004.
[291] Wittig, G.; Geissler, G. *Justus Liebigs Ann. Chem.* **1953**, *580*, 44.
[292] Maercker, A. *Org. React.* **1965**, *14*, 270.
[293] Maryanoff, B. E.; Reitz, A. B. *Chem. Rev.* **1989**, *89*, 863.
[294] Vedejs, E.; Peterson, M. J. *Top. Stereochem.* **1994**, *21*, 1.
[295] Horner, L.; Hoffmann, H.; Wippel, H. G. *Chem. Ber.* **1958**, *91*, 61.
[296] Clayden, J.; Warren, S. *Angew. Chem., Int. Ed. Engl.* **1996**, *35*, 241.
[297] Wadsworth, W. S., Jr.; Emmons, W. D. *J. Am. Chem. Soc.* **1961**, *83*, 1733.
[298] Wadsworth, W. S. *Org. React.* **1977**, *25*, 73.
[299] Peterson, D. J. *J. Org. Chem.* **1968**, *33*, 780.
[300] Ager, D. J. *Org. React.* **1990**, *38*, 1.
[301] van Staden, L. F.; Gravestock, D.; Ager, D. J. *Chem. Soc. Rev.* **2002**, *31*, 195.
[302] Johnson, C. R.; Shanklin, J. R.; Kirchhoff, R. A. *J. Am. Chem. Soc.* **1973**, *95*, 6462.
[303] Baudin, J. B.; Hareau, G.; Julia, S. A.; Ruel, O. *Tetrahedron Lett.* **1991**, *32*, 1175.
[304] Blakemore, P. R. *J. Chem. Soc., Perkin Trans. 1* **2000**, 2563.
[305] Kocienski, P. J.; Lythgoe, B.; Roberts, D. A. *J. Chem. Soc., Perkin Trans. 1* **1978**, 834.
[306] Fabre, J.-L.; Julia, M.; Verpeaux, J.-N. *Tetrahedron Lett.* **1982**, *23*, 2469.
[307] Sano, S.; Takehisa, T.; Ogawa, S.; Yokoyama, K.; Nagao, Y. *Chem. Pharm. Bull.* **2002**, *50*, 1300.
[308] Bestmann, H. J.; Ermann, P.; Ruppel, H.; Sperling, W. *Liebigs Ann. Chem.* **1986**, 479.
[309] Still, W. C.; Gennari, C. *Tetrahedron Lett.* **1983**, *24*, 4405.
[310] Ando, K. *J. Org. Chem.* **1997**, *62*, 1934.
[311] van der Klei, A.; de Jong, R. L. P.; Lugtenburg, J.; Tielens, A. G. M. *Eur. J. Org. Chem.* **2002**, 3015.
[312] Sano, S.; Teranishi, T.; Nagao, Y. *Tetrahedron Lett.* **2002**, *43*, 9183.
[313] Bernardi, A.; Cardani, S.; Scolastico, C.; Villa, R. *Tetrahedron* **1988**, *44*, 491.

[314] Truce, W. E.; Kreider, E. M.; Brand, W. W. *Org. React.* **1970**, *18*, 99.
[315] Baudin, J. B.; Hareau, S. A.; Julia, S. A.; Lorne, R.; Ruel, O. *Bull. Soc. Chim. Fr.* **1993**, *130*, 856.
[316] Charette, A. B.; Berthelette, C.; St-Martin, D. *Tetrahedron Lett.* **2001**, *42*, 5149.
[317] Blakemore, P. R.; Cole, W. J.; Kocienski, P. J.; Morley, A. *Synlett* **1998**, 26.
[318] Kocienski, P. J.; Bell, A.; Blakemore, P. R. *Synlett* **2000**, 365.
[319] Alonso, D. A.; Nájera, C.; Varea, M. *Tetrahedron Lett.* **2004**, *45*, 573.
[320] Alonso, D. A.; Fuensanta, M.; Nájera, C.; Varea, M. *Phosphorus, Sulfur Silicon Relat. Elem.* **2005**, *180*, 1119.
[321] Alonso, D. A.; Fuensanta, M.; Nájera, C.; Varea, M. *J. Org. Chem.* **2005**, *70*, 6404.
[322] Mirk, D.; Grassot, J.-M.; Zhu, J. *Synlett* **2006**, 1255.
[323] Marti, C.; Carreira, E. M. *J. Am. Chem. Soc.* **2005**, *127*, 11505.
[324] Chang, S.-K.; Paquette, L. A. *Synlett* **2005**, 2915.
[325] Chatterjee, A. K. In *Handbook of Metathesis*; Grubbs, R. H., Ed.; Wiley-VCH: Weinheim, **2003**; Vol. II, pp 246–295.
[326] Connon, S. J.; Blechert, S. *Angew. Chem. Int. Ed.* **2003**, *42*, 1900.
[327] Grubbs, R. H.; Trnka, T. M. In *Ruthenium in Organic Synthesis*; Murahashi, S. I., Ed.; Wiley-VCH: Weinheim, **2004**; pp 153–177.
[328] Toste, F. D.; Chatterjee, A. K.; Grubbs, R. H. *Pure Appl. Chem.* **2002**, *74*, 7.
[329] Engelhardt, F. C.; Schmitt, M. J.; Taylor, R. E. *Org. Lett.* **2001**, *3*, 2209.
[330] Choi, T.-L.; Chatterjee, A. K.; Grubbs, R. H. *Angew. Chem. Int. Ed.* **2001**, *40*, 1277.
[331] Chatterjee, A. K.; Grubbs, R. H. *Org. Lett.* **1999**, *1*, 1751.
[332] Shessard, S.; Stoltz, S. *Org. Lett.* **2002**, *4*, 1943.
[333] Chatterjee, A. K.; Choi, T.-L.; Sanders, D. P.; Grubbs, R. H. *J. Am. Chem. Soc.* **2003**, *125*, 11360.
[334] Michrowska, A.; Bujok, R.; Harutyunyan, S.; Sashuk, V.; Dolgonos, G.; Grela, K. *J. Am. Chem. Soc.* **2004**, *126*, 9318.
[335] Netscher, T. *J. Organomet. Chem.* **2006**, *691*, 5155.
[336] Evers, E. C.; Young, A. E., II; Panson, A. J. *J. Am. Chem. Soc.* **1957**, *79*, 5118.
[337] Blomquist, A. T.; Hiscock, B. F.; Harpp, D. N. *J. Org. Chem.* **1966**, *31*, 4121.
[338a] Fieser, L. S.; Fieser, M. *Reagents for Organic Synthesis*; Wiley: New York, **1967**; Vol. I, pp 1030–1033.
[338b] Caubère, P. *Angew. Chem. Int. Ed. Engl.* **1983**, *22*, 599.
[339] Kato, M.; Watanabe, M.; Vogler, B.; Awen, B. Z.; Masuda, Y.; Tooyama, Y.; Yoshikoshi, A. *J. Org. Chem.* **1991**, *56*, 7071.
[340] Cunningham, A. F., Jr.; Kündig, E. P. *J. Org. Chem.* **1988**, *53*, 1823.
[341] Nakamura, T.; Shiozaki, M. *Tetrahedron* **2001**, *57*, 9087.
[342] White, J. D.; Hanselmann, R.; Jackson, R. W.; Porter, W. J.; Ohba, Y.; Tiller, T.; Wang, S. *J. Org. Chem.* **2001**, *66*, 5217.
[343] Enders, D.; Jandeleit, B.; Prokopenko, O. F. *Tetrahedron* **1995**, *51*, 6273.
[344] Smith, A. B., III; Condon, S. M.; McCauley, J. A.; Leazer, J. L., Jr.; Leahy, J. W.; Maleczka, R. E., Jr. *J. Am. Chem. Soc.* **1997**, *119*, 947.
[345] Fernández-Mateos, A.; Pascual Coca, G.; Pérez Alonso, J. J.; Rubio González, R.; Simmonds, M. S. J.; Blaney, W. M. *Tetrahedron* **1998**, *54*, 14989.
[346] Caturla, F.; Nájera, C. *Tetrahedron* **1998**, *54*, 11255.
[347] Kurosawa, S.; Mori, K. *Eur. J. Org. Chem.* **2000**, 955.
[348] Trost, B. M.; Weber, L.; Strege, P.; Fullerton, T. J.; Dietsche, T. J. *J. Am. Chem. Soc.* **1978**, *100*, 3426.
[349] Carr, R. V. C.; Williams, R. V.; Paquette, L. A. *J. Org. Chem.* **1983**, *48*, 4976.
[350] Carr, R. V. C.; Paquette, L. A. *J. Am. Chem. Soc.* **1980**, *102*, 853.
[351] Posner, G. H.; Brunelle, D. J. *J. Org. Chem.* **1973**, *38*, 2747.
[352] De Chirico, G.; Fiandanese, V.; Marchese, G.; Naso, F.; Sciacovelli, O. *J. Chem. Soc., Chem. Commun.* **1981**, 523.
[353] Padwa, A.; Murphree, S. S.; Yeske, P. E. *J. Org. Chem.* **1990**, *55*, 4241.
[354] Bödeker, C.; de Waard, E. R.; Huisman, H. O. *Tetrahedron* **1981**, *37*, 1233.

[355] Desrosiers, J.-N.; Charette, A. B. *Angew. Chem. Int. Ed.* **2007**, *46*, 5955.
[356] Ballini, R.; Bosica, G.; Mecozzi, T. *Tetrahedron* **1997**, *53*, 7341.
[357] Heathcock, C. H.; Brown, R. C. D.; Norman, T. C. *J. Org. Chem.* **1998**, *63*, 5013.
[358] Adam, W.; Gogonas, E. P.; Hadjiarapoglou, L. P. *J. Org. Chem.* **2003**, *68*, 9155.
[359] Trost, B. M.; Verhoeven, T. R. *J. Am. Chem. Soc.* **1980**, *102*, 4743.
[360] Bhat, V.; Cookson, R. C. *J. Chem. Soc., Chem. Commun.* **1981**, 1123.
[361] Adam, W.; Gogonas, E. P.; Hadjiarapoglou, L. P. *Synlett* **2003**, 1165.
[362] Ballini, R.; Palmieri, A.; Petrini, M.; Torregiani, E. *Org. Lett* **2006**, *8*, 4093.
[363] Bamford, S. J.; Luker, T.; Speckamp, W. N.; Hiemstra, H. *Org. Lett.* **2000**, *2*, 1157.
[364] Basil, L. F.; Meyers, A. I.; Hassner, A. *Tetrahedron* **2002**, *58*, 207.
[365] Shirai, Y.; Seki, M.; Mori, K. *Eur. J. Org. Chem.* **1999**, 3139.
[366] Simas, A. B. C.; Furtado, L. F. O.; Costa, P. R. R. *Tetrahedron Lett.* **2002**, *43*, 6893.
[367] Nemoto, H.; Ando, M.; Fukumoto, K. *Tetrahedron Lett.* **1990**, *31*, 6205.
[368] Bakkestuen, A. K.; Gundersen, L.-L. *Tetrahedron* **2003**, *59*, 115.
[369] Taber, D. F.; Jiang, Q.; Chen, B.; Zhang, W.; Campbell, C. L. *J. Org. Chem.* **2002**, *67*, 4821.
[370] Takabe, K.; Hashimoto, H.; Sugimoto, H.; Nomoto, M.; Yoda, H. *Tetrahedron: Asymmetry* **2004**, *15*, 909.
[371] Capdevila, A.; Prasad, A. R.; Quero, C.; Petschen, I.; Bosch, M. P.; Guerrero, A. *Org. Lett.* **1999**, *1*, 845.
[372] Clive, D. L. J.; Yeh, V. S. C. *Synth. Commun.* **2000**, *30*, 3267.
[373] Tsunoda, T.; Uemoto, K.; Ohtani, T.; Kaku, H.; Itô, S. *Tetrahedron Lett.* **1999**, *40*, 7359.
[374] Hansen, T. V.; Skattebøl, L. *Tetrahedron Lett.* **2004**, *45*, 2809.
[375] Balnaves, A. S.; McGowan, G.; Shapland, P. D. P.; Thomas, E. J. *Tetrahedron Lett.* **2003**, *44*, 2713.
[376] Fürstner, A.; Gastner, T.; Rust, J. *Synlett* **1999**, 29.
[377] Trost, B. M.; Verhoeven, T. R. *J. Am. Chem. Soc.* **1978**, *100*, 3435.
[378] Mulzer, J.; Mantoulidis, A.; Öhler, E. *J. Org. Chem.* **2000**, *65*, 7456.
[379] Heathcock, C. H.; Finkelstein, B. L.; Jarvi, E. T.; Radel, P. A.; Hadley, C. R. *J. Org. Chem.* **1988**, *53*, 1922.
[380] Morita, A.; Matsuyama, S.; Oguma, Y.; Kuwahara, S. *Biosci. Biotechnol. Biochem.* **2005**, *69*, 1620.
[381] Miyaoka, H.; Kajiwara, Y.; Hara, Y.; Yamada, Y. *J. Org. Chem.* **2001**, *66*, 1429.
[382] Domon, K.; Takikawa, H.; Mori, K. *Eur. J. Org. Chem.* **1999**, 981.
[383] Magee, D. I.; Leach, J. D.; Mallais, T. C. *Tetrahedron Lett.* **1997**, *38*, 1289.
[384] Fürstner, A.; Gastner, T. *Org. Lett.* **2000**, *2*, 2467.
[385] Miyaoka, H.; Baba, T.; Mitome, H.; Yamada, Y. *Tetrahedron Lett.* **2001**, *42*, 9233.
[386] Kutner, A.; Perlman, K. L.; Lago, A.; Sicinski, R. R.; Schnoes, H. K.; DeLuca, H. F. *J. Org. Chem.* **1988**, *53*, 3450.
[387] Evano, G.; Schaus, J. V.; Panek, J. S. *Org. Lett.* **2004**, *6*, 525.
[388] Panek, J. S.; Masse, C. E. *J. Org. Chem.* **1997**, *62*, 8290.
[389] Prakash, G. K. S.; Wang, Y.; Hu, J.; Olah, G. A. *J. Fluorine Chem.* **2005**, *126*, 1361.
[390] Kim, D. Y.; Suh, K. H. *Synth. Commun.* **1998**, *28*, 83.
[391] Prakash, G. K. S.; Hu, J.; Wang, Y.; Olah, G. A. *Eur. J. Org. Chem.* **2005**, 2218.
[392] Li, Y.; Hu, J. *Angew. Chem. Int. Ed.* **2005**, *44*, 5882.
[393] Hollingworth, G. J.; Dinnell, K.; Dickinson, L. C.; Elliott, J. M.; Kulagowski, J. J.; Swain, C. J.; Thomson, C. G. *Tetrahedron Lett.* **1999**, *40*, 2633.
[394] Yu, J.; Cho, H.-S.; Chandrasekhar, S.; Falck, J. R. *Tetrahedron Lett.* **1994**, *35*, 5437.
[395] Ni, C.; Liu, J.; Zhang, L.; Hu, J. *Angew. Chem. Int. Ed.* **2007**, *46*, 786.
[396] Prakash, G. K. S.; Hu, J.; Wang, Y.; Olah, G. A. *Org. Lett.* **2004**, *6*, 4315.
[397] Kündig, E. P.; Cunningham, A. F., Jr. *Tetrahedron* **1988**, *44*, 6855.
[398] Julia, M.; Uguen, D.; Zhang, D. *Aust. J. Chem.* **1995**, *48*, 279.
[399] Mossé, S.; Alexakis, A. *Org. Lett.* **2005**, *7*, 4361.
[400] Prakash, G. K. S.; Chacko, S.; Alconcel, S.; Stewart, T.; Mathew, T.; Olah, G. A. *Angew. Chem. Int. Ed.* **2007**, *46*, 4933.

[401] Mizuta, S.; Shibata, N.; Goto, Y.; Furukawa, T.; Nakamura, S.; Toru, T. *J. Am. Chem. Soc.* **2007**, *129*, 6394.
[402] Fukuzumi, T.; Shibata, N.; Sugiura, M.; Yasui, H.; Nakamura, S.; Toru, T. *Angew. Chem. Int. Ed.* **2006**, *45*, 4973.
[403] Hoffmann, H. M.; Brandes, A. *Tetrahedron* **1995**, *51*, 155.
[404] Doi, T.; Iijima, Y.; Takasaki, M.; Takahashi, T. *J. Org. Chem.* **2007**, *72*, 3667.
[405] Liu, J.; Li, Y.; Hu, J. *J. Org. Chem.* **2007**, *72*, 3119.
[406] Gómez, G.; Rivera, H.; García, I.; Estévez, L.; Fall, Y. *Tetrahedron Lett.* **2005**, *46*, 5819.
[407] Suzuki, M.; Doi, H.; Kato, K.; Björkman, M.; Långström, B.; Watanabe, Y.; Noyori, R. *Tetrahedron* **2000**, *56*, 8263.
[408] Mirsadeghi, S.; Rickborn, B. *J. Org. Chem.* **1985**, *50*, 4340.
[409] Svatoš, A.; Huňková, Z.; Koen, V.; Hoskovec, M.; Šaman, D.; Valterová, I.; Vrkoč, J.; Koutek, B. *Tetrahedron: Asymmetry* **1996**, *7*, 1285.
[410] Ku, Y.-Y.; Patel, R. R.; Roden, B. A.; Sawick, D. P. *Tetrahedron Lett.* **1994**, *35*, 6017.
[411] Green, D. L. C.; Thompson, C. M. *Tetrahedron Lett.* **1991**, *32*, 5051.
[412] Green, D. L. C.; Kiddle, J. J.; Thompson, C. M. *Tetrahedron* **1995**, *51*, 2865.
[413] Chang, Y.-H.; Pinnick, H. W. *J. Org. Chem.* **1978**, *43*, 373.
[414] Caturla, F.; Nájera, C. *Tetrahedron Lett.* **1997**, *38*, 3789.
[415] Sagnard, I.; Sasaki, N. A.; Chiaroni, A.; Riche, C.; Potier, P. *Tetrahedron Lett.* **1995**, *36*, 3149.
[416] Gaoni, Y.; Tomažič, A. *J. Org. Chem.* **1985**, *50*, 2948.
[417] Wang, Q.; Sasaki, N. A.; Potier, P. *Tetrahedron Lett.* **1998**, *39*, 5755.
[418] Wang, Q.; Sasaki, N. A.; Potier, P. *Tetrahedron* **1998**, *54*, 15759.
[419] Sasaki, N. A.; Sagnard, I. *Tetrahedron* **1994**, *50*, 7093.
[420] Pauly, R.; Sasaki, N. A.; Potier, P. *Tetrahedron Lett.* **1994**, *35*, 237.
[421] Hodgson, D. M.; Bebbington, M. W. P.; Willis, P. *Org. Biomol. Chem.* **2003**, *1*, 3787.
[422] Kiddle, J. J.; Green, D. L. C.; Thompson, C. M. *Tetrahedron* **1995**, *51*, 2851.
[423] Llamas, T.; Gómez Arrayás, R.; Carretero, J. C. *Org. Lett.* **2006**, *8*, 1795.
[424] Coldham, I.; Burrell, A. J. M.; White, L. E.; Adams, H.; Oram, N. *Angew. Chem. Int. Ed.* **2007**, *46*, 6159.
[425] Berry, M. B.; Craig, D.; Jones, P. S.; Rowlands, G. J. *Chem. Commun.* **1997**, 2141.
[426] Nicolaou, K. C.; Harrison, S. T. *J. Am. Chem. Soc.* **2007**, *129*, 429.
[427] Ravindran, B.; Deshpande, S. G.; Pathak, T. *Tetrahedron* **2001**, *57*, 1093.
[428] Thompson, C. M.; Green, D. L. C.; Kubas, R. *J. Org. Chem.* **1988**, *53*, 5389
[429] Díez, D.; Beneitez, M. T.; Marcos, I. S.; Garrido, N. M.; Basabe, P.; Urones, J. G. *Synlett* **2001**, 655.
[430] Moreno-Vargas, A. J.; Schütz, C.; Scopelliti, R.; Vogel, P. *J. Org. Chem.* **2003**, *68*, 5632.
[431] Sasaki, N. A.; Hashimoto, C.; Potier, P. *Tetrahedron Lett.* **1987**, *28*, 6069.
[432] Buser, S.; Vasella, A. *Helv. Chim. Acta* **2005**, *88*, 3151.
[433] Liu, X.; Rainier, J. D. *Org. Lett.* **2006**, *8*, 459.
[434] Jung, M. E.; Truc, V. C. *Tetrahedron Lett.* **1988**, *29*, 6059.
[435] Takaki, K.; Maeda, T.; Ishikawa, M. *J. Org. Chem.* **1989**, *54*, 58.
[436] Díez, D.; García, P.; Moro, R. F.; Marcos, I. S.; Basabe, P.; Garrido, N. M.; Broughton, H. B.; Urones, J. G. *Synthesis* **2005**, 3327.
[437] Díez, D.; García, P.; Marcos, I. S.; Basabe, P.; Garrido, N. M.; Broughton, H. B.; Urones, J. G. *Tetrahedron* **2005**, *61*, 11641.
[438] Kimura, T.; Nakata, T. *Tetrahedron Lett.* **2007**, *48*, 43.
[439] Clive, D. L. J.; Yeh, V. S. C. *Tetrahedron Lett.* **1999**, *40*, 8503.
[440] Zai, H.; Parvez, M.; Back, T. G. *J. Org. Chem.* **2007**, *72*, 3853.
[441] Back, T. G.; Nakajima, K. *Tetrahedron Lett.* **1997**, *38*, 989.
[442] Caldwell, J. J.; Craig, D. *Angew. Chem. Int. Ed.* **2007**, *46*, 2631.
[443] Barco, A.; Benetti, S.; De Risi, C.; Marchetti, P.; Pollini, G. P.; Zanirato, V. *Tetrahedron Lett.* **1998**, *39*, 1973.
[444] Craig, D.; Jones, P. S.; Rowlands, G. J. *Synlett* **1997**, 1423.
[445] Akiyama, E.; Hirama, M. *Synlett* **1996**, 100.

[446] Schmittberger, T.; Uguen, D. *Tetrahedron Lett.* **1997**, *38*, 2837.
[447] Carretero, J. C.; Dominguez, E. *J. Org. Chem.* **1992**, *57*, 3867.
[448] Bonete, P.; Nájera, C. *Tetrahedron* **1996**, *52*, 4111.
[449] Masaki, Y.; Serizawa, Y.; Nagata, K.; Kaki, K. *Chem. Lett.* **1984**, 2105.
[450] Collins, M. A.; Jones, D. N. *Tetrahedron* **1996**, *52*, 8795.
[451] García Ruano, J. L.; Alonso de Diego, S. A.; Martín, M. R.; Torrente, E.; Martín Castro, A. M. *Org. Lett.* **2004**, *6*, 4945.
[452] Julia, M.; Badet, B. *Tetrahedron Lett.* **1974**, *15*, 1363.
[453] Adrio, J.; Carretero, J. C. *Tetrahedron* **1998**, *54*, 1601.
[454] Löfström, C. M. G.; Ericsson, A. M.; Bourrinet, L.; Juntunen, S. K.; Bäckvall, J.-E. *J. Org. Chem.* **1995**, *60*, 3586.
[455] d'Angelo, J.; Revial, G.; Costa, P. R. R.; Castro, R. N.; Antunes, O. A. C. *Tetrahedron: Asymmetry* **1991**, *2*, 199.
[456] Kondo, K.; Saito, E.; Tunemoto, D. *Tetrahedron Lett.* **1975**, *16*, 2275.
[457] François, D.; Lallemand, M.-C.; Selkti, M.; Tomas, A.; Kunesch, N.; Husson, H. P. *Angew. Chem. Int. Ed.* **1998**, *37*, 104.
[458] Bäckvall, J.-E.; Plobeck, N. A.; Juntunen, S. K. *Tetrahedron Lett.* **1989**, *30*, 2589.
[459] Bäckvall, J.-E.; Plobeck, N. A. *J. Org. Chem.* **1990**, *55*, 4528.
[460] Trost, B. M.; Bernstein, P. R.; Funfschilling, P. C. *J. Am. Chem. Soc.* **1979**, *101*, 4378.
[461] Barrero, A. F.; Alvarez-Manzaneda, E. J.; Chahboun, R.; Rivas, A. R.; Palomino, P. L. *Tetrahedron Lett.* **2000**, *56*, 6099.
[462] Najdi, S.; Kurth, M. J. *Tetrahedron Lett.* **1990**, *31*, 3279.
[463] Klenk, M.; Suter, C. M.; Archer, S. *J. Am. Chem. Soc.* **1948**, *70*, 3846.
[464] Lautens, M.; Ren, Y. *J. Am. Chem. Soc.* **1996**, *118*, 10668.
[465] Díez, D.; García, P.; Marcos, I. S.; Garrido, N. M.; Basabe, P.; Broughton, H. B.; Urones, J. G. *Org. Lett.* **2003**, *5*, 3687.
[466] Karoyan, P.; Chassaing, G. *Tetrahedron Lett.* **2002**, *43*, 1221.
[467] Karoyan, P.; Quancard, J.; Vaissermann, J.; Chassaing, G. *J. Org. Chem.* **2003**, *68*, 2256.
[468] Paquette, L. A.; Fischer, J. W.; Browne, A. R.; Doecke, C. W. *J. Am. Chem. Soc.* **1985**, *107*, 686.
[469] Solas, D.; Wolinsky, J. *J. Org. Chem.* **1983**, *48*, 1988.
[470] Barrero, A. F.; Alvarez-Manzaneda, E. J.; Herrador, M. M.; Alvarez-Manzaneda, R.; Quilez, J.; Chahboun, R.; Palomino, P. L.; Rivas, A. R. *Tetrahedron Lett.* **1999**, *40*, 8273.
[471] Mori, K.; Harada, H.; Zagatti, P.; Cork, A.; Hall, D. R. *Liebigs Ann. Chem.* **1991**, 259.
[472] Shapiro, G.; Buechler, D.; Marzi, M.; Schmidt, K.; Gomez-Lor, B. *J. Org. Chem.* **1995**, *60*, 4978.
[473] Díez, D.; García, P.; Marcos, I. S.; Garrido, N. M.; Basabe, P.; Broughton, H. B.; Urones, J. G. *Tetrahedron* **2005**, *61*, 699.
[474] Domon, L.; Uguen, D. *Tetrahedron Lett.* **2000**, *41*, 5501.
[475] Kimura, T.; Carlson, D. A.; Mori, K. *Eur. J. Org. Chem.* **2001**, 3385.
[476] Craig, D.; McCague, R.; Potter, G. A.; Williams, M. R. V. *Synlett* **1998**, 58.
[477] White, J. D.; Kawasaki, M. *J. Am. Chem. Soc.* **1990**, *112*, 4991.
[478] Ishii, Y.; Nagumo, S.; Arai, T.; Akuzawa, M.; Kawahara, N.; Akita, H. *Tetrahedron* **2006**, *62*, 716.
[479] Norley, M.; Kocienski, P.; Faller, A. *Synlett* **1996**, 900.
[480] Sasaki, M.; Tsukano, C.; Tachibana, K. *Org. Lett.* **2002**, *4*, 1747.
[481] Li, S.; Xu, R.; Bai, D. *Tetrahedron Lett.* **2000**, *41*, 3463.
[482] Larsson, M.; Galandrin, E.; Högberg, H.-E. *Tetrahedron* **2004**, *60*, 10659.
[483] Bull, J. R.; de Koning, P. D. *J. Chem. Soc., Perkin Trans. 1* **2000**, 1003.
[484] Majewski, M.; Clive, D. L. J.; Anderson, P. C. *Tetrahedron Lett.* **1984**, *25*, 2101.
[485] Larcheveque, M.; Sanner, C.; Azerad, R.; Buisson, D. *Tetrahedron* **1988**, *44*, 6407.
[486] Ley, S. V.; Norman, J.; Pinel, C. *Tetrahedron Lett.* **1994**, *35*, 2095.
[487] Baldwin, I. R.; Whitby, R. J. *Chem. Commun.* **2003**, 2786.
[488] Díez, D.; Moro, R. F.; Marcos, I. S.; Sánchez López, J. M.; Urones, J. G. *Synlett* **2000**, 794.
[489] Tanner, D.; Somfai, P. *Tetrahedron* **1987**, *43*, 4395.

[490] Hutchinson, D. K.; Fuchs, P. L. *J. Am. Chem. Soc.* **1987**, *109*, 4755.
[491] Spencer, T. A.; Li, D.; Russel, J. S.; Tomkinson, N. C. O.; Willson, T. M. *J. Org. Chem.* **2000**, *65*, 1919.
[492] Maehr, H.; Uskokovic, M. R.; Adorini, L.; Reddy, S. *J. Med. Chem.* **2004**, *47*, 6476.
[493] Ferraboschi, P.; Reza-Elahi, S.; Verza, E.; Santaniello, E. *Tetrahedron: Asymmetry* **1998**, *9*, 2193.
[494] Ferraboschi, P.; Pecora, F.; Reza-Elahi, S.; Santaniello, E. *Tetrahedron: Asymmetry* **1999**, *10*, 2497.
[495] Mori, K.; Takahashi, Y. *Liebigs Ann. Chem.* **1991**, 1057.
[496] Hioki, H.; Hamano, M.; Kubo, M.; Uno, T.; Kodama, M. *Chem. Lett.* **2001**, 898.
[497] Marshall, J. A.; Cleary, D. G. *J. Org. Chem.* **1986**, *51*, 858.
[498] Baba, T.; Huang, G.; Isobe, M. *Tetrahedron* **2003**, *59*, 6851.
[499] Baba, T.; Isobe, M. *Synlett* **2003**, 547.
[500] Miyaoka, H.; Shida, H.; Yamada, N.; Mitome, H.; Yamada, Y. *Tetrahedron Lett.* **2002**, *43*, 2227.
[501] Yang, Y.-L.; Manna, S.; Falck, J. R. *J. Am. Chem. Soc.* **1984**, *106*, 3811.
[502] Hayward, M. M.; Roth, R. M.; Duffy, K. J.; Dalko, P. I.; Stevens, K. L.; Guo, J.; Kishi, Y. *Angew. Chem. Int. Ed.* **1998**, *37*, 192.
[503] Tsuboi, K.; Ichikawa, Y.; Isobe, M. *Synlett* **1997**, 713.
[504] Smith, A. B., III; Condon, S. M.; McCauley, J. A.; Leahy, J. W.; Leazer, J. L., Jr.; Maleczka, R. E., Jr. *Tetrahedron Lett.* **1994**, *35*, 4907.
[505] Deng, L.-S.; Huang, X.-P.; Zhao, G. *J. Org. Chem.* **2006**, *71*, 4625.
[506] Park, P.; Broka, C. A.; Johnson, B. F.; Kishi, Y. *J. Am. Chem. Soc.* **1987**, *109*, 6205.
[507] Liu, T.-Z.; Kirschbaum, B.; Isobe, M. *Synlett* **2000**, 587.
[508] Martynow, J. G.; Jóźwik, J.; Szelejewski, W.; Achmatowicz, O.; Kutner, A.; Wioniewski, K.; Winiarski, J.; Zegrocka-Stendel, O.; Golębiewski, P. *Eur. J. Org. Chem.* **2007**, 689.
[509] Naito, H.; Kawahara, E.; Maruta, K.; Maeda, M.; Sasaki, S. *J. Org. Chem.* **1995**, *60*, 4419.
[510] Brittain, D. E.; Griffiths-Jones, C. M.; Linder, M. R.; Smith, M. D.; McCusker, C.; Barlow, J. S.; Akiyama, R.; Yasuda, K.; Ley, S. V. *Angew. Chem. Int. Ed.* **2005**, *44*, 2732.
[511] Holton, R. A.; Crouse, D. J.; Williams, A. D.; Kennedy, R. M. *J. Org. Chem.* **1987**, *52*, 2317.
[512] Gundermann, K.-D.; Huchting, R. *Chem. Ber.* **1959**, *92*, 415.
[513] Guo, H.; Ye, S.; Wang, J.; Zhang, Y. *J. Chem. Research (S)* **1997**, 114.
[514] House, H. O.; Larson, J. K. *J. Org. Chem.* **1968**, *33*, 61.
[515] Molander, G. A.; Hahn, G. *J. Org. Chem.* **1986**, *51*, 1135.
[516] Robin, S.; Huet, F.; Fauve, A.; Veschambre, H. *Tetrahedron: Asymmetry* **1993**, *4*, 239.
[517] Jacobs, H. K.; Mueller, B. H.; Gopalan, A. S. *Tetrahedron* **1992**, *48*, 8891.
[518] Carretero, J. C.; Rojo, J. *Tetrahedron Lett.* **1992**, *33*, 7407.
[519] Rojo, J.; García, M.; Carretero, J. C. *Tetrahedron* **1993**, *49*, 9787.
[520] Satyamurthi, N.; Singh, J.; Aidhen, I. S. *Synthesis* **2000**, 375.
[521] Thomsen, M. W.; Handwerker, B. M.; Katz, S. A.; Fisher, S. A. *Synth. Commun.* **1988**, *18*, 1433.
[522] Trost, B. M.; Verhoeven, T. R. *J. Am. Chem. Soc.* **1979**, *101*, 1595.
[523] Martín, T.; Rodríguez, C. M.; Martín, V. S. *Tetrahedron: Asymmetry* **1995**, *6*, 1151.
[524] Weichert, A.; Hoffmann, M. R. *J. Org. Chem.* **1991**, *56*, 4098.
[525] White, J. D.; Blakemore, P. R.; Milicevic, S. *Org. Lett.* **2002**, *4*, 1803.
[526] Craig, D.; Hyland, C. J. T.; Ward, S. E. *Synlett* **2006**, 2142.
[527] Takeda, H.; Watanabe, H.; Nakada, M. *Tetrahedron* **2006**, *62*, 8054.
[528] Kondo, K.; Tunemoto, D. *Tetrahedron Lett.* **1975**, *16*, 1397.
[529] Cavicchioli, S.; Savoia, D.; Trombini, C.; Umani-Ronchi, A. *J. Org. Chem.* **1984**, *49*, 1246.
[530] Mussatto, M. C.; Savoia, D.; Trombini, C.; Umani-Ronchi, A. *J. Org. Chem.* **1980**, *45*, 4002.
[531] Muñoz, L.; Rosa, E.; Bosch, M. P.; Guerrero, A. *Tetrahedron Lett.* **2005**, *46*, 3311.
[532] Hernandez-Juan, F. A.; Xiong, X.; Brewer, S. E.; Buchanan, D. J.; Dixon, D. J. *Synthesis* **2005**, 3283.
[533] Brimble, M. A.; Officer, D. L.; Williams, G. M. *Tetrahedron Lett.* **1988**, *29*, 3609.
[534] De Blas, J.; Carretero, J. C.; Domínguez, E. *Tetrahedron: Asymmetry* **1995**, *6*, 1035.
[535] Santhosh, K. C.; Balasubramanian, K. K. *Tetrahedron Lett.* **1991**, *32*, 7727.
[536] Santhosh, K. C.; Balasubramanian, K. K. *J. Chem. Soc., Chem. Commun.* **1992**, 224.

[537] Adrio, J.; Rivero, M. R.; Carretero, J. C. *Angew. Chem. Int. Ed.* **2000**, *39*, 2906.
[538] Porta, A.; Vidari, G.; Zanoni, G. *J. Org. Chem.* **2005**, *70*, 4876.
[539] Colobert, F.; Genet, J. P. *Tetrahedron Lett.* **1985**, *26*, 2779.
[540] Pavri, N. P.; Trudell, M. L. *Tetrahedron Lett.* **1997**, *38*, 7993.
[541] Zhang, C.; Ballay, C. J., II; Trudell, M. L. *J. Chem. Soc., Perkin Trans. 1* **1999**, 675.
[542] Llamas, T.; Gómez Arrayás, R.; Carretero, J. C. *Angew. Chem. Int. Ed.* **2007**, *46*, 3329.
[543] Sengupta, S.; Das, D.; Mondal, S. *Synlett* **2001**, 1464.
[544] Moreno-Vargas, A. J.; Vogel, P. *Tetrahedron Lett.* **2003**, *14*, 3173.
[545] Yakura, T.; Tanaka, K.; Iwamoto, M.; Nameki, M.; Ikeda, M. *Synlett* **1999**, 1313.
[546] Carretero, J. C.; Domínguez, E. *J. Org. Chem.* **1993**, *58*, 1596.
[547] Davies, M. J.; Moody, C. J.; Taylor, R. J. *J. Chem. Soc., Perkin Trans. 1* **1991**, 1.
[548] Davies, M. J.; Moody, C. J. *J. Chem. Soc., Perkin Trans. 1* **1991**, 9.
[549] Spino, C.; Deslongchamps, P. *Tetrahedron Lett.* **1990**, *31*, 3969.
[550] Nicolaou, K. C.; Bunnage, M. E.; McGarry, D. G.; Shi, S.; Somers, P. K.; Wallace, P. A.; Chu, X.-J.; Agrios, K. A.; Gunzner, J. L.; Yang, Z. *Chem. Eur. J.* **1999**, *5*, 599.
[551] Sasaki, N. A.; Pauly, R.; Fontaine, C.; Chiaroni, A.; Riche, C.; Potier, P. *Tetrahedron Lett.* **1994**, *35*, 241.
[552] Kajiwara, Y.; Scott, A. I. *Tetrahedron Lett.* **2002**, *43*, 8795.
[553] Lygo, B. *Synlett* **1992**, 793.
[554] McAllister, L. A.; Brand, S.; de Gentile, R.; Procter, D. J. *Chem. Commun.* **2003**, 2380.
[555] Isobe, M.; Ichikawa, Y.; Bai, D.-L.; Masaki, H.; Kawai, T.; Goto, T. *Tetrahedron* **1987**, *43*, 4767.
[556] Harmata, M.; Wacharasindhu, S. *Synthesis* **2007**, 2365.
[557] Buddhsukh, D.; Magnus, P. *J. Chem. Soc., Chem. Commun.* **1975**, 952.
[558] Larsson, M.; Högberg, H.-E. *Tetrahedron* **2001**, *57*, 7541.
[559] Sengupta, S.; Sarma, D. S.; Mondal, S. *Tetrahedron* **1998**, *54*, 9791.
[560] Caricato, G.; Savoia, D. *Synlett* **1994**, 1015.
[561] Norman, B. H.; Gareau, Y.; Padwa, A. *J. Org. Chem.* **1991**, *56*, 2154.
[562] Tsai, M.-R.; Chen, B.-F.; Cheng, C.-C.; Chang, N.-C. *J. Org. Chem.* **2005**, *70*, 1780.
[563] Kende, A. S.; Kaldor, I.; Aslanian, R. *J. Am. Chem. Soc.* **1988**, *110*, 6265.
[564] Jones, D. N.; Maybury, M. W. J.; Swallow, S.; Tomkinson, N. C. O. *Tetrahedron Lett.* **1993**, *34*, 8553.
[565] Jacobs, H. K.; Gopalan, A. S. *J. Org. Chem.* **1994**, *59*, 2014.
[566] Lai, S. M. F.; Orchison, J. J. A.; Whiting, D. A. *Tetrahedron* **1989**, *45*, 5895.
[567] Back, T. G.; Wulff, J. E. *Chem. Commun.* **2002**, 1710.
[568] Back, T. G.; Parvez, M.; Wulff, J. E. *J. Org. Chem.* **2003**, *68*, 2223.
[569] Mota, A. J.; Chiaroni, A.; Langlois, N. *Eur. J. Org. Chem.* **2003**, 4187.
[570] Poss, C. S.; Rychnovsky, S. D.; Schreiber, S. L. *J. Am. Chem. Soc.* **1993**, *115*, 3360.
[571] Posner, G. H.; Kinter, C. M. *J. Org. Chem.* **1990**, *55*, 3967.
[572] Suh, Y.-G.; Jung, J. K.; Seo, S.-Y.; Min, K. H.; Shin, D. Y.; Lee, Y. S.; Kim, S. H.; Park, H.-J. *J. Org. Chem.* **2002**, *67*, 4127.
[573] Magnus, P.; Ujjainwalla, F.; Westwood, N.; Lynch, V. *Tetrahedron* **1998**, *54*, 3069.
[574] Magnus, P.; Booth, J.; Magnus, N.; Tarrant, J.; Thom, S.; Ujjainwalla, F. *Tetrahedron Lett.* **1996**, *36*, 5331.
[575] Lygo, B.; Rudd, C. N. *Tetrahedron Lett.* **1995**, *36*, 3577.
[576] Patra, A.; Parhari, P.; Ray, S.; Mal, D. *J. Org. Chem.* **2005**, *70*, 9017.
[577] Chen, H.-W.; Hsu, R.-T.; Chang, M.-Y.; Chang, N.-C. *Org. Lett.* **2006**, *8*, 3033.
[578] Miyata, O.; Ozawa, Y.; Ninomiya, I.; Naito, T. *Synlett* **1997**, 275.
[579] Takeda, K.; Nakajima, A.; Yoshii, E. *Synlett* **1995**, 249.
[580] Dallavalle, S.; Nannei, R.; Merlini, L.; Bava, A.; Nasini, G. *Synlett* **2005**, 2676.
[581] Fulmer, T. D.; Bryson, T. A. *J. Org. Chem.* **1989**, *54*, 3496.
[582] Mushti, C. S.; Kim, J.-H.; Corey, E. J. *J. Am. Chem. Soc.* **2006**, *128*, 14050.
[583] Marshall, J. A.; Andrews, R. C.; Lebioda, L. *J. Org. Chem.* **1987**, *52*, 2378.
[584] Hu, T.; Corey, E. J. *Org. Lett.* **2002**, *4*, 2441.
[585] Paquette, L. A.; Lin, H.-S.; Coghlan, M. J. *Tetrahedron Lett.* **1987**, *28*, 5017.

[586] Chang, M.-Y.; Sun, P. P.; Chen, S. T.; Chang, N. C. *Tetrahedron Lett.* **2003**, *44*, 5271.
[587] Lebsack, A. D.; Overman, L. E.; Valentekovich, R. J. *J. Am. Chem. Soc.* **2001**, *123*, 4851.
[588] Ichikawa, Y.; Isobe, M.; Masaki, H.; Kawai, T.; Goto, T. *Tetrahedron* **1987**, *43*, 4759.
[589] Solladié, G.; Colobert, A. A. F. *Synlett* **1992**, 167.
[590] Trost, B. M.; Verhoeven, T. R. *J. Am. Chem. Soc.* **1976**, *98*, 630.
[591] White, J. D.; Avery, M. A.; Choudhry, S. C.; Dhingra, O. P.; Kang, M.; Whittle, A. J. *J. Am. Chem. Soc.* **1983**, *105*, 6517.
[592] White, J. D.; Choudhry, S. C.; Kang, M. *Tetrahedron Lett.* **1984**, *25*, 3671.
[593] Godleski, S. A.; Villhauer, E. B. *J. Org. Chem.* **1986**, *51*, 486.
[594] Araki, K.; Saito, K.; Arimoto, H.; Uemura, D. *Angew. Chem. Int. Ed.* **2004**, *43*, 81.
[595] Gandula, S. R. V.; Kumar, P. *Tetrahedron* **2006**, *62*, 9942.
[596] Chun, J.; Li, G.; Byun, H.-S.; Bittman, R. *J. Org. Chem.* **2002**, *67*, 2600.
[597] Chun, J.; Li, G.; Byun, H.-S.; Bittman, R. *Tetrahedron Lett.* **2002**, *43*, 375.
[598] Solladié, G.; Maestro, M. C.; Rubio, A.; Pedregal, C.; Carreño, M. C.; Ruano, J. L. G. *J. Org. Chem.* **1991**, *56*, 2317.
[599] Kobayashi, Y.; Taguchi, T.; Kanuma, N.; Ikekawa, N.; Oshida, J. *Tetrahedron Lett.* **1981**, *22*, 4309.
[600] Genet, J. P.; Gaudin, J. M. *Tetrahedron* **1987**, *43*, 5315.
[601] Grieco, P. A.; Masaki, Y.; Boxler, D. *J. Org. Chem.* **1975**, *40*, 2261.
[602] Chavan, S. P.; Praveen, C. *Tetrahedron Lett.* **2004**, *45*, 421.
[603] Suenaga, K.; Araki, K.; Sengoku, T.; Uemura, D. *Org. Lett.* **2001**, *3*, 527.
[604] Takahashi, T.; Kataoka, H.; Tsuji, J. *J. Am. Chem. Soc.* **1983**, *105*, 147.
[605] Jia, Y.; Li, X.; Wang, P.; Wu, B.; Zhao, X.; Tu, Y. *J. Chem. Soc., Perkin Trans. 1* **2002**, 560.
[606] Blakemore, P. R.; Browder, C. C.; Hong, J.; Lincoln, C. M.; Nagornyy, P. A.; Robarge, L. A.; Wardrop, D. J.; White, J. D. *J. Org. Chem.* **2005**, *70*, 5449.
[607] Chakraborty, T. K.; Suresh, V. R. *Tetrahedron Lett.* **1998**, *39*, 9109.
[608] Nicolaou, K. C.; Pihko, P. M.; Bernal, F.; Frederick, M. O.; Qian, W.; Uesaka, N.; Diedrichs, N.; Hinrichs, J.; Koftis, T. V.; Loizidou, E.; Petrovic, G.; Rodríguez, M.; Sarlah, D.; Zou, N. *J. Am. Chem. Soc.* **2006**, *128*, 2244.
[609] Hikage, N.; Furukawa, H.; Takao, K.; Kobayashi, S. *Tetrahedron Lett.* **1998**, *39*, 6237.
[610] Takeda, K.; Kawanishi, E.; Nakamura, H.; Yoshii, E. *Tetrahedron Lett.* **1991**, *32*, 4925.
[611] Carter, R. G.; Bourland, T. C.; Zhou, X.-T.; Gronemeyer, M. A. *Tetrahedron* **2003**, *59*, 8963.
[612] Sugimoto, T.; Ishihara, J.; Murai, A. *Tetrahedron Lett.* **1997**, *38*, 7379.
[613] Martin, S. F.; Naito, S. *J. Org. Chem.* **1998**, *63*, 7592.
[614] Jia, Y. X.; Li, X.; Wu, B.; Zhao, X. Z.; Tu, Y. Q. *Tetrahedron* **2002**, *58*, 1697.
[615] White, J. D.; Ohba, Y.; Porter, W. J.; Wang, S. *Tetrahedron Lett.* **1997**, *38*, 3167.
[616] Halim, R.; Brimble, M. A.; Merten J. *Org. Biomol. Chem.* **2006**, *4*, 1387.
[617] Boeckman, R. K., Jr.; Shao, P.; Wrobleski, S. T.; Boehmler, D. J.; Heintzelman, G. R.; Barbosa, A. J. *J. Am. Chem. Soc.* **2006**, *128*, 10572.
[618] Evans, D. A.; Kværnø, L.; Mulder, J. A.; Raymer, B.; Dunn, T. B.; Beauchemin, A.; Olhava, E. J.; Juhl, M.; Kagechika, K. *Angew. Chem. Int. Ed.* **2007**, *46*, 4693.
[619] Zhou, X.-T.; Carter, R. G. *Angew. Chem. Int. Ed.* **2006**, *45*, 1787.
[620] Paterson, I.; Feβner, K.; Finlay, M. R. V. *Tetrahedron Lett.* **1997**, *38*, 4301.
[621] Hanessian, S.; Grillo, T. A. *J. Org. Chem.* **1998**, *63*, 1049.
[622] Lautens, M.; Colucci, J. T.; Hiebert, S.; Smith, N. D.; Bouchain, G. *Org. Lett.* **2002**, *4*, 1879.
[623] Baldwin, J. E.; Adlington, R. M.; Lowe, C.; O'Neil, I. A.; Sanders, G. L.; Schofield, C. J.; Sweeney, J. B. *J. Chem. Soc., Chem. Commun.* **1988**, 1030.
[624] Taylor, N. H.; Thomas, E. J. *Tetrahedron* **1999**, *55*, 8757.
[625] Ono, N.; Miyake, H.; Kamimura, A.; Hamamoto, I.; Tamura, R.; Kaji, A. *Tetrahedron* **1985**, *41*, 4013.
[626] Baldwin, J. E.; Adlington, R. M.; Ichikawa, Y.; Kneale, C. J. *J. Chem. Soc., Chem. Commun.* **1988**, 702.
[627] Padwa, A.; Muller, C. L.; Rodríguez, A.; Watterson, S. H. *Tetrahedron* **1998**, *54*, 9651.
[628] Murphree, S. S.; Muller, C. L.; Padwa, A. *Tetrahedron Lett.* **1990**, *31*, 6145.

[629] Alonso, D. A.; Nájera, C.; Sansano, J. M. *Tetrahedron* **1994**, *50*, 6603.
[630] Yoda, H.; Shirakawa, K.; Takabe, K. *Chem. Lett.* **1989**, 1391.
[631] Edwards, G. L.; Sinclair, D. J.; Wasiowych, C. D. *Synlett* **1997**, 1285.
[632] Edwards, G. L.; Sinclair, D. J. *Synthesis* **2005**, 3613.
[633] Wada, E.; Yasuoka, H.; Kanemasa, S. *Chem. Lett.* **1994**, 145.
[634] Kumareswaran, R.; Balasubramanian, T.; Hassner, A. *Tetrahedron Lett.* **2000**, *41*, 8157.
[635] Balasubramanian, T.; Hassner, A. *Tetrahedron: Asymmetry* **1998**, *9*, 2201.
[636] Crowley, P. J.; Fawcett, J.; Griffith, G. A.; Moralee, A. C.; Percy, J. M.; Salafia, V. *Org. Biomol. Chem.* **2005**, *3*, 3297.
[637] Pearson, W. H.; Kropf, J. E.; Choy, A. L.; Lee, I. Y.; Kampf, J. W. *J. Org. Chem.* **2007**, *72*, 4135.
[638] Hodgson, D. M.; Hachisu, S.; Andrews, M. D. *Org. Lett.* **2005**, *7*, 815.
[639] Mohanty, S. S.; Uebelhart, P.; Eugster, C. H. *Helv. Chim. Acta* **2000**, *83*, 2036.
[640] Trost, B. M.; Weber, L. *J. Org. Chem.* **1975**, *40*, 3619.
[641] Lipshutz, B. H.; Bulow, G.; Fernandez-Lazaro, F.; Kim, S.-K.; Lowe, R.; Mollard, P.; Stevens, K. L. *J. Am. Chem. Soc.* **1999**, *121*, 11664.
[642] Jones, D. N.; Peel, M. R. *J. Chem. Soc., Chem. Commun.* **1986**, 216.
[643] Crich, D.; Natarajan, S.; Crich, J. Z. *Tetrahedron* **1997**, *3*, 7139.
[644] Uemoto, K.; Kawahito, A.; Matsushita, N.; Sakamoto, I.; Kaku, H.; Tsunoda, T. *Tetrahedron Lett.* **2001**, *42*, 905.
[645] Williams, D. R.; Coleman, P. J. *Tetrahedron Lett.* **1995**, *36*, 35.
[646] Miyaoka, H.; Isaji, Y.; Mitome, H.; Yamada, Y. *Tetrahedron* **2003**, *59*, 61.
[647] Boukouvalas, J.; Robichaud, J.; Maltais, F. *Synlett* **2006**, 2480.
[648] Yue, X.; Lan, J.; Li, J.; Liu, Z.; Lin, Y. *Tetrahedron* **1999**, *55*, 133.
[649] Hayakawa, K.; Nishiyama, H.; Kanematsu, K. *J. Org. Chem.* **1985**, *50*, 512.
[650] Grieco, P. A.; Masaka, Y. *J. Org. Chem.* **1974**, *39*, 2135.
[651] Nickel, A.; Maruyama, T.; Tang, H.; Murphy, P. D.; Greene, B.; Yusuff, N.; Wood, J. L. *J. Am. Chem. Soc.* **2004**, *126*, 16300.
[652] Tong, R.; Valentine, J. C.; McDonald, F. E.; Cao, R.; Fang, X.; Hardcastle, K. I. *J. Am. Chem. Soc.* **2007**, *129*, 1050.
[653] Zheng, Y. F.; Dodd, D. S.; Oehlschlager, A. C. *Tetrahedron* **1995**, *51*, 5255.
[654a] Trost, B. M.; Machacek, M. R.; Tsui, H. C. *J. Am. Chem. Soc.* **2005**, *127*, 7014.
[654b] Marshall, J. A.; Markwalder, J. A. *Tetrahedron Lett.* **1988**, *29*, 4811.
[655] Bouzbouz, S.; Kirschleger, B. *Synthesis* **1994**, 714.
[656] Sum, F. W.; Weiler, L. *J. Am. Chem. Soc.* **1979**, *101*, 4401.
[657] Trost, B. M.; Dong, G.; Vance, J. A. *J. Am. Chem. Soc.* **2007**, *129*, 4540.
[658] Eren, D.; Keinan, E. *J. Am. Chem. Soc.* **1988**, *110*, 4356.
[659] Min, J.-H.; Lee, J.-S.; Yang, J.-D.; Koo, S. *J. Org. Chem.* **2003**, *68*, 7925.
[660] Gao, Y.; Nan, F.; Xu, X. *Tetrahedron Lett.* **2000**, *41*, 4811.
[661] Tanimoto, H.; Oritani, T. *Tetrahedron* **1997**, *53*, 3527.
[662] Takanashi, S.; Mori, K. *Liebigs Ann./Recl.* **1997**, 825.
[663] Chênevert, R.; Courchesne, G. *Tetrahedron Lett.* **2002**, *43*, 7971.
[664] Brioche, J. C. R.; Goodenough, K. M.; Whatrup, D. J.; Harrity, J. P. A. *Org. Lett.* **2007**, *9*, 689.
[665] Zhang, T.; Liu, Z.; Li, Y. *Synthesis* **2001**, 393.
[666] Tanada, Y.; Mori, K. *Eur. J. Org. Chem.* **2003**, 848.
[667] Terao, S.; Kato, K.; Shiraishi, M.; Marimoto, H. *J. Chem. Soc., Perkin Trans. 1* **1978**, 1101.
[668] Yu, X.-J.; Chen, F.-E.; Dai, H.-F.; Chen, X.-X.; Kuang, Y.-Y.; Xie, B. *Helv. Chim. Acta* **2005**, *88*, 2575.
[669] Dai, H.-F.; Chen, F. E.; Yu, X. J. *Helv. Chim. Acta* **2006**, *89*, 1317.
[670] Grassi, D.; Lippuner, V.; Aebi, M.; Brunner, J.; Vasella, A. *J. Am. Chem. Soc.* **1997**, *119*, 10992.
[671] Antonio, Y.; De La Cruz, M. E.; Galeazzi, E.; Guzman, A.; Bray, B. L.; Greenhouse, R.; Kurz, L. J.; Lustig, D. A.; Maddox, M. L.; Muchowski, J. M. *Can. J. Chem.* **1994**, *15*, 15.
[672] Watanabe, Y.; Ueno, Y.; Araki, T.; Endo, T.; Okawara, M. *Tetrahedron Lett.* **1986**, *27*, 215.
[673] Girniene, J.; Tardy, S.; Tatibouët, A.; Sačkus, A.; Rollin, P. *Tetrahedron Lett.* **2004**, *45*, 6443.

[674] Mozingo, R.; Wolf, D. E.; Harris, S. A.; Folkers, K. *J. Am. Chem. Soc.* **1943**, *65*, 1013.
[675] Conreaux, D.; Bossharth, E.; Monteiro, N.; Desbordes, P.; Balme, G. *Tetrahedron Lett.* **2005**, *46*, 7917.
[676] Cossu, S.; De Luchi, O.; Durr, R.; Fabris, F. *Synth. Commun.* **1996**, *26*, 211.
[677] Simpkins, N. S. *Tetrahedron Lett.* **1987**, *28*, 989.
[678] Holmes, A. B.; Pooley, G. R. *Tetrahedron* **1992**, *48*, 7775.
[679] Moiseenkov, A. M.; Czeskis, B. A.; Ivanova, N. M.; Nefedov, O. M. *J. Chem. Soc., Perkin Trans. 1* **1991**, 2639.
[680] Ohnuma, T.; Hata, N.; Fujiwara, H.; Ban, Y. *J. Org. Chem.* **1982**, *47*, 4713.
[681] Belloch, J.; Virgili, M.; Moyano, A.; Pericàs, M. A.; Riera, A. *Tetrahedron Lett.* **1991**, *32*, 4579.
[682] Clayton, S. C.; Regan, A. C. *Tetrahedron Lett.* **1993**, *34*, 7493.
[683] Liang, F.; Navarro, H. A.; Abraham, P.; Kotian, P.; Ding, Y.-S.; Fowler, J.; Volkow, N.; Kuhar, M. J.; Carroll, F. I. *J. Med. Chem.* **1997**, *40*, 2293.
[684] Ibáñez, P. L.; Nájera, C. *Tetrahedron Lett.* **1993**, *34*, 2003.
[685] Arjona, O.; Borrallo, C.; Iradier, F.; Medel, R.; Plumet, J. *Tetrahedron Lett.* **1998**, *39*, 1977.
[686] Leung-Toung, R.; Liu, Y.; Muchowski, J. M.; Wu, Y.-L. *J. Org. Chem.* **1998**, *63*, 3235.
[687] Carpino, L. A.; Lin, Y.-Z. *J. Org. Chem.* **1990**, *55*, 247.
[688] Huang, D. F.; Shen, T. Y. *Tetrahedron Lett.* **1993**, *34*, 4477.
[689] Metcalf, B. W.; Bonilavri, E. *J. Chem. Soc., Chem. Commun.* **1978**, 914.
[690] Steglich, W.; Wegmann, H. *Synthesis* **1980**, 481.
[691] Chéry, F.; Desroses, M.; Tatibouët, A.; De Lucchi, O.; Rollin, P. *Tetrahedron* **2003**, *59*, 4563.
[692] Aceña, J. L.; Arjona, O.; Fernández de la Pradilla, R.; Plumet, J.; Viso, A. *J. Org. Chem.* **1992**, *57*, 1945.
[693] Sheehan, S. M.; Padwa, A. *J. Org. Chem.* **1997**, *62*, 438.
[694] Porta, A.; Re, S.; Zanoni, G.; Vidari, G. *Tetrahedron* **2007**, *63*, 3989.
[695] Balasubramanian, T.; Hassner, A. *Tetrahedron Lett.* **1996**, *37*, 5755.
[696] Enjo, J.; Castedo, L.; Tojo, G. *Org. Lett.* **2001**, *3*, 1343.
[697] Barry, W. J.; Finar, I. L. *J. Chem. Soc.* **1954**, 138.
[698] Castedo, L.; Delamano, J.; Enjo, J.; Fernández, J.; Grávalos, D. G.; Leis, R.; López, C.; Marcos, C. F.; Ríos, A.; Tojo, G. *J. Am. Chem. Soc.* **2001**, *123*, 5102.
[699] Schmit, C. *Synlett* **1994**, 241.
[700] Chéry F.; Pillard, C.; Tatibouët, A.; De Lucchi, O.; Rollin, P. *Tetrahedron* **2006**, *62*, 5141.
[701] Yu, S.; Pu, X.; Cheng, T.; Wang, R.; Ma, D. *Org. Lett.* **2006**, *8*, 3179.
[702] Zeng, Z.; Xu, X. *Tetrahedron Lett.* **2000**, *41*, 3459.
[703] Back, T. G.; Proudfoot, J. R.; Djerassi, C. *Tetrahedron Lett.* **1986**, *27*, 2187.
[704] Horvath, R. F.; Linde, R. G., II; Hayward, C. M.; Joglar, J.; Yohannes, D.; Danishefsky, S. L. *Tetrahedron Lett.* **1993**, *34*, 3993.
[705] Chen, S.-H.; Horvath, R. F.; Joglar, J.; Fisher, M. J.; Danishefsky, S. L. *J. Org. Chem.* **1991**, *56*, 5834.
[706] Carter, R. G.; Graves, D. E.; Gronemeyer, M. A.; Tschumper, G. S. *Org. Lett.* **2002**, *4*, 2181.
[707] Cabianca, E.; Chéry, F.; Rollin, P.; Tatibouët, A.; De Lucchi, O. *Tetrahedron Lett.* **2002**, *43*, 585.
[708] Thompson, C. M.; Frick, J. A. *J. Org. Chem.* **1989**, *54*, 890.
[709] Carretero, J. C.; Díaz, N.; Molina, M. L.; Rojo, J. *Tetrahedron Lett.* **1996**, *37*, 3179.
[710] Ono, N.; Miyake, H.; Tanura, R.; Hamamoto, I.; Kaji, A. *Chem. Lett.* **1981**, 1139.
[711] Arjona, O.; Iradier, F.; Medel, R.; Plumet, J. *Heterocycles* **1999**, *50*, 653.
[712] Lee, G. H.; Lee, H. K.; Choi, E. B.; Kim, B. T.; Pak, C. S. *Tetrahedron Lett.* **1995**, *31*, 5607.
[713] Molander, G. A.; Eastwood, P. R. *J. Org. Chem.* **1995**, *60*, 8382.
[714] Fernández-Mayoralas, A.; Marra, A.; Trumtel, M. *Tetrahedron Lett.* **1989**, *30*, 2537.
[715] Pontikis, R.; Wolf, J.; Monneret, C.; Florent, J.-C. *Tetrahedron Lett.* **1995**, *36*, 3523.
[716] Burks, J. E., Jr.; Crandall, J. K. *J. Org. Chem.* **1984**, *49*, 4663.
[717] Ono, N.; Kamimura, A.; Kaji, A. *Tetrahedron Lett.* **1986**, *27*, 1595.
[718] Lacrampe, F.; Léost, F.; Doutheau, A. *Tetrahedron Lett.* **2000**, *41*, 4773.
[719] Ono, N.; Kamimura, A.; Kaji, A. *J. Org. Chem.* **1988**, *53*, 251.
[720] Morita, Y.; Tokuyama, H.; Fukuyama, T. *Org. Lett.* **2005**, *7*, 4337.

[721] Mann, S.; Carillon, S.; Breyne, O.; Duhayon, C.; Hamon, L.; Marquet, A. *Eur. J. Org. Chem.* **2002**, 736.

[722] Davidson, A. H.; Eggleton, N.; Wallace, I. H. *J. Chem. Soc., Chem. Commun.* **1991**, 378.

[723] Roush, W. R.; Russo-Rodriguez, S. *J. Org. Chem.* **1985**, *50*, 5465.

[724] Kraus, G. A.; Jeon, I. *Tetrahedron* **2005**, *61*, 2111.

[725] Ichikawa, Y. *J. Chem. Soc., Perkin Trans. 1* **1992**, 2135.

[726] Hodgson, D. M.; Hachisu, S.; Andrews, M. D. *J. Org. Chem.* **2005**, *70*, 8866.

[727] Asakura, N.; Usuki, Y.; Lio, H.; Tanaka, T. *J. Fluorine Chem.* **2006**, *127*, 800.

[728] Kitano, Y.; Ito, T.; Suzuki, T.; Nogata, Y.; Shinshima, K.; Yoshimura, E.; Chiba, K.; Tada, M.; Sakaguchi, I. *J. Chem. Soc., Perkin Trans. 1* **2002**, 2251.

[729] Hirama, M.; Uei, M. *J. Am. Chem. Soc.* **1982**, *104*, 4251.

[730] Kazmaier, U.; Wesquet, A. *Synlett* **2005**, 1271.

[731] Adjé, N.; Domon, L.; Vogeleisen-Mutterer, F.; Uguen, D. *Tetrahedron Lett.* **2000**, *41*, 5495.

[732] Takadoi, M.; Katoh, T.; Ishiwata, A.; Terashima, S. *Tetrahedron* **2002**, *58*, 9903.

[733] Galkina, A.; Buff, A.; Schulz, E.; Hennig, L.; Findeisen, M.; Reinhard, G.; Oehme, R.; Welzel, P. *Eur. J. Org. Chem.* **2003**, 4640.

[734] Ermolenko, L.; Sasaki, N. A.; Potier, P. *Tetrahedron Lett.* **1999**, *40*, 5187.

[735] Takeda, K.; Sato, M.; Yoshii, E. *Tetrahedron Lett.* **1986**, *27*, 3903.

[736] Cox, C. M.; Whiting, D. A. *J. Chem. Soc., Perkin Trans. 1* **1991**, 1907.

[737] Shimamura, H.; Sunazuka, T.; Izuhara, T.; Hirose, T.; Shiomi, K.; Omura, S. *Org. Lett.* **2007**, *9*, 65.

[738] Demont, E.; Lopez, R.; Férézou, J.-P. *Synlett* **1998**, 1223.

[739] Oddon, G.; Uguen, D.; De Cian, A.; Fischer, J. *Tetrahedron Lett.* **1998**, *39*, 1149.

[740] Jenmalm, A.; Berts, W.; Li, Y.-L.; Luthman, K.; Csöregh, I.; Hacksell, U. *J. Org. Chem.* **1995**, *60*, 1026.

[741] Wiktelius, D.; Berts, W.; Jenmalm, A.; Gullbo, J.; Saitton, S.; Csöregh, I.; Luthman, K. *Tetrahedron* **2006**, *62*, 3600.

[742] Jenmalm, A.; Berts, W.; Li, Y.-L.; Luthman, K.; Csöregh, I.; Hacksell, U. *J. Org. Chem.* **1994**, *59*, 1139.

[743] Spaltenstein, A.; Carpino, P. A.; Miyake, F.; Hopkins, P. B. *J. Org. Chem.* **1987**, *52*, 3759.

[744] Trost, B. M.; Matsumura, Y. *J. Org. Chem.* **1977**, *42*, 2036.

[745] O'Connor, S. J.; Williard, P. G. *Tetrahedron Lett.* **1989**, *30*, 4637.

[746] Trost, B. M.; Lynch, J.; Renaut, P.; Steinman, D. H. *J. Am. Chem. Soc.* **1986**, *108*, 284.

[747] Poupon, J.-C.; Demont, E.; Prunet, J.; Férézou, J.-P. *J. Org. Chem.* **2003**, *68*, 4700.

[748] Toyooka, N.; Okumura, M.; Takahata, H.; Remoto, H. *Tetrahedron* **1999**, *55*, 10673.

[749] Hoemann, M. Z.; Agrios, K. A.; Aubé, J. *Tetrahedron* **1997**, *53*, 11087.

[750] Roush, W. R.; Peseckis, S. M. *Tetrahedron Lett.* **1982**, *23*, 4879.

[751] Hird, N. W.; Lee, T. V.; Leigh, A. J.; Maxwell, J. R.; Peakman, T. M. *Tetrahedron Lett.* **1989**, *30*, 4867.

[752] Tabuchi, H.; Hamamoto, T.; Miki, S.; Tejima, T.; Ichihara, A. *J. Org. Chem.* **1994**, *59*, 4749.

[753] Raghavan, S.; Rajender, A. *J. Org. Chem.* **2003**, *68*, 7094.

[754] Arjona, O.; Iradier, F.; Mañas, R. M.; Plumet, J. *Tetrahedron Lett.* **1998**, *39*, 8335.

[755] Zanoni, G.; Porta, A.; Vidari, G. *J. Org. Chem.* **2002**, *67*, 4346.

[756] Morzycki, J. W.; Schnoes, H. K.; DeLuca, H. F. *J. Org. Chem.* **1984**, *49*, 2148.

[757] Villalobos, A.; Danishefsky, S. J. *J. Org. Chem.* **1989**, *54*, 12.

[758] De Laszlo, S. E.; Ford, M. J.; Ley, S. V.; Maw, G. N. *Tetrahedron Lett.* **1990**, *31*, 5525.

[759] Wardrop, D. J.; Fritz, J. *Org. Lett.* **2006**, *8*, 3659.

[760] Knight, D. W.; Sibley, A. W. *Tetrahedron Lett.* **1993**, *34*, 6607.

[761] Kim, G.; Chu-Moyer, M. Y.; Danishefsky, S. J.; Schulte, G. K. *J. Am. Chem. Soc.* **1993**, *115*, 30.

[762] de Gaeta, L. S. L.; Czarniecki, M. *J. Org. Chem.* **1989**, *54*, 4004.

[763] Kubota, T.; Tsuda, M.; Kobayashi, J. *Tetrahedron* **2003**, *59*, 1613.

[764] Kirk, D. N.; Varley, M. J.; Makin, H. L. J.; Trafford, D. J. H. *J. Chem. Soc., Perkin Trans. 1* **1983**, 2563.

[765] Zhu, J.; Ma, D. *Angew. Chem. Int. Ed.* **2003**, *42*, 5348.
[766] Shimizu, S.; Nakamura, S.; Nakada, M.; Shibasaki, M. *Tetrahedron* **1996**, *52*, 13363.
[767] Mendlik, M. T.; Cottard, M.; Rein, T.; Helquist, P. *Tetrahedron Lett.* **1997**, *38*, 6375.
[768] Kocienski, P. J. *J. Org. Chem.* **1980**, *45*, 2037.
[769] Patel, D. V.; VanMiddlesworth, F.; Donaubauer, J.; Gannett, P.; Sih, C. J. *J. Am. Chem. Soc.* **1986**, *108*, 4603.
[770] Guanti, G.; Banfi, L.; Schmid, G. *Tetrahedron Lett.* **1994**, *35*, 4239.
[771] Tani, K.; Naganawa, A.; Ishida, A.; Egashira, H.; Odagaki, Y.; Miyazaki, T.; Hasegawa, T.; Kawanaka, Y.; Sagawa, K.; Harada, H.; Ogawa, M.; Maruyama, T.; Nakai, H.; Ohuchida, S.; Kondo, K.; Toda, M. *Bioorg. Med. Chem.* **2002**, *10*, 1883.
[772] Nemoto, H.; Kurobe, H.; Fukumoto, K.; Kametani, T. *J. Org. Chem.* **1986**, *51*, 5311.
[773] Sodeoka, M.; Satoh, S.; Shibasaki, M. *J. Am. Chem. Soc.* **1988**, *110*, 4823.
[774] Rao, A. V. R.; Gurjar, M. K.; Pal, S.; Pariza, R. J.; Chorghade, S. M. *Tetrahedron Lett.* **1995**, *36*, 2505.
[775] Wang, Q.; Sasaki, A. *J. Org. Chem.* **2004**, *69*, 4767.
[776] Kocienski, P.; Todd, M. *J. Chem. Soc., Perkin Trans. 1* **1983**, 1783.
[777] Suenaga, K.; Miya, S.; Kuroda, T.; Handa, T.; Kanematsu, K.; Sakakura, A.; Kigoshi, H. *Tetrahedron Lett.* **2004**, *45*, 5383.
[778] Suenaga, K.; Kimura, T.; Kuroda, T.; Matsui, K.; Miya, S.; Kuribayashi, S.; Sakakura, A.; Kigoshi, H. *Tetrahedron* **2006**, *62*, 8278.
[779] Berberich, S. M.; Cherney, R. J.; Colucci, J.; Courillon, C.; Geraci, L. S.; Kirkland, T. A.; Marx, M. A.; Schneider, M. F.; Martin, S. F. *Tetrahedron* **2003**, *59*, 6819.
[780] Kende, A. S.; Mendoza, J. S.; Fujii, Y. *Tetrahedron* **1993**, *49*, 8015.
[781] Kocienski, P. J.; Lythgoe, B. *J. Chem. Soc., Perkin Trans. 1* **1980**, 1400.
[782] Trost, B. M.; Calkins, T. L.; Bochet, C. G. *Angew. Chem. Int. Ed.* **1997**, *36*, 2632.
[783] Achmatowicz, B.; Gorobets, E.; Marczak, S.; Przezdziecka, A.; Steinmeyer, A.; Wicha, J.; Zügel, U. *Tetrahedron Lett.* **2001**, *42*, 2891.
[784] Choudhry, S. C.; Belica, P. S.; Coffen, D. L.; Focella, A.; Maehr, H.; Manchand, P. S.; Serico, L.; Yang, R. T. *J. Org. Chem.* **1993**, *58*, 1496.
[785] Perlman, K. L.; DeLuca, H. F. *Tetrahedron Lett.* **1992**, *33*, 2937.
[786] Yamamoto, K.; Shimizu, M.; Yamada, S. Iwata, S.; Hocino, O. *J. Org. Chem.* **1992**, *57*, 33.
[787] Yamada, S.; Nakayama, K.; Takayama, H. *Tetrahedron Lett.* **1981**, *22*, 2591.
[788] Edwards, M. P.; Ley, S. V.; Lister, S. G.; Palmer, B. D.; Williams, D. J. *J. Org. Chem.* **1984**, *49*, 3503.
[789] Barret, A. G. M.; Carr, R. A. E.; Attwood, S. V.; Richardson, G.; Walshe, N. D. A. *J. Org. Chem.* **1986**, *51*, 4840.
[790] Kigoshi, H.; Ojiva, M.; Suenaga, K.; Mutuo, T.; Hirano, J.; Sakakura, A.; Ogawa, T.; Nisiwaki, M.; Yamada, K. *Tetrahedron Lett.* **1994**, *35*, 1247.
[791] Kigoshi, H.; Suenaga, K.; Takagi, M.; Asao, A.; Kanematsu, K.; Kamei, N.; Okugawa, Y.; Yamada, K. *Tetrahedron* **2002**, *58*, 1075.
[792] Ghosh, A. K.; Wang, Y. *J. Am. Chem. Soc.* **2000**, *122*, 11027.
[793] Shimizu, A.; Nishiyama, S. *Synlett* **1998**, 1209.
[794] Seebach, D.; Maestro, M. A.; Sefkow, M.; Neidlein, A.; Sternfeld, F.; Adam, G.; Sommerfeld, T. *Helv. Chim. Acta* **1991**, *74*, 2112.
[795] Chen, A.; Nelson, A.; Tanikkul, N.; Thomas, E. J. *Tetrahedron Lett.* **2001**, *42*, 1251.
[796] Kozikowski, A. P.; Sorgi, K. L. *Tetrahedron Lett.* **1984**, *25*, 2085.
[797] Abel, S.; Faber, D.; Hüter, O.; Giese, B. *Synthesis* **1999**, 188.
[798] Matsuda, F.; Tomiyoshi, N.; Yanagiya, M.; Matsumoto, T. *Tetrahedron* **1990**, *46*, 3469.
[799] Danishefsky, S. J.; Selnick, H. G.; Zelle, R. E.; DeNinno, M. P. *J. Am. Chem. Soc.* **1988**, *110*, 4368.
[800] Mori, Y.; Asai, M.; Kawade, J.; Furukawa, H. *Tetrahedron* **1995**, *51*, 5315.
[801] Baker, R.; O'Mahony, M. J.; Swain, C. *J. Chem. Soc., Chem. Commun.* **1985**, 1326.
[802] Horigome, M.; Motoyoshi, H.; Watanabe, H.; Kitahara, T. *Tetrahedron Lett.* **2001**, *42*, 8207.

[803] White, J. D.; Bolton, G. L.; Dantanarayana, A. P.; Fox, C. M. J.; Hiner, R. N.; Jackson, R. W.; Sakuma, K.; Warrier, U. S. *J. Am. Chem. Soc.* **1995**, *117*, 1908.

[804] Anthony, N. J.; Armstrong, A.; Ley, S. V.; Madin, A. *Tetrahedron Lett.* **1989**, *30*, 3209.

[805] Greck, C.; Grice, P.; Jones, A. B.; Ley, S. V. *Tetrahedron Lett.* **1987**, *28*, 5759.

[806] Hikota, M.; Tone, H.; Horita, K.; Yonemitsu, O. *Tetrahedron* **1990**, *46*, 4613.

[807] Tanimoto, N.; Gerritz, S. W.; Sawabe, A.; Noda, T.; Filla, S. A.; Masamune, S. *Angew. Chem. Int. Ed. Engl.* **1994**, *33*, 673.

[808] Ohmori, K.; Ogawa, Y.; Obitsu, T.; Ishikawa, Y.; Nishiyama, S.; Yamamura, S. *Angew. Chem. Int. Ed.* **2000**, *39*, 2290.

[809] Hale, K. J.; Frigerio, M.; Manavizar, S.; Hummersone, M. G.; Fillingham, I. J.; Barsukov, I. G.; Damblon, C. F.; Gescher, A.; Roberts, G. C. K. *Org. Lett.* **2003**, *5*, 499.

[810] de Vicente, J.; Huckins, J. R.; Rychnovsky, S. D. *Angew. Chem. Int. Ed.* **2006**, *45*, 7258.

[811] Kageyama, M.; Tamura, T. Nantz, M. H.; Roberts, J. C.; Somfai, P.; Whritenour, D. C.; Masamune, S. *J. Am. Chem. Soc.* **1990**, *112*, 7407.

[812] Evans, D. A.; Kaldor, S. W.; Jones, T. K.; Clardy, J.; Stout, T. J. *J. Am. Chem. Soc.* **1990**, *112*, 7001.

CUMULATIVE CHAPTER TITLES BY VOLUME

Volume 1 (1942)

1. **The Reformatsky Reaction:** Ralph L. Shriner

2. **The Arndt-Eistert Reaction:** W. E. Bachmann and W. S. Struve

3. **Chloromethylation of Aromatic Compounds:** Reynold C. Fuson and C. H. McKeever

4. **The Amination of Heterocyclic Bases by Alkali Amides:** Marlin T. Leffler

5. **The Bucherer Reaction:** Nathan L. Drake

6. **The Elbs Reaction:** Louis F. Fieser

7. **The Clemmensen Reduction:** Elmore L. Martin

8. **The Perkin Reaction and Related Reactions:** John R. Johnson

9. **The Acetoacetic Ester Condensation and Certain Related Reactions:** Charles R. Hauser and Boyd E. Hudson, Jr.

10. **The Mannich Reaction:** F. F. Blicke

11. **The Fries Reaction:** A. H. Blatt

12. **The Jacobson Reaction:** Lee Irvin Smith

Volume 2 (1944)

1. **The Claisen Rearrangement:** D. Stanley Tarbell

2. **The Preparation of Aliphatic Fluorine Compounds:** Albert L. Henne

3. **The Cannizzaro Reaction:** T. A. Geissman

4. **The Formation of Cyclic Ketones by Intramolecular Acylation:** William S. Johnson

5. **Reduction with Aluminum Alkoxides (The Meerwein-Ponndorf-Verley Reduction):** A. L. Wilds

6. **The Preparation of Unsymmetrical Biaryls by the Diazo Reaction and the Nitrosoacetylamine Reaction:** Werner E. Bachmann and Roger A. Hoffman

2. **Cyanoethylation:** Herman L. Bruson
3. **The Diels-Alder Reaction: Quinones and Other Cyclenones:** Lewis L. Butz and Anton W. Rytina
4. **Preparation of Aromatic Fluorine Compounds from Diazonium Fluoborates: The Schiemann Reaction:** Arthur Roe
5. **The Friedel and Crafts Reaction with Aliphatic Dibasic Acid Anhydrides:** Ernst Berliner
6. **The Gattermann-Koch Reaction:** Nathan N. Crounse
7. **The Leuckart Reaction:** Maurice L. Moore
8. **Selenium Dioxide Oxidation:** Norman Rabjohn
9. **The Hoesch Synthesis:** Paul E. Spoerri and Adrien S. DuBois
10. **The Darzens Glycidic Ester Condensation:** Melvin S. Newman and Barney J. Magerlein

Volume 6 (1951)

1. **The Stobbe Condensation:** William S. Johnson and Guido H. Daub
2. **The Preparation of 3,4-Dihydroisoquinolines and Related Compounds by the Bischler-Napieralski Reaction:** Wilson M. Whaley and Tutucorin R. Govindachari
3. **The Pictet-Spengler Synthesis of Tetrahydroisoquinolines and Related Compounds:** Wilson M. Whaley and Tutucorin R. Govindachari
4. **The Synthesis of Isoquinolines by the Pomeranz-Fritsch Reaction:** Walter J. Gensler
5. **The Oppenauer Oxidation:** Carl Djerassi
6. **The Synthesis of Phosphonic and Phosphinic Acids:** Gennady M. Kosolapoff
7. **The Halogen-Metal Interconversion Reaction with Organolithium Compounds:** Reuben G. Jones and Henry Gilman
8. **The Preparation of Thiazoles:** Richard H. Wiley, D. C. England, and Lyell C. Behr
9. **The Preparation of Thiophenes and Tetrahydrothiophenes:** Donald E. Wolf and Karl Folkers
10. **Reductions by Lithium Aluminum Hydride:** Weldon G. Brown

Volume 7 (1953)

1. **The Pechmann Reaction:** Suresh Sethna and Ragini Phadke
2. **The Skraup Synthesis of Quinolines:** R. H. F. Manske and Marshall Kulka
3. **Carbon–Carbon Alkylations with Amines and Ammonium Salts:** James H. Brewster and Ernest L. Eliel

4. **The von Braun Cyanogen Bromide Reaction:** Howard A. Hageman

5. **Hydrogenolysis of Benzyl Groups Attached to Oxygen, Nitrogen, or Sulfur:** Walter H. Hartung and Robert Simonoff

6. **The Nitrosation of Aliphatic Carbon Atoms:** Oscar Touster

7. **Epoxidation and Hydroxylation of Ethylenic Compounds with Organic Peracids:** Daniel Swern

Volume 8 (1954)

1. **Catalytic Hydrogenation of Esters to Alcohols:** Homer Adkins

2. **The Synthesis of Ketones from Acid Halides and Organometallic Compounds of Magnesium, Zinc, and Cadmium:** David A. Shirley

3. **The Acylation of Ketones to Form β-Diketones or β-Keto Aldehydes:** Charles R. Hauser, Frederic W. Swamer, and Joe T. Adams

4. **The Sommelet Reaction:** S. J. Angyal

5. **The Synthesis of Aldehydes from Carboxylic Acids:** Erich Mosettig

6. **The Metalation Reaction with Organolithium Compounds:** Henry Gilman and John W. Morton, Jr.

7. **β-Lactones:** Harold E. Zaugg

8. **The Reaction of Diazomethane and Its Derivatives with Aldehydes and Ketones:** C. David Gutsche

Volume 9 (1957)

1. **The Cleavage of Non-enolizable Ketones with Sodium Amide:** K. E. Hamlin and Arthur W. Weston

2. **The Gattermann Synthesis of Aldehydes:** William E. Truce

3. **The Baeyer-Villiger Oxidation of Aldehydes and Ketones:** C. H. Hassall

4. **The Alkylation of Esters and Nitriles:** Arthur C. Cope, H. L. Holmes, and Herbert O. House

5. **The Reaction of Halogens with Silver Salts of Carboxylic Acids:** C. V. Wilson

6. **The Synthesis of β-Lactams:** John C. Sheehan and Elias J. Corey

7. **The Pschorr Synthesis and Related Diazonium Ring Closure Reactions:** DeLos F. DeTar

Volume 10 (1959)

1. **The Coupling of Diazonium Salts with Aliphatic Carbon Atoms:** Stanley J. Parmerter

2. **The Japp-Klingemann Reaction:** Robert R. Phillips

3. **The Michael Reaction:** Ernst D. Bergmann, David Ginsburg, and Raphael Pappo

Volume 11 (1960)

1. **The Beckmann Rearrangement:** L. Guy Donaruma and Walter Z. Heldt

2. **The Demjanov and Tiffeneau-Demjanov Ring Expansions:** Peter A. S. Smith and Donald R. Baer

3. **Arylation of Unsaturated Compounds by Diazonium Salts:** Christian S. Rondestvedt, Jr.

4. **The Favorskii Rearrangement of Haloketones:** Andrew S. Kende

5. **Olefins from Amines: The Hofmann Elimination Reaction and Amine Oxide Pyrolysis:** Arthur C. Cope and Elmer R. Trumbull

Volume 12 (1962)

1. **Cyclobutane Derivatives from Thermal Cycloaddition Reactions:** John D. Roberts and Clay M. Sharts

2. **The Preparation of Olefins by the Pyrolysis of Xanthates. The Chugaev Reaction:** Harold R. Nace

3. **The Synthesis of Aliphatic and Alicyclic Nitro Compounds:** Nathan Kornblum

4. **Synthesis of Peptides with Mixed Anhydrides:** Noel F. Albertson

5. **Desulfurization with Raney Nickel:** George R. Pettit and Eugene E. van Tamelen

Volume 13 (1963)

1. **Hydration of Olefins, Dienes, and Acetylenes via Hydroboration:** George Zweifel and Herbert C. Brown

2. **Halocyclopropanes from Halocarbenes:** William E. Parham and Edward E. Schweizer

3. **Free Radical Addition to Olefins to Form Carbon–Carbon Bonds:** Cheves Walling and Earl S. Huyser

4. **Formation of Carbon–Heteroatom Bonds by Free Radical Chain Additions to Carbon–Carbon Multiple Bonds:** F. W. Stacey and J. F. Harris, Jr.

Volume 14 (1965)

1. **The Chapman Rearrangement:** J. W. Schulenberg and S. Archer

2. **α-Amidoalkylations at Carbon:** Harold E. Zaugg and William B. Martin

3. **The Wittig Reaction:** Adalbert Maercker

Volume 15 (1967)

1. **The Dieckmann Condensation:** John P. Schaefer and Jordan J. Bloomfield

2. **The Knoevenagel Condensation:** G. Jones

Volume 16 (1968)

1. **The Aldol Condensation:** Arnold T. Nielsen and William J. Houlihan

Volume 17 (1969)

1. **The Synthesis of Substituted Ferrocenes and Other π-Cyclopentadienyl-Transition Metal Compounds:** Donald E. Bublitz and Kenneth L. Rinehart, Jr.

2. **The γ-Alkylation and γ-Arylation of Dianions of β-Dicarbonyl Compounds:** Thomas M. Harris and Constance M. Harris

3. **The Ritter Reaction:** L. I. Krimen and Donald J. Cota

Volume 18 (1970)

1. **Preparation of Ketones from the Reaction of Organolithium Reagents with Carboxylic Acids:** Margaret J. Jorgenson

2. **The Smiles and Related Rearrangements of Aromatic Systems:** W. E. Truce, Eunice M. Kreider, and William W. Brand

3. **The Reactions of Diazoacetic Esters with Alkenes, Alkynes, Heterocyclic, and Aromatic Compounds:** Vinod Dave and E. W. Warnhoff

4. **The Base-Promoted Rearrangements of Quaternary Ammonium Salts:** Stanley H. Pine

Volume 19 (1972)

1. **Conjugate Addition Reactions of Organocopper Reagents:** Gary H. Posner

2. **Formation of Carbon–Carbon Bonds via π-Allylnickel Compounds:** Martin F. Semmelhack

3. **The Thiele-Winter Acetoxylation of Quinones:** J. F. W. McOmie and J. M. Blatchly

4. **Oxidative Decarboxylation of Acids by Lead Tetraacetate:** Roger A. Sheldon and Jay K. Kochi

Volume 20 (1973)

1. **Cyclopropanes from Unsaturated Compounds, Methylene Iodide, and Zinc-Copper Couple:** H. E. Simmons, T. L. Cairns, Susan A. Vladuchick, and Connie M. Hoiness

2. **Sensitized Photooxygenation of Olefins:** R. W. Denny and A. Nickon

3. **The Synthesis of 5-Hydroxyindoles by the Nenitzescu Reaction:** George R. Allen, Jr.

4. **The Zinin Reaction of Nitroarenes:** H. K. Porter

Volume 21 (1974)

1. **Fluorination with Sulfur Tetrafluoride:** G. A. Boswell, Jr., W. C. Ripka, R. M. Scribner, and C. W. Tullock

2. **Modern Methods to Prepare Monofluoroaliphatic Compounds:** Clay M. Sharts and William A. Sheppard

Volume 22 (1975)

1. **The Claisen and Cope Rearrangements:** Sara Jane Rhoads and N. Rebecca Raulins

2. **Substitution Reactions Using Organocopper Reagents:** Gary H. Posner

3. **Clemmensen Reduction of Ketones in Anhydrous Organic Solvents:** E. Vedejs

4. **The Reformatsky Reaction:** Michael W. Rathke

Volume 23 (1976)

1. **Reduction and Related Reactions of α,β-Unsaturated Compounds with Metals in Liquid Ammonia:** Drury Caine

2. **The Acyloin Condensation:** Jordan J. Bloomfield, Dennis C. Owsley, and Janice M. Nelke

3. **Alkenes from Tosylhydrazones:** Robert H. Shapiro

Volume 24 (1976)

1. **Homogeneous Hydrogenation Catalysts in Organic Solvents:** Arthur J. Birch and David H. Williamson

2. **Ester Cleavages via S_N2-Type Dealkylation:** John E. McMurry

3. **Arylation of Unsaturated Compounds by Diazonium Salts (The Meerwein Arylation Reaction):** Christian S. Rondestvedt, Jr.

4. **Selenium Dioxide Oxidation:** Norman Rabjohn

Volume 25 (1977)

1. **The Ramberg-Bäcklund Rearrangement:** Leo A. Paquette

2. **Synthetic Applications of Phosphoryl-Stabilized Anions:** William S. Wadsworth, Jr.

3. **Hydrocyanation of Conjugated Carbonyl Compounds:** Wataru Nagata and Mitsuru Yoshioka

Volume 26 (1979)

1. **Heteroatom-Facilitated Lithiations:** Heinz W. Gschwend and Herman R. Rodriguez

2. **Intramolecular Reactions of Diazocarbonyl Compounds:** Steven D. Burke and Paul A. Grieco

Volume 27 (1982)

1. **Allylic and Benzylic Carbanions Substituted by Heteroatoms:** Jean-François Biellmann and Jean-Bernard Ducep

2. **Palladium-Catalyzed Vinylation of Organic Halides:** Richard F. Heck

Volume 28 (1982)

1. **The Reimer-Tiemann Reaction:** Hans Wynberg and Egbert W. Meijer

2. **The Friedländer Synthesis of Quinolines:** Chia-Chung Cheng and Shou-Jen Yan

3. **The Directed Aldol Reaction:** Teruaki Mukaiyama

Volume 29 (1983)

1. **Replacement of Alcoholic Hydroxy Groups by Halogens and Other Nucleophiles via Oxyphosphonium Intermediates:** Bertrand R. Castro

2. **Reductive Dehalogenation of Polyhalo Ketones with Low-Valent Metals and Related Reducing Agents:** Ryoji Noyori and Yoshihiro Hayakawa

3. **Base-Promoted Isomerizations of Epoxides:** Jack K. Crandall and Marcel Apparu

Volume 30 (1984)

1. **Photocyclization of Stilbenes and Related Molecules:** Frank B. Mallory and Clelia W. Mallory

2. **Olefin Synthesis via Deoxygenation of Vicinal Diols:** Eric Block

Volume 31 (1984)

1. **Addition and Substitution Reactions of Nitrile-Stabilized Carbanions:** Siméon Arseniyadis, Keith S. Kyler, and David S. Watt

Volume 32 (1984)

1. **The Intramolecular Diels-Alder Reaction:** Engelbert Ciganek

2. **Synthesis Using Alkyne-Derived Alkenyl- and Alkynylaluminum Compounds:** George Zweifel and Joseph A. Miller

Volume 33 (1985)

1. **Formation of Carbon–Carbon and Carbon–Heteroatom Bonds via Organoboranes and Organoborates:** Ei-Ichi Negishi and Michael J. Idacavage

2. **The Vinylcyclopropane-Cyclopentene Rearrangement:** Tomáš Hudlický, Toni M. Kutchan, and Saiyid M. Naqvi

Volume 34 (1985)

1. **Reductions by Metal Alkoxyaluminum Hydrides:** Jaroslav Málek

2. **Fluorination by Sulfur Tetrafluoride:** Chia-Lin J. Wang

Volume 35 (1988)

1. **The Beckmann Reactions: Rearrangements, Elimination-Additions, Fragmentations, and Rearrangement-Cyclizations:** Robert E. Gawley

2. **The Persulfate Oxidation of Phenols and Arylamines (The Elbs and the Boyland-Sims Oxidations):** E. J. Behrman

3. **Fluorination with Diethylaminosulfur Trifluoride and Related Aminofluorosulfuranes:** Miloš Hudlický

Volume 36 (1988)

1. **The [3 + 2] Nitrone-Olefin Cycloaddition Reaction:** Pat N. Confalone and Edward M. Huie

2. **Phosphorus Addition at sp^2 Carbon:** Robert Engel

3. **Reduction by Metal Alkoxyaluminum Hydrides. Part II. Carboxylic Acids and Derivatives, Nitrogen Compounds, and Sulfur Compounds:** Jaroslav Málek

Volume 37 (1989)

1. **Chiral Synthons by Ester Hydrolysis Catalyzed by Pig Liver Esterase:** Masaji Ohno and Masami Otsuka

2. **The Electrophilic Substitution of Allylsilanes and Vinylsilanes:** Ian Fleming, Jacques Dunoguès, and Roger Smithers

Volume 38 (1990)

1. **The Peterson Olefination Reaction:** David J. Ager

2. **Tandem Vicinal Difunctionalization: β-Addition to α,β-Unsaturated Carbonyl Substrates Followed by α-Functionalization:** Marc J. Chapdelaine and Martin Hulce

3. **The Nef Reaction:** Harold W. Pinnick

Volume 39 (1990)

1. **Lithioalkenes from Arenesulfonylhydrazones:** A. Richard Chamberlin and Steven H. Bloom

2. **The Polonovski Reaction:** David Grierson

3. **Oxidation of Alcohols to Carbonyl Compounds via Alkoxysulfonium Ylides: The Moffatt, Swern, and Related Oxidations:** Thomas T. Tidwell

Volume 40 (1991)

1. **The Pauson-Khand Cycloaddition Reaction for Synthesis of Cyclopentenones:** Neil E. Schore
2. **Reduction with Diimide:** Daniel J. Pasto and Richard T. Taylor
3. **The Pummerer Reaction of Sulfinyl Compounds:** Ottorino DeLucchi, Umberto Miotti, and Giorgio Modena
4. **The Catalyzed Nucleophilic Addition of Aldehydes to Electrophilic Double Bonds:** Hermann Stetter and Heinrich Kuhlmann

Volume 41 (1992)

1. **Divinylcyclopropane-Cycloheptadiene Rearrangement:** Tomáš Hudlický, Rulin Fan, Josephine W. Reed, and Kumar G. Gadamasetti
2. **Organocopper Reagents: Substitution, Conjugate Addition, Carbo/Metallo-cupration, and Other Reactions:** Bruce H. Lipshutz and Saumitra Sengupta

Volume 42 (1992)

1. **The Birch Reduction of Aromatic Compounds:** Peter W. Rabideau and Zbigniew Marcinow
2. **The Mitsunobu Reaction:** David L. Hughes

Volume 43 (1993)

1. **Carbonyl Methylenation and Alkylidenation Using Titanium-Based Reagents:** Stanley H. Pine
2. **Anion-Assisted Sigmatropic Rearrangements:** Stephen R. Wilson
3. **The Baeyer-Villiger Oxidation of Ketones and Aldehydes:** Grant R. Krow

Volume 44 (1993)

1. **Preparation of α,β-Unsaturated Carbonyl Compounds and Nitriles by Selenoxide Elimination:** Hans J. Reich and Susan Wollowitz
2. **Enone Olefin [2 + 2] Photochemical Cyclizations:** Michael T. Crimmins and Tracy L. Reinhold

Volume 45 (1994)

1. **The Nazarov Cyclization:** Karl L. Habermas, Scott E. Denmark, and Todd K. Jones
2. **Ketene Cycloadditions:** John Hyatt and Peter W. Raynolds

Volume 46 (1994)

1. **Tin(II) Enolates in the Aldol, Michael, and Related Reactions:** Teruaki Mukaiyama and Shū Kobayashi

2. **The [2,3]-Wittig Reaction:** Takeshi Nakai and Koichi Mikami

3. **Reductions with Samarium(II) Iodide:** Gary A. Molander

Volume 47 (1995)

1. **Lateral Lithiation Reactions Promoted by Heteroatomic Substituents:** Robin D. Clark and Alam Jahangir

2. **The Intramolecular Michael Reaction:** R. Daniel Little, Mohammad R. Masjedizadeh, Olof Wallquist (in part), and Jim I. McLoughlin (in part)

Volume 48 (1995)

1. **Asymmetric Epoxidation of Allylic Alcohols: The Katsuki-Sharpless Epoxidation Reaction:** Tsutomu Katsuki and Victor S. Martin

2. **Radical Cyclization Reactions:** B. Giese, B. Kopping, T. Göbel, J. Dickhaut, G. Thoma, K. J. Kulicke, and F. Trach

Volume 49 (1997)

1. **The Vilsmeier Reaction of Fully Conjugated Carbocycles and Heterocycles:** Gurnos Jones and Stephen P. Stanforth

2. **[6 + 4] Cycloaddition Reactions:** James H. Rigby

3. **Carbon–Carbon Bond-Forming Reactions Promoted by Trivalent Manganese:** Gagik G. Melikyan

Volume 50 (1997)

1. **The Stille Reaction:** Vittorio Farina, Venkat Krishnamurthy and William J. Scott

Volume 51 (1997)

1. **Asymmetric Aldol Reactions Using Boron Enolates:** Cameron J. Cowden and Ian Paterson

2. **The Catalyzed α-Hydroxyalkylation and α-Aminoalkylation of Activated Olefins (The Morita–Baylis–Hillman Reaction):** Engelbert Ciganek

3. **[4 + 3] Cycloaddition Reactions:** James H. Rigby and F. Christopher Pigge

Volume 52 (1998)

1. **The Retro–Diels–Alder Reaction. Part I. C–C Dienophiles:** Bruce Rickborn

2. **Enantioselective Reduction of Ketones:** Shinichi Itsuno

Volume 53 (1998)

1. **The Oxidation of Alcohols by Modified Oxochromium(VI)-Amine Complexes:** Frederick A. Luzzio

2. **The Retro-Diels-Alder Reaction. Part II. Dienophiles with One or More Heteroatoms:** Bruce Rickborn

Volume 54 (1999)

1. **Aromatic Substitution by the $S_{RN}1$ Reaction:** Roberto Rossi, Adriana B. Pierini, and Ana N. Santiago

2. **Oxidation of Carbonyl Compounds with Organohypervalent Iodine Reagents:** Robert M. Moriarty and Om Prakash

Volume 55 (1999)

1. **Synthesis of Nucleosides:** Helmut Vorbrüggen and Carmen Ruh-Pohlenz

Volume 56 (2000)

1. **The Hydroformylation Reaction:** Iwao Ojima, Chung-Ying Tsai, Maria Tzamarioudaki, and Dominique Bonafoux

2. **The Vilsmeier Reaction. 2. Reactions with Compounds Other Than Fully Conjugated Carbocycles and Heterocycles:** Gurnos Jones and Stephen P. Stanforth

Volume 57 (2001)

1. **Intermolecular Metal-Catalyzed Carbenoid Cyclopropanations:** Huw M. L. Davies and Evan G. Antoulinakis

2. **Oxidation of Phenolic Compounds with Organohypervalent Iodine Reagents:** Robert M. Moriarty and Om Prakash

3. **Synthetic Uses of Tosylmethyl Isocyanide (TosMIC):** Daan van Leusen and Albert M. van Leusen

Volume 58 (2001)

1. **Simmons-Smith Cyclopropanation Reaction:** André B. Charette and André Beauchemin

2. **Preparation and Applications of Functionalized Organozinc Compounds:** Paul Knochel, Nicolas Millot, Alain L. Rodriguez, and Charles E. Tucker

Volume 59 (2001)

1. **Reductive Aminations of Carbonyl Compounds with Borohydride and Borane Reducing Agents:** Ellen W. Baxter and Allen B. Reitz

Volume 60 (2002)

1. **Epoxide Migration (Payne Rearrangement) and Related Reactions:** Robert M. Hanson

2. **The Intramolecular Heck Reaction:** J. T. Link

Volume 61 (2002)

1. **[3 + 2] Cycloaddition of Trimethylenemethane and Its Synthetic Equivalents:** Shigeru Yamago and Eiichi Nakamura

AUTHOR INDEX, VOLUMES 1–72

Volume number only is designated in this index

Organic Reactions, Vol. 72, Edited by Scott E. Denmark et al.

CHAPTER AND TOPIC INDEX, VOLUMES 1–72

Many chapters contain brief discussions of reactions and comparisons of alternative synthetic methods related to the reaction that is the subject of the chapter. These related reactions and alternative methods are not usually listed in this index. In this index, the volume number is in **boldface**, the chapter number is in ordinary type.

Organic Reactions, Vol. 72, Edited by Scott E. Denmark et al.